Statistical Analysis of
Designed Experiments

WILEY SERIES IN PROBABILITY AND STATISTICS

Established by WALTER A. SHEWHART and SAMUEL S. WILKS

A complete list of the titles in this series appears at the end of this volume.

Statistical Analysis of Designed Experiments

Theory and Applications

AJIT C. TAMHANE
Northwestern University

WILEY
A JOHN WILEY & SONS, INC., PUBLICATION

Published by John Wiley & Sons, Inc., Hoboken, New Jersey.
Published simultaneously in Canada

For general information on our other products and services or for technical support, please contact our Customer Care Department within the United States at 877-762-2974, outside the United States at 317-572-3993 or fax 317-572-4002.

Wiley also publishes its books in a variety of electronic formats. Some content that appears in print may not be available in electronic formats. For more information about Wiley products, visit our web site at www.wiley.com.

Library of Congress Cataloging-in-Publication Data:

Tamhane, Ajit C.
Statistical analysis of designed experiments : theory and applications / Ajit C. Tamhane.
p. cm.
Includes bibliographical references and index.
ISBN 978-0-471-75043-7 (cloth)
1. Experimental design. I. Title.
QA279.T36 2008
519.5′7—dc22

2008009432

10 9 8 7 6 5 4 3 2 1

To All My Teachers —
From Grade School to Grad School

Contents

Preface

There are many excellent books on design and analysis of experiments, for example, Box, Hunter, and Hunter (2005), Montgomery (2005), and Wu and Hamada (2000), so one may ask, why another book? The answer is largely personal. An instructor who teaches any subject over many years necessarily develops his or her own perspective of how the subject should be taught. Specifically, in my teaching of DOE (a popular abbreviation for design of experiments that I will use below for convenience), I have felt it necessary to put equal emphasis on theory and applications. Also, I have tried to motivate the subject by using real data examples and exercises drawn from a range of disciplines, not just from engineering or medicine, for example, since, after all, the principles of DOE are applicable everywhere. Therefore I wanted to present the subject according to my personal preferences and because this mode of presentation has worked for students in my classes over the years. Accordingly, the primary goal of this book is to provide a balanced coverage of the underlying theory and applications using real data. The secondary goal is to demonstrate the versatility of the DOE methodology by showing applications to wide-ranging areas, including agriculture, biology, education, engineering, marketing, medicine, and psychology. The book is mainly intended for seniors and first-year graduate students in statistics and those in applied disciplines with the necessary mathematical and statistical prerequisites (calculus and linear algebra and a course in statistical methods covering distribution theory, confidence intervals and hypothesis tests, and simple and multiple linear regression). It can also serve as a reference for practitioners who are interested in understanding the "whys" of the designs and analyses that they use and not just the "hows."

As the title indicates, the main focus of the book is on the analysis; design and planning, although equally if not more important, require discussion of many practical issues, some of which are application-specific, and hence are not emphasized to the same degree. The book provides an in-depth coverage of most of the standard topics in a first course on DOE. Many advanced topics such as nonnormal responses, generalized linear models, unbalanced or missing data, complex aliasing, and optimal designs are not covered. An extensive coverage of these topics would require a separate volume. The readers interested in these topics are referred to a more advanced book such as Wu and Hamada (2000). However, in the future I hope to add short reviews of these topics as supplementary materials on the book's website. Additional test problems as well as data sets for all the examples and exercises in the book will also be posted on this website.

A model-based approach is followed in the book. Discussion of each new design and its analysis begins with the specification of the underlying model and assumptions. This is followed by inference methods, for example, analysis of variance (ANOVA), confidence intervals and hypothesis tests, and residual analyses for model diagnostics. Derivations of the more important formulas and technical results are given in Chapter Notes at the end of each chapter. All designs are illustrated by fully worked out real data examples. Appropriate graphics accompany each analysis. The importance of using a statistical package to perform computations and for graphics cannot be overemphasized, but some calculations are worked out by hand as, in my experience, they help to enhance understanding of the methods. Minitab® is the main package used for performing analyses as it is one of the easiest to use and thus allows a student to focus on understanding the statistical concepts rather than learning the intricacies of its use. However, any other package would work equally well if the instructor and students are familiar with it. Because of the emphasis on using a statistical package, I have not provided catalogs of designs since many of the standard designs are now available in these packages, particularly those that specialize in DOE.

The book is organized as follows. Chapter 1 introduces the basic concepts and a brief history of DOE, Chapter 2 gives a review of elementary statistical methods through multiple regression. This background is assumed as a prerequisite in the course that I teach, although some instructors may want to go over this material, especially multiple regression using the matrix approach. Chapter 3 discusses the simplest single-factor experiments (one-way layouts) using completely randomized designs with and without covariates. Chapter 4 introduces multiple comparison and selection procedures for one-way layouts. These procedures provide practical alternatives to ANOVA F-tests of equality of treatment means. Chapter 5 returns to the single-factor setup but with randomization restrictions necessitated by blocking over secondary (noise) factors in order to evaluate the robustness of the effects of primary (treatment) factors of interest or to eliminate the biasing effects of secondary factors. This gives rise to randomized block designs (including balanced incomplete block designs), Latin squares, Youden squares and Graeco-Latin squares. Chapter 6 covers two-factor and three-factor experiments. Chapter 7 covers 2^p factorial experiments in which each of the $p \geq 2$ factors is at two levels. These designs are intended for screening purposes, but become impractically large very quickly as the number of runs increases exponentially with p. To economize on the number of runs without sacrificing the ability to estimate the important main effects and interactions of the factors, 2^{p-q} fractional factorial designs are used which are studied in Chapter 8. This chapter also discusses other fractional factorial designs including Plackett-Burman and Hadamard designs. A common thread among these designs is their orthogonality property. Orthogonal arrays, which provide a general mathematical framework for these designs, are also covered in this chapter. Chapter 9 discusses three-level and mixed-level full and fractional factorial experiments using orthogonal arrays. Response optimization using DOE is discussed in Chapter 10. The methodologies covered include response surface exploration and optimization, mixture experiments and the Taguchi method for robust design of products and processes. In Chapter 11 random and mixed effects models are introduced for single-factor and crossed factors designs. These are extended to nested and crossed-nested factors designs in Chapter 12. This chapter also discusses split-plot factorial designs. These designs are commonly employed in practice because a complete randomization with respect to all factors is not possible since some factors are harder to change than others, and so randomization is done in stages. Chapter 13 introduces repeated measures designs in which observations are taken over time on the same experimental units given different treatments. In addition to the effects due to time trends, the time-series nature of the data introduces correlations among observations which must be taken into account in the analysis. Both univariate and multivariate analysis methods are presented. Finally, Chapter 14 gives a review of the theory of fixed-effects linear models, which underlies the designs discussed in Chapters 3 through 10. Chapters 11 through 13 cover designs that involve random factors, but their general theory is not covered in this book. The reader is referred to the book by Searle, Casella and McCulloch (1992) for this theory.

There are three appendices. Appendix A gives a summary of results about vector-valued random variables and distribution theory of quadratic forms under multivariate normality. This appendix supplements the theory of linear models covered in Chapter 14. Appendix B gives three case studies. Two of these case studies are taken from student projects. These case studies illustrate the level of complexity encountered in real-life experiments that students taking a DOE course based on this book may be expected to deal with. They also illustrate two interesting modern applications in medical imaging and stem cell research. Appendix C contains some useful statistical tables.

Some of the exercises are also based on student projects. The exercises in each chapter are divided by sections and within each section are further divided into theoretical and applied. It is hoped that this will enable both the instructor and the student to choose exercises to suit their theoretical/applied needs. Answers to selected exercises are included at the end. A solution manual will be made available to the instructors from the publisher upon adopting the book in their courses.

Some of the unique features of the book are as follows. Chapter 4 gives a modern introduction to multiple comparison procedures and also to ranking and selection procedures, which are not covered in most DOE texts. Chapter 13 discusses repeated-measures designs, a topic of significant practical importance that is not covered in many texts. Chapter 14 gives a survey of linear model theory (along with the associated distribution theory in Appendix A) that can serve as a concise introduction to the topic in a more theoretically oriented course.

Finally, the case studies discussed in Appendix B should give students a taste of complexities of practical experiments, including constraints on randomization, unbalanced data, and outliers.

There is obviously far more material in the book than can be covered in a term-long course. Therefore the instructor must pick and choose the topics. Chapter 1 must, of course, be covered in any course. Chapter 2 is mainly for review and reference; the sections on simple and multiple regression using matrix notation may be covered if students do not have this background. In a two-term graduate course on linear models and DOE, this material can be supplemented with Chapter 14 at a more mathematical depth but also at a greater investment of time. From the remaining chapters, for a one-term course, I suggest Chapters 3, 4, 5, 6, 7, 8, and 11. For a two-term course, Chapters 9, 10, 12, and 13 can be added in the second term. Not all sections from each chapter can be covered in the limited time, so choices will need to be made by the instructor.

As mentioned at the beginning, there are several excellent books on DOE which I have used over the years and from which I have learned a lot. Another book that I have found very stimulating and useful for providing insights into various aspects of DOE is the collection of short articles written for practitioners (many from the Quality Quandaries column in *Quality Engineering*) by Box and Friends (2006). I want to acknowledge the influence of these books on the present volume. Most examples and exercises use data sets from published sources, which I have tried to cite wherever possible. I am grateful to all publishers who gave permission to use the data sets or figures from their copyrighted publications. I am especially grateful to Pearson Education, Inc. (Prentice Hall) for giving permission without fee for reproducing large portions of material from my book *Statistics and Data Analysis: From Elementary to Intermediate* with Dorothy Dunlop. Unfortunately, in a few cases, I have lost the original references and I offer my apologies for my inability to cite them. I have acknowledged the students whose projects are used in exercises and case studies individually in appropriate places.

I am grateful to three anonymous reviewers of the book who pointed out many errors and suggested improvements in earlier drafts of the book. I especially want to thank one reviewer who offered very detailed comments, criticisms, and suggestions on the pre–final draft of the book which led to significant revision and rearrangement of some chapters. This reviewer's comments on the practical aspects of design, analysis, and interpretations of the data sets in examples were particularly useful and resulted in substantial rewriting.

I want to thank Professor Bruce Ankenman of my department at Northwestern for helpful discussions and clarifications about some subtle points. I am also grateful to my following graduate students who helped with collection of data sets for examples and exercises, drawing figures and carrying out many computations: Kunyang Shi, Xin (Cindy) Wang, Jiaxiao Shi, Dingxi Qiu, Lingyun Liu, and Lili Yao. Several generations of students in my DOE classes struggled through early drafts of the manuscript and pointed out many errors, ambiguous explanations, and so on; I thank them all. Any remaining errors are my responsibility.

Finally, I take this opportunity to express my indebtedness to all my teachers — from grade school to grad school — who taught me the value of inquiry and knowledge. This book is dedicated to all of them.

AJIT C. TAMHANE

Department of Industrial Engineering & Management Sciences
Northwestern University, Evanston, IL

Abbreviations

ANCOVA	Analysis of covariance
ANOVA	Analysis of variance
BB	Box–Behnken
BIB	Balanced incomplete block
BLUE	Best linear unbiased estimator
BTIB	Balanced-treatment incomplete block
CC	Central composite
c.d.f.	Cumulative distribution function
CI	Confidence interval
CR	Completely randomized
CWE	Comparisonwise error rate
d.f.	Degrees of freedom
E(MS)	Expected mean square
FCC	Face-centered cube
FDR	False discovery rate
FWE	Familywise error rate
GLS	Generalized least squares
GLSQ	Graeco–Latin square
iff	If and only if
i.i.d.	Independent and identically distributed
IQR	Interquartile range
LFC	Least favorable configuration
LS	Least squares
LSD	Least significant difference
LSQ	Latin square
MANOVA	Multivariate analysis of variance
MCP	Multiple comparison procedure
ML	Maximum likelihood
MOLSQ	Mutually orthogonal Latin square
MS	Mean square
MVN	Multivariate normal
n.c.p.	Noncentrality parameter

OA	Orthogonal array
OC	Operating characteristic
OME	Orthogonal main effect
PB	Plackett–Burman
P(CS)	Probability of correct selection
p.d.f	Probability distribution function
PI	Prediction interval
PSE	Pseudo standard error
QQ	Quantile–quantile
RB	Randomized block
REML	Restricted maximum likelihood
RM	Repeated measures
R&R	Reproducibility and repeatability
RSM	Response surface methodology
RSP	Ranking and selection procedure
r.v.	Random variable
SCC	Simultaneous confidence coefficient
SCI	Simultaneous confidence interval
SD	Standard deviation
SE	Standard error
SPC	Statistical process control
SS	Sum of squares
s.t.	Such that
STP	Simultaneous test procedure
UI	Union–intersection
WLS	Weighted least squares
w.r.t.	With respect to
YSQ	Youden square

CHAPTER 1

Introduction

Humans have always been curious about nature. Since prehistoric times, they have tried to understand how the universe around them operates. Their curiosity and ingenuity have led to innumerable scientific discoveries that have fundamentally changed our lives for the better. This progress has been achieved primarily through careful observation and experimentation. Even in cases of serendipity, for example, Alexander Fleming's discovery of penicillin when a petri dish in which he was growing cultures of bacteria had a clear area (because the bacteria were killed) where a bit of mold had accidentally fallen (Roberts, 1989, pp. 160–161) or Charles Goodyear's discovery of vulcanization of rubber when he inadvertently allowed a mixture of rubber and sulfur to touch a hot stove (Roberts, 1989, p. 53), experimental confirmation of a discovery is a must. This book is about how to design experiments and analyze the data obtained from them to draw useful conclusions. In this chapter we introduce the basic terminology and concepts of experimentation.

The outline of the chapter is as follows. Section 1.1 contrasts observational studies with experimental studies. Section 1.2 gives a brief history of the subject. Section 1.3 defines the basic terminology and concepts followed by a discussion of principles in Section 1.4. Section 1.5 gives a summary of the chapter.

1.1 OBSERVATIONAL STUDIES AND EXPERIMENTS

Observational studies and experiments are the two primary methods of scientific inquiry. In an **observational study** the researcher is a passive observer who records variables of interest (often categorized as **independent/explanatory variables** or **factors** and **dependent/response variables**) and draws conclusions about associations between them. In an **experiment** the researcher actively manipulates the factors and evaluates their effects on the response variables.

Statistical Analysis of Designed Experiments: Theory and Applications By Ajit C. Tamhane

For example, an observational study may find that people who exercise regularly live healthier lives. But is it the exercise that makes people healthy or is it something else that makes people exercise regularly and also makes them healthy? After all, there are many other variables such as diet, sleep, and use of medication that can affect a person's health. People who exercise regularly are likely to be more disciplined in their dietary and sleep habits and hence may be healthy. These variables are not controlled in an observational study and hence may **confound** the outcome. Only a controlled experiment in which people are randomly assigned to different exercise regimens can establish the effect of exercise on health.

An observational study can only show *association*, not *causation*, between the factors of interest (referred to as **treatment factors**) and the response variable. This is because of possible confounding caused by all other factors that are not controlled (referred to as **noise factors**) and are often not even recognized to be important to be observed (hence referred to as **lurking variables**). Any conclusion about cause–effect relationships is further complicated by the fact that some noise factors may affect not only the response variable but also the treatment factors. For example, lack of good diet or sleep may cause a person to get tired quickly and hence not exercise.

Epidemiological studies are an important class of observational studies. In these studies the suspected **risk factors** of a disease are the treatment factors, and the objective is to find out whether they are associated with the disease. These studies are of two types. In **prospective studies**, subjects with and without risk factors are followed forward in time and their disease outcome (yes or no) is recorded. In **retrospective studies** (also called **case–control studies**), subjects with and without disease are followed backward in time and their exposure to suspected risk factors (yes or no) is recorded. Retrospective studies are practically easier, but their results are more likely to be invalidated or at least more open to question because of uncontrolled lurking variables. This is also a problem in prospective studies, but to a lesser extent. For instance, if a study establishes association between obesity and hypertension, one could argue that both may be caused by a common gene rather than obesity causing hypertension. This general phenomenon of a lurking variable influencing both the predictor variable and the response variable is depicted diagrammatically in Figure 1.1. An even more perplexing possibility is that the roles of "cause" and "effect" may be reversed. For example, a person may choose not to exercise because of poor health.

On the other hand, an experiment can establish causation, that is, a cause–effect relationship between the treatment factors that are actively changed and the response variable. This is because the treatment factors are controlled by the investigator and so cannot be affected by uncontrolled and possibly unobserved noise factors. Furthermore, selected noise factors may be controlled for experimental purposes to remove their confounding effects, and the effects of the others can be averaged out using randomization; see Section 1.4.

In addition to establishing causation, another advantage of experimentation is that by active intervention in the causal system we can try to improve its

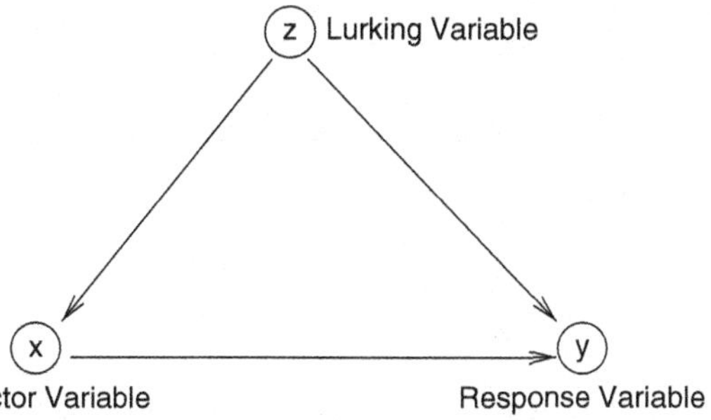

Figure 1.1 Lurking variable z influences both predictor variable x and response variable y.

performance rather than wait for serendipity to act. Even if an improvement occurs due to serendipity, we are left to guess as to which input variables actually caused the improvement.

The general goal of any experiment is knowledge and discovery about the phenomenon under study. By knowledge we mean a better understanding of the phenomenon; for example, which are the key factors that affect the outcomes of the phenomenon and how. This knowledge can then be used to discover how to make improvements by tuning the key design factors. This process is often iterative or sequential since, as our knowledge base expands, we can make additional adjustments and improvements. The sequential nature of experimentation is discussed in Section 1.4.2.

Some specific goals of an experiment include the following:

(a) Screen the treatment factors to identify the important ones.
(b) Determine the factor space, that is, the ranges of values of the treatment factors (current or new ones suggested by previous experiments), for follow-up experiments.
(c) Select the best combination of the treatment factor settings to optimize the response.
(d) Select the best combination of treatment factor settings to make the response robust (i.e., insensitive) to variations in noise factors.
(e) Fit a model that can be used to make predictions and/or to adjust the treatment factors so as to keep the response on target.
(f) Determine or expand the scope of applicability of the treatment factors and the predictive model based on them.

Statistics plays a crucial role in the design and analysis of experiments and of observational studies. The design of an experiment involves many practical considerations. Statistics is especially useful in determining the appropriate combinations of factor settings and the necessary sample sizes. This book focuses mainly on the statistical analyses of data collected from designed experiments. Often the same methods of data analysis are used for observational studies and experiments, but as explained above, stronger conclusions are possible from experiments.

1.2 BRIEF HISTORICAL REMARKS

The field of statistical design and analysis of experiments was founded by Sir Ronald A. Fisher (1890–1962) in the 1920s and 1930s while he was working at the Rothamsted Agricultural Experimental Station in England. Fisher was an intellectual giant who made seminal contributions to statistics and genetics. In design of experiments he invented many important basic ideas (e.g., randomization), experimental designs (e.g., Latin squares), and methods of analysis (e.g., analysis of variance) and wrote the first book on the subject (Fisher 1935). Figure 1.2 shows a picture of Fisher in his younger days taken from his excellent biography by his daughter, Joan Fisher-Box (1978). Fisher was followed in his position at Rothamsted by Sir Frank Yates (1902–1994), who proposed novel block designs and factorial designs and their methods of analysis.

In the 1940s and 1950s, Sir George Box, while working at the Imperial Chemical Industry, developed response surface methodology (Box and Wilson, 1953) as a statistical method for process optimization. There are some crucial differences

Figure 1.2 Sir Ronald A. Fisher in 1924. (*Source*: Fisher-Box, 1978, Plate 4).

between agricultural experimentation, the original setting of the subject, and industrial experimentation, the setting in which Box and his co-workers extended the subject in new directions:

(a) Agricultural experiments can be performed only once or twice a year, and data do not become available until the growing and harvesting seasons are over. Industrial experiments, on the other hand, are usually much shorter in duration and data often become immediately available. Therefore agricultural experiments tend to be a few in number but large in size, while several small and sequential experiments are feasible (and preferable) in industrial settings.
(b) Many industrial experiments are performed online and hence are likely to disrupt an ongoing production process. Therefore it is preferable to conduct them sequentially with several small experiments rather than one large experiment.
(c) In agricultural experiments the focus is on comparisons between crop varieties or fertilizers. As a result, analysis of variance techniques with the associated significance tests of equality of means are common. On the other hand, in industrial experiments the focus is on process modeling, optimization, and quality improvement.

In the 1950s, a mathematical theory of construction of experimental designs based on combinatorial analysis and group theory was developed by Raj Chandra Bose (1901–1987) and others. Later a theory of optimal designs was proposed by Jack Kiefer (1923–1980).

Around the same time, A. Bradford Hill (1897–1991) promoted randomized assignments of patients in clinical trials. Psychology, education, marketing, and other disciplines also witnessed applications of designed experiments. A random assignment of human subjects is not always ethical and sometimes not even practical in social and medical experiments. This led to the development of quasi-experiments in the fields of psychology and education by Donald Campbell (1916–1996) and Julian Stanley.

The most recent infusion of new ideas in design of experiments came from engineering applications, in particular designing quality into manufactured products. The person primarily responsible for this renaissance is the Japanese engineer Genichi Taguchi, who proposed that a product or a process should be designed so that its performance is insensitive to factors that are not easily controlled, such as variations in manufacturing conditions or field operating conditions. The resulting methodology of planning and analysis of experiments is called **robust design**.

1.3 BASIC TERMINOLOGY AND CONCEPTS OF EXPERIMENTATION

In designed experiments the factors whose effects on the response variable are of primary interest are referred to as treatment factors or **design factors**. The different settings of a treatment factor are called its **levels**. Because the experimenter can set the levels of the treatment factors, they are said to be **controllable factors**. In the health–exercise example, exercise (yes or no) is the treatment factor, whose effect on the subjects' health is evaluated by comparing a group that follows a prescribed exercise regimen with another group that does not exercise. The other factors that may also possibly affect the response variable can be broadly divided into two categories: noise factors and **blocking factors**. These are discussed in more detail later.

In this book we restrict discussion to a single response variable but possibly multiple treatment factors. A **qualitative factor** has categorical (nominal or ordinal) levels, while a **quantitative factor** has numerical levels. For example, the type of a drug (e.g., three analgesics: aspirin, tylenol, and ibuprofen) is a qualitative factor, while the dose of a drug is a quantitative factor. A particular combination of factor levels is called a **treatment combination** or simply a **treatment**. (If there is a single factor, then its levels are the treatments.)

The treatments are generally applied to physical entities (e.g., subjects, items, animals, plots of land) whose responses are then observed. An entity receiving an *independent* application of a treatment is called an **experimental unit**. An experimental **run** is the process of "applying" a particular treatment combination to an experimental unit and recording its response. A **replicate** is an independent run carried out on a different experimental unit under the same conditions. The importance of independent application of a treatment is worth emphasizing for estimation of **replication error** (see the next section for a discussion of different errors). If an experimental unit is further subdivided into smaller units on which measurements are made, then they do not constitute replicates, and the sample variance among those measurements does not provide an estimate of replication error. As an example, if a batch of cookie dough is made according to a certain recipe (treatment) from which many cookies are made and are scored by tasters, then the batch would be an experimental unit—not the cookies. To obtain another replicate, another batch of dough must be prepared following the same recipe.

A **repeat measurement** is another measurement of the same response of a given experimental unit; it is not an independent replicate. Taste scores on cookies made from the same batch and assigned by the same taster can be viewed as repeat measurements assuming that the cookies are fairly homogeneous and the only variation is caused by variation in the taster's perception of the taste. Sample variance among repeat measurements estimates **measurement error**—not the replication error that is needed to compare the differences between different recipes for dough. The measurement error is generally smaller than the replication error (as can be seen from the cookie example). If it is incorrectly used to compare recipes, then it may falsely find nonexisting or negligible differences between recipes as significant.

All experimental units receiving the same treatment form a **treatment group**. Often, an experiment includes a **standard** or a **control** treatment, which is used as a benchmark for comparison with other, so-called **test treatments**. For example, in a clinical trial a new therapy is compared to a standard therapy (called an **active control**) if one exists or a therapy that contains no medically active ingredient, called a **placebo** or a **passive control** (e.g., the proverbial "sugar pill"). All experimental units receiving a control treatment form a **control group**, which forms a basis for comparison for the treatment group.

Let us now turn to noise and blocking factors. These factors differ from the treatment factors in that they represent intrinsic attributes of the experimental units or the conditions of the experiment and are not externally "applied." For example, in the exercise experiment the age of a subject (young or old) may be an important factor, as well as diet, medications, and amount of sleep that a subject gets. The noise factors are not controlled or generally not even measured in observational studies. On the other hand, blocking factors are controlled in an experiment because their effects and especially their interactions with the treatment factors (e.g., consistency or lack thereof of the effects of the treatment factors across different categories of experimental units) are of interest since they determine the scope and robustness of the applicability of the treatments. For example, different varieties of a crop (treatment factor) may be compared in an agricultural experiment across different fields (blocking factor) having different growing conditions (soils, weather, etc.) to see whether there is a universal winner with the highest yield in all growing conditions. In designed experiments some noise factors may be controlled and used as blocking factors mainly for providing uniform conditions for comparing different treatments. This use of blocking to reduce the variation or bias caused by noise factors is discussed in the next section.

Example 1.1 (Heat Treatment of Steel: Treatment and Noise Factors)

A metallurgical engineer designing a heat treatment method wants to study the effects of furnace temperature (high or low) and quench oil bath temperature (high or low) on the surface hardness of steel, which is the response variable. The treatment factors are furnace temperature and quench bath temperature, each with two levels. This gives $2 \times 2 = 4$ treatments. Noise factors include deviations from constant furnace and quench oil bath temperatures, variations between steel samples, and so on.

Suppose that 20 steel samples are available for experimentation. In order to regard them as experimental units, each sample must receive an independent application of furnace heating followed by a quench bath, and the temperatures of each should be independently set in a random order (subject to the condition that all four treatments are replicated five times to have a balanced design). But this may not be feasible in practice. If the engineer can assure us that the furnace and quench bath temperatures are perfectly controllable, then a simpler experiment can be conducted in which 10 samples are heated together in the furnace at one temperature followed by the remaining 10 samples at the other temperature. Each group of 10 samples is then randomly divided into two subgroups of five samples each, which are then quenched at two different temperatures. If replication error is estimated from the samples, it will underestimate the true replication error if the assumption of perfect controllability of furnace and quench bath temperatures is not correct. In this case different methods of analyses are required. ■

As mentioned above, often multiple treatment factors are studied in a single experiment. An experiment in which the factors are simultaneously varied (in a random order) is called a **factorial experiment**. In contrast, in a **one-factor-at-a-time experiment** only one factor is varied at a time, keeping the levels of the other factors fixed. In a **full factorial experiment** all factor-level combinations are studied, while in a **fractional factorial experiment** only a subset of them are studied.

In a factorial experiment each factor can be classified as **fixed** or **random**. The levels of a **fixed factor** are chosen because of specific a priori interest in comparing them. For example, consider a clinical trial to compare three different therapies to treat breast cancer: mastectomy, chemotherapy, and radiation therapy. The therapy is then a fixed factor. The levels of a **random factor** are chosen at random from the population of all levels of that factor. The purpose generally is not to compare the specific levels chosen but rather (i) to estimate the variability of the responses over the population of all levels and (ii) to assess the generalizability of the results to that population. For example, consider an experiment to compare the mean assembly times using two types of fixtures. Suppose three different operators, Tom, Dick, and Harry, are chosen to participate in the experiment. Clearly, the fixture is a fixed factor. The operator would be a fixed factor if Tom, Dick, and Harry are chosen because the experimenter was specifically interested in comparing them or because they are the only operators in the factory. If there are many operators in the factory from whom these three are chosen at random, then the operator would be a random factor. In this latter case, there would be less interest in comparing Tom, Dick, and Harry with each other since they simply *happened* to be chosen. However, the variability among these three can be used to estimate the variability that could be expected across all operators in the factory. In practice, however, comparisons will and should be made between the chosen operators if there are large differences between them to determine the causes for the differences.

The parameters that quantify how the mean response depends on the levels of a factor are called its **effects**. For a fixed factor, the effects are fixed quantities and are called **fixed effects**; for a random factor they are random quantities and are called **random effects**. The corresponding mathematical models for the response variable are called the **fixed-effects model** (also called **model I**) and the **random-effects model** (also called **model II**). In the former case the goal is to estimate the fixed effects while in the latter case the goal is to estimate the *variances* of the random effects. Some experiments have a combination of fixed and random factors. The corresponding mathematical model is then called the **mixed-effects model** (also called **model III**). Chapters 1–10 cover designs which use fixed-effects models. Chapters 11–13 cover designs which use random and mixed-effects models.

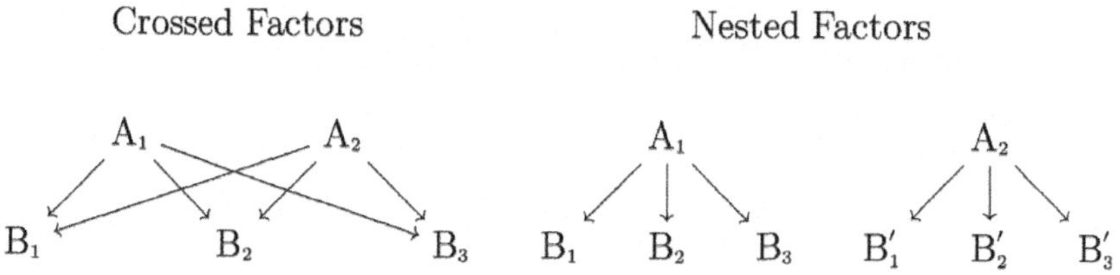

Figure 1.3 Schematic of crossed and nested factors.

Any two factors in a factorial experiment may be **crossed** with each other or one may be **nested** within the other. If the same levels of two factors are used in combination, then the factors are said to be crossed. This is possible if the levels of the factors can be set independently of each other. In the fixture–operator example above, the factors are crossed since each operator works with each fixture. Suppose the fixture experiment is carried out in two manufacturing plants and three operators are chosen from each plant. Thus the operators in plant 1 are not the same as the operators in plant 2. In this case the operators are said to be nested within the plants. More generally, if there are two factors, A and B, and *different* levels of B are observed in combination with each level of A, then B is said to be nested within A. In this case the levels of the factors cannot be set independently of each other because the levels of B are different for each level of A. A schematic of these two factorial designs is shown in Figure 1.3.

1.4 BASIC PRINCIPLES OF EXPERIMENTATION

1.4.1 How to Minimize Biases and Variability?

As noted in the previous section, the effects of treatment factors are of primary interest in an experiment. However, the treatment effects may be distorted by biases or masked by variability in the responses of the experimental units. Therefore biases and variability need to be minimized in order to detect practically important factor effects with high probability. In other words, we need to maximize the **signal-to-noise ratio**. To understand how to do this, let us look at the main components of the "noise":

- **Systematic biases/errors** are caused by systematic differences between the experimental units in different treatment groups. For example, suppose a new teaching method is to be compared with a standard method by offering them in two separate sections of a course. If which section a student registers in is voluntary, then students in the two classes are likely to differ systematically; for example, intellectually more adventurous ones may register in the new teaching method section, thus biasing the results. The noise factors on which the experimental units differ are said to **confound** or **bias** the treatment comparisons.

- **Replication or random errors** are caused by the inherent variability in the responses of similar experimental units given the same treatment. They manifest in the variation between replicate observations.
- **Measurement errors** are caused by imprecise measuring instruments or inspectors. They manifest in the variation between repeat measurements.

Replication and measurement errors are together referred to as **experimental errors**. In subsequent chapters, experimental errors are used synonymously with replication errors as measurement errors are often a much smaller part.

We first discuss how to account for the effects of noise factors in order to reduce systematic biases. As seen in Example 1.1, not all noise factors are specified or even known. Noise factors that are specified and measured may or may not be controllable. Strategies to reduce systematic biases differ depending on which type of noise factor we are dealing with. There are two common strategies that are used to design out the effects of noise factors:

- **Blocking:** Use of blocking factors for evaluating the consistency of the effects of treatment factors has been discussed in the previous section. But a noise factor may also be used as a blocking factor if it is controllable or its value for each experimental unit is known before the experiment begins. In that case, its effect can be separated from the treatment effects by dividing the experimental units into homogeneous groups (i.e., having the same or similar values of the noise factor). These groups are called **blocks**. The treatments are assigned at random to the experimental units within each block; thus the effect of the noise factor is the same across all treatments within each block and cancels out when the responses to the treatments are compared to each other.

  In Example 1.1, suppose that the 20 steel samples come from five different batches of steel, four per batch. The differences between the batches are not of primary interest. Since the batches can be identified before the experiment begins, they can be used as a blocking factor; see Example 1.3.
- **Randomization:** What about other noise factors, whether known or unknown? The experimental units may differ on these factors as well, thus biasing the results. To make the treatment groups equal on every single noise factor is impossible. Fisher's brilliant solution to this problem was to make the treatment groups *probabilistically* equal on *all* noise factors by *randomly assigning* experimental units to the different treatments. Note that randomization does not imply equality of experimental units for all treatments but that no treatment is favored.

  The chance element is introduced intentionally to reduce bias. Another important consequence of random assignment is that it makes statistical inference possible because the statistics calculated from the resulting data have well-defined sampling distributions. If assignment is selective, then such distributions may exist but cannot be derived or simulated in general.

In summary, the basic strategy for dealing with noise factors at the *design stage* is as follows: *Block over those noise factors that can be easily controlled; randomize over the rest*.

Borrowing from the widely publicized quote by former U.S. Secretary of Defense, Donald Rumsfeld, block over known knowns (known noise factors) and randomize over known and unknown unknowns (unknown noise factors).

If the noise factors are observed, then their effects can also be taken into account at the *analysis stage* by using regression methods or by forming the blocks postrandomization and data collection. Normally, only a few noise factors can be used for blocking. Some noise factors cannot be used for blocking because they are not observed before assignment of experimental units to the treatments or their levels cannot be fixed. Such noise factors are called **covariates**. In Example 1.1 percentage of carbon content of steel samples may be regarded as a covariate. A regression analysis that includes effects of both the treatment factors and the covariates is called **analysis of covariance** (see Section 3.8).

The effect of random errors can be reduced by replicating the experimental runs, that is, by making multiple independent runs under identical treatment conditions. The effect of measurement errors can be reduced by making repeat measurements. In both cases some average is used to smooth out the variability in the individual measurements. Another benefit of replicates and repeat measurements is that they can be used to estimate the corresponding error variances.

To summarize, the following four strategies are useful to improve the precision of an experiment:

(a) Blocking
(b) Randomization
(c) Replication
(d) Repeat measurements.

The first two are smart strategies because they do not require larger samples. The last two are costly strategies, but they are necessary if random and measurement errors must be estimated and controlled.

Example 1.2 (Lady Tasting Tea Experiment)

Fisher (1935) illustrated the principles of experimentation by using the following actual episode that took place at the Rothamsted Experimental Station as described in Fisher-Box (1978, p. 134). In fact, the experiment is so well-known in statistical literature that there is a book with the title *The Lady Tasting Tea: How Statistics Revolutionized Science in the Twentieth Century* by Salsburg (2001) and the experiment is described in detail in Chapter 1 of the book.

Dr. B. Muriel Bristol, an algologist, claimed that she could tell by tasting a cup of tea made with milk whether brewed tea infusion or milk was added to the cup first. It was suggested to test her claim. Fisher (1935) formulated this as a hypothesis testing problem with the **null hypothesis** (H_0) being that the lady has no discriminating ability and picks one of the two choices at random, each with probability $\frac{1}{2}$. Rejection of H_0 would be taken as a proof of the lady's claim.

One of the key decisions when planning an experiment such as this is the sample size, that is, how many cups to test. The other decisions include: in what order to offer the cups, whether to block over some factors of interest, how to control for noise factors, and the decision rule for rejecting H_0. We assume that an equal number of cups are made using each method and the lady is told of this fact. To illustrate statistical considerations used to determine the sample size, first suppose that six cups of tea are tested in a random order, three made by adding milk first (method A) and three made by adding tea infusion first (method B). The lady classifies them into two groups (A and B) of three each. There are $\binom{6}{3} = 20$ different ways of classifying the cups. Under H_0, all 20 classifications are equally likely. If correct classification of all six cups is taken as a proof of the lady's discriminating ability, then the **type I error probability**, which is the probability of rejecting a true null hypothesis, is $\frac{1}{20} = 0.05$. This is exactly equal to the commonly used standard, called the **level of significance**, or α, that the type I error probability is required not to exceed. Therefore the experimenter may want to use eight cups. A similar calculation shows that the type I error probability for correct classification of all eight cups is $\frac{1}{70} = 0.014$, which may be deemed to be acceptable. In addition to such statistical considerations, practical considerations (e.g., time and other resources available for the experiment, how many cups can be tested in one sitting, etc.) are also used to determine the sample size.

The above probability calculation would be different if a different design is used, for example, if the A and B cups were offered in pairs for tasting with a random order within each pair. This design improves the precision of the experiment because A and B are directly compared with each other, which allows the lady to better discriminate between them. Assuming that H_0 is true, the probability of classifying all four pairs correctly is $(\frac{1}{2})^4 = \frac{1}{16} = 0.063$. This illustrates how the statistical analysis of an experiment depends on its design.

The conditions for making and tasting the tea cups made by the two methods should be as similar as possible, so that any taste difference can be attributed solely to the method of preparation. If the conditions are not the same, then that could introduce bias; for example, if sugar is added to all the A cups but not to the B cups, then the lady's discriminating ability could be impaired by the sweetness of sugar. Besides introducing bias, this could change the null distribution of the number of correct classifications. In the extreme case, the classification of cups could depend solely on whether sugar is added or not; thus there will not be 20 classifications, but only two (one correct, one wrong) each with probability $\frac{1}{2}$ under H_0.

Some factors are relatively easy to control, for example, the type of cup used and the amount of sugar added to each cup. Other factors are not so easy to control, for example, the temperature of tea and the aftertaste effect. It is difficult, if not impossible, to make the conditions exactly equal for the two methods of tea preparation with respect to all such noise factors. Fisher suggested to resolve this problem by randomizing the order in which the cups are offered to the lady to taste. Note that randomization not only helps to avoid the bias but also makes the null distribution of the test statistic (i.e., all classifications equally likely) valid.

A need for blocking may arise in this experiment if a fully randomized order of tasting is not feasible or desirable for some reason. Offering A and B cups in pairs to improve the precision of the experiment is an example of blocking. Another example would be if it was of interest to find out if the lady's discriminating ability ranges over different brands of tea, for example, Earl Grey and English Breakfast. In that case, the brand would be a blocking factor with an equal number of A and B cups of each brand. ■

We now introduce two basic designs that result from the principles of randomization and blocking. The first is called a **completely randomized (CR) design**, in which *all* experimental units are assigned at random to the treatments without any restriction. This design is generally used when the treatments are applied to a single group of essentially homogeneous experimental units. Any heterogeneity among these units is averaged out by random assignment of treatments to them.

Heterogeneity among experimental units can be more explicitly accounted for by designating selected measurable noise factors as blocking factors and forming blocks of units which are similar with regard to those factors. Treatments are then randomly assigned to the units within each block. Thus randomization is done subject to a blocking restriction. The resulting design is called a **randomized block (RB) design**. This design enables more precise estimates of the treatment differences because comparisons between treatments are made among homogeneous experimental units in each block.

Example 1.3 (Heat Treatment of Steel: CR and RB Designs)

Refer to Example 1.1. Suppose that 20 samples come from five different batches, four per batch. There are four treatment combinations: $A =$ (Heating Temp.: Low, Quench Temp.: Low), $B =$ (Heating Temp.: High, Quench Temp.: Low), $C =$ (Heating Temp.: Low, Quench Temp.: High), $D =$ (Heating Temp.: High, Quench Temp.: High). A CR design consists of randomly assigning the 20 steel samples to A, B, C, D without regard to the batches that the samples came from; see Figure 1.4*a*. It is not required to assign an equal number to each treatment, although randomization is often done under that restriction.

The steel samples from the same batch are likely to be more similar to each other than are samples from different batches. Therefore we can form blocks of four samples from each of five batches and randomly assign them to the four treatments. We then have one replicate per treatment from each batch. The resulting RB design is shown in Figure 1.4*b*. ■

(a) Completely Randomized Design

Batch 1	Batch 2	Batch 3	Batch 4	Batch 5
A	C	D	D	B
B	A	A	C	D
C	D	B	C	A
B	C	D	B	A

(b) Randomized Block Design

Batch 1	Batch 2	Batch 3	Batch 4	Batch 5
A	D	C	D	B
C	A	B	C	C
B	C	D	B	A
D	B	A	A	D

Figure 1.4 Completely randomized and randomized block designs for heat treatment of steel samples.

Example 1.4 (Chemical Process Yield: CR and RB Designs)

The yield of a chemical process is to be compared under three different operating conditions: A, B, and C. Suppose that six runs can be made in a day (three in the morning, three in the afternoon) allowing two runs per treatment. The CR design would randomize the order of the runs so as to neutralize any confounding effect of the time of day. One possible random order is $\{A, C, A, B, B, C\}$. A practical difficulty in using a completely random order is that the operating conditions may not be easy to change. Then it may be necessary to make the two runs for each process condition consecutively; the resulting design is *not* a CR design but a **split-plot design** (discussed in Chapter 13).

Now suppose it is known from past experience that due to temporal trends the morning (AM) runs are different from the afternoon (PM) runs in terms of the yield obtained. Then the CR design given above is not balanced with respect to this noise factor since both A runs are made in the morning while both B runs are made in the afternoon. Therefore the A-versus-B difference is confounded with the AM-versus-PM difference. An RB design that uses the AM and PM runs as blocks provides the necessary balance. An example is $\{\underbrace{B, C, A}_{\text{AM}}, \underbrace{C, B, A}_{\text{PM}}\}$. ■

1.4.2 Sequential Experimentation

Knowledge and discovery represent an iterative *exploratory* process in which learning occurs through induction–deduction cycles; see Box, Hunter, and Hunter (2005, Chapter 1) for a thorough discussion. One-shot experiments are used mainly for *confirmatory* purposes. In the exploratory phase the investigator generally begins with a theory or a hypothesis or a model (either suggested by previously collected data or derived from some basic scientific principles). For example, the theory might state that certain factors have positive effects on the process and certain other factors have negative effects. At the beginning of an investigation, the phenomenon may be very poorly understood with little good-quality data; therefore the theory may be rather fuzzy or almost nonexistent. To arrive at a testable theory, first an exploratory experiment must be conducted. At each later stage of the investigation, the data from the experiment must be checked if they conform with the predictions made by the current theory under test. In rare cases the theory is confirmed and no further experimentation is necessary. More often, the theory is modified or a new theory is proposed because the original theory is refuted. This leads to a new cycle of experiments and a further modification of the theory.

In practice, as one learns from data, not only may the theory be modified but also the experimental conditions may be altered, as the following examples illustrate:

(a) Some of the factors used in earlier experiments may be found unsuitable and other more promising factors may replace them.
(b) The ranges of the factors may be changed because the previously used ranges may not have produced desired improvements or may have produced undesirable side effects.
(c) The response variable may itself be changed or redefined to better capture the outcome that the investigator wants to measure.
(d) The design may itself be changed, for example, from a CR design to an RB design.

There is no general methodology for designing sequential experiments that would deal with all of the above and possibly other changes dictated by data and any external scientific knowledge that becomes available during the course of the study. Methodologies have been developed for some specific applications. Two specific methodologies are sequential assemblies of fractional factorial designs (follow-up designs) to resolve confounding patterns (see Section 8.6) and the other is response surface optimization (see Section 10.1). The rest of this book deals with single experiments which are used as building blocks of sequential experiments.

When conducting experiments to search for the optimum factor-level combination to achieve the best response, the importance of a final **confirmation experiment** cannot be overemphasized. This experiment consists of a few runs to verify that the optimizing factor-level combination indeed meets the design objectives.

1.5 CHAPTER SUMMARY

(a) In an observational study an investigator passively observes potential predictor variables and the response variable. In an experimental study an investigator actively manipulates potential predictor variables to study their effects on the response variable. An observational study can only establish association between the predictor variables and the response variable. A controlled experiment can also establish a cause–effect relationship.
(b) A basic goal in any experiment is to identify the effects of the treatment factors on the response variable unconfounded from those of any noise factors. Blocking is used to evaluate the consistency of the effectiveness

of the treatments across experimental units having wide-ranging characteristics. It is also used to minimize the biasing effects of the noise factors that are easily controllable and on which data are available before experimentation begins. Randomization is used to minimize the biasing effects of all other noise factors. Completely randomized (CR) and randomized block (RB) designs are two basic designs based on these two techniques. The variation caused by random and measurement errors can be minimized through replication and repeat measurements, respectively.

EXERCISES

Section 1.1 (Observational Studies and Experiments)

1.1 Tell in each of the following instances whether the study is experimental or observational:

(a) Two computing algorithms are compared in terms of the CPU times required to do the same six test problems.

(b) A psychologist measures the response times of subjects under two stimuli; each subject is observed under both stimuli in a random order.

(c) A group of smokers and a group of nonsmokers are followed for 10 years and the numbers of subjects getting cancer over this period are noted.

(d) An advertising agency has come up with two different TV commercials for a household detergent. To compare which one is more effective, a pretest is conducted in which a sample of 50 adults is randomly divided into two groups. Each group is shown a different commercial, and the people in the group are asked to score the commercial.

(e) Two different school districts are compared in terms of the scores of the students on a standardized test.

Section 1.3 (Basic Terminology and Concepts of Experimentation)

1.2 A drilling experiment was conducted to study the effects of the bit size (0.25 and 0.5 in.), rotational speed (1000 and 2000 rpm), and feed rate (0.1 and 0.2 in./sec) on the surface finish of drilled holes. All combinations of these factors were run with three holes drilled in each case by the same operator in pieces of aluminum cut from the same stock. The coolant temperature was not controlled because its effect was not of interest and is difficult to control in practice.

(a) Which are the treatment and noise factors?

(b) How many levels of each treatment factor are studied? How many treatment combinations are studied?

(c) What are the experimental units? How many replicates are observed per treatment?
(d) Are the treatment factors crossed or nested?

1.3 A bottling plant has four beverage fillers, each with eight filling heads. To see what part of the variation in the filled amounts is due to the differences between the fillers and between the filling heads (with the rest being random variation), a random sample of five bottles is taken from two randomly selected filling heads of each filler (for a total of 8 filling heads and 40 bottles) and their contents are measured.

(a) Which are the treatment factors? Are they fixed or random? Are they crossed or nested?
(b) What are the experimental units? How many replicates are observed per treatment?

1.4 For a statistics project a student compared three methods of ripening raw bananas at home. The methods were: keep bananas in (i) open air, (ii) a brown paper bag, (iii) a basket covered with apples. A dozen bananas were ripened using these methods (four bananas per method) for one week, at the end of which the amount of ripening was measured by counting the number of black dots on the banana skins.

(a) Identify the treatments. Is there a control treatment?
(b) What are some of the noise factors in this experiment?
(c) What are the experimental units? How many replicates are observed per treatment?
(d) What is the response variable?

Section 1.4 (Basic Principles of Experimentation)

1.5 Refer to Exercise 1.2. Suppose that the drill operator randomized the sequence of the eight settings of the drill bit size, speed, and feed rate but made all three holes for each setting one after the other because it was too much work to change the settings after drilling each hole. Why is this a violation of the randomization principle? What systematic errors could this introduce? How should the randomization principle be correctly applied?

1.6 Refer to Exercise 1.4. What are the possible sources of systematic and measurement errors? What precautions would you take to minimize them?

1.7 An experiment is to be conducted to study the effects of having sugar (in the form of a donut) and caffeine (in the form of coffee) for breakfast on the alertness of students in a morning class. Sixteen students are selected to participate in the experiment. Alertness will be measured by reaction times to a certain standardized stimulus. Outline how you would conduct a completely randomized experiment. Specify the treatment factors, treatments, experimental units, three possible noise factors, and replicates.

1.8 Refer to the previous exercise. The sixteen students are recruited so that there are four from each class: freshman, sophomore, junior, and senior. Also, there are two men and two women from each class. Which of these factors (class or gender) would be more appropriate for blocking? Outline a randomized block design using the chosen blocking factor.

1.9 (From Mason, Gunst, and Hess 2003, Chapter 4, Exercise 1. Reprined by permission of John Wiley & Sons, Inc.) A test program was conducted to evaluate how the quality of epoxy–glass–fiber pipes is affected by operating condition (normal vs. severe) and water temperature (150 vs. 175°F). The test program required 16 pipes, half of which were manufactured at each of two manufacturing plants. Table 1.1 lists the test conditions that constituted the experimental protocol. Identify which of the following features are included in this design and what they are:

Table 1.1 Experimental Protocol for Testing Epoxy–Glass–Fiber Pipes

Run No.	Plant	Operating Condition	Water Temperature (°F)
1	1	Normal	175
2	1	Normal	150
3	1	Severe	150
4	2	Severe	175
5	1	Normal	175
6	1	Normal	150
7	2	Normal	150
8	1	Severe	175
9	2	Normal	175
10	2	Severe	150
11	2	Normal	150
12	1	Severe	175
13	2	Severe	175
14	2	Severe	150
15	1	Severe	150
16	2	Normal	175

Source: Mason, Gunst, and Hess (2003, p. 133). Reprinted by permission of John Wiley & Sons, Inc.

(a) treatment factors, (b) treatments, (c) blocking factors, (d) experimental units, (e) replications, (f) randomization.

1.10 Consider the lady tasting tea experiment.

(a) Calculate the type I error probability if eight cups of tea are tested but one misclassification is allowed in the decision rule, that is, H_0 is rejected if there is more than one misclassification.

(b) What is the minimum number of cups that must be tested if the type I error probability for this decision rule must not exceed $\alpha = 0.05$?

CHAPTER 2

Review of Elementary Statistics

In this chapter we review some elementary normal theory statistical methods that are useful in the remainder of the book. Section 2.1 studies the one-sample problem, which deals with the evaluation of the performance of a single treatment. Experiments for estimating the parameters of a single treatment or testing them against prescribed standards are known as **noncomparative experiments** since they do not involve comparisons between several treatments. Section 2.2 studies the two-sample problem, which involves comparisons between two treatments using **comparative experiments**. Finally, Section 2.3 discusses the topic of fitting equations to data, in particular, the simple and multiple linear regression problems. Since this is a review chapter, the coverage of the topics is necessarily succinct; we mainly emphasize results that are used in later chapters.

2.1 EXPERIMENTS FOR A SINGLE TREATMENT

The simplest type of an experiment (or an observational study) involves a single treatment (or a group). Such an experiment is generally noncomparative because the goal is to evaluate the performance of the treatment and not compare it with another treatment (except possibly with a known standard or a target). Such experiments are used to estimate the mean performance of a process or a system, for example, by sampling from a manufacturing process to estimate the fraction defective or by doing a computer simulation experiment to estimate the mean waiting time in a queuing network. Since only a single treatment is under study, no randomization of experimental units is involved. However, to avoid biases, they should be randomly selected from the population of interest. For example, a random sample of items produced by a manufacturing process should be taken to estimate the fraction defectives produced over a specified time period.

2.1.1 Summary Statistics and Graphical Plots

Denote the data on a treatment by $y_1, y_2, \ldots, y_n$. The commonly used summary statistics for the center and spread of the data are, respectively, the sample mean and sample standard deviation (SD):

$$\bar{y} = \frac{1}{n}\sum_{i=1}^{n} y_i \qquad \text{and} \qquad s = \sqrt{\frac{1}{n-1}\sum_{i=1}^{n}(y_i - \bar{y})^2}.$$

These measures of the center and spread of the data are appropriate for symmetrically distributed data. Also, these measures are quite sensitive to extreme observations or **outliers**. The sample median (the second quartile, Q_2, also denoted by $\tilde{y}$) is better suited as a measure of center for skewed data; it is also resistant to outliers. The corresponding measure of spread is the interquartile range (IQR), which is the difference between the third quartile (Q_3) and the first quartile (Q_1). The three quartiles together with the minimum and maximum data values, $\{y_{\min}, Q_1, Q_2, Q_3, y_{\max}\}$, is called the **five-number summary** of the data.

A **box plot** is often used to graphically represent the center and the variation of the data around the center. The interquartile range from Q_1 to Q_3 is displayed by a box with a line drawn at the sample median. Two lines, called **whiskers**, are drawn from each end of the box up to the extreme observations that are within the so-called **fences**, which are defined as

$$\text{Lower fence} = Q_1 - 1.5 \times \text{IQR},$$
$$\text{Upper fence} = Q_3 + 1.5 \times \text{IQR}.$$

Any data points falling outside the fences are regarded as outliers and are denoted by asterisks in the box plot. Example 2.1 gives an illustration.

Standard normal theory methods for formal inferences assume that $y_1, y_2, \ldots, y_n$ are independent and identically distributed (i.i.d.) observations from a $N(\mu, \sigma^2)$ population. This assumption can be checked by making a normal plot or by applying one of the goodness-of-fit tests for normality.

A **normal plot** is a plot of the percentiles of the standard normal distribution versus the corresponding sample percentiles [for more details, see Tamhane and Dunlop (2000), pp. 123–125]. Often, the standard normal quantile axis is marked in terms of the respective normal distribution probabilities and the plot is called a **normal probability plot**. If the data follow a normal distribution, then the plot will be roughly linear for a sufficiently large sample size.

Among the class of goodness-of-fit tests based on comparing the empirical (sample) distribution function of the standardized data values with the distribution function of the standard normal distribution (see D'Agostino and Stephens 1986, Chapter 4), we focus on the **Anderson–Darling test**. We first calculate standardized scores $z_i = (y_i - \bar{y})/s$ for the data and order them $z_{(1)} \le z_{(2)} \le \cdots \le z_{(n)}$.

Table 2.1 Selected Critical Values of Anderson–Darling Statistic

α	0.50	0.25	0.15	0.10	0.05	0.025	0.01	0.005
A^2	0.341	0.470	0.561	0.631	0.752	0.873	1.035	1.159

Next we calculate $w_{(i)} = \Phi(z_{(i)})$ $(1 \le i \le n)$, where $\Phi(\cdot)$ is the standard normal cumulative distribution function (c.d.f.). Table C.1 in Appendix C gives the values of $\Phi(z)$ for $-3.49 \le z \le 3.49$. Then the Anderson–Darling statistic equals

$$A^2 = \left(1 + \frac{0.75}{n} + \frac{2.25}{n^2}\right)\left[-n - \frac{1}{n}\sum_{i=1}^{n}(2i-1)\ln w_{(i)} + \ln\{1 - w_{(n+1-i)}\}\right],$$

where the factor in the front is a correction factor. The statistic A^2 gets larger as the normal plot deviates more and more from a straight line; thus large values of A^2 are significant. A few selected critical values are listed in Table 2.1.

It is important to note that the statistics and graphical techniques discussed above do not provide useful summaries if there are time trends in the data. For instance, if there is an increasing or cyclic trend, then computing the mean or median and making a box plot or a histogram tend to mask underlying temporal patterns. A useful graphical technique in this case is a **run chart**. We do not focus on time-series data in this book.

Example 2.1 (Roof Shingle Breaking Strength: Test for Normality)

As part of a gage R&R study (see Example 11.3 in Section 11.2.1) breaking strength data were collected on roof shingles. A force (measured in grams) was applied to a piece of shingle in a testing machine until the piece broke. Since shingles are brittle, the measured breaking strengths are highly variable. Therefore the average of five pieces from a shingle was used as a measure of the breaking strength of that shingle. Data for 20 shingles are shown in Table 2.2. (The full data set is given in Exercise 11.11.) It is desired to determine if the lot from which this sample of 20 shingles is taken meets the specification that the mean breaking strength is at least 1700 grams (see Example 2.2 for this test).

A quick eyeballing of the data indicates that earlier shingles have generally lower breaking strengths than the later shingles. Figure 2.1 shows a run chart of the data which confirms this observation. However, we have no information on the sequence in which samples were taken, so we will simply note this trend (which should be further investigated in a real application).

The mean and standard deviation (SD) of the data are $\bar{y} = 1886.4$ grams and $s = 399.6$ grams. The five-number summary is

$$y_{\min} = 1084.9, \quad Q_1 = 1680.2, \quad Q_2 = 1825.1, \quad Q_3 = 2183.4,$$
$$y_{\max} = 2553.9.$$

Table 2.2 Mean Breaking Strengths of Roof Shingles (grams)

No.	Breaking Strength	No.	Breaking Strength
1	1660.0	11	1827.9
2	1312.0	12	1738.9
3	1313.6	13	1818.7
4	1084.9	14	1822.2
5	1491.9	15	2198.8
6	1751.5	16	2349.5
7	2095.8	17	2126.4
8	2137.3	18	2523.9
9	1831.3	19	2305.0
10	1813.9	20	2494.2

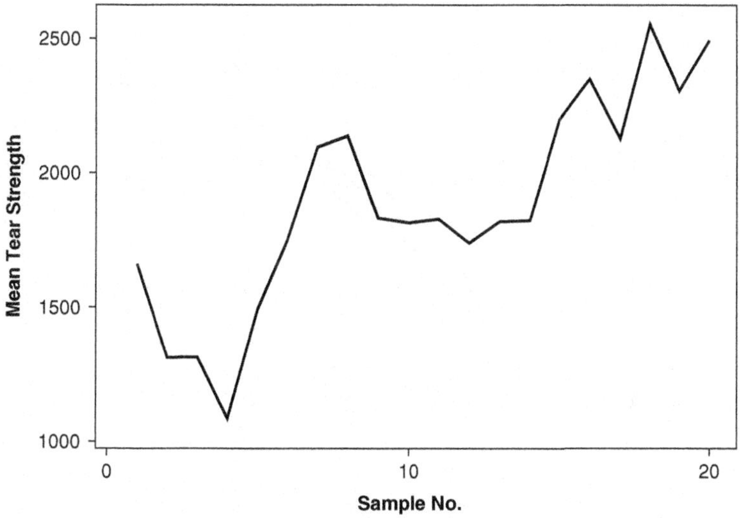

Figure 2.1 Run chart of shingle breaking strength data.

A higher mean than the median and a greater separation between Q_3 and Q_2 than between Q_1 and Q_2 suggest a right-skewed distribution. These observations are supported by the box plot shown in Figure 2.2. The right skew in the data can be seen from the asymmetric shape of the box and the whiskers. The box plot does not show any outliers. We can check this independently by computing the fences. First compute

$$\text{IQR} = Q_3 - Q_1 = 2183.4 - 1680.2 = 703.2.$$

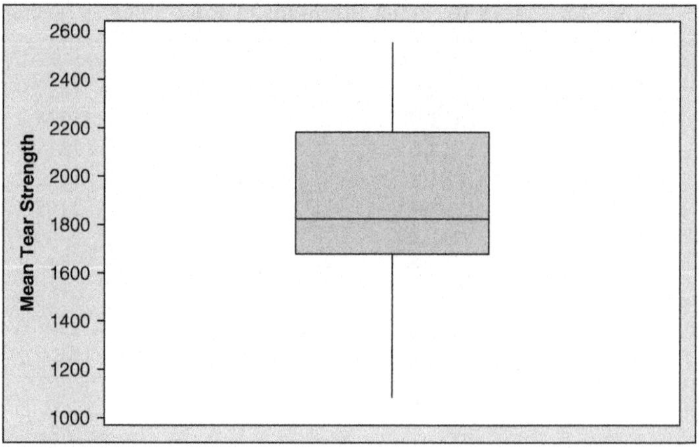

Figure 2.2 Box plot of shingle mean breaking strength data.

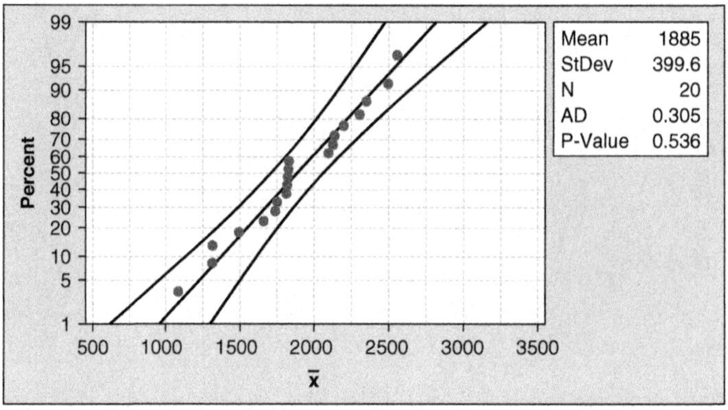

Figure 2.3 Normal plot of shingle breaking strength data.

Then the fences are

$$\text{Lower fence} = 1680.2 - 1.5 \times 703.2 = 625.4,$$

$$\text{Upper fence} = 2183.4 + 1.5 \times 703.2 = 3238.2.$$

None of the observations falls outside these fences.

We next check the normality of the data by making a normal plot, shown in Figure 2.3. Although the plot is not very linear, all the points fall within the 95% confidence band shown, indicating that the normality assumption is acceptable. This is confirmed by the Anderson–Darling test with $A^2 = 0.305$ ($p = 0.536$). ■

2.1.2 Confidence Intervals and Hypothesis Tests

When data plots and goodness-of-fit tests support the normality assumption [i.e., the data can be regarded as a random sample from a $N(\mu, \sigma^2)$ distribution], further analyses can be conducted on μ and σ^2, for example, testing the agreement of the data with specified values of these parameters or confidence intervals on them. For the data sampled from a normal distribution, they are based on the result that $\overline{y}$ and s^2 are distributed (denoted by "$\sim$") independently of each other as

$$\overline{y} \sim N\left(\mu, \frac{\sigma^2}{n}\right), \qquad s^2 \sim \frac{\sigma^2 \chi^2_{n-1}}{n-1}. \tag{2.1}$$

Here χ^2_{n-1} denotes the **chi-square distribution** with $n-1$ **degrees of freedom (d.f.)**. The sample estimate $s/\sqrt{n}$ of $\text{SD}(\overline{y}) = \sigma/\sqrt{n}$ is called the **estimated standard error** of $\overline{y}$; we will simply refer to it as the **standard error (SE)** and write $\text{SE}(\overline{y}) = s/\sqrt{n}$. More generally, we will use this notation for the estimated standard deviation of any statistic.

From (2.1) it follows that

$$t = \frac{\overline{y} - \mu}{s/\sqrt{n}} \sim t_{n-1} \qquad \text{and} \qquad \chi^2 = \frac{(n-1)s^2}{\sigma^2} \sim \chi^2_{n-1},$$

where t_{n-1} denotes the t-**distribution** with $n-1$ d.f. The above random variables (r.v.'s) are known as the pivotal quantities for μ and σ^2, respectively. A **pivotal quantity** is a function of the data and the unknown parameter for which inference is desired but of no other unknown parameters and whose distribution is free of all unknown parameters. For example, t is a function of the data through $\overline{y}$ and s and of the unknown parameter μ of interest but not of the unknown parameter σ. Furthermore, the distribution of t is free of both μ and σ.

For specified α, let $t_{\nu,\alpha/2}$ denote the upper $\alpha/2$ **critical point** of the t_ν-distribution, that is, the abscissa point that cuts off an area $\alpha/2$ in the upper tail of the distribution; these critical points are tabulated in Table C.2. A two-sided $100(1-\alpha)\%$ **confidence interval (CI)** for the unknown mean μ is given by

$$\mu \in \left[\overline{y} \pm t_{\nu,\alpha/2} \frac{s}{\sqrt{n}}\right]. \tag{2.2}$$

Let μ_0 denote a specified standard or target. A **test of hypothesis** of

$$H_0 : \mu = \mu_0 \quad \text{vs.} \quad H_1 : \mu \neq \mu_0$$

with **level of significance** α rejects H_0 if μ_0 is not included in the above 100 $(1-\alpha)\%$ CI for μ, which is equivalent to rejecting H_0 if

$$|t| = \frac{|\overline{y} - \mu_0|}{s/\sqrt{n}} > t_{n-1,\alpha/2}. \tag{2.3}$$

The **type I error probability**, which is the probability of rejecting H_0 when it is true, of this test is controlled at level α. One-sided CIs and tests are obtained by the usual modifications in the above formulas and replacing the critical point $t_{\nu,\alpha/2}$ by $t_{\nu,\alpha}$.

The above two-sided test can be also applied by computing the p-**value**, which is twice the area in the upper tail of the t_{n-1}-distribution beyond the observed $|t|$-statistic. Then we can reject H_0 at level α if $p < \alpha$. The p-value can be interpreted as the smallest α at which H_0 can be rejected. Thus a smaller p-value represents a more significant result in terms of rejecting H_0.

Since the sample mean obeys the **central limit theorem** (see Tamhane and Dunlop 2000, p. 170) under fairly mild regularity conditions, the procedures based on the t-distribution are **robust** against nonnormality. However, they are not **resistant** to outliers because the sample mean is sensitive to outliers.

A $100(1-\alpha)\%$ CI for σ^2 is given by

$$\sigma^2 \in \left[\frac{(n-1)s^2}{\chi^2_{n-1,\alpha/2}}, \frac{(n-1)s^2}{\chi^2_{n-1,1-\alpha/2}}\right], \tag{2.4}$$

where $\chi^2_{n-1,\alpha/2}$ is the upper $\alpha/2$ critical point and $\chi^2_{n-1,1-\alpha/2}$ is the lower $\alpha/2$ critical point of the χ^2_{n-1}-distribution. These critical points are tabulated in Table C.3. An α-level test of the null hypothesis $H_0 : \sigma^2 = \sigma_0^2$, where σ_0^2 is specified rejects if

$$\chi^2 = \frac{(n-1)s^2}{\sigma_0^2} > \chi^2_{n-1,\alpha/2} \quad \text{or} \quad < \chi^2_{n-1,1-\alpha/2}. \tag{2.5}$$

One-sided intervals and tests use the appropriate (upper or lower) α critical point of the χ^2_{n-1}-distribution. The two-sided p-value equals 2 times the minimum of the upper and lower tail areas of the χ^2_{n-1}-distribution from the observed value of the χ^2-statistic. These tests and CIs for σ^2 are neither robust to nonnormality nor resistant to outliers.

Example 2.2 (Roof Shingle Breaking Strength: Confidence Interval and Hypothesis Test)

Refer to Example 2.1 and the data in Table 2.2. We want to test if these data support the hypothesis that the mean breaking strength is at least 1700 grams. If μ denotes the mean breaking strength, then we need to test one-sided hypotheses, $H_0 : \mu \leq 1700$ vs. $H_1 : \mu > 1700$. The sample mean of the data is $\bar{y} = 1886.4$ and the sample SD is $s = 399.6$. Therefore the test statistic equals

$$t = \frac{(1886.4 - 1700)\sqrt{20}}{399.6} = 2.09,$$

which has 19 d.f. If using $\alpha = 0.05$, the critical value is $t_{19,.05} = 1.729$. We can thus reject H_0 at the 0.05 level and conclude that the mean breaking strength for the lot meets the minimum specification of 1700 grams; the p-value can be computed to be 0.025.

It is a common practice to give a two-sided CI for μ. For a 95% CI we need the critical point $t_{19,.025}$, which equals 2.093. Therefore the 95% CI equals

$$\left[1886.4 \pm (2.093)\frac{399.6}{\sqrt{20}}\right] = [1699.4, 2073.4].$$

Thus the specification limit of 1700 grams is just outside the boundary of the CI. In other words, a two-sided test of $H_0 : \mu = 1700$ would barely fail to reject at the 0.05 level.

Next we calculate a two-sided 95% CI for the lot standard deviation σ. We will use the upper confidence limit as a "conservative" prior estimate of σ to calculate the sample size in Example 2.3. We need the critical points $\chi^2_{19,.975} = 8.906$ and $\chi^2_{19,.025} = 32.852$. Using (2.4) we obtain the desired CI as

$$\left[\sqrt{\frac{18}{32.852}}(399.6), \sqrt{\frac{18}{8.906}}(399.6)\right] = [295.79, 568.09]. \qquad \blacksquare$$

2.1.3 Power and Sample Size Calculation

An important part of designing an experiment is the determination of the sample size and evaluation of its **power**, which is the probability of rejecting H_0 when H_1 is true. To calculate the necessary sample size, we first formulate the appropriate hypotheses and then fix the level of significance α and the power $1 - \beta$ for detecting a specified practically significant deviation $\delta > 0$ from the null hypothesis. Note that β is the **type II error probability**. We shall illustrate this calculation for the one-sided test of $H_0 : \mu \le \mu_0$ vs. $H_1 : \mu > \mu_0$ for the case of assumed *known* σ^2. (The case of *unknown* σ^2 involves the noncentral t-distribution; see Example 14.6 for this calculation.) In this case, an α-level test of H_0 rejects if

$$z = \frac{(\bar{y} - \mu_0)\sqrt{n}}{\sigma} > z_\alpha,$$

where z_α is the upper α critical point of the standard normal distribution. The power function of this test is given by

$$\pi(\mu) = \Phi\left[-z_\alpha + \frac{(\mu - \mu_0)\sqrt{n}}{\sigma}\right]. \tag{2.6}$$

Equating

$$\pi(\mu_0 + \delta) = \Phi\left[-z_\alpha + \frac{\delta\sqrt{n}}{\sigma}\right] = 1 - \beta = \Phi(z_\beta)$$

and solving for n, we get

$$n = \left[\frac{(z_\alpha + z_\beta)\sigma}{\delta}\right]^2. \tag{2.7}$$

For a two-sided test, z_α in the above formula should be replaced by $z_{\alpha/2}$. In practice, a t-test would be used which would require a slightly larger n to guarantee the same power; see Section 14.3.

Example 2.3 (Roof Shingle Breaking Strength: Sample Size Calculation) Consider the one-sided hypothesis testing problem in Example 2.2. Suppose that the manufacturer believes that the shingles exceed the specification limit by at least 200 grams (i.e., $\mu \geq 1900$) and would like to demonstrate this with a 0.05-level one-sided test having 80% power. The question is how many shingles must be sampled from each lot? For design purposes, we take the upper 95% confidence limit of 568.09 on σ calculated in Example 2.2 and round up to 570 as a "conservative" *known* (from a pilot study of 20 shingles) prior estimate of σ. Then substituting $z_{0.05} = 1.645$, $z_{0.20} = 0.8416$, $\sigma = 570$, and $\delta = 200$ in (2.7), we get

$$n = \left[\frac{(1.645 + 0.8416)(570)}{200}\right]^2 = 50.05 \approx 50.$$

Thus at least 50 shingles must be tested from each lot with five measurements on each shingle. ■

2.2 EXPERIMENTS FOR COMPARING TWO TREATMENTS

The CR design for comparing two treatments is usually referred to as the **independent samples design**, so-called because the data from the two treatments are mutually independent. The RB design for comparing two treatments is referred to as the **matched pairs design**, so-called because each block consists of two similar experimental units. If the same experimental unit is used as its own match with the order of application of the treatments being randomized, then the observations under the two treatments are generally correlated.

2.2.1 Independent Samples Design

A convenient graphical technique to compare the locations and dispersions of the data from two samples is the **side-by-side box plot**. Another technique is the **QQ (quantile–quantile) plot** in which the same quantiles of the two samples are plotted against each other. The QQ plot is useful for determining whether observations from one sample *tend* to be larger or smaller than those from the other sample.

Example 2.4 (Cloud-Seeding Experiment: Box Plots and Normal Plots)

An experiment was conducted to evaluate the effect of seeding clouds on rainfall with silver nitrate. This was done by randomly seeding 26 of 52 clouds, the other 26 clouds being left unseeded to serve as controls. There are many questions that one could ask about this experiment. For example: How were the clouds selected to be the experimental units? How was the randomization done? How was seeding done? How was the rainfall measured? These details of the experiment can be found in the original data source, Simpson, Olsen, and Eden (1975), and the references cited therein. We do not discuss these design issues here.

The rainfalls measured in both cases (to the nearest acre-feet) are shown in Table 2.3. The side-by-side box plot for the rainfall data in the upper panel of Figure 2.4 shows that the seeded clouds result in more rainfall than unseeded clouds. The extreme skewness in both data sets can be removed by log transformation as can be seen from the plots in the lower panel. The normal plots in Figure 2.5 show a serious lack of normality as well as outliers in both data sets. The Anderson–Darling statistics for rainfalls for seeded and unseeded clouds are 3.391 ($p < 0.005$) and 3.805 ($p < 0.005$), respectively, confirming the nonnormality seen in the plots. The log transformation helps to fix this problem as well, as can be seen from the normal plots in Figure 2.6. Here the Anderson–Darling statistics for rainfalls for seeded and unseeded clouds are 0.363 ($p = 0.415$) and 0.241 ($p = 0.749$), respectively, confirming the normality seen in the plots. All formal inferences will be made on the log-transformed data. ■

We next discuss formal inference procedures for comparing the means and variances of two independent samples. Denote the data from the two samples by

$$y_{11}, y_{12}, \ldots, y_{1n_1} \quad \text{and} \quad y_{21}, y_{22}, \ldots, y_{2n_2}.$$

Table 2.3 Rainfall from Seeded and Unseeded Clouds (acre-feet)

Seeded	2746	1698	1656	978	703	489	430	334	303	275	275	255	243
	201	199	130	119	118	115	92	41	33	31	18	8	4
Unseeded	1203	830	372	346	321	244	163	148	95	87	81	69	47
	41	37	29	29	26	26	24	23	17	12	5	5	1

Source: Simpson, Olsen and Eden (1975).

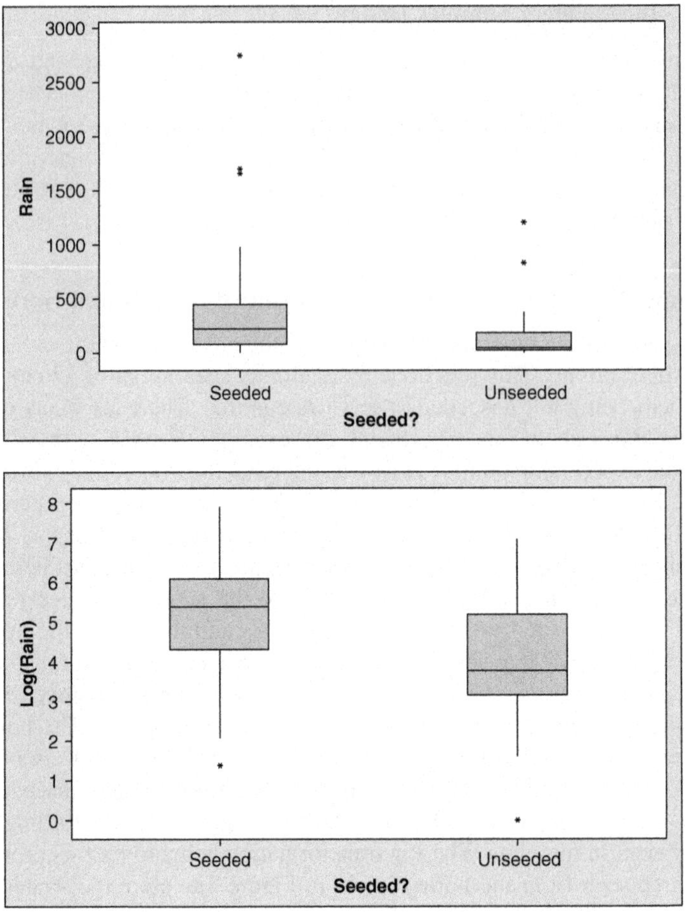

Figure 2.4 Side-by-side Box plots of rainfall data (raw and log-transformed).

Let us first assume that these are independent random samples from normal populations, $N(\mu_1, \sigma_1^2)$ and $N(\mu_2, \sigma_2^2)$, respectively. We first consider the problem of comparing the means μ_1 and μ_2. If the data are clearly nonnormal and cannot be transformed to normality, then a nonparametric method such as the Mann-Whitney (Wilcoxon rank-sum) test should be employed; see the books by Conover (1999) and Hollander and Wolfe (1998). If the data cannot be assumed to be random samples from well-defined populations, then a **randomization test** discussed in Section 2.2.1.5 should be employed.

Under the normality assumption, different procedures are used to compare μ_1 and μ_2 depending on whether one assumes $\sigma_1^2 = \sigma_2^2$ (called the **homoscedasticity assumption**) or $\sigma_1^2 \neq \sigma_2^2$ (called the **heteroscedasticity assumption**). These two procedures are discussed, respectively, in Sections 2.2.1.1 and 2.2.1.3. A test of the assumption $\sigma_1^2 = \sigma_2^2$ is given in Section 2.2.1.4. If in doubt about the

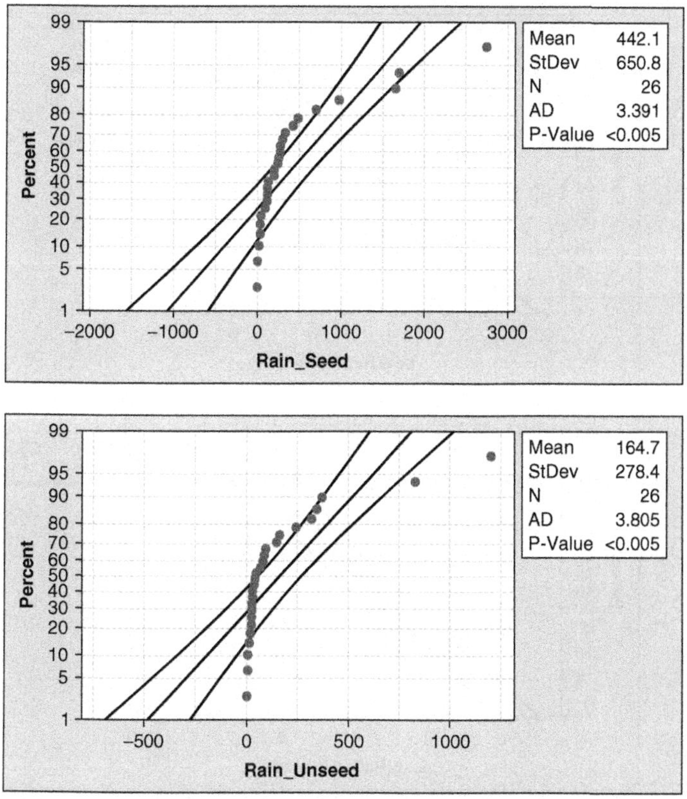

Figure 2.5 Normal plots of raw rainfall data for seeded and unseeded clouds.

homoscedasticity assumption, it is safer to use the procedure that is valid under heteroscedasticity.

2.2.1.1 Pooled Variances t-Interval and Test for Homoscedastic Data

Denote the sample means and variances of the two samples by

$$\overline{y}_i = \frac{1}{n_i}\sum_{j=1}^{n_i} y_{ij} \qquad \text{and} \qquad s_i^2 = \frac{1}{n_i - 1}\sum_{j=1}^{n_i}(y_{ij} - \overline{y}_i)^2 \qquad (i = 1, 2).$$

Under the homoscedasticity assumption, $\sigma_1^2 = \sigma_2^2 = \sigma^2$ (say), the common σ^2 can be estimated by pooling s_1^2 and s_2^2 weighted by their respective d.f., $n_1 - 1$ and $n_2 - 1$:

$$s^2 = \frac{(n_1 - 1)s_1^2 + (n_2 - 1)s_2^2}{n_1 + n_2 - 2} = \frac{\sum_{j=1}^{n_1}(y_{1j} - \overline{y}_1)^2 + \sum_{j=1}^{n_2}(y_{2j} - \overline{y}_2)^2}{n_1 + n_2 - 2}. \quad (2.8)$$

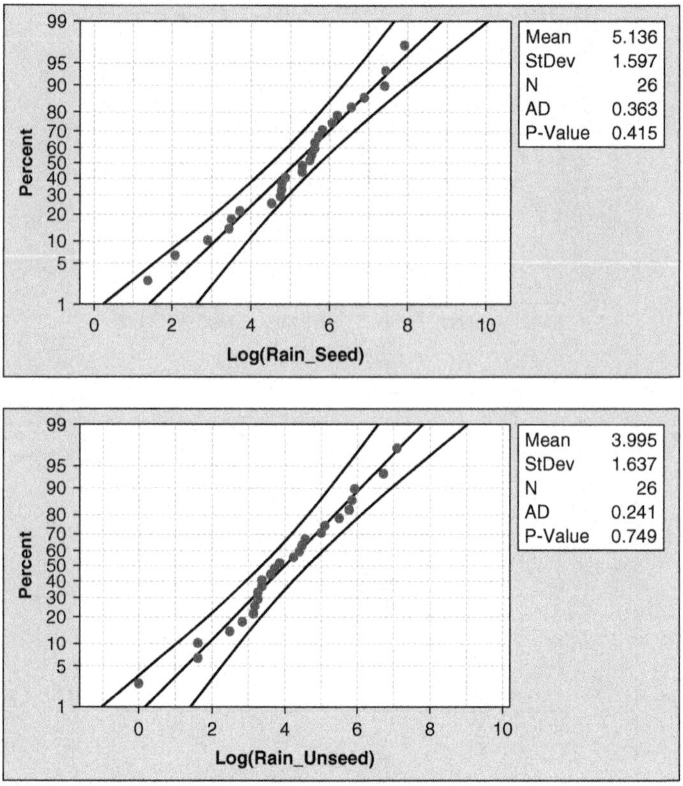

Figure 2.6 Normal plots of log-transformed rainfall data for seeded and unseeded clouds.

This estimator has $\nu = n_1 + n_2 - 2$ d.f. The **estimated standard error (SE)** of $\overline{y}_1 - \overline{y}_2$ is $s\sqrt{1/n_1 + 1/n_2}$. The pivotal quantity for $\mu_1 - \mu_2$ is

$$t = \frac{\overline{y}_1 - \overline{y}_2 - (\mu_1 - \mu_2)}{s\sqrt{1/n_1 + 1/n_2}}, \tag{2.9}$$

which has a t-distribution with $\nu = n_1 + n_2 - 2$ d.f. A two-sided $100(1-\alpha)\%$ CI on $\mu_1 - \mu_2$ is given by

$$\mu_1 - \mu_2 \in \left[\overline{y}_1 - \overline{y}_2 \pm t_{\nu,\alpha/2}\, s\sqrt{\frac{1}{n_1} + \frac{1}{n_2}}\right]. \tag{2.10}$$

An α-level test of the hypothesis $H_0 : \mu_1 = \mu_2$ vs. $H_1 : \mu_1 \neq \mu_2$ rejects H_0 if

$$|t| = \frac{|\overline{y}_1 - \overline{y}_2|}{s\sqrt{1/n_1 + 1/n_2}} > t_{\nu,\alpha/2}. \tag{2.11}$$

Example 2.5 (Cloud-Seeding Experiment: Pooled Variances t-Interval and Test)

Refer to the rainfall data in Table 2.3 after log transformation. The sample means and standard deviations are as follows:

$$\text{Seeded clouds:}\quad \bar{y}_1 = 5.136, \qquad s_1 = 1.597,$$
$$\text{Unseeded clouds:}\quad \bar{y}_2 = 3.995, \qquad s_2 = 1.637.$$

The sample sizes for both groups are equal: $n_1 = n_2 = n = 26$. Suppose we make an a priori assumption that $\sigma_1^2 = \sigma_2^2 = \sigma^2$. The pooled estimate of common σ equals

$$s = \sqrt{\frac{(25)(1.597)^2 + (25)(1.637)^2}{50}} = 1.604$$

with $26 + 26 - 2 = 50$ d.f. Using $t_{50,0.025} = 2.009$, a 95% CI for $\mu_1 - \mu_2$ equals

$$\left[5.136 - 3.995 \pm (2.009)(1.604)\sqrt{\tfrac{2}{26}}\right] = [0.247, 2.035].$$

Thus seeded clouds have between 0.247 and 2.035 units of more average rainfall on the log scale than unseeded clouds. The t-statistic equals

$$\left(\frac{5.136 - 3.995}{1.604}\right)\sqrt{\frac{26}{2}} = 2.859,$$

which is highly significant (two-sided $p = 0.006$). ■

2.2.1.2 *Power and Sample Size Calculation*

Next consider the problem of determining the sample size to guarantee a specified power requirement for the one-sided test of $H_0 : \mu_1 - \mu_2 \le 0$ versus $H_1 : \mu_1 - \mu_2 > 0$. Assume a common sample size n per group. For simplicity, let us further assume that $\sigma_1^2 = \sigma_2^2 = \sigma^2$, where σ^2 is *known*.[1] In that case an α-level test of H_0 is the usual z-test that rejects H_0 if

$$z = \left(\frac{\bar{y}_1 - \bar{y}_2}{\sigma}\right)\sqrt{\frac{n}{2}} > z_\alpha,$$

[1]If σ^2 is unknown, an upper bound on it may be assumed which would yield a conservative sample size. If σ^2 is unknown then the sample size can be calculated from the power function of the t-test. This problem is discussed in Examples 14.5 and 14.6.

where z_α is the upper α critical point of the standard normal distribution. The **power function** (the power regarded as a function of the true difference $\mu_1 - \mu_2$) of this test can be shown to be

$$\pi(\mu_1 - \mu_2) = \text{P(Test rejects } H_0 | H_1 : \mu_1 - \mu_2 > 0) = \Phi\left(-z_\alpha + \frac{\mu_1 - \mu_2}{\sigma}\sqrt{\frac{n}{2}}\right),$$

where $\Phi(\cdot)$ is the c.d.f. of the standard normal distribution.

If the practically significant difference between μ_1 and μ_2 that is worth detecting is $\delta > 0$, then the sample size n necessary to guarantee that

$$\pi(\mu_1 - \mu_2) \geq 1 - \beta \quad \text{if } \mu_1 - \mu_2 \geq \delta \tag{2.12}$$

is obtained by setting $\pi(\delta) = 1 - \beta$. The solution to this equation is

$$n = 2\left[\frac{\sigma(z_\alpha + z_\beta)}{\delta}\right]^2.$$

For the two-sided testing problem z_α in the above formula should be replaced by $z_{\alpha/2}$. When analyzing the data, the pooled estimate of σ^2 and the corresponding t-test must be used. This would increase the sample size slightly for specified power.

Example 2.6 (Cloud-Seeding Experiment: Sample Size Calculation)

Let y_1 denote the rainfall from a seeded cloud and y_2 from an unseeded cloud. As in Example 2.5, we assume that $\ln y_1 \sim N(\mu_1, \sigma^2)$ and $\ln y_2 \sim N(\mu_2, \sigma^2)$. In other words, y_1 and y_2 have **lognormal distributions** with

$$E(y_1) = e^{\mu_1 + (1/2)\sigma^2}, \qquad E(y_2) = e^{\mu_2 + (1/2)\sigma^2}, \qquad \frac{E(y_1)}{E(y_2)} = e^{\mu_1 - \mu_2}.$$

Suppose we want to test $H_0 : \mu_1 - \mu_2 = 0$ versus $H_1 : \mu_1 - \mu_2 \neq 0$ at level $\alpha = 0.05$ and want to guarantee power at least equal to $1 - \beta = 0.80$ if the seeded clouds produce three times the average amount of rainfall than unseeded clouds. Then the practically significant difference δ between μ_1 and μ_2 is given by

$$e^\delta = 3 \quad \text{or} \quad \delta = \ln 3 = 1.099.$$

For design purposes, let us assume $\sigma = 2.0$. Then the necessary sample size is given by

$$n = 2\left[\frac{\sigma(z_{.025} + z_{.20})}{\delta}\right]^2 = 2\left[\frac{2.0(1.960 + 0.842)}{1.099}\right]^2 = 52. \qquad \blacksquare$$

2.2.1.3 *Separate Variances t-Interval and Test for Heteroscedastic Data*

If $\sigma_1^2 \neq \sigma_2^2$, then their estimates cannot be pooled as in Eq. (2.8) and a pivotal quantity for $\mu_1 - \mu_2$ does not exist. Still, analogous to (2.9), approximate inferences can be based on the r.v.

$$t = \frac{\overline{y}_1 - \overline{y}_2 - (\mu_1 - \mu_2)}{\sqrt{s_1^2/n_1 + s_2^2/n_2}}$$

where $\sqrt{s_1^2/n_1 + s_2^2/n_2}$ is the standard error of $\overline{y}_1 - \overline{y}_2$. This is not a pivotal quantity because its distribution depends on the unknown variances σ_1^2 and σ_2^2 through their ratio. However, this distribution can be well-approximated by a t-distribution with d.f. estimated by

$$\nu = \frac{(s_1^2/n_1 + s_2^2/n_2)^2}{(s_1^2/n_1)^2/(n_1 - 1) + (s_2^2/n_2)^2/(n_2 - 1)}.$$

This is known as the **Welch–Satterthwaite approximation**. Notice that ν is in general fractional; it may be rounded down to the next lower integer. The formulas for CIs and hypothesis tests given for $\sigma_1^2 = \sigma_2^2$ are modified by using $\sqrt{s_1^2/n_1 + s_2^2/n_2}$ for the standard error of $\overline{y}_1 - \overline{y}_2$ in place of $s\sqrt{1/n_1 + 1/n_2}$. Furthermore, the above formula for ν must be used for the t critical points.

Example 2.7 (Failure Times of Capacitors: Comparison of Means)

Gibbons (1998) gave the data shown in Table 2.4 on the failure times of 18 capacitors, 8 of which were tested under normal operating conditions (control group) and 10 were tested under thermally stressed conditions (stressed group). We want to compare the mean failure times of the two groups using a two-sided test at $\alpha = 0.10$.

Table 2.4 Failure Times of Capacitors (Units Unknown)

Control Group		Stressed Group	
5.2	17.1	1.1	7.2
8.5	17.9	2.3	9.1
9.8	23.7	3.2	15.2
12.3	29.8	6.3	18.3
		7.0	21.1
$\overline{y}_1 = 15.538$		$\overline{y}_2 = 9.080$	
$s_1 = 8.262$		$s_2 = 6.886$	

Source: Gibbons (1998). Reprinted by permission of The McGraw-Hill Companies.

The normal plots (not shown here) indicate that the data are approximately normally distributed. The sample standard deviations s_1 and s_2 are not significantly different (see Example 2.8 for a formal test of significance), but we will not assume homoscedasticity. The t-statistic for testing the equality of the two mean failure times μ_1 and μ_2 is

$$t = \frac{15.538 - 9.080}{\sqrt{(8.262)^2/8 + (6.886)^2/10}} = 1.778.$$

The d.f. for this t-statistic are

$$\nu = \frac{(8.262)^2/8 + (6.886)^2/10)^2}{(8.262)^4/(7 \times 8^2) + (6.886)^4/(9 \times 10^2)} = 13.66,$$

which we will round down to 13. Since $t_{13,.05} = 1.771$, the difference is barely significant at $\alpha = 0.10$.

If we assume homoscedasticity, then the pooled sample variance is

$$s^2 = \frac{7(8.262)^2 + 9(6.886)^2}{16} = 56.536,$$

and $s = \sqrt{56.553} = 7.519$. Then the t-statistic equals

$$t = \frac{15.538 - 9.080}{7.519\sqrt{1/8 + 1/10}} = 1.811,$$

which is compared with $t_{16,0.05} = 1.746$. So the difference is slightly more significant ($p = 0.089$). ■

2.2.1.4 *Test of Homoscedasticity Assumption*

The homoscedasticity assumption is checked by testing $H_0 : \sigma_1^2 = \sigma_2^2$ versus $H_1 : \sigma_1^2 \neq \sigma_2^2$. The test statistic is

$$F = \frac{s_1^2}{s_2^2},$$

which has an F-**distribution** under H_0 with $\nu_1 = n_1 - 1$ and $\nu_2 = n_2 - 1$ d.f. Therefore an α-level test rejects the assumption of equal variances if

$$F > f_{\nu_1,\nu_2,\alpha/2} \qquad \text{or} \qquad F < f_{\nu_1,\nu_2,1-\alpha/2},$$

where $f_{\nu_1,\nu_2,\alpha/2}$ denotes the upper $\alpha/2$ critical point of the F-distribution with ν_1 and ν_2 d.f. These critical points are tabulated in Table C.4. Since $f_{\nu_1,\nu_2,1-\alpha/2} = 1/f_{\nu_2,\nu_1,\alpha/2}$, the above test can also be written as $F = s_{\text{max}}^2/s_{\text{min}}^2 > f_{u,v,\alpha/2}$, where u and v are the d.f. associated with s_{max}^2 and s_{min}^2, respectively.

If this test is nonsignificant with a sufficiently high p-value (e.g., $p > 0.25$), then the homoscedasticity assumption is generally made, although we know that failure to reject H_0 does not prove it. Another important point is that this test is highly sensitive to the normality assumption. For nonnormal data, a more robust test is the **Levene test**, which is discussed in Section 3.4.1 in the context of testing equality of several variances.

Example 2.8 (Failure Times of Capacitors: Test of Homoscedasticity)

Using the summary statistics from Table 2.4 we calculate $F = s^2_{\max}/s^2_{\min} = (8.262)^2/(6.886)^2 = 1.437$ with $\nu_1 = 7$ and $\nu_2 = 9$ d.f. For $\alpha = 0.10$, the critical constant with which this F-statistic must be compared is $f_{7,9,.05} = 3.29$. Thus the homoscedasticity hypothesis $H_0 : \sigma_1^2 = \sigma_2^2$ cannot be rejected (the p-value is 0.30). ■

2.2.1.5 Randomization Tests

In many applications the data cannot be assumed to be random samples from some well-defined populations. For example, in the cloud-seeding example, there are no infinite populations of homogeneous clouds from which the actual samples of seeded and unseeded clouds are drawn. The only population we have is the population of all possible randomizations of 52 eligible clouds into the seeded group and unseeded group of 26 each. What makes the application of a normal theory t-test to compare the means valid in this case? The answer lies in the fact that a test can be based on the **randomization distribution** under the null hypothesis over the population of all randomizations and this test, called the **randomization test**, can be well approximated by a normal theory t-test.

The randomization test treats the $N = n_1 + n_2$ data values as fixed. The treatment labels 1 and 2 can be assigned to these data values with n_1 to treatment 1 and n_2 to treatment 2 in $N!/(n_1!n_2!)$ ways. Under the null hypothesis of no treatment effect, all these assignments of treatment labels are equally likely. For each assignment we can calculate an appropriate statistic, for example, $\overline{y}_1 - \overline{y}_2$ or the t-statistic given by (2.9) if we are testing $\mu_1 = \mu_2$. The distribution of the statistic induced by randomization is called its randomization distribution. This can be used as a reference distribution to evaluate the p-value of the observed value of the statistic, which corresponds to the actual assignment of treatment labels. If n_1 and n_2 are large, then the task of enumerating the randomization distribution becomes computationally burdensome. But it is exactly in this case that the p-value calculated using the normal theory t-test provides a good approximation to the p-value calculated from the randomization distribution which provides a justification for using the t-test.

Example 2.9 (Walking Age of Infants: Randomization Test)

Fisher and van Belle (1993, Example 5.5) used the data from an archival source to compare the ages at which infants in two groups started to walk. The treatment

Table 2.5 Ages of Infants (months) When Beginning to Walk

Control group ($n_1 = 5$)	13.25, 11.50, 12.00, 13.50, 11.50	$\bar{y}_1 = 12.350, s_1 = 0.962$
Exercise group ($n_2 = 6$)	9.00, 9.50, 9.75, 10.00, 13.00, 9.50	$\bar{y}_2 = 10.125, s_2 = 1.447$

Source: Fisher and van Belle (1993, Example 5.5). Reprinted by permission of John Wiley & Sons, Inc.

group of newborn infants was given "walking exercises" consisting of holding them under arms and letting their bare feet touch a flat surface, which is supposed to stimulate a walking reflex. The control group was not given any such exercise. The exercises were continued for only eight weeks because the walking reflex disappears after that period. Subsequently, the ages at which the infants started to walk were reported by their mothers. There were $n_1 = 5$ infants in the control group (one observation was missing) and $n_2 = 6$ in the exercise group. The data are shown in Table 2.5.

Suppose we want to test $H_0 : \mu_1 - \mu_2 = 0$ versus $H_1 : \mu_1 - \mu_2 > 0$, where μ_1 and μ_2 are the mean walking ages for the populations of infants corresponding to the two groups. Assuming homoscedasticity, the pooled sample t-test yields $t = 2.9285$ with a one-sided $p = 0.008$. However, one might question the use of a normal theory test in this case, particularly since the number of observations is very small, making it virtually impossible to check their normality. A randomization test requires a total of $11!/(5!6!) = 462$ assignments of treatment labels to the 11 observations. A histogram of t-statistics for these 462 assignments is shown in Figure 2.7. The actual observed $t = 2.9285$ is also shown. The proportion of t-statistics that are ≥ 2.9285 is $5/462 = 0.0108$, which is the randomization p-value. Even for such a small data set, the p-value for the t-test is a relatively close approximation to the randomization p-value. ■

2.2.2 Matched Pairs Design

Denote the data on n matched pairs by

$$\begin{pmatrix} y_{11} \\ y_{21} \end{pmatrix}, \begin{pmatrix} y_{12} \\ y_{22} \end{pmatrix}, \ldots, \begin{pmatrix} y_{1n} \\ y_{2n} \end{pmatrix},$$

where y_{1j} and y_{2j} are the observations on the jth matched pair for treatments 1 and 2, respectively. If we assume the model

$$E(y_{ij}) = \mu_i + \beta_j \qquad (i = 1, 2, j = 1, 2, \ldots, n)$$

where μ_i is the effect of the ith treatment and β_j is the effect of the jth matched pair (e.g., the jth experimental unit if the same unit is used as its own match), then we can work with the differences $d_j = y_{1j} - y_{2j}$ $(1 \leq j \leq n)$ since $E(d_j) = \mu_d = \mu_1 - \mu_2$.

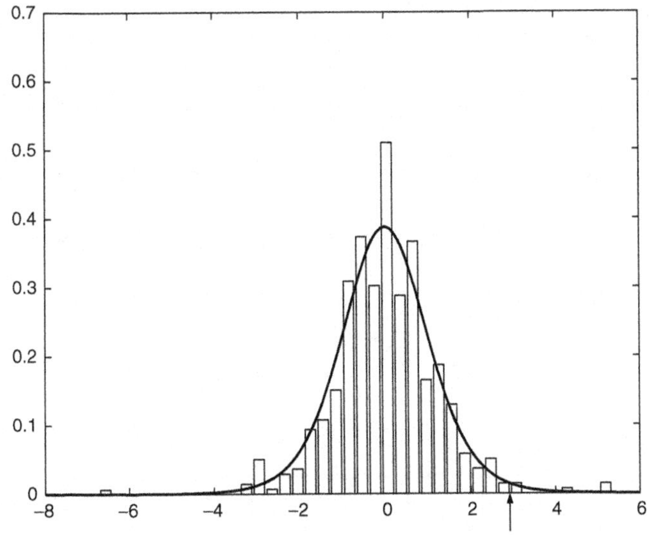

Figure 2.7 Histogram of randomization distribution of t-statistic superimposed with density function of t-distribution with nine degrees of freedom.

A graphical comparison can be made by making a scatter plot of the pairs (y_{1j}, y_{2j}) and checking if the plot falls mostly above or below the 45° line. Other plots are also possible, for example, a stem-and-leaf plot of the d_j. For a formal comparison we assume that the d_j's are i.i.d. $\mathrm{N}(\mu_d, \sigma_d^2)$ r.v.'s where $\mu_d = \mu_1 - \mu_2$.

Inferences on μ_d can be obtained using the single sample t procedures discussed in Section 2.1. Let

$$\bar{d} = \frac{\sum_{j=1}^{n} d_j}{n} \quad \text{and} \quad s_d^2 = \frac{\sum_{j=1}^{n}(d_j - \bar{d})^2}{n-1}.$$

An unbiased estimator of μ_d is $\bar{d}$ with a standard error of $s_d/\sqrt{n}$ with $n-1$ d.f. A $100(1-\alpha)\%$ CI on μ_d is given by

$$\mu_d \in \left[\bar{d} \pm t_{n-1,\alpha/2}\frac{s_d}{\sqrt{n}}\right].$$

An α-level test of $H_0 : \mu_1 = \mu_2$ versus $H_1 : \mu_1 \neq \mu_2$ rejects if

$$|t| = \frac{|\bar{d}|\sqrt{n}}{s_d} > t_{n-1,\alpha/2}. \tag{2.13}$$

One-sided CIs and tests are obtained by the usual modifications of the above formulas and replacing the critical point $t_{n-1,\alpha/2}$ by $t_{n-1,\alpha}$.

Table 2.6 Corn Yield (lb/acre)

No.	Regular Seed	Kiln-Dried Seed	Difference
1	1903	2009	−106
2	1935	1915	20
3	1910	2011	−101
4	2496	2463	33
5	2108	2180	−72
6	1961	1925	36
7	2060	2122	−62
8	1444	1482	−38
9	1612	1542	70
10	1316	1443	−127
11	1511	1535	−24
Mean	1841.46	1875.18	−33.73
SD	342.74	332.85	66.17

Source: Gosset (1908). Reprinted by permission of Oxford University Press.

Example 2.10 (Comparison of Corn Yields: *t*-Test)

William Sealy Gosset (1876–1937) introduced the t-distribution (named after Gosset's pseudoname "Student") in a 1908 classic paper from which this example is taken. Eleven pairs of adjacent plots were planted with regular seeds and kiln-dried seeds of corn to see which gives a higher yield. The data are shown in Table 2.6. Notice that the individual corn yields are quite variable ($s_1 = 342.74$, $s_2 = 332.85$), but the differences are much more consistent ($s_d = 66.17$).

An inspection of the differences in Table 2.6 indicates that there are more negative differences and they are larger in magnitude. However, the average difference is not significantly different from zero, as can be checked by calculating the paired t-statistic:

$$t = \frac{\bar{d}\sqrt{n}}{s_d} = \frac{-33.73\sqrt{11}}{66.17} = -1.691$$

with $11 - 1 = 10$ d.f. (two-sided p-value $= 0.122$). The small sample size is probably the reason for this nonsignificant result. ■

Since the analysis of a matched pairs design is the same as that of a single-treatment design, the same formulas (2.6) and (2.7) can be used for power and sample size calculation with μ_d and σ_d standing in for μ and σ, respectively. The following example illustrates this calculation.

Example 2.11 (Comparison of Corn Yields: Sample Size Determination)

Suppose that the experiment referred to in Example 2.10 was a pilot experiment and that we want to design a larger experiment based on the data obtained from it. Further suppose that, for the larger experiment, a two-sided test with $\alpha = 0.05$ should have power at least equal to $1 - \beta = 0.80$ to detect a difference of 30 lb/acre between the mean yields of the two seeds. We assume $\sigma_d = 70$ (based on $s_d = 66.17$). Then

$$n = \left[\frac{(z_{.025} + z_{.20})\sigma_d}{\delta}\right]^2 = \left[\frac{(1.960 + 0.84)70}{30}\right]^2 = 42.68.$$

Thus 43 pairs of adjacent plots should be planted with the two seeds, the seeds being assigned at random to the two plots in each pair. See Example 14.6 where it is shown how to find the sample size for the paired t-test for this example; the necessary sample size would be slightly greater than 43. ∎

2.3 LINEAR REGRESSION

Regression analysis is a primary tool used for empirical model building. The general goal is to fit a functional relationship between the expected value of a **response variable** y and a set of **predictor variables** $x_1, x_2, \ldots, x_p$ from data on these variables on N observational or experimental units. If the functional relationship is parametric, then the problem of model fitting reduces to estimating a set of unknown parameters. If it is linear in unknown parameters, then the model is called a **linear model**; otherwise it is called a **nonlinear model**. Specifically, a linear model has the form

$$y = \beta_0 + \beta_1 x_1 + \beta_2 x_2 + \cdots + \beta_p x_p + e, \tag{2.14}$$

where $\beta_0, \beta_1, \ldots, \beta_p$ are unknown fixed parameters and e is a **random error**. Here we will assume that $e \sim \mathrm{N}(0, \sigma^2)$. The parameter β_0 is referred to as the **constant term** or **intercept** and the β_j's $(1 \le j \le p)$ as **regression coefficients**. The predictor variables are assumed to be fixed by design in case of experimental data or by conditioning in case of observational data. This assumption implies that $y \sim \mathrm{N}(\mu, \sigma^2)$, where $\mu = \mathrm{E}(y) = \beta_0 + \beta_1 x_1 + \beta_2 x_2 + \cdots + \beta_p x_p$.

Note that a linear model can be nonlinear in the x's. Thus, for example, a quadratic model in a single predictor variable x, namely, $y = \beta_0 + \beta_1 x + \beta_2 x^2 + e$, is a linear model and can be fitted using the methods of linear regression. More generally, a polynomial model in one or more predictor variables is a linear model. As another example, consider two predictor variables x_1 and x_2. Their joint effect on the mean of y (called **interaction**) is often included as a product term $x_1 x_2$, and the following model is fitted: $y = \beta_0 + \beta_1 x_1 + \beta_2 x_2 + \beta_3 x_1 x_2 + e$. This is also a linear model. The general theory of linear models is given in Chapter 14, where the proofs of most of the results cited in this section can be found. Nonlinear models are more difficult to fit and their theory is more complicated. We refer the reader to the book by Bates and Watts (1988) for details.

It should be noted that the predictor variables are not restricted to be numerical; categorical variables can be included in the model by defining appropriate **indicator** or **dummy variables** as follows. If a predictor variable is binary (e.g., gender), then we need only one indicator variable, x, to represent it. For example, $x = 0$ for males and $x = 1$ for females. The coefficient of x then denotes the difference in the mean response between a female and a male keeping other predictor variables fixed. If a predictor variable has $c > 2$ categories, then we need $c - 1$ indicator variables, $x_1, \ldots, x_{c-1}$, where $x_j = 0$ if the jth category is absent and $x_j = 1$ if the jth category is present $(1 \le j \le c - 1)$; all $x_j = 0$ if the cth category is present. Thus coefficient β_j of x_j is the difference in the mean response between the jth and the cth category, which is taken as the "base" category from which the changes in the mean response of other categories are measured.

Simple linear regression, which is a special case of the linear model (2.14) for a single predictor variable, x, is discussed in Section 2.3.1. The general case of multiple linear regression is discussed in Section 2.3.2.

2.3.1 Simple Linear Regression

2.3.1.1 Least Squares Estimation

Suppose we have N pairs of observations (x_i, y_i), $i = 1, 2, \ldots, N$. We want to fit the model

$$y_i = \beta_0 + \beta_1 x_i + e_i, \tag{2.15}$$

where the e_i are i.i.d. $N(0, \sigma^2)$ r.v.'s, and hence the y_i are independent $N(\mu_i, \sigma^2)$ r.v.'s with $\mu_i = \beta_0 + \beta_1 x_i$. Thus we assume that the errors e_i are (i) normal, (ii) homoscedastic, (iii) independent and (iv) have zero means (which implies that the assumed linear model is correct; also there are no individual observations, called **outliers**, that significantly violate this model). We want to estimate β_0 and β_1 as well as the error variance σ^2 under these assumptions. Later we will check these assumptions using some **model diagnostics**.

The **least squares (LS) method** is commonly used to estimate β_0 and β_1. The LS estimators $\widehat{\beta}_0$ and $\widehat{\beta}_1$ minimize the sum of squared differences between the observed y_i and their expected values under the model (2.15), that is, they minimize

$$Q = \sum_{i=1}^{N} e_i^2 = \sum_{i=1}^{N} [y_i - (\beta_0 + \beta_1 x_i)]^2. \tag{2.16}$$

The following notation will be useful in what follows:

$$S_{xy} = \sum_{i=1}^{N} (x_i - \overline{x})(y_i - \overline{y}), \qquad S_{xx} = \sum_{i=1}^{N} (x_i - \overline{x})^2, \qquad S_{yy} = \sum_{i=1}^{N} (y_i - \overline{y})^2. \tag{2.17}$$

Then the LS estimators are given by (see Example 2.13 for the derivation)

$$\widehat{\beta}_0 = \overline{y} - \widehat{\beta}_1 \overline{x} \quad \text{and} \quad \widehat{\beta}_1 = \frac{S_{xy}}{S_{xx}}. \tag{2.18}$$

The **LS fitted equation** is $\widehat{y} = \widehat{\beta}_0 + \widehat{\beta}_1 x$. The corresponding **fitted values** and **residuals** are given by

$$\widehat{y}_i = \widehat{\beta}_0 + \widehat{\beta}_1 x_i \quad \text{and} \quad \widehat{e}_i = y_i - \widehat{y}_i \quad (1 \le i \le N).$$

An unbiased estimator of σ^2 can be computed from the residuals using

$$s^2 = \frac{\sum_{i=1}^{N} \widehat{e}_i^2}{N-2}.$$

The residuals are especially useful for checking the model assumptions such as the straight-line relationship between y and x, independence, normality and homoscedasticity of random errors, and detection of outliers. Specifically, the normality is checked by making a **normal plot** of the residuals and homoscedasticity is checked by plotting the residuals against the fitted values (called the **fitted-values plot**). Some theoretical background behind the residual plots is discussed in Section 3.4. In the case of simple linear regression, since the $\widehat{y}_i$ are linearly related to the x_i, the fitted-values plot also shows if there is a nonlinear component to the relationship between y and x since the residuals show what is left over after the linear part of the relationship is filtered out. In the case of multiple regression, one must plot residuals against the different predictor variables to see if there are nonlinear terms that need be added to the model. These uses of residuals are illustrated in Examples 2.12 and 2.15.

2.3.1.2 *Confidence Intervals and Hypothesis Tests*

For making inferences on β_0 and β_1 and more generally on any linear function of them, we need the sampling distributions of $\widehat{\beta}_0$ and $\widehat{\beta}_1$. It can be shown that they are unbiased, that is, $\mathrm{E}(\widehat{\beta}_0) = \beta_0$ and $\mathrm{E}(\widehat{\beta}_1) = \beta_1$, and are jointly normally distributed with the following variances and covariance:

$$\mathrm{Var}(\widehat{\beta}_0) = \frac{\sigma^2 \sum x_i^2}{N S_{xx}}, \quad \mathrm{Var}(\widehat{\beta}_1) = \frac{\sigma^2}{S_{xx}}, \quad \mathrm{Cov}(\widehat{\beta}_0, \widehat{\beta}_1) = -\frac{\sigma^2 \overline{x}}{S_{xx}}. \tag{2.19}$$

Next, s^2 is distributed independently of $\widehat{\beta}_0$ and $\widehat{\beta}_1$ as

$$s^2 \sim \frac{\sigma^2 \chi^2_{N-2}}{N-2}.$$

Hence

$$\frac{\widehat{\beta}_0 - \beta_0}{s\sqrt{\sum x_i^2 / N S_{xx}}} \quad \text{and} \quad \frac{\widehat{\beta}_1 - \beta_1}{s/\sqrt{S_{xx}}}$$

have t-distributions with $N-2$ d.f. Inferences on β_0 and β_1 can be based on these pivotal quantities. For example, $100(1-\alpha)\%$ CIs on β_0 and β_1 are

$$\beta_0 \in \left[\widehat{\beta}_0 \pm t_{N-2,\alpha/2}\, s\sqrt{\frac{\sum x_i^2}{N S_{xx}}}\right] \quad \text{and} \quad \beta_1 \in \left[\widehat{\beta}_1 \pm t_{N-2,\alpha/2}\frac{s}{\sqrt{S_{xx}}}\right].$$

A particular linear combination of β_0 and β_1 of interest is the expected value of $y = y^*$ when $x = x^*$, that is, $\mathrm{E}(y^*) = \mu^* = \beta_0 + \beta_1 x^*$. For example, suppose y is the viscosity of a paint which is modeled as a linear function of the percentage (x) of one of its ingredients. Then it would be of interest to estimate the mean viscosity in terms of a confidence interval if x is set at a specified target value x^*. The LS estimator of μ^* is

$$\widehat{\mu}^* = \widehat{\beta}_0 + \widehat{\beta}_1 x^* \sim \mathrm{N}\left(\mu^*, \sigma^2\left[\frac{1}{N} + \frac{(x^* - \overline{x})^2}{S_{xx}}\right]\right).$$

Hence a $100(1-\alpha)\%$ CI on μ^* is given by

$$\mu^* \in \left[\widehat{\beta}_0 + \widehat{\beta}_1 x^* \pm t_{N-2,\alpha/2}\, s\sqrt{\frac{1}{N} + \frac{(x^* - \overline{x})^2}{S_{xx}}}\right]. \tag{2.20}$$

If one is interested in *predicting* y^* (e.g., the viscosity of a particular batch of paint) instead of *estimating* $\mathrm{E}(y^*)$, then we require a $100(1-\alpha)\%$ **prediction interval (PI)**, which is given by

$$y^* \in \left[\widehat{\beta}_0 + \widehat{\beta}_1 x^* \pm t_{N-2,\alpha/2}\, s\sqrt{1 + \frac{1}{N} + \frac{(x^* - \overline{x})^2}{S_{xx}}}\right]. \tag{2.21}$$

The significance of the linear relationship between $\mathrm{E}(y)$ and x is assessed by testing $H_0 : \beta_1 = 0$ versus $H_1 : \beta_1 \neq 0$. The test statistic is

$$t = \frac{\widehat{\beta}_1}{s/\sqrt{S_{xx}}}.$$

The α-level test rejects H_0 if

$$|t| > t_{N-2,\alpha/2} \quad \text{or equivalently} \quad F = t^2 = \frac{\widehat{\beta}_1^2 S_{xx}}{s^2} > f_{1,N-2,\alpha}, \tag{2.22}$$

where $f_{\nu_1,\nu_2,\alpha}$ is the upper α critical point of the F-distribution with ν_1 and ν_2 d.f.

The proportion of the total variation in the y_i's accounted for by a regression model is used as one of the measures of its goodness of fit. The total sum of squares (SS), $\sum_{i=1}^{N}(y_i - \overline{y})^2$, denoted by $\mathrm{SS}_{\mathrm{tot}}$, can be partitioned into two components, **regression sum of squares ($\mathbf{SS_{reg}}$)** and **error sum of squares ($\mathbf{SS_e}$)**, as follows:

$$\underbrace{\sum_{i=1}^{N}(y_i - \overline{y})^2}_{\mathrm{SS}_{\mathrm{tot}}} = \underbrace{\sum_{i=1}^{N}(\widehat{y}_i - \overline{y})^2}_{\mathrm{SS}_{\mathrm{reg}}} + \underbrace{\sum_{i=1}^{N}(y_i - \widehat{y}_i)^2}_{\mathrm{SS}_e}. \tag{2.23}$$

This is called the **analysis of variance (ANOVA) identity**. Note that $\mathrm{SS}_e = \sum_{i=1}^{N}\widehat{e}_i^2$, and it can be shown that $\mathrm{SS}_{\mathrm{reg}} = \widehat{\beta}_1^2 S_{xx}$ (see Example 14.3 for the derivation). The proportion of variation in the y_i's accounted for by regression on the x_i's equals

$$R^2 = \frac{\mathrm{SS}_{\mathrm{reg}}}{\mathrm{SS}_{\mathrm{tot}}}, \tag{2.24}$$

which is called the **coefficient of determination**. Its square root, R, is the **correlation coefficient** between x and y. Using $\mathrm{SS}_{\mathrm{reg}} = \widehat{\beta}_1^2 S_{xx} = S_{xy}^2/S_{xx}$ and $\mathrm{SS}_{\mathrm{tot}} = S_{yy}$, we get

$$R = \frac{S_{xy}}{\sqrt{S_{xx}S_{yy}}}.$$

Note that $0 \le R^2 \le 1$ and $-1 \le R \le 1$. The sign of R is that of S_{xy}, that is, that of $\widehat{\beta}_1$.

Corresponding to the ANOVA identity (2.23), there is a partitioning of the degrees of freedom:

$$\underbrace{N-1}_{\text{Total d.f.}} = \underbrace{1}_{\text{Regression d.f.}} + \underbrace{N-2}_{\text{Error d.f.}}.$$

A sum of squares divided by its d.f. is referred to as a mean square (MS). Thus, **mean square regression** equals $\mathrm{MS}_{\mathrm{reg}} = \mathrm{SS}_{\mathrm{reg}}/1$ and **mean square error** equals $\mathrm{MS}_e = \mathrm{SS}_e/(N-2) = s^2$. Note that the F-statistic used in the test of $H_0 : \beta_1 = 0$ above equals $F = \mathrm{MS}_{\mathrm{reg}}/\mathrm{MS}_e$. These calculations are presented in the form of an **ANOVA table** shown in Table 2.7.

Example 2.12 (Salary Data: Simple Linear Regression)

Table 2.8 gives the data on the salaries of 46 employees of a company along with six predictor variables, which are listed in Table 2.9. Preliminary plots showed the Salary variable to be heteroscedastic; hence we made a log (to the base 10) transformation to stabilize its variance. In this example we will use only Education, which is most highly correlated with log(Salary) (correlation

Table 2.7 ANOVA Table for Simple Linear Regression

Source	SS	d.f.	MS	F
Regression	$SS_{reg} = \widehat{\beta}_1^2 S_{xx}$	1	MS_{reg}	MS_{reg}/MS_e
Error	$SS_e = \sum \widehat{e}_i^2$	$N-2$	MS_e	
Total	$SS_{tot} = \sum(y_i - \overline{y})^2$	$N-1$		

coefficient is $R = 0.782$) to fit a straight-line relationship, ignoring the remaining variables. The plot of the data with the LS fitted line is shown in Figure 2.8. The Minitab output for the LS fit is shown in Display 2.1. The LS fitted line is $\widehat{\log(\text{Salary})} = 4.428 + 0.0310 \times \text{Education}$.

Suppose we want to estimate the mean salary of employees with four years of post–high school education. Letting $y = \log(\text{Salary})$ and $x = \text{Education}$ and substituting $x = x^* = 4$ in the above equation, we get $\widehat{\mu}^* = 4.552$, so the estimated mean salary is $10^{4.552} = 35645$. To calculate a 95% CI on the mean salary we require $S_{xx} = (N-1)s_x^2 = 362.70$, $s = \sqrt{MS_e} = 0.0709$, and $t_{45,.025} = 2.014$. Substituting these values in (2.20), the desired interval on μ^* is [4.530.4.575], so the corresponding interval on the mean salary is $[10^{4.530}, 10^{4.575}] = [33884, 37583]$.

The predicted salary of a future employee with four years of post–high school education is also \$35,645. The prediction interval on y^* calculated using (2.21) is [4.408, 4.697]. Hence the PI on the predicted salary is $[10^{4.408}, 10^{4.697}] = [25568, 49785]$, which is much wider than the CI on the mean salary calculated above. This is quite typical; here the discrepancy is exaggerated because of the antilog transformation.

We see that $\widehat{\beta}_1 = 0.0310$ is highly significant ($p = 0.000$), but Education alone only explains $R^2 = 61.2\%$ of the variation in log(Salary). Figure 2.9 shows the normal and fitted-values plots of the residuals from this regression. The normal plot appears satisfactory, but the fitted-values plot shows that there is still some heteroscedasticiy left; also there appears to be a quadratic component to the relationship which is not accounted for by this linear regression. We could try fitting a quadratic regression (which is a special case of multiple regression discussed in the next section) and see if these problems with the fitted-values plot go away; we will not pursue this further exploration here. In Example 2.15 we will fit a multiple regression to these data to take into account all available predictor variables. We will see there that the resulting residual plots are quite satisfactory. ■

2.3.1.3 *Pure Error and Lack-of-Fit Test*

It is important to note that MS_e in the ANOVA table (Table 2.7) gives an unbiased estimator of σ^2 only if the assumed linear model (2.15) is correct. If, in fact, the true model is different, for example, if it is a quadratic model,

Table 2.8 Salary Data

No.	Salary	YrsEm	PriorYr	Education	Super	Gender	Dept
1	38985	18	7	9	5	F	S
2	32920	15	3	9	4	M	S
3	29548	5	6	1	0	F	S
4	24749	6	2	0	1	M	S
5	41889	22	16	7	7	F	S
6	31528	3	11	3	6	F	S
7	38791	21	4	5	9	F	S
8	39828	18	6	5	5	M	S
9	28985	0	1	4	4	M	S
10	32782	0	1	7	0	F	S
11	43674	6	9	4	2	F	P
12	35467	3	6	6	3	M	P
13	29876	2	0	3	5	M	P
14	36431	9	4	4	2	M	P
15	56326	12	3	8	6	F	P
16	36571	6	1	4	2	F	P
17	35468	9	5	4	5	M	P
18	26578	0	6	2	2	M	P
19	47536	15	5	6	4	F	A
20	23654	0	0	0	2	M	A
21	37548	19	9	4	6	F	A
22	36578	4	4	8	8	F	A
23	54679	20	3	6	4	M	A
24	53234	25	0	6	3	F	A
25	31425	7	6	5	6	M	A
26	39743	9	6	5	1	M	E
27	26452	1	3	2	0	M	E
28	34632	5	4	4	0	F	E
29	35631	6	4	4	2	F	E
30	46211	14	5	6	5	M	E
31	34231	6	2	6	3	F	E
32	26548	5	1	0	2	F	E
33	36512	6	6	4	2	M	E
34	34869	7	5	4	1	M	E
35	41255	9	4	6	4	F	E
36	39331	9	3	6	1	M	E
37	35487	8	2	2	2	M	E
38	36487	6	5	2	3	F	E
39	68425	25	2	12	1	F	E
40	69246	22	3	10	45	F	E
41	65487	27	0	12	44	M	E
42	48695	6	19	8	40	F	E
43	51698	18	6	6	1	F	E
44	46184	20	3	4	1	F	E
45	34987	9	6	2	3	M	E
46	54899	12	5	8	0	M	E

Source: McKenzie and Goldman (1999, Temco.mtw Data Set).

Table 2.9 Salary Data Variables

Variable	Explanation
Salary	Annual salary in dollars
YrsEm	No. of years employed with company
PriorYr	No. of years of prior experience
Education	No. of years of education after high school
Super	No. of people supervised
Gender	M = Male, F = Female
Dept	Department of employment
	A = Advertising
	E = Engineering
	M = Marketing
	S = Sales

Source: McKenzie and Goldman (1999, p. ED-32).

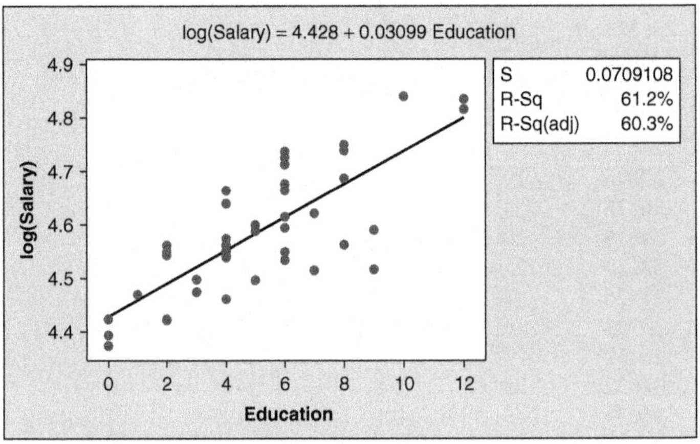

Figure 2.8 Fitted-line plot of log(Salary) versus Education.

$y_i = \beta_0 + \beta_1 x_i + \beta_2 x_i^2 + e_i$ $(1 \le i \le N)$, and we fit a linear model, then MS_e will be positively biased. In practice, we never know what the true model is; in fact, any mathematical model is an approximation to reality. The only sure way to get an unbiased estimator of σ^2 is to take i.i.d. observations on y at the same settings of x and pool the sample variances (under the homoscedasticity assumption); the resulting estimator is called a **pure-error estimator**. Not only does this method give an unbiased estimator of σ^2, but it also enables us to test for goodness of fit of the assumed linear model. Therefore, when designing an experiment for regression, it is recommended to make at least some **repeat runs**.

Consider an experiment in which n_i i.i.d. observations y_{ij} $(1 \le j \le n_i)$ are taken at each x_i $(1 \le i \le m)$, where $n_i \ge 2$ and $\sum_{i=1}^{m} n_i = N$. We assume the model (2.15) with obvious changes in notation. The LS estimators $\widehat{\beta}_0$ and $\widehat{\beta}_1$ can

```
The regression equation is
log(Salary) = 4.43 + 0.0310 Education

Predictor       Coef   SE Coef       T      P
Constant     4.42847   0.02156  205.39  0.000
Education   0.030986  0.003723    8.32  0.000

S = 0.0709108   R-Sq = 61.2%   R-Sq(adj) = 60.3%

Analysis of Variance

Source          DF       SS       MS      F      P
Regression       1  0.34834  0.34834  69.28  0.000
Residual Error  44  0.22125  0.00503
Total           45  0.56959
```

Display 2.1 Minitab output for straight-line fit of log(Salary) versus Education.

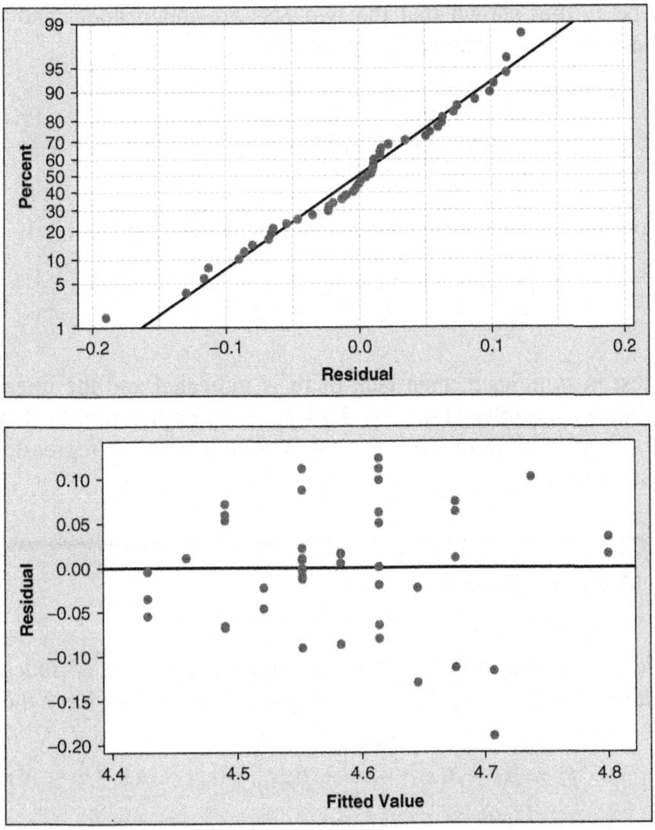

Figure 2.9 Normal (top) and fitted-values (bottom) plots of residuals from simple linear regression of log(Salary) versus Education.

be computed as before and from them the fitted values: $\widehat{y}_{ij} = \widehat{\beta}_0 + \widehat{\beta}_1 x_i$ ($1 \leq i \leq m, 1 \leq j \leq n_i$). Then SS_e can be partitioned into two terms:

$$SS_e = \sum_{i=1}^{m}\sum_{j=1}^{n_i}(y_{ij} - \widehat{y}_{ij})^2 = \underbrace{\sum_{i=1}^{m}\sum_{j=1}^{n_i}(\overline{y}_i - \widehat{y}_{ij})^2}_{SS_{\text{lof}}} + \underbrace{\sum_{i=1}^{m}\sum_{j=1}^{n_i}(y_{ij} - \overline{y}_i)^2}_{SS_{\text{pe}}},$$

where SS_{lof} represents the **lack-of-fit sum of squares** and SS_{pe} represents the **pure-error sum of squares**. The corresponding decomposition of the d.f. is given by

$$\underbrace{N-2}_{\text{Total-error d.f.}} = \underbrace{m-2}_{\text{Lack-of-fit d.f.}} + \underbrace{N-m}_{\text{Pure-error d.f.}}.$$

It can be further shown that the two SSs are independent and under the null hypothesis that the assumed linear model is correct,

$$F_{\text{lof}} = \frac{MS_{\text{lof}}}{MS_{\text{pe}}} = \frac{SS_{\text{lof}}/(m-2)}{SS_{\text{pe}}/(N-m)} \sim F_{m-2,N-m}. \quad (2.25)$$

Therefore the above null hypothesis can be rejected at level α if

$$F_{\text{lof}} > f_{m-2,N-m,\alpha}.$$

If the test is significant, then lack of fit is indicated and the linear model needs to be suitably modified. Exercise 2.14 gives an application of this test. The test can be extended in an obvious manner to multiple linear regression discussed in the next section.

2.3.2 Multiple Linear Regression

2.3.2.1 Model

Consider the linear model (2.14) and suppose we have N independent vector observations on $(x_1, \ldots, x_p, y)$. Analogous to (2.15), we have the model

$$y_i = \beta_0 + \beta_1 x_{i1} + \cdots + \beta_p x_{ip} + e_i \qquad (1 \leq i \leq N), \quad (2.26)$$

where the e_i's are assumed to be i.i.d. $N(0, \sigma^2)$ r.v.'s. and hence the y_i's are independent $N(\mu_i, \sigma^2)$ r.v.'s where $\mu_i = \beta_0 + \beta_1 x_{i1} + \cdots + \beta_p x_{ip}$.

To represent this model in matrix form, we introduce the following notation[2]:

$$y = \begin{bmatrix} y_1 \\ y_2 \\ \vdots \\ y_N \end{bmatrix}, \quad X = \begin{bmatrix} 1 & x_{11} & \cdots & x_{1p} \\ 1 & x_{21} & \cdots & x_{2p} \\ \vdots & \vdots & \ddots & \vdots \\ 1 & x_{N1} & \cdots & x_{Np} \end{bmatrix}, \quad \boldsymbol{\beta} = \begin{bmatrix} \beta_0 \\ \beta_1 \\ \vdots \\ \beta_p \end{bmatrix},$$

$$\boldsymbol{e} = \begin{bmatrix} e_1 \\ e_2 \\ \vdots \\ e_N \end{bmatrix}.$$

Then model (2.26) can be written in matrix notation as

$$\boldsymbol{y} = \boldsymbol{X}\boldsymbol{\beta} + \boldsymbol{e}, \tag{2.27}$$

where $\boldsymbol{X}$ is known as the **model matrix** and $\boldsymbol{e}$ has a multivariate normal (MVN) distribution (see Section A.3 for a discussion of the MVN distribution) with mean vector $\boldsymbol{0}$ (a null vector) and covariance matrix $\sigma^2\boldsymbol{I}$; here $\boldsymbol{I}$ represents an $N \times N$ identity matrix. We denote this by $\boldsymbol{e} \sim \text{MVN}(\boldsymbol{0}, \sigma^2\boldsymbol{I})$. It follows that $\boldsymbol{y} \sim \text{MVN}(\boldsymbol{\mu}, \sigma^2\boldsymbol{I})$, where $\boldsymbol{\mu} = \boldsymbol{X}\boldsymbol{\beta} = (\mu_1, \mu_2, \ldots, \mu_N)'$.

2.3.2.2 Least Squares Estimation

The LS method minimizes

$$Q = \sum_{i=1}^{N} [y_i - (\beta_0 + \beta_1 x_{i1} + \cdots + \beta_p x_{ip})]^2 = (\boldsymbol{y} - \boldsymbol{X}\boldsymbol{\beta})'(\boldsymbol{y} - \boldsymbol{X}\boldsymbol{\beta})$$

with respect to (w.r.t.) $\boldsymbol{\beta}$. The resulting LS estimator $\widehat{\boldsymbol{\beta}}$ satisfies the following set of **normal equations**:

$$(\boldsymbol{X}'\boldsymbol{X})\widehat{\boldsymbol{\beta}} = \boldsymbol{X}'\boldsymbol{y}. \tag{2.28}$$

If $\boldsymbol{X}$ is a full column rank matrix, then $\boldsymbol{X}'\boldsymbol{X}$ is invertible and there exists a unique $\widehat{\boldsymbol{\beta}}$ given by

$$\widehat{\boldsymbol{\beta}} = (\boldsymbol{X}'\boldsymbol{X})^{-1}\boldsymbol{X}'\boldsymbol{y}. \tag{2.29}$$

As noted before, derivations of the LS estimators and the associated theory of linear models are given in Chapter 14.

[2]All vectors are assumed to be column vectors. A prime on a vector or a matrix denotes its transpose. Vectors and matrices are denoted by bold letters. Usually, we suppress the dimension of any vector or a matrix if it is clear from the context. When necessary, we indicate the dimension by a subscript.

If X is a not a full column rank matrix, that is, if the columns of X, which represent the predictor variables, are linearly dependent, then one needs to add suitable linear constraints on the β_j's to obtain unique LS estimators $\widehat{\beta}_j$'s. This is commonly done in linear models for multifactor experiments, as we shall see in later chapters; also see Example 14.1.

The vector of fitted values is given by

$$\widehat{\boldsymbol{y}} = \widehat{\boldsymbol{\mu}} = \begin{bmatrix} \boldsymbol{x}'_1\widehat{\boldsymbol{\beta}} \\ \boldsymbol{x}'_2\widehat{\boldsymbol{\beta}} \\ \vdots \\ \boldsymbol{x}'_N\widehat{\boldsymbol{\beta}} \end{bmatrix} = \boldsymbol{X}\widehat{\boldsymbol{\beta}}, \tag{2.30}$$

and the vector of residuals is given by

$$\widehat{\boldsymbol{e}} = \boldsymbol{y} - \widehat{\boldsymbol{y}}. \tag{2.31}$$

It can be shown that the dot product of the $\widehat{\boldsymbol{e}}$ vector with every column vector of the $\boldsymbol{X}$ matrix is zero, that is, $\widehat{\boldsymbol{e}}$ is orthogonal to every column of $\boldsymbol{X}$; in particular,

$$\sum_{i=1}^{N} \widehat{e}_i = 0. \tag{2.32}$$

Therefore only $\nu = N - (p+1)$ of the $\widehat{e}_i$ are linearly independent. The **error sum of squares** $\text{SS}_e = \widehat{\boldsymbol{e}}'\widehat{\boldsymbol{e}} = \sum_{i=1}^{N} \widehat{e}_i^2$ divided by the error d.f. ν is called the **mean square error**, which gives an unbiased estimator of σ^2:

$$s^2 = \text{MS}_e = \frac{\text{SS}_e}{N - (p+1)}. \tag{2.33}$$

We now apply these results to simple linear regression and derive the expressions used in the previous section.

Example 2.13 (Simple Linear Regression: LS Estimation)

For the simple linear regression model (2.15), $\boldsymbol{y}, \boldsymbol{X}$, and $\boldsymbol{\beta}$ are as follows:

$$\boldsymbol{y} = \begin{bmatrix} y_1 \\ y_2 \\ \vdots \\ y_N \end{bmatrix}, \quad \boldsymbol{X} = \begin{bmatrix} 1 & x_1 \\ 1 & x_2 \\ \vdots & \vdots \\ 1 & x_N \end{bmatrix}, \quad \boldsymbol{\beta} = \begin{bmatrix} \beta_0 \\ \beta_1 \end{bmatrix}.$$

Therefore

$$\boldsymbol{X}'\boldsymbol{X} = \begin{bmatrix} N & \sum x_i \\ \sum x_i & \sum x_i^2 \end{bmatrix}.$$

If the x_i's are not all equal, then $\boldsymbol{X}$ is full column rank and $\boldsymbol{X}'\boldsymbol{X}$ is invertible, giving

$$(\boldsymbol{X}'\boldsymbol{X})^{-1} = \frac{1}{N\sum x_i^2 - (\sum x_i)^2}\begin{bmatrix} \sum x_i^2 & -\sum x_i \\ -\sum x_i & N \end{bmatrix} = \frac{1}{S_{xx}}\begin{bmatrix} \sum x_i^2/N & -\overline{x} \\ -\overline{x} & 1 \end{bmatrix}$$

where $S_{xx} = \sum x_i^2 - (\sum x_i)^2/N = \sum(x_i - \overline{x})^2$. Also,

$$\boldsymbol{X}'\boldsymbol{y} = \begin{bmatrix} \sum y_i \\ \sum x_i y_i \end{bmatrix}.$$

Hence the LS estimator of $\boldsymbol{\beta}$ is

$$\widehat{\boldsymbol{\beta}} = \begin{bmatrix} \widehat{\beta}_0 \\ \widehat{\beta}_1 \end{bmatrix} = (\boldsymbol{X}'\boldsymbol{X})^{-1}\boldsymbol{X}'\boldsymbol{y} = \frac{1}{S_{xx}}\begin{bmatrix} \sum x_i^2/N & -\overline{x} \\ -\overline{x} & 1 \end{bmatrix}\begin{bmatrix} \sum y_i \\ \sum x_i y_i \end{bmatrix}$$
$$= \frac{1}{S_{xx}}\begin{bmatrix} \overline{y}(\sum x_i^2) - \overline{x}\sum x_i y_i \\ -N\overline{x}\overline{y} + \sum x_i y_i \end{bmatrix}.$$

With some algebraic simplification, we get

$$\widehat{\beta}_1 = \frac{\sum x_i y_i - N\overline{x}\ \overline{y}}{S_{xx}} = \frac{\sum(x_i - \overline{x})(y_i - \overline{y})}{S_{xx}} = \frac{S_{xy}}{S_{xx}}.$$

Next,

$$\begin{aligned}
\widehat{\beta}_0 &= \frac{\overline{y}\sum x_i^2 - \overline{x}\sum x_i y_i}{S_{xx}} \\
&= \frac{\overline{y}\sum x_i^2 - \overline{x}(S_{xy} + N\overline{x}\ \overline{y})}{S_{xx}} \\
&= \frac{\overline{y}(\sum x_i^2 - N\overline{x}^2) - \overline{x}S_{xy}}{S_{xx}} \\
&= \frac{\overline{y}S_{xx} - \overline{x}\widehat{\beta}_1 S_{xx}}{S_{xx}} \\
&= \overline{y} - \widehat{\beta}_1\overline{x}.
\end{aligned}$$

In case of multiple regression this formula generalizes to

$$\widehat{\beta}_0 = \overline{y} - \widehat{\beta}_1\overline{x}_1 - \widehat{\beta}_2\overline{x}_2 - \cdots - \widehat{\beta}_p\overline{x}_p. \tag{2.34}$$

The error variance is estimated by

$$s^2 = \text{MS}_e = \frac{\sum[y_i - (\widehat{\beta}_0 + \widehat{\beta}_1 x_i)]^2}{N-2},$$

where the error d.f. is $\nu = N - 2$. ■

A case of special interest is when the columns of $\boldsymbol{X}$ are mutually orthogonal so that $\boldsymbol{X}'\boldsymbol{X}$ is diagonal. This not only makes the computation of $\widehat{\boldsymbol{\beta}}$ easy, but it also affords the following advantages:

(a) The $\widehat{\beta}_j$ are numerically independent in the sense that if any x_j term is dropped from the model by setting the corresponding $\beta_j = 0$, the other $\widehat{\beta}_k$ for $k \neq j$ remain unchanged. See Section 14.1.3 for further details.
(b) From (2.36) below, we see that the covariance matrix of $\widehat{\boldsymbol{\beta}}$ is $\sigma^2(\boldsymbol{X}'\boldsymbol{X})^{-1}$, and hence $\text{Cov}(\widehat{\beta}_j, \widehat{\beta}_k) = 0$ for all $j \neq k$, that is, the estimators $\widehat{\beta}_j$ are uncorrelated. Under the normality assumption this means that they are statistically independent.
(c) In the ANOVA, the total variation among the y_i's, SS_{tot}, can be partitioned into SSs which can be attributed independently to the individual x_j's and whose sum equals SS_{tot}. If any x_j is dropped from the model, the contributions due to the other x_j's to SS_{tot} is unchanged. The SS due to the dropped x_j gets pooled with SS_e along with its d.f.

Designs with a model matrix having this property are called **orthogonal designs**. The following provides a simple example.

Example 2.14 (A 2^2 Factorial Experiment)

A metal-cutting experiment is conducted to study the effects of cutting speed (factor A) and depth of cut (factor B) on the surface finish (y). Two levels (which we will call "low" and "high") of each factor are employed in the experiment resulting in a total of four treatment combinations. This is referred to as a 2^2 experiment because each factor has two levels and there are two factors. If there are $p \geq 2$ factors, each at two levels, then the experiment is referred to as a 2^p experiment. Let x_1 denote the level of A and x_2 denote the level of B. It is a common practice to use the coding $x_j = -1$ for the low level and $x_j = +1$ for the high level of the jth factor.

Suppose that the experiment is replicated once with $N = 4$ runs. The settings for the four runs of the experiment can be written in the form of a so-called **design matrix**:

$$\boldsymbol{D} = \begin{array}{c} \begin{matrix} x_1 & x_2 \end{matrix} \\ \begin{bmatrix} -1 & -1 \\ +1 & -1 \\ -1 & +1 \\ +1 & +1 \end{bmatrix} \end{array}.$$

This design with some made-up data is shown below:

		Factor B	
		Low	High
Factor A	Low	$y_{11} = 25$	$y_{12} = 35$
	High	$y_{21} = 30$	$y_{22} = 50$

Suppose the following model is fitted to the data:

$$E(y) = \beta_0 + \beta_1 x_1 + \beta_2 x_2 + \beta_{12} x_1 x_2.$$

Denote the response vector by $\boldsymbol{y} = (y_{11}, y_{21}, y_{12}, y_{22})' = (25, 30, 35, 50)'$. In matrix notation the model to be fitted is $E(\boldsymbol{y}) = \boldsymbol{X}\boldsymbol{\beta}$, where

$$\boldsymbol{X} = \begin{array}{c} \begin{array}{cccc} 1 & x_1 & x_2 & x_1x_2 \end{array} \\ \begin{bmatrix} +1 & -1 & -1 & +1 \\ +1 & +1 & -1 & -1 \\ +1 & -1 & +1 & -1 \\ +1 & +1 & +1 & +1 \end{bmatrix} \end{array} \quad \text{and} \quad \boldsymbol{\beta} = \begin{bmatrix} \beta_0 \\ \beta_1 \\ \beta_2 \\ \beta_{12} \end{bmatrix}. \tag{2.35}$$

The model matrix is orthogonal as can be verified by computing

$$\boldsymbol{X}'\boldsymbol{X} = \begin{bmatrix} 4 & 0 & 0 & 0 \\ 0 & 4 & 0 & 0 \\ 0 & 0 & 4 & 0 \\ 0 & 0 & 0 & 4 \end{bmatrix}.$$

Therefore,

$$\widehat{\boldsymbol{\beta}} = (\boldsymbol{X}'\boldsymbol{X})^{-1}\boldsymbol{X}'\boldsymbol{y} = \frac{1}{4}\begin{bmatrix} +25 + 30 + 35 + 50 \\ -25 + 30 - 35 + 50 \\ -25 - 30 + 35 + 50 \\ +25 - 35 - 35 + 50 \end{bmatrix} = \begin{bmatrix} 35 \\ 5 \\ 7.5 \\ 3.5 \end{bmatrix}.$$

It is readily checked that if any term is dropped from the model (e.g., the x_1x_2 term), the LS estimates of the remaining terms (e.g., β_0, β_1, and β_2) remain unchanged.

Let us interpret the estimates that we just computed. First, $\widehat{\beta}_0 = 35$ is the grand mean of the four y-values. Next, $\widehat{\beta}_1$, being the slope of x_1, equals the average change in y for a unit change in x_1. Thus, as x_1 is changed from -1 to $+1$ by two units when A is changed from low to high, the average response changes from $(35 + 25)/2$ to $(50 + 30)/2$; hence

$$\widehat{\beta}_1 = \frac{1}{2}\left[\frac{50 + 30}{2} - \frac{35 + 25}{2}\right] = 5.$$

An alternative way of expressing $\widehat{\beta}_1$, namely,

$$\widehat{\beta}_1 = \frac{1}{2}\left[\frac{(50-35)+(30-25)}{2}\right] = 5,$$

shows that it is proportional to the average of the differences between the y-values at high and low levels of A at the two levels of B. Thus $\widehat{\beta}_1$ is the average of the effects of changing A from low to high levels at the two levels of B. Hence $\widehat{\beta}_1$ is said to measure the average effect, called the main effect of A. Similarly,

$$\widehat{\beta}_2 = \frac{1}{2}\left[\frac{50+35}{2} - \frac{30+25}{2}\right] = \frac{1}{2}\left[\frac{(50-30)+(35-25)}{2}\right] = 7.5$$

measures the main effect of B. Finally,

$$\widehat{\beta}_{12} = \frac{1}{2}\left[\frac{(50-35)-(30-25)}{2}\right] = \frac{1}{2}\left[\frac{(50-30)+(35-25)}{2}\right] = 2.5$$

compares the change in response (namely, 50 – 35) when A is changed from low to high level at the high level of B versus the same change (namely, 30 – 25) at the low level of B. Hence it is called the **interaction** between A and B. Note that the estimated interaction is symmetrically defined in A and B, and it equals zero if $y_{22} - y_{12} = y_{21} - y_{11}$, that is, the changes in the response as A is changed from low to high levels at the two levels of B are equal. ■

2.3.2.3 *Confidence Intervals and Hypothesis Tests*

The sampling distribution of $\widehat{\boldsymbol{\beta}}$ can be shown to be multivariate normal (see Section A.3) with

$$\mathrm{E}(\widehat{\boldsymbol{\beta}}) = \boldsymbol{\beta} \quad \text{and} \quad \mathrm{Cov}(\widehat{\boldsymbol{\beta}}) = \sigma^2(\boldsymbol{X}'\boldsymbol{X})^{-1} = \sigma^2\boldsymbol{V} \quad \text{(say)}. \tag{2.36}$$

Furthermore,

$$s^2 = \mathrm{MS}_e \sim \frac{\sigma^2\chi^2_\nu}{\nu}$$

independently of $\widehat{\boldsymbol{\beta}}$. It follows that a $100(1-\alpha)\%$ CI on β_j is given by

$$\beta_j \in \left[\widehat{\beta}_j \pm t_{\nu,\alpha/2}s\sqrt{v_{jj}}\right], \tag{2.37}$$

where v_{jj} is the diagonal entry of $\boldsymbol{V} = (\boldsymbol{X}'\boldsymbol{X})^{-1}$ corresponding to $\widehat{\beta}_j$. Significance of any $\widehat{\beta}_j$ can be assessed by testing $H_{0j} : \beta_j = 0$, which is rejected at level α if

$$|t_j| = \frac{|\widehat{\beta}_j|}{s\sqrt{v_{jj}}} > t_{\nu,\alpha/2} \quad \text{or equivalently} \quad F_j = t_j^2 > f_{1,\nu,\alpha}. \tag{2.38}$$

Next consider the problem of estimating the expected value μ^* of $y = y^*$ when $\boldsymbol{x} = \boldsymbol{x}^* = (1, x_1^*, \ldots, x_p^*)'$, that is, $\mathrm{E}(y^*) = \mu^* = \boldsymbol{x}^{*'}\boldsymbol{\beta}$. The LS estimator of μ^* is $\widehat{\mu}^* = \widehat{y}^* = \boldsymbol{x}^{*'}\widehat{\boldsymbol{\beta}}$, which is normally distributed with

$$\mathrm{E}(\widehat{y}^*) = \mu^* \quad \text{and} \quad \mathrm{Var}(\widehat{y}^*) = \sigma^2(\boldsymbol{x}^{*'}\boldsymbol{V}\boldsymbol{x}^*). \tag{2.39}$$

Therefore the standard error of $\widehat{\mu}^* = \widehat{y}^*$ equals

$$\mathrm{SE}(\widehat{y}^*) = s\sqrt{\boldsymbol{x}^{*'}\boldsymbol{V}\boldsymbol{x}^*}. \tag{2.40}$$

Hence a $100(1-\alpha)\%$ CI on μ^* is given by

$$\mu^* \in \left[\boldsymbol{x}^{*'}\widehat{\boldsymbol{\beta}} \pm t_{\nu,\alpha/2}s\sqrt{\boldsymbol{x}^{*'}\boldsymbol{V}\boldsymbol{x}^*}\right].$$

Analogous to (2.21), a $100(1-\alpha)\%$ PI on y^* is given by

$$y^* \in \left[\boldsymbol{x}^{*'}\widehat{\boldsymbol{\beta}} \pm t_{\nu,\alpha/2}s\sqrt{1 + \boldsymbol{x}^{*'}\boldsymbol{V}\boldsymbol{x}^*}\right].$$

As in simple linear regression, the overall goodness of fit of the model may be assessed by R^2 defined in (2.24). Significance of goodness of fit can be tested by a test of the overall null hypothesis $H_0 : \beta_1 = \cdots = \beta_p = 0$ versus $H_1 :$ At least one $\beta_j \neq 0$. If H_0 cannot be rejected, then none of the predictor variables has a significant effect on $\mathrm{E}(y)$ and so the conclusion would be that the proposed model does not provide a good fit to the data and $\mathrm{E}(y)$ is essentially constant regardless of the values of the predictor variables. This hypothesis testing problem is a generalization of the problem $H_0 : \beta_1 = 0$ versus $H_1 : \beta_1 \neq 0$ in the case of simple linear regression, and its test derived below is a generalization of the F-test in (2.22).

The test of H_0 can be derived using the **extra sum of squares method**, which is discussed in more generality in Section 14.2.5. The method is as follows:

(a) Fit the **full model** (2.26) and compute the associated error sum of squares, $\mathrm{SS}_e = \sum_{i=1}^N \widehat{e}_i^2$, which has $\nu = N - (p+1)$ d.f.

(b) Fit the **reduced model** under H_0, which is $y_i = \beta_0 + e_i$ $(1 \le i \le N)$, and compute the associated error sum of squares, $\mathrm{SS}_{e0} = \sum_{i=1}^N \sum_{i=1}^N (y_i - \overline{y})^2$ (since $\widehat{\beta}_0 = \overline{y}$ under the reduced model), which has $\nu = N - 1$ d.f.

(c) The F-statistic for testing H_0 is given by

$$F = \frac{\mathrm{SS}_{H_0}/H_0 \text{ d.f.}}{\mathrm{SS}_e/\nu} = \frac{\mathrm{MS}_{H_0}}{\mathrm{MS}_e},$$

where $\mathrm{SS}_{H_0} = \mathrm{SS}_{e0} - \mathrm{SS}_e$ is the **hypothesis sum of squares** and the H_0 d.f. equals the number of linearly independent restrictions imposed by H_0 which is p. The H_0 d.f. also equals the difference in the error d.f. under the two models, namely, $N - 1 - [N - (p+1)] = p$.

Table 2.10 ANOVA Table for Multiple Linear Regression

Source	SS	d.f.	MS	F
Regression	$\text{SS}_{\text{reg}} = \widehat{\boldsymbol{\beta}}'\boldsymbol{X}'\boldsymbol{y} - N\overline{y}^2$	p	MS_{reg}	$\text{MS}_{\text{reg}}/\text{MS}_e$
Error	$\text{SS}_e = \widehat{\boldsymbol{e}}'\widehat{\boldsymbol{e}}$	$N-(p+1)$	MS_e	
Total	$\text{SS}_{\text{tot}} = \boldsymbol{y}'\boldsymbol{y} - N\overline{y}^2$	$N-1$		

(d) Since $\text{SS}_{e_0} = \text{SS}_{\text{tot}}$, it follows from the ANOVA identity (2.23) that $\text{SS}_{H_0} = \text{SS}_{e0} - \text{SS}_e = \text{SS}_{\text{tot}} - \text{SS}_e = \text{SS}_{\text{reg}}$. Therefore

$$F = \frac{\text{SS}_{\text{reg}}/p}{\text{SS}_e/\nu} = \frac{\text{MS}_{\text{reg}}}{\text{MS}_e},$$

which has an $F_{p,\nu}$-distribution under H_0 and hence the α-level F-test in (2.22) generalizes to the following:

$$\text{Reject } H_0 : \beta_1 = \cdots = \beta_p = 0 \quad \text{if } F = \frac{\text{SS}_{\text{reg}}/p}{\text{SS}_e/\nu} = \frac{\text{MS}_{\text{reg}}}{\text{MS}_e} > f_{p,\nu,\alpha}. \tag{2.41}$$

It is shown in Theorem 14.2 that $\text{SS}_{\text{reg}} = \widehat{\boldsymbol{\beta}}'\boldsymbol{X}'\boldsymbol{y} - N\overline{y}^2$.

These calculations can be summarized in the form of an ANOVA table shown in Table 2.10.

Finally, it may be noted that the extra sum of squares method can be used to also test partial null hypotheses of the form $H_0 : \beta_1 = \cdots = \beta_q = 0$ for $q < p$, resulting in F-tests; the d.f. of the F-statistic will be q and ν. In fact, the F-test in (2.38) for $H_{0j} : \beta_j = 0$ can be derived in this way with $q = 1$. The same principle underlies the partial F-tests used in stepwise regression (see Draper and Smith, 1998).

Example 2.15 (Salary Data: Multiple Linear Regression)

In Example 2.12 we fitted a simple linear regression model to log(Salary) using Education as the only predictor variable. Here we fit a multiple linear regression model using all the given predictor variables in Table 2.8, including the categorical variables Gender and Department, which are coded as follows:

Female = 1	if Gender = Female
Advertising = 1	if Department = Advertising
Engineering = 1	if Department = Engineering
Purchase = 1	if Department = Purchase
Female = 0	if Gender = Male

Advertising = 0	if Department ≠ Advertising
Engineering = 0	if Department ≠ Engineering
Purchase = 0	if Department ≠ Purchase

Minitab output for the fitted regression model that includes all the predictor variables is shown in Display 2.2. We see that PriorYr and Super are highly nonsignificant ($p = 0.395$ and 0.631, respectively). The indicator variable Female is also nonsignificant ($p = 0.115$), indicating no significant difference between male and female salaries keeping other variables fixed. We drop these nonsignificant variables and refit the model; the resulting output is shown in Display 2.3. We see that all predictor variables are significant in this fitted model, so we keep this as the final model.

It should be remarked that we have used a rather naive strategy of dropping nonsignificant variables from the model. In fact, model fitting is an exercise in finding the best subset of predictor variables, and there are methods available for this purpose, such as **stepwise regression** and **best-subsets regression**; see Section 2.3.2.4 for references on these methods. Also, different criteria may be used to choose the best model, for example, predictive ability. Lastly, model diagnostics based on residual plots and other analyses are an essential part of model fitting. We have not pursued these various lines of inquiry here because

```
The regression equation is
log(Salary) = 4.34 + 0.00748 YrsEm + 0.00168 PriorYr + 0.0170
              Education + 0.000390 Super + 0.0231 Female + 0.0550
              Advertising + 0.0880 Engineering + 0.0938 Purchase

Predictor          Coef      SE Coef        T       P
Constant        4.33502      0.02273   190.74   0.000
YrsEm          0.007479     0.001193     6.27   0.000
PriorYr        0.001684     0.001957     0.86   0.395
Education      0.017034     0.003336     5.11   0.000
Super         0.0003901    0.0008056     0.48   0.631
Female          0.02307      0.01429     1.61   0.115
Advertising     0.05500      0.02301     2.39   0.022
Engineering     0.08805      0.01806     4.88   0.000
Purchase        0.09378      0.02257     4.15   0.000

S = 0.0458640   R-Sq = 86.3% R-Sq(adj) = 83.4%

Analysis of Variance

Source          DF        SS        MS       F      P
Regression       8  0.491757  0.061470   29.22  0.000
Residual Error  37  0.077830  0.002104
Total           45  0.569587
```

Display 2.2 Minitab output for regression of log(Salary) on all predictor variables.

```
The regression equation is
log(Salary) = 4.35 + 0.00766 YrsEm + 0.0184 Education + 0.0511
              Advertising + 0.0851 Engineering + 0.0876 Purchase

Predictor          Coef    SE Coef       T      P
Constant        4.35141    0.01960  221.97  0.000
YrsEm          0.007660   0.001208    6.34  0.000
Education      0.018371   0.003124    5.88  0.000
Advertising     0.05110    0.02318    2.20  0.033
Engineering     0.08509    0.01800    4.73  0.000
Purchase        0.08759    0.02274    3.85  0.000

S = 0.0467652   R-Sq = 84.6% R-Sq(adj) = 82.7%

Analysis of Variance

Source          DF        SS        MS      F      P
Regression       5  0.482107  0.096421  44.09  0.000
Residual Error  40  0.087479  0.002187
Total           45  0.569587
```

Display 2.3 Minitab output for regression of log(Salary) on all predictor variables.

we did not cover these techniques in this brief review of multiple regression. See Draper and Smith (1998) for an exhaustive coverage of these techniques.

In the final model we see that YrsEm and Education have significant effects on Salary. The significant coefficients of Advertising, Engineering, and Purchase mean that these department salaries differ significantly from those of Sales. The residual plots from this multiple regression are shown in Figure 2.10; we see that both the normal and fitted-values plots of the residuals are quite satisfactory and so there are no gross violations of the model assumptions. ■

2.3.2.4 Additional Remarks

Multiple regression is a vast subject; in this brief review we have only scratched its surface. Two excellent references on the subject are the books by Draper and Smith (1998) and Montgomery, Peck, and Vining (2006). Here we make a few remarks about some important aspects of regression model fitting that are not covered in our brief review:

(a) In practice, one normally has a choice of many candidate predictor variables to select from for fitting the model. These include the original variables plus any additional ones created by making transformations to account for possible nonlinearities in the relationships, interactions between the variables, and so on. Stepwise regression and best-subsets regression algorithms are commonly used to select the variables to enter the model.

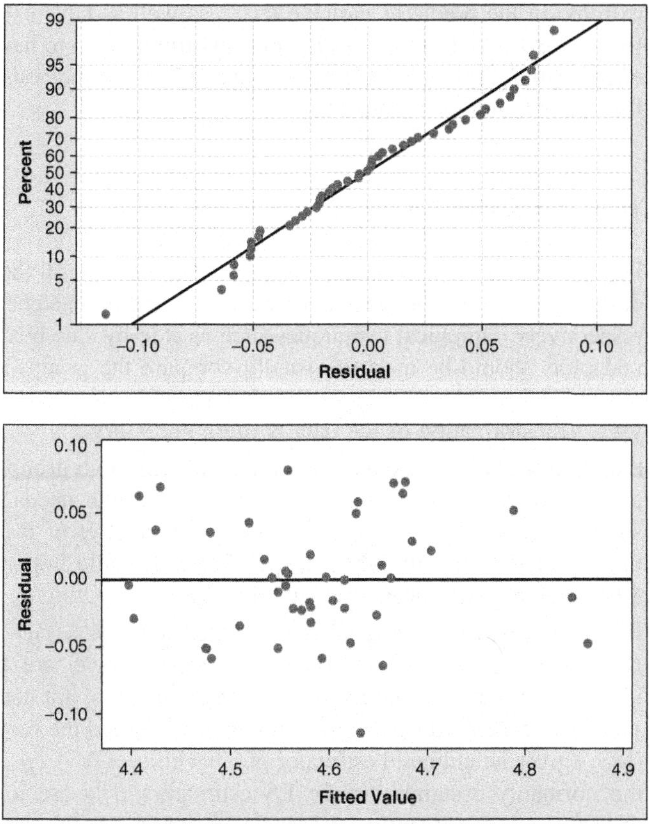

Figure 2.10 Normal (top) and fitted-values (bottom) plots of residuals from multiple linear regression of log(Salary) versus YrsEm, Education, and department.

(b) A frequent problem, especially when dealing with large data sets, is that there are approximate linear dependencies among the predictor variables. As a result, the columns of the model matrix $\boldsymbol{X}$ are approximately linearly dependent, making the $\boldsymbol{X}'\boldsymbol{X}$ matrix difficult to invert. This is called the **multicollinearity problem**. Methods such as **ridge regression** are available to deal with this problem.

Exact linear dependencies can exist if the predictor variables are constrained by a linear relationship as occurs in **mixture experiments** (discussed in Chapter 10) where the predictor variables are the proportions of different ingredients in a product which add up to 1. Specialized models are used to account for this linear dependency.

(c) Another important problem that arises in regression analysis is that of **outliers** and **influential observations**. Outlier observations have y-values that deviate significantly from the fitted $\widehat{y}$-values according to the given model and are identified by large standardized residuals. Influential observations exert undue influence on the model fit. For example, observations that are extreme in the predictor variable space as well as have extreme y-values (hence also called **high-leverage observations**), tend to have undue influence on the slope coefficients. Robust regression methods may be used for dealing with such data.

2.4 CHAPTER SUMMARY

(a) For comparing two treatment groups, the analogs of the CR and RB designs are the independent samples design and the matched pairs design, respectively. Graphical techniques such as side-by-side box plots and normal plots should be used to visually compare the groups and detect any nonnormality or outliers in the data before subjecting them to formal analyses. Transformation of the data is often necessary.

(b) Standard normal theory methods for comparing two groups are the t-test and t-interval. To determine the sample size, one needs to specify the significance level (type I error) α and the type II error β or equivalently the power, $1 - \beta$, with which a specified practically important difference δ between the two means must be detected.

(c) In linear regression, the goal is to fit the model $y_i = \beta_0 + \beta_1 x_{i1} + \cdots + \beta_p x_{ip} + e_i$ $(1 \leq i \leq N)$ where the random errors e_i are assumed to be i.i.d. $N(0, \sigma^2)$ r.v.'s. The model is linear in the β's, not necessarily in the x's. The LS method is used to estimate the β_j's and the mean square error (MS_e) gives an unbiased estimator of σ^2 with $\nu = N - (p + 1)$ d.f. Under the normality assumptions the LS estimators $\widehat{\beta}_j$'s are jointly normally distributed independently of $MS_e = s^2$, which is chi-square distributed with ν d.f. This provides the statistical basis for inferences on the β_j's based on the t- or F-distribution. The theory of linear models and the derivations of the results used in this chapter are given in Chapter 14.

EXERCISES

Section 2.1 (Experiments for a Single Treatment)

2.1 (From Tamhane and Dunlop, 2000, Example 7.7. Reprinted by permission of Pearson Education, Inc.) Henry Cavendish, a British scientist, made 29 measurements on the density of earth (expressed in grams per cubic centimeters) in 1798. They are given in Table 2.11.

(a) Make a box plot and a normal plot of the data and comment on their shapes. Does the normality assumption seem reasonable?

(b) If the normality assumption seems reasonable, calculate a 95% CI for the mean density of earth.

Table 2.11 Cavendish's Measurements on Density of Earth

5.50	5.61	4.88	5.07	5.26	5.55	5.36	5.29	5.58	5.65
5.57	5.53	5.62	5.29	5.44	5.34	5.79	5.10	5.27	5.39
5.42	5.47	5.63	5.34	5.46	5.30	5.75	5.86	5.85	

Source: Tamhane and Dunlop (2000, p. 250). Reprinted by permission of Pearson Education, Inc.

Table 2.12 Tear Strength in PSI of Silicone Rubber

33.74	34.40	32.62	32.57	34.69	33.78	36.76	34.31
37.61	33.78	35.43	33.22	33.53	33.68	33.24	32.98

Source: Tamhane and Dunlop (2000, p. 251). Reprinted by permission of Pearson Education, Inc.

2.2 (From Tamhane and Dunlop, 2000, Example 7.8. Reprinted by permission of Pearson Education, Inc.) Table 2.12 gives data on tear strengths of 16 sample sheets of silicone rubber used in a high-voltage transformer.

(a) Make a box plot and a normal plot of the data and comment on their shapes. Are there outliers in the data? Does the normality assumption seem reasonable?

(b) Delete any outliers, and if the normality assumption now seems reasonable, calculate a 95% CI for the mean tear strength .

2.3 (From Tamhane and Dunlop, 2000, Exercise 7.17. Reprinted by permission of Pearson Education, Inc.) A new home glucose monitor used by diabetic patients is tested for accuracy (lack of bias) and precision (low variance) by having 25 persons measure the glucose level of a sample having a known glucose concentration of 118 mg/dl. The new monitor is preferred to the current monitor if it is accurate (has negligible bias) and if its standard deviation σ is less than 10 mg/dl. Table 2.13 gives the 25 readings.

(a) Make a normal plot of the data. Does the normality assumption seem reasonable?

(b) Check if the new monitor is preferable to the current monitor by testing if its mean deviates significantly from the nominal value of 118 mg/dl and if its standard deviation $\sigma < 10$ mg/dl. Do both tests at $\alpha = 0.05$.

Table 2.13 Glucose Level Readings on Test Sample

125	123	117	123	115	112	128	118	124	111
116	109	125	120	113	123	112	118	121	118
122	115	105	118	131					

Source: Tamhane and Dunlop (2000, p. 265). Reprinted by permission of Pearson Education, Inc.

Table 2.14 Hospitalization Cost Data (dollars)

Control group	478	605	626	714	818	1,203	1,204	1,323	2,150	2,700
	2,969	3,151	3,565	3,626	3,739	4,148	4,382	4,576	6,953	6,963
	7,062	7,284	7,829	8,681	9,319	12,664	17,539	23,237	26,677	32,913
Treatment group	528	650	912	1,130	1,289	1,651	1,706	1,708	1,882	1,998
	2,391	2,432	2,697	2,829	3,039	3,186	3,217	3,345	3,590	4,153
	5,928	6,018	6,267	6,790	7,365	8,635	9,615	13,890	15,225	19,083

Source: Tamhane and Dunlop (2000, Table 8.1). Reprinted by permission of Pearson Education, Inc.

2.4 Refer to the previous exercise. Suppose we are designing this experiment and want to know how many monitors to test. The experiment is to be designed so that if there is bias of 4 mg/dl or more in the monitor, it will be detected with at least 80% probability using a 0.05-level two-sided test of the hypothesis that the bias is zero, that is, the true mean reading of the monitor is 118 mg/dl. For design purposes, assume $\sigma = 8$ mg/dl. Determine the number of monitors to be tested.

Section 2.2 (Experiments for Two Treatments)

2.5 Tell in each of the instances of Exercise 1.1 whether the study uses an independent samples design or a matched pairs design

2.6 (From Tamhane and Dunlop, 2000, Example 8.3. Reprinted by permission of Pearson Education, Inc.) Table 2.14 gives hospitalization cost data (in dollars) for geriatric patients cared for by two different methods.

(a) Make side-by-side box plots of the data. What do you conclude?

(b) Make normal plots of the data for the two groups. Is the normality assumption satisfied?

(c) Log-transform the data and make the normal plots. Is the normality assumption now satisfied?

2.7 Refer to the previous exercise. Use the log-transformed cost data for the following tests. You may assume normality.

(a) Test $H_0 : \sigma_1^2 = \sigma_2^2$ versus $H_1 : \sigma_1^2 \neq \sigma_2^2$ using $\alpha = 0.10$. Is the homoscedasticity assumption consistent with the data?

(b) Test $H_0 : \mu_1 = \mu_2$ versus $H_1 : \mu_1 \neq \mu_2$ at $\alpha = 0.05$ assuming $\sigma_1^2 = \sigma_2^2$. What is your conclusion?

2.8 (From Tamhane and Dunlop, 2000, Exercise 8.11. Reprinted by permission of Pearson Education, Inc.) Two methods of measuring the atomic weight

Table 2.15 Atomic Weights of Carbon Determined by Two Methods

Method 1	12.0129	12.0072	12.0064	12.0054	12.0016
	11.9853	11.9949	11.9985	12.0077	12.0061
Method 2	12.0318	12.0246	12.0069	12.0006	12.0075

Source: Best and Rayner (1987).

Table 2.16 Heights of Plants Fertilized by Two Methods

Plant Pair	Cross-Fertilized	Self-Fertilized
1	23.5	17.4
2	12.0	20.4
3	21.0	20.0
4	22.0	20.0
5	19.1	18.4
6	21.5	18.6
7	22.1	18.6
8	20.4	15.3
9	18.3	16.5
10	21.6	18.0
11	23.3	16.3
12	21.0	18.0
13	22.1	12.8
14	23.0	15.5
15	12.0	18.0

Source: Hand et al. (1994, data set 3). Reprinted by permission of Taylor & Francis.

of carbon (the nominal atomic weight is 12) yielded the results shown in Table 2.15.

(a) Test $H_0 : \sigma_1^2 = \sigma_2^2$ versus $H_1 : \sigma_1^2 \neq \sigma_2^2$ using $\alpha = 0.10$. Is the homoscedasticity assumption consistent with the data?

(b) Test $H_0 : \mu_1 = \mu_2$ versus $H_1 : \mu_1 \neq \mu_2$ at $\alpha = 0.05$ assuming $\sigma_1^2 = \sigma_2^2$. What is your conclusion?

(c) Repeat (b) without assuming $\sigma_1^2 = \sigma_2^2$. Compare the results.

2.9 Fifteen pairs of seedlings, one produced by cross-fertilization and the other produced by self-fertilization, were grown together under nearly identical conditions (Darwin, 1876). The data shown in Table 2.16 are the final heights of plants after a fixed period of time. The aim of the experiment was to show that the cross-fertilized plants grow bigger. Do appropriate analyses and draw conclusions.

2.10 (From Tamhane and Dunlop, 2000, Exercise 8.26. Reprinted by permission of Pearson Education, Inc.) Consider the problem of testing H_0 : $\mu_1 = \mu_2$ versus $H_1 : \mu_1 > \mu_2$ based on the independent random samples from two normal populations, $N(\mu_1, \sigma_1^2)$ and $N(\mu_2, \sigma_2^2)$. For design purposes, assume that σ_1^2 and σ_2^2 are known . Let n_1 and n_2 be the sample sizes and $\overline{y}_1$ and $\overline{y}_2$ be the sample means. The α-level test of H_0 rejects if

$$z = \frac{\overline{y}_1 - \overline{y}_2}{\sqrt{\sigma_1^2/n_1 + \sigma_2^2/n_2}} > z_\alpha.$$

(a) Show that the power of the α-level test as a function of $\mu_1 - \mu_2$ is given by

$$\pi(\mu_1 - \mu_2) = \Phi\left(-z_\alpha + \frac{\mu_1 - \mu_2}{\sqrt{\sigma_1^2/n_1 + \sigma_2^2/n_2}}\right).$$

(b) For detecting a specified difference $\mu_1 - \mu_2 = \delta > 0$, show that for a fixed total sample size $n_1 + n_2 = N$ the power is maximized when

$$n_1 = \frac{\sigma_1}{\sigma_1 + \sigma_2}N \quad \text{and} \quad n_2 = \frac{\sigma_2}{\sigma_1 + \sigma_2}N,$$

that is, the optimum values of the n_i are proportional to the σ_i (ignoring the integer restrictions on the n_i).

(c) Show that the smallest total sample size N required to guarantee at least $1 - \beta$ power when $\mu_1 - \mu_2 = \delta > 0$ is given by

$$N = \left[\frac{(z_\alpha + z_\beta)(\sigma_1 + \sigma_2)}{\delta}\right]^2.$$

(d) Calculate the sample sizes n_1 and n_2 for $\alpha = 0.05, \beta = 0.10$, $\delta = 2.0, \sigma_1 = 2.0, \sigma_2 = 4.0$.

Section 2.3 (Linear Regression)

2.11 Show that for fitting a regression line through the origin, that is, for fitting the straight-line model $y = \beta x + e$, based on N pairs of observations, $(x_1, y_1), (x_2, y_2), \ldots, (x_N, y_N)$, the LS estimator of the slope coefficient is

$$\widehat{\beta} = \frac{\sum x_i y_i}{\sum x_i^2}.$$

2.12 Derive formula (2.34) for $\widehat{\beta}_0$.

2.13 Consider the following design problem for fitting a simple linear regression model: Choose the levels $x_1, x_2, \ldots, x_N$ over a specified range $[L, U]$ where the response variable y should be observed in order to minimize the variance of the slope estimator $\widehat{\beta}_1$.

(a) Show that the optimum choice of the x_i's is to take $N/2$ observations at L and $N/2$ observations at U. The corresponding minimum $\text{Var}(\widehat{\beta}_1) = 4\sigma^2/N(U-L)^2$.

(b) If you are not certain that a linear relationship holds between y and x over the interval $[L, U]$, would you take observations only at L and U? If not, how would you spread your observations? What information would you need for deciding the spacings of the observations?

2.14 (From Tamhane and Dunlop, 2000, Exercise 10.20. Reprinted by permission of Pearson Education, Inc.) Table 2.17 gives data on stopping distance y at different speeds. Two measurements were made at each speed.

(a) Fit a straight line to these data. Plot the residuals against speed. Which model assumption is violated?

(b) Use the lack-of-fit test from (2.25) to check if a straight line provides a good fit to these data. Compare the result with the residual plot from part (a).

(c) Fit a quadratic equation of stopping distance versus speed squared. Plot the residuals against speed. Do the residuals show any pattern? Are the model assumptions satisfied?

(d) Provide a physical explanation for the quadratic relationship between stopping distance and speed.

2.15 To obtain numerically stable computation of regression coefficients it is often a good idea to first standardize all variables as follows:

$$y_i^* = \frac{y_i - \bar{y}}{s_y} \quad \text{and} \quad x_{ij}^* = \frac{x_{ij} - \bar{x}_j}{s_{x_j}} \quad (1 \le i \le N, 1 \le j \le p),$$

where $\bar{y}$ and $\bar{x}_j$ are the sample means and s_y and s_{x_j} are the sample standard deviations of the corresponding variables. Denote the estimated

Table 2.17 Stopping Distance of a Car at Different Speeds

Speed x, mph	20	20	30	30	40	40	50	50	60	60
Distance y, ft	16.3	26.7	39.2	63.5	65.7	98.4	104.1	155.6	217.2	160.8

Source: Mosteller et al. (1983, Chapter 12, Example 2A).

regression coefficients (called the standardized regression coefficients) for these data by $\widehat{\beta}_j^*$ $(0 \le j \le p)$. Note that the $\widehat{\beta}_j^*$ are unitless and can be compared with each other to assess the relative importance of the x_j's in their strength of linear relationship of y in the presence of other predictor variables.

(a) Use formula (2.34) to show that $\widehat{\beta}_0^* = 0$. Hence we may define the model matrix $\boldsymbol{X}^*$ for the standardized data to be an $N \times p$ matrix by excluding the first column of 1's; similarly let $\widehat{\boldsymbol{\beta}}^* = (\widehat{\beta}_1^*, \ldots, \widehat{\beta}_p^*)'$ to be a $p \times 1$ vector.

(b) Show that the correlation matrix $\boldsymbol{R} = \{r_{x_j x_k}\}$ among the x_j's and the correlation vector $\boldsymbol{r} = (r_{yx_1}, \ldots, r_{yx_p})'$ among y and x_j for $j = 1, 2, \ldots, p$ are given by

$$\boldsymbol{R} = \frac{1}{N-1}\boldsymbol{X}^{*'}\boldsymbol{X}^* \quad \text{and} \quad \boldsymbol{r} = \frac{1}{N-1}\boldsymbol{X}^{*'}\boldsymbol{y}^*.$$

(c) Show that the vector of standardized regression coefficients is given by $\widehat{\boldsymbol{\beta}}^* = \boldsymbol{R}^{-1}\boldsymbol{r}$; thus it is sufficient to know only the correlation coefficients among all the variables to compute $\widehat{\boldsymbol{\beta}}^*$.

(d) Show that the unstandardized regression coefficients can be obtained from the standardized ones as follows:

$$\widehat{\beta}_j = \widehat{\beta}_j^* \left(\frac{s_y}{s_{x_j}}\right) \qquad (1 \le j \le p)$$

and $\widehat{\beta}_0$ is obtained from (2.34). Thus, to compute $\widehat{\boldsymbol{\beta}}$, we only need to know the sample means and the sample standard deviations of all the variables, besides $\boldsymbol{R}$ and $\boldsymbol{r}$.

2.16 Consider a 2^3 experiment in which there are three factors, A, B, and C, each at two levels. Suppose $N = 8$ runs are made at each of the eight treatment combinations, and the following model is fitted:

$$\begin{aligned} \mathrm{E}(y) = {} & \beta_0 + \beta_1 x_1 + \beta_2 x_2 + \beta_3 x_3 \\ & + \beta_{12} x_1 x_2 + \beta_{13} x_1 x_3 + \beta_{23} x_2 x_3 + \beta_{123} x_1 x_2 x_3, \end{aligned}$$

where $x_j = -1$ or $+1$ depending on whether the jth factor is set at the low or high level $(j = 1, 2, 3)$.

(a) Write the design matrix for this experiment.

(b) Write the model matrix for the above model.

Table 2.18 Industrial Sales Data

No.	x_1	x_2	y
1	31	1.85	4.20
2	46	2.80	7.28
3	40	2.20	5.60
4	49	2.85	8.12
5	38	1.80	5.46
6	49	2.80	7.42
7	31	1.85	3.36
8	38	2.30	5.88
9	33	1.60	4.62
10	42	2.15	5.88

Source: Tamhane and Dunlop (2000, Table 11.4). Reprinted by permission of Pearson Education, Inc.

2.17 (From Tamhane and Dunlop, 2000, Example 11.7. Reprinted by permission of Pearson Education, Inc.) Table 2.18 gives data for 10 sales territories of an industrial products company on the sales revenue (y) in millions of dollars, number of salespersons (x_1), and amount of total sales expenditures (including salaries, advertising, etc.) (x_2) in millions of dollars. It is desired to build a predictive model for sales revenue based on the two predictor variables.

(a) Compute the correlation matrix between x_1, x_2, and y. What do you conclude?

(b) Fit the regression model $E(y) = \beta_0 + \beta_1 x_1 + \beta_2 x_2$. Which one of x_1 and x_2 has a significant linear relationship with y using $\alpha = 0.05$? Explain the results of regression based on the correlations computed in part (a).

(c) Drop the nonsignificant predictor variable and fit a simple linear regression model in the other predictor variable.

CHAPTER 3

Single Factor Experiments: Completely Randomized Designs

The CR design introduced in Section 1.4 is used when the experimenter has decided to apply the treatments to a single group of essentially homogeneous experimental units—any heterogeneity among them being controlled by their random assignment to the treatments. The simplest CR design involves a single treatment factor; later chapters discuss factorial experiments that use a CR design. If there is no random assignment, then we have an observational study. Both CR experimental designs and observational studies with a single treatment factor are referred to as **one-way layouts** because the data can be classified according to the levels of a single factor. As mentioned in Chapter 1, the same methods of analysis are used for both types of studies, but the conclusions that can be drawn from experimental studies are stronger. Generally, a key part of the analysis consists of a one-way ANOVA. If there are covariates present, they can be taken into account by a combination of ANOVA and regression analysis, called the **analysis of covariance** (**ANCOVA**).

The outline of the chapter is as follows. Section 3.1 introduces simple exploratory analyses (summary statistics and graphical plots) for one-way layout data. Section 3.2 gives a linear model for one-way layout data. Section 3.3 discusses methods of inference, including estimation of parameters and ANOVA. Section 3.4 covers diagnostic methods for detecting model violations and outliers. These methods are based on residual analyses and plots and are applicable to more complex designs. Section 3.5 discusses data transformations that are useful for correcting some of the model violations. Section 3.6 shows how the power analysis of the ANOVA F-test can be used for sample size determination. Section 3.7 studies the experiments where the treatment factor is quantitative and its effects on the response variable can be partitioned into various polynomial terms (e.g., linear, quadratic). Section 3.8 studies the

Statistical Analysis of Designed Experiments: Theory and Applications By Ajit C. Tamhane

Table 3.1 Data from One-Way Layout Experiment

Treatment	Data	Mean	SD
1	$y_{11}, y_{12}, \ldots, y_{1n_1}$	$\overline{y}_1$	s_1
2	$y_{21}, y_{22}, \ldots, y_{2n_2}$	$\overline{y}_2$	s_2
$\vdots$	$\vdots$	$\vdots$	$\vdots$
a	$y_{a1}, y_{a2}, \ldots, y_{an_a}$	$\overline{y}_a$	s_a

one-way ANCOVA design. Section 3.9 gives some mathematical details and derivations and Section 3.10 gives a chapter summary.

3.1 SUMMARY STATISTICS AND GRAPHICAL DISPLAYS

We will first introduce the basic notation. Let $a \geq 2$ denote the number of treatments, which are the levels of a single treatment factor. Let N denote the total number of experimental units. In a CR design, the experimental units are randomly assigned to the treatments with n_i to the ith treatment subject to $\sum_{i=1}^{a} n_i = N$. In clinical trials this is known as a **parallel-arms study** with different therapies being the treatment arms and the patients being the experimental units. Let y_{ij} denote the observed response of the jth experimental unit in the ith treatment group $(1 \leq i \leq a, 1 \leq j \leq n_i)$. The data may be presented in a tabular form as shown in Table 3.1.

The first step in the analysis of one-way layout data is to compute some measures of the center and dispersion for each treatment and make graphical plots. Usually, the sample means $\overline{y}_i$ and the sample SDs s_i are used:

$$\overline{y}_i = \frac{\sum_{j=1}^{n_i} y_{ij}}{n_i} \quad \text{and} \quad s_i = \sqrt{\frac{\sum_{j=1}^{n_i}(y_{ij} - \overline{y}_i)^2}{n_i - 1}} \quad (1 \leq i \leq a). \tag{3.1}$$

Side-by-side box plots allow a convenient visual comparison of the treatments. The following example illustrates these techniques.

Example 3.1 (Anorexia Data: Summary Statistics and Graphical Displays)

An experiment was conducted to compare three therapies for treating anorexia in young girls to help them gain weight. The therapies were control therapy, cognitive behavioral therapy, and family therapy. We refer to the three groups as control, behavioral, and family.

There were unequal numbers of girls in each group, but for the purpose of this example we dropped some observations from the control and behavioral groups to have an equal number, $n = 17$, per group. Table 3.2 gives the baseline weights and weight gains (posttreatment weight minus pretreatment weight) of the girls.

Table 3.2 Baseline Weights and Weight Gains (lb) of Anorexia Patients

	Control Group (1)		Behavioral Group (2)		Family Group (3)	
Patient No.	Baseline Weight	Weight Gain	Baseline Weight	Weight Gain	Baseline Weight	Weight Gain
1	80.7	−0.5	80.5	1.7	83.8	11.4
2	89.4	−9.3	84.9	0.7	83.3	11.0
3	91.8	−5.4	81.5	−0.1	86.0	5.5
4	74.0	12.3	82.6	−0.7	82.5	9.4
5	78.1	−2.0	79.9	−3.5	86.7	13.6
6	88.3	−10.2	88.7	14.9	79.6	−2.9
7	87.3	−12.2	94.9	3.5	76.9	−0.1
8	75.1	11.6	76.3	17.1	94.2	7.4
9	80.6	−7.1	81.0	−7.6	73.4	21.5
10	78.4	6.2	80.5	1.6	80.5	−5.3
11	77.6	−0.2	85.0	11.7	81.6	−3.8
12	88.7	−9.2	89.2	6.1	82.1	13.4
13	81.3	8.3	81.3	1.1	77.6	13.1
14	78.1	3.3	76.5	−4.0	83.5	9.0
15	70.5	11.3	70.0	20.9	89.9	3.9
16	77.3	0.0	80.4	−9.1	86.0	5.7
17	85.2	−1.0	83.3	2.1	87.3	10.7

Source: From Hand et al. (1994, data set 285). Reprinted by permission of Taylor & Francis.

Here the variable of interest is weight gain and the baseline weight is a possible covariate. To compare the average weight gains of the three groups in an unbiased manner, we must first check if weight gains and baseline weights are correlated. This is done by making a groupwise scatter plot of weight gain versus baseline weight with separately fitted regression lines in Figure 3.1. We see that in each group there is a negative trend between the two variables with the trend being strongest in the control group. The plot shows that the weight gains are more consistent (regardless of the baseline weight) in the family group. On the other hand, in the control group heavier girls lost weights while the lighter girls did not lose as much or slightly gained weights. This is probably because heavier girls are more prone to lose while the lighter girls are not. In any case, as a first step, it is necessary to check whether the random assignment has made the three groups comparable w.r.t. the baseline weights. (A more sophisticated approach would be to adjust for the effects of baseline weights by using ANCOVA; see Exercise 3.29.) Next we would like to know (i) whether there are significant differences between the groups, (ii) whether the behavioral and family therapies are better than the control therapy, and (iii) which therapy is best. To begin answering these questions, we give the summary statistics in Table 3.3 and side-by-side box plots in Figure 3.2 for baseline weights and weight gains.

We see from the summary statistics and box plots that the three therapy groups are quite comparable w.r.t. the baseline weights. Thus randomization appears to

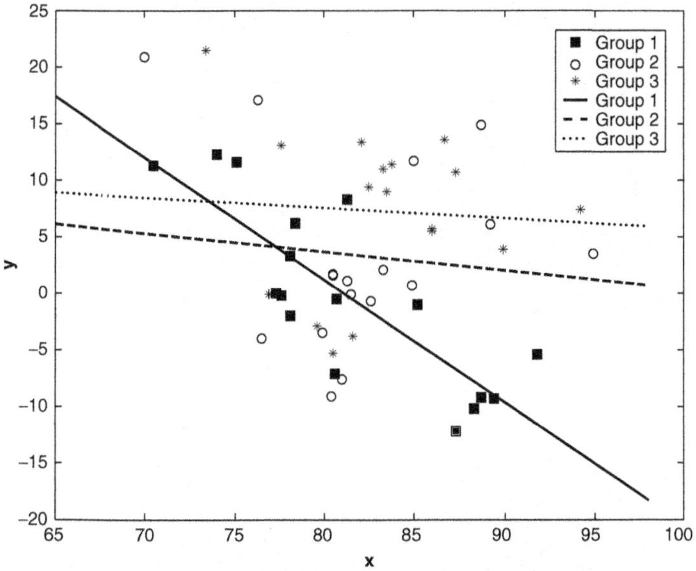

Figure 3.1 Plot of weight gains versus baseline weights for three therapy groups.

Table 3.3 Summary Statistics for Anorexia Data

Therapy	Baseline Weight		Weight Gain	
Group	Mean	SD	Mean	SD
Control	81.318	6.117	−0.241	8.061
Behavior	82.147	5.602	3.318	8.409
Family	83.229	5.017	7.265	7.157

have been effective in equalizing the groups. Also, there are differences in weight gains, suggesting that the behavioral and family therapies had the desired effect of a positive weight gain. There are two outliers in the baseline weight data of the behavioral therapy group, but without a specific reason for their invalidity, we do not exclude them from analysis. ■

3.2 MODEL

The standard model for a one-way layout assumes that the y_{ij} are independently distributed as $N(\mu_i, \sigma^2)$ r.v.'s. We write this model as

$$y_{ij} = \mu_i + e_{ij} \qquad (1 \le i \le a, 1 \le j \le n_i), \tag{3.2}$$

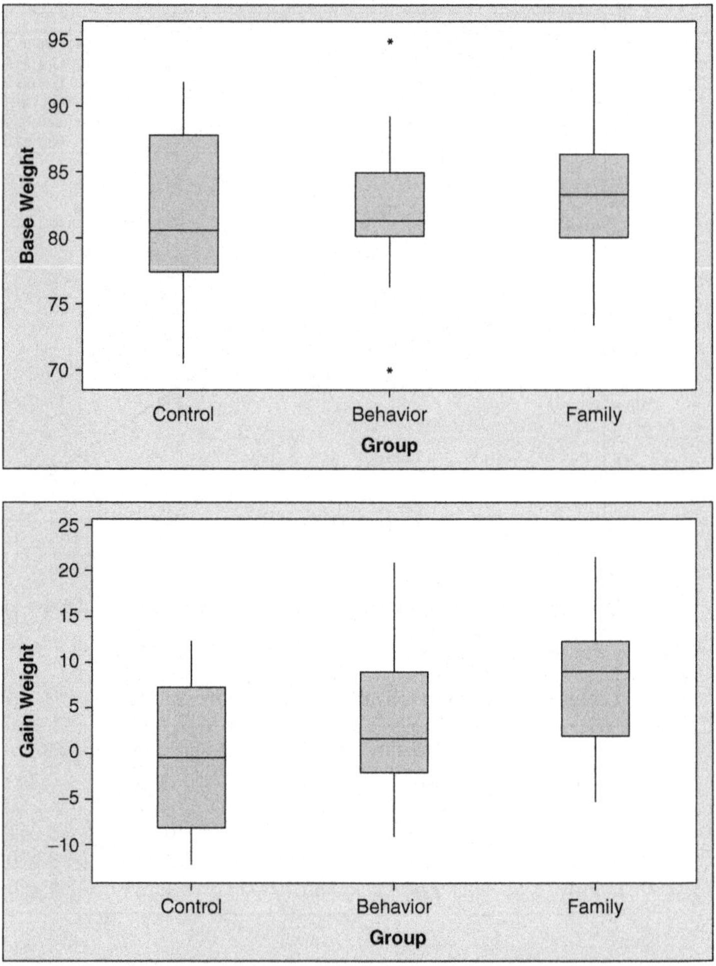

Figure 3.2 Side-by-side box plots of baseline weights and weight gains.

where the μ_i are unknown *fixed* treatment means and the e_{ij} are i.i.d. $N(0, \sigma^2)$ experimental errors. The assumption of a common error variance for all treatments is known as the **homoscedasticity assumption**.

Often, the above model is rewritten as

$$y_{ij} = \mu + \alpha_i + e_{ij} \qquad (1 \leq i \leq a, 1 \leq j \leq n_i), \tag{3.3}$$

so that the $\alpha_i = \mu_i - \mu$ are deviations of the treatment means μ_i from some overall mean μ and are referred to as **treatment effects**. Clearly, to define the parameters of this model uniquely, we need to specify a constraint on them. We will employ the constraint $\sum n_i\alpha_i = 0$, which implies that

$$\mu = \frac{\sum_{i=1}^{a} n_i\mu_i}{\sum_{i=1}^{a} n_i} \qquad \text{and} \qquad \alpha_i = \mu_i - \mu. \tag{3.4}$$

Alternately, one may specify the constraint $\sum \alpha_i = 0$, which leads to $\mu = \sum \mu_i/a$ being the unweighted average of the μ_i.

3.3 STATISTICAL ANALYSIS

3.3.1 Estimation

It is easy to see that the LS estimators of the treatment means μ_i are $\widehat{\mu}_i = \overline{y}_i$. Although the definition of μ as the weighted average of the μ_i, being dependent on the sample sizes n_i, is rather nonintuitive, the LS estimators

$$\widehat{\mu} = \frac{\sum_{i=1}^{a} n_i\overline{y}_i}{\sum_{i=1}^{a} n_i} = \overline{\overline{y}} \qquad \text{and} \qquad \widehat{\alpha}_i = \overline{y}_i - \overline{\overline{y}} \qquad (1 \le i \le a),$$

where $\overline{\overline{y}}$ is the grand mean of the data, are quite intuitive. If the alternate constraint is used, then $\widehat{\mu} = \sum \overline{y}_i/a$ and $\widehat{\alpha}_i = \overline{y}_i - \sum \overline{y}_i/a$. In either case, as noted above, $\widehat{\mu}_i = \widehat{\mu} + \widehat{\alpha}_i = \overline{y}_i$ $(1 \le i \le a)$.

The fitted values equal

$$\widehat{y}_{ij} = \widehat{\mu}_i = \overline{y}_i$$

and the residuals equal

$$\widehat{e}_{ij} = y_{ij} - \widehat{y}_{ij} = y_{ij} - \overline{y}_i. \tag{3.5}$$

An unbiased estimator of σ^2 is given by

$$s^2 = \text{MS}_e = \frac{\text{SS}_e}{N-a} = \frac{\sum_{i=1}^{a}\sum_{j=1}^{n_i} \widehat{e}_{ij}^{\,2}}{N-a},$$

where SS_e is the **error sum of squares**, MS_e is the **mean square error**, and $\nu = N - a$ is the **error degrees of freedom (d.f.)**, being the total number of observations minus a, the number of independent parameters (namely, the μ_i's) in the model. Note that s^2 can be expressed as

$$s^2 = \frac{\sum_{i=1}^{a}(n_i - 1)s_i^2}{\sum_{i=1}^{a}(n_i - 1)},$$

which is the pooled sample variance from all treatments. The weights used in pooling are the d.f., $n_i - 1$, of the individual sample variances s_i^2, and the total error d.f. is the sum of the individual d.f.

3.3.2 Analysis of Variance

Formal comparisons between the treatment means usually begin with a test of the **overall null hypothesis** of equality of all means:

$$H_0 : \mu_1 = \mu_2 = \cdots = \mu_a \quad \text{vs.} \quad H_1 : \mu_i \neq \mu_j \quad \text{for some } i \neq j \tag{3.6}$$

or, equivalently,

$$H_0 : \alpha_1 = \alpha_2 = \cdots = \alpha_a = 0 \quad \text{vs.} \quad H_1 : \alpha_i \neq 0 \quad \text{for some } i. \tag{3.7}$$

The F-test of H_0 can be derived by using the **extra sum of squares method** following the steps in Section 2.3.2.3; for details see Example 14.4. The basic result used to derive the F-test is the ANOVA identity:

$$\underbrace{\sum_{i=1}^{a}\sum_{j=1}^{n_i}(y_{ij} - \bar{\bar{y}})^2}_{\text{SS}_{\text{tot}}} = \underbrace{\sum_{i=1}^{a} n_i(\bar{y}_i - \bar{\bar{y}})^2}_{\text{SS}_{\text{trt}}} + \underbrace{\sum_{i=1}^{a}\sum_{j=1}^{n_i}(y_{ij} - \bar{y}_i)^2}_{\text{SS}_{\text{e}}}, \tag{3.8}$$

where SS_{tot} is the **total sum of squares** and SS_{trt} corresponds to SS_{reg} in (2.23) and is referred to as the **treatment sum of squares**. Define the **treatment mean square** ($\mathbf{MS_{trt}}$) as the ratio of SS_{trt} to $a - 1$; here $a - 1$ is the d.f. associated with the hypothesis H_0 of (3.6) since H_0 places $a - 1$ linearly independent restrictions on the parameters. Then using the general results from Section 14.2.5 it follows that under H_0

$$F = \frac{\text{MS}_{\text{trt}}}{\text{MS}_e} = \frac{\text{SS}_{\text{trt}}/(a-1)}{\text{SS}_e/\nu} \sim F_{a-1,\nu},$$

where $\nu = N - a$. An α-level F-test of H_0 rejects if

$$F = \frac{\text{MS}_{\text{trt}}}{\text{MS}_e} > f_{a-1,\nu,\alpha}, \tag{3.9}$$

where $f_{a-1,\nu,\alpha}$ is the upper α critical point of the central F-distribution with $a - 1$ and ν d.f. The ANOVA given in Table 3.4 summarizes these calculations.

Table 3.4 ANOVA Table for One-Way Layout

Source	SS	d.f.	MS	F
Treatments	$\text{SS}_{\text{trt}} = \sum n_i(\bar{y}_i - \bar{\bar{y}})^2$	$a - 1$	MS_{trt}	$\text{MS}_{\text{trt}}/\text{MS}_e$
Error	$\text{SS}_e = \sum\sum(y_{ij} - \bar{y}_i)^2$	$N - a$	MS_e	
Total	$\text{SS}_{\text{tot}} = \sum\sum(y_{ij} - \bar{\bar{y}})^2$	$N - 1$		

The above derivation of the F-test is based on the assumption of normality of the data. There is an alternative justification based on the randomization distribution of this statistic, which is discussed in Section 3.9.1.

Example 3.2 (Anorexia Data: Analysis of Variance)

Refer to the data in Table 3.2. First we make the ANOVA table for baseline weights using Minitab (see Display 3.1) to see if the three treatment groups are comparable.

It is evident that the treatment groups do not differ significantly in terms of the baseline weights ($F = 0.50$, $p = 0.610$).

Next we make the ANOVA table for weight gains (see Display 3.2) to see if the three treatment groups are significantly different. Here it is evident that the treatment groups differ significantly in terms of the weight gains ($F = 3.85$, $p = 0.028$). Both these conclusions agree with our initial impressions from the summary statistics and the box plots given in Example 3.1. ■

```
Source    DF       SS     MS      F      P
Group      2     31.2   15.6   0.50  0.610
Error     48   1503.5   31.3
Total     50   1534.7
S = 5.597   R-Sq = 2.04%   R-Sq(adj) = 0.00%
                                 Individual 95%  CIs For Mean Based on
                                 Pooled StDev
Level       N    Mean   StDev   -------+---------+---------+---------+--
Behavior   17  82.147   5.602       (-------------*------------)
Control    17  81.318   6.117   (-------------*------------)
Family     17  83.229   5.017            (------------*-------------)
                                -------+---------+---------+---------+--
                                    80.0      82.0      84.0      86.0
Pooled StDev = 5.597
```

Display 3.1 Minitab ANOVA output for baseline weights of anorexia patients.

```
Source    DF       SS     MS      F      P
Group      2    479.3  239.7   3.85  0.028
Error     48   2990.6   62.3
Total     50   3469.9
S = 7.893   R-Sq = 13.81%   R-Sq(adj) = 10.22%
                                 Individual 95%  CIs For Mean Based on Pooled
                                 StDev
Level       N    Mean   StDev      +---------+---------+---------+---------
Behavior   17   3.318   8.409              (--------*---------)
Control    17  -0.241   8.061      (--------*---------)
Family     17   7.265   7.157                      (--------*---------)
                                   +---------+---------+---------+---------
                                 -4.0       0.0       4.0       8.0
Pooled StDev = 7.893
```

Display 3.2 Minitab ANOVA output for weight gains of anorexia patients.

3.3.3 Confidence Intervals and Hypothesis Tests

If the overall null hypothesis is rejected, then more detailed comparisons between the treatment means are required to answer questions such as which treatments are different from each other or which treatments are "better" than others or is there one treatment which is clearly the "best." Such inferences involve making **multiple comparisons**, which entail the risk of inflating the type I error probability. Proper control of this error probability requires special procedures, which are discussed in Chapter 4. Here we give procedures from elementary statistics which treat each comparison separately.

First consider inferences on individual treatment means μ_i. The usual single-sample t-intervals and t-tests can be used in this case, with the only change being that the estimator of σ^2 is the pooled MS_e based on all the data. Thus $100(1-\alpha)\%$ separate CIs on the μ_i are given by

$$\mu_i \in \left[\overline{y}_i \pm t_{\nu,\alpha/2}\frac{s}{\sqrt{n_i}}\right] \qquad (1 \le i \le a), \tag{3.10}$$

where $t_{\nu,\alpha/2}$ is the upper $\alpha/2$ critical point of the t-distribution with $\nu = N - a$ d.f.

The **least significant difference (LSD) procedure** (see Section 4.2.1) compares any pair of treatment means, say μ_i and μ_j, by the usual two-sample t-interval or t-test. In this procedure separate $100(1-\alpha)\%$ CIs on the pairwise differences $\mu_i - \mu_j$ are given by

$$\mu_i - \mu_j \in \left[\overline{y}_i - \overline{y}_j \pm t_{\nu,\alpha/2}s\sqrt{\frac{1}{n_i} + \frac{1}{n_j}}\right] \qquad (1 \le i < j \le a). \tag{3.11}$$

Example 3.3 (Anorexia Data: Comparisons of Means)

To compare the mean weight gains for the three therapies, we compare the three therapies with each other by computing separate 95% CIs for the three pairwise differences using (3.11):

Control vs. Behavioral:

$$\left[(-0.241 - 3.318) \pm (2.011)(7.893)\sqrt{\frac{2}{17}}\right] = [-9.002, 1.884],$$

Control vs. Family:

$$\left[(-0.241 - 7.265) \pm (2.011)(7.893)\sqrt{\frac{2}{17}}\right] = [-12.949, -2.063],$$

Behavioral vs. Family:

$$\left[(3.318 - 7.265) \pm (2.011)(7.893)\sqrt{\frac{2}{17}}\right] = [-9.390, 1.496].$$

There is a significant difference only between the control therapy and family therapy with the family therapy resulting in a higher weight gain. ■

3.4 MODEL DIAGNOSTICS

Four main assumptions that underlie model (3.2) are as follows:

(a) the errors e_{ij} are normally distributed;
(b) the errors e_{ij} are homoscedastic;
(c) the errors e_{ij} are independently distributed; and
(d) $E(e_{ij}) = 0$ or equivalently $E(y_{ij}) = \mu_i$.

The validity of the analyses described in the previous section depends on these assumptions, which must therefore be carefully checked. Outliers can also vitiate the analyses.

Graphical plots or formal tests of significance based on residuals are the key tools used to check the above assumptions. We describe both methods in this section. Tests on the residuals are based on the following distributional result: Under the given model, the $\widehat{e}_{ij}$ are jointly normally distributed with zero means, and

$$\begin{aligned} \text{Var}(\widehat{e}_{ij}) &= \sigma^2 \left(\frac{n_i - 1}{n_i} \right), \\ \text{Corr}(\widehat{e}_{ij}, \widehat{e}_{ik}) &= -\frac{1}{n_i - 1} \quad \text{for} j \neq k, \\ \text{Corr}(\widehat{e}_{ij}, \widehat{e}_{k\ell}) &= 0 \quad \text{for } i \neq k; \end{aligned} \tag{3.12}$$

see Exercise 3.7 for a proof of this result.

It must be mentioned that, if the sample sizes are small, then the plots of residuals may be difficult to interpret and can even be misleading. As an example of what can go wrong for small sample sizes, recall that the residuals are subject to the restriction $\sum_{j=1}^{n_i} \widehat{e}_{ij} = 0$ for each treatment $i = 1, 2, \ldots, a$. The consequence of this fact is that the $\widehat{e}_{ij}$ and $\widehat{e}_{ik}$ are negatively correlated as given in (3.12). For $n_i = 2$, the correlation equals -1.

The plots are also difficult to interpret if the data are granular, that is, if there are only a few distinct values. Suppose that a particular data value, say y^*, occurs once in three different treatment groups. Then the residuals for these three values, namely $y^* - \overline{y}_1$, $y^* - \overline{y}_2$, and $y^* - \overline{y}_3$, will be perfectly negatively correlated with the corresponding treatment means $\overline{y}_i$. Therefore, if the residuals are plotted versus $\overline{y}_i$ (as is done to check the homoscedasticity assumption; see Section 3.4.1 below), the three points will fall on a negatively sloping line. For other distinct data values there will be similar lines and these lines will be parallel to each other. Thus the residual plot will appear nonrandom with a parallel streak pattern.

3.4.1 Checking Homoscedasticity

The t- and F-tests are not robust to lack of homoscedasticity (Box, 1954). Therefore this assumption should always be checked first and appropriate corrective actions should be taken if the assumption is not satisfied, for example, by applying a variance-stabilizing transformation (see Section 3.5) or by using the modified F-test given in Section 3.9.2 for heteroscedastic variances. If the assumption holds, then the spread of the residuals should be roughly the same for each treatment. This can be checked by plotting the residuals against the treatment means $\overline{y}_i$. This plot should exhibit a parallel band. The reason for plotting the residuals against the $\overline{y}_i = \widehat{\mu}_i$ is that, if the treatment variances σ_i^2 are not constant, then they often are functions of the μ_i. In interpreting the plot one should bear in mind that the spread of the residuals is proportional to the standard deviations. Thus, if the spread appears to increase linearly with the $\overline{y}_i$, then the likely relation is $\sigma_i \propto \mu_i$. An alternative way of seeing the relationship between the σ_i and the μ_i is to plot ln s_i versus ln $\overline{y}_i$. We will see an application of this in Section 3.5.

There are several formal tests of homoscedasticity which test

$$H_0 : \sigma_1^2 = \sigma_2^2 = \cdots = \sigma_a^2 \quad \text{vs.} \quad H_1 : \sigma_i^2 \neq \sigma_j^2 \quad \text{for some } i \neq j. \tag{3.13}$$

The most common among these is the **Bartlett test**. The Bartlett statistic equals

$$B = \frac{2}{c}\left[\nu \ln s - \sum_{i=1}^{a} \nu_i \ln s_i\right], \tag{3.14}$$

where c is a correction factor given by

$$c = 1 + \frac{1}{3(a-1)}\left[\sum_{i=1}^{a} \frac{1}{\nu_i} - \frac{1}{\nu}\right],$$

$\nu_i = n_i - 1$ and $\nu = \sum \nu_i = N - a$. If the n_i's are equal (balanced one-way layout), then this formula simplifies; see Exercise 3.8.

For large n_i's, the Bartlett statistic has a chi-square distribution with $a - 1$ d.f. under H_0. Thus an α-level test rejects H_0 if $B > \chi^2_{a-1,\alpha}$. For $a = 2$, the Bartlett statistic reduces to the ratio of sample variances s_1^2/s_2^2 which, as seen in Section 2.2.1.4, is F distributed under H_0. A strike against the Bartlett test is that it is highly nonrobust in case of nonnormal data and therefore should not be used if the normality assumption is suspect.

An alternative test that is robust against nonnormality is the **Levene test**. It tests the equality of the means of the deviations of the y_{ij} from some measure of the center of the data for each treatment. Commonly, the sample median, denoted by $\widetilde{y}_i$, is used as a measure of the center. Let

$$z_{ij} = |y_{ij} - \widetilde{y}_i| \qquad (1 \le i \le a, 1 \le j \le n_i)$$

be the absolute deviations from the sample median. Then the Levene test statistic (denoted by L) is simply the ANOVA F-statistic applied to the z_{ij}.

3.4.2 Checking Normality

Normality can be checked by using normal plots of the data introduced in Section 2.1.1. If the homoscedasticity assumption is in doubt, then separate normal plots of the data from each group should be made. On the other hand, if the homoscedasticity assumption is not contradicted by the tests of the previous section, then a common normal plot for all the data can be made, but the data must first be centered by subtracting the sample means $\overline{y}_i$ from the observations y_{ij}, that is, the residuals $\widehat{e}_{ij}$ must be used.

In general, normality is less of a concern than homoscedasticity unless the sample sizes are small because the t- and F-tests used in the analysis are quite robust to nonnormality. The reason is that they are functions of the sample means $\overline{y}_i$, which are approximately normally distributed under relatively mild regularity conditions for modestly large sample sizes, even if the raw data y_{ij} are nonnormal, thanks to the central limit theorem.

3.4.3 Checking Independence

To check independence, one must have some prior notion about how the assumption might be violated. Usually, dependence arises because of the time order of data collection. The observations that are close in time tend to be more correlated than those that are farther apart. Alternatively, there could be periodic spikes in correlations. If the time order of data collection is known, the residuals should be plotted versus that order (called the **time sequence plot** or **run chart**) to see if any temporal patterns are present. If the independence assumption holds, then this plot should be quite random. Positive serial correlations result in a few cycles with relatively long periods, while negative serial correlations result in a large number of cycles with short periods (sawtooth pattern). A **runs test** (see, e.g., Hollander and Wolfe, 1998) may be used to check if the number of cycles is too few or too many. In interpreting the time sequence plot, bear in mind the warning given at the beginning of this section about the residuals not being independent (especially when the sample sizes are small) even if the random errors are.

3.4.4 Checking Outliers

Outliers can be detected using one of the graphical plots described above or by a formal test of significance. In graphical plots, outliers stand out as isolated points. For example, in a normal plot of residuals, outliers may be identified as points in either tail that significantly deviate from the general linear pattern of the plot.

Using the formula for $\text{Var}(\widehat{e}_{ij})$ from (3.12), a formal test to determine whether a particular observation y_{ij} is an outlier can be based on a standardized residual:

$$\widehat{e}_{ij}^{*} = \frac{\widehat{e}_{ij}}{s\sqrt{1 - 1/n_i}}. \tag{3.15}$$

In an **internally standardized residual**, s is the usual pooled mean square error estimate computed from all data, whereas in an **externally standardized residual**, s is replaced by $s_{(ij)}$, which is the pooled mean square error estimate computed by omitting the observation y_{ij}. The reason for omitting this observation under test is that, if it is an outlier, then its inclusion would inflate the error estimate and thereby understate its extremeness. The test declares y_{ij} to be an outlier if $|\widehat{e}_{ij}^*|$ is large, for example, if $|\widehat{e}_{ij}^*| > 2$.

How to deal with outliers is not an easy problem. If they are wrong observations and the sources of error can be identified then they should be deleted. If they can be suitably corrected then one should do that. Otherwise, one should use inferential methods that are resistant to them (called **resistant** or **robust methods**), e.g., methods based on medians or trimmed means or Winsorized means; see the book by Wilcox (2005).

Example 3.4 (Anorexia Data: Checking of Assumptions)

Refer to the anorexia data in Table 3.2. To check the homoscedasticity assumption, the residuals are plotted against the fitted values for the baseline weights and for weight gains in Figure 3.3. Both plots show fairly random distributions of residuals with approximately equal spreads. Thus the homoscedasticity assumption appears to be satisfied. This conclusion is confirmed by the Bartlett and Levene tests conducted below.

Bartlett and Levene Tests for Baseline Weights From the summary statistics in Table 3.3 we have

$$s_1 = 6.117, \qquad s_2 = 5.602, \qquad s_3 = 5.017, \qquad s = 5.597.$$

Therefore

$$2(N-a)\ \ln s\ -2\sum_{i=1}^{a}(n_i-1)\ \ln s_i = 96\ \ln(5.597) - 32[\ln(6.117) + \ln(5.602) + \ln(5.017)] = 0.626.$$

The correction factor is

$$c = 1 + \frac{1}{(3)(2)}\left[\frac{3}{16} - \frac{1}{48}\right] = 1.028.$$

So the Bartlett statistic equals

$$B = \frac{0.626}{1.028} = 0.608,$$

which is clearly nonsignificant ($p = 0.738$) when compared with the chi-square distribution with two d.f. The Levene test conducted using Minitab gives $L = 0.572$ ($p = 0.588$).

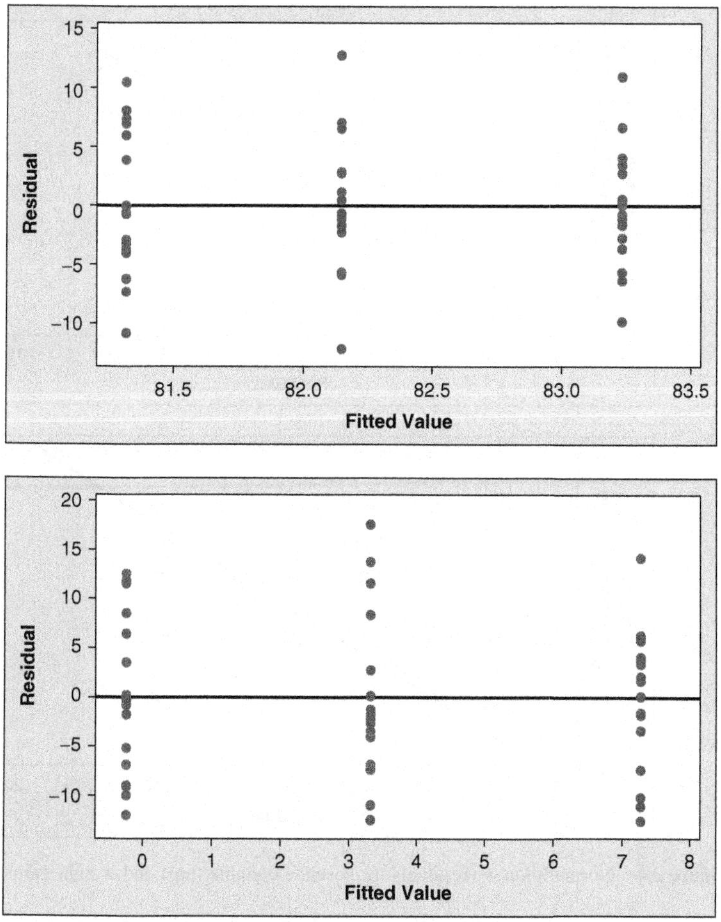

Figure 3.3 Fitted-value plots of residuals for baseline weights (top) and weight gains (bottom).

Bartlett and Levene Tests for Weight Gains From the summary statistics in Table 3.3 we have

$$s_1 = 8.061, \qquad s_2 = 8.409, \qquad s_3 = 7.157, \qquad s = 7.893.$$

Therefore

$$2(N-a)\ \ln s\ -2\sum_{i=1}^{a}(n_i-1)\ \ln s_i = 96\ \ln(7.893) - 32[\ln(8.061) + \ln(8.409) + \ln(7.157)] = 0.412.$$

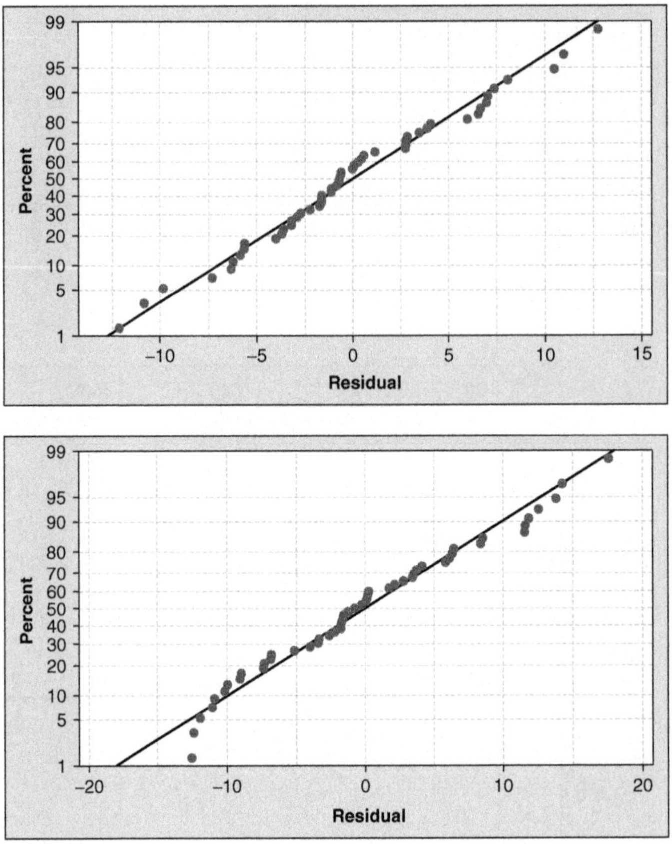

Figure 3.4 Normal plots of residuals for baseline weights (top) and weight gains (bottom).

The correction factor is the same as before. So the Bartlett statistic equals

$$B = \frac{0.412}{1.028} = 0.401,$$

which is clearly nonsignificant ($p = 0.818$) when compared with the chi-square distribution with two d.f. The Levene test gives $L = 0.133$ ($p = 0.876$).

We next check the normality assumption for both variables. Since the homoscedasticity assumption is not rejected, we do not need to make separate groupwise normal plots. Instead we can pool the residuals from all three groups for each variable and make normal plots of the residuals. Normal plots for both variables are shown in Figure 3.4. Both plots are reasonably linear, suggesting no gross violation of the normality assumption. The Anderson–Darling statistics for residuals for baseline weights and weight gains are $A^2 = 0.405$ ($p = 0.342$) and $A^2 = 0.387$ ($p = 0.375$), respectively. Thus this goodness-of-fit test confirms the conclusions drawn from the normal plots. ■

3.5 DATA TRANSFORMATIONS

If some of the model assumptions listed at the beginning of the previous section are violated, then there are usually two courses of action. The first is to use alternative methods of analysis that do not require those assumptions. For example, if the normality assumption is not satisfied, then one may use a nonparametric method such as the Kruskal–Wallis test as an alternative to the F-test. Alternatively, the randomization test described in Section 3.9.1 may be used.

The second course of action is to transform the raw data so that the necessary assumptions are at least approximately met. Standard normal theory methods of this chapter can then be applied to the transformed data. In this section we discuss **variance-stabilizing transformations** for heteroscedastic data.

As noted in Section 3.4.1, if the variance of the response variable is not constant, it often depends on the mean. In other words, if $\mathrm{E}(y) = \mu$, then $\mathrm{SD}(y) = g(\mu)$ for some nonnegative function $g(\cdot)$. In a linear model, $\mathrm{E}(y)$ is a function of the values of the predictor variables. Therefore, as predictor variable values change, $\mathrm{Var}(y)$ changes with $\mathrm{E}(y)$ and is not constant as required by the homoscedasticity assumption. We want to find a transformation $h(y)$ so that $\mathrm{Var}[h(y)]$ is approximately constant.

Suppose $g(\cdot)$ is a known smooth function. The so-called **delta method** approximates $\mathrm{Var}[h(y)]$ in terms of $\mathrm{Var}(y) = g^2(\mu)$ by using the first-order Taylor series expansion as follows. Let

$$h(y) \approx h(\mu) + (y - \mu)h'(\mu),$$

where $h'(\mu)$ is the first derivative of $h(\cdot)$ evaluated at μ. Therefore

$$\mathrm{Var}[h(y)] \approx [h'(\mu)]^2\mathrm{Var}(y) = [h'(\mu)]^2 g^2(\mu).$$

To make $\mathrm{Var}[h(y)]$ approximately constant, we set $[h'(\mu)]^2 g^2(\mu) = 1$ and solve for $h(\cdot)$, resulting in the transformation

$$h(y) = \int \frac{dy}{g(y)}. \tag{3.16}$$

Some important cases of this formula follow from the special case $\mathrm{SD}(y) \propto \mu^\alpha$. Then $g(y) = cy^\alpha$, where $c > 0$ is a proportionality constant which may be ignored. The application of formula (3.16) yields the variance-stabilizing transformation as

$$h(y) = \int \frac{dy}{y^\alpha} = \frac{1}{1-\alpha} y^{1-\alpha}.$$

Put $\lambda = 1 - \alpha$. Furthermore, to avoid the discontinuity at $\lambda = 0$, define

$$h(y) = \begin{cases} \dfrac{y^\lambda - 1}{\lambda} & \text{for } \lambda \neq 0, \\ \ln y & \text{for } \lambda = 0. \end{cases}$$

This is called the family of **power transformations**. Note that

$$\lim_{\lambda \to 0} \frac{y^\lambda - 1}{\lambda} = \ln y.$$

Three power transformations of particular interest are as follows:

Square Root Transformation For $\alpha = \frac{1}{2}$, $h(y) = \sqrt{y}$ (example: Poisson data).

Logarithmic Transformation For $\alpha = 1$, $h(y) = \ln y$ (example: exponential or lognormal data).

Inverse Transformation For $\alpha = 2$, $h(y) = 1/y$. ■

To apply the power transformation, one needs to estimate α and then set $\lambda = 1 - \alpha$. This estimation can be done by plotting $\ln s_i$ versus $\ln \overline{y}_i$ for $i = 1, 2, \ldots, a$. If the plot is roughly linear, then the slope of the line gives an estimate of α.

Example 3.5 (Aerosol Paint Data: Variance-Stabilizing Power Transformation)

Nelson (1998) gave the data shown in Table 3.5 on the amounts of gloss white aerosol paints required by repeatedly spraying a hiding chart to hide the light to dark-gray stripes on its white background. Three brands of paints were compared in the study with four samples of each brand.

It is clear that the variances are increasing with the means. The plot of $\ln s_i$ vs. $\ln \overline{y}_i$ is shown in Figure 3.5. The slope of the fitted line is $\alpha = 1.286$. Hence $\lambda = -0.286 \approx -0.30$. Table 3.6 gives the transformed data and associated means and SDs. Notice that the transformation has stabilized the variances. ■

A variance-stabilizing transformation often helps to make the distribution of the response variable more nearly normal, but not always. For example,

Table 3.5 Amounts (oz/ft^2) of Three Brands of Paints to Cover a Hiding Chart

Brand 1	Brand 2	Brand 3
2.1	4.7	6.4
1.9	3.6	8.5
1.8	3.9	7.9
2.2	3.8	8.8
$\overline{y}_1 = 2.0$	$\overline{y}_2 = 4.0$	$\overline{y}_3 = 7.9$
$s_1^2 = 0.0333$	$s_2^2 = 0.2333$	$s_3^2 = 1.1400$
$\ln(\overline{y}_1) = 0.6931$	$\ln(\overline{y}_2) = 1.3863$	$\ln(\overline{y}_3) = 2.0669$
$\ln(s_1) = -1.7005$	$\ln(s_2) = -0.7277$	$\ln(s_3) = 0.0655$

Source: Nelson (1998, Table 15.8,). Reprinted by permission of The McGraw-Hill Companies.

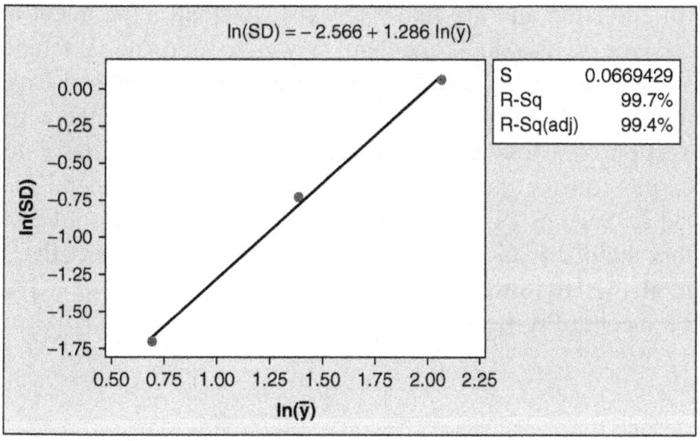

Figure 3.5 Plot of ln s_i versus ln $\bar{y}_i$.

Table 3.6 Transformed Amounts ($y_{ij} \rightarrow 10y_{ij}^{-0.3}$) of Three Brands of Paints to Cover a Hiding Chart

Brand 1	Brand 2	Brand 3
8.005	6.286	5.730
8.248	6.809	5.379
8.383	6.648	5.262
7.894	6.700	5.208
$\bar{y}_1 = 8.133$	$\bar{y}_2 = 6.611$	$\bar{y}_3 = 5.395$
$s_1^2 = 0.0497$	$s_2^2 = 0.0514$	$s_3^2 = 0.0550$

as noted above, the square-root transformation stabilizes the variance of Poisson-distributed data, but the normalizing transformation is $y \rightarrow y^{2/3}$.

Box and Cox (1964) proposed a variant of the power transformation in (3.17) called the **Box–Cox transformation**:

$$h(y) = \begin{cases} \dfrac{y^\lambda - 1}{\lambda \tilde{y}^{\lambda-1}} & \text{for } \lambda \neq 0, \\ \tilde{y} \ln y & \text{for } \lambda = 0, \end{cases}$$

where $\tilde{y}$ is the geometric mean of all y_{ij}. They proposed the maximum likelihood method for estimating λ; for further details, see Draper and Smith (1998, pp. 280–282).

3.6 POWER OF F-TEST AND SAMPLE SIZE DETERMINATION

A basic tenet of experimentation is that the sample sizes should be large enough to detect, with sufficiently high probability (power), the treatment differences that are worth detecting and are likely to exist. A small experiment may be a waste of effort since it may lack adequate power to find the differences of interest. Therefore it is important to assess the power of a designed experiment and to determine the minimum sample size necessary to guarantee a specified power. In this section we show how to do these calculations for the F-test (3.9).

An expression for the power of the F-test for the general linear hypothesis is derived in Section 14.3. We now specialize that result to the one-way layout. Recall that the distribution of the F-statistic under the alternative hypothesis is a **noncentral F-distribution** [denoted by $F_{a-1,\nu}(\lambda^2)$] with $a-1$ and $\nu = N-a$ d.f. and noncentrality parameter (n.c.p.) λ^2, where

$$\lambda^2 = \frac{\sum_{i=1}^{a} n_i \alpha_i^2}{\sigma^2}. \tag{3.17}$$

Note that $\lambda^2 = 0$ only under $H_0 : \alpha_1 = \alpha_2 = \cdots = \alpha_a = 0$. Furthermore,

$$\mathrm{E}(\mathrm{MS}_{\mathrm{trt}}) = \frac{\sigma^2}{a-1}(a-1+\lambda^2) = \sigma^2 + \frac{1}{a-1}\sum_{i=1}^{a} n_i \alpha_i^2.$$

The power of the α-level F-test under H_1 is given by

$$\mathrm{P}\left\{F_{a-1,\nu}(\lambda^2) > f_{a-1,\nu,\alpha}\right\} = 1-\beta. \tag{3.18}$$

As discussed in Section 14.3, the **Pearson–Hartley charts** give the plots of the power curves of the F-test as a function of the d.f. of F (denoted in general by ν_1 and ν_2; here $\nu_1 = a-1$ and $\nu_2 = N-a$), the significance level α, and the quantity

$$\phi = \frac{\lambda}{\sqrt{\nu_1+1}} = \sqrt{\frac{\sum_{i=1}^{a} n_i \alpha_i^2}{a\sigma^2}}. \tag{3.19}$$

As can be seen from Figure 14.4, these charts are difficult to read. Therefore the power and sample size calculations based on them are not very accurate. Now many computer programs such as JMP, Minitab, and SAS are available which give exact results. We recommend use of these programs for power and sample size calculation.

A question that an experimenter is often interested in is: "What is the minimum power if at least two μ_i's differ by a practically significant amount δ?" The dual problem is to find the minimum sample size n per treatment to guarantee a specified power $1-\beta$ to detect a practically significant difference δ between any two treatment means. For example, in the anorexia experiment the researcher may be interested in knowing the minimum power of the F-test if any two therapies differ in their mean weight gains by at least 5 lb. We now explain how to find this minimum power. For convenience, we will assume equal sample sizes, $n_i \equiv n$.

It can be shown and is also intuitively clear that the power function defined in (3.18) is increasing in λ^2 and hence in ϕ. To find the minimum power, we need to find the minimum λ^2 or equivalently the minimum of $\sum_{i=1}^{a} \alpha_i^2$ subject to $\max |\mu_i - \mu_j| \geq \delta$. This minimum is attained when exactly two μ_i's are δ apart and all other μ_i's are at the midpoint of the two. In other words, the **least favorable configuration (LFC)** for minimum power is

$$\alpha_1 = -\tfrac{1}{2}\delta, \qquad \alpha_2 = \cdots = \alpha_{a-1} = 0, \qquad \alpha_a = +\tfrac{1}{2}\delta. \tag{3.20}$$

Then the minimum values of λ^2 and ϕ are

$$\lambda^2_{\min} = \frac{n\delta^2}{2\sigma^2} \quad \text{and} \quad \phi_{\min} = \frac{\delta}{\sigma}\sqrt{\frac{n}{2a}}. \tag{3.21}$$

To calculate $\phi_{\min}$ for specified n, it is necessary to have a prior estimate of σ, which may be an educated guess or may be taken from a pilot study or historical data. In some applications it may be easier to specify a *relative* difference δ/σ that is practically significant rather than an *absolute* difference δ; in that case a prior estimate of σ is not needed.

Example 3.6 (Anorexia Data: Power and Sample Size Calculations)

Suppose a researcher is interested in detecting mean differences exceeding 5 lb, that is, a weight gain of 5 lb is regarded as practically significant. To calculate the power and sample size, we assume $\sigma = 8$ lb based on $s = 7.893$ lb calculated in Example 3.2. Then for comparing $a = 3$ therapies with sample sizes $n_1 = n_2 = n_3 = n = 17$, the minimum power of the 0.05-level F-test is calculated to be 0.3329 using Minitab.

It is common to design experiments so that they achieve a minimum power of about 0.70–0.90. If we want the minimum power to be 0.70, then using Minitab, the sample size per therapy group is calculated to be 41. If we want the minimum power to be 0.90, then the sample size per therapy group is calculated to be 66.

The above calculations can be done using Pearson–Hartley charts. To calculate the power for $n = 17$, we use

$$\phi_{\min} = \frac{\delta}{\sigma}\sqrt{\frac{n}{2a}} = \frac{5}{8}\sqrt{\frac{17}{(2)(3)}} = 1.052.$$

The sample size calculation requires trial and error since an increase in n increases both $\phi_{\min}$ and $\nu_2 = a(n-1)$. ■

3.7 QUANTITATIVE TREATMENT FACTORS

In many experiments the treatments are numerical, for example, different levels of temperature. An appropriate analysis in this case is regression analysis. A special form of regression analysis called **orthogonal polynomial regression** is especially useful if the levels of the quantitative factor are equispaced and the design is balanced.

As an example, consider an experiment to study the dependence of process yield y on the reaction temperature by setting the temperature at three equispaced levels x_1, x_2, x_3. Let μ_1, μ_2, μ_3 be the mean yields at the three temperatures which are estimated by the sample means $\overline{y}_1$, $\overline{y}_2$, $\overline{y}_3$ based on n replicate observations at each temperature. The one-way ANOVA provides a test of the null hypothesis: $H_0 : \mu_1 = \mu_2 = \mu_3$. However, this test is not useful in revealing the nature of the relationship between the yield and temperature. A quadratic regression model of the form $\mathrm{E}(y) = \beta_0 + \beta_1 x + \beta_2 x^2$ can be fitted to the data. A more useful way which is related to orthogonal polynomial regression is to estimate the **linear effect** of temperature on yield by

$$(\overline{y}_3 - \overline{y}_2) + (\overline{y}_2 - \overline{y}_1) = \overline{y}_3 - \overline{y}_1$$

and the **quadratic effect** of temperature on yield by the difference of two linear effects:

$$(\overline{y}_3 - \overline{y}_2) - (\overline{y}_2 - \overline{y}_1) = \overline{y}_3 - 2\overline{y}_2 + \overline{y}_1.$$

If there is no quadratic effect, then $\overline{y}_3 - \overline{y}_2$ and $\overline{y}_2 - \overline{y}_1$ should be roughly equal and their difference should cancel out.

The above two effects are linear functions of the sample means $\overline{y}_i$ with the property that the coefficients of each linear function add up to zero. Such linear functions of the sample means are called **contrasts**. In particular, $\overline{y}_3 - \overline{y}_1 = \overline{y}_3 + 0\overline{y}_2 - \overline{y}_1$ is called a **linear contrast** and $\overline{y}_3 - 2\overline{y}_2 + \overline{y}_1$ is called a **quadratic contrast**. The coefficient vectors of these two contrasts are $(-1, 0, +1)$ and $(+1, -2, +1)$. Their dot product is zero, and so the two contrasts are **orthogonal**. Orthogonal contrasts can be shown to be uncorrelated for balanced designs; see Section 4.4.

In general, one can fit a polynomial of degree $p \leq a - 1$ when there are $a \geq 2$ distinct settings of the quantitative factor. Suppose the settings are equispaced at levels $x_1, x_2, \ldots, x_a$ that are distance $d > 0$ apart and there are n replicate observations at each setting. Let $N = an$ be the total sample size. It is convenient to normalize the levels x_i to

$$z_i = \frac{x_i - \overline{x}}{d} \qquad (1 \leq i \leq a).$$

Note that if a is even, say $a = 2m$, then $z_i = i - (m+1)/2$, and if a is odd, say $a = 2m + 1$, then $z_i = i - (m + 1)$. For example, if $a = 3$, then $z_1 = -1, z_2 =$

0, $z_3 = +1$, and if $a = 4$, then $z_1 = -\frac{3}{2}$, $z_2 = -\frac{1}{2}$, $z_3 = +\frac{1}{2}$, $z_4 = +\frac{3}{2}$. With this rescaling, one can fit an orthogonal polynomial of degree $p \le a - 1$ of the form

$$\begin{aligned} y_{ij} &= \beta_0 \xi_0(z_i) + \beta_1 \xi_1(z_i) + \beta_2 \xi_2(z_i) + \cdots \\ &\quad + \beta_p \xi_p(z_i) + e_{ij} \quad (1 \le i \le a,\ 1 \le j \le n), \end{aligned} \tag{3.22}$$

where $\xi_k(z_i)$ is a kth-degree polynomial with the property that

$$\begin{aligned} \sum_{i=1}^{a} \xi_k(z_i) &= 0 \quad \text{for } k = 1, 2, \ldots, p, \\ \sum_{i=1}^{a} \xi_k(z_i)\xi_\ell(z_i) &= 0 \quad \text{for all } k \ne \ell. \end{aligned} \tag{3.23}$$

The first five orthogonal polynomials are

$$\begin{aligned} \xi_0(z) &= 1, \\ \xi_1(z) &= \lambda_1 z, \\ \xi_2(z) &= \lambda_2 \left[z^2 - \left(\frac{a^2 - 1}{12} \right) \right], \\ \xi_3(z) &= \lambda_3 \left[z^3 - z \left(\frac{3a^2 - 7}{20} \right) \right], \\ \xi_4(z) &= \lambda_4 \left[z^4 - z^2 \left(\frac{3a^2 - 13}{14} \right) + \frac{3(a^2 - 1)(a^2 - 9)}{560} \right], \end{aligned}$$

where the λ_k are constants chosen so that the polynomial values are integers. Table 3.7 gives the orthogonal polynomial coefficients for $a = 2(1)5$. As an example, if $a = 3$, then

$$\begin{aligned} &\xi_0(z_1) = 1, \qquad \xi_0(z_2) = 1, \quad \xi_0(z_3) = 1, \\ &\xi_1(z_1) = -\lambda_1, \quad \xi_1(z_2) = 0, \quad \xi_1(z_3) = +\lambda_1, \\ &\xi_2(z_1) = \lambda_2 \left[1 - \tfrac{1}{12}(9 - 1)\right] = \tfrac{1}{3}\lambda_2, \ \xi_2(z_2) = \lambda_2 \left[0 - \tfrac{1}{12}(9 - 1)\right] = -\tfrac{2}{3}\lambda_2, \\ &\xi_2(z_3) = \lambda_2 \left[1 - \tfrac{1}{12}(9 - 1)\right] = \tfrac{1}{3}\lambda_2. \end{aligned}$$

These orthogonal polynomials take integer values if we set $\lambda_1 = 1$ and $\lambda_2 = 3$. Thus the orthogonal polynomials are $\xi_0(z) = 1$, $\xi_1(z) = z$, and $\xi_2(z) = 3z^2 - 2$. Then we have $\xi_1(z_1) = -1$, $\xi_1(z_2) = 0$, $\xi_1(z_3) = +1$ and $\xi_2(z_1) = +1$, $\xi_2(z_2) = -2$, $\xi_2(z_3) = +1$. These are exactly the linear and quadratic coefficient vectors given earlier.

Table 3.7 Coefficients of Orthogonal Polynomials

	$a=2$	$a=3$		$a=4$			$a=5$			
x_i	ξ_1	ξ_1	ξ_2	ξ_1	ξ_2	ξ_3	ξ_1	ξ_2	ξ_3	ξ_4
1	−1	−1	1	−3	1	−1	−2	2	−1	1
2	1	0	−2	−1	−1	3	−1	−1	2	−4
3		1	1	1	−1	−3	0	−2	0	6
4				3	1	1	1	−1	−2	−4
5							2	2	1	1
$\sum_{i=1}^{a} \xi_k^2(z_i)$	2	2	6	20	4	20	10	14	10	70
λ_k	1	1	3	2	1	10/3	1	1	5/6	35/12

The LS estimators of the β_k are (the derivations are given in Section 3.9.3)

$$\widehat{\beta}_0 = \frac{\sum\sum y_{ij}}{N} = \overline{\overline{y}} \quad \text{and}$$

$$\widehat{\beta}_k = \frac{\sum\sum \xi_k(z_i) y_{ij}}{n \sum \xi_k^2(z_i)} = \frac{\sum \xi_k(z_i)\overline{y}_i}{\sum \xi_k^2(z_i)} \qquad (1 \le k \le p). \tag{3.24}$$

It is shown in Section 14.1.3 that addition or deletion of terms in the orthogonal polynomial leaves the estimators of the coefficients of the remaining terms unaltered.

The variances and covariances of the $\widehat{\beta}_k$ are (the derivations are given in Section 3.9.3)

$$\text{Var}(\widehat{\beta}_0) = \frac{\sigma^2}{N},$$

$$\text{Var}(\widehat{\beta}_k) = \frac{\sigma^2}{n \sum_{i=1}^{a} \xi_k^2(z_i)}, \qquad \text{Cov}(\widehat{\beta}_k, \widehat{\beta}_\ell) = 0 \quad \text{for } k \neq \ell. \tag{3.25}$$

Thus the estimated coefficients of an orthogonal polynomial are uncorrelated, and because of the normality assumption, they are independent. The estimator of σ^2 is the usual pooled mean square error estimator $s^2 = \text{MS}_e$ with $\nu = N - (p+1)$ d.f.

The t-statistic for testing the significance of $\widehat{\beta}_k$ is

$$t_k = \frac{\widehat{\beta}_k \sqrt{n \sum_{i=1}^{a} \xi_k^2(z_i)}}{s}$$

with $\nu = N - (p+1)$ d.f. The corresponding F-statistic is

$$F_k = t_k^2 = \frac{\widehat{\beta}_k^2 \{n \sum_{i=1}^{a} \xi_k^2(z_i)\}}{s^2}$$

with 1 and $\nu = N - (p+1)$ d.f. The numerator of the F-statistic can be thought of as the mean square for the kth-order effect with one d.f. Hence it is the same as the sum of squares for the kth-order effect, denoted by SS_k, where

$$SS_k = \widehat{\beta}_k^2 \left\{ n \sum_{i=1}^{a} \xi_k^2(z_i) \right\} \qquad (1 \le k \le p). \tag{3.26}$$

Since the $\widehat{\beta}_k$ coefficients are independent, it follows that their sums of squares SS_k are also independent. Furthermore, it can be shown that

$$SS_1 + SS_2 + \cdots + SS_{a-1} = SS_{\text{trt}},$$

which has $a-1$ d.f. Thus orthogonal polynomial regression analysis enables us to partition SS_{trt} into $a-1$ independent sums of squares, each SS having one d.f., representing linear, quadratic, cubic, and so on, terms in the model. Generally, it is not necessary to fit a full model of degree $a-1$. If a pth-degree (with $p < a-1$) model is fitted, the sums of squares for the remaining $a-1-p$ terms can be pooled with SS_e to yield a new SS_e with $\nu = N-(p+1)+(a-1-p) = N+a-2(p+1)$ d.f.

Example 3.7 (Tear Factor of Paper: Fitting an Orthogonal Polynomial)
Table 3.8 gives the tear factor of paper (the percentage of the standard force necessary to tear the paper) for five batches made at different pressures, with four sheets tested from each batch. The pressures are equally spaced on the log scale. We will fit an orthogonal fourth-degree polynomial to these data to illustrate how the sum of squares for the batches with four d.f. can be partitioned into independent contributions from the four terms of the polynomial, although, in general, it is not advisable to fit more than a cubic polynomial, particularly with observations taken only at five points.

First we treat this as a one-way layout and compute the ANOVA table shown in Display 3.3. There are significant ($p = 0.016$) differences between the mean tear factors under different applied pressures. The pooled estimate of σ is $s = 5.727$ with 15 d.f.

Table 3.8 Tear Factor of Paper Made at Different Pressures

Pressure	Tear Factor				Sample Mean
35.0	112	119	117	113	115.25
49.5	108	99	112	118	109.25
70.0	120	106	102	109	109.25
99.0	110	101	99	104	103.50
140.0	100	102	96	101	99.75

Source: Williams (1959, Table 2.5).

```
One-way ANOVA: Tear versus Pressure
Source    DF      SS     MS     F      P
Pressure   4   568.8  142.2  4.34  0.016
Error     15   492.0   32.8
Total     19  1060.8
S = 5.727   R-Sq = 53.62%   R-Sq(adj) = 41.25%
```

Display 3.3 Minitab ANOVA output for the tear factor data.

Before fitting an orthogonal polynomial, we plot the tear factor values versus the logarithm of pressure in Figure 3.6 to see the nature of the relationship. We see that the relationship is roughly linear with a negative slope. Next we compute the linear, quadratic, cubic, and quartic terms in the fourth-degree orthogonal polynomial using the coefficients from Table 3.7:

$$\widehat{\beta}_0 = \tfrac{1}{4}(115.25 + 109.25 + 109.25 + 103.50 + 99.75) = 107.4,$$

$$\widehat{\beta}_1 = \tfrac{1}{10}[(-2)(115.25) + (-1)(109.25) + (0)(109.25) + (1)(103.50) + (2)(99.75)] = -3.675,$$

$$\widehat{\beta}_2 = \tfrac{1}{14}[(2)(115.25) + (-1)(109.25) + (-2)(109.25) + (-1)(103.50) + (2)(99.75)] = -0.089,$$

$$\widehat{\beta}_3 = \tfrac{1}{10}[(-1)(115.25) + (2)(109.25) + (0)(109.25) + (-2)(103.50) + (1)(99.75)] = -0.400,$$

$$\widehat{\beta}_4 = \tfrac{1}{70}[(1)(115.25) + (-4)(109.25) + (6)(109.25) + (-4)(103.50) + (1)(99.75)] = 0.279.$$

The t-statistics for the $\widehat{\beta}_j$ $(1 \le j \le 4)$ are

$$t_1 = \frac{-3.675}{5.727/\sqrt{(4)(10)}} = -4.056,$$

$$t_2 = \frac{-0.089}{5.727/\sqrt{(4)(14)}} = -0.117,$$

$$t_3 = \frac{-0.400}{5.727/\sqrt{(4)(10)}} = -0.442,$$

$$t_4 = \frac{0.279}{5.727/\sqrt{(4)(70)}} = 0.814.$$

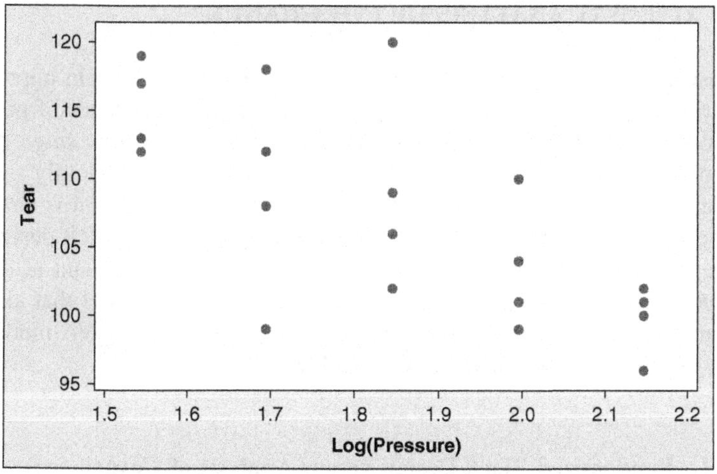

Figure 3.6 Plot of tear factor versus logarithm of pressure.

We see that only the linear term is statistically significant, as was evident from Figure 3.6. We conclude that the relationship between the tear factor and log(pressure) is essentially linear and is given by $\widehat{y} = 107.4 - 3.675z$, where z is on the coded scale taking values -2 when log(pressure) = log(35.0) = 1.544 and +2 when log(pressure) = log(140.0) = 2.145.

The sums of squares for the four effects can be computed using formula (3.26):

$$\text{SS}_1 = (40)(3.675)^2 = 540.23, \text{SS}_2 = (56)(-0.089)^2 = 0.44,$$

$$\text{SS}_3 = (40)(-0.4)^2 = 6.40, \text{SS}_4 = (280)(-0.279)^2 = 21.80.$$

Note that $\text{SS}_1 + \text{SS}_2 + \text{SS}_3 + \text{SS}_4 = \text{SS}_{\text{trt}} = 568.87$. The resulting ANOVA is shown in Table 3.9. ■

Table 3.9 ANOVA for Tear Factor Data Using Orthogonal Polynomials

Source	SS	d.f.	MS	F	p
Pressure	568.87	4	142.22	4.34	0.016
Linear Effect	540.23	1	540.23	16.47	0.001
Quadratic Effect	0.44	1	0.44	0.013	0.911
Cubic Effect	6.40	1	6.40	0.195	0.665
Quartic Effect	21.80	1	21.80	0.665	0.428
Error	492.00	15	32.80		
Total	1060.87	19			

3.8 ONE-WAY ANALYSIS OF COVARIANCE

Consider an experiment to compare different treatments for pain improvement of arthritic patients. The age of a patient is an important predictor of pain relief. It is possible to take this noise factor into account at the *design stage* by dividing the patients into blocks with roughly equal ages and then randomly assigning the patients in each block to the study drugs so that the allocation is balanced with respect to age. As discussed in Section 1.4, this would be an RB design with age as a blocking factor. It is also possible to take age into account at the *analysis stage* by using it as a predictor variable in a regression model that also includes dummy predictor variables for treatments. This is the ANCOVA model with age as a **covariate** or a **concomitant variable**.

3.8.1 Randomized Block Design versus Analysis of Covariance

Both the RB design and ANCOVA adjust for the effect of the covariate, thereby rendering the comparisons between the treatments more precise. Which of these two approaches to use in a given application depends on the nature of the covariate. If the values of the covariate for the experimental units are known beforehand, then either approach can be used. If the values become known only during or after the experiment and are not affected by the treatments or if the decision to use the data on the covariate is made post hoc, then only the ANCOVA approach is possible. An example of the former situation is the muzzle velocity of a bullet (which can be measured only when the bullet is fired) in a ballistic experiment to compare different shell materials in terms of their resistance to bullet penetration.

There are pros and cons associated with each approach. The RB design is easier to explain to a user and does not assume any particular functional form for the relationship between the response variable and the covariate. However, it ignores the actual numerical values of the covariate and treats it as a nominal variable. The RB design is also more difficult to conduct in practice because of the randomization restrictions. Another problem with the RB design is that it may be difficult to form blocks if there are several covariates. On the other hand, the ANCOVA is a very flexible technique which treats covariates as quantitative and in which a linear or a nonlinear relationship with one or more covariates can be fitted.

3.8.2 Model

In this section we present the standard ANCOVA model for a single covariate which is assumed to be linearly related to the response variable. The basic concepts are best explained in this simple setting. As mentioned above, it is straightforward to model several covariates with complex relationships by using a suitable statistical software.

Consider the same setup as in Section 3.1, that is, a single treatment factor with N experimental units randomly allocated to $a \geq 2$ treatments with n_i units to the ith treatment. But now, in addition to the response y_{ij}, we also observe a covariate x_{ij} on the jth experimental unit in the ith treatment group. We assume the model

$$y_{ij} = \mu_i + \beta(x_{ij} - \overline{\overline{x}}) + e_{ij} \qquad (1 \leq i \leq a, 1 \leq j \leq n_i), \tag{3.27}$$

where the e_{ij} are i.i.d. $N(0, \sigma^2)$ r.v.'s, μ_i is the ith adjusted (for the covariate) treatment mean, β is the slope coefficient which is assumed to be common for all treatments, and $\overline{\overline{x}}$ is the grand mean of the x_{ij}. As in the case of the one-way layout model (3.3), the μ_i are commonly expressed as $\mu_i = \mu + \alpha_i$ where the α_i are subject to the constraint $\sum n_i\alpha_i = 0$.

The x_{ij} are assumed to be fixed; if in fact they are random, then the inference is conditional on their observed values, as in the case of multiple regression (see Section 2.3). Thus the ANCOVA model can be regarded as a combination of the ANOVA model (3.2) and the simple linear regression model (2.15).

Note that the common slope coefficient β implies that the regression lines for the treatments are parallel. Therefore the effect of the covariate on the response variable is the same across all treatments. We say that there is no treatment by covariate interaction. In this case the difference between the mean responses of any two treatments i and j for any fixed covariate value x is independent of x and equals $\mu_i - \mu_j$. This assumption is the same as the no treatment–block interaction for the RB design that we make in Section 5.1. A test for the common slopes assumption is discussed in Section 3.8.3.4 on model diagnostics.

Figure 3.7 shows how misleading conclusions can result if the effect of the covariate is ignored. To fix ideas, suppose that three different teaching methods are compared in a study where x is the pretest score of students and y is the posttest score. First look at the top panel in which the methods do not differ, but because the pretest scores of the students in each group are very different (e.g., due to self-selection bias), the posttest scores are also very different. As a result, the null hypothesis of equal method effects will be rejected with a high probability. Now look at the bottom in which the methods are different, but the methods with low effects are assigned to students with high pretest scores and vice versa. Since the slopes of the regression lines are positive (as would be the case for a typical pretest–posttest study), the low method effects are compensated by contributions from the high pretest scores and vice versa, resulting in roughly equal mean posttest scores. Therefore the null hypothesis of equal method effects will be accepted with a high probability even though it is false. Hence it is extremely important to eliminate the possible confounding effects of the covariates via ANCOVA so as to obtain unbiased comparisons between the treatments.

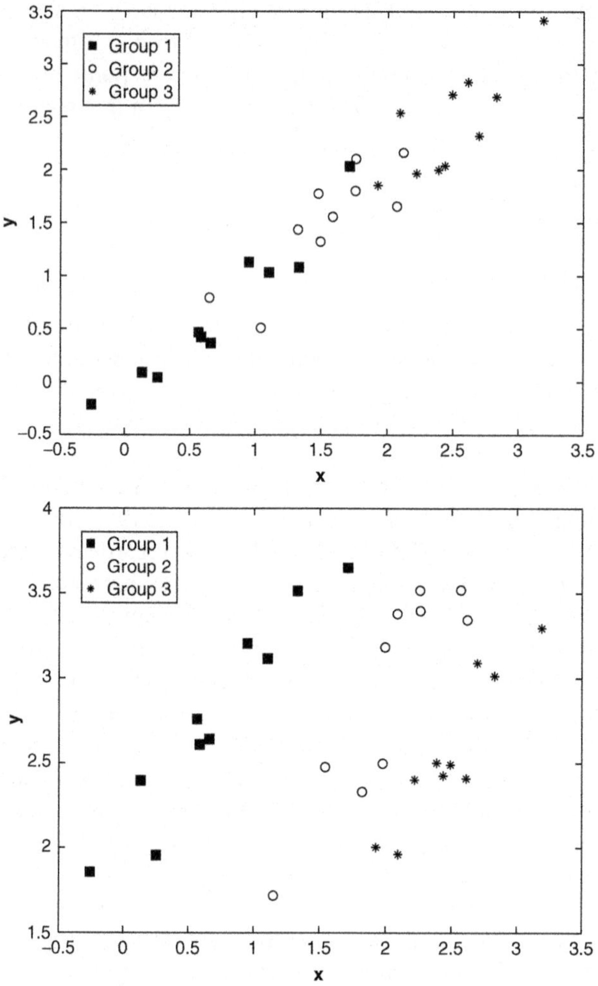

Figure 3.7 Two situations where ignoring the covariate leads to a wrong decision on $H_0 : \mu_1 = \mu_2 = \mu_3$.

3.8.3 Statistical Analysis

3.8.3.1 Estimation

To give the necessary formulas, we define the following sums of cross-products and squares:

$$R_{xy} = \sum_{i=1}^{a} n_i(\overline{x}_i - \overline{\overline{x}})(\overline{y}_i - \overline{\overline{y}}), \qquad R_{xx} = \sum_{i=1}^{a} n_i(\overline{x}_i - \overline{\overline{x}})^2,$$

$$R_{yy} = \sum_{i=1}^{a} n_i(\overline{y}_i - \overline{\overline{y}})^2,$$

$$S_{xy} = \sum_{i=1}^{a}\sum_{j=1}^{n_i}(x_{ij} - \overline{x}_i)(y_{ij} - \overline{y}_i), \qquad S_{xx} = \sum_{i=1}^{a}\sum_{j=1}^{n_i}(x_{ij} - \overline{x}_i)^2,$$

$$S_{yy} = \sum_{i=1}^{a}\sum_{j=1}^{n_i}(y_{ij} - \overline{y}_i)^2,$$

$$T_{xy} = \sum_{i=1}^{a}\sum_{j=1}^{n_i}(x_{ij} - \overline{\overline{x}})(y_{ij} - \overline{\overline{y}}), \qquad T_{xx} = \sum_{i=1}^{a}\sum_{j=1}^{n_i}(x_{ij} - \overline{\overline{x}})^2,$$

$$T_{yy} = \sum_{i=1}^{a}\sum_{j=1}^{n_i}(y_{ij} - \overline{\overline{y}})^2.$$

Note that the quantities, R, S, and T are the between, within, and total sums of cross products and squares of the x's and the y's, respectively. It follows from the one-way ANOVA identity (3.8) that $R_{xy} + S_{xy} = T_{xy}$, $R_{xx} + S_{xx} = T_{xx}$, and $R_{yy} + S_{yy} = T_{yy}$.

The LS estimators of the parameters in the ANCOVA model (3.27) can be shown to be (see Section 3.9.4)

$$\widehat{\mu}_i = \overline{y}_i - \widehat{\beta}(\overline{x}_i - \overline{\overline{x}}) \qquad (1 \le i \le a) \qquad \text{and} \qquad \widehat{\beta} = \frac{S_{xy}}{S_{xx}}. \qquad (3.28)$$

The $\widehat{\mu}_i$ are called the estimated **adjusted treatment means**. Note that $\widehat{\beta}$ has exactly the same form as the corresponding slope estimator $\widehat{\beta}_1$ in simple linear regression; see Eq. (2.13). One can view $\widehat{\beta}$ as a pooled estimator of the common slope coefficient of $a \ge 2$ parallel regression lines.

3.8.3.2 Analysis of Covariance

In analogy with the ANOVA for simple linear regression given in Table 2.7, it is not difficult to show that the error sum of squares for this model is

$$\text{SS}_e = S_{yy} - \frac{S_{xy}^2}{S_{xx}}$$

with $\nu = N - a - 1$ d.f. It follows that an unbiased estimator of σ^2 is given by

$$s^2 = \text{MS}_e = \frac{\text{SS}_e}{N - a - 1},$$

which is distributed as $\sigma^2\chi_\nu^2/\nu$ independently of the $\widehat{\mu}_i$ and $\widehat{\beta}$. Compared to the one-way ANOVA, the error d.f. is less by 1 because of the additional parameter β in the ANCOVA model. Also, in analogy with Table 2.7, the sum of squares for regression equals

$$\text{SS}_{\text{reg}} = \frac{S_{xy}^2}{S_{xx}},$$

with one d.f. and $\text{SS}_{\text{tot}} = T_{yy}$.

The treatment sum of squares, SS_{trt}, cannot be obtained simply by subtraction (i.e., $SS_{trt} \neq SS_{tot} - SS_{reg} - SS_e$) because this is not an orthogonal design. In particular, the column corresponding to the β term in the $\boldsymbol{X}$ matrix for the design is not orthogonal to the columns corresponding to the μ_i terms. However, SS_{trt} can be obtained by applying the extra sum of squares method as follows. Under the hypothesis $H_0 : \mu_1 = \mu_2 = \cdots = \mu_a$, the model (3.27) reduces to

$$y_{ij} = \mu + \beta(x_{ij} - \bar{\bar{x}}) + e_{ij} \qquad (1 \le i \le a, 1 \le j \le n_i).$$

Because the same regression line holds for all treatments (the same intercept μ and the same slope coefficient β), the data from all the treatments can be pooled, yielding the following LS estimators under H_0:

$$\widehat{\mu}_0 = \bar{\bar{y}} \qquad \text{and} \qquad \widehat{\beta}_0 = \frac{T_{xy}}{T_{xx}}.$$

The error sum of squares for this reduced model is

$$SS_{e0} = T_{yy} - \frac{T_{xy}^2}{T_{xx}}.$$

Therefore, by the extra sum of squares method,

$$SS_{trt} = SS_{e_0} - SS_e = \left[T_{yy} - \frac{T_{xy}^2}{T_{xx}}\right] - \left[S_{yy} - \frac{S_{xy}^2}{S_{xx}}\right]$$

with $a - 1$ d.f. These and the associated calculations of the mean squares and F-ratios are shown in the ANCOVA given in Table 3.10. An α-level test of $H_0 : \mu_1 = \mu_2 = \cdots = \mu_a$ rejects if

$$F = \frac{MS_{trt}}{MS_e} > f_{a-1,\nu,\alpha},$$

and an α-level test of $H_0 : \beta = 0$ rejects if

$$F = \frac{MS_{reg}}{MS_e} > f_{1,\nu,\alpha}.$$

Example 3.8 (Home Heating: Analysis of Covariance)

The average outside temperature and heating gas consumption were observed for a house in England for 26 weeks before and 30 weeks after installation of insulation. The house thermostat was set at 20°C throughout. Table 3.11 gives a subset of these data obtained by omitting alternate observations in each group resulting in the sample sizes of $n_1 = 13$ (before insulation) and $n_2 = 15$ (after insulation).

Table 3.10 ANCOVA Table for One-Way Layout with Covariate

Source	SS	d.f.	MS	F
Regression	$SS_{reg} = \frac{S_{xy}^2}{S_{xx}}$	1	MS_{reg}	MS_{reg}/MS_e
Treatments	$SS_{trt} = \left[T_{yy} - \frac{T_{xy}^2}{T_{xx}}\right] - \left[S_{yy} - \frac{S_{xy}^2}{S_{xx}}\right]$	$a-1$	MS_{trt}	MS_{trt}/MS_e
Error	$SS_e = S_{yy} - \frac{S_{xy}^2}{S_{xx}}$	$N-a-1$	MS_e	
Total	$SS_{tot} = T_{yy}$	$N-1$		

Table 3.11 Home Heating Data

	Before Insulation		After Insulation	
	Average Outside Temperature (°C)	Gas Consumption (1000 cf)	Average Outside Temperature (°C)	Gas Consumption (1000 cf)
	−0.8	7.2	−0.7	4.8
	0.4	6.4	1.0	4.7
	2.9	5.8	1.5	4.2
	3.6	5.6	2.3	4.1
	4.2	5.8	2.5	3.5
	5.4	4.9	3.9	3.9
	6.0	4.3	4.0	3.7
	6.2	4.5	4.3	3.5
	6.9	3.7	4.7	3.5
	7.4	4.2	4.9	3.7
	7.5	3.9	5.0	3.6
	8.0	4.0	6.2	2.8
	9.1	3.1	7.2	2.8
			8.0	2.7
			8.8	1.3
Mean	5.138	4.877	4.240	3.520
SD	2.966	1.195	2.648	0.874

Source: Data adapted from Hand et al. (1994, data set 88). Reprinted by permission of Taylor & Francis.

We want to determine if insulation saved on the average gas consumption. In making this determination we must take into account the outside temperature effect. We first plot the data in Figure 3.8. The after-insulation plot is systematically offset below the before-insulation plot, indicating definite energy savings due to insulation. The regression lines fitted to the two plots are roughly parallel, although one could argue based on the theory of heat transfer that the rate of gas consumption per degree change in outside temperature should be different with and without insulation. For simplicity, we will proceed with the assumption of equal slopes but test this assumption later in Example 3.10.

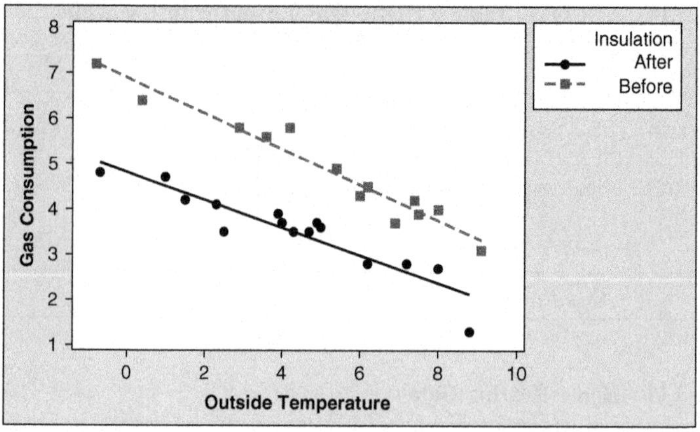

Figure 3.8 Plot of gas consumption against outside temperature before and after insulation.

Although calculations for ANCOVA are routinely performed on a computer, here we will carry them out using the given formulas for illustration purposes. First calculate

$$\overline{x}_1 = 5.138, \qquad \overline{x}_2 = 4.240, \qquad \overline{\overline{x}} = 4.657,$$
$$\overline{y}_1 = 4.877, \qquad \overline{y}_2 = 3.520, \qquad \overline{\overline{y}} = 4.150,$$

and

$$S_{xy} = -71.260, \qquad S_{xx} = 203.728, \qquad S_{yy} = 27.830,$$
$$T_{xy} = -62.775, \qquad T_{xx} = 209.358, \qquad T_{yy} = 40.662.$$

This gives

$$\widehat{\beta} = \frac{S_{xy}}{S_{xx}} = \frac{-71.260}{203.728} = -0.350.$$

Therefore the average gas consumption decreases by 0.35 units (i.e., 350 ft^3) for each 1°C increase in the average outside temperature. Furthermore,

$$\widehat{\mu}_1 = \overline{y}_1 - \widehat{\beta}(\overline{x}_1 - \overline{\overline{x}}) = 4.877 - (-0.350)(5.138 - 4.657) = 5.045$$

and

$$\widehat{\mu}_2 = \overline{y}_2 - \widehat{\beta}(\overline{x}_2 - \overline{\overline{x}}) = 3.520 - (-0.350)(4.240 - 4.657) = 3.374,$$

and the difference between the two intercepts is $5.045 - 3.374 = 1.671$.

Table 3.12 ANCOVA for Home Heating Data

Source	SS	d.f.	MS	F	p
Outside temperature	24.925	1	24.925	214.38	0.000
Insulation	18.934	1	18.934	162.94	0.000
Error	2.905	25	0.116		
Total	40.662	27			

The sums of squares in Table 3.12 are calculated as follows:

$$\mathrm{SS}_{\mathrm{reg}} = \frac{S_{xy}^2}{S_{xx}} = \frac{(-71.260)^2}{203.728} = 24.925,$$

$$\mathrm{SS}_e = S_{yy} - \mathrm{SS}_{\mathrm{reg}} = 27.830 - 24.925 = 2.905,$$

$$\mathrm{SS}_{\mathrm{trt}} = \left[T_{yy} - \frac{T_{xy}^2}{T_{xx}}\right] - \mathrm{SS}_e = \left[40.662 - \frac{(-62.775)^2}{209.358}\right] - 2.905 = 18.934,$$

$$\mathrm{SS}_{\mathrm{tot}} = T_{yy} = 40.662.$$

From Table 3.12 we see that both the insulation and outside temperature effects are highly significant. ■

3.8.3.3 *Confidence Intervals and Hypothesis Tests*

The adjusted treatment means μ_i are the primary parameters of interest in the ANCOVA problem. The distribution of their LS estimators $\widehat{\mu}_i$ can be derived using the following results from simple linear regression (see, e.g., Tamhane and Dunlop, 2000, pp. 358–360) extended to the present ANCOVA setup:

(a) $\widehat{\beta}$ is normally distributed with mean β and variance σ^2/S_{xx}.

(b) $\overline{y}_i$'s are independently and normally distributed with means $\mu_i + \beta(\overline{x}_i - \overline{\overline{x}})$ and variances σ^2/n_i.

(c) $\widehat{\beta}$ and $\overline{y}_i$ are independently distributed.

From these results it follows that

$$\widehat{\mu}_i = \overline{y}_i - \widehat{\beta}(\overline{x}_i - \overline{\overline{x}}) \sim \mathrm{N}\left(\mu_i, \sigma^2\left[\frac{1}{n_i} + \frac{(\overline{x}_i - \overline{\overline{x}})^2}{S_{xx}}\right]\right) \tag{3.29}$$

and

$$\widehat{\mu}_i - \widehat{\mu}_j = \overline{y}_i - \overline{y}_j - \widehat{\beta}(\overline{x}_i - \overline{x}_j) \sim \mathrm{N}\left(\mu_i - \mu_j, \sigma^2\left[\frac{1}{n_i} + \frac{1}{n_j} + \frac{(\overline{x}_i - \overline{x}_j)^2}{S_{xx}}\right]\right). \tag{3.30}$$

Inferences on the adjusted treatment means can be based on these two results. For example, a $100(1-\alpha)\%$ CI on the pairwise difference $\mu_i - \mu_j$ is given by

$$\mu_i - \mu_j \in \left[\widehat{\mu}_i - \widehat{\mu}_j \pm t_{\nu,\alpha/2} s \sqrt{\frac{1}{n_i} + \frac{1}{n_j} + \frac{(\overline{x}_i - \overline{x}_j)^2}{S_{xx}}}\right], \tag{3.31}$$

where $\nu = N - a - 1$.

Example 3.9 (Home Heating: Effect of Insulation)

We will compare the adjusted treatment means before and after insulation by finding a 95% CI on their difference. The sample estimates of the adjusted treatment means are $\widehat{\mu}_1 = 5.045$ and $\widehat{\mu}_2 = 3.374$. Also, $s = \sqrt{\text{MS}_e} = \sqrt{0.116} = 0.341$ with 25 d.f. From (3.31), we calculate a 95% CI on $\mu_1 - \mu_2$ (using $t_{25,0.025} = 2.060$) as

$$\left[5.045 - 3.374 \pm (2.060)(0.341)\sqrt{\frac{1}{13} + \frac{1}{15} + \frac{(5.138 - 4.240)^2}{203.728}}\right]$$
$$= [1.671 \pm 0.270] = [1.401, 1.941].$$

The corresponding t-statistic for testing $H_0 : \mu_1 = \mu_2$ is

$$t = \frac{5.045 - 3.374}{0.341\sqrt{\frac{1}{13} + \frac{1}{15} + \frac{(5.138 - 4.240)^2}{203.728}}} = 12.749.$$

Note that $t^2 = (12.749)^2 = 214.38$, which is the F-statistic given in Table 3.12. Thus the t- and F-statistics are equivalent and have the same p-values. ■

3.8.3.4 Model Diagnostics

The assumption of equal slopes can be tested by including the treatment–covariate interaction term in the model and testing its significance. This test is equivalent to the extra sum of squares F-test (see Section 2.3.2.3) in which the full model that includes the interaction term implies unequal slopes and the partial model is the ANCOVA model (3.27) that assumes equal slopes. A significant interaction means that the assumption of equal slopes is not valid.

Residual plots as described in Section 3.4 can be used to check the normality and homoscedasticity assumptions. The fitted values and the residuals are given by

$$\widehat{y}_{ij} = \widehat{\mu}_i + \widehat{\beta}(x_{ij} - \overline{\overline{x}}) = \overline{y}_i + \widehat{\beta}(x_{ij} - \overline{x}_i) \quad \text{and} \quad \widehat{e}_{ij} = y_{ij} - \widehat{y}_{ij}.$$

The residuals can be plotted against the x_{ij} to check if the assumption of linearity with respect to the covariate is valid. The normality assumption can be checked by making a normal plot of the residuals and a goodness-of-fit test.

Most statistical software packages use standardized residuals to detect outliers. The standard error of a residual is given by (see Exercise 3.25 for a derivation)

$$\mathrm{SE}(\widehat{e}_{ij}) = s\sqrt{1 - \left[\frac{1}{n_i} + \frac{(x_{ij} - \overline{x}_i)^2}{S_{xx}}\right]}. \tag{3.32}$$

Thus, the observation (x_{ij}, y_{ij}) is regarded as an outlier if the standardized residual

$$\widehat{e}^{*}_{ij} = \frac{\widehat{e}_{ij}}{\mathrm{SE}(\widehat{e}_{ij})}$$

is large in magnitude. The assumption of parallel regression lines or equal slopes can be tested by using the extra sum of squares method.

Example 3.10 (Home Heating: Model Diagnostics)

We shall first check the equal-slopes assumption. The scatter plot in Figure 3.8 suggests approximately parallel regression lines. To more clearly see the differences in the slopes, we plot the residuals from the equal-slopes model identified by the treatment (before or after insulation) in Figure 3.9. The plot shows that

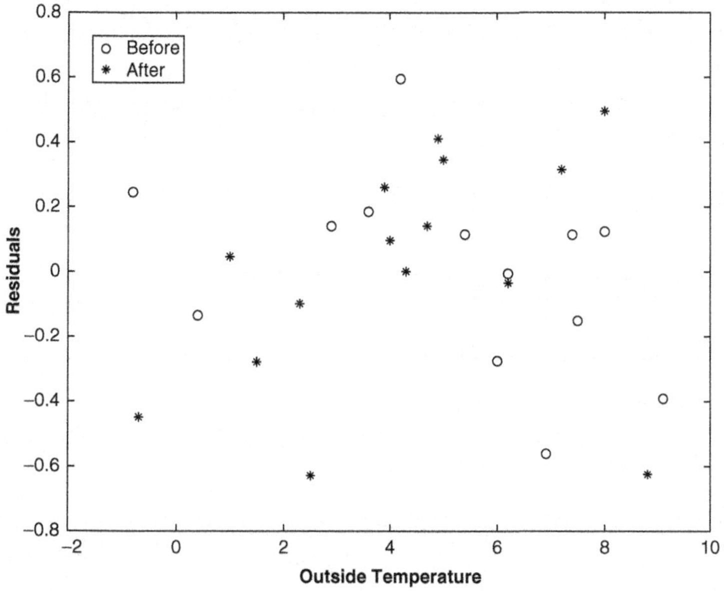

Figure 3.9 Labeled scatter plot of residuals from equal-slopes model identified by before and after insulation.

```
Analysis of Variance for Gas Consumption, using Adjusted SS for Tests
Source                           DF   Seq SS   Adj SS   Adj MS        F      P
Insulation                        1   12.823    7.516    7.516    71.94  0.000
Outside Temperature               1   24.926   24.664   24.664   236.05  0.000
Insulation*Outside Temperature    1    0.394    0.394    0.394     3.77  0.064
Error                            24    2.508    2.508    0.104
Total                            27   40.650
S = 0.323243    R-Sq = 93.83%    R-Sq(adj) = 93.06%
Term                           Coef   SE Coef        T      P
Constant                     5.8509    0.1228    47.66  0.000
Outside Temp               -0.34817   0.02266   -15.36  0.000
Outside Temp*Insulation
                After       0.04398   0.02266     1.94  0.064
```

Display 3.4 Minitab output for ANCOVA model with unequal slopes.

the before-insulation residuals trend downward while the after-insulation residuals trend upward, indicating that the after-insulation slope is steeper than the before-insulation slope.

Display 3.4 shows the Minitab output for the ANCOVA model that includes the insulation–outside temperature interaction. We see that the interaction term has $p = 0.064$ and is thus not significant at $\alpha = 0.05$. Nonetheless, it is sufficiently on the borderline of significance to cause concern since slight nonsignificance may be caused by one or two outlier observations. Note, however, that if the unequal-slopes model is assumed, then the effect of insulation depends upon outside temperature.

A normal plot and a fitted-values plot of the residuals are shown in Figure 3.10. Both plots are satisfactory and do not indicate violation of the normality and homoscedasticity assumptions. The Anderson–Darling statistic for the normality test equals $A^2 = 0.361$ ($p = 0.422$) and the F-statistic for the equality of the variances equals 1.416 ($p = 0.552$), confirming the conclusions from the plots. ■

3.9 CHAPTER NOTES

3.9.1 Randomization Distribution of F-Statistic

The randomization test for the two independent samples t-statistic was introduced in Section 2.2.1.5. In this section we generalize it to the F-test for the one-way ANOVA. As in the case of the t-statistic, the **randomization distribution** of the F-statistic is obtained by treating the observed data values as fixed and considering all possible assignments of treatment labels to these data values and calculating the F-statistic for each assignment. There are a total of

$$C(n_1, n_2, \ldots, n_a) = \frac{N!}{n_1! n_2! \cdots n_a!}$$

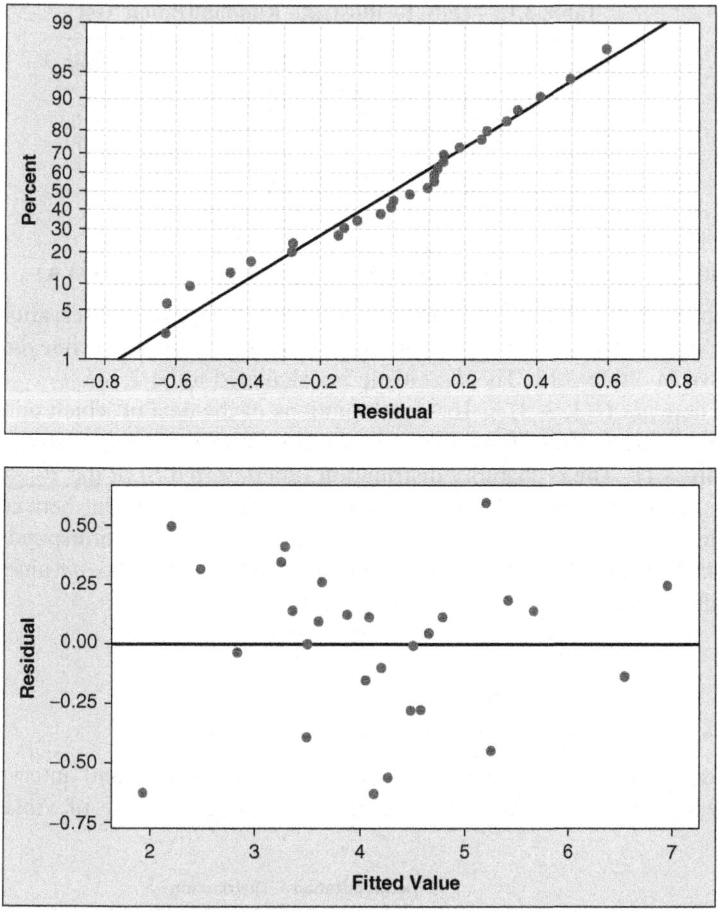

Figure 3.10 Normal (top) and fitted-values (bottom) plots of residuals from ANCOVA of home heating data.

possible assignments and under the overall null hypothesis H_0 they are equally likely. The upper tail area of this distribution from the actual observed F-statistic gives its p-value.

Note that the probability distribution is introduced not by the randomness of the data (which are assumed fixed) but by the act of randomization. Pitman (1937) showed that for large n_i's this distribution is well approximated by the F-distribution with a-1 and N-a d.f. which justifies using the normal theory ANOVA F-test, although the normality assumption is not used in this derivation.

Even for small n_i's, the number of possible assignments, $C(n_1, n_2, \ldots, n_a)$, is extremely large and even modern fast computers are unable to cope with the task of enumerating all of them. So we will illustrate the randomization F-test by a small example.

Table 3.13 Data to Illustrate Randomization Test

Group 1	Group 2	Group 3
2	1	5
4	6	8
3	7	9

Example 3.11 (Randomization F-Test for One-Way ANOVA)

Suppose that we have $a = 3$ treatment groups with $n = 3$ observations on each. The data (which for the sake of illustration are taken to be integers 1–9) are shown in Table 3.13. The F-statistic is calculated to be 2.745.

There are $C(3, 3, 3) = 1680$ permutations of the data of which only 280 yield distinct values of the F-statistic. The histogram of these F-statistics is shown in Figure 3.11. The probability distribution function (p.d.f.) of the $F_{2,6}$-distribution is superimposed on the histogram. Note the close agreement between the two. From the randomization distribution we find the p-value corresponding to $F = 2.745$ to be 0.1393. If referred to the $F_{2,6}$-distribution, the p-value is 0.1424, again confirming the closeness of approximation. ■

3.9.2 F-Test for Heteroscedastic Treatment Variances

As stated earlier, the F-test and the associated t-tests and intervals are not very robust if the treatment variances are unequal. In case of violation of the

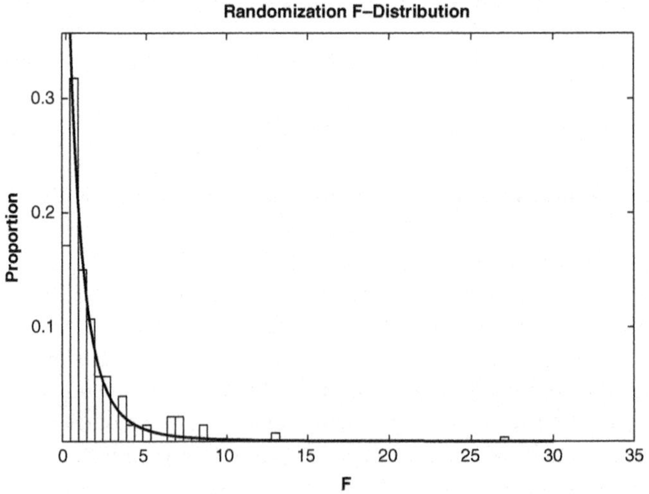

Figure 3.11 Histogram of F-statistics under randomization distribution superimposed with $F_{2,6}$-distribution.

homoscedasticity assumption, Brown and Forsythe (1974) proposed a modification of the F-test which is a generalization of the Welch–Satterthwaite approximate two-sample t-test discussed in Section 2.2.1.3. This modification involves approximating the distribution of an F-like statistic by an F-distribution with estimated d.f.

The F-like statistic is computed using the formula

$$F^* = \frac{\sum_{i=1}^a n_i(\bar{y}_i - \bar{\bar{y}})^2}{\sum_{i=1}^a (1 - n_i/N)s_i^2}, \tag{3.33}$$

where s_i^2 are the sample variances of the treatments. Note that for a balanced one-way layout ($n_i = n$ for all i) F^* equals the usual F-statistic. Brown and Forsythe (1974) proposed approximating the null distribution of F^* by an F_{ν_1,ν_2}-distribution with $\nu_1 = a - 1$ and ν_2 obtained by the moment-matching method, which yields

$$\nu_2 = \frac{\left[\sum_{i=1}^a (1 - n_i/N)s_i^2\right]^2}{\sum_{i=1}^a \dfrac{(1 - n_i/N)^2 s_i^4}{n_i - 1}}.$$

Mehrotra (1997a) improved this approximation by also estimating ν_1 from the data using the same moment matching method, which yields

$$\nu_1 = \frac{\left(\sum_{i=1}^a s_i^2 - \sum_{i=1}^a n_i s_i^2/N\right)^2}{\sum_{i=1}^a s_i^4 + \left(\dfrac{\sum_{i=1}^a n_i s_i^2}{N}\right)^2 - \dfrac{2\sum_{i=1}^a n_i s_i^4}{N}}.$$

Finally, an approximate α-level test rejects $H_0 : \mu_1 = \mu_2 = \cdots = \mu_a$ if $F^* > f_{\nu_1,\nu_2,\alpha}$.

Example 3.12 (HIV Data: Heterascedastic F-Test)

This example is taken from Mehrotra (1997b). The summary data in Table 3.14 come from a three-parallel-arm clinical trial for a vaccine against HIV infection. The response variable is the change in baseline CD4 cell count at 24 weeks after start of the treatment. The overall mean equals

$$\bar{\bar{y}} = \frac{(11)(53.9) + (17)(54.3) + (12)(49.2)}{40} = 52.66.$$

Therefore the treatment sum of squares equals

$$\text{SS}_{\text{trt}} = 11(53.9 - 52.66)^2 + 17(54.3 - 52.66)^2 + 12(49.2 - 52.66)^2 = 206.30.$$

Table 3.14 Summary Statistics for Change in Baseline CD4 Cell Counts for Three Treatment Groups

Treatment Group	n	Mean	SD
1	11	53.9	7.9
2	17	54.3	3.1
3	12	49.2	2.2

Source: Mehrotra (1977b).

Also,

$$s^2 = \mathrm{MS}_e = \frac{(10)(7.9)^2 + (16)(3.1)^2 + (11)(2.2)^2}{37} = 22.462.$$

Hence

$$F = \frac{206.30/2}{22.462} = 4.592$$

with p-value $= 0.0165$. However, the variance homogeneity assumption is not valid as can be checked using the Bartlett test; see Exercise 3.11. Therefore we apply Mehrotra's modification of the Brown–Forsythe test: Compute

$$F^* = \frac{206.30}{(1 - 11/40)(7.9)^2 + (1 - 17/40)(3.1)^2 + (1 - 12/40)(2.2)^2} = 3.809.$$

The d.f. calculated using the given formulas are $\nu_1 = 1.63$ and $\nu_2 = 14.12$. The corresponding p-value is 0.04, which is more than twice that obtained using the usual F-test. ■

3.9.3 Derivations of Formulas for Orthogonal Polynomials

The model (3.22) can be written in matrix notation of Section 2.3.2 as $\mathrm{E}(\boldsymbol{y}) = \boldsymbol{X}\boldsymbol{\beta}$, where

$$\boldsymbol{y} = \begin{bmatrix} y_{11} \\ \vdots \\ y_{1n} \\ y_{21} \\ \vdots \\ y_{2n} \\ \vdots \\ y_{a1} \\ \vdots \\ y_{an} \end{bmatrix}, \quad \boldsymbol{X} = \begin{bmatrix} 1 & \xi_1(z_1) & \ldots & \xi_p(z_1) \\ \vdots & \vdots & \vdots & \vdots \\ 1 & \xi_1(z_1) & \ldots & \xi_p(z_1) \\ 1 & \xi_1(z_2) & \ldots & \xi_p(z_2) \\ \vdots & \vdots & \ddots & \vdots \\ 1 & \xi_1(z_2) & \ldots & \xi_p(z_2) \\ \vdots & \vdots & \ddots & \vdots \\ 1 & \xi_1(z_a) & \ldots & \xi_p(z_a) \\ \vdots & \vdots & \ddots & \vdots \\ 1 & \xi_1(z_a) & \ldots & \xi_p(z_a) \end{bmatrix}, \quad \boldsymbol{\beta} = \begin{bmatrix} \beta_0 \\ \beta_1 \\ \vdots \\ \beta_p \end{bmatrix}.$$

Since the columns of $\boldsymbol{X}$ are orthogonal because of the conditions (3.23), it follows that

$$\boldsymbol{X}'\boldsymbol{X} = \begin{bmatrix} N & 0 & \dots & 0 \\ 0 & n\sum \xi_1^2(z_i) & \dots & 0 \\ \vdots & \vdots & \ddots & \vdots \\ 0 & 0 & \dots & n\sum \xi_p^2(z_i) \end{bmatrix}.$$

Furthermore,

$$\boldsymbol{X}'\boldsymbol{y} = \begin{bmatrix} \sum\sum y_{ij} \\ \sum\sum \xi_1(z_i)y_{ij} \\ \vdots \\ \sum\sum \xi_p(z_i)y_{ij} \end{bmatrix}.$$

Using (2.29), the LS estimator of $\boldsymbol{\beta}$ is given by

$$\begin{aligned}
\widehat{\boldsymbol{\beta}} &= (\boldsymbol{X}'\boldsymbol{X})^{-1}\boldsymbol{X}'\boldsymbol{y} \\
&= \begin{bmatrix} N^{-1} & 0 & \dots & 0 \\ 0 & [n\sum \xi_1^2(z_i)]^{-1} & \dots & 0 \\ \vdots & \vdots & \ddots & \vdots \\ 0 & 0 & \dots & [n\sum \xi_p^2(z_i)]^{-1} \end{bmatrix} \begin{bmatrix} \sum\sum y_{ij} \\ \sum\sum \xi_1(z_i)y_{ij} \\ \vdots \\ \sum\sum \xi_p(z_i)y_{ij} \end{bmatrix} \\
&= \begin{bmatrix} \dfrac{\sum\sum y_{ij}}{N} \\ \dfrac{\sum\sum \xi_1(z_i)y_{ij}}{n\sum \xi_1^2(z_i)} \\ \vdots \\ \dfrac{\sum\sum \xi_p(z_i)y_{ij}}{n\sum \xi_p^2(z_i)} \end{bmatrix}.
\end{aligned}$$

This gives the estimators in (3.24).

Next, from Eq. (14.18) we have

$$\begin{aligned}
\mathrm{Cov}(\widehat{\boldsymbol{\beta}}) &= \sigma^2(\boldsymbol{X}'\boldsymbol{X})^{-1} \\
&= \sigma^2 \begin{bmatrix} N^{-1} & 0 & \dots & 0 \\ 0 & [n\sum_{i=1}^{a} \xi_1^2(z_i)]^{-1} & \dots & 0 \\ \vdots & \vdots & \ddots & \vdots \\ 0 & 0 & \dots & [n\sum_{i=1}^{a} \xi_p^2(z_i)]^{-1} \end{bmatrix}.
\end{aligned}$$

This gives the variances and covariances of the estimators in (3.25).

3.9.4 Derivation of LS Estimators for One-Way Analysis of Covariance

We can write the ANCOVA model (3.27) in matrix notation of Section 2.3.2 as follows. The vector of parameters is $\boldsymbol{\beta} = (\mu_1, \ldots, \mu_a, \beta)'$ and the vector of response variables is $\boldsymbol{y} = (y_{11}, \ldots, y_{1n_1}, \ldots, y_{a1}, \ldots, y_{an_a})'$. The $\boldsymbol{X}$ matrix is given as

$$\boldsymbol{X} = \begin{bmatrix} 1 & & & x_{11} - \bar{\bar{x}} \\ \vdots & & & \vdots \\ 1 & & & x_{1n_1} - \bar{\bar{x}} \\ & 1 & & x_{21} - \bar{\bar{x}} \\ & \vdots & & \vdots \\ & 1 & & x_{2n_2} - \bar{\bar{x}} \\ & & \ddots & \vdots \\ & & 1 & x_{a1} - \bar{\bar{x}} \\ & & \vdots & \vdots \\ & & 1 & x_{an_a} - \bar{\bar{x}} \end{bmatrix}.$$

Hence we get

$$\boldsymbol{X}'\boldsymbol{X} = \begin{bmatrix} n_1 & & & \sum(x_{1j} - \bar{\bar{x}}) \\ & \ddots & & \vdots \\ & & n_a & \sum(x_{aj} - \bar{\bar{x}}) \\ \sum(x_{1j} - \bar{\bar{x}}) & \cdots & \sum(x_{aj} - \bar{\bar{x}}) & \sum\sum(x_{ij} - \bar{\bar{x}})^2 \end{bmatrix}.$$

The normal equations [see (2.28)] are

$$\begin{aligned} n_1\mu_1 + \cdots + n_a\mu_a &= \sum\sum y_{ij}, \\ n_1\mu_1 + \beta\sum(x_{1j} - \bar{\bar{x}}) &= \sum y_{1j}, \\ &\vdots \\ n_a\mu_a + \beta\sum(x_{aj} - \bar{\bar{x}}) &= \sum y_{aj}, \\ \mu_1\sum(x_{1j} - \bar{\bar{x}}) + \cdots + \mu_a\sum(x_{aj} - \bar{\bar{x}}) + \beta\sum\sum(x_{ij} - \bar{\bar{x}})^2 & \\ &= \sum\sum(x_{ij} - \bar{\bar{x}})y_{ij}. \end{aligned}$$

If we define $\mu = \sum n_i \mu_i / N$ as the overall (weighted) mean of the μ_i's, then the first equation gives $\widehat{\mu} = \overline{\overline{y}}$. The next a equations then simplify to

$$\mu_i = \frac{1}{n_i}\left[\sum_j y_{ij} - \beta \sum_j (x_{ij} - \overline{\overline{x}})\right] = \overline{y}_i - \beta(\overline{x}_i - \overline{\overline{x}}) \qquad (1 \le i \le a). \tag{3.34}$$

The left-hand side of the last equation simplifies to

$$\sum_j (x_{1j} - \overline{\overline{x}})\left[\overline{y}_1 - \beta(\overline{x}_1 - \overline{\overline{x}})\right] + \cdots + \sum_j (x_{aj} - \overline{\overline{x}})\left[\overline{y}_a - \beta(\overline{x}_a - \overline{\overline{x}})\right]$$

$$+\beta \sum_i \sum_j (x_{ij} - \overline{\overline{x}})^2 = \sum_i n_i(\overline{x}_i - \overline{\overline{x}})\overline{y}_i$$

$$+\beta\left[\sum_i \sum_j (x_{ij} - \overline{\overline{x}})^2 - \sum_i n_i(\overline{x}_i - \overline{\overline{x}})^2\right] = R_{xy} + \beta(T_{xx} - R_{xx})$$

and the right-hand side simplifies to

$$\sum_i \sum_j (x_{ij} - \overline{\overline{x}}) y_{ij} = \sum_i \sum_j (x_{ij} - \overline{\overline{x}})(y_{ij} - \overline{\overline{y}}) = T_{xy}.$$

Thus from the last equation we obtain

$$\widehat{\beta} = \frac{T_{xy} - R_{xy}}{T_{xx} - R_{xx}} = \frac{S_{xy}}{S_{xx}}.$$

Substituting this in (3.34) we obtain

$$\widehat{\mu}_i = \overline{y}_i - \widehat{\beta}(\overline{x}_i - \overline{\overline{x}}) \qquad (1 \le i \le a).$$

This completes the derivation of the formulas in (3.28) for both $\widehat{\mu}_i$ and $\widehat{\beta}$.

3.10 CHAPTER SUMMARY

(a) A one-way layout consists of data classified by a single treatment or grouping factor. Such data may be obtained through a CR experimental design or an observational study. The same methods of analysis are used in both cases.

(b) Graphical comparisons consisting of side-by-side box plots and summary statistics such as the means and standard deviations are useful first steps in the analysis of data. More formal inferences often begin with an analysis of variance followed by confidence intervals and tests on the individual treatment means and pairwise differences among them.

(c) The normal theory model assumptions, in particular, normality and homoscedasticity, must be checked through normal and fitted-values plots and Bartlett or Levine tests. Sometimes model violations can be corrected by making data transformations. In other cases, an alternative method, such as a distribution-free test or a randomization test, must be used. Outliers should also be identified and suitably accounted for (either by correcting, modifying, or deleting them or by using inferential methods that are resistant to them).

(d) The power of the ANOVA F-test involves a noncentral F-distribution; its n.c.p. is a measure of the deviation of the alternative hypothesis from the null hypothesis (3.7). Sample size calculations can be done to guarantee a specified power requirement using charts or programs based on this distribution.

(e) If the treatment factor is quantitative and its levels are equispaced on some scale, then the treatment sum of squares and its d.f. can be partitioned into independent (orthogonal) contributions due to various effects such as linear, quadratic, and so on. These effects can then be independently estimated and tested.

(f) If data on some covariates are available on each experimental unit in a one-way layout, then the treatment effects can be estimated and tested after adjusting for the effects of the covariates using ANCOVA, which is a combination of ANOVA and regression. Treatment effects are meaningful only if there is no treatment by covariate interaction. In other words, the regression lines with respect to each covariate for different treatments are parallel.

EXERCISES

Section 3.3 (Statistical Analysis)

Theoretical Exercises

3.1 (From Tamhane and Dunlop, 2000, Exercise 12.30. Reprinted by permission of Pearson Education, Inc.) Derive the ANOVA identity (3.8) by expressing the three sums of squares as follows:

$$\mathrm{SS}_{\mathrm{tot}} = \sum_{i=1}^{a}\sum_{j=1}^{n_i} y_{ij}^2 - N\bar{\bar{y}}^2, \qquad \mathrm{SS}_{\mathrm{trt}} = \sum_{i=1}^{a} n_i \bar{y}_i^2 - N\bar{\bar{y}}^2,$$

$$\mathrm{SS}_e = \sum_{i=1}^{a}\sum_{j=1}^{n_i} y_{ij}^2 - \sum_{i=1}^{a} n_i \bar{y}_i^2.$$

3.2 (From Tamhane and Dunlop, 2000, Exercise 12.33. Reprinted by permission of Pearson Education, Inc.) For $a = 2$ show that the one-way ANOVA F-test is equivalent to the two-sample t-test of $H_0 : \mu_1 = \mu_2$ versus $H_1 : \mu_1 \neq \mu_2$ by carrying out the following steps:

(a) Show that

$$\text{MS}_{\text{trt}} = \text{SS}_{\text{trt}} = \frac{n_1 n_2}{n_1 + n_2}(\overline{y}_1 - \overline{y}_2)^2.$$

(b) Show that

$$F = \frac{\text{MS}_{\text{trt}}}{\text{MS}_e} = \left(\frac{\overline{y}_1 - \overline{y}_2}{s\sqrt{1/n_1 + 1/n_2}} \right)^2 = t^2.$$

(c) Finally show that the α-level F-test of H_0 rejects whenever the α-level two-sided t-test of H_0 rejects by showing that

$$F > f_{1,\nu,\alpha} \Longleftrightarrow |t| > t_{\nu,\alpha/2},$$

where $\nu = n_1 + n_2 - 2$. [*Hint*: Use the result (A.10) that $t^2_{\nu,\alpha/2} = f_{1,\nu,\alpha}$.]

Applied Exercises

3.3 Table 3.15 provides data on wood glue strengths. Eight wood joints were tested per glue. Calculate the ANOVA table and test if there are significant differences at $\alpha = 0.01$ between the mean strengths of the glues. Assume normality and homoscedasticity.

3.4 (From Tamhane and Dunlop, 2000, Examples 12.1 and 12.3. Reprinted by permission of Pearson Education, Inc.) Table 3.16 gives data on the weights of plastic bottles produced at six different stations on an injection molding machine.

(a) Make side-by-side box plots of the data to identify any differences (in mean weights and standard deviations) between the molding stations.

Table 3.15 Sample Means and Standard Deviations for Wood Joint Strength Data (kN)

Glue no.	1	2	3	4	5	6	7	8
Mean	1821	1968	1439	616	1354	1424	1694	1669
SD	214	435	243	205	135	191	225	551

Source: Pellicane (1990, Table 1). Copyright ASTM INTERNATIONAL. Reprinted with permission.

Table 3.16 Plastic Bottle Weights (grams)

	Station 1	Station 2	Station 3	Station 4	Station 5	Station 6
	51.28	51.46	51.07	51.70	51.82	52.12
	51.63	51.15	51.44	51.69	51.70	52.29
	51.06	51.21	50.91	52.12	51.25	51.42
	51.66	51.07	51.11	51.23	51.68	51.88
	52.20	51.84	50.77	51.51	51.76	52.00
	51.27	51.46	51.86	52.02	51.63	51.84
	52.31	51.50	51.22	51.35	51.61	51.57
	51.87	50.99	51.54	51.36	52.14	51.74
Sample mean	51.660	51.335	51.240	51.623	51.699	51.858
Sample SD	0.450	0.281	0.357	0.322	0.247	0.284

Source: Tamhane and Dunlop (2000, Table 12.2). Reprinted by permission of Pearson Education Inc.

(b) Calculate the ANOVA table. Are the mean differences between the stations significant at the 0.01 level?

3.5 (From Tamhane and Dunlop, 2000, Exercise 12.5. Reprinted by permission of Pearson Education, Inc.) Hemoglobin levels were measured on patients with three different types of sickle cell disease: sickle cell disease with two SS genes (HB SS), the combined problem of sickle cell trait with thalassemia (HB S/thalassemia), and the variant of sickle cell disease which has one S and one C gene (HB SC). The purpose was to investigate whether the mean hemoglobin levels differ with the type of disease. The data are given in Table 3.17.

(a) Make side-by-side box plots for the three disease types. Are any obvious differences indicated by the plots?

(b) Make an ANOVA table. Test the null hypothesis of equal mean hemoglobin levels.

3.6 A dose–response study was conducted to investigate the thyroid iodide transport mechanism in rats, in particular, to determine if the effect of

Table 3.17 Hemoglobin Levels in Sickle Cell Disease Patients

HB SS	7.2	7.7	8.0	8.1	8.3	8.4	8.4	8.5
	8.6	8.7	9.1	9.1	9.8	10.1	10.3	
HB S/thalassemia	8.1	9.2	10.0	10.4	10.6	10.9	11.1	11.9
	12.0	12.1						
HB SC	10.7	11.3	11.5	11.6	11.7	11.8	12.0	12.1
	12.3	12.6	12.6	13.3	13.3	13.8	13.9	

Source: Hand et al. (1994, data set 310). Reprinted by permission of Taylor & Francis.

Table 3.18 T/S Ratio as Function of Iodide Level (micrograms) and with or without PTU

	0 μg	0.25 μg	2.5 μg	45 μg
	Without PTU			
	184	87	92	146
	137	128	138	162
	207	172	117	119
	166	172	83	68
	276	81	152	141
	145	167	115	139
	197	130	99	164
Mean	187.4	133.9	113.7	134.1
SD	46.8	38.8	24.8	32.9
	With PTU			
	215	204	245	260
	169	145	132	359
	274	317	291	196
	116	315	354	228
	209	282	629	222
	127	241	204	90
	214	141	249	475
Mean	189.1	235.0	300.6	261.4
SD	55.5	74.5	160.4	123.5

Source: Dunnett (1970). Reprinted by permission of M.I.T. Press.

iodide, if any, can be prevented by propylthiouracil (PTU). Eight groups of seven rats each were administered with four different doses of iodide (including a zero dose to serve as a control), the first four groups without PTU and the remaining four groups with PTU. Table 3.18 gives the ratio of iodide concentrations in the thyroid and in the serum (T/S ratio). You will analyze these data in detail in several exercises in the present and later chapters. In this exercise you will begin by doing a simple one-way ANOVA by ignoring the two-factor structure.

Calculate the ANOVA table and do the F-test for the equality of the means of the eight dose groups. What is the p-value of the test? Are the differences significant at the 0.05 level?

Section 3.4 (Model Diagnostics)

Theoretical Exercises

3.7 Show the result (3.12) using elementary methods. (*Hint*: Write $\widehat{e}_{ij} = y_{ij} - \overline{y}_i$ and use the results from Appendix A concerning the distributions of linear functions of normally distributed r.v.'s and the formula for the covariance between two linear functions of r.v.'s.)

3.8 Show that for a balanced one-way layout with n observations per treatment the Bartlett statistic (3.14) simplifies to

$$B = \frac{a(n-1)\ln(s^2/\widetilde{s^2})}{1+(a+1)/3a(n-1)},$$

where s^2 is the pooled sample variance, which in this case is simply the arithmetic average of the s_i^2 and $\widetilde{s^2} = \left(\prod_{i=1}^{a} s_i^2\right)^{1/a}$ is the geometric average. This form of the statistic makes clear that $\ln(s^2/\widetilde{s^2})$ is a measure of discrepancies between the s_i^2. This measure increases as the s_i^2 deviate from each other and has a minimum value of zero when the s_i^2 are all equal, in which case $s^2 = \widetilde{s^2}$.

Applied Exercises

3.9 Refer to the data in Table 3.15. Show that the homoscedasticity assumption is rejected by the Bartlett test at $\alpha = 0.10$.

3.10 Refer to the iodide data in Table 3.18. Do the Bartlett and Levene tests for homoscedasticity on the raw data and log-transformed data. What do you conclude?

3.11 Refer to the data from Example 3.12. Perform the Bartlett test and show that the homoscedasticity assumption does not hold for these data.

3.12 Refer to Exercise 3.6. Perform the F^*-test given in Section 3.9.2 for equality of treatment means assuming heteroscedastic variances. Compare the result with that obtained assuming homoscedastic variances.

3.13 Toxicity of a crop protection compound was studied by feeding it to rodents through their diet over a 90-day period at the end of which they were sacrificed and their kidney weights and body weights were measured. Three doses of the compound were compared with a zero-dose control. The response variable was the ratio of the kidney weight to the body weight—the larger the ratio, the more toxic the dose. The summary statistics are given in Table 3.19.

(a) Do the Bartlett test to show that the variances are significantly different from each other using $\alpha = 0.05$.

(b) Perform the F^*-test given in Section 3.9.2 for equality of dose means assuming heteroscedastic variances.

Table 3.19 Kidney Weight/Body Weight ($\times 10^3$)

	Dose			
	0	1	2	3
Mean	6.5606	6.9975	7.6778	9.2606
SD	0.5094	0.5755	0.5949	1.0052
n	18	20	19	18

Source: Tamhane and Logan (2004). Reprinted by permission of Taylor & Francis.

3.14 Bacterial killing ability of a fungicide was studied using four different solvents. The fungicide mixture was sprayed on the fungi and the percentage of fungi killed was measured. Fifteen samples were run using each solvent and the data are shown in Table 3.20.

(a) Make the normal and the fitted-value plots of the residuals. Are the model assumptions satisfied? Are there any outliers in the data?

(b) Do the Bartlett and Levene tests for homoscedasticity. What is your conclusion?

(c) Perform the F^*-test given in Section 3.9.2 for heteroscedastic variances.

Table 3.20 Percentage of Bacteria Killed

	Solvent 1	Solvent 2	Solvent 3	Solvent 4
	96.44	93.63	93.58	97.18
	96.87	93.99	93.02	97.42
	97.24	94.61	93.86	97.65
	95.41	91.69	92.90	95.90
	95.29	93.00	91.43	96.35
	95.61	94.17	92.68	97.13
	95.58	94.67	92.65	96.71
	98.20	95.28	95.31	98.11
	98.29	95.13	95.33	98.38
	98.30	95.68	95.17	98.35
	98.65	97.52	98.59	98.05
	98.43	97.52	98.00	98.25
	98.41	97.37	98.79	98.12
Mean	96.7300	94.4943	94.0686	97.2764
SD	1.6420	1.6796	2.1763	0.8880

Source: Bishop and Dudewicz (1978, Table 2A).

Section 3.5 (Data Transformations)

Theoretical Exercises

3.15 Recall that for a binomial proportion $\widehat{p}$ based on a sample of size n we have $\mathrm{E}(\widehat{p}) = p$ and $\mathrm{Var}(\widehat{p}) = p(1-p)/n$. Show that the variance-stabilizing transformation of $\widehat{p}$ is $2\sqrt{n}\,\sin^{-1}\sqrt{\widehat{p}}$. This is called the **arcsin square-root transformation**.

3.16 An r.v. y is said to have a lognormal distribution if $\ln y$ is $\mathrm{N}(\mu, \sigma^2)$ for some μ and σ^2. The mean and the variance of a lognormal distribution are given by

$$\mathrm{E}(y) = e^{\mu+(1/2)\sigma^2} \quad \text{and} \quad \mathrm{Var}(y) = e^{2\mu+\sigma^2}\left(e^{\sigma^2} - 1\right).$$

Show that the variance-stabilizing transformation of y is $h(y) = \ln y$, which is also the exact normalizing transformation.

Applied Exercises

3.17 Refer to the iodide data in Table 3.18.

(a) Make the normal and the fitted-values plots of the residuals. Are the model assumptions satisfied?

(b) Plot $\ln s_i$ versus $\ln \bar{y}_i$. Explain why an inverse transformation is suggested by this plot.

(c) Make inverse and logarithmic transformations of the data and do the ANOVA F-test for both. Make the normal and fitted-values plots of the residuals for both transformations and decide for which transformation the model assumptions are better satisfied.

3.18 Lifetime tests were conducted on light bulbs at four different voltage settings with six bulbs per setting. The failure times in minutes are shown in Table 3.21. These data are obviously not normally distributed. Try a suitable power transformation and make plots to check if the homoscedasticity and normality assumptions are satisfied.

Section 3.6 (Power of F-Test and Sample Size Determination)

Theoretical Exercise

3.19 Prove that the configuration (3.20) minimizes the power. Specifically, prove that, subject to

$$\sum_{i=1}^{a} \alpha_i = 0 \quad \text{and} \quad \max_{1 \le i < j \le a} |\alpha_i - \alpha_j| \ge \delta,$$

Table 3.21 Failure Times (min.) of Light Bulbs for Four Voltage Settings

3.25 V	3.50 V	4.00 V	4.70 V
8,668	2,481	634	95
9,200	2,557	679	101
14,010	2,640	733	102
15,022	4,219	931	129
19,277	5,496	984	133
22,041	6,661	1,036	145

Source: Timmer (2000). Reprinted by permission of Taylor & Francis.

$\sum_{i=1}^{a} \alpha_i^2$ is minimized under the configuration (3.20).

Applied Exercise

3.20 Suppose that five hypertension drugs are to be compared in terms of their ability to reduce systolic blood pressure (BP). The standard deviation of repeat measurements is 3 mm of Hg. It is desired to test the null hypothesis of equality of the five drugs using an F-test at $\alpha = 0.05$ and reject it with probability at least 0.90 if in fact any two drugs differ in their mean reduction in systolic BP by 5 mm of Hg or more. How many subjects should be tested per drug? If 10 subjects are tested per drug, what is the power of the F-test?

Section 3.7 (Quantitative Treatment Factors)

Theoretical Exercise

3.21 Compute the orthogonal polynomials for $a = 4$.

Applied Exercises

3.22 To study the effect of engine size on the gas mileage, four models of large passenger cars having eight cylinder engines with engine sizes of 300, 350, 400, and 450 cubic inches were compared. Four cars of each model were tested with the results shown in Table 3.22.

(a) Fit a cubic orthogonal polynomial to these data and characterize the nature of the relationship between gas mileage and engine size. Use $\alpha = 0.05$ to test the significance of the coefficients.

(b) Compute the sums of squares corresponding to the linear, quadratic, and cubic terms and check that they add up to the sum of squares for engine size.

Table 3.22 Gas Mileage as Function of Engine Size in Cubic Inches

300 in.^{3}	350 in.^{3}	400 in.^{3}	450 in.^{3}
16.6	14.4	12.4	11.5
16.9	14.9	12.7	12.8
15.8	14.2	13.3	12.1
15.5	14.1	13.6	12.0

3.23 Refer to Exercise 3.18. The voltage settings are not equispaced, but for the purpose of this exercise you may treat them as such. Fit a cubic orthogonal polynomial to the transformed data. Assess the goodness of fit. Which effects (linear, quadratic or cubic) are significant at the 0.05 level? Give the final fitted regression equation for predicting failure time from voltage setting.

3.24 Fluidity of molten iron was measured as a function of silicon content. Three replicate measurements were made for each level of silicon content. The data are shown in Table 3.23.

(a) Plot the data as a function of silicon content. What relationship do you see?

(b) Fit a cubic orthogonal polynomial to the data. Assess the goodness of fit. Which effects (linear, quadratic or cubic) are significant at the 0.05 level? Give the final fitted regression equation for predicting fluidity from silicon content.

Section 3.8 (One-Way Analysis of Covariance)

Theoretical Exercise

3.25 Derive formula (3.32) by carrying out the following steps:

Table 3.23 Fluidity of Molten Iron as Function of Silicon Content in Percent

1.25%	1.50%	1.75%	2.00%	2.25%
47.5	60.0	65.0	72.5	77.5
55.0	55.0	67.5	75.0	85.0
37.5	50.0	70.0	75.0	75.0

Source: Hamaker (1955). Reprinted by permission of the International Biometric Society.

(a) Show that

$$\mathrm{Var}(\widehat{y}_{ij}) = \sigma^2\left[\frac{1}{n_i} + \frac{(x_{ij} - \overline{x}_i)^2}{S_{xx}}\right].$$

(b) Next show that

$$\mathrm{Cov}(y_{ij}, \widehat{y}_{ij}) = \sigma^2\left[\frac{1}{n_i} + \frac{(x_{ij} - \overline{x}_i)^2}{S_{xx}}\right].$$

(c) Finally, using $\widehat{e}_{ij} = y_{ij} - \widehat{y}_{ij}$, derive $\mathrm{Var}(\widehat{e}_{ij})$, which gives the formula (3.32).

Applied Exercises

3.26 Steel and Torrie (1980, p. 412) give data on the ascorbic acid content (measured in milligrams per 100 grams of dry weight) for 11 varieties of lima beans planted in five fields. Consider the data only for the first six varieties. From past experience it was known that the ascorbic acid content was negatively correlated with maturity of the plants at harvest. Percentage of dry matter (from 100 grams of freshly harvested beans) was measured as an index of maturity and used as a covariate. The data are shown in Table 3.24.

(a) This is an RB design with percentage of dry matter as a covariate and fields as blocks. Write an ANCOVA model with an additional term for block effects (see Section 5.1 for details about the RB design and its model).

(b) Calculate the ANCOVA table. Are there significant differences among the varieties? Use $\alpha = 0.05$.

3.27 A study reported in Fisher and van Belle (1993) was conducted to compare the effects of general anesthesia (treatment 1) versus local anesthesia

Table 3.24 Percentage of Dry matter (x) and Ascorbic Acid Content (y) of Lima Beans

	Field 1		Field 2		Field 3		Field 4		Field 5	
Variety	x	y	x	y	x	y	x	y	x	y
1	34.0	93.0	33.4	94.8	34.7	91.7	38.9	80.8	36.1	80.2
2	39.6	47.3	39.8	51.5	51.2	33.3	52.0	27.2	56.2	20.6
3	31.7	81.4	30.1	109.0	33.8	71.6	39.6	57.5	47.8	30.1
4	37.7	66.9	38.2	74.1	40.3	64.7	39.4	69.3	41.3	63.2
5	24.9	119.5	24.0	128.5	24.9	125.6	23.5	129.0	25.1	126.2
6	30.3	106.6	29.1	111.4	31.7	99.0	28.3	126.1	34.2	95.6

Source: Steel and Torrie (1980, Table 17.2). Reprinted by permission of Prof. David Dickey.

(treatment 2) for surgeries performed on trauma victims. The response variable (y) was the percent depression of lymphocyte transformation following surgery. The covariate (x) was the trauma classification (0–4), a higher number implying more severe trauma. The summary statistics were as follows:

General Anesthesia	$n_1 = 35$	$\bar{x}_1 = 2.371$	$\bar{y}_1 = 25.600$
Local Anesthesia	$n_2 = 42$	$\bar{x}_2 = 1.262$	$\bar{y}_2 = 6.738$
$\bar{\bar{x}} = 1.766$	$\bar{\bar{y}} = 15.312$	$S_{xx} = 64.0$	$S_{yy} = 49388.15$
	$S_{xy} = 542.59$		

(a) First compare the two treatments ignoring the covariate by calculating a two-sample t-statistic. Is the difference significant at $\alpha = 0.05$?

(b) Calculate the adjusted treatment means and the standard error of the difference between the adjusted treatment means.

(c) Calculate the t-statistic for testing the difference between the adjusted treatment means. Is the difference significant at $\alpha = 0.05$?

(d) Explain why different results are obtained in the two comparisons made in parts (a) and (c).

3.28 Table 3.25 gives sales for a chain of electronics stores.

(a) Plot Sales Volume (y) versus number of Households (x) using different plotting symbols for the three locations and plot the three regression lines. Are the regression lines approximately parallel? What do you conclude about the differences between the three locations?

(b) Prepare the ANCOVA table. Are there significant differences between the locations? Use $\alpha = 0.01$.

(c) Do pairwise comparisons between the locations by computing 90% CIs. Which locations are significantly different from each other?

3.29 Refer to the anorexia data in Table 3.2.

(a) Perform an ANCOVA of the data with the weight gain as the response variable and the baseline weight as the covariate. Make pairwise comparisons between groups using $\alpha = 0.05$. Do the conclusions differ from those obtained in Example 3.3 where the covariate is ignored.

(b) Looking at the groupwise scatter plots with fitted regression lines in Figure 3.1, does the assumption of common slopes appear valid? Check the assumption by including the group–covariate interaction term in the model and testing its significance. Use $\alpha = 0.10$.

(c) If the equal-slopes assumption is not consistent with the data, are the comparisons between the groups made in part (a) meaningful? Why or why not?

Table 3.25 Electronic Stores Sales

Store	Number of Households (in thousands)	Location	Sales Volume ($, in thousands)
1	161	Street	157.27
2	99	Street	93.28
3	135	Street	136.81
4	120	Street	123.79
5	164	Street	153.51
6	221	Mall	241.74
7	179	Mall	201.54
8	204	Mall	206.71
9	214	Mall	229.78
10	101	Mall	135.22
11	231	Downtown	224.71
12	206	Downtown	195.29
13	248	Downtown	242.16
14	107	Downtown	115.21
15	205	Downtown	197.82

Source: Bowerman and O'Connell (1997, Table 12.10).

3.30 Refer to Example 3.8. Calculate the fitted value, residual, and standardized residual for the observation $(x_{2,15}, y_{2,15}) = (8.8, 1.30)$. Show that the standardized residual exceeds 2.0 in absolute value and therefore that observation is a likely outlier.

CHAPTER 4

Single-Factor Experiments: Multiple Comparison and Selection Procedures

The previous chapter focused on estimation of the treatment means and the F-test of the overall null hypothesis of equality of all treatment means in a one-way layout. The overall null hypothesis is *global* in nature in the sense that its rejection does not pinpoint where the *local* differences lie. To determine which local differences between the treatment means result in the rejection of the overall null hypothesis, detailed **multiple comparisons** between the treatment means are necessary. For example, consider an experiment to compare different brands of paper towels in terms of their water absorbency. If the null hypothesis of equal mean absorbencies is rejected, the investigator would want to know which brands differ from each other and if there is a clear winner. If a test brand is compared with current brands on the market, then the investigator would want to know which current brands differ (are better or worse) from the test brand. In many instances, an experimenter is primarily interested in these detailed comparisons regardless of the outcome of the test of the overall null hypothesis.

Multiple comparisons entail the risk of increased chance of false significances with increase in the number of comparisons. For example, if one makes 100 tests, each at the 5% level, and if all 100 null hypotheses are true, then the expected number of type I errors is 5. Furthermore, if the tests are independent, then the probability of at least one type I error is nearly 1 (see Exercise 4.1). Therefore a suitable **multiplicity adjustment** is necessary to control the type I error rate correctly. The present chapter discusses **multiple comparison procedures (MCPs)** for this purpose. Hochberg and Tamhane (1987) is a comprehensive reference on MCPs. Other key references include Miller (1981), Westfall and Young (1993), Hsu (1996), and Westfall et al. (1999).

Statistical Analysis of Designed Experiments: Theory and Applications By Ajit C. Tamhane

A related problem is how to select a subset of "good" treatments or a single "best" treatment. For example, in the paper towel experiment the investigator may want to select the most absorbent towel or rank the top three towels. The last section of this chapter discusses **ranking and selection procedures (RSPs)** designed for such goals. References on this topic include Gibbons, Olkin, and Sobel (1977), Gupta and Panchapakesan (1979), and Bechhofer, Santner, and Goldsman (1995).

The chapter outline is as follows. Section 4.1 elucidates the basic concepts of multiple comparisons, including families, a **familywise error rate (FWE)**, and some basic methods for controlling the FWE. Section 4.2 discusses procedures for all pairwise comparisons among means, including the well-known Tukey procedure. Section 4.3 discusses procedures for all comparisons of treatment means with a control mean, including the well-known Dunnett procedure. Section 4.4 discusses procedures for general contrasts, including the well-known Scheffé procedure. Procedures for ranking and selection are covered in Section 4.5, which includes two different approaches: the indifference-zone approach due to Bechhofer (1954) and the subset selection approach due to Gupta (1956). Section 4.5.3 considers the problem of multiple comparisons with the best treatment (having the largest mean) which bridges the two areas of MCPs and RSPs. Section 4.6 gives a summary of the chapter.

4.1 BASIC CONCEPTS OF MULTIPLE COMPARISONS

4.1.1 Family

In confirmatory studies multiple inferences need to be considered *jointly* rather than *separately* if they are contextually related and some common decisions follow from them. For example, suppose a new drug will be recommended over a control drug or a placebo if it is shown to be better on at least one of $m \geq 2$ outcome variables. In exploratory studies the goal is generally not decision making but rather identification of promising leads to pursue in future studies. Many hypotheses are tested for this purpose. Again, these hypotheses need to be considered jointly in order to control the risk of too many false leads. The collection of such contextually related multiple inferences (hypothesis tests or confidence intervals) or equivalently the collection of the parameters on which these inferences are made is called a **family**.

The concept of a family is at the core of the multiple comparisons approach. It should be noted that the family includes not only the inferences that are actually made but also those other similar inferences that could potentially have been made, had the data turned out differently. For example, in a one-way layout experiment, a researcher in the hunt of a significant finding may compare treatments producing $\bar{y}_{\max}$ and $\bar{y}_{\min}$ to see if they differ significantly. Here, the particular treatments are compared because of the way the data turned out—not because that pair was of a priori interest. Had another pair of treatments produced the two extreme sample means, they would have been compared instead. Therefore the family comprises of *all* $\binom{a}{2}$ pairwise comparisons between the treatment means. As we shall see, defining the family in this way permits us to make any comparison from that family even a posteriori in light of the data.

4.1.2 Familywise Error Rate

The FWE (also called the **experimentwise error rate**) is defined as

$$\text{FWE} = \text{P}\{\text{Reject at least one true null hypothesis in a family}\}. \quad (4.1)$$

It is clear that if each hypothesis is tested at level α, as in a single hypothesis test [this is called controlling the **comparisonwise error rate (CWE)**], the FWE will exceed α in general. The primary goal in the multiple comparisons approach is to control the FWE at a specified α level in order to limit the occurrences of false significance.

Recently, Benjamini and Hochberg (1995) have introduced another error rate, called the **false discovery rate (FDR)**, which has found wide applications in problems involving a very large number of tests. An example is microarray analysis, which typically involves several thousand gene expression comparisons. If R denotes the number of rejected null hypotheses (termed as "discoveries") and if V denotes the number of type I errors (falsely rejected null hypotheses or false discoveries) among them, then

$$\text{FDR} = \text{E}\left(\frac{V}{R}\right),$$

where the ratio V/R is defined to be zero if $R = 0$. It can be shown that FDR $\leq$ FWE, so controlling the FDR is a less stringent requirement (if FWE is controlled below α, then FDR is also controlled below α), which is appropriate for exploratory studies. On the other hand, controlling the FWE is appropriate for confirmatory studies. In this chapter we will focus on controlling the FWE.

The FWE is a function of the underlying true configuration of the parameters. For example, if all null hypotheses are false, then the FWE equals zero since there can be no type I errors in this case. We want to control the FWE under *all* configurations, regardless of which and how many null hypotheses are true; that is, we want to control $\max \text{FWE} \leq \alpha$ where the maximum is taken over all configurations. This is referred to as **strong FWE control**. Some procedures control the FWE only under certain configurations, for example, only under the overall null hypothesis. This is referred to as **weak FWE control**. We restrict to procedures that control the FWE strongly.

Simultaneous confidence intervals (SCIs) are complementary to multiple hypothesis tests on the parameters in a family. In this case the quantity complementary to the FWE is the **simultaneous confidence coefficient (SCC)**, defined as

$$\text{SCC} = \text{P}\{\text{All confidence intervals include their respective true parameters}\}. \tag{4.2}$$

We require that SCC be at least equal to a specified level $1-\alpha$ for all parameter configurations. If we have $(1-\alpha)$-level SCIs, then they can be used to test hypotheses on the corresponding parameters with the FWE controlled at level α.

4.1.3 Bonferroni Method

One of the easiest and most general methods to control the FWE is the Bonferroni method, which takes its name from the **Bonferroni inequality**, stated as follows: Let $A_1, A_2, \ldots, A_m$ be $m \geq 2$ some specified random events. Then

$$\text{P}\left\{\bigcup_{i=1}^{m} A_i\right\} \leq \sum_{i=1}^{m} \text{P}\{A_i\}. \tag{4.3}$$

Suppose that we are simultaneously testing $m \geq 2$ hypotheses, $H_{01}, H_{02}, \ldots, H_{0m}$. Identify event A_i with the rejection of H_{0i}. If each hypothesis is tested at level α/m and if $m_0 \leq m$ null hypotheses are true, then using the Bonferroni inequality, it follows that

$$\text{FWE} = \text{P}\{\text{At least one true } H_{0i} \text{ is rejected}\} \leq \frac{m_0\alpha}{m} \leq \alpha.$$

Intuitively, the Bonferroni method divides the allowable type I error rate α among the m hypotheses and tests each hypothesis at level α/m. For example, if $\alpha = 0.05$ and $m = 10$, then each comparison is made at the $0.05/10 = 0.005$ level. (More generally, H_{0i} may be tested at level α_i such that $\sum_{i=1}^{m} \alpha_i = \alpha$.)

If the p-values for the hypotheses H_{0i} are available (denoted by p_i), then the Bonferroni method rejection rule is

$$\text{Reject } H_{0i} \text{ if } p_i < \frac{\alpha}{m} \qquad (1 \leq i \leq m). \tag{4.4}$$

Since rejection of the global null hypothesis $H_0 = \bigcap_{i=1}^{m} H_{0i}$ (i.e., all H_{0i} are true) is implied by rejection of any H_{0i}, the above rejection rule is equivalent to

$$\text{Reject } H_0 \text{ if } p_{\min} = \min_{1 \leq i \leq m} p_i < \frac{\alpha}{m}. \tag{4.5}$$

Although the Bonferroni method provides a general solution to a large class of multiple comparison problems, it is often too conservative, especially when the number of comparisons is large and/or the test statistics are highly correlated. In such cases the inequality (4.3) is not very sharp. As a result, the actual maximum

FWE is much less than α and the power is not as high as it could be. More accurate methods are available for specific problems, some of which are described in the sequel. The Bonferroni method is in fact a conservative solution to a more general method which is discussed next.

4.1.4 Union–Intersection Method

Let $\{H_{0i},\ i \in \mathcal{I}\}$ be an arbitrary family of hypotheses where $\mathcal{I}$ is a finite or an infinite index set. Let $H_0 = \bigcap_{i\in\mathcal{I}} H_{0i}$ be the overall null hypothesis. Roy (1953) proposed the following method for testing H_0. Suppose a test of each H_{0i} is available. Then reject H_0 if and only if at least one H_{0i} is rejected. If $\mathcal{R}_i$ denotes the rejection region of H_{0i}, then the rejection region of H_0 is $\mathcal{R} = \bigcup_{i\in\mathcal{I}} \mathcal{R}_i$. Thus the rejection region for the *intersection* of elementary hypotheses H_{0i} is the *union* of their rejection regions, hence the name **union–intersection (UI) method**.

As a typical application of the UI method, suppose the rejection region $\mathcal{R}_i$ is of the form $\{t_i > c\}$ where t_i is a test statistic for H_{0i} and c is a common critical constant. Then $\mathcal{R} = \bigcup_{i\in\mathcal{I}} \mathcal{R}_i$ equals $\bigcup_{i\in\mathcal{I}}\{t_i > c\} = \{t_{\max} > c\}$. To provide an α-level test of H_0, the critical constant c must be chosen to be the upper α critical point of the null distribution of $t_{\max}$; that is, it is the solution to the equation $\mathrm{P}_{H_0}\{t_{\max} > c\} = \alpha$. Since the p-values are inversely related to the test statistics t_i, if d is the critical constant with which the p_i are compared, then $\mathcal{R}_i$ is of the form $\{p_i < d\}$ and $\mathcal{R}$ is of the form $\{p_{\min} < d\}$.

Our goal, of course, is not to test the *single* overall hypothesis H_0 but to test the *multiple* component hypotheses H_{0i}. Since rejection of H_0 is equivalent to rejecting at least one H_{0i}, the MCP that rejects any H_{0i} for which $t_i > c$ controls the FWE at level α. These tests can be inverted in the usual manner to obtain $100(1-\alpha)\%$ SCIs on the parameters under test.

The Bonferroni method is a conservative approximation to the UI method. If the t_i's have the same *marginal* distribution under H_0, then it chooses c to be the upper α/m critical point of that distribution. Control of the FWE at level α follows from the Bonferroni inequality:

$$\mathrm{P}_{H_0}\{t_{\max} > c\} = \mathrm{P}_{H_0}\left\{\bigcup_{i=1}^{m}(t_i > c)\right\} \le \sum_{i=1}^{m}\mathrm{P}_{H_0}\{t_i > c\} = \sum_{i=1}^{m}\frac{\alpha}{m} = \alpha.$$

Thus the Bonferroni method avoids the difficulty of dealing with the distribution of $t_{\max}$, which involves the *joint* distribution of the t_i's.

As discussed above, the UI method results in an MCP that uses the same critical constant c with which all test statistics are compared. Therefore the tests can be carried out simultaneously without reference to each other. Such MCPs are called **single-step MCPs** or **simultaneous test procedures (STPs)**. These STPs have SCIs associated with them, derived by inverting the multiple tests and which use the same critical constant c. Better power performance can be achieved if the hypotheses can be tested in a stepwise manner, adjusting the critical constant at each step as hypotheses are rejected in earlier steps to account for the reduced number of hypotheses that remain to be tested. Such a stepwise MCP can be constructed using the closure method discussed in the next section. In general, stepwise MCPs cannot be easily inverted to obtain the corresponding SCIs.

4.1.5 Closure Method

Let $\{H_{0i},\ (1 \le i \le m)\}$ be an arbitrary finite family of hypotheses. The closure method of constructing multiple tests for all H_{0i} while controlling the FWE strongly at level α is due to Marcus, Peritz and Gabriel (1976). First consider a larger family consisting of all nonempty intersections of H_{0i}. This is called the **closure** of the original family. For example, if the original family had three hypotheses H_{01}, H_{02}, H_{03}, then the closure has seven hypotheses:

$$H_{01}, H_{02}, H_{03}, H_{01} \bigcap H_{02}, H_{01} \bigcap H_{03}, H_{02} \bigcap H_{03}, \text{ and } H_{01} \bigcap H_{02} \bigcap H_{03}.$$

The closure method rejects any H_{0i} $(1 \le i \le m)$ at the familywise level α if and only if all intersection hypotheses containing H_{0i} are rejected at the individual level α. Thus, in the above example, H_{01} will be rejected if and only if $H_{01} \bigcap H_{02} \bigcap H_{03}$, $H_{01} \bigcap H_{02}$, $H_{01} \bigcap H_{03}$, and H_{01} are all significant at individual level α. More generally, the closure method rejects any intersection hypothesis if and only if all intersection hypotheses that contain it are rejected at individual level α. Thus, $H_{01} \bigcap H_{02}$ will be rejected at level α if and only if both $H_{01} \bigcap H_{02} \bigcap H_{03}$ and $H_{01} \bigcap H_{02}$ are significant at level α. It can be shown that this multiple test procedure controls the FWE strongly at level α.

To apply the closure method we need an α-level test of each intersection hypothesis. Then, we need a step-down procedure for testing all hierarchically ordered intersection hypotheses, beginning with the overall intersection of all m hypotheses, next intersections of all subsets of $m - 1$ hypotheses, and so on. Note that the overall intersection implies subset intersections. For example, the overall intersection $H_{01} \bigcap H_{02} \bigcap H_{03}$ implies the subset intersections $H_{01} \bigcap H_{02}$, $H_{01} \bigcap H_{03}$, and $H_{02} \bigcap H_{03}$. Similarly, H_{01} and H_{02} are implied by $H_{01} \bigcap H_{02}$.

Subset hypotheses of an intersection hypothesis are tested if and only if the latter is rejected at level α; otherwise all subset hypotheses are accepted without testing them. In the above example, $H_{01} \bigcap H_{02} \bigcap H_{03}$ is tested first at level α. If it is rejected, then $H_{01} \bigcap H_{02}$, $H_{01} \bigcap H_{03}$, and $H_{02} \bigcap H_{03}$ are each tested at the individual level α; if it is accepted, then all hypotheses in the closure family implied by it (which are all the remaining hypotheses) are accepted and testing stops. The procedure continues in this step-down manner as long as rejections occur. Figure 4.1 shows a tree diagram of this step-down test procedure.

Although this procedure may appear unwieldy, in many cases there is a shortcut to it which turns out to be a simple and powerful step-down procedure. Here is an example.

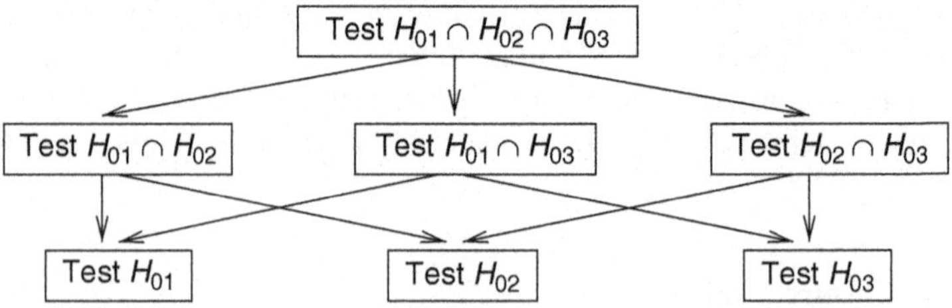

Figure 4.1 Tree diagram of closed procedure for testing three hypotheses.

Example 4.1 (Holm Procedure)

Consider m hypotheses $H_{01}, H_{02}, \ldots, H_{0m}$ with p-values $p_1, p_2, \ldots, p_m$. Let $p_{(1)} \leq p_{(2)} \leq \cdots \leq p_{(m)}$ be the ordered p-values, and let $H_{0(1)}, H_{0(2)}, \ldots, H_{0(m)}$ be the corresponding hypotheses. Suppose we use the Bonferroni test (4.5) for each intersection hypothesis. Thus we first test the overall intersection hypothesis $H_0 = \bigcap_{i=1}^{m} H_{0i}$ and reject it if $p_{\min} = p_{(1)} < \alpha/m$. The next step in the closure method is to test all intersections of size $m-1$ using the Bonferroni test, which uses the critical value $\alpha/(m-1)$. Any intersection that includes $H_{0(1)}$ has $p_{\min} = p_{(1)}$ and $p_{(1)} \leq \alpha/m < \alpha/(m-1)$; thus all those intersections will be rejected. This holds for every intersection that includes $H_{0(1)}$ (since $p_{(1)} \leq \alpha/m < \alpha/i$ for all $i < m$). Hence $H_{0(1)}$ can be rejected at the first step itself if $p_{(1)} < \alpha/m$.

This argument can be extended to $H_{0(2)}, H_{0(3)}, \ldots$, resulting in the following step-down procedure which was proposed by Holm (1979): At step 1, reject $H_{0(1)}$ if $p_{(1)} < \alpha/m$ and go to step 2; otherwise stop testing and accept all hypotheses. At step 2, reject $H_{0(2)}$ if $p_{(2)} < \alpha/(m-1)$ and go to step 3; otherwise stop testing and accept all the remaining hypotheses, and so on. Notice that, while the Bonferroni procedure uses the same critical value α/m against which all p-values are tested, the Holm procedure uses increasingly larger critical values, $\alpha/(m-i+1)$, for $i = 1, 2, \ldots, m$. Hence the Holm procedure is uniformly more powerful than the Bonferroni procedure since it rejects all the hypotheses rejected by the latter and possibly more.

To illustrate the Holm procedure, consider the following numerical example. Suppose $m = 5$ and the ordered p-values are $0.008, 0.012, 0.015, 0.028$, and 0.048. All the p-values are less than 0.05; so if the CWE is controlled at $\alpha = 0.05$, then all five hypotheses can be rejected. However, to control the FWE at $\alpha = 0.05$ using the Bonferroni procedure, we can only reject $H_{0(1)}$ since only $p_{(1)} = 0.008 < 0.05/5 = 0.01$. Using the Holm procedure, we can reject not only $H_{0(1)}$ but also $H_{0(2)}$ and $H_{0(3)}$ since $p_{(2)} = 0.012 < 0.05/4 = 0.0125$ and $p_{(3)} = 0.015 < 0.05/3 = 0.0167$. The procedure stops at step 4 and accepts $H_{0(4)}$ (and hence also $H_{0(5)}$) since $p_{(4)} = 0.028 > 0.05/2 = 0.025$. ■

4.2 PAIRWISE COMPARISONS

Consider the one-way layout setup of Section 3.2 with $a \geq 2$ treatments. As a follow-up to the ANOVA F-test, it is often of interest to determine which treatments differ from each other. This entails making $m = \binom{a}{2}$ pairwise comparisons between the treatment means which constitute the family. For example, one may want to test the hypotheses

$$H_{0ij} : \mu_i - \mu_j = 0 \quad \text{vs.} \quad H_{1ij} : \mu_i - \mu_j \neq 0 \qquad (1 \le i < j \le a) \tag{4.6}$$

with FWE $\le \alpha$ or calculate $100(1-\alpha)\%$ SCIs on $\mu_i - \mu_j$ $(1 \le i < j \le a)$. Only two-sided inferences are meaningful here unless there is an a priori ordering among the treatments. The pivotal quantities for $\mu_i - \mu_j$ are

$$t_{ij} = \frac{\bar{y}_i - \bar{y}_j - (\mu_i - \mu_j)}{s\sqrt{1/n_i + 1/n_j}} \qquad (1 \le i < j \le a), \tag{4.7}$$

which are marginally t-distributed with $\nu = N - a$ d.f.

4.2.1 Least Significant Difference and Bonferroni Procedures

The LSD procedure consists of performing separate α-level pairwise t-tests as given in Section 3.3.3, that is, reject H_{0ij} if $|t_{ij}| > t_{\nu,\alpha/2}$ $(1 \le i < j \le a)$. The corresponding SCIs are the usual pairwise t-intervals:

$$\mu_i - \mu_j \in \left[\bar{y}_i - \bar{y}_j \pm t_{\nu,\alpha/2} s \sqrt{\frac{1}{n_i} + \frac{1}{n_j}} \right] \qquad (1 \le i < j \le a). \tag{4.8}$$

Obviously, this procedure does not control the FWE. Fisher (1935) suggested that the LSD procedure be used only if the ANOVA F-test is significant at level α. This is referred to as the **protected LSD procedure** because it protects against excessive FWE by providing the *screen* of the preliminary F-test. Although this procedure controls the FWE under the overall null configuration $\mu_1 = \mu_2 = \cdots = \mu_a$, it does not do so under all configurations. To illustrate this point, consider a partial null configuration with $\mu_1 = \mu_2 = \cdots = \mu_{a-1}$ but μ_a infinitely different from the rest. Then the F-test rejects with probability tending to 1. Since pairwise comparisons among $\mu_1, \mu_2, \ldots, \mu_{a-1}$ are made using the LSD procedure at level α, it is clear that FWE $> \alpha$ (except for $a = 3$ in which case there is only one comparison between μ_1 and μ_2). Thus even the protected LSD procedure does not control the FWE strongly and therefore should not be used. For opposing views, see Rothman (1990).

The Bonferroni procedure for pairwise comparisons replaces the critical constant $t_{\nu,\alpha/2}$ used in (4.8) for the LSD procedure by the critical constant $t_{\nu,\alpha/2m}$ where $m = \binom{a}{2}$. The resulting intervals are wider but guarantee SCC $\ge 1 - \alpha$; equivalently, the corresponding tests strongly control the FWE $\le \alpha$.

4.2.2 Tukey Procedure for Pairwise Comparisons

Tukey's (1953) procedure is a UI procedure obtained by expressing

$$\{H_0 : \mu_1 = \mu_2 = \cdots = \mu_a\} = \bigcap_{1 \le i < j \le a} \{H_{0ij} : \mu_i = \mu_j\}.$$

For **balanced one-way layouts** (i.e., equal n_i's with $n_i \equiv n$), the Tukey procedure involves the distribution of

$$\max_{1 \le i < j \le a} |t_{ij}| = \max_{1 \le i < j \le a} \frac{|\bar{y}_i - \bar{y}_j - (\mu_i - \mu_j)|}{s\sqrt{2/n}}.$$

Tukey derived this distribution in terms of the r.v.

$$\begin{aligned} Q &= \sqrt{2} \max_{1 \le i < j \le a} |t_{ij}| \\ &= \max_{1 \le i < j \le a} \frac{|(\bar{y}_i - \bar{y}_j) - (\mu_i - \mu_j)|}{s/\sqrt{n}} \\ &= \max_{1 \le i < j \le a} \frac{|(\bar{y}_i - \mu_i)\sqrt{n}/\sigma - (\bar{y}_j - \mu_j)\sqrt{n}/\sigma|}{s/\sigma} \\ &= \max_{1 \le i < j \le a} \frac{|z_i - z_j|}{s/\sigma} \\ &= \frac{z_{\max} - z_{\min}}{s/\sigma}, \end{aligned}$$

where the $z_i = (\bar{y}_i - \mu_i)\sqrt{n}/\sigma$ are i.i.d. N(0, 1) r.v.'s independent of $s/\sigma \sim \sqrt{\chi^2_\nu/\nu}$. Thus Q equals the range of a i.i.d. N(0, 1) r.v.'s divided by an independently distributed $\sqrt{\chi^2_\nu/\nu}$ r.v. The distribution of Q is known as the **Studentized range distribution**. It is free of all unknown parameters and depends only on the number of treatments a and the error d.f. ν.

Let $q_{a,\nu,\alpha}$ denote the upper α critical point of this distribution. These critical points are tabulated in Table C.5. Then it is a simple matter to show that $P\{Q \le q_{a,\nu,\alpha}\} = 1 - \alpha$ implies that

$$P\left\{\mu_i - \mu_j \in \left[\bar{y}_i - \bar{y}_j \pm q_{a,\nu,\alpha}\frac{s}{\sqrt{n}}\right] \quad (1 \le i < j \le a)\right\} = 1 - \alpha. \qquad (4.9)$$

Thus the intervals in the above expression are $100(1-\alpha)\%$ SCIs for all pairwise differences.

Example 4.2 (Comparisons of Barley Varieties: Tukey Procedure)

The sample mean yields of seven varieties of barley from a field experiment are shown in Table 4.1.

Table 4.1 Mean Yields of Varieties of Barley (bushels per acre)

Variety	A	B	C	D	E	F	G
Sample Mean	49.6	71.2	67.6	61.5	71.3	58.1	61.0

Source: Duncan (1955). Reprinted by permission of the International Biometric Society.

These sample means are based on six replicates from each variety. An RB design was used in which each of the seven varieties was replicated once in each of the six different fields. The corresponding ANOVA yields $\text{MS}_e = 79.64$ with 30 d.f. We want to find which varieties differ from each other using the Tukey procedure with $\alpha = 0.10$.

The standard error of each $\bar{y}_i$ is $s/\sqrt{n} = \sqrt{79.64/6} = 3.643$. Any difference $|\bar{y}_i - \bar{y}_j|$ that exceeds

$$q_{7,30,0.10}\frac{s}{\sqrt{n}} = (4.016)(3.643) = 14.630$$

is significant at the 0.10 level using the Tukey procedure. The following three mean differences exceed this critical value:

$$E - A = 71.3 - 49.6 = 21.7, \qquad B - A = 71.2 - 49.6 = 21.6,$$
$$C - A = 67.8 - 49.6 = 18.2.$$

Thus variety A differs significantly from (and has lower mean yield than) varieties E, B, and C. The results are summarized in Display 4.1. A subset of means connected by a common line is called a homogeneous subset since any pair of means in that subset are not significantly different from each other. On the other hand, any two means not connected by a common line are significantly different. Note that A and C, B, E are not connected by a common line. ■

For **unbalanced one-way layouts**, the SCIs (4.9) can be modified to

$$\mu_i - \mu_j \in \left[\bar{y}_i - \bar{y}_j \pm \frac{q_{a,\nu,\alpha}}{\sqrt{2}} s\sqrt{\frac{1}{n_i} + \frac{1}{n_j}}\right] \qquad (1 \le i < j \le a). \tag{4.10}$$

A	F	G	D	C	B	E
49.6	58.1	61.0	61.5	67.6	71.2	71.3

Display 4.1 Diagram showing homogeneous subsets of barley varieties.

Hayter (1984) has shown that these intervals are in fact conservative; that is, their coverage probability is $>1-\alpha$. The corresponding t-statistic is

$$t_{ij} = \frac{\bar{y}_i - \bar{y}_j}{s\sqrt{1/n_i + 1/n_j}} \qquad (1 \le i < j \le a).$$

The Tukey procedure rejects any pairwise null hypothesis H_{0ij} with FWE $\le \alpha$ if $\sqrt{2}|t_{ij}| > q_{a,\nu,\alpha}$. Obviously, the Tukey procedure is more powerful than the Bonferroni. The LSD intervals in (4.8) are shorter than Tukey's, but they do not guarantee that SCC $\ge 1-\alpha$.

4.2.2.1 *Extension of the Tukey Procedure to General Designs*

The Tukey procedure is applicable to designs more general than one-way layouts, for example, ANCOVA designs (see Section 3.8) and block designs (see Section 5.1). In these designs we have the following setup. The treatment means of interest are $\mu_1, \mu_2, \ldots, \mu_a$ and $\widehat{\mu}_1, \widehat{\mu}_2, \ldots, \widehat{\mu}_a$ are their least squares estimators. The $\widehat{\mu}_i$ are jointly normally distributed with means μ_i, variances $\sigma^2 v_{ii}$, and covariances $\sigma^2 v_{ij}$. Finally, s^2 is an unbiased estimator of σ^2 based on ν d.f. such that $\nu s^2/\sigma^2 \sim \chi^2_\nu$ independently of the $\widehat{\mu}_i$. If the design is **pairwise variance-balanced**, that is, if

$$\mathrm{Var}(\widehat{\mu}_i - \widehat{\mu}_j) = \sigma^2(v_{ii} + v_{jj} - 2v_{ij}) = \sigma^2 v \qquad (1 \le i < j \le a),$$

then the following Tukey intervals are *exact* $100(1-\alpha)\%$ SCIs:

$$\mu_i - \mu_j \in \left[\widehat{\mu}_i - \widehat{\mu}_j \pm q_{a,\nu,\alpha} s\sqrt{\tfrac{1}{2}v}\right] \qquad (1 \le i < j \le a). \tag{4.11}$$

An example of a pairwise variance-balanced design is the balanced incomplete block (BIB) design studied in Section 5.2. Example 5.4 in that section gives an application of these Tukey intervals. If a design is not pairwise variance-balanced and the $\widehat{\mu}_i$ are correlated, then the following intervals may be used as approximate $100(1-\alpha)\%$ SCIs:

$$\mu_i - \mu_j \in \left[\widehat{\mu}_i - \widehat{\mu}_j \pm \frac{q_{a,\nu,\alpha}}{\sqrt{2}} \mathrm{SE}(\widehat{\mu}_i - \widehat{\mu}_j)\right] \qquad (1 \le i < j \le a),$$

where $\mathrm{SE}(\widehat{\mu}_i - \widehat{\mu}_j) = s\sqrt{v_{ii} + v_{jj} - 2v_{ij}}$.

4.2.3 Step-Down Procedures for Pairwise Comparisons

Both the Bonferroni and Tukey procedures can be used to obtain SCIs as well tests for the differences between pairs of treatment means. If only tests are required, then a more powerful method is to do the testing in a step-down manner. The theory of these procedures is based on the closure method discussed in Section 4.1.5.

We now discuss these step-down testing procedures for the balanced one-way layout setup. Step-down procedures in the unbalanced case are more complicated and are described in Hochberg and Tamhane (1987, pp. 116–121).

Order the sample means: $\bar{y}_{(1)} \leq \bar{y}_{(2)} \leq \cdots \leq \bar{y}_{(a)}$. At the first step, the significance of the difference between the largest and the smallest sample means, $\bar{y}_{(a)}$ and $\bar{y}_{(1)}$, is tested. If this difference is not significant, then all a sample means are declared not significantly different from each other by implication and testing stops. Otherwise, $\bar{y}_{(a)}$ and $\bar{y}_{(1)}$ are declared significantly different. At the next step, the differences between the next pairs of extreme means, $(\bar{y}_{(a)}, \bar{y}_{(2)})$ and $(\bar{y}_{(a-1)}, \bar{y}_{(1)})$, are tested. Testing continues in this manner. If at any step the difference between the two extremes of any group of adjacent sample means is found nonsignificant, then the intermediate sample means in that group are also declared to be not significantly different and testing stops on that group of means. For a graphical display, the means are written in ascending order and a line is drawn connecting those means that are not significantly different from each other. No further tests are made within this homogeneous subset of the means. The critical constants used to test the significance of successively smaller ranges of means become successively smaller at each step, which results in higher power.

In general, the significance level used to test the range of p sample means (e.g., the first step tests the range of $p = a$ sample means, the second step tests the range of $p = a - 1$ sample means, etc.) depends on p; we denote it by α_p, called the p-**mean nominal significance level**. The corresponding critical constant, denoted by c_p, equals q_{p,ν,α_p}, which is the upper α_p critical point of the Studentized range distribution with parameter p and d.f. ν. The critical constants are adjusted to be monotone: $c_2 \leq c_3 \leq \cdots \leq c_a$ if they are not already so; that is, if $c_{p-1} > c_p$, then we set $c_p = c_{p-1}$.

At the first step, $(\bar{y}_{(a)} - \bar{y}_{(1)})\sqrt{n}/s$ is compared to c_a; at the second step, $(\bar{y}_{(a-1)} - \bar{y}_{(1)})\sqrt{n}/s$ and $(\bar{y}_{(a)} - \bar{y}_{(2)})\sqrt{n}/s$ are compared to c_{a-1}; and so on. If a significant difference is not found at any step, then no further tests are made within that p-mean range.

Some choices for α_p resulting in different step-down procedures are as follows:

Newman–Keuls Procedure: $\alpha_p = \alpha \quad (2 \leq p \leq a)$.

Duncan Procedure: $\alpha_p = 1 - (1 - \alpha)^{p-1} \quad (2 \leq p \leq a)$.

Tukey–Ryan Procedure: $\alpha_p = 1 - (1 - \alpha)^{p/a} (2 \leq p \leq a - 2), \alpha_p = \alpha$ $(p = a - 1, a)$. The original Ryan (1960) procedure uses a slightly less powerful choice: $\alpha_p = p\alpha/a \ (2 \leq p \leq a)$.

It can be shown that the Newman–Keuls and Duncan procedures do not control the FWE at level α; the Duncan procedure is especially lax in this regard. The Tukey–Ryan procedure controls the FWE and is therefore recommended.

Example 4.3 (Comparisons of Barley Varieties: Tukey–Ryan Procedure)

Refer to the barley yield data from Example 4.2. We now apply the Tukey–Ryan procedure to determine which varieties differ from each other at $\alpha = 0.10$.

Table 4.2 Calculation of Critical Constants for Tukey–Ryan Procedure

p	2	3	4	5	6	7
α_p	0.0297	0.0442	0.0584	0.0725	0.10	0.10
$q_{p,30,\alpha_p}$	3.213	3.562	3.743	3.860	3.851	4.016
$c_p = q_{p,30,\alpha_p} s/\sqrt{n}$	11.705	12.976	13.636	14.062	14.062	14.630

From Table C.5, the critical constants for $p = 7$ and $p = 6$ equal $q_{7,30,0.10} = 4.016$ and $q_{6,30,0.10} = 3.851$, respectively. Hence $c_7 = (4.016)(3.643) = 14.630$ and $c_6 = (3.851)(3.643) = 14.029$. The critical constants for $p < 6$ need to be found by interpolation in α because of the nonstandard α_p-values. This interpolation is more accurate if done linearly in $-\log_{10} \alpha$. Thus, for example, for $p = 5$ we have $\alpha_5 = 1 - (1 - 0.10)^{5/7} = 0.0725$, but the Studentized range critical points are tabulated for $\alpha = 0.05, 0.10$. We have $-\log_{10}(0.05) = 1.30$, $-\log_{10}(0.10) = 1.00$, and $-\log_{10}(0.0725) = 1.14$. From Table C.5 we find $q_{5,30,0.05} = 4.102$ and $q_{5,30,0.10} = 3.648$. Therefore,

$$q_{5,30,0.0725} \approx 3.648 + (4.102 - 3.648)\left(\frac{1.14 - 1.00}{1.30 - 1.00}\right) = 3.860,$$

which gives $c_5 = (3.648)(3.643) = 14.062$. Since $c_5 > c_6$, to maintain monotonicity, we set $c_6 = 14.062$. Other critical constants are interpolated in the same manner. The results are summarized in Table 4.2.

To apply the procedure, begin by writing the ordered sample means as follows:

A	F	G	D	C	B	E
49.6	58.1	61.0	61.5	67.6	71.2	71.3

Step 1: E and A are significantly different since $71.3 - 49.6 = 21.7 > c_7 = 14.630$.

Step 2: (a) B and A are significantly different since $71.2 - 49.6 = 21.6 > c_6 = 14.062$.
(b) E and F are not significantly different since $71.3 - 58.1 = 13.2 < c_6 = 14.062$. Draw a line connecting E and F. Hence varieties $\{F, G, D, C, B, E\}$ form a homogeneous subset and no further tests are conducted within this subset.

Step 3: C and A are significantly different since $67.6 - 49.6 = 18.0 > c_5 = 14.062$.

Step 4: D and A are not significantly different since $61.5 - 49.6 = 11.9 < c_4 = 13.636$. Draw a line connecting D and A. Hence varieties $\{A, F, G, D\}$ form another homogeneous subset and no further tests are conducted within this subset.

This concludes the test procedure. Significant differences are found between A and E, A and B, and A and C. These results are identical to those found by the Tukey procedure. ■

4.3 COMPARISONS WITH A CONTROL

The importance of including a **control group** in a comparative experiment was emphasized in Chapter 1. In some experiments the comparisons of interest are those between the new treatments (called the **test treatments**) and the control. For example, the control may be an inactive placebo or an active standard drug and the test treatments may be new drugs, one or more of which are to be recommended for further testing and eventual use. In a dose–response study, the control may be the zero dose of a drug and it may be of interest to compare the higher doses with it to determine the minimum effective dose. Such comparisons are known as comparisons with a control or **many-to-one comparisons**.

Label the control treatment as 0 and the test treatments as $1, 2, \ldots, a$. Assume the same one-way layout setup as in Section 3.2. The parameters of interest are the mean differences, $\mu_i - \mu_0 \ (1 \le i \le a)$.

Consider the following two families of hypothesis testing problems:

$$H_{0i} : \mu_i - \mu_0 = 0 \quad \text{vs.} \quad H_{1i} : \mu_i - \mu_0 > 0 \qquad (1 \le i \le a) \quad \text{(one-sided)} \tag{4.12}$$

and

$$H_{0i} : \mu_i - \mu_0 = 0 \quad \text{vs.} \quad H_{1i} : \mu_i - \mu_0 \neq 0 \qquad (1 \le i \le a) \quad \text{(two-sided).} \tag{4.13}$$

There are corresponding one-sided and two-sided SCIs on $\mu_i - \mu_0 \ (1 \le i \le a)$.

The pivotal quantities for $\mu_i - \mu_0$ are

$$t_i = \frac{\bar{y}_i - \bar{y}_0 - (\mu_i - \mu_0)}{s\sqrt{1/n_i + 1/n_0}} \qquad (1 \le i \le a), \tag{4.14}$$

which are marginally t-distributed with ν d.f. where $\nu = \sum_{i=0}^{a} n_i - (a+1) = N - (a+1)$.

4.3.1 Dunnett Procedure for Comparisons with a Control

Dunnett's (1955) procedure is a UI procedure obtained by expressing

$$\{H_0 : \mu_0 = \mu_1 = \cdots = \mu_a\} = \bigcap_{1 \le i \le a} \{H_{0i} : \mu_i = \mu_0\},$$

where $H_{0i} : \mu_i - \mu_0 = 0$ are the individual null hypotheses.

For one-sided inferences we need the critical points of $\max_{1 \le i \le a} t_i$ and for two-sided inferences we need the critical points of $\max_{1 \le i \le a} |t_i|$. The *marginal* distribution of each t_i is Student's t, but to evaluate the necessary critical points, we need the *joint* distribution of the t_i's. This joint distribution is known as the **multivariate t-distribution** (discussed in Section A.6) with ν d.f. and correlation matrix $\{\rho_{ij}\}$ with elements

$$\rho_{ij} = \text{Corr}(\overline{y}_i - \overline{y}_0, \overline{y}_j - \overline{y}_0) = \sqrt{\frac{n_i n_j}{(n_i + n_0)(n_j + n_0)}} \qquad (1 \le i < j \le a); \tag{4.15}$$

for a derivation of this formula, see Exercise 4.10.

If $n_1 = n_2 = \cdots = n_a = n$ (say) but n_0 and n are not necessarily equal, then the one-way layout is said to be balanced w.r.t. the test treatments. In that case, all ρ_{ij} equal $\rho = n/(n + n_0)$ and the associated distribution is referred to as an **equicorrelated multivariate** t. If, in addition, $n_0 = n$, then the one-way layout is said to be fully balanced and $\rho = \frac{1}{2}$.

When using a one-way layout balanced w.r.t. the test treatments, the following design question arises: What is the optimum ratio between n_0 and n subject to a fixed total sample size $N = n_0 + an$? An asymptotically (as $N \to \infty$) optimum solution is obtained by minimizing the common variance of $\overline{y}_i - \overline{y}_0$ $(1 \le i \le a)$. The solution can be shown to be $n_0/n = \sqrt{a}$; see Exercise 4.11. For example, if there are four test treatments, then twice as many observations should be taken on the control as on each test treatment. This is known as the **square-root allocation rule**.

Let $t_{a,\nu,\{\rho_{ij}\},\alpha}$ and $|t|_{a,\nu,\{\rho_{ij}\},\alpha}$ denote the upper α critical points of $\max_{1 \le i \le a} t_i$ and $\max_{1 \le i \le a} |t_i|$, respectively, where the ρ_{ij} are given by (4.15). Similarly, let $t_{a,\nu,\rho,\alpha}$ and $|t|_{a,\nu,\rho,\alpha}$ denote the corresponding critical points for the equicorrelated multivariate t-distribution with common correlation ρ. Comprehensive tables of the critical points are available in the equicorrelated case for selected values of ρ in Bechhofer and Dunnett (1988). Tables C.6 and C.7 give values of the critical points for $\rho = \frac{1}{2}$ and Table C.8 and C.9 for $\rho = 0$. It is worth noting that the critical points $t_{a,\nu,\{\rho_{ij}\},\alpha}$ and $|t|_{a,\nu,\{\rho_{ij}\},\alpha}$ can be well approximated by the critical points of the equicorrelated t-distribution, with common correlation $\overline{\rho}$, the arithmetic average of the ρ_{ij}. The critical point for $\overline{\rho}$ can be approximated by linear interpolation w.r.t. $1/(1 - \rho)$ between the critical points for $\rho = 0$ and $\rho = \frac{1}{2}$ if $\overline{\rho} \leqslant \frac{1}{2}$.

The one-sided Dunnett procedure rejects $H_{0i} : \mu_i - \mu_0 \le 0$ if

$$t_i = \frac{\overline{y}_i - \overline{y}_0}{s\sqrt{1/n_i + 1/n_0}} > t_{a,\nu,\{\rho_{ij}\},\alpha} \qquad (1 \le i \le a). \tag{4.16}$$

The corresponding $100(1 - \alpha)\%$ one-sided SCIs on $\mu_i - \mu_0$ are

$$\mu_i - \mu_0 \ge \overline{y}_i - \overline{y}_0 - t_{a,\nu,\{\rho_{ij}\},\alpha} s \sqrt{\frac{1}{n_i} + \frac{1}{n_0}} \qquad (1 \le i \le a). \tag{4.17}$$

The two-sided Dunnett procedure rejects $H_{0i} : \mu_i - \mu_0 = 0$ if

$$|t_i| = \frac{|\bar{y}_i - \bar{y}_0|}{s\sqrt{1/n_i + 1/n_0}} > |t|_{a,\nu,\{\rho_{ij}\},\alpha} \qquad (1 \le i \le a). \tag{4.18}$$

The corresponding $100(1-\alpha)\%$ two-sided SCIs on $\mu_i - \mu_0$ are

$$\mu_i - \mu_0 \in \left[\bar{y}_i - \bar{y}_0 \pm |t|_{a,\nu,\{\rho_{ij}\},\alpha} s\sqrt{\frac{1}{n_i} + \frac{1}{n_0}}\right] \qquad (1 \le i \le a). \tag{4.19}$$

Example 4.4 (Weight Gain Therapies for Anorexia Patients: Comparisons with Control Group)

Refer to the data from Example 3.1. The ANOVA for weight gains in Display 3.2 shows that there are significant differences between the three therapy groups. Suppose we want to find out if either of the treatment groups (behavioral therapy and family therapy) significantly improves the weight gain compared to the control group. To make these comparisons, we use Dunnett's procedure. The summary statistics are shown in Table 4.3.

First we make one-sided comparisons by calculating 95% simultaneous lower confidence bounds on $\mu_1 - \mu_0$ and $\mu_2 - \mu_0$. The required multivariate t critical point is $t_{2,48,1/2,0.05} \approx t_{2,50,1/2,0.05} = 1.959$. Therefore

$$\mu_1 - \mu_0 \ge 3.318 - (-0.241) - (1.959)(7.893)\sqrt{\tfrac{2}{17}} = 3.559 - 5.304 = -1.545$$

and

$$\mu_2 - \mu_0 \ge 7.265 - (-0.241) - (1.959)(7.893)\sqrt{\tfrac{2}{17}} = 7.506 - 5.304 = 2.202.$$

If two-sided comparisons are to be made, then the required multivariate t critical point is $|t|_{2,48,1/2,0.05} \approx |t|_{2,50,1/2,0.05} = 2.276$. Therefore

$$\begin{aligned}\mu_1 - \mu_0 &\in \left[3.318 - (-0.241) \pm (2.276)(7.893)\sqrt{\tfrac{2}{17}}\right] = [3.559 \pm 6.162] \\ &= [-2.603, 9.721]\end{aligned}$$

Table 4.3 Means and Standard Deviations for Anorexia Weight Gain Data

Control group (0)	$\bar{y}_0 = -0.241$	$n_0 = 17$
Behavioral group (1)	$\bar{y}_1 = 3.318$	$n_1 = 17$
Family group (2)	$\bar{y}_2 = 7.265$	$n_2 = 17$
Pooled SD	$s = 7.893$	d.f. $= 48$

and

$$\mu_2 - \mu_0 \in \left[7.265 - (-0.241) \pm (2.276)(7.893)\sqrt{\tfrac{2}{17}}\right] = [7.506 \pm 6.162]$$
$$= [1.344, 13.668].$$

In both cases the family therapy shows a significant improvement in weight gain over the control therapy group but the behavioral therapy group fails to show a significant improvement.

Note that if Tukey's procedure is used to make all three pairwise two-sided comparisons, then the critical value for each comparison would be

$$q_{3,48,0.05}\frac{s}{\sqrt{n}} \approx q_{3,40,0.05}\frac{s}{\sqrt{n}} = (3.442)\frac{7.893}{\sqrt{17}} = 6.589.$$

This critical value is naturally larger than the critical value 6.162 used by the two-sided Dunnett procedure. ■

4.3.2 Step-Down Procedures for Comparisons with a Control

Dunnett's procedure gives multiple tests as well as SCIs. If only hypothesis tests are desired, then higher power can be achieved by testing the hypotheses in a step-down manner. The procedure is easy to state when the design is balanced[1] with respect to the test treatments with $n_1 = n_2 = \cdots = n_a = n$. Let $\rho = n/(n + n_0)$. For the one-sided testing problem (4.12), order the test statistics t_i from (4.16) as $t_{(1)} \le t_{(2)} \le \cdots \le t_{(a)}$ and denote the corresponding hypotheses as $H_{0(1)}, H_{0(2)}, \ldots, H_{0(a)}$. Begin by comparing $t_{(a)}$ against the critical constant $t_{a,\nu,\rho,\alpha}$. If $t_{(a)} \le t_{a,\nu,\rho,\alpha}$, then accept all H_{0i} and stop testing. Otherwise reject $H_{0(a)}$ and proceed to test $H_{0(a-1)}$ by comparing $t_{(a-1)}$ against the critical constant $t_{a-1,\nu,\rho,\alpha}$. Proceed in this manner as long as rejection occurs. If at any step $t_{(i)} \le t_{i,\nu,\rho,\alpha}$, then accept all the remaining hypotheses $H_{0(j)}$ $(1 \le j \le i)$ and stop testing. The two-sided test procedure is exactly the same except that it uses the ordered test statistics $|t|_{(i)}$ from (4.18) and the critical constants $|t|_{i,\nu,\rho,\alpha}$ $(1 \le i \le a)$. Both these step-down procedures are clearly more powerful than the corresponding single-step procedures since they use smaller critical constants for $i < a$.

Example 4.5 (Estrogen Activity in Mice: Comparisons with a Control Using a Step-Down Procedure)

Table 4.4 gives uterine weights of mice obtained from an estrogen assay of a control solution and six test solutions that had been subjected to an in vitro activation technique. Uterine weights are used as measures of estrogen activity

[1]See Dunnett and Tamhane (1991) for the procedure in the unbalanced case.

Table 4.4 Uterine Weights of Mice (mg)

	Solutions						
	0	1	2	3	4	5	6
	89.8	84.4	64.4	75.2	88.4	56.4	65.6
	93.8	116.0	79.8	62.4	90.2	83.2	79.4
	88.4	84.0	88.0	62.4	73.2	90.4	65.6
	112.6	68.6	69.4	73.8	87.8	85.6	70.2
$\bar{y}_i$	96.15	88.25	75.40	68.45	84.90	78.90	70.20
s_i	11.20	19.91	10.57	7.01	7.87	15.30	6.51

Source: Steel and Torrie (1980, data for Exercise 7.3.1). Reprinted by permission of Prof. David Dickey.

Table 4.5 t-Statistics for the Differences in Means Between Control Solution and Test Solutions, and Multivariate t Critical Constants

p	1	2	3	4	5	6
t_p	0.925	2.430	3.244	1.317	2.020	3.039
$t_{(p)}$	0.925	1.317	2.020	2.430	3.039	3.244
$t_{p,20,1/2,.10}$	1.325	1.642	1.813	1.927	2.013	2.081

(lower weights correspond to higher estrogen activity). The design is a balanced one-way layout with $n = 4$ mice per group. We want to determine which test solutions have significantly lower mean uterine weights and hence higher estrogen activity compared to the control solution using $\alpha = 0.10$.

The pooled $s = \sqrt{\text{MS}_e} = \sqrt{145.84} = 12.08$ with 21 d.f. The ANOVA F-statistic equals 2.76 with p-value $= 0.039$, which indicates significant differences between the solutions. The t-statistics for comparing each test solution with the control solution along with their ordered values are shown in Table 4.5. The ordered t-statistics are compared with the critical constants $t_{p,20,1/2,.10}$ (we use 20 d.f., which is tabled, instead of 21 d.f.), which are also shown in the same table.

Begin by comparing $t_{(6)} = t_3 = 3.244$ with $t_{6,20,1/2,0.10} = 2.081$. Since $3.244 > 2.081$, we conclude a significant difference between the control solution and the test solution 3. Proceeding in this way, we find that $t_{(5)} = t_6 = 3.039 > t_{5,20,1/2,0.10} = 2.013$, $t_{(4)} = t_2 = 2.430 > t_{4,20,1/2,0.10} = 1.927$, $t_{(3)} = t_5 = 2.020 > t_{3,20,1/2,0.10} = 1.813$, but $t_{(2)} = t_4 = 1.317 < t_{2,20,1/2,0.10} = 1.642$. Testing stops at this step with the conclusion that the test solutions 3, 6, 2, and 5 differ significantly from the control solution (have higher estrogen activity) but the test solutions 1 and 4 do not differ significantly. Note that Dunnett's procedure compares all t-statistics with a common critical constant $t_{6,20,1/2,.10} = 2.081$ and thus fails to find the test solution 5 to be significantly different from the control solution. ■

4.4 GENERAL CONTRASTS

In the previous two sections we discussed procedures for pairwise comparisons and comparisons with a control. These are special cases of a more general type of comparison, called a **contrast**, which is a linear function of the treatment means with constant coefficients:

$$c_1\mu_1 + c_2\mu_2 + \cdots + c_a\mu_a = \sum_{i=1}^{a} c_i\mu_i,$$

where $\sum_{i=1}^{a} c_i = 0$. The c_i's are called the **contrast coefficients**. A pairwise comparison or a comparison with a control is an **elementary contrast** of the type $\mu_i - \mu_j$, which corresponds to all contrast coefficients equal to zero except $c_i = +1$ and $c_j = -1$ ($j = 0$ when the comparison is with a control). The LS estimator of the above contrast is $\sum_{i=1}^{a} c_i\overline{y}_i$. Its variance equals

$$\text{Var}\left(\sum_{i=1}^{a} c_i\overline{y}_i\right) = \sum_{i=1}^{a} c_i^2\text{Var}(\overline{y}_i) = \sigma^2\sum_{i=1}^{a}\frac{c_i^2}{n_i}.$$

Now consider a family of m contrasts ($2 \le m \le a-1$):

$$\sum_{i=1}^{a} c_{ij}\mu_i \qquad (1 \le j \le m). \tag{4.20}$$

The pivotal quantities for these contrasts are

$$t_j = \frac{\sum_{i=1}^{a} c_{ij}(\overline{y}_i - \mu_i)}{s\sqrt{\sum_{i=1}^{a}(c_{ij}^2/n_i)}} \qquad (1 \le j \le m). \tag{4.21}$$

The t_j are marginally t-distributed with ν d.f. To make simultaneous inferences on these m contrasts, we need the joint distribution of $(t_1, t_2, \ldots, t_m)$. This is an m-variate t-distribution (discussed in Section A.6) with $\nu = N - a$ d.f. and correlation matrix $\{\rho_{jk}\}$ where

$$\rho_{jk} = \text{Corr}\left(\sum_{i=1}^{a} c_{ij}\overline{y}_i, \sum_{i=1}^{a} c_{ik}\overline{y}_i\right) = \frac{\sum_{i=1}^{a}(c_{ij}c_{ik}/n_i)}{\sqrt{[\sum_{i=1}^{a}(c_{ij}^2/n_i)][\sum_{i=1}^{a}(c_{ik}^2/n_i)]}}. \tag{4.22}$$

For a balanced one-way layout (i.e., $n_i = n$ for all i), the above expression simplifies to

$$\rho_{jk} = \frac{\sum_{i=1}^{a} c_{ij}c_{ik}}{\sqrt{(\sum_{i=1}^{a} c_{ij}^2)(\sum_{i=1}^{a} c_{ik}^2)}} = \frac{\boldsymbol{c}_j'\boldsymbol{c}_k}{\sqrt{(\boldsymbol{c}_j'\boldsymbol{c}_j)(\boldsymbol{c}_k'\boldsymbol{c}_k)}}, \tag{4.23}$$

where $\boldsymbol{c}_j = (c_{1j}, c_{2j}, \ldots, c_{aj})'$ $(1 \le j \le m)$. A special case of interest is when $\rho_{jk} \equiv 0$, which occurs for a balanced one-way layout when $\boldsymbol{c}'_j \boldsymbol{c}_k = 0$ for all $j \ne k$, that is, when the contrast vectors are orthogonal. **Orthogonal contrasts** are uncorrelated and hence are independent under normality. We have seen an example of orthogonal contrasts in Section 3.7 when fitting orthogonal polynomials. Here are two more examples: For $a = 3$, if treatment 3 is a control, two contrasts of interest might be $0.5\mu_1 + 0.5\mu_2 - \mu_3$, which is a comparison between the average of the two test treatments with the control, and $\mu_1 - \mu_2$, which is a comparison between the two test treatments. The contrast vectors $(0.5, 0.5, -1)$ and $(1, -1, 0)$ are orthogonal. Another example is the main effects and interactions in balanced two-level factorial experiments (discussed in Chapter 6) among the means of treatment combinations.

4.4.1 Tukey Procedure for Orthogonal Contrasts

Consider the family of orthogonal contrasts (4.20) for a balanced one-way layout. The pivotal quantities (4.21) have an m-variate t-distribution with ν d.f. and common correlation $\rho = 0$. Consider testing one-sided hypotheses

$$H_{0j} : \sum_{i=1}^{a} c_{ij}\mu_i = 0 \quad \text{vs.} \quad H_{1j} : \sum_{i=1}^{a} c_{ij}\mu_i > 0 \qquad (1 \le j \le m).$$

The test statistics are

$$t_j = \frac{\sqrt{n}\sum_{i=1}^{a} c_{ij}\bar{y}_i}{s\sqrt{\sum_{i=1}^{a} c_{ij}^2}}.$$

The null distribution (under the overall null hypothesis $H_0 : \mu_1 = \cdots = \mu_a$) of $\max_{1\le j\le m} t_j$ is known as the **Studentized maximum distribution**; its critical points are $t_{m,\nu,0,\alpha}$, which are denoted for convenience as $M_{m,\nu,\alpha}$ and are tabulated in Table C.8. The procedure rejects H_{0j} if

$$t_j > M_{m,\nu,\alpha} \qquad (1 \le j \le m).$$

For testing two-sided hypotheses,

$$H_{0j} : \sum_{i=1}^{a} c_{ij}\mu_i = 0 \quad \text{vs.} \quad H_{1j} : \sum_{i=1}^{a} c_{ij}\mu_i \ne 0 \qquad (1 \le j \le m),$$

we need the null distribution of $\max_{1\le j\le m} |t_j|$. This is known as the **Studentized maximum-modulus distribution**; its critical points are $|t|_{m,\nu,0,\alpha}$, which are denoted for convenience as $|M|_{m,\nu,\alpha}$ and are tabulated in Table C.9. The procedure rejects H_{0j} if

$$|t_j| > |M|_{m,\nu,\alpha} \qquad (1 \le j \le m).$$

The associated SCIs and step-down procedures are exactly analogous to those for comparisons with a control except that here $\rho = 0$.

Example 4.6 (Anorexia Data: Simultaneous Confidence Intervals on Treatment Means)

The above procedure can be applied to compute SCIs on treatment means in a one-way layout. Although the sample means $\overline{y}_i$ are not contrasts, they are uncorrelated, which is all that is required for this procedure to be applicable. For an illustration, refer to Example 3.3, which gives 95% separate CIs for the means of the control, behavioral, and family groups. To calculate 95% SCIs, we need the critical constant $|M|_{3,48,0.05}$, which equals 2.476 using the entry for $\nu = 45$ from Table C.9 as a conservative approximation. The corresponding intervals are

$$\text{Control:} \quad \left[-0.241 \pm \frac{(2.476)(7.893)}{\sqrt{17}}\right] = [-4.969, 4.507],$$

$$\text{Behavioral:} \quad \left[3.318 \pm \frac{(2.476)(7.893)}{\sqrt{17}}\right] = [-1.430, 8.066],$$

$$\text{Family:} \quad \left[7.265 \pm \frac{(2.476)(7.893)}{\sqrt{17}}\right] = [2.517, 12.013].$$

Once again, only the family therapy is demonstrated to be effective in causing a positive weight gain. ■

For another illustration of this procedure, refer to Example 3.7, where we fitted an orthogonal polynomial of fourth degree to the tear factor data. The t-statistics for the linear, quadratic, cubic, and quartic coefficients were

$$t_1 = -4.056, \qquad t_2 = -0.117, \qquad t_3 = -0.442, \qquad t_4 = 0.814,$$

each with 15 d.f. As we saw in Section 3.7, these coefficients are uncorrelated. To test their significance simultaneously at the 0.05 level, the t-statistics are compared with $|M|_{4,15,0.05} = 2.805$ using Table C.9. Once again, we see that only the linear coefficient is significant.

4.4.2 Scheffé Procedure for All Contrasts

The three families of contrasts studied in previous sections, namely pairwise, many-to-one, and orthogonal, are useful when the type of comparisons of interest may be specified a priori. There are situations, especially in exploratory studies, where comparisons are selected a posteriori, and they could be quite general. This requires a procedure that allows for **data snooping**, that is, testing *any* contrast that may catch an experimenter's eye upon examining the data and still control the FWE. Scheffé's (1953) procedure meets this requirement because it addresses the infinite family of *all* contrasts.

The Scheffé procedure can be derived as a UI procedure by expressing the overall null hypothesis as an infinite intersection of null hypotheses on all contrasts:

$$\{H_0 : \mu_1 = \cdots = \mu_a\} = \bigcap_{c \in C} \{H_{0c} : \boldsymbol{c}'\boldsymbol{\mu} = 0\}, \tag{4.24}$$

where $\boldsymbol{c} = (c_1, c_2, \ldots, c_a)'$ is a **contrast vector**, $\boldsymbol{\mu} = (\mu_1, \mu_2, \ldots, \mu_a)'$ is a vector of the treatment means, and

$$\mathcal{C} = \left\{ \boldsymbol{c} \,\middle|\, \sum_{i=1}^{a} c_i = 0 \right\}$$

is the **contrast space**, which has dimension $a - 1$. The equivalence (4.24) follows from the fact that all treatment means are equal if and only if every contrast among them equals zero.

The test statistic for testing H_{0c} is

$$t_c = \frac{\sum_{i=1}^{a} c_i \bar{y}_i}{s\sqrt{\sum_{i=1}^{a}(c_i^2/n_i)}}.$$

Therefore the test statistic for testing H_0 is the maximum (over all contrasts) of $|t_c|$ or equivalently of t_c^2. It can be shown (see Exercise 4.17) that

$$\max_{c \in C} t_c^2 = \max_{c \in C} \left\{ \frac{(\sum_{i=1}^{a} c_i \bar{y}_i)^2}{s^2 \sum_{i=1}^{a}(c_i^2/n_i)} \right\} = \frac{\sum_{i=1}^{a} n_i(\bar{y}_i - \bar{\bar{y}})^2}{s^2} = (a-1)F, \tag{4.25}$$

where F is the ANOVA F-statistic from (3.9). Therefore the maximum of t_c^2 is distributed as $(a-1)F_{a-1,\nu}$ under H_0. Hence the Scheffé procedure rejects any $H_{0c} : \boldsymbol{c}'\boldsymbol{\mu} = 0$ at FWE $\leq \alpha$ if

$$t_c^2 > (a-1) f_{a-1,\nu,\alpha} \qquad \text{or equivalently} \qquad |t_c| > \sqrt{(a-1) f_{a-1,\nu,\alpha}}. \tag{4.26}$$

The corresponding $(1 - \alpha)$-level SCIs on all contrasts are given by

$$\sum_{i=1}^{a} c_i \mu_i \in \left[\sum_{i=1}^{a} c_i \bar{y}_i \pm \{(a-1) f_{a-1,\nu,\alpha}\}^{1/2} s \sqrt{\sum_{i=1}^{a} \left(\frac{c_i^2}{n_i} \right)} \right]. \tag{4.27}$$

Equation (4.25) shows that there is a one-to-one correspondence between the ANOVA F-test and the Scheffé procedure (4.26) for all contrasts. The F-test is significant at level α if and only if there is at least one contrast, $\sum c_i \bar{y}_i$, that is significant at level α. This contrast can be shown to be $c_i = n_i(\bar{y}_i - \bar{\bar{y}})$ $(1 \leq i \leq a)$; however, it may not be of any practical interest.

Example 4.7 (Comparisons of Barley Varieties: Scheffé Procedure)

Refer to Example 4.2, where we used the Tukey procedure to make pairwise comparisons between the seven barley varieties. As seen there, any difference $|\overline{y}_i - \overline{y}_j|$ that exceeds

$$q_{7,30,0.10}\frac{s}{\sqrt{n}} = (4.016)(3.643) = 14.630$$

is significant at the 0.10 level. If we were to use the Scheffé procedure, then the critical value would be

$$\sqrt{6 f_{6,30,0.10} s^2 \left(\frac{2}{n}\right)} = \sqrt{(6 \times 1.98)(79.64)\left(\tfrac{2}{6}\right)} = 17.759.$$

Thus the Scheffé critical value is larger than the Tukey critical value by a factor of $17.759/14.630 = 1.214$. The Scheffé procedure is more conservative because it addresses the family of all contrasts. Therefore it should not be used if only pairwise or many-to-one contrasts are of interest. In this particular example, however, both procedures lead to the same significant differences, namely,

$$E - A = 71.3 - 49.6 = 21.7,$$

$$B - A = 71.2 - 49.6 = 21.6,$$

$$C - A = 67.8 - 49.6 = 18.2.$$

■

4.5 RANKING AND SELECTION PROCEDURES

Ranking and selection procedures were developed to address the goal of selecting the best treatment, for example, the treatment having the largest mean. The implicit premise underlying this goal is that the treatment means are not equal and so it is not of interest to test the overall null hypothesis $H_0 : \mu_1 = \mu_2 = \cdots = \mu_a$. Two formulations of this goal have been principally used: Bechhofer's (1954) indifference-zone formulation and Gupta's (1956, 1965) subset selection formulation. They are discussed in the following two sections. Section 4.5.3 discusses Hsu's (1984) multiple comparisons with the best formulation which provides a unified framework.

4.5.1 Indifference-Zone Formulation

This formulation is geared toward designing an experiment, specifically, determining the sample size. As before, we assume that the observations from the ith treatment (referred to as a *population* in the ranking and selection literature and denoted by Π_i) are i.i.d. $N(\mu_i, \sigma^2)$ r.v.'s $(1 \le i \le a)$.

Denote the ordered means by $\mu_{[1]} \leq \mu_{[2]} \leq \cdots \leq \mu_{[a]}$ and the corresponding treatments by $\Pi_{[1]}, \Pi_{[2]}, \ldots, \Pi_{[a]}$. This ordering is of course unknown. The treatment $\Pi_{[a]}$ associated with the largest mean is referred to as the **"best" treatment** (assumed to be unique). The *goal* of the experiment is to select the best treatment. This is called a **correct selection (CS)**.

We require the **probability of a correct selection [P(CS)]** of our procedure to be at least equal to a preassigned value, $1-\alpha$. However, it is easy to see that this requirement will not be met if the mean of the best treatment is close to that of the second best. In fact, if all the μ_i's become equal, then $\mathrm{P(CS)} \rightarrow 1/a$. On the other hand, in this case it does not matter which treatment is selected. This suggests that the requirement $\mathrm{P(CS)} \geq 1-\alpha$ need only be satisfied when the best treatment is better than the rest by a practically significant amount, denoted by $\delta > 0$. For instance, in the paper towel example mentioned at the beginning of the chapter, it may be regarded as a practically significant improvement if one brand absorbs additional $\delta = 1$ ml of water on the average than another brand. Therefore we specify the **probability requirement** as follows:

$$\mathrm{P}_{\boldsymbol{\mu}}(\mathrm{CS}) \geq 1-\alpha \quad \text{ if } \mu_{[a]} - \mu_{[i]} \geq \delta \qquad (1 \leq i \leq a-1)$$

where $\boldsymbol{\mu} = (\mu_1, \mu_2, \ldots, \mu_a)'$. The subset of the parameter space where this requirement must be met, that is, where

$$\mu_{[a]} - \mu_{[i]} \geq \delta \;\; (1 \leq i \leq a-1) \Longleftrightarrow \mu_{[a]} - \mu_{[a-1]} \geq \delta \tag{4.28}$$

is called the **preference zone [PZ(δ)]**. The complementary subset where $\mu_{[a]} - \mu_{[a-1]} < \delta$ holds is called the **indifference zone [IZ(δ)]** because in that case we are indifferent to whether the best treatment is selected or not. Then the above probability requirement can be expressed as

$$\min_{\boldsymbol{\mu} \in \mathrm{PZ}(\delta)} \mathrm{P}_{\boldsymbol{\mu}}(\mathrm{CS}) \geq 1-\alpha. \tag{4.29}$$

Note that this requirement is analogous to the power requirement in hypothesis testing; see Eq. (2.12). For the procedures discussed here, this minimum is attained at the LFC given by

$$\mu_{[1]} = \mu_{[2]} = \cdots = \mu_{[a-1]} = \mu_{[a]} - \delta. \tag{4.30}$$

If σ^2 is known, then the probability requirement (4.29) can be met using a **single-stage procedure** given in Section 4.5.1.1 that takes a fixed-size sample from each population. If σ^2 is unknown, then a procedure that uses more than one stage of sampling must be used; a **two-stage procedure** is given in Section 4.5.1.2.

4.5.1.1 *Single-Stage Procedure*

Consider a natural selection procedure that takes a random sample of size n from each treatment and selects the treatment that yields the largest sample mean $\overline{y}_{\max}$ as the best.

To write the P(CS) of this procedure, let $\overline{y}_{[i]}$ denote the sample mean for the treatment with mean $\mu_{[i]}$, that is, $E(\overline{y}_{[i]}) = \mu_{[i]}$. Note that $\overline{y}_{[i]}$ is not necessarily the ith largest sample mean. Then P(CS) is given by

$$\begin{aligned}
P_{\mu}(CS) &= P_{\mu}\{\overline{y}_{[a]} > \overline{y}_{[i]} \ (1 \le i \le a-1)\} \\
&= P_{\mu}\left\{\left(\frac{\overline{y}_{[i]} - \overline{y}_{[a]} + \mu_{[a]} - \mu_{[i]}}{\sigma}\right)\sqrt{\frac{n}{2}}\right. \\
&\qquad \left. < \left(\frac{\mu_{[a]} - \mu_{[i]}}{\sigma}\right)\sqrt{\frac{n}{2}} \ (1 \le i \le a-1)\right\} \\
&= P_{\mu}\left\{z_i < \left(\frac{\mu_{[a]} - \mu_{[i]}}{\sigma}\right)\sqrt{\frac{n}{2}} \ (1 \le i \le a-1)\right\} \qquad (4.31)
\end{aligned}$$

where

$$z_i = \left(\frac{\overline{y}_{[i]} - \overline{y}_{[a]} + \mu_{[a]} - \mu_{[i]}}{\sigma}\right)\sqrt{\frac{n}{2}} \qquad (1 \le i \le a-1)$$

are standard normal r.v.'s with $\mathrm{Corr}(z_i, z_j) = \frac{1}{2}$ for $i \ne j$. In other words, $z_1, z_2, \ldots, z_{a-1}$ have a multivariate standard normal distribution (see Section A.3) with common correlation $\frac{1}{2}$.

Since the above P(CS) expression is an increasing function of each $\mu_{[a]} - \mu_{[i]}$, its minimum over $\boldsymbol{\mu} \in PZ(\delta)$, where $PZ(\delta)$ is defined in (4.28), is attained when the differences $\mu_{[a]} - \mu_{[i]}$ are minimized, that is, at the LFC given by (4.30). Denoting this LFC by $\boldsymbol{\mu}(\delta)$, we have

$$\begin{aligned}
\min_{\boldsymbol{\mu} \in PZ(\delta)} P_{\boldsymbol{\mu}}(CS) &= P_{\boldsymbol{\mu}(\delta)}\left\{z_i < \frac{\delta}{\sigma}\sqrt{\frac{n}{2}} \ (1 \le i \le a-1)\right\} \\
&= P_{\boldsymbol{\mu}(\delta)}\left\{\max_{1 \le i \le a-1} z_i < \frac{\delta}{\sigma}\sqrt{\frac{n}{2}}\right\}. \qquad (4.32)
\end{aligned}$$

To meet the requirement (4.37), we need to make the right-hand side of the above equation equal to $1 - \alpha$. This is achieved by setting

$$\frac{\delta}{\sigma}\sqrt{\frac{n}{2}} = z_{a-1,1/2,\alpha},$$

where $z_{a-1,1/2,\alpha}$ is the upper α critical point of $\max_{1 \le i \le a-1} z_i$. These critical points are tabulated in Table C.6. From the above equation we get the following formula for the sample size per treatment:

$$n = 2\left(\frac{z_{a-1,1/2,\alpha}\sigma}{\delta}\right)^2. \tag{4.33}$$

Example 4.8 (How Many Paper Towels to Test?)

Consider an experiment to compare five brands of paper towel in terms of their water absorbency. An increase in the mean absorbency by 1 ml is regarded as practically significant. The standard deviation σ in the absorbency measurements is assumed to be 2 ml. If the P(CS) must be at least 0.95, how many towels of each brand must be tested?

For $a = 5$ and $\alpha = 0.05$, we find $z_{a-1,1/2,\alpha} = z_{4,1/2,0.05} = 2.160$ from Table C.6. Substituting this in Eq. (4.33) we find

$$n = 2\left[\frac{(2.160)(2)}{1}\right]^2 = 18.66.$$

Therefore 19 towels of each brand must be tested. ■

4.5.1.2 *Two-Stage Procedure*

In practice, σ^2 is generally unknown. In that case the single-stage procedure of the previous section can only be used if δ is specified in relative terms as a multiple of σ. For instance, δ may be specified as 0.5σ. The disadvantage of this approach is that the magnitude of δ depends on the unknown σ; if $\sigma = 1$ then $\delta = 0.5$, whereas if $\sigma = 10$ then $\delta = 5$. Alternatively, if there is a known upper bound on σ then it can be used in (4.33) to find an upper bound on the necessary sample size n. If either of these options is not feasible then the following two-stage procedure proposed by Bechhofer, Dunnett, and Sobel (1954) may be used. In this procedure the first-stage sample estimate of σ determines the second-stage sample size.

Stage 1: Take a random sample of size $n_1 \geq 2$ from each treatment. Calculate the sample means $\bar{y}_i^{(1)}$ and the pooled sample variance:

$$s^2 = \frac{1}{a(n_1-1)}\left[\sum_{i=1}^{a}\sum_{j=1}^{n_1}\left(y_{ij} - \bar{y}_i^{(1)}\right)^2\right]$$

with $\nu = a(n_1 - 1)$ d.f. Calculate

$$N = \max\left\{n_1, 2\left(\frac{t_{a-1,\nu,1/2,\alpha}s}{\delta}\right)^2\right\}, \tag{4.34}$$

where $t_{a-1,\nu,1/2,\alpha}$ is the upper α critical point of the $(a-1)$-variate t-distribution with $\nu = a(n_1 - 1)$ d.f. and common correlation $\frac{1}{2}$. If $N = n_1$

[i.e., if $2(t_{a-1,\nu,1/2,\alpha}s/\delta)^2 \le n_1$], then stop sampling and select the treatment yielding $\max_{1\le i\le a} \bar{y}_i^{(1)}$ as the best. If $N > n_1$, then go to the second stage.

Stage 2: Take an additional random sample of size $n_2 = N - n_1$ from each treatment. Calculate the overall sample means

$$\bar{y}_i = \frac{1}{N}\sum_{j=1}^{N} y_{ij} = \frac{n_1\bar{y}_i^{(1)} + n_2\bar{y}_i^{(2)}}{N} \qquad (1 \le i \le a),$$

where $\bar{y}_i^{(1)}$ and $\bar{y}_i^{(2)}$ are the first- and second-stage sample means, respectively, of the ith treatment. Select the treatment yielding $\max_{1\le i\le a} \bar{y}_i$ as the best.

Note that the total sample size N per treatment is an r.v., and it is roughly proportional to s^2.

The P(CS) of the two-stage procedure can be written as follows:

$$\begin{aligned}
P_{\boldsymbol{\mu}}(\text{CS}) &= P_{\boldsymbol{\mu}}\{\bar{y}_{[a]} > \bar{y}_{[i]} \ (1 \le i \le a-1)\}\\
&= P_{\boldsymbol{\mu}}\left\{\left(\frac{\bar{y}_{[i]} - \bar{y}_{[a]} + \mu_{[a]} - \mu_{[i]}}{s}\right)\sqrt{\frac{N}{2}}\right.\\
&\qquad \left. < \left(\frac{\mu_{[a]} - \mu_{[i]}}{s}\right)\sqrt{\frac{N}{2}} \ (1 \le i \le a-1)\right\}\\
&\ge P_{\boldsymbol{\mu}(\delta)}\left\{t_i < \frac{\delta}{s}\sqrt{\frac{N}{2}} \ (1 \le i \le a-1)\right\} \quad (\text{ since } \mu_{[a]} - \mu_{[i]} \ge \delta),
\end{aligned}$$

where the r.v.'s

$$t_i = \left(\frac{\bar{y}_{[i]} - \bar{y}_{[a]} + \mu_{[a]} - \mu_{[i]}}{s}\right)\sqrt{\frac{N}{2}} \qquad (1 \le i \le a-1) \tag{4.35}$$

have an $(a-1)$-variate t-distribution[2] with $\nu = a(n_1 - 1)$ d.f. and common correlation $\frac{1}{2}$; the inequality in the last step is obtained by using the fact that the minimum of P(CS) over $\boldsymbol{\mu} \in \text{PZ}(\delta)$ is attained at the LFC (4.30). Now from (4.34) we have

$$N \ge 2\left(\frac{t_{a-1,\nu,1/2,\alpha}s}{\delta}\right)^2 \iff \frac{\delta}{s}\sqrt{\frac{N}{2}} \ge t_{a-1,\nu,1/2,\alpha}.$$

[2]This result is not as straightforward as it seems since the sample size N is an r.v., being a function of s. It can be proved by conditioning on s and hence on N.

Therefore a further lower bound on P(CS) over $\boldsymbol{\mu} \in \mathrm{PZ}(\delta)$ is given by

$$P_{\boldsymbol{\mu}}(\mathrm{CS}) \geq \mathrm{P}\left\{ \max_{1 \leq i \leq a-1} t_i < t_{a-1,\nu,1/2,\alpha} \right\} = 1 - \alpha, \tag{4.36}$$

which proves that the two-stage procedure meets the probability requirement (4.29).

Example 4.9 (Airline Reservation Systems: Two-Stage Simulation Experiment to Select the Best System)

Goldsman, Nelson, and Schmeiser (1991) discuss how to compare four different airline reservation systems using simulation. Each system consists of two computers in parallel. The four systems differ in the parameters that affect the time-to-failure (TTF). The system with the largest E(TTF) is regarded as the best. Because computer failures are rare, E(TTF)'s are very large—of the order of 100,000 minutes (about 70 days).

Each system was simulated for 20 replications (each replication consisted of 20 microreplications whose TTF values were averaged to yield a single observation). The sample means and the sample SDs are shown in Table 4.6. Because each y_{ij} is an average of 20 observations, they are approximately normal, thanks to the central limit theorem. Therefore we may check the homoscedasticity assumption using the Bartlett test (see Section 3.4.1). The Bartlett statistic equals 1.056 on 3 d.f., which is nonsignificant. So we may pool the separate variances to compute the pooled SD,

$$s = \sqrt{\frac{(2.916)^2 + (2.429)^2 + (2.532)^2 + (2.081)^2}{4}} = 2.507$$

with $\nu = 76$ d.f.

Suppose that this constitutes the first stage of the simulation experiment. We now want to design the second stage to select the best system with P(CS) ≥ 0.95 if E(TTF) for the best system exceeds E(TTF) for the other systems by at least 3000 min, that is, the threshold δ that defines the preference zone PZ(δ) equals 0.3. Hence the total number of replications from each system is calculated using

Table 4.6 Means and Standard Deviations of Simulated TTF Values ($\times 10^4$ min)

	System 1	System 2	System 3	System 4
$\bar{y}_i$	10.829	10.769	9.617	8.975
s_i	2.916	2.429	2.532	2.081

Source: Goldsman, Nelson, and Schmeiser (1991, Table 2).

(4.34) with $t_{3,76,1/2,0.05} = 2.095$ and $s = 2.507$ as

$$N = \max\left\{20, 2\left(\frac{(2.095)(2.507)}{0.3}\right)^2\right\} = 613.$$

Thus an additional $613 - 20 = 593$ replications (each consisting of 20 microreplications) must be made for each system in the second stage. The system that will produce the largest overall sample mean TTF will be selected as the best with 95% confidence. ■

4.5.2 Subset Selection Formulation

Instead of focusing on selecting the single best treatment, the subset selection formulation focuses on selecting a subset of "good" treatments. A **correct selection (CS)** is defined as selection of any subset that includes the best treatment. This goal is relevant when **screening** a large number of treatments to keep good ones. The **probability requirement** is

$$\mathrm{P}_{\boldsymbol{\mu}}(\mathrm{CS}) \geq 1 - \alpha \quad \text{for all } \boldsymbol{\mu}. \tag{4.37}$$

Note that there is no preference zone in this formulation, and the probability requirement must be guaranteed over the entire parameter space.

An advantage of the subset selection formulation is that a single-stage procedure can be used even if σ^2 is unknown regardless of the sample sizes because it aims to control the type I error rate. On the other hand, the indifference-zone formulation requires a two-stage procedure because it aims to control the type II error rate (or power), which involves determination of the sample size. Thus the subset selection formulation is geared toward analysis of the already collected data, while the indifference-zone formulation is geared toward design of an experiment.

Assume a balanced one-way layout with n observations per treatment. Let $\bar{y}_i$ be the sample means $(1 \leq i \leq a)$ and let s^2 be the pooled sample variance (MS_e) based on ν d.f. Then Gupta's (1956, 1965) selection procedure selects the following subset of treatments:

$$S = \left\{\Pi_i : \bar{y}_i > \bar{y}_{\max} - t_{a-1,\nu,1/2,\alpha} s\sqrt{\frac{2}{n}}\right\}. \tag{4.38}$$

To show that this procedure guarantees the probability requirement (4.37), write its P(CS) as follows:

$$\begin{aligned}\mathrm{P}_{\boldsymbol{\mu}}(\mathrm{CS}) &= \mathrm{P}_{\boldsymbol{\mu}}\left\{\Pi_{[a]} \in S\right\} \\ &= \mathrm{P}_{\boldsymbol{\mu}}\left\{\bar{y}_{[a]} > \bar{y}_{[i]} - t_{a-1,\nu,1/2,\alpha} s\sqrt{\frac{2}{n}} \ (1 \leq i \leq a-1)\right\}\end{aligned}$$

$$= \mathrm{P}_{\mu}\left\{\left(\frac{\bar{y}_{[i]} - \bar{y}_{[a]} + \mu_{[a]} - \mu_{[i]}}{s}\right)\sqrt{\frac{n}{2}}\right.$$

$$\left. < \left(\frac{\mu_{[a]} - \mu_{[i]}}{s}\right)\sqrt{\frac{n}{2}} + t_{a-1,\nu,1/2,\alpha} \quad (1 \le i \le a-1)\right\}$$

$$\ge \mathrm{P}_{\mu}\left\{\max_{1 \le i \le a-1}\left(\frac{\bar{y}_{[i]} - \bar{y}_{[a]} + \mu_{[a]} - \mu_{[i]}}{s}\right)\sqrt{\frac{n}{2}} < t_{a-1,\nu,1/2,\alpha}\right\}$$

$$= \mathrm{P}\left\{\max_{1 \le i \le a-1} t_i < t_{a-1,\nu,1/2,\alpha}\right\}$$

$$= 1 - \alpha. \tag{4.39}$$

In the above, the final equality follows because the t_i have an $(a-1)$-variate distribution with ν d.f. and common correlation $\rho = \frac{1}{2}$. Note that the LFC where the P(CS) of the subset selection procedure is minimized is the equal-means configuration $\mu_{[1]} = \mu_{[2]} = \cdots = \mu_{[a]}$.

Example 4.10 (Airline Reservation Systems: Subset Selection)

Refer to the data from Example 4.11. Suppose that the goal is to select a subset of the four systems that includes the best system with P(CS) ≥ 0.95. Then the systems with

$$\bar{y}_i > \bar{y}_{\max} - t_{a-1,\nu,1/2,\alpha}s\sqrt{\frac{2}{n}} = 10.829 - (2.095)(2.507)\sqrt{\frac{2}{20}} = 9.168$$

will be included in the subset. Thus only system 4 will be excluded from the subset. ■

4.5.3 Multiple Comparisons with the Best

In this section we will study SCIs for comparing the mean of the best treatment with the others. It is convenient to work with $\mu_i - \max_{j \ne i} \mu_j$ $(1 \le i \le a)$. Note that if $\mu_i - \max_{j \ne i} \mu_j > 0$, then $\mu_i = \mu_{\max}$ and hence the ith treatment is the best. On the other hand, if $\mu_i - \max_{j \ne i} \mu_j < 0$, then $\mu_i < \mu_{\max}$ and hence the ith treatment is nonbest. The lower and upper confidence limits on $\mu_i - \max_{j \ne i} \mu_j$ can be used to check these inequalities, respectively.

Let us assume a balanced one-way layout with a common sample size of n per treatment. Hsu (1981, 1984) showed that $100(1-\alpha)\%$ SCIs for $\mu_i - \max_{j \ne i} \mu_j$ $(1 \le i \le a)$ are given by

$$\mu_i - \max_{j \ne i} \mu_j \in \left[\ell_i = \left(\bar{y}_i - \max_{j \ne i} \bar{y}_j - t_{a-1,\nu,1/2,\alpha}s\sqrt{\frac{2}{n}}\right)^{-},\right.$$

$$\left. u_i = \left(\bar{y}_i - \max_{j \ne i} \bar{y}_j + t_{a-1,\nu,1/2,\alpha}s\sqrt{\frac{2}{n}}\right)^{+}\right]. \tag{4.40}$$

Here $x^- = x$ if $x < 0$ and $x^- = 0$ if $x \geq 0$. Similarly, $x^+ = x$ if $x > 0$ and $x^+ = 0$ if $x \leq 0$.

Note that the lower limits ℓ_i of the above SCIs are constrained to be ≤ 0 and the upper limits u_i are constrained to be ≥ 0. If $\ell_i = 0$ (there can be at most one such i), it means that $\mu_i \geq \max_{j \neq i} \mu_j$. Discounting the possibility of tied μ's, it means that the ith treatment is the unique best treatment. On the other hand, if $u_i = 0$, then we can conclude that $\mu_i \leq \max_{j \neq i} \mu_j$, so the ith treatment is nonbest (again discounting the possibility of tied μ's). If $\ell_i < 0$ and $u_i > 0$, then the ith treatment is a possible candidate for being the best.

Example 4.11 (Airline Reservation Systems: Comparisons with the Best System)

The critical constant required to apply Hsu's procedure at the 95% confidence level equals $t_{a-1,\nu,1/2,\alpha} = t_{3,76,1/2,0.05} = 2.095$. Therefore the 95% SCI's are:

$$\mu_1 - \max_{i \neq 1} \mu_i : \quad \left[\left(10.829 - 10.769 - (2.095)(2.507)\sqrt{\frac{2}{20}}\right)^-, \left(10.829 - 10.769 + (2.095)(2.507)\sqrt{\frac{2}{20}}\right)^+\right] = [-1.601, 1.721],$$

$$\mu_2 - \max_{i \neq 2} \mu_i : \quad \left[\left(10.769 - 10.829 - (2.095)(2.507)\sqrt{\frac{2}{20}}\right)^-, \left(10.769 - 10.829 + (2.095)(2.507)\sqrt{\frac{2}{20}}\right)^+\right] = [-1.721, 1.601],$$

$$\mu_3 - \max_{i \neq 3} \mu_i : \quad \left[\left(9.617 - 10.829 - (2.095)(2.507)\sqrt{\frac{2}{20}}\right)^-, \left(9.617 - 10.829 + (2.095)(2.507)\sqrt{\frac{2}{20}}\right)^+\right] = [-2.873, 0.449],$$

$$\mu_4 - \max_{i \neq 4} \mu_i : \quad \left[\left(8.975 - 10.829 - (2.095)(2.507)\sqrt{\frac{2}{20}}\right)^-, \left(8.975 - 10.829 + (2.095)(2.507)\sqrt{\frac{2}{20}}\right)^+\right] = [-3.515, 0].$$

Thus systems 1, 2, and 3 are not significantly different from the best of the remaining systems, but system 4 is nonbest. ■

4.5.4 Connection between Multiple Comparisons with Best and Selection of Best Treatment

Both the indifference-zone selection and subset selection inferences are implied by Hsu's SCIs (4.40) for multiple comparisons with the best. First consider indifference-zone selection. Denote the selected treatment by Π^* (which produced $\overline{y}_{\max}$). We know that if $\boldsymbol{\mu} \in \text{PZ}(\delta)$, then this treatment is guaranteed to be the best with probability at least $1 - \alpha$; in other words,

$$\text{P}_{\boldsymbol{\mu}}\left\{\mu^* = \mu_{[a]}\right\} \geq 1 - \alpha \qquad \text{for } \boldsymbol{\mu} \in \text{PZ}(\delta),$$

where μ^* is the mean of the selected treatment. But what if $\boldsymbol{\mu} \in \text{IZ}(\delta)$? Fabian (1962) provided an answer to this question by showing that

$$\text{P}_{\boldsymbol{\mu}}\left\{\mu^* \geq \mu_{[a]} - \delta\right\} \geq 1 - \alpha \qquad \text{for all } \boldsymbol{\mu}.$$

Thus, if we define any treatment with mean within $\delta > 0$ of $\mu_{[a]}$ as good, then the treatment selected using the indifference-zone selection procedures (single-stage or two-stage) is guaranteed to be good with $100(1 - \alpha)\%$ confidence.

Hsu's SCIs (4.40) provide stronger inference in that they imply that *all* treatments in the set

$$\widehat{G} = \{\Pi_i : \ell_i \geq -\delta\} = \left\{\Pi_i : \overline{y}_i \geq \max_{j \neq i} \overline{y}_j + t_{a-1,\nu,1/2,\alpha} s\sqrt{\frac{2}{n}} - \delta\right\} \tag{4.41}$$

are good with probability at least $1 - \alpha$. In other words, if

$$G = \left\{\Pi_i : \mu_i \geq \mu_{[a]} - \delta\right\}$$

denotes the set of all good treatments, then

$$\text{P}_{\boldsymbol{\mu}}\left\{\widehat{G} \subseteq G\right\} \geq 1 - \alpha.$$

For the set $\widehat{G}$ to be nonempty, we must have

$$t_{a-1,\nu,1/2,\alpha} s\sqrt{\frac{2}{n}} - \delta \leq 0 \iff n \geq 2\left(\frac{t_{a-1,\nu,1/2,\alpha} s}{\delta}\right)^2,$$

so that at least $\overline{y}_{\max}$ satisfies the condition in (4.41) and $\Pi^* \in \widehat{G}$. If σ is unknown, then this condition cannot be met using a single-stage procedure (because n is fixed while s can be arbitrarily large); however, the two-stage procedure sample size N meets this condition, as can be seen from (4.34). If σ is known, then this condition is met using the single-stage procedure from Section 4.5.1.1, as can be seen from (4.33), because in that case σ replaces s and $z_{a-1,1/2,\alpha}$ replaces $t_{a-1,\nu,1/2,\alpha}$.

Whereas the connection of indifference-zone selection with the SCIs (4.40) is through the lower confidence limits ℓ_i, the connection with subset selection is through the upper confidence limits u_i. As noted in Section 4.5.3, $u_i > 0$ implies that Π_i is a possible candidate for the best treatment. Note that the subset selection procedure selects subset S of treatments for which $u_i > 0$ since

$$u_i > 0 \iff \overline{y}_i > \overline{y}_{\max} - t_{a-1,\nu,1/2,\alpha} s \sqrt{\frac{2}{n}}.$$

4.6 CHAPTER SUMMARY

(a) When multiple comparisons (a priori planned or a posteriori selected) are performed on the same set of data, the probability of finding false significances increases with the number of comparisons. A systematic approach to address this problem is through defining a family of inferences and then controlling the probability of making at least one type I error in the family, called the familywise error rate (FWE).

(b) The UI method is useful for constructing single-step MCPs. The closure method is useful for constructing step-down MCP's. Single-step MCPs can be used to perform hypothesis tests as well as to compute SCIs. Step-down MCPs are primarily useful for testing and are more powerful than single-step MCPs. The choice between the two depends on whether one is interested only in tests or also in SCIs.

(c) The Bonferroni procedure is an omnibus but conservative single-step MCP that can be used in almost any problem with a finite family of a priori specified inferences. In a testing application, it only requires p-values for all the hypotheses; therefore it can be used even when different test statistics, for example, t or χ^2 or nonparametric, are used for different hypotheses. The Holm procedure is a step-down and more powerful version of the Bonferroni procedure.

(d) For specific families more powerful MCPs are available that should be used if the required assumptions, such as normality and homoscedasticity, are satisfied. The Bonferroni procedure is useful for finite number of arbitrary prespecified contrasts, the Tukey procedure is useful for pairwise comparisons, and the Dunnett procedure is useful for comparisons with a control. These procedures also have more powerful step-down versions. The Scheffé procedure is useful for general data-snooped contrasts.

(e) For selecting the best treatment (with the largest population mean), there are two approaches. The indifference-zone approach addresses the design problem of determining the sample size necessary to guarantee the probability requirement that the best treatment is selected with probability at least $1 - \alpha$ if the best population mean exceeds the rest by at least a prespecified amount $\delta > 0$. The subset selection approach addresses the data analysis problem of selecting a subset of treatments that contains the best treatment with probability at least $1 - \alpha$.

(f) The choice between MCPs and RSPs depends on the experimenter's inferential goal. The MCPs are useful for detailed and possibly data-snooped comparisons between treatment means. The goals in RSPs are more structured, for example, select the best treatment or select a subset containing the best treatment. Among the MCPs, the choice depends, first of all, on the family of inferences. Next, one needs to decide whether to use one of the normal theory procedures or a p-value based procedure such as the Bonferroni or Holm. The choice between single-step and step-down procedures depends on the trade-off between the desirability of having SCIs against more power for testing.

EXERCISES

Section 4.1 (Basic Concepts of Multiple Comparisons)

Theoretical Exercises

4.1 Consider testing $m \geq 2$ null hypotheses, each at level α, and suppose that the tests are independent. Show that if all null hypotheses are true, then

$$\text{FWE} = 1 - (1 - \alpha)^m > \alpha.$$

Calculate the FWE for $m = 100$ and $\alpha = 0.05$.

4.2 Show that FWE $\geq$ FDR with equality holding if and only if all null hypotheses in the family are true. Therefore control of the FWE is more stringent than control of the FDR.

4.3 Draw the decision tree for testing four hypotheses using a closed procedure.

Section 4.2 (Pairwise Comparisons)

Theoretical Exercises

4.4 (a) Show that $q_{2,\nu,\alpha}/\sqrt{2} = t_{\nu,\alpha/2}$. Check this relation numerically for $\nu =$ 30 d.f. and $\alpha = 0.05$.

(b) Show that the Tukey procedure for comparing two treatments reduces to the independent-samples two-sided t-test.

4.5 Show that the c.d.f. of the Studentized range r.v. Q is given by

$$\mathrm{P}\{Q \leq x\} = a \int_0^{\infty} \left\{ \int_{-\infty}^{\infty} \left[\Phi(z) - \Phi\left(z - x\sqrt{\frac{u}{\nu}} \right) \right]^{a-1} \phi(z)\, dz \right\} g_\nu(u)\, du,$$

where $\Phi(z)$ and $\phi(z)$ are the standard normal c.d.f. and p.d.f., respectively, and $g_\nu(u)$ is the p.d.f. of the χ^2_ν r.v. Hence the upper α critical point $q_{a,\nu,\alpha}$ of this distribution is the solution in x to the equation obtained by setting the above integral to $1 - \alpha$.

Applied Exercises

4.6 Find the critical constant $t_{\nu,\alpha/2m}$ used by the Bonferroni procedure, $t_{\nu,\alpha/2}$ used by the LSD procedure, and $q_{a,\nu,\alpha}/\sqrt{2}$ used by the Tukey procedure for making pairwise comparisons at $\alpha = 0.05$ between $a = 5$ treatment groups with error d.f. $\nu = 30$. Check that the following inequalities hold:

$$t_{\nu,\alpha/2} < q_{a,\nu,\alpha}/\sqrt{2} < t_{\nu,\alpha/2m}.$$

4.7 Refer to the plastic bottle weights data from Exercise 3.4. Apply the LSD, Bonferroni, and Tukey procedures at $\alpha = 0.05$ to determine which stations differ from each other. Use the pooled SD as an estimate of the error SD.

4.8 Refer to the previous exercise. Apply the Tukey–Ryan step-down procedure at $\alpha = 0.05$ to determine which stations differ from each other. Compare the results with those obtained by applying the Tukey procedure.

4.9 Refer to the hemoglobin data from Exercise 3.5. Perform Bonferroni and Tukey pairwise comparisons [using the modification (4.10) for unbalanced one-way layouts] at $\alpha = 0.05$. Which disease groups are significantly different from each other using these two procedures?

Section 4.3 (Comparisons with a Control)

Theoretical Exercises

4.10 Derive formula (4.15):

$$\rho_{ij} = \sqrt{\frac{n_i n_j}{(n_i + n_0)(n_j + n_0)}}.$$

4.11 Consider comparing $a \geq 2$ treatments, each with n replicates, with a control having n_0 replicates. Suppose that the total sample size $N = n_0 + an$

is fixed. Ignoring the integer restrictions on the n's, show that the common variance of $\bar{y}_i - \bar{y}_0$ $(1 \le i \le a)$ is minimized by the **square-root allocation rule**: $n_0/n = \sqrt{a}$.

Applied Exercises

4.12 Spurrier and Solorzano (2004) report data from an experiment to study adverse effects of a pollutant (OP) and estrogen (E) on the reproductive organs of male rats. A 2^2 factorial design obtained by combining two doses of OP (20 and 80 mg) and E (0.8 and 8 mg) was employed. A corn oil vehicle was used to inject each OP–E combination into rats. For comparison purposes two controls were used, no injection (passive control) and corn oil injection (active control). A one-way layout design with $n = 6$ rats per group was used. A number of outcome variables were measured; here we consider weights of pituitary glands one month after the treatment (a higher weight indicates an adverse effect). The summary statistics are given in Table 4.7.

(a) Compute 95% two-sided SCIs between the four treatments and the passive control. Which treatments are significantly different (and increase the risk of elevated pituitary gland weight) from the passive control? Repeat for active control.

(b) Suppose we pool the means of the passive and active controls because there is not a significant difference between the two, as can be checked. If we regard the pooled controls as a single control with twice the sample size, what is the common correlation between the four comparisons?

(c) Calculate 95% two-sided SCIs between the four treatments and the pooled control. Approximate the required critical point by linearly interpolating w.r.t. $1/(1-\rho)$ between the critical points for $\rho = 0$ and $\rho = \frac{1}{2}$ in Tables C.9 and C.7, respectively.

Table 4.7 Mean Pituitary Gland Weights

Treatment	Mean Weight (mg)
Passive control	7.19
Active control	6.70
20 mg OP	6.96
0.8 mg E	7.83
80 mg OP	9.02
8 mg E	14.12
Mean-square error	1.0626

Source: Spurrier and Solorzano (2004). Reprinted by permission of the Institute of Mathematical Statistics.

Table 4.8 Blood Cell Counts of Animals (millions per cubic millimeter)

Control	Drug *A*	Drug *B*
7.40	9.76	12.80
8.50	8.80	9.68
7.20	7.68	12.16
8.24	9.36	9.20
9.84		10.55
8.32		

Source: Dunnett (1955).

4.13 Consider the data on blood cell counts of three groups of animals shown in Table 4.8. It is desired to compute 95% two-sided SCIs to compare the mean cell counts of the two treated groups with the control group.

(a) Calculate the correlation coefficient ρ between the two contrasts that compare the two drugs with control.

(b) Calculate the two CIs. Approximate the required critical point using the same method as in Exercise 4.12(d). Is either drug significantly different from the control?

4.14 Refer to the iodide data from Exercise 3.6. Model diagnostic tests made in Exercise 3.10 suggested log transformation of the data. The means and SDs for the log-transformed data (using natural logs) are shown in Table 4.9. Use the common pooled estimate $s = \sqrt{\text{MS}_e} = 0.3567$ with 48 d.f. in the following calculations.

(a) Do the Dunnett test on the No-PTU data to see which iodide doses differ significantly from the zero dose. Use $\alpha = 0.05$.

(b) Repeat part (a) for the PTU data.

Table 4.9 Means and Standard Deviations for Log-Transformed Iodide Data

	By Iodide Level			
	0 (μg)	0.25 (μg)	2.5 (μg)	45 (μg)
No PTU				
Mean	5.2086	4.8561	4.7134	4.8647
SD	0.2370	0.3173	0.2172	0.3040
PTU				
Mean	5.2028	5.4117	5.6009	5.4596
SD	0.3105	0.3434	0.4832	0.5233

4.15 Refer to the previous exercise. Redo the two tests using the stepwise Dunnett procedure from Section 4.3.2.

Section 4.4 (General Contrasts)

Theoretical Exercises

4.16 Derive formula (4.22):

$$\rho_{jk} = \text{Corr}\left(\sum_{i=1}^{a} c_{ij}\bar{y}_i, \sum_{i=1}^{a} c_{ik}\bar{y}_i\right) = \frac{\sum_{i=1}^{a}(c_{ij}c_{ik}/n_i)}{\sqrt{[\sum_{i=1}^{a}(c_{ij}^2/n_i)][\sum_{i=1}^{a}(c_{ik}^2/n_i)]}}.$$

4.17 Show the result in (4.25) that

$$\max_{c\in C} t_c^2 = \max_{c\in C}\left\{\frac{(\sum_{i=1}^{a} c_i\bar{y}_i)^2}{s^2\sum_{i=1}^{a}(c_i^2/n_i)}\right\} = \frac{\sum_{i=1}^{a} n_i(\bar{y}_i - \bar{\bar{y}})^2}{s^2}.$$

What are the maximizing values of the contrast coefficients c_i? (*Hint*: Write

$$\sum_{i=1}^{a} c_i\bar{y}_i = \sum_{i=1}^{a}\left(\frac{c_i}{\sqrt{n_i}}\right)[\sqrt{n_i}(\bar{y}_i - g)],$$

where g is an arbitrary constant. Without loss of generality, fix $\sum_{i=1}^{a}(c_i^2/n_i) = h$, where $h > 0$ is another constant. Then apply the **Cauchy–Schwarz inequality**, which states that

$$\left(\sum_{i=1}^{n} u_i v_i\right)^2 \le \left(\sum_{i=1}^{n} u_i^2\right)\left(\sum_{i=1}^{n} v_i^2\right)$$

for any real vectors $(u_1, u_2, \ldots, u_n)$ and $(v_1, v_2, \ldots, v_n)$ with equality holding if and only if the vectors are collinear. Finally show that $g = \bar{\bar{y}}$ in order for the maximizing values of the c_i to be contrasts.)

Applied Exercises

4.18 Refer to Exercise 4.14. An important question of a priori interest is whether there is a PTU effect.

(a) Suppose it was proposed in the protocol that the PTU effect will be evaluated by comparing the average of the responses of *all* the four doses with PTU with those of the four doses without PTU. Is any multiplicity adjustment needed for this comparison? Do the comparison using $\alpha = 0.05$.

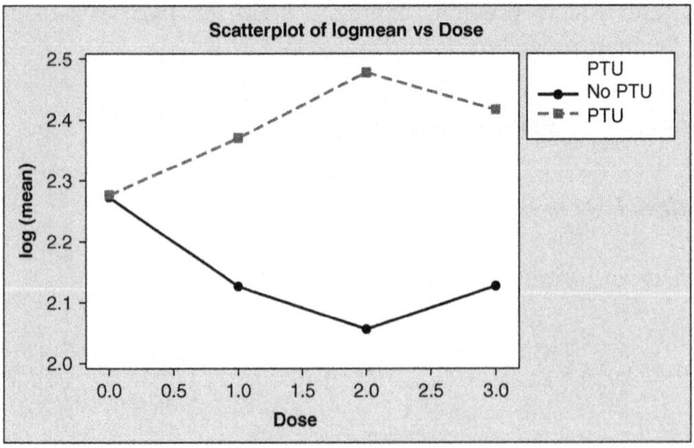

Figure 4.2 Plot of means of log(T/S) versus dose levels for no-PTU and PTU data.

(b) Suppose it was also proposed in the protocol that the PTU effect will be evaluated by comparing the average response of *each* of the four doses with PTU with the corresponding average response of each of the four doses without PTU. Is any multiplicity adjustment needed for these comparisons? Do the comparisons using $\alpha = 0.05$.

4.19 Refer to Exercise 4.14. The sample means of log(T/S) versus the dose levels are plotted separately for No PTU and PTU in Figure 4.2. We see that the dose response is quadratic in both cases but in opposite directions. This was an unanticipated finding, and we would like to formally test if PTU changes the nature of the quadratic response (i.e., is there a PTU–quadratic effect interaction). The contrast vector $(+1, -1, -1, +1)$ estimates the quadratic effect among the four doses.

(a) Estimate the quadratic effects for the No-PTU and PTU data. Find the difference between the two quadratic effects. Give the contrast vector for this difference.

(b) Calculate the standard error of the above contrast.

(c) Which multiple comparison method should be used to test the significance of this contrast and why? Do the corresponding significance test at $\alpha = 0.05$.

Section 4.5 (Ranking and Selection Procedures)

Theoretical Exercises

4.20 Suppose we have $a = 2$ populations with $N(\mu_1, \sigma_1^2)$ and $N(\mu_2, \sigma_2^2)$ distributions where μ_1 and μ_2 are unknown means and σ_1^2 and σ_2^2 are known variances. To select the population with the larger mean, we take random samples of sizes n_1 and n_2 from the two populations and select the one producing the larger sample mean $\overline{y}_i$.

(a) Show that for this procedure

$$\mathrm{P(CS)} = \Phi\left(\frac{\mu_{[2]} - \mu_{[1]}}{\sqrt{\sigma_1^2/n_1 + \sigma_2^2/n_2}}\right),$$

where $\mu_{[1]} \le \mu_{[2]}$ are the ordered values of μ_1, μ_2.

(b) Show that P(CS) is maximized subject to fixed total sample size $N = n_1 + n_2$ by choosing $n_i \propto \sigma_i$, that is,

$$\frac{\sigma_1}{n_1} = \frac{\sigma_2}{n_2}.$$

Thus, if the integer restrictions on the n_i are ignored, then

$$n_1 = \frac{\sigma_1}{\sigma_1 + \sigma_2}N \qquad \text{and} \qquad n_2 = \frac{\sigma_2}{\sigma_1 + \sigma_2}N.$$

(c) Hence show that, to guarantee the probability requirement (4.29), the minimum total sample size is given by

$$N = \left[\frac{z_\alpha(\sigma_1 + \sigma_2)}{\delta}\right]^2.$$

(d) Calculate the sample sizes n_1 and n_2 for $1 - \alpha = 0.95$, $\delta = 2.0$, $\sigma_1 = 2.0$, $\sigma_2 = 4.0$. Note that n_1 and n_2 should be rounded up to the nearest integers.

4.21 Show that the multivariate normal probability in (4.31) can be alternately written as

$$\mathrm{P}_{\mu}(\mathrm{CS}) = \int_{-\infty}^{\infty} \prod_{i=1}^{a-1} \Phi\left[z + \frac{(\mu_{[a]} - \mu_{[i]})\sqrt{n}}{\sigma}\right] \phi(z)\, dz.$$

Hence under the LFC (4.30) we have an equivalent expression for (4.32):

$$\mathrm{P}_{\mu(\delta)}(\mathrm{CS}) = \int_{-\infty}^{\infty} \left\{\Phi\left[z + \frac{\delta\sqrt{n}}{\sigma}\right]\right\}^{a-1} \phi(z)\, dz.$$

Further show that if this integral is equated to $1 - \alpha$ and by putting $c = \delta\sqrt{n}/\sigma$ solved for c, then the solution equals $c = \sqrt{2}z_{a-1,1/2,\alpha}$. This solution is tabulated in Bechhofer (1954). [*Hint*: In the first step of (4.31) use the standardization

$$z_i = \frac{(\overline{y}_{[i]} - \mu_{[i]})\sqrt{n}}{\sigma} \qquad (1 \le i \le a);$$

then condition on $z_a = z$ and finally uncondition on z.]

4.22 Consider the generalization of the setup of Exercise 4.20 for $a \ge 2$ normal populations $N(\mu_i, \sigma_i^2)$, $i = 1, 2, \ldots, a$. Suppose that n_i observations are taken from the ith population and the one producing the largest sample mean is selected as the best. Show that if we put $\tau_i = \sigma_i/\sqrt{n_i}$, then the P(CS) of this rule under the LFC (4.30) equals

$$P_{\mu(\delta)}(CS) = \int_{-\infty}^{\infty} \prod_{i=1}^{a-1} \Phi\left[z\left(\frac{\tau_{[a]}}{\tau_{[i]}}\right) + \frac{\delta}{\tau_{[i]}}\right] \phi(z)\,dz,$$

where $\tau_{[i]}$ is associated with the population having mean $\mu_{[i]}$ $(1 \le i \le a)$. Hence show that if we allocate $n_i \propto \sigma_i^2$, that is, make the $\text{Var}(\overline{y}_i)$ equal,

$$\frac{\sigma_1^2}{n_1} = \frac{\sigma_2^2}{n_2} = \cdots = \frac{\sigma_a^2}{n_a},$$

then the n_i satisfying the probability requirement (4.29) can be found from the equation

$$n_i = \left(\frac{c\sigma_i}{\delta}\right)^2,$$

where $c = \sqrt{2}z_{a-1,1/2,\alpha}$ is the critical constant defined in Exercise 4.21. However, this allocation is not optimal, as can be seen from Exercise 4.20 for the $a = 2$ case.

4.23 Show that the critical constant $t_{a-1,\nu,1/2,\alpha}$ occurring in (4.36) and (4.39) is the solution in t to the equation

$$\int_0^{\infty} \left\{ \int_{-\infty}^{\infty} \left\{ \Phi\left[z + t\sqrt{\frac{u}{\nu}}\right] \right\}^{a-1} \phi(z)\,dz \right\} g_\nu(u)\,du = 1 - \alpha,$$

where $g_\nu(u)$ is the density function of a χ_ν^2 r.v. (*Hint*: Write the multivariate t r.v.'s, $t_i = z_i/\sqrt{\chi_\nu^2/\nu}$, where the z_i are multivariate standard normal r.v.'s with common correlation $\frac{1}{2}$ and are independent of χ^2. Condition on $\chi^2 = u$ and write the probability in terms of the z_i using the integral representation derived in Exercise 4.21 and then uncondition on χ^2.)

Applied Exercises

4.24 Refer to Exercise 3.3. Suppose that the experiment was to be designed to compare the eight different formulations of wood glue to select the best one having the highest mean bonding strength. For design purposes assume homoscedasticity (this assumption is checked in Exercise 3.9) and the common σ is 300 kilo-newtons (kN). The best glue must be selected with probability at least 0.95 if its mean bonding strength exceeds that of the second best by at least 200 kN. How many wood joints must be tested using each glue?

4.25 Refer to the previous exercise, but now suppose that the common σ is unknown and the data given in Table 3.15 is the first stage of the experiment with $n_1 = 8$ is used to estimate it. Using the pooled sample variance from this first stage to estimate the common σ^2, calculate the second-stage sample size required per glue to guarantee the same probability requirement as in the previous exercise.

4.26 Refer to Exercise 4.24. Assume homoscedasticity and use the pooled sample variance to estimate the common σ^2.

(a) Apply Gupta's procedure to select a subset of the glues that contains the best glue (having the highest mean bonding strength) with probability at least 0.95. Which glues are excluded from this subset?
(b) Apply Hsu's procedure to compute 95% SCIs for the differences between the mean bonding strengths of the "best" glue and the others. Explain how these SCIs are related to the subset selected in part (a).

4.27 Refer to the data in Table 4.1 which were obtained from a field experiment conducted using an RB design with six blocks. The MS_e from the ANOVA was 79.64 with 30 d.f. Apply Hsu's procedure with $\alpha = 0.10$ to classify the varieties as nonbest and possible candidates for the best. Is there one variety that is clearly the best?

CHAPTER 5

Randomized Block Designs and Extensions

In Chapter 1 we saw that blocking is a useful technique for improving the precision of comparisons between treatments by minimizing the biasing effects of selected noise factors. Even if those noise factors do not turn out to have significant effects on the response variable, blocking on them is a good insurance policy, at least in preliminary phases of experimentation, to protect against any challenge to the results of the experiment (e.g., the treatment differences being shown significant) as being caused by possible confounding due to noise factors. Blocking is equally useful for evaluating the consistency of the effectiveness of the treatments over a range of categories of experimental units. For example, in a clinical trial it is of interest to determine not only if a new treatment is effective but also if its effectiveness is consistent across different patient subgroups, that is, whether there is a treatment by patient subgroup interaction. Similarly, in an experiment to compare several manufacturing processes, if the raw materials come from different suppliers, then it would be of interest to study if the differences in the processes depend on the suppliers.

In an RB design, blocking factors are used to divide the experimental units into relatively homogeneous groups, called **blocks**, and the treatments are randomly assigned to the experimental units within each block. Thus blocks introduce a restriction on randomization. RB designs are discussed in Section 5.1. Sometimes not all treatments can be accommodated in each block, that is, the block size is less than the number of treatments. A **balanced incomplete-block (BIB) design** may be used in this case. BIB designs are discussed in Section 5.2.

RB and BIB designs eliminate heterogeneity caused by a single blocking factor (and hence are called **one-way elimination of heterogeneity designs**). If there are two or more blocking factors, then they are combined into a single one. For example, in a clinical trial, patients may be grouped into six blocks

Statistical Analysis of Designed Experiments: Theory and Applications By Ajit C. Tamhane

based on their gender and three age categories. If the individual effects of gender and age are of interest, then experimental units must be blocked separately by gender and age group. **Youden square (YSQ)** (discussed in Section 5.3) and **Latin square (LSQ)** designs (discussed in Section 5.4) allow blocking on two factors (and hence are called **two-way elimination of heterogeneity designs**). They may also be used when there are two treatment factors and one blocking factor. **Graeco–Latin square (GLSQ)** designs, which are extensions of Latin squares, are discussed in Section 5.4.5. Section 5.5 covers some additional topics and proofs of the results used in the chapter. Section 5.6 gives a summary of the chapter.

5.1 RANDOMIZED BLOCK DESIGNS

5.1.1 Model

RB designs were first used in agricultural trials for comparing different varieties of a crop or different fertilizers. For example, trials are often conducted in different fields, which are the blocks. Each field is divided into **plots** which are the experimental units. Thus the plots within each field are relatively homogeneous compared to those in different fields. By assigning the treatments (e.g., varieties of a crop) at random to the plots within each field, more precise comparisons between the treatments can be made.

In some applications, the treatments are applied to the same experimental unit in a random or specified order if there are no long-lasting carryover effects. For example, in a matched pairs experiment, the same subject may be given two different treatments in a random order. Strictly speaking, this is a **repeated measures (RM) design**—not an RB design since the treatments are not independently applied to different experimental units within the same block. This aspect is generally ignored in practice, especially since these designs are small and there is not enough data to estimate the correlations among the observations on the same experimental unit; so the data are analyzed as in an RB design. The same practice is used in other extensions of RB designs covered in this chapter, for example, BIB, YSQ, and LSQ designs.

Consider a treatment factor with $a \geq 2$ levels (treatments) and a blocking factor with $b \geq 2$ levels (blocks). Each block is assumed to be **complete**, that is, it consists of a plots. Thus each treatment is replicated only once ($n = 1$) in each block. More generally, one can have $n \geq 1$ replicates of each treatment within each block, in which case the block size is na; this design is discussed in Section 5.1.3.

Both the treatments and blocks are assumed to be fixed. The analysis methods given here apply equally well if the blocks are random since the block effects are eliminated when comparing the treatment effects. The same is true in case of BIB designs studied in the next section. In that case, however, it is possible to recover extra information about the treatment effects if the blocks are random. This method of analysis is discussed in Section 5.2.2.

Let y_{ij} denote the response observed on a plot in the jth block that is given the ith treatment. The y_{ij} are assumed to be independent $N(\mu_{ij}, \sigma^2)$ r.v.'s. The usual model for an RB design is

$$y_{ij} = \mu + \alpha_i + \beta_j + e_{ij} \qquad (1 \le i \le a, 1 \le j \le b), \tag{5.1}$$

where μ is the "overall" mean, α_i is the ith treatment effect, β_j is the jth block effect, and the e_{ij} are i.i.d. $N(0, \sigma^2)$ r.v.'s. Thus $E(y_{ij}) = \mu_{ij} = \mu + \alpha_i + \beta_j$. Only $a - 1$ of the α_i's and $b - 1$ of the β_j's are independent because of the constraints

$$\sum_{i=1}^{a} \alpha_i = \sum_{j=1}^{b} \beta_j = 0. \tag{5.2}$$

Therefore this model contains $1 + (a - 1) + (b - 1) = a + b - 1$ independent parameters leaving

$$\nu = ab - [1 + (a - 1) + (b - 1)] = (a - 1)(b - 1)$$

d.f. for estimating the error.

The model (5.1) is referred to as an **additive model** because it postulates that, apart from the overall mean, the mean response is the sum of the treatment and block effects. Therefore the difference between the means of any two treatments, i and i', is the same across all blocks since

$$\mu_{ij} - \mu_{i'j} = (\mu + \alpha_i + \beta_j) - (\mu + \alpha_{i'} + \beta_j) = \alpha_i - \alpha_{i'}. \tag{5.3}$$

If the treatment means μ_{ij} are plotted against blocks (or against treatments), then the **mean profiles** are parallel, as shown in Figure 5.1.

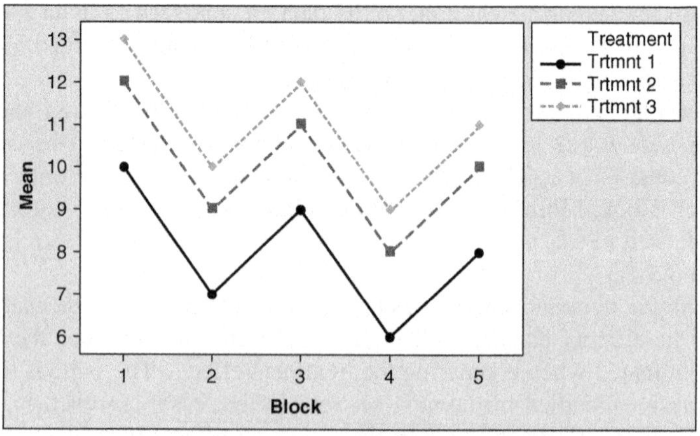

Figure 5.1 Parallel mean profiles when model is additive.

Any effects (positive or negative) over and above the additive effects of the treatments and blocks, namely

$$(\alpha\beta)_{ij} = \mu_{ij} - (\mu + \alpha_i + \beta_j), \tag{5.4}$$

are called **interactions**. It is easily seen that interactions cannot be estimated or tested if $n = 1$ since in that case the total number of parameters equals the total number of observations, namely, ab, and so $\nu = 0$. Under a more restrictive model in which the interactions are assumed to follow a **multiplicative model**, that is, $(\alpha\beta)_{ij} = \gamma\alpha_i\beta_j$ (so that there is only one parameter for interactions, namely, γ), Tukey (1949) gave a formal test of $H_{0AB} = \gamma = 0$, which is discussed in Section 6.2.5. This test can be used for an RB design with $n = 1$. However, for the general model, we need $n > 1$ replicates of each treatment in each block. The test for interaction for this design is given in Section 5.1.3.

It should be emphasized that an RB design is not a factorial design with two crossed factors. In an RB design the blocking factor is used to group the experimental units into blocks, while the treatments are applied to them using restricted randomization within each block. Thus only the levels of the treatment factor are randomly assigned to the experimental units. On the other hand, in a two-factor design (discussed in Chapter 6), both factors are treatment factors, and the levels of both factors are randomly assigned to the experimental units. It should be noted that a factorial experiment may be run as an RB design or any of its extensions discussed in this chapter.

5.1.2 Statistical Analysis

5.1.2.1 Estimation

The LS estimators of μ, the α_i's and the β_j's, minimize

$$\sum_{i=1}^{a}\sum_{j=1}^{b}[y_{ij} - (\mu + \alpha_i + \beta_j)]^2.$$

They can readily shown to be equal to

$$\widehat{\mu} = \overline{y}_{..}, \qquad \widehat{\alpha}_i = \overline{y}_{i.} - \overline{y}_{..} \quad (1 \le i \le a), \qquad \widehat{\beta}_j = \overline{y}_{.j} - \overline{y}_{..} \quad (1 \le j \le b),$$

where

$$\overline{y}_{i.} = \frac{1}{b}\sum_{j=1}^{b} y_{ij}, \qquad \overline{y}_{.j} = \frac{1}{a}\sum_{i=1}^{a} y_{ij}, \qquad \overline{y}_{..} = \frac{1}{ab}\sum_{i=1}^{a}\sum_{j=1}^{b} y_{ij}.$$

This **dot notation** is used throughout the rest of the text, where dots in the subscript mean that the given quantity is averaged (if there is a bar on the symbol) or summed (if there is no bar) over the subscripts replaced by the dots keeping all other indexes fixed.

Note that the estimators $\widehat{\alpha}_i$'s and $\widehat{\beta}_j$'s satisfy the respective constraints (5.2) on the α_i's and β_j's. The fitted values of the y_{ij} are given by

$$\widehat{y}_{ij} = \widehat{\mu}_{ij} = \widehat{\mu} + \widehat{\alpha}_i + \widehat{\beta}_j = \overline{y}_{..} + (\overline{y}_{i.} - \overline{y}_{..}) + (\overline{y}_{.j} - \overline{y}_{..}) = \overline{y}_{i.} + \overline{y}_{.j} - \overline{y}_{..}. \tag{5.5}$$

Hence the residuals equal

$$\widehat{e}_{ij} = y_{ij} - \widehat{y}_{ij} = y_{ij} - \overline{y}_{i.} - \overline{y}_{.j} + \overline{y}_{..}. \tag{5.6}$$

An unbiased estimator of σ^2 is given by

$$s^2 = \text{MS}_e = \frac{\text{SS}_e}{(a-1)(b-1)} = \frac{\sum_{i=1}^{a}\sum_{j=1}^{b}\widehat{e}_{ij}^2}{(a-1)(b-1)},$$

where $(a-1)(b-1) = \nu$ is the error d.f. Note that this estimator is unbiased only under the additive model.

5.1.2.2 Analysis of Variance and Multiple Comparisons

Represent each y_{ij} as a sum of four components by substituting $\widehat{\mu}, \widehat{\alpha}_i, \widehat{\beta}_j$, and $\widehat{e}_{ij}$ in (5.1):

$$y_{ij} = \overline{y}_{..} + (\overline{y}_{i.} - \overline{y}_{..}) + (\overline{y}_{.j} - \overline{y}_{..}) + \widehat{e}_{ij}.$$

Square both sides and sum over $i = 1, 2, \ldots, a$ and $j = 1, 2, \ldots, b$. All six cross products on the right-hand side add to zero when summed over i and j because of orthogonality between the treatment and block effects, thus yielding

$$\begin{aligned}\sum_{i=1}^{a}\sum_{j=1}^{b} y_{ij}^2 &= N\overline{y}_{..}^2 + \sum_{i=1}^{a}\sum_{j=1}^{b}(\overline{y}_{i.} - \overline{y}_{..})^2 + \sum_{i=1}^{a}\sum_{j=1}^{b}(\overline{y}_{.j} - \overline{y}_{..})^2 + \sum_{i=1}^{a}\sum_{j=1}^{b}\widehat{e}_{ij}^2 \\ &= N\overline{y}_{..}^2 + \underbrace{b\sum_{i=1}^{a}\widehat{\alpha}_i^2}_{\text{SS}_{\text{trt}}} + \underbrace{a\sum_{j=1}^{b}\widehat{\beta}_j^2}_{\text{SS}_{\text{blk}}} + \underbrace{\sum_{i=1}^{a}\sum_{j=1}^{b}\widehat{e}_{ij}^2}_{\text{SS}_e}.\end{aligned} \tag{5.7}$$

Transferring $N\overline{y}_{..}^2$ to the left-hand side and noting that $\sum\sum y_{ij}^2 - N\overline{y}_{..}^2 = \sum\sum(y_{ij} - \overline{y}_{..})^2 = \text{SS}_{\text{tot}}$, we get the **ANOVA identity** for an RB design:

$$\text{SS}_{\text{tot}} = \text{SS}_{\text{trt}} + \text{SS}_{\text{blk}} + \text{SS}_e.$$

The corresponding partition of the d.f. is

$$\underbrace{ab-1}_{\text{total d.f.}} = \underbrace{a-1}_{\text{treatments d.f.}} + \underbrace{b-1}_{\text{blocks d.f.}} + \underbrace{(a-1)(b-1)}_{\text{error d.f.}}.$$

The mean squares are computed in the usual way by dividing the sums of squares by their d.f.:

$$\text{MS}_{\text{trt}} = \frac{\text{SS}_{\text{trt}}}{a-1}, \qquad \text{MS}_{\text{blk}} = \frac{\text{SS}_{\text{blk}}}{b-1}, \qquad \text{MS}_e = \frac{\text{SS}_e}{(a-1)(b-1)}.$$

The ratio $F_{\text{trt}} = \text{MS}_{\text{trt}}/\text{MS}_e$ can be used to test the null hypothesis of no treatment effects, $H_{0A} : \alpha_1 = \cdots = \alpha_a = 0$, as follows: Reject H_{0A} at level α if

$$F_{\text{trt}} = \frac{\text{MS}_{\text{trt}}}{\text{MS}_e} > f_{a-1,\nu,\alpha}, \tag{5.8}$$

where $\nu = (a-1)(b-1)$. This test can be derived by using the extra sum of squares method. The ANOVA is given in Table 5.1.

There is some controversy about the validity of the analogous test on block effects. Anderson and McLean (1974) argue that because the levels of the blocking factor are not randomly assigned to the experimental units, the F-test of equality of block effects is not valid. To show this, they introduced a different model which incorporates a **restriction error** due to restricted randomization. This model is discussed in Section 5.5.1. Under this model, the restriction error is unique to each block and cannot be separately estimated. Anderson and McLean do point out, however, that the F-test is valid for a broader null hypothesis of no block effects and no restriction error, which translates in practical terms to the hypothesis that blocking is not effective in reducing the error variance. We shall interpret the null hypothesis in this broader sense, implicitly assuming that the restriction error variance is zero. This is in accord with the common practice that many books and practitioners follow of using the magnitude of $F_{\text{blk}} = \text{MS}_{\text{blk}}/\text{MS}_e$ at least informally, to decide whether blocking has been effective.

For making multiple comparisons between the treatments, an RB design can be treated as a balanced one-way layout with b observations per treatment, but with $s = \sqrt{\text{MS}_e}$ based on $\nu = (a-1)(b-1)$ d.f. Thus, for example, the $100(1-\alpha)\%$ Tukey SCIs for pairwise differences $\alpha_i - \alpha_j$ between the treatment effects are

$$\alpha_i - \alpha_j \in \left[\bar{y}_{i\cdot} - \bar{y}_{j\cdot} \pm q_{a,\nu,\alpha}\frac{s}{\sqrt{b}}\right] \qquad (1 \le i < j \le a),$$

where $q_{a,\nu,\alpha}$ is the upper α critical point of the Studentized range distribution (see Section 4.2.2) with parameter a and error d.f. ν.

Example 5.1 (Chick Growth Promoter Study: ANOVA and Multiple Comparisons)

Snee (1985a) gave data from a study to evaluate the effect of a growth promoter on the weights of baby chicks. Three levels of the growth promoter were compared: zero dose (control), low dose, and high dose. The experiment was conducted in eight different locations (blocks) in a bird house. The response variable was the average weight per chick in ounces. The data are given in Table 5.2.

From the plot of the data in Figure 5.2 we see that the average weight of chicks is generally increasing with the dose level of the growth promoter except in locations 2 and 5, the latter location showing a more flagrant violation of the monotone dose response observed in other locations. Overall, we see that the growth promoter is effective in promoting growth, and the majority of the mean profiles are roughly parallel, indicating no significant interaction between the growth promoter and location.

From the ANOVA shown in Display 5.1 we see that the growth promoter has a highly significant effect ($F = 17.84$, $p = 0.000$), whereas the location effect

Table 5.1 ANOVA Table for RB Design

Source	SS	d.f.	MS	F
Treatments	$SS_{trt} = b\sum(\bar{y}_{i\cdot} - \bar{y}_{\cdot\cdot})^2$	$a-1$	MS_{trt}	MS_{trt}/MS_e
Blocks	$SS_{blk} = a\sum(\bar{y}_{\cdot j} - \bar{y}_{\cdot\cdot})^2$	$b-1$	MS_{blk}	MS_{blk}/MS_e
Error	SS_e = by subtraction	$(a-1)(b-1)$	MS_e	
Total	$SS_{tot} = \sum\sum(y_{ij} - \bar{y}_{\cdot\cdot})^2$	$ab-1$		

Table 5.2 Chick Weights (oz)

	Growth Promoter			
Location	Zero Dose (0)	Low Dose (1)	High Dose (2)	Row Mean
1	3.93	3.99	4.08	4.000
2	3.78	3.96	3.94	3.893
3	3.88	3.96	4.02	3.953
4	3.93	4.03	4.06	4.007
5	3.84	4.10	3.94	3.960
6	3.75	4.02	4.09	3.953
7	3.98	4.06	4.17	4.070
8	3.84	3.92	4.12	3.960
Column mean	3.866	4.005	4.053	3.974

Source: Snee (1985a, Table 1).

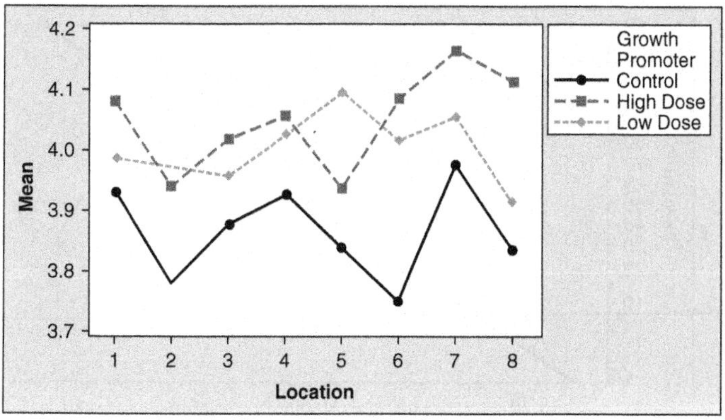

Figure 5.2 Plot of chick weight data.

```
Analysis of Variance for Weight

Source               DF        SS        MS       F       P
Growth Promoter       2  0.149858  0.074929   17.84   0.000
Location              7  0.056129  0.008018    1.91   0.144
Error                14  0.058808  0.004201
Total                23  0.264796
S = 0.0648120       R-Sq = 77.79%    R-Sq(adj) = 63.51%
```

Display 5.1 Minitab ANOVA output for chick weight data.

is nonsignificant ($F = 1.91$, $p = 0.144$). Therefore blocking by locations could be dispensed with in future experiments based on what is learned from this experiment. However, at least in this experiment it has served the purpose of an insurance policy mentioned at the beginning of this chapter.

The normal plot and the fitted-values plot of residuals are shown in Figure 5.3. The normal plot is reasonably linear (the Anderson–Darling statistic is 0.250 with $p = 0.716$), and the fitted-values plot is also fairly random, so the normality and the homoscedasticity assumptions are acceptable. There are no obvious outliers in the data.

The two doses can be compared with the control using Dunnett's one-sided procedure [see Eq. (4.17)] assuming that a priori the growth promoter was expected to have a positive effect. The one-sided multivariate t critical point for two treatment comparisons with the control using $\alpha = 0.05$, $\rho = \frac{1}{2}$ [see Eq. (4.15); here $n_0 = n_1 = n_2 = 8$], and error d.f. $= 14$ is $t_{2,14,1/2,0.05} = 2.079$ from Table C.6. The ANOVA estimate of σ is $s = 0.0648$. Therefore the lower one-sided 95% SCIs on the differences between the doses and the control are

$$\alpha_1 - \alpha_0 \geq 4.005 - 3.866 - (2.079)(0.0648)\sqrt{\frac{2}{8}} = 0.072,$$

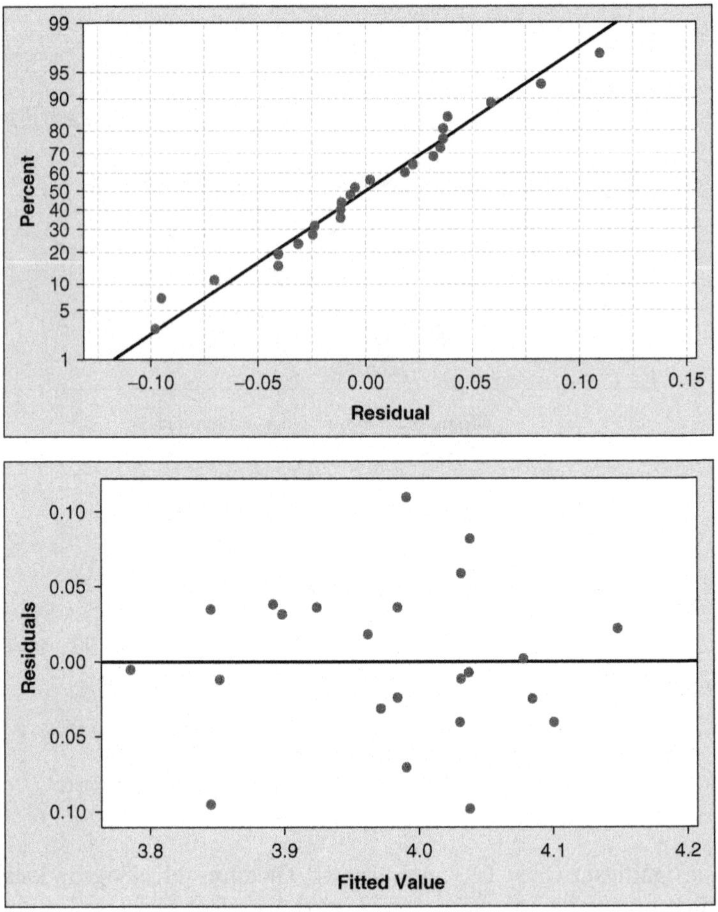

Figure 5.3 Normal (top) and fitted-values (bottom) plots of residuals for chick weight data.

and

$$\alpha_2 - \alpha_0 \geq 4.053 - 3.866 - (2.079)(0.0648)\sqrt{\tfrac{2}{8}} = 0.120.$$

Thus both doses demonstrate significant increases in weight gains over the control dose. That leads to the question of whether the high dose is significantly better than the low dose. The difference in their mean weight gains is $4.053 - 4.005 = 0.048$ oz, which is highly nonsignificant using a two-sample t-test, that is, even without any adjustment for multiple testing. Thus the extra cost of the high dose of the growth promoter may not be worthwhile. This question may need further investigation using additional data and a trade-off between the marginal weight gain with the high dose and its extra cost. ■

5.1.3 Randomized Block Designs with Replicates

As we saw in previous sections, an RB design with a single replicate observation on each treatment per block does not permit inferences on treatment-block interactions, which are of primary interest when consistency of the treatment effects across different blocks of experimental units is to be assessed. There are many situations in practice where multiple replicate observations using an RB design are readily available. For instance, in the clinical trial example mentioned at the beginning of this chapter, we would certainly require multiple patients in each block who receive the same treatment to obtain sufficiently high power to detect treatment differences, if not to estimate the treatment-block interactions. If we have a treatments, then we can form blocks (based on relevant blocking factors, e.g., gender and age group) of na patients each for a total of $N = abn$ patients. The treatments would then be assigned randomly to the patients in each block subject to the balance restriction of n patients per treatment.

In some applications the experimenter has a choice of a trade-off between having more blocks or more replications within blocks. The first choice should be used if it is desired to test the applicability of the treatments over a wider population of the experimental units. The second choice should be used if it is more important to be able to estimate and test treatment–block interaction.

5.1.3.1 Model

The model for this design is a simple extension of (5.1) that includes the interaction terms $(\alpha\beta)_{ij}$ defined in (5.4):

$$\begin{aligned} y_{ijk} = \mu_{ij} + e_{ijk} &= \mu + \alpha_i + \beta_j \\ &+ (\alpha\beta)_{ij} + e_{ijk} \qquad (1 \le i \le a, 1 \le j \le b, 1 \le k \le n), \end{aligned} \tag{5.9}$$

where the $(\alpha\beta)_{ij}$ are subject to the constraints $\sum_{i=1}^{a}(\alpha\beta)_{ij} = 0$ for all j and $\sum_{j=1}^{b}(\alpha\beta)_{ij} = 0$ for all i and other terms have the same meanings and satisfy the same constraints as before. Thus there are $(a-1)(b-1)$ linearly independent interaction terms which account for $(a-1)(b-1)$ d.f. In the single-replicate case, these d.f. were assigned to the error. In the present case, there are a total of $N = abn$ observations and ab parameters, thus leaving $abn - ab = ab(n-1)$ d.f. for estimating the error.

5.1.3.2 Analysis of Variance

The LS estimators of the parameters in this model are given by

$$\begin{aligned} &\widehat{\mu} = \overline{y}_{\dots}, \qquad \widehat{\alpha}_i = \overline{y}_{i\cdot\cdot} - \overline{y}_{\dots}, \qquad \widehat{\beta}_j = \overline{y}_{\cdot j\cdot} - \overline{y}_{\dots}, \\ &\widehat{(\alpha\beta)}_{ij} = \overline{y}_{ij\cdot} - \overline{y}_{i\cdot\cdot} - \overline{y}_{\cdot j\cdot} + \overline{y}_{\dots}, \end{aligned} \tag{5.10}$$

where the dot notation is as defined in Section 5.1.2.1. Thus the fitted values equal $\widehat{y}_{ijk} = \widehat{\mu} + \widehat{\alpha}_i + \widehat{\beta}_j + \widehat{(\alpha\beta)}_{ij} = \overline{y}_{ij\cdot}$ and the residuals equal $\widehat{e}_{ijk} = y_{ijk} - \widehat{y}_{ijk} =$

Table 5.3 ANOVA Table for RB Design with Replicates

Source	SS	d.f.	MS	F
Treatments (A)	$SS_A = bn\sum(\overline{y}_{i..} - \overline{y}_{...})^2$	$a-1$	MS_A	MS_A/MS_e
Blocks (B)	$SS_B = an\sum(\overline{y}_{.j.} - \overline{y}_{...})^2$	$b-1$	MS_B	MS_B/MS_e
AB Interaction	$SS_{AB} = n\sum\sum(\overline{y}_{ij.} - \overline{y}_{i..} - \overline{y}_{.j.} + \overline{y}_{...})^2$	$(a-1)(b-1)$	MS_{AB}	MS_{AB}/MS_e
Error	$SS_e = \sum\sum\sum(y_{ijk} - \overline{y}_{ij.})^2$	$ab(n-1)$	MS_e	
Total	$SS_{tot} = \sum\sum\sum(y_{ijk} - \overline{y}_{...})^2$	$abn-1$		

Table 5.4 Plant Heights of Four Cultivars

		Plant Height	
Cultivar	Bench	1	2
A	1	19.3	17.2
	2	16.7	15.5
	3	17.7	19.8
B	1	20.1	19.4
	2	21.2	20.8
	3	21.0	21.9
C	1	17.4	16.6
	2	14.4	13.6
	3	15.8	17.4
D	1	16.6	15.7
	2	13.5	12.9
	3	12.8	14.7

$y_{ijk} - \overline{y}_{ij.}$. The total sum of squares can be partitioned into four orthogonal components as $SS_{tot} = SS_A + SS_B + SS_{AB} + SS_e$, where A denotes the treatment factor, B denotes the blocking factor, and AB denotes their interaction. The resulting ANOVA is shown in Table 5.3.

Example 5.2 (Effect of Cultivar on Plant Heights: Analysis of Variance)

An experiment was conducted to compare the differences in growth among four different cultivars of a house plant. The greenhouse had three benches in three different locations which form natural blocks. Two pots of each cultivar were randomly assigned to each bench for a total of six pots per cultivar and eight pots per bench. Thus we have $a = 4$, $b = 3$, and $n = 2$ in this experiment. The block size is $c = 8$. The data are shown in Table 5.4.

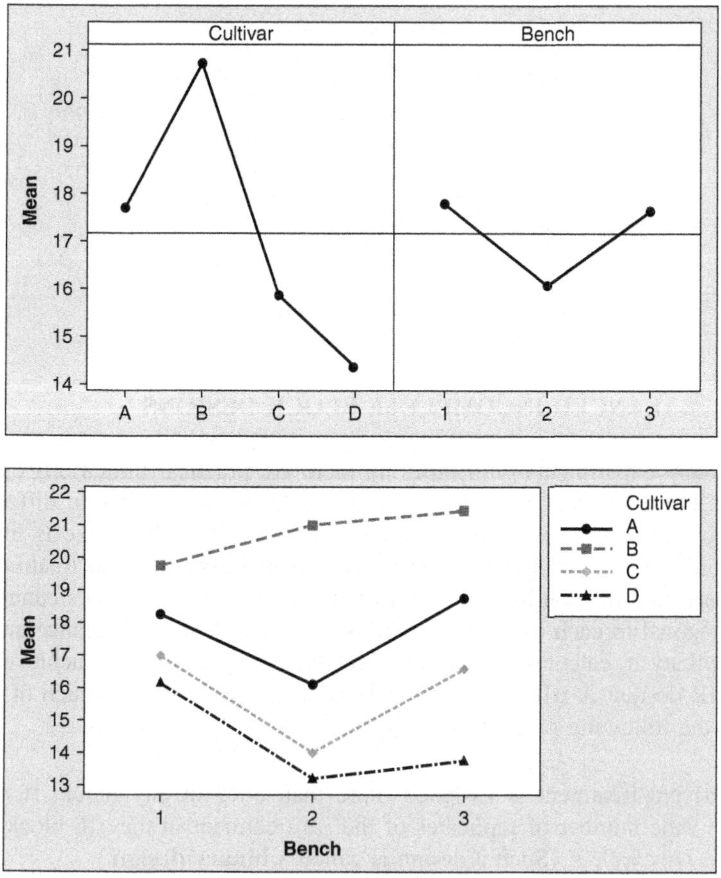

Figure 5.4 Main-effects (top) and interaction (bottom) plots for cultivar data.

Figure 5.4 shows the main-effect and interaction plots for the cultivar data. We see that the average plant heights of cultivars are ordered at all three locations with *B* having the highest height followed by *A*, then *C*, with *D* coming last. The mean profile plots are reasonably parallel except the profile for *B* is concave and increasing, while the profiles for the other three cultivars are convex with minimums at bench 2 (keep in mind, however, that the benches are not ordered in any way), suggesting that there is a possible interaction. The main effects of cultivars are obviously large, but those of benches are large too. These visual impressions are confirmed by the ANOVA shown in Display 5.2. The cultivar–bench interaction is also significant ($p = 0.035$). The conclusion to be drawn from this analysis is that, unlike the chick weight example, blocking by bench (location) is important and should be adhered to in future experiments. ■

```
Analysis of Variance for Height

Source           DF       SS       MS       F       P
Cultivar          3  135.213   45.071   53.18   0.000
Bench             2   14.391    7.195    8.49   0.005
Cultivar*Bench    6   17.099    2.850    3.36   0.035
Error            12   10.170    0.847
Total            23  176.873
S = 0.920598    R-Sq = 94.25%    R-Sq(adj) = 88.98%
```

Display 5.2 Minitab ANOVA output for cultivar data.

5.2 BALANCED INCOMPLETE BLOCK DESIGNS

In many experiments with blocking there are practical limitations on the block size. For example, suppose we want to compare more than four different brands of tire and cars are to be used as blocks. Then the block size is limited to 4. In such cases, the blocks are said to be **incomplete** since all treatments cannot be assigned in each block. Therefore the treatments and blocks cannot be made orthogonal to each other. Nonetheless, a certain degree of balance and resulting simplicity in calculations and interpretation of results can be achieved by using a BIB design. A BIB design for a treatments with b blocks, each of size $c < a$, has the following properties:

(a) No treatment is assigned more than once in any block. If n_{ij} denotes the number of replicates of the ith treatment in the jth block, then each $n_{ij} = 0, 1$. (Such a design is called a **binary design**.)

(b) Each treatment is replicated the same number of times. This number equals

$$n = \sum_{j=1}^{b} n_{ij} = \frac{bc}{a}.$$

(c) Each treatment occurs with every other treatment in the same block the same number of times, say λ. To get an expression for λ, fix any treatment, say 1. This treatment occurs in n blocks. The remaining $c - 1$ plots of these blocks are assigned to the remaining $a - 1$ treatments, each of which occurs in λ blocks. Therefore we have $n(c - 1) = \lambda(a - 1)$. Hence

$$\lambda = \frac{n(c-1)}{(a-1)} = \frac{bc(c-1)}{a(a-1)}. \tag{5.11}$$

Although a BIB design is not an orthogonal design, explicit expressions for treatment and block effects and their sums of squares can be obtained because of the above balance properties. A BIB design with $a = b$ is called a **symmetric BIB design**.

In the remainder of this chapter we will denote the treatments by letters with the obvious correspondence $A = 1$, $B = 2$, and so on.

Example 5.3 (BIB Design with $a = 7$ and $c = 4$)

For the problem of comparing seven brands of tire, the following symmetric BIB design with seven cars can be used:

Car						
1	2	3	4	5	6	7
A	E	E	E	E	G	G
B	F	G	G	A	A	F
C	C	B	A	F	F	B
D	D	D	C	B	D	C

For this design we have $a = b = 7, c = 4, n = 4$, and $\lambda = 2$. ■

Note that for $a \geq 2$ treatments and block size $c < a$ one can always construct a BIB design by forming $b = \binom{a}{c}$ blocks of all combinations of c out of a treatments. Thus, for $a = 4, c = 2$ there exists a BIB design with $b = \binom{4}{2} = 6$ blocks, for $a = 5, c = 2$ there exists a BIB design with $b = \binom{5}{2} = 10$ blocks, and so on. However, in many cases, a BIB design with a smaller number of blocks exists. For instance, for $a = 7, c = 4$ we have $b = \binom{7}{4} = 35$, but the BIB design given in the above example uses only seven blocks.

For given a and c, the following conditions are necessary (but not sufficient) for a BIB design with b blocks to exist:

$$b \geq a: \qquad n = \frac{bc}{a} \text{ and } \lambda = \frac{bc(c-1)}{a(a-1)} \text{ must be integers.}$$

In this example, $a = 7$ and $c = 4$, so

$$n = \frac{4b}{7} \quad \text{and} \quad \lambda = \frac{2b}{7},$$

and the smallest value of $b \geq a$ for which n and λ are integers is 7. Tables of BIB designs with minimum number of blocks for selected combinations of (a, c) are given in Fisher and Yates (1953).

5.2.1 Statistical Analysis

The model for a BIB design is the same additive model (5.1) for an RB design:

$$y_{ij} = \mu + \alpha_i + \beta_j + e_{ij} \qquad (1 \le i \le a, 1 \le j \le b) \tag{5.12}$$

except that only a subset of all treatment–block combinations (i, j) are observed.

5.2.1.1 Estimation

To present the formulas for the LS estimators of the model parameters and the associated tests, we first introduce some notation. Derivations of all the formulas are given in Section 5.5.2. Denote

$$v_i = y_{i\cdot} = \sum_{j=1}^{b} n_{ij} y_{ij}, \qquad w_j = y_{\cdot j} = \sum_{i=1}^{a} n_{ij} y_{ij},$$

$$v_j^* = \sum_{i=1}^{a} n_{ij} v_i, \qquad w_i^* = \sum_{j=1}^{b} n_{ij} w_j. \tag{5.13}$$

These quantities have the following interpretations: v_i is the ith treatment total, w_j is the jth block total, v_j^* is the sum of all treatment totals for those treatments assigned in the jth block, and w_i^* is the sum of all block totals for those blocks in which the ith treatment is assigned. The ith **adjusted treatment total** is defined as

$$u_i = v_i - \frac{1}{c} w_i^* \qquad (1 \le i \le a). \tag{5.14}$$

The LS estimators of the model parameters are given by

$$\begin{aligned} \widehat{\mu} &= \overline{y}_{\cdot\cdot}, \qquad \widehat{\alpha}_i = \frac{c}{a\lambda} u_i \qquad (1 \le i \le a), \\ \widehat{\beta}_j &= \frac{1}{c} w_j - \overline{y}_{\cdot\cdot} - \frac{1}{a\lambda} v_j^* \qquad (1 \le j \le b). \end{aligned} \tag{5.15}$$

It can be shown that $\sum_{i=1}^{a} \widehat{\alpha}_i = \sum_{j=1}^{b} \widehat{\beta}_j = 0$. The fitted values and the residuals are given by $\widehat{y}_{ij} = \widehat{\mu} + \widehat{\alpha}_i + \widehat{\beta}_j$ and $\widehat{e}_{ij} = y_{ij} - \widehat{y}_{ij}$.

5.2.1.2 Analysis of Variance and Multiple Comparisons

The sums of squares for testing the hypotheses of no treatment effects, $H_{0A}: \alpha_1 = \alpha_2 = \cdots = \alpha_a = 0$, and no block effects, $H_{0B}: \beta_1 = \beta_2 = \cdots = \beta_b = 0$, respectively equal (see Section 5.5.2 for the derivation)

$$\text{SS}_{\text{trt(adj)}} = \frac{c}{a\lambda} \sum_{i=1}^{a} u_i^2 \quad \text{and} \quad \text{SS}_{\text{blk(adj)}} = \frac{c}{a\lambda} \sum_{i=1}^{a} u_i^2 - \frac{1}{n} \sum_{i=1}^{a} v_i^2 + \frac{1}{c} \sum_{j=1}^{b} w_j^2,$$

Table 5.5 ANOVA Table for a BIB Design

Source	SS	d.f.	MS	F
Treatments,				
(adjusted)	$SS_{trt(adj)} = (c/a\lambda)\sum u_i^2$	$a-1$	$MS_{trt(adj)}$	$MS_{trt(adj)}/MS_e$
Blocks				
(unadjusted)	$SS_{blk(unadj)} = c\sum(\bar{y}_{\cdot j \cdot} - \bar{y}_{\cdot\cdot})^2$			
(adjusted)	$SS_{blk(adj)} = (c/a\lambda)\sum u_i^2 - (1/n)\sum v_i^2 + (1/c)\sum w_j^2$	$b-1$	$MS_{blk(adj)}$	$MS_{blk(adj)}/MS_e$
Error	$SS_e = SS_{tot} - SS_{trt(adj)} - SS_{blk(unadj)}$	$bc-a-b+1$	MS_e	
Total	$SS_{tot} = \sum\sum(y_{ij} - \bar{y}_{\cdot\cdot})^2$	$bc-1$		

where adj stands for adjusted for the other factor effects. The unadjusted sum of squares for blocks, $SS_{blk(unadj)} = c\sum_{j=1}^{b}(\bar{y}_{\cdot j} - \bar{y}_{\cdot\cdot})^2$, can be used to calculate SS_e by subtraction using the identity $SS_{tot} = SS_{trt(adj)} + SS_{blk(unadj)} + SS_e$.

The d.f. for the treatments, blocks, and error are $a-1, b-1$, and $bc-a-b+1$, respectively. The mean squares and the F-statistics are computed in the usual way. Table 5.5 gives the ANOVA table.

The variance of the difference between any pair of estimated treatment effects is given by

$$\text{Var}(\widehat{\alpha}_i - \widehat{\alpha}_j) = \frac{2c\sigma^2}{a\lambda} \qquad (1 \le i < j \le a). \tag{5.16}$$

This means that BIB designs are **pairwise variance balanced**. Therefore, as noted in Section 4.2.2, the exact Tukey procedure can be applied resulting in the following $100(1-\alpha)\%$ SCIs for all pairwise differences:

$$\alpha_i - \alpha_j \in \left[\widehat{\alpha}_i - \widehat{\alpha}_j \pm q_{a,\nu,\alpha} s\sqrt{\frac{c}{a\lambda}}\right] \qquad (1 \le i < j \le a), \tag{5.17}$$

where $\nu = bc - a - b + 1$.

Example 5.4 (Access Times of Disk Drives: BIB Design)

Nelson (1998) gave data on average access times (in milliseconds) of five brands of fixed disk drives. The computers that the disk drives were installed in are the blocks. However, only four drives could be installed in each computer. As a result, a BIB design was employed with $a = 5, b = 5, c = 4$, and $\lambda = 3$. We assume that the the disk drives are randomly assigned to the slots in each computer. The design and the data are shown in Table 5.6.

The ANOVA computed using Minitab is shown in Display 5.3. We see that the differences between the brands are significant at $\alpha = 0.05$ ($F = 3.58$,

Table 5.6 Average Access Times (msec) of Five Brands of Disk Drives

Disk Drive Brand (Treatment)	Computer (Block) 1	2	3	4	5	Row Total
A	35	41	—	32	40	148
B	42	45	40	—	38	165
C	31	—	42	33	35	141
D	30	32	33	35	—	130
E	—	40	39	36	37	152
Column total	138	158	154	136	150	736

Source: Nelson (1998, Table 1).

```
Source      DF    Seq SS    Adj SS    Adj MS      F       P
Brand        4   168.700   139.200   34.800    3.58   0.042
Computer     4    65.700    65.700   16.425    1.69   0.222
Error       11   106.800   106.800    9.709
Total       19   341.200
S = 3.11594    R-Sq = 68.70%    R-Sq(adj) = 45.93%
```

Display 5.3 Minitab ANOVA output for disc access time data (BIB design).

Table 5.7 Calculation of Brand Effect Estimates

i	v_i	w_i^*	u_i	$\widehat{\alpha}_i$
A	148	582	2.5	0.667
B	165	600	15.0	4.000
C	141	578	−3.5	−0.933
D	130	586	−16.5	−4.400
E	152	598	2.5	0.667

$p = 0.042$), but the differences between the computers are not ($F = 1.69$, $p = 0.222$). The error estimate is $s = 3.116$.

Next we make pairwise comparisons between the brands using the Tukey procedure to see which brands differ from each other. First we need to calculate the brand effects $\widehat{\alpha}_i$. These calculations (shown in Table 5.7) use the row (treatment) totals v_i and the column (block) totals w_j. As an example of the calculation, $w_A^* = w_1 + w_2 + w_4 + w_5 = 138 + 158 + 136 + 150 = 582$, which gives $u_A = 148 - (582/4) = 2.5$ and so $\widehat{\alpha}_A = (4/15)(2.5) = 0.667$.

Using (5.17), the common allowance for 95% Tukey SCIs is

$$q_{5,11,0.05}s\sqrt{\frac{c}{a\lambda}} = (4.574)(3.116)\sqrt{\frac{4}{(5)(3)}} = 7.360.$$

We see that only the difference between brand B and brand D, $4.000 - (-4.400) = 8.400$, exceeds this allowance and hence is significant at the 5% level. The 95% SCI for $\alpha_B - \alpha_D$ is $[8.400 \pm 7.360] = [1.040, 15.760]$.

Brand	D	C	E	A	B
$\widehat{\alpha}_i$	-4.400	-0.933	0.667	0.667	4.000

As in Example 4.2, the above diagram summarizes the results of pairwise comparisons. ■

In a BIB design, all treatments are regarded symmetrically. If one treatment plays the role of a control to which other treatments are compared, then a design needs to take this into account by increasing replications on the control but at the same time maintaining the symmetry between treatment–control comparisons. Toward this end, Bechhofer and Tamhane (1981) proposed a class of designs called **balanced treatment incomplete block (BTIB)** designs. Here is an example of a BTIB design for comparing four treatments (labeled A, B, C, D) with a control (labeled 0) in seven blocks of size 3 each. Note that the control is replicated nine times while each of the treatment is replicated three times. Every treatment occurs with every other treatment one time but occurs with the control three times. Also note that this design is not binary since the control is replicated twice in blocks 4, 5, and 6.

Block						
1	2	3	4	5	6	7
0	0	0	0	0	0	A
A	B	C	0	0	0	B
C	C	D	A	B	D	D

5.2.2 Interblock Analysis

The method of analysis of a BIB design given above is called **intrablock analysis** as it eliminates the block effects (whether fixed or random) when comparing the treatment effects. However, Yates (1940) showed that if the block effects β_j are assumed to be i.i.d. $N(0, \sigma_B^2)$ (this is the same assumption made in the random effects model in Chapter 11 for two-factor experiments), then additional information about the treatment effects can be recovered via so-called **interblock analysis**. Essentially, we can compute LS estimators $\widetilde{\alpha}_i$ of the α_i that are independent of the $\widehat{\alpha}_i$ given by (5.15). We call the $\widehat{\alpha}_i$ **intrablock estimators** and the $\widetilde{\alpha}_i$ **interblock estimators**. Thus more precise estimators of the α_i can be obtained by pooling these two estimators.

We begin by combining the random terms β_j and e_{ij} in the BIB design model (5.12). Summing the observations in the jth block and using the notation introduced in Eq. (5.13), we obtain the following model:

$$w_j = y_{\cdot j} = c\mu + \sum_{i=1}^{a} n_{ij}\alpha_i + \left(c\beta_j + \sum_{i=1}^{a} n_{ij}e_{ij}\right) \qquad (1 \le j \le b).$$

In this model, the error term $e_{ij}^* = c\beta_j + \sum_{i=1}^{a} e_{ij} \sim \mathrm{N}(0, c^2\sigma_B^2 + c\sigma^2)$. The LS estimators of the fixed parameters μ and the α_i are obtained by minimizing

$$Q = \sum_{j=1}^{b}\left[w_j - \left(c\mu + \sum_{i=1}^{a} n_{ij}\alpha_i\right)\right]^2.$$

The normal equations can be shown to be

$$N\mu + n\sum_{i=1}^{a}\alpha_i = y_{\cdot\cdot},$$

$$cn\mu + (n-\lambda)\alpha_i = \sum_{j=1}^{b} n_{ij}w_j = w_i^* \qquad (1 \le i \le a).$$

Using the constraint $\sum_{i=1}^{a}\alpha_i = 0$, we get the following intrablock estimators as the solutions to these equations:

$$\widetilde{\mu} = \overline{y}_{\cdot\cdot} \quad \text{and} \quad \widetilde{\alpha}_i = \frac{w_i^* - cn\overline{y}_{\cdot\cdot}}{n-\lambda} \qquad (1 \le i \le a). \tag{5.18}$$

Note that $\widetilde{\mu} = \widehat{\mu}$ but $\widetilde{\alpha}_i \neq \widehat{\alpha}_i$. In fact, it can be shown that $\widehat{\alpha}_i$ and $\widetilde{\alpha}_i$ are independent of each other; both are normal with a common mean α_i (i.e., they are both unbiased) and

$$\tau_1^2 = \mathrm{Var}(\widehat{\alpha}_i) = \frac{c(a-1)}{a^2\lambda}\sigma^2 \quad \text{and} \quad \tau_2^2 = \mathrm{Var}(\widetilde{\alpha}_i) = \frac{c(a-1)}{a(n-\lambda)}(\sigma_B^2 + \sigma^2).$$

The formula for $\mathrm{Var}(\widehat{\alpha}_i)$ is derived in Section 5.5.2 and the formula for $\mathrm{Var}(\widetilde{\alpha}_i)$ can be derived similarly.

It is well-known that if we have two independent unbiased estimators of the same parameter, then the (minimum variance) unbiased estimator is given by a weighted average of the two estimators weighted inversely by their variances. Thus the optimal estimator of α_i and its variance are given by

$$\widehat{\alpha}_i^* = \frac{\tau_2^2\widehat{\alpha}_i + \tau_1^2\widetilde{\alpha}_i}{\tau_1^2 + \tau_2^2} \quad \text{and} \quad \mathrm{Var}(\widehat{\alpha}_i^*) = \frac{\tau_1^2\tau_2^2}{\tau_1^2 + \tau_2^2}.$$

This optimal estimator cannot be used directly because τ_1^2 and τ_2^2, which depend on σ^2 and σ_B^2, are unknown. Their unbiased estimators are

$$\widehat{\sigma}^2 = \text{MS}_e \qquad \text{and} \qquad \widehat{\sigma}_B^2 = \frac{\left(\text{MS}_{\text{blk(adj)}} - \text{MS}_e\right)(b-1)}{a(n-1)}.$$

If $\text{MS}_{\text{blk(adj)}} < \text{MS}_e$, then we set $\widehat{\sigma}_B^2 = 0$ (which introduces a slight bias in the estimator).

Substituting $\widehat{\sigma}^2$ and $\widehat{\sigma}_B^2$ in the formulas for τ_1^2 and τ_2^2, we can obtain estimators $\widehat{\tau}_1^2$ and $\widehat{\tau}_2^2$, which can then be used to compute the (approximately) optimal estimators $\widehat{\alpha}_i^*$. The following example illustrates these calculations.

Example 5.5 (Access Times of Disk Drives: Interblock Analysis of BIB Design)

In Example 5.4 we calculated the intrablock estimates of the α_i. We now calculate interblock estimates and then find the pooled estimates. The following quantities are useful in these calculations: $n = 4$, $\lambda = 3$, and $\overline{y}_{..} = 736/20 = 36.8$. Also, $\text{MS}_e = 9.709$ and $\text{MS}_{\text{blk(adj)}} = 16.425$ from the ANOVA table in Display 5.3. The weights used for pooling are obtained as follows:

$$\widehat{\sigma}^2 = 9.709 \qquad \text{and} \qquad \widehat{\sigma}_B^2 = \frac{(16.425 - 9.709)(4)}{(5)(3)} = 1.791,$$

which yield

$$\widehat{\tau}_1^2 = \left(\frac{4 \times 4}{3 \times 25}\right) 9.709 = 2.071 \qquad \text{and}$$

$$\widehat{\tau}_2^2 = \left(\frac{4 \times 4}{5 \times 1}\right) (1.791 + 9.709) = 36.800.$$

Therefore the optimal estimated weights on $\widehat{\alpha}_i$ and $\widetilde{\alpha}_i$ are, respectively,

$$\frac{36.800}{2.071 + 36.800} = 0.947 \qquad \text{and} \qquad \frac{2.071}{2.071 + 36.800} = 0.053.$$

We see that most of the weight is placed on the intrablock estimates; as a result, the pooled estimates are not very different from intrablock estimates. Table 5.8 shows the calculations using the results from Table 5.7. The standard error of the

Table 5.8 Calculation of Interblock Estimates of Brand Effects

i	w_i^*	$\widehat{\alpha}_i$	$\widetilde{\alpha}_i$	$\widehat{\alpha}_i^*$
A	582	0.667	−6.8	0.271
B	600	4.000	11.2	4.382
C	578	−0.933	−10.8	−1.456
D	586	−4.400	−2.8	−4.315
E	598	0.667	9.2	1.119

pooled estimates is given by

$$\text{SE}(\widehat{\alpha}_i^*) = \sqrt{\frac{\widehat{\tau}_1^2\widehat{\tau}_2^2}{\widehat{\tau}_1^2 + \widehat{\tau}_2^2}} = \sqrt{\frac{(2.071)(36.800)}{2.071 + 36.800}} = 1.400,$$

which is slightly smaller than $\text{SE}(\widehat{\alpha}_i) = \sqrt{2.071} = 1.439$. ■

5.3 YOUDEN SQUARE DESIGNS

Consider Example 5.3, which gives a symmetric BIB design for testing seven brands of tires on seven cars using four of each brand. That design assumes that the tires are assigned at random to the four wheel positions in each car. Instead of a random assignment, a balanced assignment can be done so that a tire of each brand is mounted in each wheel position the same number of times (once). This can be done by simply rearranging the treatments in each column of the BIB design resulting in the following design:

Wheel Position	Car 1	2	3	4	5	6	7
1	A	F	B	C	D	E	G
2	B	C	D	E	G	A	F
3	C	D	E	G	A	F	B
4	D	E	G	A	F	B	C

This is called a Youden square design, named after William J. Youden (1900–1971), who proposed this design to study the effects of seven treatments on the tobacco mosaic virus. The treatments were applied to top, middle, and bottom leaves (three positions) on seven tobacco plants (blocks) and balance with respect to both the leaf positions and plants was sought. The resulting design can be obtained as a dual of the above design consisting of missing treatments in each block.

In general, a YSQ design can always be constructed from a symmetric BIB design by rearranging the treatments so that each treatment is assigned to each position the same number of times. Similarly, a YSQ design is always a Latin square (LSQ) design (studied in the next section) from which one or more rows (or columns or diagonals) have been deleted; however, deleting rows arbitrarily from an LSQ design does not always result in a YSQ design.

5.3.1 Statistical Analysis

The model for a YSQ design is the same as that for a BIB design but with an additional term for the position effect. Thus an observation y_{ijk} on the ith treatment in the jth block and the kth position within that block is modeled as

$$y_{ijk} = \mu + \alpha_i + \beta_j + \gamma_k + e_{ijk} \qquad (1 \le i \le a, 1 \le j \le b, 1 \le k \le c), \quad (5.19)$$

where the parameters μ, α_i, β_j have the same meanings as before, γ_k is the kth position effect, and the e_{ijk} are i.i.d. $N(0, \sigma^2)$ r.v.'s. Note that not all (i, j, k) combinations are observed. The parameters of this additive model are subject to the constraints

$$\sum_{i=1}^{a} \alpha_i = \sum_{j=1}^{b} \beta_j = \sum_{k=1}^{c} \gamma_k = 0.$$

5.3.1.1 Estimation

The LS estimators of $\mu, \alpha_i,$ and β_j are the same as those given in (5.15) for the BIB design. In addition, the LS estimators of the position effects are

$$\widehat{\gamma}_k = \bar{y}_{\cdot\cdot k} - \bar{y}_{\cdot\cdot\cdot} \qquad (1 \le k \le c).$$

Note that, although the treatment and block effects are not orthogonal to each other, the position effects are orthogonal to both since every treatment–position and block–position combination is observed the same number of times. This is also clear from the fact that if the position effects are dropped from the model, the LS estimators of the treatment and block effects remain unchanged and are the same as those obtained for the BIB design.

5.3.1.2 Analysis of Variance and Multiple Comparisons

The ANOVA table for a YSQ design is given in Table 5.9. It has an additional entry corresponding to the position effects. The error sum of squares and the error d.f. are reduced accordingly.

Example 5.6 (Access Times of Disk Drives: YSQ Design)

Refer to the data from Example 5.4, but now suppose that, instead of assigning the disk drives randomly to the slots in the computers (positions), their assignment is balanced so that each drive is assigned to each of the four slots in the four of the

Table 5.9 ANOVA Table for Youden Square Design

Source	SS	d.f.	MS	F
Treatments, (adjusted)	$SS_{trt(adj)} = \frac{c}{a\lambda} \sum u_i^2$	$a - 1$	$MS_{trt(adj)}$	$MS_{trt(adj)}/MS_e$
Blocks				
(unadjusted)	$SS_{blk(unadj)} = c \sum (\overline{y}_{.j.} - \overline{y}_{..})^2$			
(adjusted)	$SS_{blk(adj)} = (c/a\lambda) \sum u_i^2 - (1/n) \sum v_i^2 + (1/c) \sum w_j^2$	$b - 1$	$MS_{blk(adj)}$	$MS_{blk(adj)}/MS_e$
Positions	$SS_{pos} = b \sum (\overline{y}_{..k} - \overline{y}_{...})^2$	$c - 1$	MS_{pos}	MS_{pos}/MS_e
Error	$SS_e = SS_{tot} - SS_{trt(adj)} - SS_{blk(unadj)} - SS_{pos}$	$bc - a - b - c + 2$	MS_e	
Total	$SS_{tot} = \sum \sum \sum (y_{ijk} - \overline{y}_{...})^2$	$bc - 1$		

Table 5.10 Average Access Times (msec) of Five Brands of Disk Drives

Disk Drive Position	Computer (Block) 1	2	3	4	5	Position Total
1	*A* (35)	*B* (45)	*C* (42)	*D* (35)	*E* (37)	194
2	*B* (42)	*A* (41)	*D* (33)	*E* (36)	*C* (35)	187
3	*C* (31)	*D* (32)	*E* (39)	*A* (32)	*B* (38)	172
4	*D* (30)	*E* (40)	*B* (40)	*C* (33)	*A* (40)	183
Block Total	138	158	154	136	150	736

Source: Nelson (1998, Table 1).

```
Source     DF    Seq SS    Adj SS    Adj MS      F      P
Brand       4   168.700   148.584    37.146   5.06  0.025
Computer    4    65.700    73.084    18.271   2.49  0.127
Position    3    48.089    48.089    16.030   2.18  0.168
Error       8    58.711    58.711     7.339
Total      19   341.200
```

Display 5.4 Minitab ANOVA output for disc access time data (Youden square design).

five computers in which it is tested. The resulting design is a YSQ design, and one such assignment (chosen merely for illustration purposes) is shown in Table 5.10. Note that the brands are denoted by letters *A* through *E* in this example, and the access time is shown in the parentheses against the brand letter.

The ANOVA for these data calculated using Minitab is shown in Display 5.4. We see that, compared to the ANOVA for the BIB design in Display 5.3, the SS_e and the error d.f. are reduced by the respective amounts accounted for by the positions, but the MS_e is reduced as well (even though the position effects are nonsignificant). As a result, the brand effects are shown to be more significant ($p = 0.025$) despite the reduced error d.f.

The brand effect calculations are exactly the same as for the BIB design; the only quantities that change here are the MS_e and the error d.f. For example, the 95% Tukey SCI for $\alpha_B - \alpha_D$ equals

$$\left[\widehat{\alpha}_B - \widehat{\alpha}_D \pm q_{5,8,0.05} s\sqrt{\frac{c}{a\lambda}}\right] = \left[4.00 - (-4.40) \pm (4.886)\sqrt{7.339}\sqrt{\frac{4}{(5)(3)}}\right]$$
$$= [8.40 \pm 6.835] = [1.565, 15.235].$$

Thus the *B*-versus-*D* difference is significant at the simultaneous 5% level. ∎

5.4 LATIN SQUARE DESIGNS

An LSQ of size $a \geq 3$ is an $a \times a$ array of a Latin letters such that each letter occurs exactly once in each row and in each column. Typically, the letters represent the treatments, and the rows and columns represent the levels of two blocking factors. Note that all three factors must have the same number a of levels. If two of the factors are blocking factors, then the LSQ design is a BIB design with a^2 blocks, but only one experimental unit and hence only one treatment is assigned in each block. If all three factors are treatment factors, then it is a $(1/a)$th fraction of a complete a^3 factorial. Because of this high degree of fractionation, all interactions must be ignored and only an additive model can be entertained.

The balance property that each factor level occurs with every other factor level the same number of times makes the LSQ an **orthogonal design**. As a result, the treatment, row, and column factor effects can be estimated independently of each other. Figure 5.5 shows a graphic picture of a 5×5 LSQ for a field trial designed by R. A. Fisher to study the effect of exposure on five different types of evergreens.

Here is another example of an LSQ design (see also Exercise 5.21): Suppose that four brands of tire (A, B, C, D) are to be compared in terms of tire wear. Four tires of each brand and four cars are available. To eliminate the differences between the cars (and their drivers), one should block by cars, that is, allocate the four brands of tire to each car and mount them on four wheel positions (1: left front, 2: right front, 3: left rear, 4: right rear). If the tires are assigned at random to the wheel positions, then this would be an RB design. But there is also the position effect that should be balanced out. The LSQ design shown in Display 5.5 does this by assigning each brand of tire exactly once to each car and to each position.

A final example is a sudoku puzzle, which involves completing a 9×9 Latin square (in which numbers 1–9 are used instead of letters) where a few entries are provided from which the missing entries have to be found. There is an additional restriction that the nine 3×3 blocks in which the square is divided each contain all nine digits (which reduces the number of possible Latin squares tremendously, but it is still a humongous number; see Peterson 2005). Table 5.11 shows an example of a solution to a sudoku puzzle.

5.4.1 Choosing a Latin Square

How should we choose a particular LSQ to use in an experiment and where is randomization used? A systematic arrangement such as in Display 5.5 might be undesirable in some cases because the treatments are in the same order in each row, which may lead to bias. For instance, consider a taste-testing experiment with treatments as different drinks, rows as tasters, and columns as the order of tasting. Thus the first taster tastes the drinks in the order A, B, C, D; the second taster tastes the drinks in the order B, C, D, A; and so on. This design is not

Figure 5.5 A 5 × 5 Latin square for studying the effects of exposure on five types of evergreens. (*Source*: Fisher-Box, 1978, Plate 6.) Reprinted by permission of John Wiley & Sons, Inc.

balanced for aftertaste effects, which may confound the true taste differences between the drinks.

Fisher pointed out that to protect against any known or unknown systematic trends, selection of an LSQ should be done by randomization. But how many LSQs are there of each size, and how can we randomly choose one among them? To answer this question, we define a **standard Latin square**—an LSQ in which the letters in the first row and the first column are in alphabetical order. The LSQ in Display 5.5 is a standard LSQ obtained by writing the first row in alphabetical order and then writing each successive row by cyclically permuting the letters from the previous row. However, there can be more than one standard LSQ for

Wheel Position	Car			
	1	2	3	4
1	*A*	*B*	*C*	*D*
2	*B*	*C*	*D*	*A*
3	*C*	*D*	*A*	*B*
4	*D*	*A*	*B*	*C*

Display 5.5 Systematic Latin square.

Table 5.11 Solution to a Sudoku Puzzle

1 8 2	4 7 5	6 3 9
9 6 4	3 1 2	7 5 8
5 7 3	9 8 6	4 1 2
6 2 1	7 5 4	9 8 3
4 3 8	2 6 9	1 7 5
7 9 5	1 3 8	2 4 6
8 1 9	6 4 3	5 2 7
2 5 7	8 9 1	3 6 4
3 4 6	5 2 7	8 9 1

A B C D	*A B C D*	*A B C D*	*A B C D*
B C D A	*B A D C*	*B D A C*	*B A D C*
C D A B	*C D B A*	*C A D B*	*C D A B*
D A B C	*D C A B*	*D C B A*	*D C B A*

Display 5.6 Standard Latin squares for $a = 4$.

each value of a. For example, there are four standard LSQs for $a = 4$ which are listed in Display 5.6.

The numbers of distinct standard LSQs for $a = 3(1)7$ are listed in Table 5.12. By permuting the rows and columns of each standard LSQ, we can obtain distinct (nonstandard) LSQs. This is done as follows: Fix the first row and permute the remaining $a - 1$ rows; there are $(a - 1)!$ such permutations. Next permute all a columns; there are $a!$ such permutations. Thus from each standard LSQ we can generate $a!(a - 1)!$ distinct LSQs (including the original standard LSQ). The total number of LSQs for any a is therefore $a!(a - 1)!\times$ the number of standard LSQs for that a.

Table 5.12 Number of Standard Latin Squares of Different Sizes

3×3	4×4	5×5	6×6	7×7
1	4	56	9408	16,942,080

A random selection of an LSQ can be made by starting with a standard LSQ (tables of which are given in Fisher and Yates, 1953) and choosing a random permutation of $a-1$ rows and then a columns. Finally, the treatments should be assigned to the letters at random.

5.4.2 Model

Let y_{ijk} denote the observation on the ith treatment assigned to the jth row and the kth column $(1 \le i, j, k \le a)$. Note that all (i, j, k) combinations are not observed. The following additive model is postulated for y_{ijk}:

$$y_{ijk} = \mu + \alpha_i + \beta_j + \gamma_k + e_{ijk} \qquad (1 \le i, j, k \le a), \tag{5.20}$$

where μ is the overall mean, α_i is the ith treatment effect, β_j is the jth row effect, γ_k is the kth column effect, and the e_{ijk} are i.i.d. $N(0, \sigma^2)$ r.v.'s. Here,

$$\sum_{i=1}^{a} \alpha_i = \sum_{j=1}^{a} \beta_j = \sum_{k=1}^{a} \gamma_k = 0. \tag{5.21}$$

This model assumes that all interactions between treatments, rows, and columns are zero. The model contains $1 + 3(a-1) = 3a - 2$ independent parameters leaving $\nu = a^2 - (3a-2) = (a-1)(a-2)$ d.f. for estimating the error.

5.4.3 Statistical Analysis

5.4.3.1 Estimation

In what follows, all summations w.r.t. i, j, and k are over $1, 2, \ldots, a$ and are assumed to include only those (i, j, k) combinations that are actually observed. The LS estimators of μ, α_i, β_j, and γ_k can be shown to be (see Exercise 5.16)

$$\widehat{\mu} = \overline{y}_{...}, \qquad \widehat{\alpha}_i = \overline{y}_{i..} - \overline{y}_{...}, \qquad \widehat{\beta}_j = \overline{y}_{.j.} - \overline{y}_{...}, \qquad \widehat{\gamma}_k = \overline{y}_{..k} - \overline{y}_{...}. \tag{5.22}$$

The fitted values are given by

$$\widehat{y}_{ijk} = \widehat{\mu} + \widehat{\alpha}_i + \widehat{\beta}_j + \widehat{\gamma}_k = \overline{y}_{i..} + \overline{y}_{.j.} + \overline{y}_{..k} - 2\overline{y}_{...} \tag{5.23}$$

and the residuals are given by

$$\widehat{e}_{ijk} = y_{ijk} - \widehat{y}_{ijk} = y_{ijk} - \overline{y}_{i..} - \overline{y}_{.j.} - \overline{y}_{..k} + 2\overline{y}_{...}. \tag{5.24}$$

An unbiased estimator of σ^2 equals

$$s^2 = \mathrm{MS}_e = \frac{\mathrm{SS}_e}{(a-1)(a-2)} = \frac{\sum_{i,j,k} \widehat{e}_{ijk}^2}{(a-1)(a-2)}.$$

5.4.3.2 Analysis of Variance and Multiple Comparisons

Following the same steps as in Section 5.1.2.2, we get

$$\underbrace{\sum_{i,j,k}(y_{ijk} - \overline{y}_{...})^2}_{\mathrm{SS}_{\mathrm{tot}}} = \sum_{i,j,k} \widehat{\alpha}_i^2 + \sum_{i,j,k} \widehat{\beta}_j^2 + \sum_{i,j,k} \widehat{\gamma}_k^2 + \sum_{i,j,k} \widehat{e}_{ijk}^2$$

$$= \underbrace{a \sum_i \widehat{\alpha}_i^2}_{\mathrm{SS}_{\mathrm{trt}}} + \underbrace{a \sum_j \widehat{\beta}_j^2}_{\mathrm{SS}_{\mathrm{row}}} + \underbrace{a \sum_k \widehat{\gamma}_k^2}_{\mathrm{SS}_{\mathrm{col}}} + \underbrace{\sum_{i,j,k} \widehat{e}_{ijk}^2}_{\mathrm{SS}_e}. \quad (5.25)$$

The left-hand side of the above equation is the total sum of squares, the first term on the right-hand side is the sum of squares between the treatment means $\overline{y}_{i..}$, the second term is the sum of squares between the row means $\overline{y}_{.j.}$, the third term is the sum of squares between the column means $\overline{y}_{..k}$, and the last term is SS_e. Thus we get the following **ANOVA identity** for any LSQ design:

$$\mathrm{SS}_{\mathrm{tot}} = \mathrm{SS}_{\mathrm{trt}} + \mathrm{SS}_{\mathrm{row}} + \mathrm{SS}_{\mathrm{col}} + \mathrm{SS}_e.$$

The corresponding partitioning of the d.f. is

$$\underbrace{a^2 - 1}_{\text{total d.f.}} = \underbrace{a-1}_{\text{treatments d.f.}} + \underbrace{a-1}_{\text{row d.f.}} + \underbrace{a-1}_{\text{column d.f.}} + \underbrace{(a-1)(a-2)}_{\text{error d.f.}}.$$

The mean squares are computed in the usual way by dividing each sum of squares by its d.f.:

$$\mathrm{MS}_{\mathrm{trt}} = \frac{\mathrm{SS}_{\mathrm{trt}}}{a-1}, \qquad \mathrm{MS}_{\mathrm{row}} = \frac{\mathrm{SS}_{\mathrm{row}}}{a-1}, \qquad \mathrm{MS}_{\mathrm{col}} = \frac{\mathrm{SS}_{\mathrm{col}}}{a-1},$$

$$\mathrm{MS}_e = \frac{\mathrm{SS}_e}{(a-1)(a-2)}.$$

The ratio $F_{\mathrm{trt}} = \mathrm{MS}_{\mathrm{trt}}/\mathrm{MS}_e$ can be used to perform an F-test of the null hypothesis of no treatment effects, $H_{0A} : \alpha_1 = \cdots = \alpha_a = 0$: Reject H_{0A} at level α if

$$F_{\mathrm{trt}} = \frac{\mathrm{MS}_{\mathrm{trt}}}{\mathrm{MS}_e} > f_{a-1,\nu,\alpha}. \quad (5.26)$$

Table 5.13 ANOVA Table for Latin Square Design

Source	SS	d.f.	MS	F
Treatments	$\text{SS}_{\text{trt}} = a\sum(\bar{y}_{i..} - \bar{y}_{...})^2$	$a-1$	MS_{trt}	$\text{MS}_{\text{trt}}/\text{MS}_e$
Rows	$\text{SS}_{\text{row}} = a\sum(\bar{y}_{.j.} - \bar{y}_{...})^2$	$a-1$	MS_{row}	$\text{MS}_{\text{row}}/\text{MS}_e$
Columns	$\text{SS}_{\text{col}} = a\sum(\bar{y}_{..k} - \bar{y}_{...})^2$	$a-1$	MS_{col}	$\text{MS}_{\text{col}}/\text{MS}_e$
Error	$\text{SS}_e =$ by subtraction	$(a-1)(a-2)$	MS_e	
Total	$\text{SS}_{\text{tot}} = \sum\sum\sum(y_{ijk} - \bar{y}_{...})^2$	a^2-1		

The tests for the row and column effects are analogous. These tests can be derived by using the extra sum of squares method. Table 5.13 gives the ANOVA table.

The Tukey simultaneous $100(1-\alpha)\%$ SCIs for comparing any pair of treatments are given by

$$\alpha_i - \alpha_j \in \left[\bar{y}_{i..} - \bar{y}_{j..} \pm q_{a,\nu,\alpha}\frac{s}{\sqrt{a}}\right] \qquad (1 \le i < j \le a),$$

where $s = \sqrt{\text{MS}_e}$ and $\nu = (a-1)(a-2)$ is the error d.f.

Example 5.7 (Effect of Shelf Space on Hominy Sales: Analysis of Variance) This example is taken from Hoaglin, Mosteller, and Tukey (1991). The goal of the experiment was to examine the conjecture that the sales volume of a grocery item is related to the shelf space allocated to it. The experiment was conducted in six randomly selected supermarkets in Texas. The number of shelves was varied from 4 to 14 in increments of 2. The sales are known to depend on the particular week. Therefore data collection was spread over six weeks. Since a supermarket could only assign a fixed number of shelves to a given product during each week of the study, a 6×6 LSQ design was chosen with the number of shelves as the treatment factor and the supermarkets and weeks as blocking factors.[1] The sales data for one particular product, namely hominy, are shown in Table 5.14.

First we make a plot of the means for all three factors (called the **main-effects plot**) shown in Figure 5.6. We see that the shelf space (the treatment factor) has generally an increasing effect on the sales, although there is a drop-off at the maximum number of shelves. The supermarkets vary greatly in their sales (as might be expected), but there are not large differences between the weeks.

The ANOVA table for these data is shown in Display 5.7. We see that the effect of the number of shelves is nearly significant at the 0.01 level, but the differences between the supermarkets are highly significant. The differences

[1]In fact, supermarkets are experimental units to which the treatments of number of shelves are applied in a constrained random order. So this can be thought of as a repeated measures design. For the reasons mentioned in Section 5.1, we will treat supermarkets as blocks.

Table 5.14 Sales Volume for Hominy by Number of Shelves[a], Supermarket, and Week

	Week					
Supermarket	1	2	3	4	5	6
1	*F* (131)	*B* (126)	*C* (130)	*E* (188)	*A* (133)	*D* (154)
2	*E* (140)	*D* (150)	*A* (71)	*C* (111)	*B* (121)	*F* (127)
3	*A* (109)	*E* (134)	*D* (96)	*B* (123)	*F* (127)	*C* (84)
4	*C* (67)	*F* (94)	*B* (49)	*A* (93)	*D* (112)	*E* (161)
5	*B* (58)	*C* (71)	*F* (59)	*D* (62)	*E* (49)	*A* (27)
6	*D* (37)	*A* (36)	*E* (52)	*F* (58)	*C* (38)	*B* (51)

[a]Number of shelves are coded as follows: $A = 4$, $B = 6$, $C = 8$, $D = 10$, $E = 12$, $F = 14$.

Source: Hoaglin, Mosteller, and Tukey (1991). Reprinted by permission of John Wiley & Sons, Inc.

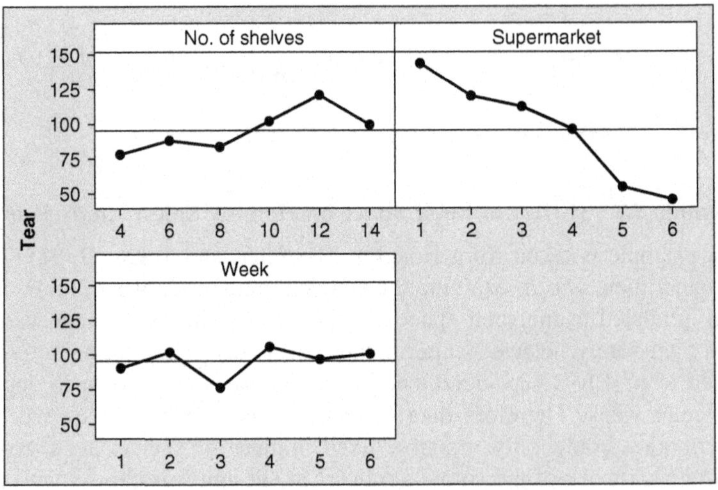

Figure 5.6 Main-effects plots for hominy sales data.

between the weeks are not significant. If we collapse the data over weeks, then this becomes a two-way design. An interaction plot between the number of shelves and supermarkets is shown in Figure 5.7. We see that this plot shows evidence of interaction, but without replicate observations over several weeks we cannot test its significance. ■

5.4.4 Crossover Designs

In some clinical trials, the same subjects are used as their own matches and each subject is given all treatments. Different subjects receive the treatments

```
Source      DF    Seq SS    Adj SS   Adj MS      F      P
Shelfno.     5    7130.9    7130.9   1426.2   4.01  0.011
Supmarket    5   44455.9   44455.9   8891.2  25.00  0.000
Week         5    3450.3    3450.3    690.1   1.94  0.132
Error       20    7113.7    7113.7    355.7
Total       35   62150.8
S = 18.8596    R-Sq = 88.55%    R-Sq(adj) = 79.97%
```

Display 5.7 Minitab ANOVA output for hominy sales data.

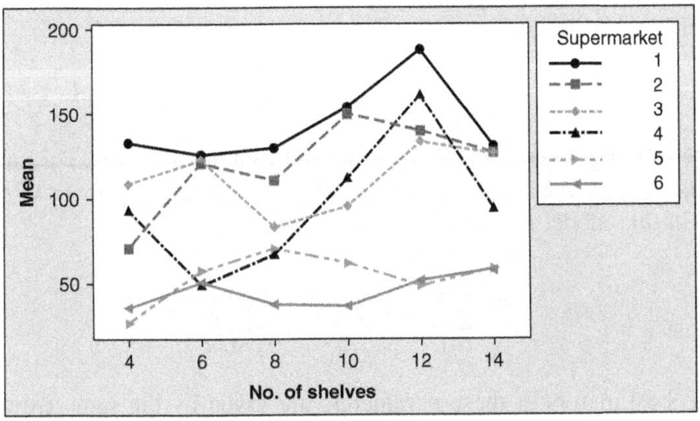

Figure 5.7 Interaction plot between number of shelves and supermarkets for hominy sales data.

in different but predetermined orders. In contrast to an RB design, the order is not randomized for each subject since the time period and the sequence of application of treatments are important factors. For example, if two drugs are compared by administering both to each subject on two separate occasions, then it may matter which drug is given first and which is given second. Crossover designs based on combinatorial ideas of LSQ designs are used to balance different sequences. Therefore we discuss these designs here as an application of LSQ designs, although they are, in fact, repeated measures designs since the same subjects are measured repeatedly under different treatments.

A crossover design takes the order of application of treatments into account as follows. Suppose $n = 2m$ subjects are available for comparing two drugs. Then the subjects are randomly divided into two groups of m each. In the first period group 1 is given drug A and group 2 is given drug B. In the second period the other drug is given to each group. The resulting design is orthogonal, and the effects of the drugs and the periods can be estimated independently of each other. The design can be represented as m replicates of 2×2 LSQs as shown below. In each LSQ the periods are represented by rows and the subjects by columns. If there is no period effect, then this design can be analyzed as a matched pairs design.

	Subject						
Period	1	2	3	4	...	$n-1$	n
1	A	B	A	B	...	A	B
2	B	A	B	A	...	B	A

More generally, suppose there are $a \geq 2$ treatments to be administered to each subject in a periods. Assume that there are $n = am$ subjects for some $m \geq 2$. Then the crossover design consists of m replicates of $a \times a$ LSQs. The additive model for this design is

$$y_{ijk} = \mu + \alpha_i + \beta_j + \gamma_k + e_{ijk} \qquad (1 \leq i, j \leq a, 1 \leq k \leq n), \tag{5.27}$$

where μ is the overall mean, α_i's are the treatment effects, β_j's are the period effects, γ_k's are the subject effects, and e_{ijk}'s are i.i.d. $N(0, \sigma^2)$ r.v.'s. The parameters in this model are subject to the constraints

$$\sum_{i=1}^{a} \alpha_i = \sum_{j=1}^{a} \beta_j = \sum_{k=1}^{n} \gamma_k = 0.$$

The LS estimators of these parameters are given by the same expressions as in (5.22). Table 5.15 gives the ANOVA.

Example 5.8 (Treating Insomnia: Crossover Design Clinical Trial)

Fisher and Wallenstein (1993) gave typical data from a clinical trial for comparing an active investigational hypnotic medication (A) for treatment of insomnia with a placebo (P). Twelve patients were randomly divided into two groups of six each. The first group took P in the first period (Monday night) and, after a two-day washout period, took A in the second period (Thursday night). The reason for the washout period is to mitigate any residual effect of the first treatment. The sequence was opposite for the second group of patients. The trial was conducted in a **double-blind** manner (i.e., neither the patient nor the evaluating clinician

Table 5.15 ANOVA Table for Crossover Design

Source	SS	d.f.	MS	F
Treatments	$SS_{trt} = n\sum(\bar{y}_{i..} - \bar{y}_{...})^2$	$a-1$	MS_{trt}	MS_{trt}/MS_e
Periods	$SS_{prd} = n\sum(\bar{y}_{.j.} - \bar{y}_{...})^2$	$a-1$	MS_{prd}	MS_{prd}/MS_e
Subjects	$SS_{sub} = a\sum(\bar{y}_{..k} - \bar{y}_{...})^2$	$n-1$	MS_{sub}	MS_{sub}/MS_e
Error	SS_e = by subtraction	$(a-1)(n-2)$	MS_e	
Total	$SS_{tot} = \sum(y_{ijk} - \bar{y}_{...})^2$	$an-1$		

Table 5.16 Duration of Sleep (Hrs.)

Group	Period	Treatment	Patient 1	2	3	4	5	6
1	1	*P*	7.00	6.75	5.90	7.20	6.25	4.85
	2	*A*	8.50	7.40	6.80	8.60	6.66	6.00
		$A - P$	1.50	0.65	0.90	1.40	0.41	1.15
2	1	*A*	6.50	6.00	8.17	7.00	8.12	7.35
	2	*P*	7.50	4.75	6.10	6.05	7.25	6.16
		$A - P$	−1.00	1.25	2.07	0.95	0.87	1.19

Source: Fisher and Wallenstein (1994, Table 10.1). Reprinted by permission of Taylor & Francis.

```
Source      DF    Seq SS     Adj SS     Adj MS       F      P
Patient     11   14.8832    14.8832     1.3530    4.44  0.013
Treatment    1    5.3582     5.3582     5.3582   17.56  0.002
Period       1    0.0193     0.0193     0.0193    0.06  0.807
Error       10    3.0506     3.0506     0.3051
Total       23   23.3112
```

Display 5.8 Minitab ANOVA output for insomnia trial using a crossover design including the period effect.

```
Source      DF    Seq SS     Adj SS     Adj MS       F      P
Patient     11   14.8832    14.8832     1.3530    4.85  0.007
Treatment    1    5.3582     5.3582     5.3582   19.20  0.001
Error       11    3.0699     3.0699     0.2791
Total       23   23.3112
```

Display 5.9 Minitab ANOVA output for insomnia trial using a crossover design excluding the period effect.

was aware of the treatment given to the patient). The variable of interest was the duration of sleep. The data are shown in Table 5.16.

The ANOVA obtained by fitting the model (5.27) is shown in Display 5.8. We see that the treatment and patient effects are highly significant ($p = 0.013$ and 0.002), but the period effect is nonsignificant ($p = 0.807$). Thus we may omit the period effect and refit the model. The resulting ANOVA is shown in Display 5.9. It should be noted that this latter analysis is equivalent to doing a simple paired t-test on the A-versus-P differences. The corresponding t-statistic equals 4.38, which is the square root of $F = 19.20$ shown in the ANOVA table. ■

In the above discussion we have not considered the possible **carryover effects** of treatments. For example, for studying three treatments, a crossover design consisting of m replicates of the standard 3×3 LSQ (obtained by cyclically permuting the letters A, B, and C) could be used as a building block and replicated

m times. However, in the resulting design A is always followed by B, B by C, and C by A (except when a treatment is the last one in the sequence, in which case there is no follow-up treatment). In this design, if the previously given treatment has a carryover effect in the following period, then these effects will be confounded with the main effects of the treatments. A washout period between treatments helps to eliminate carryover effects.

If the $a!$ sequences are replicated m times using a total of $n = (a!)m$ subjects, then an additive model for this design is

$$y_{hijk} = \mu + \alpha_i + \beta_j + \gamma_k + \delta_{hi} + e_{hijk} \qquad (1 \le h, i, j \le a, h \ne i, 1 \le k \le n),$$

where y_{hijk} is the response of the kth subject in the jth period given the ith treatment, which was preceded by the hth treatment; μ, α_i, and β_j have the same meanings as before; and δ_{hi} is the carryover effect of the hth treatment when it is followed by the ith treatment. If $j = 1$, then there is no preceding treatment and the term δ_{hi} is dropped from the model. For the analysis of this model, see John (1998, pp. 117–119).

5.4.5 Graeco–Latin Square Designs

A GLSQ of size $a \ge 4$ is an $a \times a$ array of a Latin letters and a Greek letters such that each letter appears exactly once in each row and each column; thus the two kinds of letters individually form LSQs. Furthermore, each Latin letter appears exactly once with each Greek letter. Such LSQs are said to be **mutually orthogonal**. Typically, the Latin and the Greek letters represent two treatment factors, and rows and columns represent two blocking factors. Note that all four factors have the same number a of levels. Thus there are a^4 possible factor-level combinations of which only a^2 are observed, that is, a GLSQ design of size a is a $1/a^2$ fraction of a complete factorial. Once again, only an additive model can be entertained for this design.

Because of the exact balance between the levels of the four factors, a GLSQ is an orthogonal design. Display 5.10 shows two GLSQ designs of sizes 4 and 5.

5.4.5.1 Model

The additive model for a GLSQ design is

$$y_{ijk\ell} = \mu + \alpha_i + \beta_j + \gamma_k + \delta_\ell + e_{ijk\ell} \qquad (1 \le i, j, k, \ell \le a),$$

Aα	Bβ	Cγ	Dδ
Bγ	Aδ	Dα	Cβ
Cδ	Dγ	Aβ	Bα
Dβ	Cα	Bδ	Aγ

Aα	Bβ	Cγ	Dδ	Eε
Bγ	Cδ	Dε	Eα	Aβ
Cε	Dα	Eβ	Aγ	Bδ
Dβ	Eγ	Aδ	Bε	Cα
Eδ	Aε	Bα	Cβ	Dγ

Display 5.10 A 4×4 and a 5×5 Graeco-Latin square.

where μ is the overall mean, α_i is the effect of the ith level of the Latin factor, β_j is the effect of the jth level of the Greek factor, γ_k is the effect of the kth row, and δ_ℓ is the effect of the ℓth column.[2] Note that not all (i, j, k, ℓ) combinations are observed. The α_i, β_j, γ_k, and δ_ℓ satisfy the constraints

$$\sum_{i=1}^{a} \alpha_i = \sum_{j=1}^{a} \beta_j = \sum_{k=1}^{a} \gamma_k = \sum_{\ell=1}^{a} \delta_\ell = 0.$$

Because of these constraints, only $a - 1$ of each of these parameters are independent. Thus there are a total of $1 + 4(a - 1) = 4a - 3$ independent parameters, leaving $a^2 - (4a - 3) = (a - 1)(a - 3)$ d.f. for estimating error. Hence we need $a > 3$ in order for $\nu > 0$.

5.4.5.2 *Statistical Analysis*

The LS estimators of the model parameters can be shown to be

$$\widehat{\mu} = \overline{y}_{....}, \qquad \widehat{\alpha}_i = \overline{y}_{i...} - \overline{y}_{....}, \qquad \widehat{\beta}_j = \overline{y}_{.j..} - \overline{y}_{....}, \qquad \widehat{\gamma}_k = \overline{y}_{..k.} - \overline{y}_{....},$$
$$\widehat{\delta}_\ell = \overline{y}_{...\ell} - \overline{y}_{....}.$$

The fitted values are given by

$$\widehat{y}_{ijk\ell} = \widehat{\mu} + \widehat{\alpha}_i + \widehat{\beta}_j + \widehat{\gamma}_k + \widehat{\delta}_\ell = \overline{y}_{i...} + \overline{y}_{.j..} + \overline{y}_{..k.} + \overline{y}_{...\ell} - 3\overline{y}_{....}$$

and the residuals are given by

$$\widehat{e}_{ijk\ell} = y_{ijk\ell} - \widehat{y}_{ijk\ell} = y_{ijk\ell} - \overline{y}_{i...} - \overline{y}_{.j..} - \overline{y}_{..k.} - \overline{y}_{...\ell} + 3\overline{y}_{....}.$$

An unbiased estimator of σ^2 then equals

$$s^2 = \text{MS}_e = \frac{\text{SS}_e}{(a-1)(a-3)} = \frac{\sum_{i,j,k,\ell} \widehat{e}^2_{ijk\ell}}{(a-1)(a-3)}.$$

The F-tests for all factors can be derived using the extra sum of squares method. The ANOVA including the F-statistics is given in Table 5.17.

The Tukey simultaneous $100(1 - \alpha)\%$ SCIs for comparing any pair of treatments are given by

$$\alpha_i - \alpha_j \in \left[\overline{y}_{i...} - \overline{y}_{j...} \pm q_{a,\nu,\alpha} \frac{s}{\sqrt{a}}\right] \qquad (1 \le i < j \le a),$$

where $s = \sqrt{\text{MS}_e}$ and $\nu = (a - 1)(a - 3)$ is the error d.f.

[2]The Greek letters for the parameters must not be confused with the letters for the levels of the Greek factor.

Table 5.17 ANOVA Table for Graeco–Latin Square Design

Source	SS	d.f.	MS	F
Latin treatments	$SS_{latin} = a\sum(\bar{y}_{i...} - \bar{y}_{....})^2$	$a-1$	MS_{latin}	MS_{latin}/MS_e
Greek treatments	$SS_{greek} = a\sum(\bar{y}_{.j..} - \bar{y}_{....})^2$	$a-1$	MS_{greek}	MS_{greek}/MS_e
Rows	$SS_{row} = a\sum(\bar{y}_{..k.} - \bar{y}_{....})^2$	$a-1$	MS_{row}	MS_{row}/MS_e
Columns	$SS_{col} = a\sum(\bar{y}_{...\ell} - \bar{y}_{....})^2$	$a-1$	MS_{col}	MS_{col}/MS_e
Error	$SS_e =$ by subtraction	$(a-1)(a-3)$	MS_e	
Total	$SS_{tot} = \sum\sum\sum\sum(y_{ijk\ell} - \bar{y}_{....})^2$	a^2-1		

Table 5.18 Amplitude of Signal (microvolts $\times 10^{-2}$)

	Operator			
Machine	1	2	3	4
1	$A\alpha$ (8)	$C\gamma$ (11)	$D\delta$ (2)	$B\beta$ (8)
2	$C\delta$ (7)	$A\beta$ (5)	$B\alpha$ (2)	$D\gamma$ (4)
3	$D\beta$ (3)	$B\delta$ (9)	$A\gamma$ (7)	$C\alpha$ (9)
4	$B\gamma$ (4)	$D\alpha$ (5)	$C\beta$ (9)	$A\delta$ (3)

Source: Nelson (1993, Table 4).

Example 5.9 (Comparing Substrates for Disk Drives: GLSQ Design)

This example is taken from Nelson (1993). It concerns the effects of four different substrates [aluminum (A), nickel-plated aluminum (B), type I glass (C), and type II glass (D)] on the amplitude of the signal obtained from a disk. To account for the differences between the machines, operators, and days of production, a 4×4 GLSQ design was employed with machines as columns, operators as rows, days of production as Greek letters, and substrates (the treatment factor) as Latin letters. The design and the coded data are shown in Table 5.18.

The ANOVA for these data calculated using Minitab is shown in Display 5.11. None of the factors have significant main effects. A possible reason for this is the smallness of the experiment with only three d.f. for error. ■

A Historical Note

Mutually orthogonal Latin squares (MOLSQs) were originally investigated by Euler. He was able to construct MOLSQs for $a = 4, 5, 7, 8, 9$ but not for $a = 6$ and 10. This led him to conjecture that MOLSQs do not exist for $a = 2 \pmod 4$. The conjecture was verified by Tarry (1901) for $a = 6$. It remained open until Bose, Shrikhande, and Parker (1960) showed the conjecture to be *false* for every $a = 2 \pmod 4 > 6$ and produced a pair of MOLSQs for $a = 10$ using computer search. ■

Source	DF	Seq SS	Adj SS	Adj MS	F	P
Substrat	3	61.500	61.500	20.500	2.86	0.206
Day	3	3.500	3.500	1.167	0.16	0.915
Machine	3	21.500	21.500	7.167	1.00	0.500
Operator	3	14.000	14.000	4.667	0.65	0.633
Error	3	21.500	21.500	7.167		
Total	15	122.000				

Display 5.11 Minitab ANOVA output for signal amplitude data.

5.5 CHAPTER NOTES

5.5.1 Restriction Error Model for Randomized Block Designs

To understand the restriction error model of Anderson and McLean (1974), let us go back to the CR design model (3.2) for a one-way layout and ask where the random error e_{ij} comes from. Well, it comes from the act of randomizing the experimental units to the treatments, which introduces a random error in the sense that, if another randomization is performed, different values of the y_{ij} will be observed because of the different values of the e_{ij}. In an RB design this random component is also present; in addition, because randomization is performed within each block separately, there will be a random component associated with each block regardless of whether the blocks are fixed or random. This random component arises because, as in the case of a CR design, if another randomization is performed within a given block, then different values of the y_{ij} will be observed in that block. Denote this random component by δ_j. Then the model (5.1) is modified to

$$y_{ij} = \mu + \alpha_i + \beta_j + \delta_j + e_{ij} \qquad (1 \le i \le a, 1 \le j \le b), \tag{5.28}$$

where the δ_j are are assumed to be i.i.d. N(0, σ_δ^2) r.v.'s. With this modification, the expected values of the various mean squares can be derived and are shown in Table 5.19. The SS and MS expressions are the same as those in Table 5.1 and hence are not shown in this table. There is no SS or MS available for the restriction error and its d.f. are zero. Furthermore, $Q_A = (a-1)^{-1}\sum_{i=1}^{a}\alpha_i^2$ and $Q_B = (b-1)^{-1}\sum_{j=1}^{b}\beta_j^2$.

Table 5.19 E(MS) Expressions for Restriction Error Model for RB Design

Source	d.f.	E(MS)
Treatments	$a-1$	$\sigma^2 + bQ_A$
Blocks	$b-1$	$\sigma^2 + a\sigma_\delta^2 + aQ_B$
Error	$(a-1)(b-1)$	σ^2
Restriction error	0	$\sigma^2 + a\sigma_\delta^2$

From this table one can see that under $H_{0A} : \alpha_1 = \cdots = \alpha_a = 0$ (which is equivalent to $Q_A = 0$) we have $E(MS_{trt}) = E(MS_e) = \sigma^2$; this fact along with other distributional assumptions enable us to use MS_{trt}/MS_e as the F-statistic. On the other hand, under $H_{0B} : \beta_1 = \cdots = \beta_b = 0$ (which is equivalent to $Q_B = 0$) MS_{blk} and MS_e do not have the same expectations and hence their ratio does not have a central F-distribution. The correct error term for testing H_{0B} is the mean square for restriction error; unfortunately, this term is not estimable.

Anderson and McLean (1974) point out that in many applications the investigator wants to determine if the blocks have been effective in reducing the error variance of intertreatment comparisons. The corresponding hypothesis can be stated as $H_0 : Q_B = 0$ and $\sigma_\delta^2 = 0$. This hypothesis can be tested using $F = MS_{blk}/MS_e$ in the usual manner.

5.5.2 Derivations of Formulas for BIB Design

Let us begin by considering the following simple BIB design as an example.

Block			
1	2	3	4
A	A	A	B
B	B	C	C
C	D	D	D

For this design, the model (5.1) can be written in matrix form $\boldsymbol{y} = \boldsymbol{X\theta} + \boldsymbol{e}$ of (2.27) (here we have used $\boldsymbol{\theta}$ in place of $\boldsymbol{\beta}$) as

$$\underbrace{\begin{bmatrix} y_{11} \\ y_{12} \\ y_{13} \\ y_{24} \\ y_{21} \\ y_{22} \\ y_{33} \\ y_{34} \\ y_{31} \\ y_{42} \\ y_{43} \\ y_{44} \end{bmatrix}}_{\boldsymbol{y}} = \underbrace{\begin{bmatrix} 1 & 1 & 0 & 0 & 0 & 1 & 0 & 0 & 0 \\ 1 & 1 & 0 & 0 & 0 & 0 & 1 & 0 & 0 \\ 1 & 1 & 0 & 0 & 0 & 0 & 0 & 1 & 0 \\ 1 & 0 & 1 & 0 & 0 & 0 & 0 & 0 & 1 \\ 1 & 0 & 1 & 0 & 0 & 1 & 0 & 0 & 0 \\ 1 & 0 & 1 & 0 & 0 & 0 & 1 & 0 & 0 \\ 1 & 0 & 0 & 1 & 0 & 0 & 0 & 1 & 0 \\ 1 & 0 & 0 & 1 & 0 & 0 & 0 & 0 & 1 \\ 1 & 0 & 0 & 1 & 0 & 1 & 0 & 0 & 0 \\ 1 & 0 & 0 & 0 & 1 & 0 & 1 & 0 & 0 \\ 1 & 0 & 0 & 0 & 1 & 0 & 0 & 1 & 0 \\ 1 & 0 & 0 & 0 & 1 & 0 & 0 & 0 & 1 \end{bmatrix}}_{\boldsymbol{X}} \underbrace{\begin{bmatrix} \mu \\ \alpha_1 \\ \alpha_2 \\ \alpha_3 \\ \alpha_4 \\ \beta_1 \\ \beta_2 \\ \beta_3 \\ \beta_4 \end{bmatrix}}_{\boldsymbol{\theta}} + \underbrace{\begin{bmatrix} e_{11} \\ e_{12} \\ e_{13} \\ e_{24} \\ e_{21} \\ e_{22} \\ e_{33} \\ e_{34} \\ e_{31} \\ e_{42} \\ e_{43} \\ e_{44} \end{bmatrix}}_{\boldsymbol{e}},$$

where treatments A, B, C, D are labeled as 1, 2, 3, 4, respectively. The LS estimator

$$\widehat{\boldsymbol{\theta}} = (\widehat{\mu}, \widehat{\alpha}_1, \widehat{\alpha}_2, \widehat{\alpha}_3, \widehat{\alpha}_4, \widehat{\beta}_1, \widehat{\beta}_2, \widehat{\beta}_3, \widehat{\beta}_4)'$$

is obtained by solving the normal equations $\boldsymbol{X}'\boldsymbol{X}\boldsymbol{\theta} = \boldsymbol{X}'\boldsymbol{y}$ [see (2.28)], which for the present example become

$$\begin{bmatrix} 12 & 3 & 3 & 3 & 3 & 3 & 3 & 3 & 3 \\ 3 & 3 & 0 & 0 & 0 & 1 & 1 & 1 & 0 \\ 3 & 0 & 3 & 0 & 0 & 1 & 1 & 0 & 1 \\ 3 & 0 & 0 & 3 & 0 & 1 & 0 & 1 & 1 \\ 3 & 0 & 0 & 0 & 3 & 0 & 1 & 1 & 1 \\ 3 & 1 & 1 & 1 & 0 & 3 & 0 & 0 & 0 \\ 3 & 1 & 1 & 0 & 1 & 0 & 3 & 0 & 0 \\ 3 & 1 & 0 & 1 & 1 & 0 & 0 & 3 & 0 \\ 3 & 0 & 1 & 1 & 1 & 0 & 0 & 0 & 3 \end{bmatrix} \begin{bmatrix} \mu \\ \alpha_1 \\ \alpha_2 \\ \alpha_3 \\ \alpha_4 \\ \beta_1 \\ \beta_2 \\ \beta_3 \\ \beta_4 \end{bmatrix} = \begin{bmatrix} y_{\cdot\cdot} \\ y_{1\cdot} \\ y_{2\cdot} \\ y_{3\cdot} \\ y_{4\cdot} \\ y_{\cdot 1} \\ y_{\cdot 2} \\ y_{\cdot 3} \\ y_{\cdot 4} \end{bmatrix} = \begin{bmatrix} y_{\cdot\cdot} \\ v_1 \\ v_2 \\ v_3 \\ v_4 \\ w_1 \\ w_2 \\ w_3 \\ w_4 \end{bmatrix}.$$

Generalizing from the above example, the normal equations for an arbitrary BIB design are

$$\begin{bmatrix} N & n & \cdots & n & c & \cdots & c \\ n & n & \cdots & \cdots & n_{11} & \cdots & n_{1b} \\ \vdots & & \ddots & & \vdots & \ddots & \vdots \\ n & & & n & n_{a1} & \cdots & n_{ab} \\ c & n_{11} & \cdots & n_{a1} & c & & \\ \vdots & \vdots & \ddots & \vdots & & \ddots & \\ c & n_{1b} & \cdots & n_{ab} & & & c \end{bmatrix} \begin{bmatrix} \mu \\ \alpha_1 \\ \vdots \\ \alpha_a \\ \beta_1 \\ \vdots \\ \beta_b \end{bmatrix} = \begin{bmatrix} y_{\cdot\cdot} \\ y_{1\cdot} \\ \vdots \\ y_{a\cdot} \\ y_{\cdot 1} \\ \vdots \\ y_{\cdot b} \end{bmatrix} = \begin{bmatrix} y_{\cdot\cdot} \\ v_1 \\ \vdots \\ v_a \\ w_1 \\ \vdots \\ w_b \end{bmatrix},$$

where $N = bc$ is the total number of observations.

Let $\boldsymbol{\alpha} = (\alpha_1, \ldots, \alpha_a)'$, $\boldsymbol{\beta} = (\beta_1, \ldots, \beta_b)'$, $\boldsymbol{N} = \{n_{ij}\}$ (called the **incidence matrix** of the design which is of dimension $a \times b$), $\boldsymbol{v} = (v_1, \ldots, v_a)'$, and $\boldsymbol{w} = (w_1, \ldots, w_b)'$. Also let $\mathbf{1}_a$ be a vector of 1's of dimension a and $\boldsymbol{I}_a$ be an $a \times a$ identity matrix. Then the normal equations can be written as

$$\begin{bmatrix} N & n\mathbf{1}_a' & c\mathbf{1}_b' \\ n\mathbf{1}_a & n\boldsymbol{I}_a & \boldsymbol{N} \\ c\mathbf{1}_b & \boldsymbol{N}' & c\boldsymbol{I}_b \end{bmatrix} \begin{bmatrix} \mu \\ \boldsymbol{\alpha} \\ \boldsymbol{\beta} \end{bmatrix} = \begin{bmatrix} y_{\cdot\cdot} \\ \boldsymbol{v} \\ \boldsymbol{w} \end{bmatrix}. \tag{5.29}$$

Written out in full form, the normal equations are

$$N\mu + n\mathbf{1}_a'\boldsymbol{\alpha} + c\mathbf{1}_b'\boldsymbol{\beta} = y_{\cdot\cdot},$$

$$n\mu\mathbf{1}_a + n\boldsymbol{\alpha} + \boldsymbol{N}\boldsymbol{\beta} = \boldsymbol{v},$$

$$c\mu\mathbf{1}_b + \boldsymbol{N}'\boldsymbol{\alpha} + c\boldsymbol{\beta} = \boldsymbol{w}.$$

Using the side constraints $\mathbf{1}_a'\boldsymbol{\alpha} = \sum_{i=1}^{a} \alpha_i = 0$ and $\mathbf{1}_b'\boldsymbol{\beta} = \sum_{j=1}^{b} \beta_j = 0$ in the first equation, we get

$$\widehat{\mu} = \frac{y_{..}}{N} = \overline{y}_{..}.$$

Substituting this value in the remaining equations and rearranging the terms, we get

$$\begin{aligned} n\boldsymbol{\alpha} + \boldsymbol{N}\boldsymbol{\beta} &= \boldsymbol{v} - n\overline{y}_{..}\mathbf{1}_a, \\ \boldsymbol{N}'\boldsymbol{\alpha} + c\boldsymbol{\beta} &= \boldsymbol{w} - c\overline{y}_{..}\mathbf{1}_b. \end{aligned} \tag{5.30}$$

Eliminating $\boldsymbol{\beta}$ between the two equations, we get

$$\left(n\boldsymbol{I}_a - \frac{1}{c}\boldsymbol{N}\boldsymbol{N}'\right)\boldsymbol{\alpha} = \boldsymbol{v} - \frac{1}{c}\boldsymbol{N}\boldsymbol{w} - n\overline{y}_{..}\mathbf{1}_a + \overline{y}_{..}\boldsymbol{N}\mathbf{1}_b.$$

However, since $\sum_{j=1}^{b} n_{ij} = n$ for all i, we have $\boldsymbol{N}\mathbf{1}_b = n\mathbf{1}_a$. So the last two terms of the above equation cancel, giving the result

$$\left(n\boldsymbol{I}_a - \frac{1}{c}\boldsymbol{N}\boldsymbol{N}'\right)\boldsymbol{\alpha} = \boldsymbol{v} - \frac{1}{c}\boldsymbol{N}\boldsymbol{w} = \boldsymbol{u}, \tag{5.31}$$

where $\boldsymbol{u} = (u_1, u_2, \ldots, u_a)'$ is the vector of adjusted treatment totals, u_i, defined in (5.14). Let $\boldsymbol{C} = n\boldsymbol{I}_a - (1/c)\boldsymbol{N}\boldsymbol{N}'$, which is referred to as the **coefficient matrix** or the **$\boldsymbol{C}$-matrix**. So Eq. (5.31) can be written as

$$\boldsymbol{C}\boldsymbol{\alpha} = \boldsymbol{u}. \tag{5.32}$$

Note that the ith diagonal entry of $\boldsymbol{N}\boldsymbol{N}'$ equals $\sum_{j=1}^{b} n_{ij}^2 = \sum_{j=1}^{b} n_{ij} = n$ (since $n_{ij} = 0$ or 1) and the (i, i')th off-diagonal entry equals $\sum_{j=1}^{b} n_{ij}n_{i'j} = \lambda$ (since this sum counts how many times n_{ij} and $n_{i'j}$ are both equal to 1, which is the number of concurrences). Therefore the entries of $\boldsymbol{C}$ are

$$c_{ij} = \begin{cases} n\left(1 - \dfrac{1}{c}\right) = \dfrac{n(c-1)}{c} & \text{if } i = j, \\ -\dfrac{\lambda}{c} & \text{if } i \neq j. \end{cases}$$

Each row of $\boldsymbol{C}$ sums to zero,

$$\sum_{j=1}^{b} c_{ij} = \frac{n(c-1)}{c} - \frac{\lambda(a-1)}{c} = 0,$$

by substituting for λ from (5.11). Therefore $\boldsymbol{C}$ is singular. But the system of equations (5.32) can be solved by utilizing the constraint $\sum_{i=1}^{a} \alpha_i = 0$ and hence $\sum_{j \neq i} \alpha_j = -\alpha_i$ as follows. The ith equation is

$$\frac{1}{c}\left[(c-1)n\alpha_i - \lambda \sum_{j \neq i} \alpha_j\right] = \frac{1}{c}[(c-1)n + \lambda]\alpha_i = \frac{a\lambda}{c}\alpha_i = u_i \qquad \text{and hence}$$

$$\widehat{\alpha}_i = \frac{c}{a\lambda} u_i. \tag{5.33}$$

Next, we derive the formulas for $\text{Var}(\widehat{\alpha}_i)$ and $\text{Var}(\widehat{\alpha}_i - \widehat{\alpha}_j)$. First,

$$\begin{aligned}\text{Var}(\widehat{\alpha}_i) &= \left(\frac{c}{a\lambda}\right)^2 \left[\text{Var}(v_i) + \frac{1}{c^2}\text{Var}(w_i^*) - \frac{2}{c}\text{Cov}(v_i, w_i^*)\right] \\ &= \left(\frac{c}{a\lambda}\right)^2 \left[n\sigma^2 + \frac{n}{c}\sigma^2 - \frac{2n}{c}\sigma^2\right] \\ &= \frac{c(a-1)\sigma^2}{a^2\lambda}.\end{aligned}$$

In the above, $\text{Var}(v_i) = n\sigma^2$ because v_i equals the sum of n independent y_{ij}'s, each having variance σ^2. Similarly, $\text{Var}(w_i^*) = cn\sigma^2$ because w_i^* is the sum of n block totals (in which the ith treatment occcurs) and each block has c observations. Finally, since $w_i^* = v_i +$ observations from treatments other than i, we have $\text{Cov}(v_i, w_i^*) = \text{Cov}(v_i, v_i) = \text{Var}(v_i) = n\sigma^2$.

To find $\text{Var}(\widehat{\alpha}_i - \widehat{\alpha}_j)$, we need $\text{Cov}(\widehat{\alpha}_i, \widehat{\alpha}_j)$. Since $\sum_{i=1}^{a} \widehat{\alpha}_i = 0$, we have

$$\begin{aligned}0 &= \text{Var}\left(\sum_{i=1}^{a} \widehat{\alpha}_i\right) \\ &= \sum_{i=1}^{a} \text{Var}(\widehat{\alpha}_i) + 2 \sum_{j=i+1}^{a} \sum_{i=1}^{a-1} \text{Cov}(\widehat{\alpha}_i, \widehat{\alpha}_j) \\ &= a\,\text{Var}(\widehat{\alpha}_i) + a(a-1)\,\text{Cov}(\widehat{\alpha}_i, \widehat{\alpha}_j).\end{aligned}$$

Hence

$$\text{Cov}(\widehat{\alpha}_i, \widehat{\alpha}_j) = -\frac{\text{Var}(\widehat{\alpha}_i)}{a-1} = -\frac{c}{a^2\lambda}\sigma^2.$$

Therefore

$$\begin{aligned}\text{Var}(\widehat{\alpha}_i - \widehat{\alpha}_j) &= \text{Var}(\widehat{\alpha}_i) + \text{Var}(\widehat{\alpha}_j) - 2\,\text{Cov}(\widehat{\alpha}_i, \widehat{\alpha}_j) \\ &= \frac{2c(a-1)\sigma^2}{a^2\lambda} + \frac{2c\sigma^2}{a^2\lambda} \\ &= \frac{2c\sigma^2}{a\lambda}.\end{aligned}$$

Lastly, we derive a formula for $SS_{trt(adj)}$ by the extra sum of squares method. The SS_e under the full model is given by

$$SS_e = (\boldsymbol{y} - \widehat{\boldsymbol{y}})'(\boldsymbol{y} - \widehat{\boldsymbol{y}}) = \boldsymbol{y}'\boldsymbol{y} - \widehat{\boldsymbol{y}}'\widehat{\boldsymbol{y}}.$$

Using (2.30), we can write $\widehat{\boldsymbol{y}}'\widehat{\boldsymbol{y}} = \widehat{\boldsymbol{\theta}}'\boldsymbol{X}'\boldsymbol{X}\widehat{\boldsymbol{\theta}} = \widehat{\boldsymbol{\theta}}'\boldsymbol{X}'\boldsymbol{y}$, where $\widehat{\boldsymbol{\theta}}' = (\widehat{\mu}, \widehat{\boldsymbol{\alpha}}', \widehat{\boldsymbol{\beta}}')$, the $\boldsymbol{X}$ matrix is as defined above, and from (5.29) we have $\boldsymbol{X}'\boldsymbol{y} = (y_{..}, \boldsymbol{v}', \boldsymbol{w}')'$. Therefore

$$SS_e = \boldsymbol{y}'\boldsymbol{y} - (\widehat{\mu}, \widehat{\boldsymbol{\alpha}}', \widehat{\boldsymbol{\beta}}')\begin{bmatrix} y_{..} \\ \boldsymbol{v} \\ \boldsymbol{w} \end{bmatrix} = \boldsymbol{y}'\boldsymbol{y} - \frac{y_{..}^2}{bc} - \boldsymbol{v}'\widehat{\boldsymbol{\alpha}} - \boldsymbol{w}'\widehat{\boldsymbol{\beta}}.$$

But from (5.30) we have $\widehat{\boldsymbol{\beta}} = (1/c)(\boldsymbol{w} - c\overline{y}_{..}\boldsymbol{1}_b - \boldsymbol{N}'\widehat{\boldsymbol{\alpha}})$, which when substituted in the above equation yields

$$\begin{aligned} SS_e &= \boldsymbol{y}'\boldsymbol{y} - \frac{y_{..}^2}{bc} - \boldsymbol{v}'\widehat{\boldsymbol{\alpha}} - \frac{1}{c}\boldsymbol{w}'(\boldsymbol{w} - c\overline{y}_{..}\boldsymbol{1}_b - \boldsymbol{N}'\widehat{\boldsymbol{\alpha}}) \\ &= \boldsymbol{y}'\boldsymbol{y} - \frac{y_{..}^2}{bc} - \frac{1}{c}\boldsymbol{w}'\boldsymbol{w} + \overline{y}_{..}\boldsymbol{w}'\boldsymbol{1}_b - \left(\boldsymbol{v} - \frac{1}{c}\boldsymbol{N}\boldsymbol{w}\right)'\widehat{\boldsymbol{\alpha}} \\ &= \boldsymbol{y}'\boldsymbol{y} - \frac{1}{c}\boldsymbol{w}'\boldsymbol{w} - \left(\boldsymbol{v} - \frac{1}{c}\boldsymbol{N}\boldsymbol{w}\right)'\widehat{\boldsymbol{\alpha}} \end{aligned}$$

since

$$\overline{y}_{..}\boldsymbol{w}'\boldsymbol{1}_b = \overline{y}_{..}\left(\sum_{j=1}^{b} w_j\right) = \frac{y_{..}}{bc}(y_{..}) = \frac{y_{..}^2}{bc}.$$

Next we compute SS_{e0}, the error sum of squares under the hypothesis of no treatment effects, $H_{0A} : \alpha_1 = \cdots = \alpha_a = 0$. Under H_{0A} the BIB design is equivalent to a balanced one-way layout with respect to b blocks, each with c observations. Therefore

$$SS_{e0} = \sum_{i=1}^{a}\sum_{j=1}^{b}(y_{ij} - \overline{y}_{\cdot j})^2 = \sum_{i=1}^{a}\sum_{j=1}^{b} y_{ij}^2 - \frac{1}{c}\sum_{j=1}^{b} y_{\cdot j}^2 = \boldsymbol{y}'\boldsymbol{y} - \frac{1}{c}\boldsymbol{w}'\boldsymbol{w}.$$

Hence,

$$\begin{aligned} SS_{trt(adj)} &= SS_{e0} - SS_e \\ &= \left(\boldsymbol{v} - \frac{1}{c}\boldsymbol{N}\boldsymbol{w}\right)'\widehat{\boldsymbol{\alpha}} \end{aligned}$$

$$
\begin{aligned}
&= \boldsymbol{u}'\widehat{\boldsymbol{\alpha}} \\
&= \frac{c}{a\lambda}\boldsymbol{u}'\boldsymbol{u} \\
&= \frac{c}{a\lambda}\sum_{i=1}^{a} u_i^2.
\end{aligned}
$$

5.6 CHAPTER SUMMARY

(a) If some noise factors are known and controllable or at least measurable before the start of an experiment, then the experimental units should be grouped into homogeneous blocks according to the values of those noise factors, known as blocking factors. Treatments are assigned at random within each block leading to an RB design. Since the variability in responses of the experimental units given different treatments caused by the blocking factors is thus removed (or at least reduced) by a combination of design followed by appropriate analysis, precision of comparisons between treatments is enhanced. Blocking is also used to evaluate the consistency of the effectiveness of the treatments across wide-ranging types of experimental units.

(b) The analysis of an RB design assumes an additive model because of the single replicate setup. This assumption can be checked by making an interaction plot. Inferences on the treatment effects can be carried out by regarding an RB design as a balanced one-way layout with b replicates per treatment except that MS_e is calculated differently and has $(a-1)(b-1)$ d.f. Block effects can also be tested similarly under the standard normal theory model; however, if the restriction error model of Anderson and McLean (1974) is assumed, then no valid F-test exists for the block effects.

(c) When there are more treatments than can be accommodated in each block, a balanced incomplete block (BIB) design is recommended. This design has the balance property that every treatment occurs with every other treatment in the same block the same number of times and every treatment is equally replicated. In spite of this balance property, the design is not orthogonal because not every treatment occurs in every block.

(d) YSQ and LSQ designs involve two blocking factors. A YSQ is used in the same situation as a BIB (i.e., the block size is less than the number of treatments), but the experimental units within each block are identified by their position, which is a second blocking factor (e.g., a day could be a block and morning, afternoon, and evening could be the positions). So the treatments are not randomized within each block but are assigned to achieve balance with respect to the positions. In the resulting design the treatments and the positions are orthogonal, but not the treatments and the blocks. The ANOVA accounts for the variability due to both the blocks and positions, thus enhancing the precision of between-treatment comparisons.

(e) An LSQ is an orthogonal design with the number of levels of the treatment factor and those of the two blocking factors being equal. As in other designs discussed in this chapter, an additive model is assumed in its analysis. It is used as a building block for crossover designs. A GLSQ consists of two mutually orthogonal LSQs. Thus it can be used to study two treatment factors with two blocking factors.

EXERCISES

Section 5.1 (Randomized-Block Designs)

Theoretical Exercises

5.1 Derive the F-test (5.8) of the null hypothesis of no treatment effects, $H_{0A}: \alpha_1 = \cdots = \alpha_a = 0$, using the extra sum of squares method.

5.2 (From Tamhane and Dunlop, 2000, Exercise 12.34. Reprinted by permission of Pearson Education, Inc.) For $a = 2$, show that the ANOVA F-test (5.8) for an RB design with $b = n$ is equivalent to the two-sided matched pairs t-test (2.13) by carrying out the following steps.

(a) Show that

$$\text{MS}_{\text{trt}} = \text{SS}_{\text{trt}} = \left(\frac{\overline{y}_{1\cdot} - \overline{y}_{2\cdot}}{\sqrt{2/n}} \right)^2 = \frac{n\overline{d}^2}{2},$$

where $\overline{d}$ is the mean of the $d_j = y_{1j} - y_{2j}$.

(b) Show that

$$(y_{1j} - \overline{y}_{1\cdot} - \overline{y}_{\cdot j} + \overline{y}_{\cdot\cdot})^2 = (y_{2j} - \overline{y}_{2\cdot} - \overline{y}_{\cdot j} + \overline{y}_{\cdot\cdot})^2 = \tfrac{1}{4}(d_j - \overline{d})^2.$$

Hence

$$\text{MS}_e = \frac{\sum_{i=1}^{a}\sum_{j=1}^{b}(y_{ij} - \overline{y}_{i\cdot} - \overline{y}_{\cdot j} + \overline{y}_{\cdot\cdot})^2}{(a-1)(b-1)} = \frac{\sum_{j=1}^{n}(d_j - \overline{d})^2}{2(n-1)} = \frac{s_d^2}{2},$$

where s_d is the standard deviation of the d_j's. [*Hint*: Write $\overline{y}_{\cdot j} = (y_{1j} + y_{2j})/2$ and $\overline{y}_{\cdot\cdot} = (\overline{y}_{1\cdot} + \overline{y}_{2\cdot})/2$.]

(c) Finally show that

$$F_{\text{trt}} = \frac{\text{MS}_{\text{trt}}}{\text{MS}_e} = \left(\frac{\bar{d}}{s_d/\sqrt{n}}\right)^2 = t^2.$$

Furthermore,

$$F_{\text{trt}} > f_{1,n-1,\alpha} \Longleftrightarrow |t| > t_{n-1,\alpha/2}$$

and hence the desired result follows.

Applied Exercises

5.3 The effect of temperature of the measuring bridge on the self-inductance of coils was studied in an experiment. Five coils and four temperatures were used. The temperatures for each coil were set in a random order. The data in Table 5.20 give the percentage deviations from a standard.

(a) Give the ANOVA table. What do you conclude about the differences between the temperatures and between the coils?

(b) Determine which temperatures differ from each other at $\alpha = 0.10$ using Tukey's procedure.

5.4 Table 5.21 gives data on the percentage drip loss in meat loaves (i.e., the amount of liquid that dripped out of a meat loaf during baking divided by the original weight of the loaf). The meat loaves were placed in eight different oven positions during baking. Three batches of loaves were baked in turn, each batch consisting of eight loaves. The loaves from each batch were randomly assigned to the eight positions. The goal of the experiment was to compare the eight oven positions in terms of drip loss.

(a) Give the ANOVA table. What do you conclude about the differences between the oven positions and between the batches?

Table 5.20 Effect of Temperature on Self-Inductance of Coils

	Coil				
Temperature (°C)	1	2	3	4	5
22	1.400	0.264	0.478	1.010	0.629
23	1.400	0.235	0.467	0.990	0.620
24	1.375	0.212	0.444	0.968	0.495
25	1.370	0.208	0.440	0.967	0.495

Source: Hamaker (1955, Table 7). Reprinted by permission of the International Biometric Society.

Table 5.21 Drip Loss from Meat Loaves

Oven Position	Batch 1	Batch 2	Batch 3
1	7.33	8.11	8.06
2	3.22	3.72	4.28
3	3.28	5.11	4.56
4	6.44	5.78	8.61
5	3.83	6.50	7.72
6	3.28	5.11	5.56
7	5.06	5.11	7.83
8	4.44	4.28	6.33

Source: Ryan and Joiner (2001, Table 10.18).

Table 5.22 Chronograph Velocity Measurements (m/sec)

	Chronograph		
Round	Fotobalk	Counter	Terma
1	793.8	794.6	793.2
2	793.1	793.9	793.3
3	792.4	793.2	792.6
4	794.0	794.0	793.8
5	791.4	792.2	791.6
6	792.4	793.1	791.6
7	791.7	792.4	791.6
8	792.3	792.8	792.4
9	789.6	790.2	788.5
10	794.4	795.0	794.7
11	790.9	791.6	791.3
12	793.5	793.8	793.5

Source: Grubbs (1973, Table I).

(b) Determine which positions differ significantly from each other at $\alpha = 0.05$ using Tukey's procedure.

5.5 Table 5.22 gives data on velocity measurements (in meters per second) made by three chronographs on 12 successive rounds from a 155-mm gun. Fotobalk and Counter were standard instruments, while Terma was a test instrument.

(a) Give the ANOVA table. What do you conclude about the differences between the chronographs?

(b) Compare the test instrument Terma with both standards using two-sided Dunnett's 95% SCIs. Does Terma differ significantly from either standard?

Section 5.2 (Balanced Incomplete Block Designs)

Theoretical Exercises

5.6 Compare an RB design with a BIB design in terms of $\text{Var}(\widehat{\alpha}_i - \widehat{\alpha}_j)$ when both use the same number of observations N. Which design estimates $\alpha_i - \alpha_j$ with greater precision? Why? Assume the same experimental variance σ^2 for both designs. (*Hint*: Note that if a is the number of treatments, b_{RB} and b_{BIB} are the number of blocks of the RB and BIB designs, respectively, and c is the block size of the BIB design, then $N = ab_{\text{RB}} = cb_{\text{BIB}}$.)

5.7 The previous exercise assumed the same σ^2 for both designs. The blocks for a BIB design are likely to be more homogeneous since the block size is smaller than that for an RB design. Therefore we may assume $\sigma^2_{\text{BIB}} < \sigma^2_{\text{RB}}$. Find a condition on the ratio $\sigma^2_{\text{BIB}}/\sigma^2_{\text{RB}}$ so that a BIB design will estimate $\alpha_i - \alpha_j$ with greater precision than an RB design.

5.8 Suppose $a = 8$ treatments are to be tested in blocks of size $c = 4$. A BIB design with $b = 14$ blocks and $\lambda = 3$ is shown below:

$$\begin{bmatrix} A & A & A & A & A & A & A & E & B & C & B & C & B & B \\ B & C & B & D & B & D & C & F & D & D & C & D & C & D \\ C & F & E & E & G & F & E & G & E & G & F & E & E & F \\ D & H & F & H & H & G & G & H & G & H & G & F & G & H \end{bmatrix}$$

Show that for $a = 8$ and $c = 4$, this is a BIB design with the minimum possible number of blocks. How many blocks are required for a BIB design obtained by taking all possible combinations of four treatments from eight treatments?

5.9 Suppose there are four treatments, A, B, C, D, which are to be compared in blocks of size 2. A BIB design would require six blocks, but suppose that only four blocks are available. The following **lattice square design**, which consists of two 2×2 squares, may be used in this case. Note that in this design A and D and B and C do not occur in the same block but all other pairs of treatments do occur in the same block.

Block			
1	2	3	4
A	C	A	B
B	D	C	D

(a) Show that the differences A versus D and B versus C can be estimated but with twice the variance (under the assumption of independent homoscedastic errors) than the common variance of the estimators of the other pairwise differences.
(b) How can lattice square designs be constructed if the number of treatments is a perfect square, say $a = c^2$, and there are $b = 2c$ blocks, each of size c? In particular, write the lattice square design for $a = 9$.

Applied Exercises

5.10 Natrella (1963) described an experiment to compare four geometric shapes of four resistors on the current noise. Only three resistors could be mounted on a plate at a time, so a BIB design was necessary. Table 5.23 gives the logarithms of the current noise for each combination of the shape and plate that was observed.

(a) Prepare an ANOVA table. Are there significant differences among the shapes using $\alpha = 0.05$?
(b) Make pairwise comparisons between shapes using the Tukey procedure to determine which (if any) shapes are significantly different from each other using $\alpha = 0.05$.

5.11 Refer to the previous exercise.

(a) Calculate the interblock estimates of the shape effects.
(b) Calculate the pooled estimates of the shape effects. How do they compare with the intrablock estimates?
(c) Compare the standard errors of the pooled estimates with those of the intrablock estimates.

5.12 (From Kuehl, 2000). In a process called vinylation used in industrial polymerization, acetylene under high pressure is used to convert methyl glucoside to monovinyl isomers. An experiment was done to study the

Table 5.23 Logarithms of Current Noise

	Plates			
Shapes	1	2	3	4
A	1.11	1.70	1.66	
B	—	1.22	1.11	1.22
C	0.95	—	1.52	1.54
D	0.82	0.97	—	1.18

Source: Natrella (1963, pp. 13–14).

Table 5.24 Percent Conversion of Methyl Glucoside under Five Pressures

	Pressure				
Run	250	325	400	475	550
1	16	18	—	32	
2	19	—	—	46	45
3	—	26	39	—	61
4	—	—	21	35	55
5	—	19	—	47	48
6	20	—	33	31	
7	13	13	34		
8	21	—	30	—	52
9	24	10	—	—	50
10	—	24	31	37	

Source: Kuehl (2000, Table 9.2).

effect of five selected reaction pressures on the percent conversion. Since only three high-pressure chambers were available, only three pressures could be tested in one experimental run. The chemists planning the experiment thought that there could be a large run-to-run variation because of raw material changes and changes in setups. The data for the experiment are shown in Table 5.24. In the following tests, use $\alpha = 0.05$.

(a) Prepare an ANOVA table. Are there significant differences between the pressures?
(b) Which pressures are significantly different from each other using the Tukey procedure?
(c) Is there significant run-to-run variation as expected by the chemists?

Section 5.3 (Youden Square Designs)

Theoretical Exercises

5.13 Explain why in a YSQ design treatments and blocks are not orthogonal to each other but the positions are orthogonal to both.

5.14 A taste-testing experiment is to be designed using four subjects to compare four wines. Each subject can test only three wines in a seating. To balance the effect of the order of testing, each wine must be tested exactly once in each spot, first, second, or third, in the sequence.

(a) Propose a suitable YSQ design.
(b) Write the model matrix of the YSQ design using the model (5.19).

Table 5.25 Weight Loss (mg) of Rubber Samples Tested in Martindale Wear Tester

	Run				
Position	1	2	3	4	5
α	A (268)	E (251)	D (265)	C (256)	B (240)
β	B (249)	A (233)	E (249)	D (250)	C (250)
γ	C (280)	B (231)	A (254)	E (270)	D (248)
δ	D (271)	C (291)	B (314)	A (281)	E (289)

Source: Davies (1963, Table 6.43).

(c) Show using the model matrix that the wines and subjects are not orthogonal to each other but the order of testing is orthogonal to both.

Applied Exercises

5.15 Davies (1963) described an experiment to compare abrasion resistance of five samples of rubber (labeled A through E) on a Martindale wear tester. Only four samples can be mounted on the tester at a time and the mounting positions (labeled α through δ) are a potential blocking factor. So a YSQ design shown in Table 5.25 was used; the table also gives the wear data.

(a) Prepare an ANOVA table and test if there are significant differences between the samples and positions using $\alpha = 0.05$.

(b) Make pairwise comparisons using the Tukey procedure between the positions using $\alpha = 0.05$. Which positions are significantly different from each other?

Section 5.4 (Latin Square Designs)

Theoretical Exercises

5.16 Show that the LS estimators of μ, α_i, β_j, and γ_k of the LSQ model (5.20) and (5.21) are given by

$$\widehat{\mu} = \overline{y}_{...}, \qquad \widehat{\alpha}_i = \overline{y}_{i..} - \overline{y}_{...}, \qquad \widehat{\beta}_j = \overline{y}_{.j.} - \overline{y}_{...}, \qquad \widehat{\gamma}_k = \overline{y}_{..k} - \overline{y}_{...}.$$

[*Hint*: Minimize

$$\sum_{i,j,k} [y_{ijk} - (\mu + \alpha_i + \beta_j + \gamma_k)]^2 I_{ijk},$$

where $I_{ijk} = 1$ if the factor-level combination (i, j, k) is observed and zero otherwise.]

5.17 Consider the 3×3 LSQ

$$\begin{bmatrix} A & B & C \\ B & C & A \\ C & A & B \end{bmatrix}.$$

(a) Verify that the model matrix for this design is as shown in Display 5.12.

(b) Using the relations $\sum_{i=1}^{3} \alpha_i = \sum_{j=1}^{3} \beta_j = \sum_{k=1}^{3} \gamma_k = 0$, reduce the above matrix to a full column rank matrix by eliminating α_3, β_3, and γ_3.

(c) Show that the design is orthogonal by showing that the columns in each group, $\{\mu\}, \{\alpha_1, \alpha_2\}, \{\beta_1, \beta_2\}, \{\gamma_1, \gamma_2\}$, are mutually orthogonal.

5.18 The ANOVA given in Table 5.15 for a crossover design assumes $n = ma$ different subjects and thus the column levels are different in the m replicated $a \times a$ LSQs. Give the modified ANOVA table if the row levels and column levels are the same in all LSQs. An example of such a design is the following: Four machines are to be compared in terms of their yield of good items produced, each machine can be operated by one of four workers, and a day is divided into four time periods. Thus for each day we can use a 4×4 LSQ with machines as letters, time periods as rows, and workers as columns. Now suppose that this experiment is replicated over five days (with possibly a different 4×4 LSQ each day).

5.19 A crossover trial is designed to study three drugs. Besides the effects of the drugs, patients, and periods, the effect of the sequence in which the drugs are taken is also of interest. Therefore all six sequences,

$$\begin{array}{cccccccccc} \mu & \alpha_1 & \alpha_2 & \alpha_3 & \beta_1 & \beta_2 & \beta_3 & \gamma_1 & \gamma_2 & \gamma_3 \end{array}$$
$$\begin{bmatrix} 1 & 1 & 0 & 0 & 1 & 0 & 0 & 1 & 0 & 0 \\ 1 & 0 & 1 & 0 & 1 & 0 & 0 & 0 & 1 & 0 \\ 1 & 0 & 0 & 1 & 1 & 0 & 0 & 0 & 0 & 1 \\ 1 & 0 & 1 & 0 & 0 & 1 & 0 & 1 & 0 & 0 \\ 1 & 0 & 0 & 1 & 0 & 1 & 0 & 0 & 1 & 0 \\ 1 & 1 & 0 & 0 & 0 & 1 & 0 & 0 & 0 & 1 \\ 1 & 0 & 0 & 1 & 0 & 0 & 1 & 1 & 0 & 0 \\ 1 & 1 & 0 & 0 & 0 & 0 & 1 & 0 & 1 & 0 \\ 1 & 0 & 1 & 0 & 0 & 0 & 1 & 0 & 0 & 1 \end{bmatrix}$$

Display 5.12 Model matrix for LSQ design from Exercise 5.17.

$\{ABC, ACB, BAC, BCA, CAB, CBA\}$, are observed with three patients assigned to each sequence for a total of 18 patients.

(a) Diagram the design as two LSQs (each replicated three times) with rows as periods and columns as patients.

(b) The model used to analyze the data is

$$y_{ijk\ell} = \mu + \alpha_i + \beta_j + \gamma_k + \delta_\ell + e_{ijk\ell},$$

where α_i's are the drug effects, β_j's are the period effects, γ_k's are the patient effects, δ_ℓ's are the sequence effects, and $e_{ijk\ell}$'s are i.i.d. $N(0, \sigma^2)$ random errors. The various factor effects sum to zero when summed over the levels of the corresponding factors. Give the ANOVA table with the formulas for the various sums of squares (SS_e is obtained by subtraction).

5.20 Suppose that the number of treatments to be compared in a crossover study is more than the number of periods for which the study can be conducted. Let a denote the number of treatments and c denote the number of periods with $c < a$.

(a) Suppose $a = 4$ and $c = 3$. Construct a crossover design by first constructing a BIB design and then expanding each of its columns (i.e., block) into an LSQ. This basic design can be replicated as many times as necessary. How many subjects are needed for this basic design?

(b) Explain how the ideas used to construct the design in Part (a) can be generalized to arbitrary a and c with $c < a$.

Applied Exercises

5.21 (From Hicks and Turner, 1999). Suppose that four brands of tires, labeled A, B, C, D, are to be compared in terms of tire wear after 20,000 miles of road driving. Four tires of each brand and four cars are available. The CR design, in which the 16 tires are randomly mounted on the 16 positions of the four cars, is not satisfactory since the car and the mounting position effects are not balanced out. An RB design, in which the four tires of the four brands are randomly mounted in four positions, balances out the differences between the cars with cars as blocks. The best design that also balances the differences between the four positions is a 4×4 LSQ shown in Table 5.26. The numbers in the parentheses are the amounts of tread wear in mils (thousandths of an inch).

(a) Assuming that the data are obtained using a CR design, compute the ANOVA table. Are the brand differences significant at the 0.05 level?

Table 5.26 Tire Wear (mils)

Wheel Position	Car 1	Car 2	Car 3	Car 4
1	*C* (12)	*D* (11)	*A* (13)	*B* (8)
2	*B* (14)	*C* (12)	*D* (11)	*A* (13)
3	*A* (17)	*B* (14)	*C* (10)	*D* (9)
4	*D* (13)	*A* (14)	*B* (13)	*C* (9)

Source: Hicks and Turner (1999, Chapter 4).

(b) Assuming that the data are obtained using an RB design with cars as blocks, compute the ANOVA table. Are the brand differences significant at the 0.05 level?

(c) Compute the ANOVA table by treating the design as an LSQ. Are the brand differences significant at the 0.05 level? Comment on the results obtained from the three analyses.

5.22 Halvorsen (1991) describes the following experiment to test which yogurt used as a starter makes the best home-made sour cream. The experimenter tested five brands of yogurts: Continental (A), Healthglow (B), Maya (C), Nancy's (D), and Yami (E). The sour cream was made by dividing a batch of about two pints of sweet cream in five containers, adding two teaspoons of different brands of yogurt in each container, mixing the contents thoroughly, and placing the containers in five positions of a thermostatically controlled yogurt maker for 22–23 hours. The experiment was conducted over five days using a different batch of sweet cream each day. The response variable was the acidity of the sour cream (the higher the acidity, the more the bacterial growth and the better the end product), which was measured by the amount of titrant (in milliliters) required to neutralize the acidity of 9 ml of sour cream. The data are shown in Table 5.27. Compute the ANOVA table. Which factors show significant effects on the acidity at the 0.05 level?

5.23 A tangram puzzle is an ancient Chinese game, similar to the jigsaw puzzle, in which a contestant chooses a geometric pattern and then uses seven pieces to assemble that pattern in quickest time. Among simple geometric shapes, there are two types: concave and convex. A concave pattern has at least two possible solutions, while a convex pattern has only one solution. Therefore concave patterns are generally easier to solve than convex patterns. Two students (Jun Liu and Erin Sahlsteen) in my Spring 2004 Statistics 351 class at Northwestern University performed an experiment as part of their project to compare the assembly times for a concave and a convex pattern (shown in Figure 5.8). To account for the learning effect, they used a crossover experiment in which 20 volunteer subjects were randomly divided into two groups of 10 each. One group assembled the

Table 5.27 Acidity of Sour Cream Made Using Different Starters, Batches of Sweet Cream, and Positions in a Yogurt Maker

Sweet Cream	Position				
Batch	1	2	3	4	5
1	C (8.04)	B (6.61)	D (11.99)	E (7.78)	A (8.40)
2	D (9.58)	E (6.58)	A (6.66)	C (5.34)	B (7.92)
3	E (7.98)	C (7.98)	B (8.98)	A (7.94)	D (11.32)
4	B (9.74)	A (9.46)	C (9.14)	D (12.00)	E (9.32)
5	A (9.66)	D (11.28)	E (8.04)	B (8.12)	C (6.72)

Source: Halvorsen (1991, Table 6-13). Reprinted by permission of John Wiley & Sons, Inc.

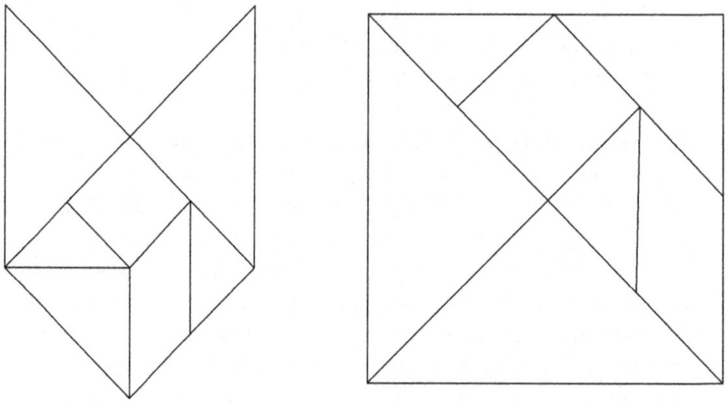

Figure 5.8 Concave and convex Tangram patterns used in crossover experiment.

concave pattern first and then the convex pattern, while the other group followed the opposite order. The response variable was the assembly time in seconds. The data are shown in Table 5.28. If a subject could not assemble the pattern in 20 min, then the trial was terminated. There was only one such censored observation, but you may ignore its censored nature.

(a) Prepare an ANOVA table. Which effects are significant at the 0.01 level? Does the concave pattern take significantly less time on average to assemble than the convex pattern?

(b) Check the data for any model violations.

5.24 (Reprinted by permission of University of Adelaide Library.) Solve the following puzzle from Fisher's (1935, pp. 89–90) book by constructing three suitable 4×4 MOLSQs: Sixteen passengers on a liner discover that they are an exceptionally representative body. Four are Englishmen, four are Scots, four are Irish, and four are Welsh. There are also four each of

Table 5.28 Time (sec) Taken to Complete a Tangram Puzzle

			Subject									
Group	Period	Pattern	1	2	3	4	5	6	7	8	9	10
1	1	Concave	178	223	661	641	176	153	153	166	199	408
	2	Convex	564	348	261	625	900	63	763	643	492	549
2	1	Convex	514	251	287	732	1200	714	793	600	463	539
	2	Concave	298	236	111	274	443	64	844	129	533	379

four different ages, 35, 45, 55, and 65, and no two of the same age are of same nationality. By profession also four are lawyers, four soldiers, four doctors, and four clergymen, and no two of the same profession are of the same age or of the same nationality.

It appears, also, that four are bachelors, four married, four widowed, and four divorced, and no two of the same marital status are of the same profession or the same age or the same nationality. Finally, four are conservatives, four liberals, four socialists, and four fascists, and no two of the same political sympathies are of the same marital status, or the same profession or the same age or the same nationality

Three of the fascists are known to be unmarried English lawyer of 65, a married Scots soldier of 55, and a widowed Irish doctor of 45. It is then easy to specify the remaining fascist.

It is further given that the Irish socialist is 35, the conservative of 45 is a Scotsman, and the Englishman of 55 is a clergyman. What do you know of the Welsh lawyer?

CHAPTER 6

General Factorial Experiments

Most experiments are run to study more than one factor. For example, agricultural trials are conducted to evaluate how the yield of a crop depends on several factors such as different strains of the seed and the type of fertilizer used in combination. In electroplating, the thickness of the deposited layer depends on the concentration of the electrolytic solution and the voltage applied across the electrodes. One strategy that investigators sometimes follow when experimentation can be done sequentially is to vary one factor at a time, keeping the others fixed, choosing the best level of that factor and keeping it fixed in subsequent runs. However, as discussed in Section 6.1 below, this is not a good strategy as it misses information on the interactions between factors and therefore may reach wrong conclusions. Another strategy is the so-called "Goldilocks design," which consists of trying different treatment combinations in a haphazard manner until the result is just right. Obviously, this is not a good strategy either. A **factorial experiment** that studies all factor-level combinations simultaneously offers a systematic and statistically valid strategy to find the best result. In this chapter we discuss factorial experiments in which all factors are fixed and crossed; experiments with random and/or nested factors are covered in Chapters 11 and 12. (See Section 1.3 for an explanation of the terminology.)

The outline of the chapter is as follows. Section 6.1 compares factorial versus one-factor-at-a-time strategies of experimentation. Section 6.2 studies equireplicated two-factor experiments (also referred to as **balanced two-way layouts**). These designs are orthogonal, which makes their analysis and interpretation particularly simple. Section 6.3 extends this to unbalanced two-way layouts, which are nonorthogonal designs. Section 6.4 gives derivations of some mathematical results and complementary material on balanced three-way layouts. Section 6.5 gives a summary of the chapter.

Statistical Analysis of Designed Experiments: Theory and Applications By Ajit C. Tamhane

6.1 FACTORIAL VERSUS ONE-FACTOR-AT-A-TIME EXPERIMENTS

When there are two or more factors (say, A, B, and C), one experimental strategy is to fix the levels of all except one factor (say, A). The levels of B and C may be fixed at their currently used values or the values that the experimenter thinks would give good results. Then A is varied to find its "best" level. Next, this best level of A is fixed and one of the remaining factors (say, B) is varied to find its best level keeping C at its original level. This process is then repeated with C. This is called a **one-factor-at-a-time experiment**. It is a common misunderstanding that this is the scientific method of experimentation.

There is a serious flaw in this strategy: The best level of A that is found can only be claimed to be the best for the chosen levels of B and C. There is no guarantee that it is the overall best level because had we chosen different levels of B and C, it is possible that we could have gotten a better response with a different level of A. This would happen if two or more factors act jointly, that is, if the change in the mean response with the change in the level of A depends on the levels of B and/or C.

To avoid this flaw, Fisher (1935) suggested that all factor-level combinations should be observed, which is a factorial experiment. Not only does this enable us to find the best overall combination, but also we can learn how the factors influence the response through their individual average effects (called the **main effects**) and through their joint effects (called the **interactions**). The two types of experiments are contrasted in the following example.

Example 6.1 (Improving the Yield of a Spring Manufacturing Process: One-Factor-at-a-Time Experiment)

Box and Bisgaard (1988) a discuss an example in which a process for making springs is to be designed to maximize the yield of springs that are free of cracks. Three treatment factors are thought to be important. Factor A is the carbon content (low level: 0.5%, high level: 0.7%), factor B is the quench oil temperature (low level: 70°F, high level: 120°F), and factor C is the temperature to which steel is heated before quenching (low level: 1450°F, high level: 1600°F). Figure 6.1 shows the percent yields of good springs that would be observed if the runs were made at different factor settings. In this figure -1 represents the low level and $+1$ represents the high level of each factor.

Let us suppose that we start with all three factors at low levels. The percent yield is 67%. Now we keep B and C fixed at low levels and change A to high level. The percent yield is 61%. Since the percent yield decreased with this change, we stay with the low level of A. Next B is changed to high level, keeping A and C at low levels. The percent yield is 59%, so again we stay with the low level of B. Finally, C is changed to high level, keeping A and B at low levels. The percent yield is 79%, so the high level of C is better. We would then conclude that A low, B low, C high is the best treatment combination. However, this strategy missed not only the overall best yield of 90% at A low, B high, C

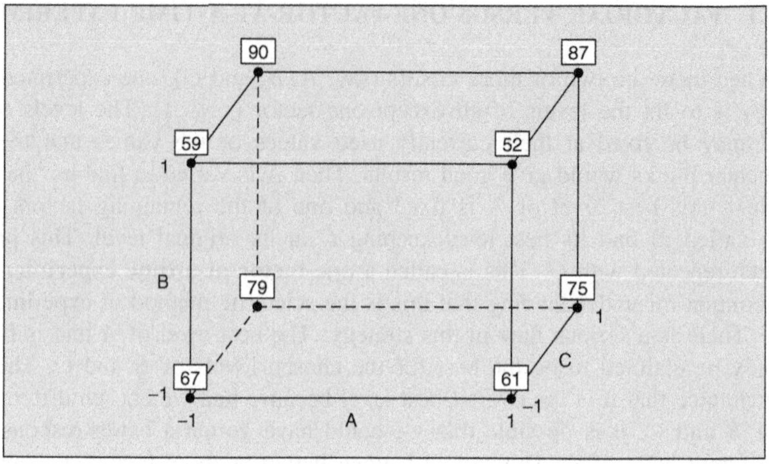

Figure 6.1 Percent yields of springs at different factor settings.

high but also the second best yield of 87% at A high, B high, C high. A factorial experiment would have identified these superior settings. ■

Factorial experiments not only yield more accurate conclusions by taking into account interactions but also use the data more efficiently. To illustrate this point, suppose we have two factors, A and B. Denote the four observations at the low–low, high–low, low–high, and high–high levels of the two factors by y_{11}, y_{21}, y_{12}, and y_{22}, respectively. In a factorial experiment all four observations are taken, while in a one-factor-at-a-time experiment only three observations are taken, say, y_{11}, y_{21}, and y_{22}. The factorial experiment estimates the main effects of A and B by taking the average of the changes when each factor is changed from its low to high level, keeping the level of the other factor fixed (see Section 7.1.1 for a more detailed discussion of estimation of factorial effects). Thus the estimators of the main effects of A and B are

$$\widehat{A} = \tfrac{1}{2}[(y_{22} - y_{12}) + (y_{21} - y_{11})] \qquad \text{and} \qquad \widehat{B} = \tfrac{1}{2}[(y_{22} - y_{21}) + (y_{12} - y_{11})].$$

In addition, the interaction between A and B can be estimated by the difference in the two changes:

$$\widehat{AB} = \tfrac{1}{2}[(y_{22} - y_{12}) - (y_{21} - y_{11})] = \tfrac{1}{2}[(y_{22} - y_{21}) - (y_{12} - y_{11})].$$

If the y_{ij} are independent, each with variance σ^2, then the variance of all three estimators, $\widehat{A}$, $\widehat{B}$, and $\widehat{AB}$, is σ^2. On the other hand, the one-factor-at-a-time experiment estimates the main effects by

$$\widehat{A} = y_{21} - y_{11} \qquad \text{and} \qquad \widehat{B} = y_{22} - y_{12}$$

but cannot estimate the interaction at all. Furthermore, these estimators are biased if interaction is present and each has variance $2\sigma^2$. We see that the former experiment is twice as efficient (yields estimators with half the variance) since effectively it uses each observation twice. If the differences in the sample sizes of the two experiments (3 for the one-factor-at-a-time and 4 for the factorial) are taken into account, then the factorial experiment is 50% more efficient.

6.2 BALANCED TWO-WAY LAYOUTS

Consider two fixed factors, A with $a \geq 2$ levels and B with $b \geq 2$ levels, so that there are ab treatment combinations. We refer to A as the **row factor** and B as the **column factor**. Suppose that a CR experiment is conducted in which $n_{ij} \geq 2$ experimental units are randomly assigned to the (i, j)th **treatment combination**. A two-factor experiment is called **balanced** if $n_{ij} = n$ for all treatment combinations (i, j) and **unbalanced** if the n_{ij} are not all equal. In this section we consider balanced two-factor experiments using a CR design. More generally, there could be some blocking factors and one of the designs from Chapter 5, for example, an RB design, may be used.

Let y_{ijk} denote the kth observation on the (i, j)th treatment combination $(1 \leq i \leq a, 1 \leq j \leq b, 1 \leq k \leq n)$ and let $N = abn$ be the total number of observations. A tabular layout of the data is shown in Table 6.1. The **cells** of the table correspond to different treatment combinations. We refer to the quantities associated with the treatment combinations as the respective cell quantities, for example, cell means, cell standard deviations, and cell sample sizes or **cell frequencies**.

6.2.1 Summary Statistics and Graphical Plots

The first step in data analysis is to explore patterns in the data by computing summary statistics and making graphical plots. This is best illustrated through an example.

Table 6.1 Data from Two-Factor Experiment

Factor A	Factor B Levels					
Levels	1	2	$\cdots$	j	$\cdots$	b
1	$y_{111}, \ldots, y_{11n}$	$y_{121}, \ldots, y_{12n}$	$\cdots$	$y_{1j1}, \ldots, y_{1jn}$	$\cdots$	$y_{1b1}, \ldots, y_{1bn}$
2	$y_{211}, \ldots, y_{21n}$	$y_{221}, \ldots, y_{22n}$	$\cdots$	$y_{2j1}, \ldots, y_{2jn}$	$\cdots$	$y_{2b1}, \ldots, y_{2bn}$
$\vdots$	$\vdots$	$\vdots$	$\vdots$	$\vdots$	$\vdots$	$\vdots$
i	$y_{i11}, \ldots, y_{i1n}$	$y_{i21}, \ldots, y_{i2n}$	$\cdots$	$y_{ij1}, \ldots, y_{ijn}$	$\cdots$	$y_{ib1}, \ldots, y_{ibn}$
$\vdots$	$\vdots$	$\vdots$	$\vdots$	$\vdots$	$\vdots$	$\vdots$
a	$y_{a11}, \ldots, y_{a1n}$	$y_{a21}, \ldots, y_{a2n}$	$\cdots$	$y_{aj1}, \ldots, y_{ajn}$	$\cdots$	$y_{ab1}, \ldots, y_{abn}$

Source: Tamhane and Dunlop (2000, Table 13.1). Reprinted by permission of Pearson Education, Inc.

Example 6.2 (Bonding Strength of Capacitors: Summary Statistics and Graphical Plots)

(From Tamhane and Dunlop, 2000, Example 13.1. Reprinted by permission of Pearson Education, Inc.) Capacitors are bonded onto a circuit board used in high-voltage electronic equipment. Engineers designed and carried out an experiment to study how the mechanical bonding strength of capacitors depends on the type of substrate and the type of bonding material. There were three types of substrates (factor A): aluminum oxide (Al_2O_3) with bracket, Al_2O_3 without bracket, and beryllium oxide (BeO) without bracket. Four types of bonding material (factor B) were used: epoxy I, epoxy II, solder I, and solder II. Thus, factor B is a nested combination of two factors, namely material, epoxy or solder, and type within material, I or II. Factor A is also a partially nested combination since Al_2O_3 is used with or without bracket.

Sixteen substrate boards of each type were randomly assigned to 16 capacitors. Each of the four bonding materials was then used to bond four capacitors to their substrates, and their bonding strengths were measured after curing the bond for a specified time. Thus this is a CR design with two treatment factors. Note that if only four substrate boards were used and four capacitors were mounted on the same board, then the resulting design would have been a split-plot design whose analysis is discussed in Chapter 13. For proprietary reasons, simulated data shown in Table 6.2 are used.

The first step in summarizing such two-way cross-classified data is to compute the cell means and standard deviations. Usually, the row and column means are also computed to see how the mean response varies with respect to each factor. These calculations are shown in Table 6.3. How the two factors affect the bonding strength can be best seen by plotting the row and column means (called the **main-effect plots**), and the individual cell means (called the **interaction plot**). These two plots are shown in Figure 6.2.

Let us look at the interaction plot first since, if interactions are present, then the main effects are not very meaningful. We see that all three substrates yield about equally stronger bonds with solder II than with solder I. On the other hand, epoxy II yields a stronger bond with Al_2O_3 (with or without bracket),

Table 6.2 Bonding Strength of Capacitors

Substrate	Bonding Material (Factor B)			
(Factor A)	Epoxy I	Epoxy II	Solder I	Solder II
Al_2O_3 without bracket	1.51, 1.96, 1.83, 1.98	2.62, 2.82, 2.69, 2.93	2.96, 2.82, 3.11, 3.11	3.67, 3.40, 3.25, 2.90
Al_2O_3 with bracket	1.63, 1.80, 1.92, 1.71	3.12, 2.94, 3.23, 2.99	2.91, 2.93, 3.01, 2.93	3.48, 3.51, 3.24, 3.45
BeO	3.04, 3.16, 3.09, 3.50	1.91, 2.11, 1.78, 2.25	3.04, 2.91, 2.48, 2.83	3.47, 3.42, 3.31, 3.76

Source: Tamhane and Dunlop (2000, Table 13.2). Reprinted by permission of Pearson Education, Inc.

Table 6.3 Cell Sample Means and Standard Deviations[a] for Capacitor Bonding Strength Data

Substrate	Bonding Material (Factor B)				Row
(Factor A)	Epoxy I	Epoxy II	Solder I	Solder II	Mean
Al_2O_3 without bracket	1.820 (0.217)	2.765 (0.138)	3.000 (0.139)	3.305 (0.321)	2.723
Al_2O_3 with bracket	1.765 (0.124)	3.070 (0.131)	2.945 (0.044)	3.420 (0.122)	2.800
BeO	3.198 (0.208)	2.013 (0.209)	2.815 (0.240)	3.490 (0.192)	2.879
Column mean	2.261	2.616	2.920	3.405	2.800

[a]The cell standard deviations are shown in parentheses.

Source: Tamhane and Dunlop (2000, Table 13.3). Reprinted by permission of Pearson Education, Inc.

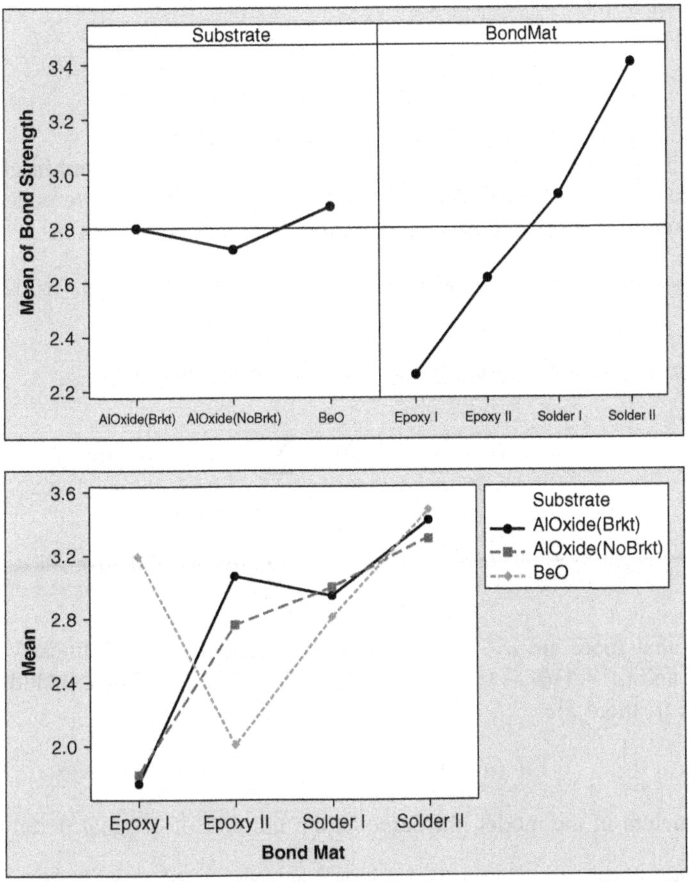

Figure 6.2 Main-effect (top) and interaction plots (bottom) for capacitor bonding strength data.

while epoxy I yields a stronger bond with BeO.[1] This means that there is no interaction between the type of solder and the type of substrate but there is interaction between the type of epoxy and the type of substrate. Because of the strong interaction between the type of substrate and epoxy, we conclude that the main effects are not very meaningful, especially the main effects of substrates, which are shown to be negligible by the flat plot (the average bonding strengths of substrates are similar). Tests for significance of these effects are given in Section 6.2.3.2. ■

6.2.2 Model

We assume that the y_{ijk} are independent $N(\mu_{ij}, \sigma^2)$ r.v.'s. or equivalently we can write

$$y_{ijk} = \mu_{ij} + e_{ijk} \qquad (1 \le i \le a, 1 \le j \le b, 1 \le k \le n), \tag{6.1}$$

where the e_{ijk} are i.i.d. $N(0, \sigma^2)$ r.v.'s. Using the "dot notation" defined in Chapter 5, let

$$\mu = \overline{\mu}_{..}, \qquad \alpha_i = \overline{\mu}_{i.} - \overline{\mu}_{..}, \qquad \beta_j = \overline{\mu}_{.j} - \overline{\mu}_{..},$$
$$(\alpha\beta)_{ij} = \mu_{ij} - \overline{\mu}_{i.} - \overline{\mu}_{.j} + \overline{\mu}_{..}. \tag{6.2}$$

Here μ is the **grand mean**, the α_i are called the A or **row main effects**, the β_j are called the B or **column main effects**, and the $(\alpha\beta)_{ij}$ are called the AB or **row–column interactions**. Then model (6.1) can be written as

$$y_{ijk} = \mu + \alpha_i + \beta_j + (\alpha\beta)_{ij} + e_{ijk} \qquad (1 \le i \le a, 1 \le j \le b, 1 \le k \le n), \tag{6.3}$$

where the parameters satisfy the following side conditions:

$$\sum_{i=1}^{a} \alpha_i = 0, \quad \sum_{j=1}^{b} \beta_j = 0, \quad \sum_{i=1}^{a} (\alpha\beta)_{ij} = 0 \quad \text{for all } j,$$
$$\sum_{j=1}^{b} (\alpha\beta)_{ij} = 0 \quad \text{for all } i. \tag{6.4}$$

Note that there are $a - 1$ linearly independent α_i's, $b - 1$ linearly independent β_j's, and $(a-1)(b-1)$ linearly independent $(\alpha\beta)_{ij}$'s. Thus, including the grand mean μ, there are

$$1 + (a-1) + (b-1) + (a-1)(b-1) = ab$$

parameters in the model, the same as the number of original parameters, μ_{ij}'s.

[1]In practice, one may want to check whether the epoxy I and epoxy II data are inadvertently exchanged for BeO.

6.2.3 Statistical Analysis

6.2.3.1 Estimation

The LS estimators of the parameters in model (6.3) are the "natural" estimators given by

$$\widehat{\mu} = \overline{y}_{\dots}, \qquad \widehat{\alpha}_i = \overline{y}_{i\cdot\cdot} - \overline{y}_{\dots}, \qquad \widehat{\beta}_j = \overline{y}_{\cdot j\cdot} - \overline{y}_{\dots},$$
$$\widehat{(\alpha\beta)}_{ij} = \overline{y}_{ij\cdot} - \overline{y}_{i\cdot\cdot} - \overline{y}_{\cdot j\cdot} + \overline{y}_{\dots}. \tag{6.5}$$

The derivation of these estimators is given in Section 6.4.1. It is easy to see that the $\widehat{\alpha}_i$, $\widehat{\beta}_j$, and $\widehat{(\alpha\beta)}_{ij}$ satisfy the linear constraints given in (6.4) on the corresponding parameters. The fitted values and residuals equal

$$\widehat{y}_{ijk} = \widehat{\mu} + \widehat{\alpha}_i + \widehat{\beta}_j + \widehat{(\alpha\beta)}_{ij} = \overline{y}_{ij\cdot} \qquad \text{and} \qquad \widehat{e}_{ijk} = y_{ijk} - \overline{y}_{ij\cdot}. \tag{6.6}$$

The sample variance for the (i, j)th cell equals

$$s_{ij}^2 = \frac{\sum_{k=1}^{n}(y_{ijk} - \overline{y}_{ij\cdot})^2}{n-1} = \frac{\sum_{k=1}^{n}\widehat{e}_{ijk}^2}{n-1},$$

which is an estimator of the error variance σ^2 for that cell. Since the model assumes homoscedasticity, that is, σ^2 is common for all cells, the cell sample variances can be averaged together to give a pooled estimator of σ^2. The weights used for averaging are the d.f., which are equal to $n-1$ for all cells. The resulting pooled estimator of σ^2 is

$$s^2 = \frac{\sum_{i=1}^{a}\sum_{j=1}^{b}(n-1)s_{ij}^2}{ab(n-1)} = \frac{\sum_{i=1}^{a}\sum_{j=1}^{b}\sum_{k=1}^{n}\widehat{e}_{ijk}^2}{N-ab}. \tag{6.7}$$

This pooled estimator is based on $\nu = ab(n-1) = N - ab$ d.f.

6.2.3.2 Analysis of Variance

The ANOVA of the balanced two-way layout data is based on the following identity:

$$\underbrace{\sum_{i=1}^{a}\sum_{j=1}^{b}\sum_{k=1}^{n}(y_{ijk} - \overline{y}_{\dots})^2}_{\text{SS}_{\text{tot}}} = \sum_{i=1}^{a}\sum_{j=1}^{b}\sum_{k=1}^{n}(\overline{y}_{i\cdot\cdot} - \overline{y}_{\dots})^2 + \sum_{i=1}^{a}\sum_{j=1}^{b}\sum_{k=1}^{n}(\overline{y}_{\cdot j\cdot} - \overline{y}_{\dots})^2$$
$$+ \sum_{i=1}^{a}\sum_{j=1}^{b}\sum_{k=1}^{n}(\overline{y}_{ij\cdot} - \overline{y}_{i\cdot\cdot} - \overline{y}_{\cdot j\cdot} + \overline{y}_{\dots})^2 + \sum_{i=1}^{a}\sum_{j=1}^{b}\sum_{k=1}^{n}(y_{ijk} - \overline{y}_{ij\cdot})^2$$
$$= \underbrace{bn\sum_{i=1}^{a}\widehat{\alpha}_i^2}_{\text{SS}_A} + \underbrace{an\sum_{j=1}^{b}\widehat{\beta}_j^2}_{\text{SS}_B} + \underbrace{n\sum_{i=1}^{a}\sum_{j=1}^{b}\widehat{(\alpha\beta)}_{ij}^2}_{\text{SS}_{AB}} + \underbrace{\sum_{i=1}^{a}\sum_{j=1}^{b}\sum_{k=1}^{n}\widehat{e}_{ijk}^2}_{\text{SS}_e}. \tag{6.8}$$

There is a corresponding decomposition of their d.f.:

$$\underbrace{N-1}_{\text{total d.f.}} = \underbrace{a-1}_{\text{A main–effects d.f.}} + \underbrace{b-1}_{\text{B main–effects d.f.}} + \underbrace{(a-1)(b-1)}_{\text{AB interactions d.f.}} + \underbrace{N-ab}_{\text{error d.f.}}.$$

The mean squares are defined in the usual way as the sums of squares divided by their d.f.:

$$\text{MS}_A = \frac{SS_A}{a-1},\ MS_B = \frac{\text{SS}_B}{b-1},\ \text{MS}_{AB} = \frac{\text{SS}_{AB}}{(a-1)(b-1)},\ \text{MS}_e = \frac{\text{SS}_e}{N-ab}.$$

Note that MS_e equals the pooled estimator s^2 of σ^2 given by (6.7). The above mean squares are used to test the null hypotheses on the A and B main effects (H_{0A} and H_{0B}) and AB interactions (H_{0AB}), stated as follows:

$$H_{0A} : \alpha_1 = \alpha_2 = \cdots = \alpha_a = 0, \qquad H_{0B} : \beta_1 = \beta_2 = \cdots = \beta_b = 0,$$
$$H_{0AB} : (\alpha\beta)_{ij} = 0 \text{ for all } i, j \qquad .$$

The F-statistics for testing the significance of these hypotheses are computed by taking the ratios of the respective mean squares to MS_e. The d.f. of the F-statistics correspond to the d.f. of the mean squares which form the ratios. Specifically, the α-level tests of H_{0A}, H_{0B}, and H_{0AB} are as follows:

$$\text{Reject } H_{0A} \text{ if } F_A = \frac{\text{MS}_A}{\text{MS}_e} > f_{a-1,\nu,\alpha},$$

$$\text{Reject } H_{0B} \text{ if } F_B = \frac{\text{MS}_B}{\text{MS}_e} > f_{b-1,\nu,\alpha},$$

$$\text{Reject } H_{0AB} \text{ if } F_{AB} = \frac{\text{MS}_{AB}}{\text{MS}_e} > f_{(a-1)(b-1),\nu,\alpha}.$$

These tests can be derived by using the extra sum of squares method; the derivation of the test for H_{0A} is given in Section 6.4.2. The above calculations are summarized in the ANOVA shown in Table 6.4.

Just as one should examine the interaction plot before the main-effects plots, similarly one should test the interaction hypothesis H_{0AB} before the main-effects

Table 6.4 ANOVA Table for Balanced Two-Way Layout

Source	SS	d.f.	MS	F
A main effect	$\text{SS}_A = bn\sum(\bar{y}_{i..} - \bar{y}_{...})^2$	$a-1$	MS_A	MS_A/MS_e
B main effect	$\text{SS}_B = an\sum(\bar{y}_{.j.} - \bar{y}_{...})^2$	$b-1$	MS_B	MS_B/MS_e
AB interaction	$\text{SS}_{AB} = n\sum\sum(\bar{y}_{ij.} - \bar{y}_{i..} - \bar{y}_{.j.} + \bar{y}_{...})^2$	$(a-1)(b-1)$	MS_{AB}	$\text{MS}_{AB}/\text{MS}_e$
Error	$\text{SS}_e = \sum\sum\sum(y_{ijk} - \bar{y}_{ij.})^2$	$N-ab$	MS_e	
Total	$\text{SS}_{tot} = \sum\sum\sum(y_{ijk} - \bar{y}_{...})^2$	$N-1$		

hypotheses H_{0A} and H_{0B}. The reason is that if there are significant interactions, tests for the significance of the main effects become moot. After all, the main effects concern the mean responses for the levels of one factor averaged over the levels of the other factor. If interactions are present, it does not follow that the given factor has no effect even if these mean responses are the same. It means that the effect of the factor depends on the level of the other factor. This point is well illustrated by the bonding strength data from Example 6.2. We saw in Figure 6.2 that the main effect of the substrate is negligible, and yet the effect of substrate (Al_2O_3 vs. BeO) is different for epoxy I and epoxy II.

Example 6.3 (Bonding Strength of Capacitors: Analysis of Variance)

The ANOVA is shown in Display 6.1. We see that the main effect of bonding material and the interaction between the bonding material and substrate are both highly significant, but the main effect of substrate is not significant at the 0.05 level. These results are in agreement with the conclusions reached from Figure 6.2. ■

6.2.3.3 Multiple Comparisons

Pairwise comparisons between the rows and the columns can be conducted by using the exact Tukey procedure (see Section 4.2.2). Note that each row mean is based on *bn* observations and each column mean is based on *an* observations. The resulting $100(1-\alpha)\%$ SCIs on the pairwise differences between the row main effects are

$$\alpha_i - \alpha_j \in \left[\bar{y}_{i..} - \bar{y}_{j..} \pm q_{a,\nu,\alpha}\frac{s}{\sqrt{bn}}\right] \qquad (1 \le i < j \le a),$$

where $s = \sqrt{\text{MS}_e}$. Similarly, $100(1-\alpha)\%$ SCIs on the pairwise differences between the column main effects are

$$\beta_i - \beta_j \in \left[\bar{y}_{.i.} - \bar{y}_{.j.} \pm q_{b,\nu,\alpha}\frac{s}{\sqrt{an}}\right] \qquad (1 \le i < j \le b).$$

More general comparisons involve **contrasts** among the cell means of the form $\sum_{i=1}^{a}\sum_{j=1}^{b} c_{ij}\mu_{ij}$, where the c_{ij} are the contrast coefficients that sum to zero.

Source	DF	SS	MS	F	P
Substrate	2	0.1953	0.0977	2.80	0.074
BondMat	3	8.4605	2.8202	80.77	0.000
Substrate*BondMat	6	7.5869	1.2645	36.21	0.000
Error	36	1.2570	0.0349		
Total	47	17.4998			

S = 0.186864 R-Sq = 92.82% R-Sq(adj) = 90.62%

Display 6.1 Analysis of variance of bonding strength data.

A special type of contrast with $\sum_{i=1}^{a} c_{ij} = 0$ for all j and $\sum_{j=1}^{b} c_{ij} = 0$ for all i is called an **interaction contrast** because it satisfies

$$\sum_{i=1}^{a}\sum_{j=1}^{b} c_{ij}\mu_{ij} = \sum_{i=1}^{a}\sum_{j=1}^{b} c_{ij}(\alpha\beta)_{ij}.$$

If such contrasts are selected by data snooping, then the Scheffé procedure (see Section 4.4.2) should be used to test for their significance. Since the dimension of the space of all interaction contrast vectors $\{c_{ij}\}$ (written as $a \times b$ arrays) is $(a-1)(b-1)$, the d.f. for the F critical value are $(a-1)(b-1)$ and $\nu = N - ab$. Thus $100(1-\alpha)\%$ SCIs on all interaction contrasts are given by

$$\sum_{i=1}^{a}\sum_{j=1}^{b} c_{ij}\mu_{ij} \in \left[\sum_{i=1}^{a}\sum_{j=1}^{b} c_{ij}\bar{y}_{ij\cdot} \pm \sqrt{(a-1)(b-1)f_{(a-1)(b-1),\nu,\alpha}}\sqrt{\frac{s^2}{n}\sum_{i=1}^{a}\sum_{j=1}^{b} c_{ij}^2}\right].$$

The following example illustrates the use of these intervals.

Example 6.4 (Bonding Strength of Capacitors: Interaction Contrasts)

(From Tamhane and Dunlop, 2000, Example 13.2. Reprinted by permission of Pearson Education, Inc.) From Figure 6.2 we have seen that there is an evident interaction between the type of substrate and the type of epoxy but not between the type of substrate and the type of solder. Let us check this visual impression by a formal significance test using the Scheffé procedure. If this impression is confirmed, as shown below, then we would want to report the results separately for epoxy and solder.

To test the type of epoxy versus the type of substrate interaction, we compare the difference between the average bonding strengths of epoxy I for the two varieties of the Al_2O_3 substrates and that for BeO with the same difference for epoxy II. This contrast equals

$$\left(\frac{\bar{y}_{11\cdot}+\bar{y}_{21\cdot}}{2} - \bar{y}_{31\cdot}\right) - \left(\frac{\bar{y}_{12\cdot}+\bar{y}_{22\cdot}}{2} - \bar{y}_{32\cdot}\right)$$
$$= \left(\frac{1.820+1.765}{2} - 3.198\right) - \left(\frac{2.765+3.070}{2} - 2.013\right)$$
$$= -2.310.$$

Note that this is an interaction contrast since the contrast coefficients

$$c_{11} = +\tfrac{1}{2}, \quad c_{21} = +\tfrac{1}{2}, \quad c_{31} = -1,$$
$$c_{12} = -\tfrac{1}{2}, \quad c_{22} = -\tfrac{1}{2}, \quad c_{32} = +1$$

sum to zero when added over either subscript. Similarly, the type of solder versus the type of substrate interaction equals

$$\left(\frac{\bar{y}_{13\cdot} + \bar{y}_{23\cdot}}{2} - \bar{y}_{33\cdot}\right) - \left(\frac{\bar{y}_{14\cdot} + \bar{y}_{24\cdot}}{2} - \bar{y}_{34\cdot}\right)$$
$$= \left(\frac{3.000 + 2.945}{2} - 2.815\right) - \left(\frac{3.305 + 3.420}{2} - 3.490\right)$$
$$= 0.285.$$

The standard error of each contrast equals

$$\sqrt{\frac{s^2}{n}\sum_{i=1}^{a}\sum_{j=1}^{b} c_{ij}^2} = \sqrt{\frac{0.0349}{4}\left\{\frac{1}{4} + \frac{1}{4} + 1 + \frac{1}{4} + \frac{1}{4} + 1\right\}} = 0.162.$$

The critical constant for 95% Scheffé SCIs equals

$$\sqrt{(a-1)(b-1)f_{(a-1)(b-1),\nu,\alpha}} = \sqrt{6 f_{6,36,0.05}} = \sqrt{(6)(2.37)} = 3.771.$$

Therefore the 95% Scheffé interval for the type of epoxy versus the type of substrate interaction equals

$$[-2.310 \pm (3.771)(0.162)] = [-2.921, -1.699]$$

and that for the type of solder versus the type of substrate interaction equals

$$[0.285 \pm (3.771)(0.162)] = [-0.326, 0.896].$$

The corresponding t-statistics are -14.262 and 1.759, respectively. Thus the type of epoxy versus the type of substrate interaction is significant, while the type of solder versus the type of substrate interaction is not significant at the 0.05 level, as was suggested by the interaction plot. ■

6.2.4 Model Diagnostics

The residuals defined in (6.6) (which are simply the differences between the data values and the corresponding cell sample means) can be used in the usual way to check the model assumptions. In particular, the plot of residuals against the fitted values can be used to check the homoscedasticity assumption and the normal plot of the residuals can be used to check the normality assumption. A run chart of the residuals can be used to check the independence assumption if the time order of data collection is available.

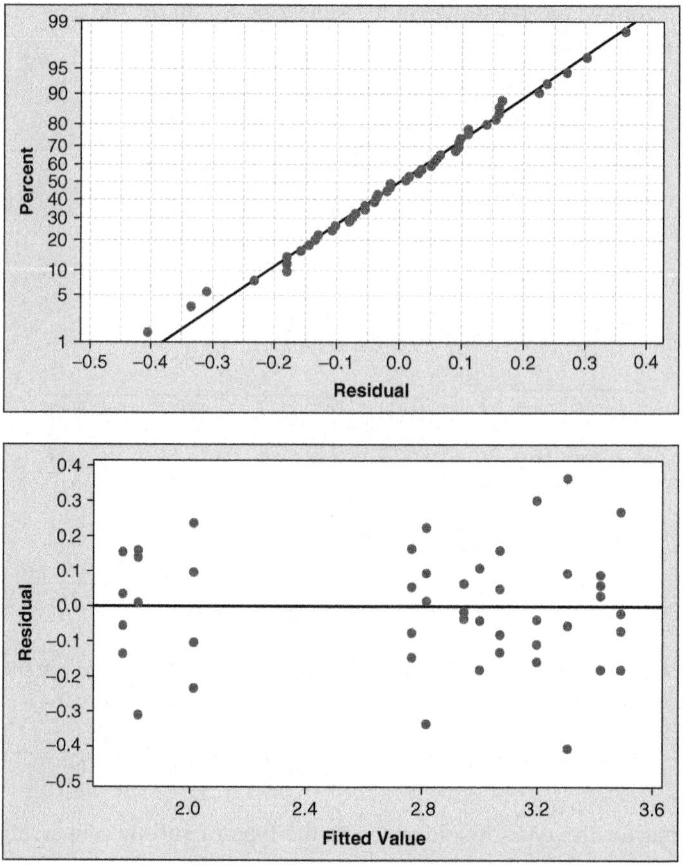

Figure 6.3 Normal (top) and fitted-values plots (bottom) of residuals for bonding strength data.

Example 6.5 (Bonding Strength of Capacitors: Model Diagnostics)

(From Tamhane and Dunlop, 2000, Example 13.3. Reprinted by permission of Pearson Education, Inc.) The normal plot and fitted-values plot of residuals are shown in Figure 6.3. Since the data are simulated using the normal, homoscedastic model, the normal plot is as linear as might be expected, and the fitted-values plot exhibits a parallel band across all fitted values. ■

6.2.5 Tukey's Test for Interaction for Singly Replicated Two-Way Layouts

As discussed in Section 5.1.1, if there is a single replicate per cell, then the full model postulated in (6.3) has the same number of parameters as the total number of observations (namely, $N = ab$). Such models are called **saturated models** because all the data are needed to estimate the unknown parameters, there being no degrees of freedom left to estimate the error term. It is for this reason that for the RB design we assumed the model (5.1) without interactions and used what would have been the interaction sum of squares as the error sum of squares with the associated $(a-1)(b-1)$ d.f. as the error d.f. Instead of ignoring the interactions altogether, Tukey (1949) assumed **multiplicative interactions**:

$$(\alpha\beta)_{ij} = \gamma\alpha_i\beta_j \quad \text{for all } i, j,$$

where γ is a fixed unknown parameter, and gave an exact test for $H_{0AB} : \gamma = 0$. Under this special form of interaction, the full model for a balanced two-way layout with $n = 1$ is

$$y_{ij} = \mu + \alpha_i + \beta_j + \gamma\alpha_i\beta_j + e_{ij} \qquad (1 \le i \le a, 1 \le j \le b), \tag{6.9}$$

where $\sum_{i=1}^{a} \alpha_i = \sum_{j=1}^{b} \beta_j = 0$. Notice that the interactions $(\alpha\beta)_{ij} = \gamma\alpha_i\beta_j$ sum to zero if summed over i for all j and if summed over j for all i. However, they depend only on one extra parameter, γ, and thus consume only one d.f. instead of $(a-1)(b-1)$. Therefore we have $\nu = (a-1)(b-1) - 1 = ab - a - b$ d.f. available to estimate the error.

The model (6.10) is not a linear model because of the multiplicative form of the interaction term. However, the LS estimators of the parameters of the model and the extra sum of squares test for H_{0AB} can be derived in the usual manner. The following motivation given by Scheffé (1959, p. 131) makes the test easy to understand. Treating the α_i and the β_j as known constants, the LS estimator of γ can be shown to be equal to

$$\widehat{\gamma} = \frac{\sum_{i=1}^{a}\sum_{j=1}^{b} \alpha_i\beta_j y_{ij}}{(\sum_{i=1}^{a} \alpha_i^2)(\sum_{j=1}^{b} \beta_j^2)}. \tag{6.10}$$

Then the LS estimator of $(\alpha\beta)_{ij}$ equals $\widehat{(\alpha\beta)}_{ij} = \widehat{\gamma}\alpha_i\beta_j$. The interaction sum of squares for testing $H_{0AB} : \gamma = 0$ for known α_i's and β_j's can then be shown, using the extra sum of squares method, to be equal to

$$\text{SS}_{AB} = \sum_{i=1}^{a}\sum_{j=1}^{b} \widehat{(\alpha\beta)}_{ij}^2 = \widehat{\gamma}^2 \left(\sum_{i=1}^{a} \alpha_i^2\right)\left(\sum_{j=1}^{b} \beta_j^2\right) = \frac{\left(\sum_{i=1}^{a}\sum_{j=1}^{b} \alpha_i\beta_j y_{ij}\right)^2}{(\sum_{i=1}^{a} \alpha_i^2)(\sum_{j=1}^{b} \beta_j^2)}.$$

Substituting $\widehat{\alpha}_i = \bar{y}_{i.} - \bar{y}_{..}$ and $\widehat{\beta}_j = \bar{y}_{.j} - \bar{y}_{..}$ for the unknown α_i and the β_j, we obtain the following expression for the interaction sum of squares:

$$\text{SS}_{AB} = \frac{\left(\sum_{i=1}^{a}\sum_{j=1}^{b} \widehat{\alpha}_i\widehat{\beta}_j y_{ij}\right)^2}{\left(\sum_{i=1}^{a} \widehat{\alpha}_i^2\right)\left(\sum_{j=1}^{b} \widehat{\beta}_j^2\right)},$$

which has one d.f. Then SS_e can be obtained by subtraction as

$$SS_e = \sum_{i=1}^{a} \sum_{j=1}^{b} (y_{ij} - \overline{y}_{i.} - \overline{y}_{.j} + \overline{y}_{..})^2 - SS_{AB}$$

with $\nu = ab - a - b$ d.f. Finally, it can be shown that under $H_{0AB} : \gamma = 0$ the statistic

$$F_{AB} = \frac{SS_{AB}/1}{SS_e/(ab - a - b)} = \frac{MS_{AB}}{MS_e} \sim F_{1,\nu},$$

and so an α-level test of H_{0AB} rejects if $F_{AB} > f_{1,\nu,\alpha}$. This test is referred to as **Tukey's single-degree-of-freedom test for additivity**. The tests for the main effects of A and B are the same as those for the RB design except for the modified error estimator and error d.f.

Example 6.6 (Resistor Data: Tukey's Test for Additivity)

Hamaker (1955) gave the data shown in Table 6.5 on the relative changes observed in a life test with resistors of different resistances and nominal powers. We want to check if there is a significant interaction between the power and resistance using Tukey's test for additivity. Table 6.6 shows the ANOVA for these data using the additive model (without interaction).

We first make an interaction plot, shown in Figure 6.4. We see that the mean profile plots cross each other, indicating the presence of interaction. The pattern of interaction becomes clearer if we look at the residuals, $\widehat{e}_{ij} = y_{ij} - \overline{y}_{i.} - \overline{y}_{.j} + \overline{y}_{..}$, after fitting the additive model shown in Table 6.7. We see that the residuals are mostly positive if the $\widehat{\alpha}_i$ and $\widehat{\beta}_j$ have the same signs and negative if the $\widehat{\alpha}_i$ and $\widehat{\beta}_j$ have opposite signs. This suggests that a multiplicative interaction,

Table 6.5 Relative Changes in Resistance (per mil)

Power	Resistance (ohms)					
(watts)	100	200	500	1000	2000	$\overline{y}_{i.}$
$\frac{1}{8}$	2.8	2.6	3.0	2.5	2.1	2.6
$\frac{1}{4}$	1.9	2.1	4.1	2.2	1.7	2.4
$\frac{1}{2}$	3.2	2.8	1.9	3.4	4.2	3.1
1	2.0	3.4	3.1	3.6	3.9	3.2
2	2.1	2.6	3.4	4.3	6.1	3.7
$\overline{y}_{.j}$	2.4	2.7	3.1	3.2	3.6	$\overline{y}_{..} = 3.0$

Note: Each value is an average of five resistors.

Source: Hamaker (1955). Reprinted by permission of the International Biometric Society.

Table 6.6 ANOVA of Resistance Data Using Additive Model

Source	SS	d.f.	MS	F	p-Value
Power	5.300	4	1.325	1.425	0.271
Resistance	4.300	4	1.075	1.156	0.367
Error	14.880	16	0.930		
Total	24.480	24			

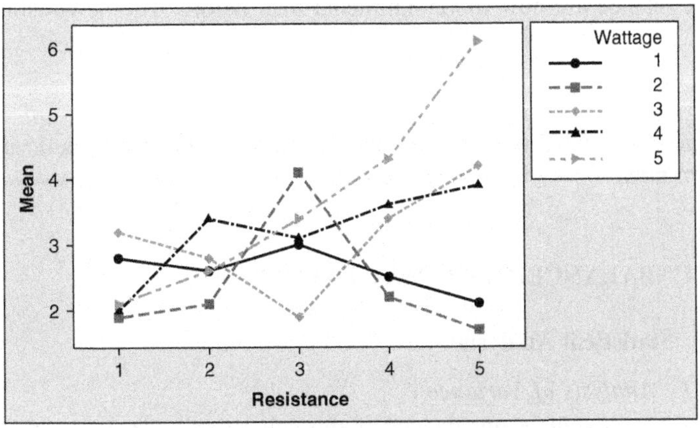

Figure 6.4 Interaction plot for resistance data.

Table 6.7 Residuals of Resistance Data from Additive Model Fit

Power (watts)	Resistance (ohms) 100	200	500	1000	2000	$\widehat{\alpha}_i = \bar{y}_{i\cdot} - \bar{y}_{\cdot\cdot}$
$\frac{1}{8}$	+0.8	+0.3	+0.3	−0.3	−1.1	−0.4
$\frac{1}{4}$	+0.1	0.0	+1.6	−0.4	−1.3	−0.6
$\frac{1}{2}$	+0.7	0.0	−1.3	+0.1	+0.5	+0.1
1	−0.6	+0.5	−0.2	+0.2	+0.1	+0.2
2	−1.0	−0.8	−0.4	+0.4	+1.8	+0.7
$\widehat{\beta}_j = \bar{y}_{\cdot j} - \bar{y}_{\cdot\cdot}$	−0.6	−0.3	+0.1	+0.2	+0.6	

$(\alpha\beta)_{ij} = \gamma\alpha_i\beta_j$ with $\gamma > 0$, may fit the data well. So we calculate

$$\widehat{\gamma} = \frac{\sum_{i=1}^{a}\sum_{j=1}^{b}\widehat{\alpha}_i\widehat{\beta}_j y_{ij}}{(\sum_{i=1}^{a}\widehat{\alpha}_i^2)(\sum_{j=1}^{b}\widehat{\beta}_j^2)} = \frac{2.367}{(1.06)(0.86)} = 2.597.$$

The corresponding sum of squares is

$$\mathrm{SS}_{AB} = \frac{\left(\sum_{i=1}^{a}\sum_{j=1}^{b}\widehat{\alpha}_i\widehat{\beta}_j y_{ij}\right)^2}{\left(\sum_{i=1}^{a}\widehat{\alpha}_i^2\right)\left(\sum_{j=1}^{b}\widehat{\beta}_j^2\right)} = \frac{(2.367)^2}{(1.06)(0.86)} = 6.146$$

with 1 d.f. Therefore $\mathrm{SS}_e = 14.880 - 6.146 = 8.734$ with 15 d.f. The revised ANOVA is shown in Table 6.8. We see that the interaction term is significant. Because of the reduction in MS_e, the F-statistics for the main effects of power and resistance are now more significant than before. The new residuals

$$\widehat{e}_{ij} = y_{ij} - \widehat{\mu} - \widehat{\alpha}_i - \widehat{\beta}_j - \widehat{\gamma}\widehat{\alpha}_i\widehat{\beta}_j$$

are shown in Table 6.9. We see that the signs of the residuals do not show any pattern now. ■

6.3 UNBALANCED TWO-WAY LAYOUTS

6.3.1 Statistical Analysis

6.3.1.1 Analysis of Variance

A two-way layout is unbalanced if the cell sample sizes n_{ij} are unequal. We will only consider the case where all $n_{ij} > 0$; the case of some empty or missing cells ($n_{ij} = 0$) is discussed in Hocking (1996, Sections 13.2 and 13.4). Denote the total sample size by $N = \sum_{i=1}^{a}\sum_{j=1}^{b} n_{ij}$.

We use the same model (6.3) for this design with the parameters defined as in (6.2) so that the side conditions (6.4) hold. The LS estimators of $\mu, \overline{\mu}_{i\cdot}$, and $\overline{\mu}_{\cdot j}$ are as follows:

$$\widehat{\mu} = \frac{1}{ab}\sum_{i=1}^{a}\sum_{j=1}^{b}\overline{y}_{ij\cdot}, \qquad \widehat{\overline{\mu}}_{i\cdot} = \frac{1}{b}\sum_{j=1}^{b}\overline{y}_{ij\cdot}, \qquad \widehat{\overline{\mu}}_{\cdot j} = \frac{1}{a}\sum_{i=1}^{a}\overline{y}_{ij\cdot}.$$

Table 6.8 ANOVA of Resistance Data Assuming Multiplicative Interaction

Source	SS	d.f.	MS	F	p
Power	5.300	4	1.325	2.276	0.109
Resistance	4.300	4	1.075	1.846	0.173
Interaction	6.146	1	6.146	10.555	0.026
Error	8.734	15	0.582		
Total	24.480	24			

Table 6.9 Residuals of Resistance Data from Multiplicative Interaction Model Fit

Power	Resistance (ohms)				
(watts)	100	200	500	1000	2000
$\frac{1}{8}$	+0.177	−0.012	+0.404	−0.092	−0.476
$\frac{1}{4}$	−0.835	−0.467	+1.756	−0.088	−0.365
$\frac{1}{2}$	+0.856	+0.078	−1.326	+0.048	+0.344
1	−0.288	+0.655	−0.252	+0.096	−0.212
2	+0.091	−0.255	−0.582	+0.036	+0.709

These estimators are known as the **LS means**; in the balanced case they equal the corresponding overall, row, and column means, namely, $\bar{y}_{...}$, $\bar{y}_{i..}$, and $\bar{y}_{.j.}$, respectively. The LS estimators of the α_i, β_j, and $(\alpha\beta)_{ij}$ are

$$\widehat{\alpha}_i = \widehat{\bar{\mu}}_{i.} - \widehat{\mu}, \qquad \widehat{\beta}_j = \widehat{\bar{\mu}}_{.j} - \widehat{\mu}, \qquad \widehat{(\alpha\beta)}_{ij} = \bar{y}_{ij.} - \widehat{\bar{\mu}}_{i.} - \widehat{\bar{\mu}}_{.j} + \widehat{\mu}.$$

The fitted values and residuals are given by the same formulas as in the balanced case, namely, $\widehat{y}_{ijk} = \bar{y}_{ij.}$ and $\widehat{e}_{ijk} = y_{ijk} - \bar{y}_{ij.}$. The error sum of squares equals

$$\mathrm{SS}_e = \sum_{i,j,k} \widehat{e}^2_{ijk} = \sum_{i,j,k} (y_{ijk} - \bar{y}_{ij.})^2. \tag{6.11}$$

Analogous to (6.7), the mean square error estimator of σ^2 is given by

$$s^2 = \mathrm{MS}_e = \frac{\mathrm{SS}_e}{N - ab}.$$

The hypotheses of interest are

$$H_{0A} : \bar{\mu}_{1.} = \bar{\mu}_{2.} = \cdots = \bar{\mu}_{a.} \text{ or } \alpha_1 = \alpha_2 = \cdots = \alpha_a = 0,$$

$$H_{0B} : \bar{\mu}_{.1} = \bar{\mu}_{.2} = \cdots = \bar{\mu}_{.b} \text{ or } \beta_1 = \beta_2 = \cdots = \beta_b = 0$$

$$H_{0AB} : \mu_{ij} - \mu_{i'j} - \mu_{ij'} + \mu_{i'j'} = 0 \text{ for all } (i, j) \neq (i', j') \text{ or}$$

$$(\alpha\beta)_{ij} = 0 \text{ for all } (i, j).$$

The sums of squares for testing these hypotheses can be computed using the extra sum of squares method. Since unequal cell sample sizes make the design nonorthogonal, these sums of squares are not statistically independent and do not add up to the total sum of squares, that is,

$$\mathrm{SS}_A + \mathrm{SS}_B + \mathrm{SS}_{AB} + \mathrm{SS}_e \neq \mathrm{SS}_{\mathrm{tot}}.$$

Example 6.7 (Bonding Strength of Capacitors: ANOVA for Unbalanced Data)

(From Tamhane and Dunlop, 2000, Example 13.4. Reprinted by permission of Pearson Education, Inc.) Consider the capacitor bonding strength data, but suppose that some of the data values are missing, resulting in an unbalanced two-way layout as shown in Table 6.10. The ANOVA table is shown in Display 6.2.

Notice that the four adjusted sums of squares do not add up to the total sum of squares:

$$SS_A + SS_B + SS_{AB} + SS_e = 0.1985 + 7.0349 + 4.9919 + 0.8871$$
$$= 13.1124 \neq SS_{\text{tot}} = 13.2802.$$

The p-values of the F-statistics are very similar to those obtained from the ANOVA of the balanced data given in Display 6.1. Thus the conclusions regarding the effects of the substrates and bonding materials are unchanged. ■

6.3.1.2 *Multiple Comparisons*

If the row–column interaction is not significant, then we may want to make comparisons between the row effects and the column effects. To obtain SCIs on the pairwise differences between the row effects, $\alpha_i - \alpha_{i'} = \overline{\mu}_{i\cdot} - \overline{\mu}_{i'\cdot}$ $(1 \leq$

Table 6.10 Bonding Strength of Capacitors (Unbalanced Data)

	Bonding Material			
Substrate	Epoxy I	Epoxy II	Solder I	Solder II
Al_2O_3 without bracket	1.51, 1.96	2.62, 2.82, 2.69	2.96, 2.82, 3.11	3.67, 3.40, 3.25, 2.90
Al_2O_3 with bracket	1.63, 1.80, 1.92, 1.71	3.12, 2.94, 3.23, 2.99	2.91, 2.93, 3.01	3.48, 3.51
BeO	3.04, 3.16, 3.09	1.91, 2.11	3.04, 2.91, 2.48, 2.83	3.47, 3.42, 3.31, 3.76

Source: Tamhane and Dunlop (2000, Table 13.8). Reprinted by permission of Pearson Education, Inc.

```
Source              DF   Seq SS   Adj SS   Adj MS      F      P
Substrt              2   0.4371   0.1985   0.0993   2.91  0.072
Bondmatl             3   6.9640   7.0349   2.3450  68.73  0.000
Substrt*Bondmatl     6   4.9919   4.9919   0.8320  24.38  0.000
Error               26   0.8871   0.8871   0.0341
Total               37  13.2802
S = 0.184717   R-Sq = 93.32%   R-Sq(adj) = 90.49%
```

Display 6.2 Minitab output for ANOVA of bonding strength data in Table 6.10.

$i < i' \le a$), we begin with the pairwise differences between the LS means, $\widehat{\alpha}_i - \widehat{\alpha}_{i'} = \widehat{\overline{\mu}}_i - \widehat{\overline{\mu}}_{i'}$, which are normally distributed with means $\alpha_i - \alpha_{i'}$ and

$$\text{Var}\,(\widehat{\alpha}_i - \widehat{\alpha}_{i'}) = \frac{\sigma^2}{b^2}\sum_{j=1}^{b}\left(\frac{1}{n_{ij}} + \frac{1}{n_{i'j}}\right). \tag{6.12}$$

Therefore $100(1-\alpha)\%$ SCIs on all $\alpha_i - \alpha_{i'}$ are given by the following extension of the Tukey procedure for unbalanced designs from Section 4.2.2:

$$\widehat{\overline{\mu}}_{i\cdot} - \widehat{\overline{\mu}}_{i'\cdot} \pm \frac{q_{a,\nu,\alpha}s}{b}\sqrt{\frac{1}{2}\sum_{j=1}^{b}\left(\frac{1}{n_{ij}} + \frac{1}{n_{i'j}}\right)} \quad (1 \le i < i' \le a).$$

In the same manner, $100(1-\alpha)\%$ SCIs on all $\beta_j - \beta_{j'}$ are given by

$$\widehat{\overline{\mu}}_{\cdot j} - \widehat{\overline{\mu}}_{\cdot j'} \pm \frac{q_{b,\nu,\alpha}s}{a}\sqrt{\frac{1}{2}\sum_{i=1}^{a}\left(\frac{1}{n_{ij}} + \frac{1}{n_{ij'}}\right)} \quad (1 \le j < j' \le b).$$

These SCIs are approximate, not exact.

Example 6.8 (Bonding Strength of Capacitors: Multiple Comparisons for Unbalanced Data)

Suppose that we wish to make pairwise comparisons between the substrates. In Minitab these comparisons can be obtained by using the general linear model option under ANOVA. The resulting output is shown in Display 6.3. The first part of the output shows the LS means for the three substrates, and the second part shows 95% SCIs for the three pairwise differences which are as follows.

Substrate 2 vs. substrate 1: $[-0.0484, 0.3317]$,

Substrate 3 vs. substrate 1: $[-0.0155, 0.3647]$,

Substrate 3 vs. substrate 2: $[-0.1542, 0.2201]$.

Since all three SCIs include zero, none of the pairwise differences is significant.

For better understanding, we show by hand calculation how the SCI for substrate 2 versus substrate 1 is computed. To calculate the LS means, we first need all the cell means, which are shown in Table 6.11. Therefore the LS means of substrates 1 and 2 are

$$\widehat{\overline{\mu}}_{1\cdot} = \tfrac{1}{4}(1.735 + 2.710 + 2.963 + 3.305) = 2.678,$$

$$\widehat{\overline{\mu}}_{2\cdot} = \tfrac{1}{4}(1.765 + 3.070 + 2.950 + 3.495) = 2.820.$$

```
Least Squares Means for Strength
Substrt    Mean   SE Mean
1         2.678  0.05496
2         2.820  0.05332
3         2.853  0.05332
Tukey 95.0% Simultaneous Confidence Intervals
Response Variable Strength
All Pairwise Comparisons among Levels of Substrt
Substrt = 1   subtracted from:

Substrt      Lower   Center    Upper      +---------+---------+---------+------

2         -0.04840  0.1417   0.3317            (-----------*------------)
3         -0.01548  0.1746   0.3647              (------------*-----------)
                                          +---------+---------+---------+------

                                       -0.15      0.00      0.15      0.30
Substrt = 2   subtracted from:

Substrt     Lower    Center    Upper      +---------+---------+---------+------

3         -0.1542   0.03292  0.2201       (-----------*------------)
                                          +---------+---------+---------+------

                                       -0.15      0.00      0.15      0.30
```

Display 6.3 Minitab output for Tukey pairwise comparisons between LS means of substrates using bonding strength data in Table 6.10.

Table 6.11 Cell Means for Bonding Strength Data in Table 6.10

	Bonding Material			
Substrate	1	2	3	4
1	1.735	2.710	2.963	3.305
2	1.765	3.070	2.950	3.495
3	3.097	2.010	2.815	3.490

The estimated standard error of $\widehat{\overline{\mu}}_{2\cdot} - \widehat{\overline{\mu}}_{1\cdot}$ is

$$\frac{s}{b}\sqrt{\sum_{j=1}^{b}\left(\frac{1}{n_{2j}}+\frac{1}{n_{1j}}\right)} = \frac{0.1847}{4}\sqrt{\left[\frac{1}{4}+\frac{1}{4}+\frac{1}{3}+\frac{1}{2}+\frac{1}{2}+\frac{1}{3}+\frac{1}{3}+\frac{1}{4}\right]} = 0.0766.$$

Also, $q_{3,26,0.05} \approx 3.532$ from Table C.5. Therefore the desired SCI for $\alpha_2 - \alpha_1$ is

$$(2.820 - 2.678) \pm \frac{(3.532)}{\sqrt{2}}(0.0766) = 0.142 \pm 0.191 = [-0.049, 0.333].$$

Other SCIs are calculated in the same manner. ■

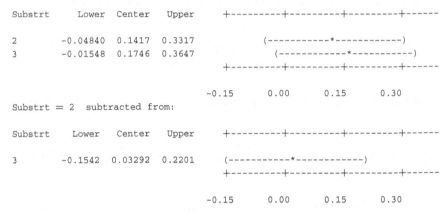

6.4 CHAPTER NOTES

6.4.1 Derivation of LS Estimators of Parameters for Balanced Two-Way Layouts

First we derive the LS estimators given in (6.5). As an example of a balanced two-way layout consider $a = 3, b = 2$, and $n = 2$. Then the model (6.3) can be written in matrix form, $\boldsymbol{y} = \boldsymbol{X}\boldsymbol{\beta} + \boldsymbol{e}$, as follows:

$$\begin{bmatrix} y_{111} \\ y_{112} \\ y_{121} \\ y_{122} \\ y_{211} \\ y_{212} \\ y_{221} \\ y_{222} \\ y_{311} \\ y_{312} \\ y_{321} \\ y_{322} \end{bmatrix} = \left[\begin{array}{c|ccc|cc|cccccc} 1 & 1 & 0 & 0 & 1 & 0 & 1 & 0 & 0 & 0 & 0 & 0 \\ 1 & 1 & 0 & 0 & 1 & 0 & 1 & 0 & 0 & 0 & 0 & 0 \\ 1 & 1 & 0 & 0 & 0 & 1 & 0 & 1 & 0 & 0 & 0 & 0 \\ 1 & 1 & 0 & 0 & 0 & 1 & 0 & 1 & 0 & 0 & 0 & 0 \\ 1 & 0 & 1 & 0 & 1 & 0 & 0 & 0 & 1 & 0 & 0 & 0 \\ 1 & 0 & 1 & 0 & 1 & 0 & 0 & 0 & 1 & 0 & 0 & 0 \\ 1 & 0 & 1 & 0 & 0 & 1 & 0 & 0 & 0 & 1 & 0 & 0 \\ 1 & 0 & 1 & 0 & 0 & 1 & 0 & 0 & 0 & 1 & 0 & 0 \\ 1 & 0 & 0 & 1 & 1 & 0 & 0 & 0 & 0 & 0 & 1 & 0 \\ 1 & 0 & 0 & 1 & 1 & 0 & 0 & 0 & 0 & 0 & 1 & 0 \\ 1 & 0 & 0 & 1 & 0 & 1 & 0 & 0 & 0 & 0 & 0 & 1 \\ 1 & 0 & 0 & 1 & 0 & 1 & 0 & 0 & 0 & 0 & 0 & 1 \end{array}\right] \begin{bmatrix} \mu \\ \alpha_1 \\ \alpha_2 \\ \alpha_3 \\ \beta_1 \\ \beta_2 \\ (\alpha\beta)_{11} \\ (\alpha\beta)_{12} \\ (\alpha\beta)_{21} \\ (\alpha\beta)_{22} \\ (\alpha\beta)_{31} \\ (\alpha\beta)_{32} \end{bmatrix} + \begin{bmatrix} e_{111} \\ e_{112} \\ e_{121} \\ e_{122} \\ e_{211} \\ e_{212} \\ e_{221} \\ e_{222} \\ e_{311} \\ e_{312} \\ e_{321} \\ e_{322} \end{bmatrix}.$$

Hence the normal equations in matrix form, $\boldsymbol{X}'\boldsymbol{X}\boldsymbol{\beta} = \boldsymbol{X}'\boldsymbol{y}$ [see (2.28)], for obtaining the LS estimator $\widehat{\boldsymbol{\beta}}$ are

$$\left[\begin{array}{c|ccc|cc|cccccc} 12 & 4 & 4 & 4 & 6 & 6 & 2 & 2 & 2 & 2 & 2 & 2 \\ \hline 4 & 4 & 0 & 0 & 2 & 2 & 2 & 2 & 0 & 0 & 0 & 0 \\ 4 & 0 & 4 & 0 & 2 & 2 & 0 & 0 & 2 & 2 & 0 & 0 \\ 4 & 0 & 0 & 4 & 2 & 2 & 0 & 0 & 0 & 0 & 2 & 2 \\ \hline 6 & 2 & 2 & 2 & 6 & 0 & 2 & 0 & 2 & 0 & 2 & 0 \\ 6 & 2 & 2 & 2 & 0 & 6 & 0 & 2 & 0 & 2 & 0 & 2 \\ \hline 2 & 2 & 0 & 0 & 2 & 0 & 2 & 0 & 0 & 0 & 0 & 0 \\ 2 & 2 & 0 & 0 & 0 & 2 & 0 & 2 & 0 & 0 & 0 & 0 \\ 2 & 0 & 2 & 0 & 2 & 0 & 0 & 0 & 2 & 0 & 0 & 0 \\ 2 & 0 & 2 & 0 & 0 & 2 & 0 & 0 & 0 & 2 & 0 & 0 \\ 2 & 0 & 0 & 2 & 2 & 0 & 0 & 0 & 0 & 0 & 2 & 0 \\ 2 & 0 & 0 & 2 & 0 & 2 & 0 & 0 & 0 & 0 & 0 & 2 \end{array}\right] \begin{bmatrix} \mu \\ \alpha_1 \\ \alpha_2 \\ \alpha_3 \\ \beta_1 \\ \beta_2 \\ (\alpha\beta)_{11} \\ (\alpha\beta)_{12} \\ (\alpha\beta)_{21} \\ (\alpha\beta)_{22} \\ (\alpha\beta)_{31} \\ (\alpha\beta)_{32} \end{bmatrix} = \begin{bmatrix} y_{\cdots} \\ y_{1\cdot\cdot} \\ y_{2\cdot\cdot} \\ y_{3\cdot\cdot} \\ y_{\cdot 1\cdot} \\ y_{\cdot 2\cdot} \\ y_{11\cdot} \\ y_{12\cdot} \\ y_{21\cdot} \\ y_{22\cdot} \\ y_{31\cdot} \\ y_{32\cdot} \end{bmatrix},$$

where the dot notation without the overbar means that the y_{ijk} are summed over the dotted subscripts; for example, $y_{\cdots} = \sum_i \sum_j \sum_k y_{ijk}$. The individual normal equations are

$$12\mu + 4\sum_{i=1}^{3} \alpha_i + 6\sum_{j=1}^{2} \beta_j + 2\sum_{i=1}^{3}\sum_{j=1}^{2}(\alpha\beta)_{ij} = y_{\cdots},$$

$$4\mu + 4\alpha_i + 2\sum_{j=1}^{2}\beta_j + 2\sum_{j=1}^{2}(\alpha\beta)_{ij} = y_{i\cdot\cdot} \qquad (1 \le i \le 3),$$

$$6\mu + 2\sum_{i=1}^{3}\alpha_i + 6\beta_j + 2\sum_{i=1}^{3}(\alpha\beta)_{ij} = y_{\cdot j\cdot} \qquad (1 \le j \le 2),$$

$$2\mu + 2\alpha_i + 2\beta_j + 2(\alpha\beta)_{ij} = y_{ij\cdot} \qquad (1 \le i \le 3, 1 \le j \le 2).$$

In general, these normal equations are

$$N\mu + bn\sum_{i=1}^{a}\alpha_i + an\sum_{j=1}^{b}\beta_j + n\sum_{i=1}^{a}\sum_{j=1}^{b}(\alpha\beta)_{ij} = y_{\cdots},$$

$$bn\mu + bn\alpha_i + n\sum_{j=1}^{b}\beta_j + n\sum_{j=1}^{b}(\alpha\beta)_{ij} = y_{i\cdot\cdot} \quad (1 \le i \le a),$$

$$an\mu + n\sum_{i=1}^{a}\alpha_i + an\beta_j + n\sum_{i=1}^{a}(\alpha\beta)_{ij} = y_{\cdot j\cdot} \quad (1 \le j \le b),$$

$$n\mu + n\alpha_i + n\beta_j + n(\alpha\beta)_{ij} = y_{ij\cdot} \quad (1 \le i \le a, 1 \le j \le b).$$

Using the side constraints (6.4), the first equation yields $N\mu = y_{\cdots}$ and hence $\widehat{\mu} = y_{\cdots}/N = \overline{y}_{\cdots}$. Substituting this value in the second equation and again using the side constraints (6.4), we get

$$\widehat{\alpha}_i = \frac{y_{i\cdot\cdot}}{bn} - \widehat{\mu} = \overline{y}_{i\cdot\cdot} - \overline{y}_{\cdots} \qquad (1 \le i \le a).$$

Similarly we get

$$\widehat{\beta}_j = \frac{y_{\cdot j\cdot}}{an} - \widehat{\mu} = \overline{y}_{\cdot j\cdot} - \overline{y}_{\cdots} \qquad (1 \le j \le b)$$

and

$$\widehat{(\alpha\beta)}_{ij} = \frac{y_{ij\cdot}}{n} - \widehat{\alpha}_i - \widehat{\beta}_j - \widehat{\mu} = \overline{y}_{ij\cdot} - \overline{y}_{i\cdot\cdot} - \overline{y}_{\cdot j\cdot} + \overline{y}_{\cdots} \qquad (1 \le i \le a, 1 \le j \le b).$$

6.4.2 Derivation of ANOVA Sums of Squares and F-Tests for Balanced Two-Way Layouts

Next we derive the expressions for the sums of squares in the ANOVA given in Table 6.4. Write

$$\begin{aligned} y_{ijk} &= \overline{y}_{\cdots} + (\overline{y}_{i\cdot\cdot} - \overline{y}_{\cdots}) + (\overline{y}_{\cdot j\cdot} - \overline{y}_{\cdots}) + (\overline{y}_{ij\cdot} - \overline{y}_{i\cdot\cdot} - \overline{y}_{\cdot j\cdot} + \overline{y}_{\cdots}) + (y_{ijk} - \overline{y}_{ij\cdot}) \\ &= \widehat{\mu} + \widehat{\alpha}_i + \widehat{\beta}_j + \widehat{(\alpha\beta)}_{ij} + \widehat{e}_{ijk}. \end{aligned}$$

Subtract $\widehat{\mu} = \overline{y}_{...}$ from both sides, square them, and then sum over i, j, k. Note that all cross products on the right-hand side sum to zero. For example,

$$\sum_{i=1}^{a}\sum_{j=1}^{b}\sum_{k=1}^{n}\widehat{\alpha}_i\widehat{\beta}_j = n\sum_{i=1}^{a}\widehat{\alpha}_i\sum_{j=1}^{b}\widehat{\beta}_j = 0$$

since $\sum_{i=1}^{a}\widehat{\alpha}_i = \sum_{j=1}^{b}\widehat{\beta}_j = 0$. Therefore we get

$$\underbrace{\sum_{i=1}^{a}\sum_{j=1}^{b}\sum_{k=1}^{n}(y_{ijk} - \overline{y}_{...})^2}_{\text{SS}_{\text{tot}}} = \underbrace{bn\sum_{i=1}^{a}\widehat{\alpha}_i^2}_{\text{SS}_A} + \underbrace{an\sum_{j=1}^{b}\widehat{\beta}_j^2}_{\text{SS}_B} + \underbrace{n\sum_{i=1}^{a}\sum_{j=1}^{b}\widehat{(\alpha\beta)}_{ij}^2}_{\text{SS}_{AB}} + \underbrace{\sum_{i=1}^{a}\sum_{j=1}^{b}\sum_{k=1}^{n}\widehat{e}_{ijk}^2}_{\text{SS}_e},$$

which is the ANOVA identity (6.8).

Consider testing the hypothesis $H_{0A} : \alpha_1 = \alpha_2 = \cdots = \alpha_a = 0$. To derive the test by the extra sum of squares method, we need to find SS_{e0} by computing the LS estimators of μ, β_j's, and $(\alpha\beta)_{ij}$'s obtained by setting all α_i's equal to zero in the model (6.3). Because of the orthogonality of the design, the LS estimators of the remaining parameters under H_{0A} are the same as in (6.5). To see this, write the objective function to be minimized in the least squares problem as

$$\begin{aligned}
Q &= \sum_{i=1}^{a}\sum_{j=1}^{b}\sum_{k=1}^{n}[y_{ijk} - (\mu + \alpha_i + \beta_j + (\alpha\beta)_{ij})]^2 \\
&= \sum_{i=1}^{a}\sum_{j=1}^{b}\sum_{k=1}^{n}[(y_{ijk} - \overline{y}_{ij.}) + (\overline{y}_{...} - \mu) + \{(\overline{y}_{i..} - \overline{y}_{...}) - \alpha_i\} \\
&\quad + \{(\overline{y}_{.j.} - \overline{y}_{...}) - \beta_j\} + \{(\overline{y}_{ij.} - \overline{y}_{i..} - \overline{y}_{.j.} + \overline{y}_{...}) - (\alpha\beta)_{ij}\}]^2 \\
&= \sum_{i=1}^{a}\sum_{j=1}^{b}\sum_{k=1}^{n}\widehat{e}_{ijk}^2 + N(\overline{y}_{...} - \mu)^2 + bn\sum_{i=1}^{a}\{(\overline{y}_{i..} - \overline{y}_{...}) - \alpha_i\}^2 \\
&\quad + an\sum_{j=1}^{b}\{(\overline{y}_{.j.} - \overline{y}_{...}) - \beta_j\}^2 + n\sum_{i=1}^{a}\sum_{j=1}^{b}\{(\overline{y}_{ij.} - \overline{y}_{i..} - \overline{y}_{.j.} + \overline{y}_{...}) - (\alpha\beta)_{ij}\}^2,
\end{aligned}$$

where the last step follows because all the cross products are zero. It is easy to see that this expression is minimized by making the last four terms equal to zero. This is achieved when the LS estimators[2] are as given in (6.5) and the resulting minimum equals

$$Q_{\min} = \sum_{i=1}^{a}\sum_{j=1}^{b}\sum_{k=1}^{n}\widehat{e}_{ijk}^2 = \text{SS}_e.$$

[2]This is an alternative way of deriving the LS estimators (6.5).

Under the reduced model with all $\alpha_i = 0$, the above objective function equals

$$Q = SS_e + N(\bar{y}_{...} - \mu)^2 + bn \sum_{i=1}^{a} (\bar{y}_{i..} - \bar{y}_{...})^2$$
$$+an \sum_{j=1}^{b} \{(\bar{y}_{.j.} - \bar{y}_{...}) - \beta_j\}^2 + n \sum_{i=1}^{a} \sum_{j=1}^{b} \{(\bar{y}_{ij.} - \bar{y}_{i..} - \bar{y}_{.j.} + \bar{y}_{...}) - (\alpha\beta)_{ij}\}^2,$$

which shows that the LS estimators of μ, the β_j, and the $(\alpha\beta)_{ij}$ are the same as before, and

$$Q_{\min} = SS_{e0} = SS_e + bn \sum_{i=1}^{a} (\bar{y}_{i..} - \bar{y}_{...})^2 = SS_e + SS_A.$$

Therefore

$$SS_{e0} - SS_e = SS_A.$$

As noted before, there are $a - 1$ d.f. associated with the hypothesis H_{0A}. Therefore the extra sum of squares F-statistic equals

$$F_A = \frac{(SS_{e0} - SS_e)/(a-1)}{SS_e/(N - ab)} = \frac{SS_A/(a-1)}{SS_e/(N - ab)} = \frac{MS_A}{MS_e} \sim F_{a-1,N-ab}$$

under H_{0A}. The statistics F_B and F_{AB} for testing H_{0B} and H_{0AB}, respectively, can be derived in the same manner.

6.4.3 Three- and Higher Way Layouts

Consider three factors A, B, and C, each with at least two levels, a, b, and c, respectively. The observations $y_{ijk\ell}$ at the treatment combination (i, j, k) are assumed to be i.i.d. $N(\mu_{ijk}, \sigma^2)$ for $\ell = 1, 2, \ldots, n_{ijk}$. We will only consider the balanced case with $n_{ijk} = n \geq 2$ for all treatment combinations. Denote the total sample size by $N = abcn$. Consider the following model:

$$y_{ijk\ell} = \mu + \alpha_i + \beta_j + \gamma_k + (\alpha\beta)_{ij} + (\alpha\gamma)_{ik} + (\beta\gamma)_{jk} + (\alpha\beta\gamma)_{ijk} + e_{ijk\ell}, \tag{6.13}$$

where the $e_{ijk\ell}$ are i.i.d. $N(0, \sigma^2)$ r.v.'s and each set of the parameters sums to zero when summed over any subscript keeping the other subscripts fixed, for example,

$$\sum_{i=1}^{a} \alpha_i = 0, \qquad \sum_{i=1}^{a} (\alpha\beta)_{ij} = 0 \quad \text{for all } j, \sum_{i=1}^{a} (\alpha\beta\gamma)_{ijk} = 0 \quad \text{for all } j, k.$$

The parameters have the same meanings as in the two-way layout case. For example, $\mu = \overline{\mu}_{...}$ is the overall mean, $\alpha_i = \overline{\mu}_{i..} - \overline{\mu}_{...}$ is the main effect of the ith level of A, $(\alpha\beta)_{ij} = \overline{\mu}_{ij.} - \overline{\mu}_{i..} - \overline{\mu}_{.j.} + \overline{\mu}_{...}$ is the two-way interaction associated with the treatment combination consisting of the ith level of A and the jth level of B, and finally

$$\begin{aligned}(\alpha\beta\gamma)_{ijk} &= \mu_{ijk} - \mu - \alpha_i - \beta_j - \gamma_k - (\alpha\beta)_{ij} - (\alpha\gamma)_{ik} - (\beta\gamma)_{jk} \\ &= \mu_{ijk} - \overline{\mu}_{ij.} - \overline{\mu}_{i.k} - \overline{\mu}_{.jk} + \overline{\mu}_{i..} + \overline{\mu}_{.j.} + \overline{\mu}_{..k} - \overline{\mu}_{...}\end{aligned}$$

is the three-way interaction associated with the treatment combination consisting of the ith level of A, jth level of B, and kth level of C.

The LS estimators of these parameters are the intuitive ones obtained by replacing the $\overline{\mu}$'s by the corresponding sample means $\overline{y}$'s. For example, $\widehat{\mu} = \overline{y}_{....}$, $\widehat{\alpha}_i = \overline{y}_{i...} - \overline{y}_{....}$, $\widehat{(\alpha\beta)}_{ij} = \overline{y}_{ij..} - \overline{y}_{i...} - \overline{y}_{.j..} + \overline{y}_{....}$, and $\widehat{(\alpha\beta\gamma)}_{ijk} = \overline{y}_{ijk.} - \overline{y}_{ij..} - \overline{y}_{i.k.} - \overline{y}_{.jk.} + \overline{y}_{i...} + \overline{y}_{.j..} + \overline{y}_{..k.} - \overline{y}_{....}$. The fitted values equal $\widehat{y}_{ijk\ell} = \widehat{\mu}_{ijk} = \overline{y}_{ijk.}$ and the residuals equal $\widehat{e}_{ijk\ell} = y_{ijk\ell} - \widehat{y}_{ijk\ell} = y_{ijk\ell} - \overline{y}_{ijk.}$. These residuals can be analyzed in the usual way to check the model assumptions.

The total sum of squares can be decomposed into its orthogonal components as follows:

$$\begin{aligned}\underbrace{\sum_{i,j,k,\ell}(y_{ijk\ell} - \overline{y}_{ijk.})^2}_{\text{SS}_{\text{tot}}} &= \underbrace{bcn\sum_i \widehat{\alpha}_i^2}_{\text{SS}_A} + \underbrace{acn\sum_j \widehat{\beta}_j^2}_{\text{SS}_B} + \underbrace{abn\sum_k \widehat{\gamma}_k^2}_{\text{SS}_C} \\ &\quad + \underbrace{cn\sum_{i,j} \widehat{(\alpha\beta)}_{ij}^2}_{\text{SS}_{AB}} + \underbrace{bn\sum_{i,k} \widehat{(\alpha\gamma)}_{ik}^2}_{\text{SS}_{AC}} + \underbrace{an\sum_{j,k} \widehat{(\beta\gamma)}_{jk}^2}_{\text{SS}_{BC}} \\ &\quad + \underbrace{n\sum_{i,j,k} \widehat{(\alpha\beta\gamma)}_{ijk}^2}_{\text{SS}_{ABC}} + \underbrace{\sum_{i,j,k,\ell} \widehat{e}_{ijk\ell}^2}_{\text{SS}_e}.\end{aligned}$$

There is a corresponding decomposition of the degrees of freedom:

$$\begin{aligned}\underbrace{abcn - 1}_{\text{total d.f.}} &= \underbrace{a-1}_{A\text{ main–effect d.f.}} + \underbrace{b-1}_{B\text{ main–effect d.f.}} + \underbrace{c-1}_{C\text{ main–effect d.f.}} \\ &\quad + \underbrace{(a-1)(b-1)}_{AB\text{ interaction d.f.}} + \underbrace{(a-1)(c-1)}_{AC\text{ interaction d.f.}} + \underbrace{(b-1)(c-1)}_{BC\text{ interaction d.f.}} \\ &\quad + \underbrace{(a-1)(b-1)(c-1)}_{ABC\text{ interaction d.f.}} + \underbrace{abc(n-1)}_{\text{error d.f.}}.\end{aligned}$$

The respective mean squares are obtained by dividing the sums of squares by their associated d.f., which can be used to perform F-tests of the null hypotheses on the corresponding parameters. The denominator for each F-statistic is $\text{MS}_e =$

$SS_e/(N - abc)$, which is an unbiased estimator of σ^2 with $\nu = N - abc$ d.f. An abbreviated ANOVA is given in Table 6.12.

6.5 CHAPTER SUMMARY

(a) A factorial experiment not only enables us to estimate the interactions between the factors but also is more efficient than a one-factor-at-a-time experiment because effectively it uses each observation twice to estimate the factor effects.

(b) A balanced two-way layout design is an orthogonal design. Unequal cell sample sizes spoil the orthogonality of the design, but otherwise the analyses are similar. In analyzing any factorial design, interactions should be tested first. Inferences on the main effects are meaningful only if the interactions are not significant. Main-effect and interaction plots are useful graphical aids to visualize the respective effects.

(c) For singly replicated factorial experiments there are not sufficient degrees of freedom available to include interactions in the linear model. Hence, often an additive model is assumed. However, in some applications it may be sensible to assume a multiplicative form of interaction, $(\alpha\beta)_{ij} = \gamma\alpha_i\beta_j$ for all i, j, which consumes only one d.f. due to γ. Tukey's single-degree-of-freedom test for nonadditivity can be used to test the hypothesis of no interaction for this model.

EXERCISES

Section 6.1 (Factorial versus One-Factor-at-a-Time Experiments)

Theoretical Exercise

6.1 In the one-factor-at-a-time experiment described in Section 6.1, only two out of three observations are used to estimate the main effects A and B.

Table 6.12 ANOVA Table for Balanced Three-Way Layout

Source	SS	d.f.
A	$SS_A = bcn \sum (\bar{y}_{i\cdots} - \bar{y}_{\cdots\cdot})^2$	$a - 1$
B	$SS_B = acn \sum (\bar{y}_{\cdot j\cdot\cdot} - \bar{y}_{\cdots\cdot})^2$	$b - 1$
C	$SS_C = abn \sum (\bar{y}_{\cdot\cdot k\cdot} - \bar{y}_{\cdots\cdot})^2$	$c - 1$
AB	$SS_{AB} = cn \sum\sum (\bar{y}_{ij\cdot\cdot} - \bar{y}_{i\cdots} - \bar{y}_{\cdot j\cdot\cdot} + \bar{y}_{\cdots\cdot})^2$	$(a-1)(b-1)$
AC	$SS_{AC} = bn \sum\sum (\bar{y}_{i\cdot k\cdot} - \bar{y}_{i\cdots} - \bar{y}_{\cdot\cdot k\cdot} + \bar{y}_{\cdots\cdot})^2$	$(a-1)(c-1)$
BC	$SS_{BC} = an \sum\sum (\bar{y}_{\cdot jk} - \bar{y}_{\cdot j\cdot\cdot} - \bar{y}_{\cdots k} + \bar{y}_{\cdots\cdot})^2$	$(b-1)(c-1)$
ABC	SS_{ABC} = By subtraction	$(a-1)(b-1)(c-1)$
Error	$SS_e = \sum\sum\sum\sum (y_{ijk\ell} - \bar{y}_{ijk\cdot})^2$	$N - abc$
Total	$SS_{tot} = \sum\sum\sum\sum (y_{ijk\ell} - \bar{y}_{\cdots\cdot})^2$	$N - 1$

Suppose the following estimators are suggested that use all three observations:

$$\widehat{A} = \tfrac{1}{2}(y_{21} + y_{22}) - y_{11} \qquad \text{and} \qquad \widehat{B} = y_{22} - \tfrac{1}{2}(y_{21} + y_{11}).$$

Assume the additive model $\mathrm{E}(y_{ij}) = \mu + \alpha_i + \beta_j (i, j = 1, 2)$ where $\alpha_1 + \alpha_2 = \beta_1 + \beta_2 = 0$ and the main effects are $A = \alpha_2 - \alpha_1 = 2\alpha_2$ and $B = \beta_2 - \beta_1 = 2\beta_2$. Show that the above estimators are biased.

Section 6.2 (Balanced Two-Way Layouts)

Theoretical Exercises

6.2 For a two-way layout with $a = b = 2$, show that there is only one linearly independent parameter of each type: the row main effect α_i, the column main effect β_j, and the row–column interaction $(\alpha\beta)_{ij}$.

6.3 Consider the balanced two-factor experiment set up but suppose that an RB design is used. Denote the blocking factor by C with $c \geq 2$ levels (blocks) and assume that each of the ab treatment combinations is replicated $n \geq 2$ times in each block. Write the full linear model for this design, including all treatment–block interactions with appropriate side constraints. How many linearly independent parameters are there in this model? How many degrees of freedom for error do you have? If $n = 1$, then how will you estimate the error and how many degrees of freedom will you have for estimating it?

6.4 Derive the LS estimator (6.10) for γ in Tukey's multiplicative interaction model assuming that the α_i and the β_j are known.

Applied Exercises

6.5 (From Tamhane and Dunlop, 2000, Exercise 13.2. Reprinted by permission of Pearson Education, Inc.) Table 6.13 gives yields of a chemical process using six different alcohol–base combinations. Four replicate observations were made for each combination.

(a) Make the main-effects and interaction plots. What do these plots suggest about the effects of alcohol and base on the percent yield?

(b) Calculate the ANOVA table. Test for the alcohol and base main effects and interaction. Use $\alpha = 0.05$.

(c) Make residual plots to check the normality and constant-variance assumptions.

6.6 Refer to the iodide data from Exercise 3.6, where the design was considered as a one-way layout with eight treatments. Now consider it as a 2×4

Table 6.13 Percent Yields for Alcohol–Base Combinations

	Base	
Alcohol	1	2
1	91.3, 89.9, 90.7, 91.4	87.3, 89.4, 91.5, 88.3
2	89.3, 88.1, 90.4, 91.4	92.3, 91.5, 90.6, 94.7
3	89.5, 87.6, 88.3, 90.3	93.1, 90.7, 91.5, 89.8

Source: Nelson (1998, Table 15.32). Reprinted by permission of The McGraw Hill Companies.

layout and calculate the corresponding ANOVA table. Are there significant effects due to PTU and iodide level? What about their interaction? Use $\alpha = 0.05$.

6.7 Hamaker (1955) described an experiment to investigate how the thickness of electrolytic deposition varies with the position of the rod and the vertical height on the rod at which the deposition is measured. Five nickel rods, 1 mm in diameter, were put in a metallic clamp and immersed in a suspension of aluminum oxide. Voltage of 100 volts was applied for a few seconds between the rods (which had negative charge) and the conducting vessel containing the suspension. As a result, oxide particles (which

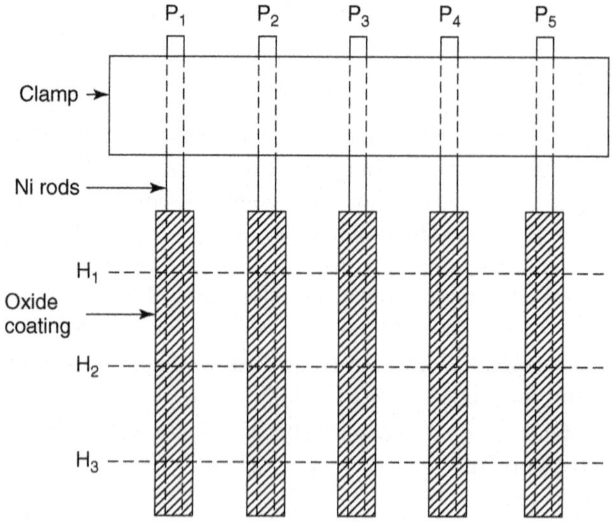

Figure 6.5 Experimental setup for measuring oxide deposit thickness (Hamaker, 1955). Reprinted by permission of the International Biometric Society.

Table 6.14 Deposition Thickness (micrometers)

	Position				
Height	P_1	P_2	P_3	P_4	P_5
H_1	125	130	128	134	143
H_2	126	150	127	124	118
H_3	130	155	168	159	138

Source: Hamaker (1955, Table 1). Reprinted by permission of the International Biometric Society.

had positive charge) were deposited on the nickel rods. Figure 6.5 shows the experimental setup. The thickness of the deposition was measured in micrometers and is given in Table 6.14.

(a) Calculate the row and column main effects.

(b) Test whether the main effects are significant at the 0.05 level assuming an additive model.

6.8 Refer to the data from Exercise 6.7.

(a) Make an interaction plot. Does it show presence of interaction?

(b) Apply Tukey's test for additivity. Do you get a significant result for interaction? Why do you think this result disagrees with the conclusion from the interaction plot?

Section 6.3 (Unbalanced Two-Way Layouts)

Theoretical Exercises

6.9 The purpose of this exercise is to show that the sums of squares for an unbalanced two-way layout can be computed by the extra sum of squares method applied to a properly formulated regression model.

(a) Show that the model

$$y = \mu + \sum_{i=1}^{a-1} \alpha_i u_i + \sum_{j=1}^{b-1} \beta_j v_j + \sum_{i=1}^{a-1} \sum_{j=1}^{b-1} (\alpha\beta)_{ij} u_i v_j + e$$

satisfies the constraints (6.4) where, for $1 \le i \le a$ and $1 \le j \le b$,

$$u_i = \begin{cases} +1 & \text{if observation is from } i\text{th row,} \\ -1 & \text{if observation is from } a\text{th row,} \\ 0 & \text{otherwise,} \end{cases}$$

$$v_j = \begin{cases} +1 & \text{if observation is from } j\text{th column,} \\ -1 & \text{if observation is from } b\text{th column,} \\ 0 & \text{otherwise.} \end{cases}$$

(b) Compute SS_A by applying the extra sum of squares method with any regression program to the data in Table 6.2 using the above model and verify that you get the same SS_A given in Display 6.2.

Applied Exercises

6.10 (From Tamhane and Dunlop, 2000, Exercise 13.14. Reprinted by permission of Pearson Education, Inc.) Table 6.15 reproduces data from Table 6.13 with some observations missing.

(a) Calculate the ANOVA table. Check that the ANOVA decomposition identity does not hold.

(b) Calculate the LS means for the two bases (columns) and their estimated standard errors.

(c) Calculate the t-statistic for comparing the LS means of the two columns. Check that this t-statistic is equivalent to the F-statistic from the ANOVA table according to the relation $t^2 = F$.

(d) Calculate the t-statistic for comparing the ordinary means of the two columns. Check that this t-statistic is *not* equivalent to the F-statistic from the ANOVA table according to the relation $t^2 = F$.

6.11 Kleinbaum, Kupper, and Muller (1988, pp. 458–459) gave hypothetical data on satisfaction scores (on a scale of 1–10) of pregnancy patients. The patients were categorized into six groups based on their worry about pregnancy (negative or positive) and their communication with physicians (low, medium, and high). This being an observational study with groups formed post hoc based on responses to questionnaires administered to

Table 6.15 Percent Yields for Alcohol–Base Combinations

	Base	
Alcohol	1	2
1	91.3, —, 90.7, 91.4	87.3, 89.4, 91.5, 88.3
2	—, 88.1, 90.4, —	92.3, 91.5, —, 94.7
3	89.5, 87.6, 88.3, 90.3	—, —, 91.5, 89.8

Source: Nelson (1998, Table 15.32) *Note:* Some values are deleted. Reprinted by permission of the McGraw Hill Companies.

Table 6.16 Satisfaction Scores of Pregnancy Patients

Communication Level	Worry: Negative	Worry: Positive
Low	8, 7, 5, 9, 9, 10, 8, 6, 8, 10 ($n_{11} = 10$) ($\bar{y}_{11\cdot} = 8$)	5, 8, 6, 6, 9, 7, 7, 8 ($n_{12} = 8$) ($\bar{y}_{12\cdot} = 7$)
Medium	4, 6, 3, 3 ($n_{21} = 4$) ($\bar{y}_{21\cdot} = 4$)	7, 7, 8, 6, 4, 9, 8, 7 ($n_{22} = 8$) ($\bar{y}_{22\cdot} = 7$)
High	2, 5, 8, 6, 2, 4, 3, 10 ($n_{31} = 8$) ($\bar{y}_{31\cdot} = 5$)	7, 5, 8, 6, 3, 5, 6, 4, 5, 6, 8, 9 ($n_{32} = 12$) ($\bar{y}_{32\cdot} = 6$)

Source: Kleinbaum, Kupper, and Muller (1988, Table 20-2).

patients and physicians, the cell sample sizes were not equal. The data are shown in Table 6.16.

(a) Make main-effects and interaction plots. What do you see?

(b) Calculate the ANOVA table. Which effects are significant at the 0.05 level?

(c) Summarize your conclusions.

6.12 Refer to the previous exercise.

(a) Calculate the LS means for the communication levels (rows).

(b) Calculate 95% Tukey SCIs on pairwise differences between the row means to determine which rows differ significantly from each other.

CHAPTER 7

Two-Level Factorial Experiments

An important class of factorial experiments consists of designs in which each factor is studied at two levels. If there are $p \geq 2$ factors, then we have a 2^p experiment. Often these experiments are used for **screening** many factors, only a few which are likely to be active. This is referred to as the **effect sparsity principle** (Box and Meyer 1985), which is analogous to the **Pareto principle** of "vital few trivial many." If the factors are quantitative and have monotonic effects over the selected ranges, then two-level factorial experiments are efficient and effective for screening purposes. Once the **active factors** are identified, a more detailed experiment with more than two levels of these factors can be carried out to obtain a more thorough understanding of their effects. If p is large, then even a 2^p design requires too many runs. In that case, a suitably chosen fraction of the full-factorial design may be used. Fractional factorial designs are discussed in the next chapter.

Two other guiding principles useful in screening experiments are the **effect heredity principle** (Hamada and Wu 1992) and the closely related **hierarchical ordering principle**. The effect heredity principle states that if a higher order effect is important, then at least one of the lower order effects must be important; thus if a two-factor interaction is important, then at least one of the main effects must be important. The hierarchical ordering principle states that often the higher order effects are smaller in magnitude and hence less important than the lower order effects such as the main-effects and two-factor interactions. Clearly, these are not universal principles, so one must be cautious in applying them.

The outline of this chapter is as follows. In Section 7.1, we begin with a discussion of estimation of effects in a 2^2 design, proceeding to a 2^3 design, and finally generalizing to a 2^p design. Section 7.2 discusses hypotheses tests and confidence interval estimation. Section 7.3 considers the case of a singly replicated design. In this case an estimator of the error variance is not available unless some interactions are ignored, so alternative methods of inference are needed. Section 7.3.3 discusses augmenting a 2^p design with some runs at the

Statistical Analysis of Designed Experiments: Theory and Applications By Ajit C. Tamhane

center point. They help in checking for curvature in the response surface and also to obtain a pure estimator of the error variance. Until this section we assume that a CR design is used, but it should be noted that a 2^p experiment can be conducted using any design, for example, an RB, a BIB, or an LSQ design. Section 7.4 considers the problem of blocking the 2^p runs in 2^q incomplete blocks of size 2^{p-q} so that no main-effects or lower order interactions of interest are confounded. Notes on additional topics and mathematical derivations are given in Section 7.5. The chapter concludes with a summary in Section 7.6.

7.1 ESTIMATION OF MAIN EFFECTS AND INTERACTIONS

7.1.1 2^2 Designs

Consider first a 2^2 design with factors A and B, each at two levels, labeled high and low. Note that for a quantitative factor, "high" and "low" do not necessarily correspond to the actual high and low levels (although it is certainly convenient). For a qualitative factor, the labels merely serve to distinguish between two nominal categories.

A new notation will be used to denote the treatment combinations in a 2^2 design:

$$\begin{aligned}(1) &= (A \text{ Low}, B \text{ Low}),\\ a &= (A \text{ High}, B \text{ Low}),\\ b &= (A \text{ Low}, B \text{ High}),\\ ab &= (A \text{ High}, B \text{ High}).\end{aligned}$$

Thus the presence of a lowercase letter in the notation for a treatment combination means that the corresponding factor is at high level, while the absence of the letter means that the factor is at low level; when both factors are at low levels, the treatment combination is denoted by (1). This notation extends to more than two treatments in an obvious way. Another notation that we will use is to represent

Table 7.1 Data from Balanced 2^2 Design

Factor		Treatment		
A	B	Combination	Data	Mean
$-$	$-$	(1)	$y_{(1)1}, \ldots, y_{(1)n}$	$\overline{y}_{(1)}$
$+$	$-$	a	$y_{a1}, \ldots, y_{an}$	$\overline{y}_{a}$
$-$	$+$	b	$y_{b1}, \ldots, y_{bn}$	$\overline{y}_{b}$
$+$	$+$	ab	$y_{ab1}, \ldots, y_{abn}$	$\overline{y}_{ab}$

Source: Tamhane and Dunlop (2000, Table 13.10). Reprinted by permission of Pearson Education, Inc.

the low level of a factor by -1 and the high level by $+1$ (or more simply by minus and plus signs).

We assume a balanced CR design with n observations per treatment combination. Denote these observations by y_{ij} where $i = (1), a, b, ab$ and $j = 1, 2, \ldots, n$. A tabular layout of the data is shown in Table 7.1. The **design matrix** is given by the two columns A and B of the table. The y_{ij} are assumed to be independent normal with means μ_i and a common variance σ^2. By the **principle of sufficiency** (see, e.g., Rice 1988, pp. 257–258), we can summarize the data in terms of the sample means $\overline{y}_i$ and $s^2 = \text{MS}_e = \sum_i \sum_j (y_{ij} - \overline{y}_i)^2/4(n-1)$. Note that the $\overline{y}_i \sim \text{N}(\mu_i, \sigma^2/n)$ for $i = (1), a, b, ab$ independently of $s^2 \sim \sigma^2\chi^2_\nu/\nu$, where $\nu = 4(n-1)$ are the error d.f. It will be convenient to take the $\overline{y}_i$ as the response variables instead of the raw data y_{ij}; the remainder of the information in the data is contained in s^2, which estimates σ^2.

We will use the regression approach to analyze 2^p designs. Toward this end, define the following indicator variables:

$$x_1 = \begin{cases} -1 & \text{if } A \text{ is low,} \\ +1 & \text{if A is high} \end{cases} \quad \text{and} \quad x_2 = \begin{cases} -1 & \text{if } B \text{ is low,} \\ +1 & \text{if } B \text{ is high.} \end{cases}$$

Consider the multiple regression model:

$$\overline{y} = \beta_0 + \beta_1 x_1 + \beta_2 x_2 + \beta_{12} x_1 x_2 + e, \tag{7.1}$$

where $e \sim \text{N}(0, \sigma^2/n)$. In the factorial experiments literature, the following notation is commonly used:

$$I = \beta_0, \qquad A = 2\beta_1, \qquad B = 2\beta_2, \qquad AB = 2\beta_{12},$$

where I is called the **grand mean effect**, A and B are called the **main effects** of factors A and B, and AB is called their **interaction**. It is easy to show that these effects are related to the treatment means as follows:

$$\begin{aligned} I &= \beta_0 = \tfrac{1}{4}[\mu_{ab} + \mu_b + \mu_a + \mu_{(1)}], \\ A &= 2\beta_1 = \tfrac{1}{2}[\mu_{ab} - \mu_b + \mu_a - \mu_{(1)}], \\ B &= 2\beta_2 = \tfrac{1}{2}[\mu_{ab} + \mu_b - \mu_a - \mu_{(1)}], \\ AB &= 2\beta_{12} = \tfrac{1}{2}[\mu_{ab} - \mu_b - \mu_a + \mu_{(1)}]. \end{aligned} \tag{7.2}$$

Thus, $I = \beta_0$ is the average of the μ_i's and A, B, AB (or $\beta_1, \beta_2, \beta_{12}$) are contrasts among the μ_i's. The coefficients of the μ_i's in the above definitions are summarized in Table 7.2. The coefficients of the μ_i's in the definitions of the β's in the regression model (7.1) are the same except they all have the multiplier $\frac{1}{4}$.

Table 7.2 Contrast Coefficients of Effects in 2^2 Design

Treatment	Effect			
Combination	I	A	B	AB
(1)	+	−	−	+
a	+	+	−	−
b	+	−	+	−
ab	+	+	+	+
Multiplier	$\frac{1}{4}$	$\frac{1}{2}$	$\frac{1}{2}$	$\frac{1}{2}$

A Note on Notation

For economy of notation, we have used A, B, and AB in three different ways. First, A and B denote the two factors. Second, A, B, and AB denote the main effects and their interaction, respectively. Third, A, B, and AB also denote the contrast vectors shown in Table 7.2. These conventions extend to more than two factors. The sense in which these notations are used in the sequel will be generally clear from the context.

The interpretation of the effects defined in (7.3) is as follows. The main effects, A and B, each equal to the difference between the mean response at the high level of the given factor versus the mean response at the low level of that factor averaged over the two levels of the other factor. The interaction AB is obtained by taking the difference of the two differences in the mean response at the high level of A versus that at the low level of A over the two levels of B. Note that $AB = 0$ if the change in the mean response from low A to high A is the same at both levels of B or equivalently if the change in the mean response from low B to high B is the same at both levels of A; in other words, if the main effect of a factor does not depend on the level of the other factor.

The **model matrix** for the regression model (7.1) is the same as Table 7.2 of coefficients (without the multipliers) and is given by

$$\boldsymbol{X} = \begin{bmatrix} 1 & -1 & -1 & 1 \\ 1 & 1 & -1 & -1 \\ 1 & -1 & 1 & -1 \\ 1 & 1 & 1 & 1 \end{bmatrix}.$$

Let $\bar{\boldsymbol{y}} = (\bar{y}_{(1)}, \bar{y}_a, \bar{y}_b, \bar{y}_{ab})'$ be the response vector. The model matrix is orthogonal with $\boldsymbol{X}'\boldsymbol{X} = 4\boldsymbol{I}$, where $\boldsymbol{I}$ is a 4×4 identity matrix. Therefore, using $\widehat{\boldsymbol{\beta}} = (\boldsymbol{X}'\boldsymbol{X})^{-1}\boldsymbol{X}'\bar{\boldsymbol{y}}$ from (2.29), the LS estimator of $\boldsymbol{\beta}$ is given by

$$\widehat{\boldsymbol{\beta}} = \begin{bmatrix} \widehat{\beta}_0 \\ \widehat{\beta}_1 \\ \widehat{\beta}_2 \\ \widehat{\beta}_{12} \end{bmatrix} = \frac{1}{4} \begin{bmatrix} \bar{y}_{ab} + \bar{y}_b + \bar{y}_a + \bar{y}_{(1)} \\ \bar{y}_{ab} - \bar{y}_b + \bar{y}_a - \bar{y}_{(1)} \\ \bar{y}_{ab} + \bar{y}_b - \bar{y}_a - \bar{y}_{(1)} \\ \bar{y}_{ab} - \bar{y}_b - \bar{y}_a + \bar{y}_{(1)} \end{bmatrix}. \tag{7.3}$$

The estimators of the effects are given by

$$\widehat{A} = 2\widehat{\beta}_1, \qquad \widehat{B} = 2\widehat{\beta}_2, \qquad \widehat{AB} = 2\widehat{\beta}_{12}.$$

From $\mathrm{Cov}(\widehat{\boldsymbol{\beta}}) = (\sigma^2/n)(\boldsymbol{X}'\boldsymbol{X})^{-1} = (\sigma^2/4n)\boldsymbol{I}$, it follows that the $\widehat{\beta}_j$ have a common variance $\sigma^2/4n$; furthermore, they are uncorrelated and, due to the normality assumption, are independent. Similarly, all the estimated main effects and interactions have a common variance σ^2/n and are independent. The common standard error of all $\widehat{\beta}_j$ is $s/\sqrt{4n}$ and that of all the estimated main effects and interactions is $s/\sqrt{n}$. These facts are used in Section 7.2 to make inferences on the β_j's. The following example shows the calculation of these effects.

Example 7.1 (Mouse Embryo Bioassay: Calculation of Main Effects and Interactions)

Gorrill et al. (1991) report experiments for comparing hamster sperm motility assay with mouse embryo bioassay as quality control tests for in vitro fertilization (IVF). In the mouse two-cell embryo bioassay part of the experiment they studied the ability of the assay to discriminate between four culture media prepared by combining each one of two solutions (modified Ham's F-10 solution, labeled as solution 1, or modified Tyrode's solution, labeled as solution 2) with each one of two types of water (tap water and MilliQ or ultrapure water). The response variable was the percentage of total embryos reaching a certain development stage. This response variable was measured by three analysts for samples of embryos taken from four female mice. Here we only analyze the average percentages for the four culture media shown in Table 7.3.

The overall sample mean equals

$$\widehat{\beta}_0 = \tfrac{1}{4}[77.53 + 54.15 + 79.25 + 55.18] = 66.5275.$$

The estimates of the main effects are

$$\widehat{\beta}_1 = \tfrac{1}{4}[(77.53 - 54.15) + (79.25 - 55.18)] = 11.6375 \quad \text{or} \quad \widehat{A} = 23.275$$

Table 7.3 Average Percentage of Mouse Two-Cell Embryos Reaching Development Stage

	Water (B)	
Solution (A)	Tap	MilliQ
1	55.18	54.15
2	79.25	77.53

and

$$\widehat{\beta}_2 = \tfrac{1}{4}[(77.53 - 79.25) + (54.15 - 55.18)] = -0.6875 \quad \text{or} \quad \widehat{B} = -1.375.$$

The estimate of the interaction effect is

$$\widehat{\beta}_{12} = \tfrac{1}{2}[(77.53 - 54.15) - (79.25 - 55.18)] = -0.1725 \quad \text{or} \quad \widehat{AB} = -0.345.$$

We see that only the solution effect is large with solution 2 yielding much higher percentages of embryos reaching development stage. The difference between the two types of water and the interaction between the solutions and the types of water are both negligible in comparison. ■

7.1.2 2^3 Designs

Consider three factors, A, B, and C, and assume that n observations are taken at each of the $2^3 = 8$ treatment combinations. Using the notation for treatment combinations introduced earlier, the data can be represented in a tabular form as shown in Table 7.4. The sample means $\overline{y}_i$ are assumed to be independent $N(\mu_i, \sigma^2)$ for $i = (1), a, b, ab, c, ac, bc, abc$.

As in the case of a 2^2 design, we can fit the following regression model to these data:

$$\begin{aligned}\overline{y} &= \beta_0 + \beta_1 x_1 + \beta_2 x_2 + \beta_3 x_3 + \beta_{12} x_1 x_2 \\ &\quad + \beta_{13} x_1 x_3 + \beta_{23} x_2 x_3 + \beta_{123} x_1 x_2 x_3 + e,\end{aligned} \tag{7.4}$$

Table 7.4 Data from Balanced 2^3 Design

Factor			Treatment		
A	B	C	Combination	Data	Mean
−	−	−	(1)	$y_{(1)1}, \ldots, y_{(1)n}$	$\overline{y}_{(1)}$
+	−	−	a	$y_{a1}, \ldots, y_{an}$	$\overline{y}_a$
−	+	−	b	$y_{b1}, \ldots, y_{bn}$	$\overline{y}_b$
+	+	−	ab	$y_{ab1}, \ldots, y_{abn}$	$\overline{y}_{ab}$
−	−	+	c	$y_{c1}, \ldots, y_{cn}$	$\overline{y}_c$
+	−	+	ac	$y_{ac1}, \ldots, y_{acn}$	$\overline{y}_{ac}$
−	+	+	bc	$y_{bc1}, \ldots, y_{bcn}$	$\overline{y}_{bc}$
+	+	+	abc	$y_{abc1}, \ldots, y_{abcn}$	$\overline{y}_{abc}$

Source: Tamhane and Dunlop (2000, Table 1). Reprinted by permission of Pearson Education, Inc.

where $e \sim N(0, \sigma^2/n)$ and

$$x_1 = \begin{cases} -1 & \text{if } A \text{ is low,} \\ +1 & \text{if } A \text{ is high,} \end{cases}$$

$$x_2 = \begin{cases} -1 & \text{if } B \text{ is low,} \\ +1 & \text{if } B \text{ is high,} \end{cases}$$

$$x_3 = \begin{cases} -1 & \text{if } C \text{ is low,} \\ +1 & \text{if } C \text{ is high.} \end{cases}$$

The regression coefficients in this model are related to the effects used in the factorial experiments literature as follows:

$$I = \beta_0, \quad A = 2\beta_1, \quad B = 2\beta_2, \quad C = 2\beta_3,$$
$$AB = 2\beta_{12}, \quad AC = 2\beta_{13}, \quad BC = 2\beta_{23}, \quad ABC = 2\beta_{123}, \tag{7.5}$$

where I is the grand mean effect, A, B, C are the main effects, AB, AC, BC are the two-factor interactions, and ABC is the three-factor interaction. It is easy to show that, except for I, the remaining effects are contrasts among the μ_i's. The coefficients of the μ_i's in the definitions of all the effects are summarized in Table 7.5. The coefficients of the μ_i's in the definitions of the β's in the regression model (7.4) are the same except they all have the multiplier $\frac{1}{8}$.

From Table 7.5 we see that the main-effect contrast for each factor has minus signs for those treatment combinations in which that factor is at the low level and plus signs for those treatment combinations in which that factor is at the high level. Any interaction contrast coefficient vector is obtained by taking the componentwise product of the main-effect contrast coefficient vectors. Thus

$$AB = A \times B, \quad BC = B \times C, \quad AC = A \times C, \quad ABC = A \times B \times C.$$

Table 7.5 Contrast Coefficients for Effects in 2^3 Design

Treatment	Effect							
Combination	I	A	B	AB	C	AC	BC	ABC
(1)	+	−	−	+	−	+	+	−
a	+	+	−	−	−	−	+	+
b	+	−	+	−	−	+	−	+
ab	+	+	+	+	−	−	−	−
c	+	−	−	+	+	−	−	+
ac	+	+	−	−	+	+	−	−
bc	+	−	+	−	+	−	+	−
abc	+	+	+	+	+	+	+	+
Multiplier	$\frac{1}{8}$	$\frac{1}{4}$	$\frac{1}{4}$	$\frac{1}{4}$	$\frac{1}{4}$	$\frac{1}{4}$	$\frac{1}{4}$	$\frac{1}{4}$

In these componentwise products, the powers of the letters follow mod 2 arithmetic. For example, $AB \times BC = AB^2C = AC$. This is because the componentwise square of any column gives the I column with all plus signs. This method can be used to obtain signs of the contrast coefficients for higher order factorial designs.

The interpretations of the effects are as follows. Each main effect, A, B, and C, is the average of the differences between the mean responses at the high level of the respective factor versus the mean responses at the low level of that factor, keeping the levels of the other factors fixed. For example, the main effect A is the average of the four differences $\mu_{abc} - \mu_{bc}$, $\mu_{ac} - \mu_c$, $\mu_{ab} - \mu_b$, and $\mu_a - \mu_{(1)}$ or, equivalently, the difference between the two average differences $\frac{1}{4}(\mu_{abc} + \mu_{ac} + \mu_{ab} + \mu_a)$ and $\frac{1}{4}(\mu_{bc} + \mu_c + \mu_b + \mu_{(1)})$ at the high level of A and the low level of A, respectively.

To understand the two-factor interactions, consider AB, for example. Using the definition of AB given for a 2^2 design previously, we have the following two expressions:

$$AB = \tfrac{1}{2}[(\mu_{abc} - \mu_{bc}) - (\mu_{ac} - \mu_c)] \quad \text{if } C \text{ is high}$$

and

$$AB = \tfrac{1}{2}[(\mu_{ab} - \mu_b) - (\mu_a - \mu_{(1)})] \quad \text{if } C \text{ is low.}$$

The overall AB interaction is defined as the average of these two interactions, whereas the three-factor interaction ABC is defined as the difference between these two interactions (or, equivalently, the difference between the AC interactions at high B and low B or the difference between the BC interactions at high A and low A).

The LS estimators of the β's or the corresponding effects can be obtained by using $\widehat{\boldsymbol{\beta}} = (\boldsymbol{X}'\boldsymbol{X})^{-1}\boldsymbol{X}'\overline{\boldsymbol{y}}$, where the model matrix $\boldsymbol{X}$ is an orthogonal 8×8 matrix whose columns are the same as the columns of the coefficients in Table 7.5 without the multipliers. Hence we get $\boldsymbol{X}'\boldsymbol{X} = 8\boldsymbol{I}$. The first element of $\boldsymbol{X}'\overline{\boldsymbol{y}}$ equals the sum of all $\overline{y}_i$'s and the remaining elements are the same mutually orthogonal contrasts among the $\overline{y}_i$'s as defined by the columns of $\boldsymbol{X}$. Thus we obtain

$$\widehat{\boldsymbol{\beta}} = \begin{bmatrix} \widehat{\beta}_0 \\ \widehat{\beta}_1 \\ \widehat{\beta}_2 \\ \widehat{\beta}_3 \\ \widehat{\beta}_{12} \\ \widehat{\beta}_{13} \\ \widehat{\beta}_{23} \\ \widehat{\beta}_{123} \end{bmatrix} = \frac{1}{8} \begin{bmatrix} \{(\overline{y}_{abc} + \overline{y}_{bc}) + (\overline{y}_{ac} + \overline{y}_c)\} + \{(\overline{y}_{ab} + \overline{y}_b) + (\overline{y}_a + \overline{y}_{(1)})\} \\ \{(\overline{y}_{abc} - \overline{y}_{bc}) + (\overline{y}_{ac} - \overline{y}_c)\} + \{(\overline{y}_{ab} - \overline{y}_b) + (\overline{y}_a - \overline{y}_{(1)})\} \\ \{(\overline{y}_{abc} - \overline{y}_{ac}) + (\overline{y}_{bc} - \overline{y}_c)\} + \{(\overline{y}_{ab} - \overline{y}_a) + (\overline{y}_b - \overline{y}_{(1)})\} \\ \{(\overline{y}_{abc} - \overline{y}_{ab}) + (\overline{y}_{bc} - \overline{y}_b)\} + \{(\overline{y}_{ac} - \overline{y}_a) + (\overline{y}_c - \overline{y}_{(1)})\} \\ \{(\overline{y}_{abc} - \overline{y}_{bc}) - (\overline{y}_{ac} - \overline{y}_c)\} + \{(\overline{y}_{ab} - \overline{y}_b) - (\overline{y}_a - \overline{y}_{(1)})\} \\ \{(\overline{y}_{abc} - \overline{y}_{bc}) - (\overline{y}_{ab} - \overline{y}_b)\} + \{(\overline{y}_{ac} - \overline{y}_c) - (\overline{y}_a - \overline{y}_{(1)})\} \\ \{(\overline{y}_{abc} - \overline{y}_{ac}) - (\overline{y}_{ab} - \overline{y}_a)\} + \{(\overline{y}_{bc} - \overline{y}_c) - (\overline{y}_b - \overline{y}_{(1)})\} \\ \{(\overline{y}_{abc} - \overline{y}_{bc}) - (\overline{y}_{ac} - \overline{y}_c)\} - \{(\overline{y}_{ab} - \overline{y}_b) - (\overline{y}_a - \overline{y}_{(1)})\} \end{bmatrix}. \tag{7.6}$$

As before, the estimators of the effects are given by

$$\widehat{A} = 2\widehat{\beta}_1, \qquad \widehat{B} = 2\widehat{\beta}_2, \qquad \widehat{C} = 2\widehat{\beta}_3,$$

$$\widehat{AB} = 2\widehat{\beta}_{12}, \qquad \widehat{AC} = 2\widehat{\beta}_{13}, \qquad \widehat{BC} = 2\widehat{\beta}_{23}, \qquad \widehat{ABC} = 2\widehat{\beta}_{123}.$$

From $\text{Cov}(\widehat{\boldsymbol{\beta}}) = (\sigma^2/n)(\boldsymbol{X}'\boldsymbol{X})^{-1} = (\sigma^2/8n)\boldsymbol{I}$, it follows that the $\widehat{\beta}_j$ have a common variance $\sigma^2/8n$; furthermore they are uncorrelated and, due to the normality assumption, are independent. Similarly, all the estimated main effects and interactions have a common variance $\sigma^2/2n$ and are independent. The common standard error of all $\widehat{\beta}_j$ is $s/\sqrt{8n}$ and that of all the estimated main effects and interactions is $s/\sqrt{2n}$, where $s^2 = \text{MS}_e = \sum_i \sum_j (y_{ij} - \overline{y}_i)^2/8(n-1)$ is an estimate of σ^2 with $\nu = 8(n-1)$ d.f. These facts are used in Section 7.2 to make inferences on the β_j's.

The order in which the treatment combinations are listed in Tables 7.4 and 7.5 is called the **standard order**. In this order, factor A levels alternate signs in singletons $(-, +, -, +, \ldots)$, factor B levels alternate signs in pairs $(-, -, +, +, \ldots)$, factor C levels alternate signs in quadruplets $(-, -, -, -, +, +, +, +, \ldots)$, and so on. Additional factors can be appended to the contrast coefficients table in an obvious manner by following the standard order. The effects, namely $I, A, B, AB, \ldots$, are also listed in the standard order in Table 7.5. It should be emphasized that the **run order** (the order in which the experimental runs are made) should always be randomized subject to any practical constraints.

Example 7.2 (Bicycle Data: Estimation of the Effects)

(From Tamhane and Dunlop 2000, Example 13.7. Reprinted by permission of Pearson Education, Inc.) Box, Hunter, and Hunter (2005, pp. 215–217) describe a student project in which a student studied the effects of bicycle seat height, generator use, and tire pressure on the time taken to make a half-block uphill run. The study was done as a 2^3 design replicated twice ($n = 2$). The levels of the factors were as follows:

Seat height (factor A):	26 in. $(-)$, 30 in. $(+)$,
Generator (factor B):	off $(-)$, on $(+)$,
Tire pressure (factor C):	40 psi $(-)$, 55 psi $(+)$.

The data are shown in Table 7.6. The estimated effects can be calculated as follows.

Main effects:

$$\begin{aligned}\widehat{A} &= \tfrac{1}{4}\{[(42.5 - 52.0) + (39.0 - 49.0)] \\ &\quad + [(43.5 - 57.0) + (42.0 - 52.5)]\} = -10.875,\end{aligned}$$

Table 7.6 Travel Times from Bicycle Experiment

Factor			Time (sec.)		
A	B	C	Run 1	Run 2	Mean
−	−	−	51	54	52.5
+	−	−	41	43	42.0
−	+	−	54	60	57.0
+	+	−	44	43	43.5
−	−	+	50	48	49.0
+	−	+	39	39	39.0
−	+	+	53	51	52.0
+	+	+	41	44	42.5

Source: Box, Hunter, and Hunter (2005, Table 5.16). Reprinted by permission of John Wiley & Sons, Inc.

$$\widehat{B} = \tfrac{1}{4}\{[(42.5 - 39.0) + (52.0 - 49.0)] + [(43.5 - 42.0) + (57.0 - 52.5)]\} = +3.125,$$

$$\widehat{C} = \tfrac{1}{4}\{[(42.5 - 43.5) + (52.0 - 57.0)] + [(39.0 - 42.0) + (49.0 - 52.5)]\} = -3.125.$$

Interactions:

$$\widehat{AB} = \tfrac{1}{4}\{[(42.5 - 52.0) - (39.0 - 49.0)] + [(43.5 - 57.0) - (42.0 - 52.5)]\} = -0.625,$$

$$\widehat{AC} = \tfrac{1}{4}\{[(42.5 - 52.0) - (43.5 - 57.0)] + [(39.0 - 49.0) - (42.0 - 52.5)]\} = +1.125,$$

$$\widehat{BC} = \tfrac{1}{4}\{[(42.5 - 39.0) - (43.5 - 42.0)] + [(52.0 - 49.0) - (57.0 - 52.5)]\} = +0.125,$$

$$\widehat{ABC} = \tfrac{1}{4}\{[(42.5 - 52.0) - (39.0 - 49.0)] - [(43.5 - 57.0) - (42.0 - 52.5)]\} = +0.875.$$

Only the main effects are large; all interactions are small in comparison. Formal statistical tests will be done in Example 7.3 to confirm these findings. The $\widehat{A}$ and $\widehat{C}$ main effects are negative, so the high levels of these factors (seat height 30 in., tire pressure 55 psi) reduce the travel time. The $\widehat{B}$ main effect is positive, so the generator turned off reduces the travel time. If there were significant interactions, then the optimal settings of the factors cannot be determined solely from the main effects. ■

7.1.3 2^p Designs

We now generalize the definitions of the main effects and interactions to $p > 3$ factors. Denote the factors by $A, B, \ldots, P$ or alternatively by $1, 2, \ldots, p$ and the treatment combinations by $\boldsymbol{i} = (i_1 i_2 \ldots i_p)$, where i_k is the kth lowercase letter if the kth factor is at high level ($x_k = +1$) and i_k is suppressed if the kth factor is at low level ($x_k = -1$). For example, for $p = 3$, if A is at high level ($x_1 = +1$), B is at low level ($x_2 = -1$), and C is at high level ($x_3 = +1$), then $i_1 = a$, i_2 is suppressed, and $i_3 = c$, which is the treatment combination $\boldsymbol{i} = ac$. If all $x_k = -1$, then we set $\boldsymbol{i} = (1)$ as before. Further let $\boldsymbol{j} = (j_1 j_2 \ldots j_p)$ be the index of an effect representing a main effect or an interaction such that $j_k = k$ if the kth factor is involved in the main effect or interaction and j_k is suppressed if the kth factor is not involved. For example, $\boldsymbol{j} = 1$ for the main effect of A, $\boldsymbol{j} = 13$ for the interaction AC, and so on. The corresponding regression coefficient is denoted by β_j (e.g., β_1 and β_{13}).

To illustrate the above notation, consider $p = 4$. In this case there are 16 treatment combinations $\boldsymbol{i} = (1), a, b, ab, \ldots, abcd$. Similarly, there are 16 effects (including the grand mean effect) $\boldsymbol{j} = 0, 1, 2, 12, \ldots, 1234$ or, equivalently, $I, A, B, AB, \ldots, ABCD$.

Suppose that n i.i.d. observations $y_{ih} \sim N(\mu_i, \sigma^2)$ $(1 \le h \le n)$ are taken at the $\boldsymbol{i}$th treatment combination. As before, we need only consider the sufficient statistics: the sample means $\overline{y}_i$ and the pooled mean square error

$$\text{MS}_e = s^2 = \frac{\sum_i \sum_{h=1}^{n} (y_{ih} - \overline{y}_i)^2}{2^p(n-1)} \tag{7.7}$$

with $\nu = 2^p(n-1)$ d.f.

The full model for this design is

$$\begin{aligned}\overline{y} = \beta_0 + \sum_k \beta_k x_k + \sum_{k<\ell} \beta_{k\ell} x_k x_\ell + \sum_{k<\ell<m} \beta_{k\ell m} x_k x_\ell x_m \\ + \cdots + \beta_{12\cdots p} x_1 x_2 \cdots x_p + e,\end{aligned} \tag{7.8}$$

where $e \sim N(0, \sigma^2/n)$. The model for all $\overline{y}_i$'s can be written in matrix form as $\overline{\boldsymbol{y}} = \boldsymbol{X\beta} + \boldsymbol{e}$. The $2^p \times 2^p$ matrix $\boldsymbol{X}'\boldsymbol{X}$ equals $2^p \boldsymbol{I}$ and $\boldsymbol{X}'\overline{\boldsymbol{y}}$ is a $2^p \times 1$ vector whose first element equals the sum of all $\overline{y}_i$'s and the remaining elements are mutually orthogonal contrasts with coefficients $c_{ij} = \pm 1$ among the $\overline{y}_i$'s. Therefore the elements of $\widehat{\boldsymbol{\beta}} = (\boldsymbol{X}'\boldsymbol{X})^{-1}\boldsymbol{X}'\overline{\boldsymbol{y}}$ are given by

$$\widehat{\beta}_0 = \frac{\sum_i \overline{y}_i}{2^p} = \overline{\overline{y}} \quad \text{and} \quad \widehat{\beta}_j = \frac{\sum_i c_{ij} \overline{y}_i}{2^p}. \tag{7.9}$$

The estimated main effects and interactions are obtained by multiplying the appropriate $\widehat{\beta}$'s by 2; thus, for example,

$$\widehat{A} = 2\widehat{\beta}_1 = \frac{\sum_i c_{i1} \overline{y}_i}{2^{p-1}}, \qquad \widehat{B} = 2\widehat{\beta}_2 = \frac{\sum_i c_{i2} \overline{y}_i}{2^{p-1}}, \qquad \text{etc.} \tag{7.10}$$

The contrast coefficients for any main effect are given by $c_{ij} = -1$ for the treatment combinations $\boldsymbol{j}$ in which that factor is at low level and $c_{ij} = +1$ for the treatment combinations $\boldsymbol{j}$ in which that factor is at high level. The contrast coefficients for any interaction are obtained by taking a componentwise product of the contrast vectors for the corresponding main effects.

From

$$\mathrm{Cov}(\widehat{\boldsymbol{\beta}}) = \left(\frac{\sigma^2}{n}\right)(\boldsymbol{X}'\boldsymbol{X})^{-1} = \left(\frac{\sigma^2}{2^p n}\right)\boldsymbol{I} = \left(\frac{\sigma^2}{N}\right)\boldsymbol{I},$$

where $N = 2^p n$ is the total sample size, it follows that the $\widehat{\beta}_j$ have a common variance σ^2/N; furthermore, they are uncorrelated and, due to the normality assumption, are independent. Similarly, all the estimated main effects and interactions have a common variance $\sigma^2/(N/4)$ and are independent. The common standard error of all $\widehat{\beta}_j$ is $s/\sqrt{N}$ and that of all the estimated main effects and interactions is $s/\sqrt{N/4}$. These facts are used in Section 7.2 to make inferences on the β_j's.

It should be noted that, in case of quantitative factors, the main and interaction effects are related to the linear effects of the factors. For example, the main effects are averaged linear effects, two-factor interactions are differences between the linear effects of one factor at the two levels of the other factor for each pair of factors, and so on. This is also clear from the regression model (7.8), which is linear in all x_j's since they occur with only their first powers. Quadratic effects of the quantitative factors cannot be estimated using 2^p designs or their fractions. Second-order designs, discussed in Section 10.1.3, are required for this purpose.

7.2 STATISTICAL ANALYSIS

7.2.1 Confidence Intervals and Hypothesis Tests

As noted above, the standard error of $\widehat{\beta}_j$ and that of the corresponding effect equal

$$\mathrm{SE}(\widehat{\beta}_j) = \frac{s}{\sqrt{N}} \quad \text{and} \quad \mathrm{SE}(\widehat{\mathrm{Effect}_j}) = \frac{s}{\sqrt{N/4}}. \tag{7.11}$$

Using the distributional results from Section 2.3.2.3, it follows that a $100(1-\alpha)\%$ CI for any β_j is given by

$$\widehat{\beta}_j \pm t_{\nu,\alpha/2}\frac{s}{\sqrt{N}}, \tag{7.12}$$

where $\nu = 2^p(n-1)$. The corresponding t-statistic for testing $H_{0j} : \beta_j = 0$ is

$$t_j = \frac{\widehat{\beta}_j}{s/\sqrt{N}} = \frac{\widehat{\beta}_j\sqrt{N}}{s}. \tag{7.13}$$

Equivalently, one can use the following F-statistic with one and ν d.f.:

$$F_j = t_j^2 = \frac{N\widehat{\beta}_j^2}{s^2} = \frac{N\left(\widehat{\text{Effect}_j}\right)^2}{4s^2}. \tag{7.14}$$

H_{0j} is rejected at level α if

$$|t_j| > t_{\nu,\alpha/2} \Longleftrightarrow F_j > f_{1,\nu,\alpha}. \tag{7.15}$$

The above test does not take into account the multiplicity of inferences on the effects. To take multiplicity into account, we must control the FWE at level α (see Chapter 4). This can be done by noting that the estimated effects are orthogonal contrasts. Using the methods of Section 4.4.1, we see that the appropriate distribution to use is the **Studentized maximum modulus distribution**; its required critical point is denoted by $|M|_{2^p-1,\nu,\alpha}$, which is tabulated in Table C.9. The maximum modulus test rejects H_{0j} at level α if $|t_j| > |M|_{2^p-1,\nu,\alpha}$.

Example 7.3 (Bicycle Data: Tests of Significance of Effects)

(From Tamhane and Dunlop, 2000, Example 13.8. Reprinted by permission of Pearson Education, Inc.) To test the significance of the individual effects calculated in Example 7.2, we first need to compute MS_e. This can be obtained by regarding the design as a one-way layout with eight treatments and computing the resulting ANOVA table shown in Display 7.1.

Using (7.11), the common standard error of all $\widehat{\beta}$'s equals $\text{SE}(\widehat{\beta}) = s/\sqrt{N} = 2.046/\sqrt{16} = 0.512$, and the common standard error of all estimated effects equals $2 \times 0.512 = 1.024$. The t-statistics are obtained by dividing the estimated effects by their common standard error and are given in Table 7.7. For separate comparisons the t-statistics are compared with $t_{8,0.05} = 1.860$. For multiple comparisons they are compared with $|M|_{7,8,0.10} = 2.96$. Using either method we find that only the main effects are significant. Therefore we may use the main-effects model for these data. ■

7.2.2 Analysis of Variance

The ANOVA in Display 7.1 ignores the **factorial structure** of the treatments. By taking this structure into account, we can partition SS_{trt} into single d.f. orthogonal

```
Source          DF       SS       MS       F       P
Bike Setting     7  560.938   80.134   19.14   0.000
Error            8   33.500    4.188
Total           15  594.438
S = 2.04634    R-Sq = 94.36%    R-Sq(adj) = 89.43%
```

Display 7.1 One-way ANOVA of bicycle data.

Table 7.7 Estimates and Test Statistics for Effects from Bicycle Data

Effect	Estimate	t
A	−10.875	−10.628
B	3.125	3.054
C	−3.125	−3.054
AB	−0.625	−0.610
AC	1.125	1.098
BC	0.125	0.122
ABC	0.875	0.855

components (each of which corresponds to a main effect or an interaction) and do a finer analysis of the data as follows.

The F-statistic of (7.14) can be viewed as the ratio of two mean squares. The numerator is

$$\text{MS}_{\text{Effect}} = \text{SS}_{\text{Effect}} = \frac{N}{4}(\widehat{\text{Effect}})^2 \tag{7.16}$$

with one d.f and the denominator is $s^2 = \text{MS}_e$ with $\nu = 2^p(n-1)$ d.f. Since the estimated effects are mutually orthogonal contrasts among the $\overline{y}$'s, the sums of squares for them add up to SS_{trt} for the 2^p treatment combinations. Furthermore, under the overall null hypothesis, the sums of squares are independently distributed as $\sigma^2\chi_1^2$ r.v.'s. These results follow from Cochran's theorem (see Section A5). If $\text{SS}_A, \text{SS}_B, \ldots, \text{SS}_{AB}, \ldots, \text{SS}_{AB\cdots P}$ denote the sums of squares for the $2^p - 1$ effects, then we have

$$\text{SS}_{\text{trt}} = n\sum_i(\overline{y}_i - \overline{\overline{y}})^2 = \text{SS}_A + \text{SS}_B + \cdots + \text{SS}_{AB\cdots P}.$$

For example, in a 2^3 design, SS_{trt} has $2^3 - 1 = 7$ d.f. and can be partitioned into seven sums of squares, each with one d.f., as follows:

$$\text{SS}_{\text{trt}} = \text{SS}_A + \text{SS}_B + \text{SS}_C + \text{SS}_{AB} + \text{SS}_{AC} + \text{SS}_{BC} + \text{SS}_{ABC}.$$

A detailed ANOVA table for the bicycle data is given in Table 7.8. Note that the sums of squares for all the effects add up to 560.9375, which is the sum of squares for the eight treatment combinations as seen from Display 7.1. The three estimated main effects are significant at $\alpha = 0.05$, while all interactions are nonsignificant.

Table 7.8 ANOVA Table for Bicycle Data

Effect	SS	d.f.	MS	F	p
A	473.0625	1	473.0625	112.97	0.000
B	39.0625	1	39.0625	9.33	0.016
C	39.0625	1	39.0625	9.33	0.016
AB	1.5625	1	1.5625	0.37	0.565
AC	5.0625	1	5.0625	1.21	0.304
BC	0.0625	1	0.0625	0.01	0.907
ABC	3.0625	1	3.0625	0.73	0.426
Error	33.5000	8	4.1875		
Total	594.9375	15			

7.2.3 Model Fitting and Diagnostics

We can summarize the results of the analysis in a final regression model in which only the significant terms are retained. Because of the orthogonal nature of the model, if any terms are dropped from the full model (7.8), the estimates of the remaining terms are unchanged; see Section 2.3.2.2.

The regression model is useful for predicting the response at any specified combination of the levels of the experimental factors. For a numerical factor, one can linearly interpolate between the low and high levels of the factor. In general, extrapolation outside the experimental range of a factor should be avoided because the fitted model may not hold. If L_i and H_i denote the low and high levels, respectively, of factor i, then the coded value corresponding to the actual value X_i is given by

$$x_i = \frac{X_i - (H_i + L_i)/2}{(H_i - L_i)/2}.$$

For a nominal factor it is meaningless to interpolate between the low and high levels. If a full model is fitted, then the predicted response for each cell equals the corresponding cell sample mean. If a reduced model is fitted, then the predicted responses will differ in general from the corresponding cell sample means. The differences between the observed responses and the corresponding predicted (fitted) responses are the residuals, which can be used for model diagnostics. We now illustrate this methodology by applying it to the bicycle data.

Example 7.4 (Bicycle Data: Fitted Regression Model and Residual Analysis)

(From Tamhane and Dunlop, 2000, Example 13.11. Reprinted by permission of Pearson Education, Inc.) The final main-effects model is

$$\widehat{y} = 47.1875 - 5.4375x_1 + 1.5625x_2 - 1.5625x_3.$$

The sums of squares for the dropped interaction terms may be pooled, namely,

$$\mathrm{SS}_{AB} + \mathrm{SS}_{AC} + \mathrm{SS}_{BC} + \mathrm{SS}_{ABC} = 1.5625 + 5.0625 + 0.0625 + 3.0625 = 9.750$$

with four d.f. and added to pure $\mathrm{SS}_e = 33.50$ with eight d.f. to obtain an overall pooled $\mathrm{SS}_e = 33.500 + 9.750 = 43.250$ with $8 + 4 = 12$ d.f. This gives pooled $\mathrm{MS}_e = 43.250/12 = 3.604$ and $s = \sqrt{3.604} = 1.898$.

Since the estimated effects of A (seat height) and C (tire pressure) are negative while that of B (generator on/off) is positive, the average travel time is minimized when A and C are set at high levels and B at low level. The minimum predicted travel time is

$$\widehat{y} = 47.1875 - 5.4375(+1) + 1.5625(-1) - 1.5625(+1) = 38.625 \text{ sec.}$$

Suppose we want to predict the travel time for the following combination: seat height 29 in., generator on, and tire pressure 45 psi. The corresponding coded values are as follows:

$$x_1 = \frac{29 - (30 + 26)/2}{(30 - 26)/2} = 0.5, \quad x_2 = 1, \quad x_3 = \frac{45 - (55 + 40)/2}{(55 - 40)/2} = -0.333.$$

The predicted travel time at these settings equals

$$\widehat{y} = 47.1875 - 5.4375(0.5) + 1.5625(1) - 1.5625(-0.333) = 46.552 \text{ sec.}$$

The standard error of this estimate can be computed using (2.40) as follows. Here $\boldsymbol{x} = (1, 0.5, 1, -0.333)'$ and $\mathrm{Cov}(\widehat{\boldsymbol{\beta}}) = (\sigma^2/16)\boldsymbol{I}$. Therefore

$$\mathrm{SE}(\widehat{y}) = 1.898\sqrt{\frac{1}{16}\boldsymbol{x}'\boldsymbol{x}} = 0.729.$$

Hence a 95% CI for the *average* travel time at these factor settings (using $t_{12,.025} = 2.179$) equals

$$[46.552 \pm (2.179)(0.729)] = [44.964, 48.140].$$

A 95% PI for *random* travel time on a ride using these factor settings equals

$$[46.552 \pm (2.179)(1.898)\sqrt{1 + \frac{1}{16}\boldsymbol{x}'\boldsymbol{x}}] = [46.552 \pm 4.430] = [42.122, 50.982].$$

Next we perform residual analyses to check model diagnostics. The normal and the fitted-value plots of the residuals are shown in Figure 7.1. Neither plot is fully satisfactory. Also, there is one outlier, namely, the travel time of 60 sec in run 2 for the treatment combination b. This observation has a residual of +4.25 with a standardized value of +2.58. The problem with the residual plots appears to be caused by this single outlier rather than the model itself, as can be checked by deleting this observation and refitting the model. ■

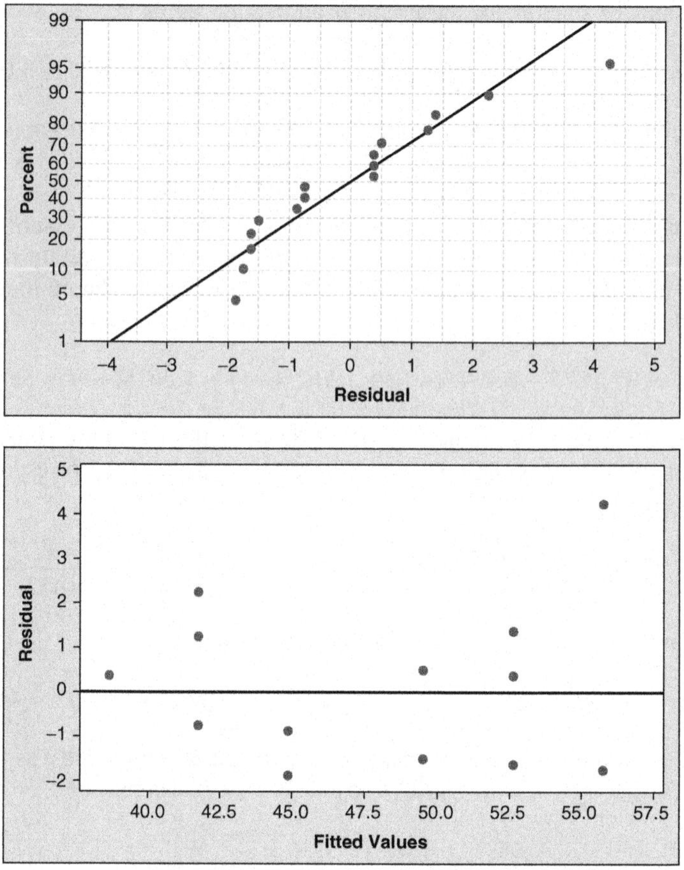

Figure 7.1 Normal (top) and fitted-values (bottom) plots of residuals for bicycle data.

7.3 SINGLE-REPLICATE CASE

If a factorial experiment is replicated only once ($n = 1$), then there are no error d.f. available for estimating σ^2 if a full model is fitted. Such a model is called a **saturated model**. In this model *exact* formal tests and confidence intervals given above cannot be used to assess the significance of the estimated effects because of the lack of an estimate of error. A graphical method based on a normal plot is usually employed instead and is discussed in Section 7.3.1. An approximate method due to Lenth (1989) may also be used and is discussed in Section 7.3.2.

7.3.1 Normal and Half-Normal Plots of Estimated Effects

When the number of effects, $2^p - 1$, is large, then according to the effect sparsity principle, a majority of them are likely to be small. Recall that the estimated effects are independent and normally distributed and have a common variance.

Therefore a majority of them can be thought of as a random sample from a normal distribution with a zero mean and a common variance. Hence, if we make their normal plot, then most of them will roughly fall along a straight line and a few large ones will appear as outliers. These outliers can be identified as significant effects. This approach may be used even for $n > 1$.

An alternative to a normal plot is a **half-normal plot** proposed by Daniel (1959). The basic idea here is that since the magnitudes and not the signs of the effects are of importance, we should plot the ordered *absolute* values of the estimated effects to assess their significance. However, they should be plotted versus the quantiles of the $|N(0, 1)|$ distribution (called the **folded normal distribution**). If we denote the $100(1 - \gamma)$th percentile of this distribution by $|z|_\gamma$ (i.e., $P\{|N(0, 1)| \leq |z|_\gamma\} = 1 - \gamma$), then $|z|_\gamma = z_{\gamma/2}$, where $z_{\gamma/2}$ is the $100(1 - \gamma/2)$th percentile of the $N(0, 1)$ distribution. For example, the 95th percentile of the folded normal distribution equals $|z|_{0.05} = z_{0.025} = 1.96$. If we are plotting $m = 2^p - 1$ absolute estimated effects, then for the ith ordered effect, $1 - \gamma$ equals $i/(m + 1) = i/2^p$. The advantage of this plot is that the significant effects show up only at the high end of the plot rather than at both ends, so the plot is easier to interpret. We recommend the half-normal plot, but if it is not available in the software, then the normal plot may be used without much difficulty. For the sake of illustration, we show both the normal and half-normal plots of the estimated effects for the bicycle data in Figure 7.2. They confirm that the three main effects are the only significant ones; all interactions are nonsignificant.

Once the significant effects are identified through a normal or a half-normal plot, one could fit a reduced model with only those effects. The sums of squares for the omitted effects may be pooled to form SS_e. Since the sums of squares of the effects are independent, each with one d.f., the error d.f. equals the number of pooled effects to form SS_e. The resulting MS_e can then be calculated by dividing the SS_e just obtained by the error d.f. and formal statistical inferences can be carried out about the retained effects. One could also begin by specifying a reduced model (e.g., a model that includes only the main effects and two-factor interactions) and use the sums of squares of the remaining effects to estimate σ^2 and carry out formal inferences. There are problems with both approaches. The first approach does not take into account type II errors that large effects may turn out to be nonsignificant due to lack of power. The problem with the second approach is that the ignored effects may not be actually negligible. In either case, the error estimate will be inflated.

Example 7.5 (Iron Concentration Data: Half-Normal Plots of the Estimated Effects)

(From Tamhane and Dunlop 2000, Example 13.13. Reprinted by permission of Pearson Education, Inc.) A student team in a quality control class at Northwestern University analyzed a process that makes a paramagnetic microparticle solution used in an immunodiagnostic testing machine for a health care products firm. The

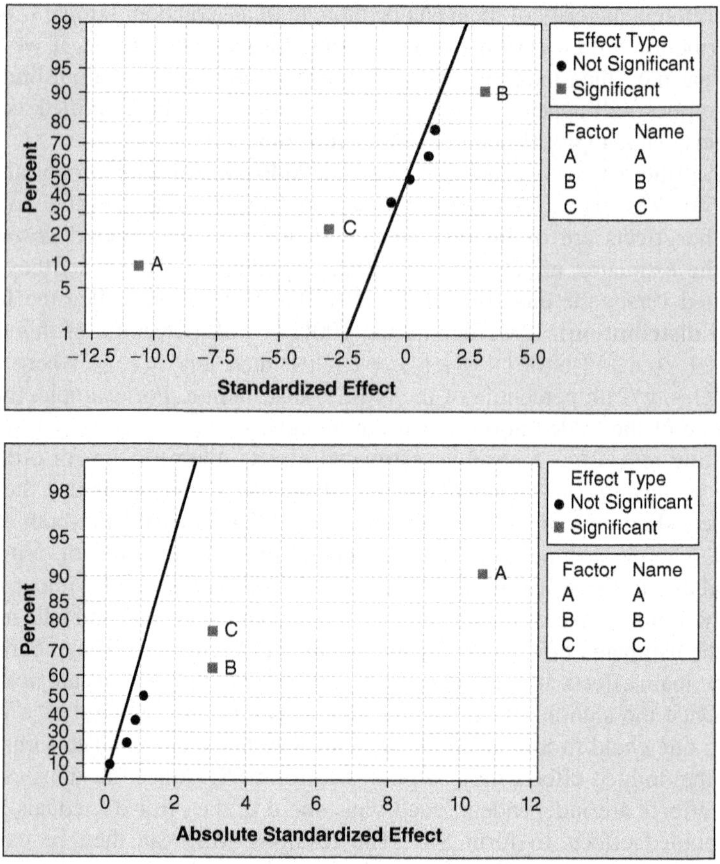

Figure 7.2 Normal (top) and half-normal (bottom) plots of effects for bicycle data.

purpose was to evaluate which of the four factors (each studied at two levels) listed in Table 7.9 affect the uniformity of the solution.

The 16 factor-level combinations were run in a random order. After each run a sample of the microparticle solution was taken from the top, middle, and bottom of the bottle using a pipette and the iron concentrations of the samples were measured. The uniformity of iron concentration was evaluated as the ratio of the maximum to the minimum of the three concentrations. The data (courtesy of Bruce Ankenman, Department of Industrial Engineering and Management Sciences at Northwestern University) are shown in Table 7.10.

When dealing with ratios it is preferable to analyze their logs. Table 7.10 lists the logs of the ratios (multiplied by 100) of the maximum to the minimum iron concentration. The effect estimates are calculated using the contrast coefficients in Table 7.11. The results are given in Table 7.12. The half-normal plot of these effect estimates is shown in Figure 7.3. From this plot we see that the $\widehat{C}$ main

Table 7.9 Factors and Their Levels for Iron Concentration Experiment

Factors	Low (−)	High (+)
Speed (A)	50 ft/min	150 ft/min
Bottle size (B)	2 L	50 L
Start % full (C)	20	85
Mix time (D)	10 min	30 min

Table 7.10 Iron Concentration Uniformity Data

Treatment Combination	Ratio of Maximum to Minimum Concentration	$100 \log_{10}$(Ratio)
(1)	1.0317	1.3553
a	1.0315	1.3469
b	1.0306	1.3090
ab	1.0491	2.0817
c	1.0275	1.1782
ac	1.0133	0.5738
bc	1.0203	0.8728
abc	1.0095	0.4106
d	1.0377	1.6072
ad	1.0431	1.8326
bd	1.0359	1.5318
abd	1.0542	2.2923
cd	1.0208	0.8941
acd	1.0402	1.7117
bcd	1.0177	0.7620
$abcd$	1.0393	1.6741

Source: Tamhane and Dunlop (2000, Table 13.20). Reprinted by permission of Pearson Education, Inc.

effect is by far the largest. The other large effects are $\widehat{A}$, $\widehat{D}$, $\widehat{AD}$, and $\widehat{ACD}$. The factor B (bottle size) does not seem to have much effect.

The sums of squares of the 10 negligible effects ($\widehat{B}$, $\widehat{AB}$, $\widehat{AC}$, $\widehat{BC}$, $\widehat{BD}$, $\widehat{CD}$, $\widehat{ABC}$, $\widehat{ABD}$, $\widehat{BCD}$, and $\widehat{ABCD}$) can be computed using formula (7.16) with $n = 1$ and $p = 4$. The pooled sum of squares equals 0.5664, which may be taken as the SS_e with 10 d.f. Hence $MS_e = 0.05664$ and $s = \sqrt{0.05664} = 0.2380$. Therefore the common standard error for testing the significance of the retained effects ($\widehat{A}$, $\widehat{C}$, $\widehat{D}$, $\widehat{AD}$, and $\widehat{ACD}$) can be calculated from (7.11) as

$$\text{SE}(\widehat{\text{Effect}}) = \frac{s}{\sqrt{N/4}} = \frac{0.2380}{\sqrt{16/4}} = 0.1190.$$

Table 7.11 Contrast Coefficients for Effects in 2^4 Design

Treatment	Effect															
Combination	I	A	B	AB	C	AC	BC	ABC	D	AD	BD	ABD	CD	ACD	BCD	$ABCD$
(1)	+	−	−	+	−	+	+	−	−	+	+	−	+	−	−	+
a	+	+	−	−	−	−	+	+	−	−	+	+	+	+	−	−
b	+	−	+	−	−	+	−	+	−	+	−	+	+	−	+	−
ab	+	+	+	+	−	−	−	−	−	−	−	−	+	+	+	+
c	+	−	−	+	+	−	−	+	−	+	+	−	−	+	+	−
ac	+	+	−	−	+	+	−	−	−	−	+	+	−	−	+	+
bc	+	−	+	−	+	−	+	−	−	+	−	+	−	+	−	+
abc	+	+	+	+	+	+	+	+	−	−	−	−	−	−	−	−
d	+	−	−	+	−	+	+	−	+	−	−	+	−	+	+	−
ad	+	+	−	−	−	−	+	+	+	+	−	−	−	−	+	+
bd	+	−	+	−	−	+	−	+	+	−	+	−	−	+	−	+
abd	+	+	+	+	−	−	−	−	+	+	+	+	−	−	−	−
cd	+	−	−	+	+	−	−	+	+	−	−	+	+	−	−	+
acd	+	+	−	−	+	+	−	−	+	+	−	−	+	+	−	−
bcd	+	−	+	−	+	−	+	−	+	−	+	−	+	−	+	−
$abcd$	+	+	+	+	+	+	+	+	+	+	+	+	+	+	+	+
Multiplier	$\frac{1}{16}$	$\frac{1}{8}$	$\frac{1}{8}$	$\frac{1}{8}$	$\frac{1}{8}$	$\frac{1}{8}$	$\frac{1}{8}$	$\frac{1}{8}$	$\frac{1}{8}$	$\frac{1}{8}$	$\frac{1}{8}$	$\frac{1}{8}$	$\frac{1}{8}$	$\frac{1}{8}$	$\frac{1}{8}$	$\frac{1}{8}$

Table 7.12 Effect Estimates for Iron Concentration Uniformity Data (Using 100 $\log_{10}$(Ratio) as Response Variable)

$\widehat{\beta}_0 = 1.3396$	$\widehat{A} = 0.3017$	$\widehat{B} = 0.0543$	$\widehat{AB} = 0.1941$
$\widehat{C} = -0.6599$	$\widehat{AC} = -0.1359$	$\widehat{BC} = -0.2139$	$\widehat{ABC} = -0.1349$
$\widehat{D} = 0.3972$	$\widehat{AD} = 0.3772$	$\widehat{BD} = -0.0007$	$\widehat{ABD} = -0.0367$
$\widehat{CD} = 0.1044$	$\widehat{ACD} = 0.3218$	$\widehat{BCD} = 0.0754$	$\widehat{ABCD} = 0.0248$

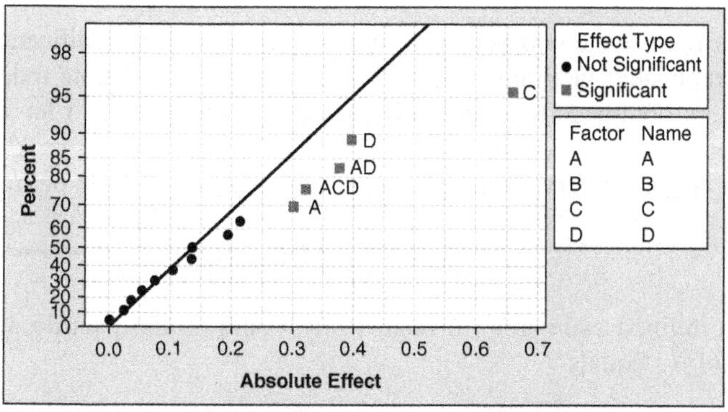

Figure 7.3 Half-normal plot of effects for iron concentration data.

We thus obtain the following t-statistics for the retained effects:

Effect	Estimate	t
A	0.3017	2.535
C	−0.6599	−5.545
D	0.3972	3.338
AD	0.3772	3.170
ACD	0.3218	2.704

Comparing the t-statistics with the critical constant $t_{10,0.05} = 1.812$, we see that all five estimated effects are significant at the $\alpha = 0.10$ level. To adjust for multiplicity of tests, if we use the Studentized maximum modulus procedure from Section 4.4.1 with the critical point $|M|_{5,10,0.10} = 2.678$, then only the effect $\widehat{A}$ fails to be significant.

The final model for predicting the uniformity of iron concentration is

$$\begin{aligned}\widehat{y} &= \overline{\overline{y}} + \frac{\widehat{A}}{2}x_1 + \frac{\widehat{C}}{2}x_3 + \frac{\widehat{D}}{2}x_4 + \frac{\widehat{AD}}{2}x_1x_4 + \frac{\widehat{ACD}}{2}x_1x_3x_4 \\ &= 1.3396 + 0.1508x_1 - 0.3300x_3 + 0.1986x_4 + 0.1886x_1x_4 + 0.1609x_1x_3x_4.\end{aligned}$$

We can check that the best combination of the factors to minimize $\widehat{y}$ (i.e., maximize the uniformity) is A low (speed = 50 ft/min), C high (start % full = 85), and D low (mix time = 10 min); the level of B is statistically unimportant and can be set based on economic and other side considerations. Note that because $x_1 = x_4 = -1$, there is a positive contribution due to the x_1x_4 interaction. However, if the levels of x_1 and x_4 are chosen to have opposite signs, then one of their main effects as well as the interaction $x_1x_3x_4$ will have positive contributions, thus failing to achieve the minimum value of $\widehat{y}$. ■

7.3.2 Lenth Method

Lenth's (1989) method provides approximate tests of significance and CIs for singly replicated factorial experiments. Let $\widehat{\theta}_1, \ldots, \widehat{\theta}_m$ denote independent effect estimators obtained from mutually orthogonal contrasts and let τ^2 denote their common variance. For an unreplicated 2^p full factorial, $m = 2^p - 1$ and $\tau^2 = \sigma^2/2^p$. A preliminary crude estimator of τ is given by (see below for an explanation)

$$\widehat{\tau}_0 = 1.5 \times \text{Median}(|\widehat{\theta}_i|).$$

A more refined estimator [referred to by Lenth as the **pseudo standard error (PSE)** of $\widehat{\theta}_i$] equals

$$\text{PSE} = 1.5 \times \text{Median}(|\widehat{\theta}_i| : |\widehat{\theta}_i| < 2.5\widehat{\tau}_0),$$

where the median of only those $|\widehat{\theta}_i|$ is taken that are less than $2.5\widehat{\tau}_0$ (again, see below for an explanation). Under $H_{0i} : \theta_i = 0$, the statistic $t_i = \widehat{\theta}_i/\text{PSE}$ can be regarded as approximately t-distributed with $m/3$ d.f. Thus each estimated effect can be tested at level α by comparing $|t_i|$ with $t_{m/3,\alpha/2}$. To adjust for multiplicity of tests, one may use the Studentized maximum modulus critical point instead.

Lenth's method is based on the following result: If a r.v. $Z \sim \text{N}(0, \tau^2)$, then the median of $|Z|$ approximately equals 0.675τ, and the 0.495th quantile approximately equals 0.665τ. Assuming initially (based on the effect sparsity principle) that all effects are null, $\widehat{\tau}_0$ is a consistent estimator of $1.5 \times 0.675\tau = 1.0125\tau$. This estimator is used to remove the potentially nonnull large effects, which exceed the two-sided 1% critical value of $2.5\widehat{\tau}_0$. This is done to make the estimator of τ robust to these "outlier" effects. The PSE is therefore an approximately consistent estimator of 1.5 times the 0.495th quantile of $|Z|$, which is $1.5 \times 0.665\tau \approx \tau$.

Example 7.6 (Iron Concentration Data: Lenth's Method)

Refer to Example 7.5 and consider the effect estimates tabulated in Table 7.12. The median of their absolute values is $|\widehat{AC}| = 0.1359$. Therefore $\widehat{\tau}_0 = 1.5 \times 0.1359 = 0.20385$. The only effect estimate with absolute value greater than $2.5 \times 0.20385 = 0.5096$ is $|\widehat{C}| = 0.6599$. The median of the absolute values of the remaining effect estimates is $(0.1349 + 0.1359)/2 = 0.1354$. Therefore $\text{PSE} = 1.5 \times 0.1354 = 0.2031$. Then the only t-statistic significant at the 0.05 level is that for $\widehat{C}$, which equals $0.6599/0.2031 = 3.249$ and exceeds the critical constant $t_{15/3,0.025} = t_{5,0.025} = 2.571$. On the other hand, the normal plot method identified additional effects $\widehat{A}$, $\widehat{D}$, $\widehat{AD}$, and $\widehat{ACD}$ as significant. ■

7.3.3 Augmenting a 2^p Design with Observations at the Center Point

As we have seen, replicate observations are needed to obtain a pure-error estimate of σ^2. Replicating the entire design can be costly if 2^p is large, and replicating only a fraction of the experiment may cause the design to lose its orthogonality property (see, however, Exercise 7.14 for an example in which 4 of the runs in a 2^3 experiment are replicated to obtain a 12-run orthogonal design). If all factors are quantitative, then a simple solution to this dilemma is to augment the factorial design with replicates at the center point of the design, which is the midpoint between the low and high levels of all factors. The level of each factor is coded as zero at this point. If we have $n_c \geq 2$ replicates at the center point, then we have $n_c - 1$ pure-error d.f. for estimating σ^2.

As an example, consider a 2^3 factorial design augmented by four center points. The design matrix is given by

$$\boldsymbol{D} = \begin{array}{c} \begin{matrix} A & B & C \end{matrix} \\ \begin{bmatrix} -1 & -1 & -1 \\ 1 & -1 & -1 \\ -1 & 1 & -1 \\ 1 & 1 & -1 \\ -1 & -1 & 1 \\ 1 & -1 & 1 \\ -1 & 1 & 1 \\ 1 & 1 & 1 \\ 0 & 0 & 0 \\ 0 & 0 & 0 \\ 0 & 0 & 0 \\ 0 & 0 & 0 \end{bmatrix} \end{array}.$$

The model matrix for fitting the model (7.4) is

$$\boldsymbol{X} = \begin{bmatrix} 1 & -1 & -1 & -1 & 1 & 1 & 1 & -1 \\ 1 & 1 & -1 & -1 & -1 & -1 & 1 & 1 \\ 1 & -1 & 1 & -1 & -1 & 1 & -1 & 1 \\ 1 & 1 & 1 & -1 & 1 & -1 & -1 & -1 \\ 1 & -1 & -1 & 1 & 1 & -1 & -1 & 1 \\ 1 & 1 & -1 & 1 & -1 & 1 & -1 & -1 \\ 1 & -1 & 1 & 1 & -1 & -1 & 1 & -1 \\ 1 & 1 & 1 & 1 & 1 & 1 & 1 & 1 \\ 1 & 0 & 0 & 0 & 0 & 0 & 0 & 0 \\ 1 & 0 & 0 & 0 & 0 & 0 & 0 & 0 \\ 1 & 0 & 0 & 0 & 0 & 0 & 0 & 0 \\ 1 & 0 & 0 & 0 & 0 & 0 & 0 & 0 \end{bmatrix}.$$

It is easy to check from this matrix that the estimates of the main effects and interactions are unaffected by the center-point replicates (which is explained by the fact that these are *linear* effects and hence depend only on the observations at the two extremes of each factor); only the estimate of the constant term is now the average of all 12 observations.

Another advantage of taking observations at the center point is that it allows testing of curvature in the model by fitting the following quadratic model:

$$y = \beta_0 + \sum_k \beta_k x_k + \sum_{k<\ell} \beta_{k\ell} x_k x_\ell + \sum_{k<\ell<m} \beta_{k\ell m} x_k x_\ell x_m$$

$$+ \cdots + \beta_{12\cdots p} x_1 x_2 \cdots x_p + \sum_k \beta_{kk} x_k^2 + e. \tag{7.17}$$

Since the x_k^2 terms are identical for all k, the individual β_{kk} terms are not estimable — only their sum. Exercise 7.9 asks you to show that $\overline{y}_f - \overline{y}_c$ is an unbiased (in fact, least squares) estimator of $\sum_k \beta_{kk}$, where $\overline{y}_f$ is the sample mean of the n_f observations at the factorial points (e.g., for a single replicate of a 2^p full factorial, $n_f = 2^p$) and $\overline{y}_c$ is the sample mean of the n_c observations at the center point.

Since $\text{Var}(\overline{y}_f - \overline{y}_{cs}) = \sigma^2(1/n_f + 1/n_c)$, a test of $H_0 : \sum_{k=1}^p \beta_{kk} = 0$ can be based on the t-statistic

$$t = \frac{\overline{y}_f - \overline{y}_c}{s\sqrt{1/n_f + 1/n_c}},$$

where $s = \sqrt{\text{MS}_e}$ is computed from the n_c replicates at the center point with $n_c - 1$ d.f. Therefore an α-level test of H_0 rejects if $|t| > t_{n_c-1,\alpha/2}$. This is called a **test for curvature**.

If some factors are qualitative, then we need to replicate the center points of the quantitative factors for each combination of the low and high levels of qualitative factors; for details see Exercises 7.10 and 7.11.

Example 7.7 (Effect of Production Factors on Viscosity of Mayonnaise: ANOVA and Test for Curvature)

Bjerke et al. (2004) report a response surface experiment to study the effects of three production factors: starch concentration (A), temperature in heat exchanger (B), and RPM in viscorotor process unit (C) on the viscosity of low-fat mayonnaise. Two storage factors, storage temperature and storing time, were also studied, but in this example we focus only on the production factors. In particular, we consider the data for mayonnaise samples stored for seven weeks in refrigerator at 4°C. The actual experimental design was a second-order design, a face-centered cube (see Section 10.1.3.1) with four center points, but for the purpose of this example we consider the data only at the factorial (vertex) points

Table 7.13 Mayonnaise Viscosity Data

A	B	C	Viscosity
−1	−1	−1	15
1	−1	−1	41
−1	1	−1	31
1	1	−1	81
−1	−1	1	30
1	−1	1	40
−1	1	1	49
1	1	1	95
0	0	0	35
0	0	0	48
0	0	0	43
0	0	0	45

Source: Bjerke et al. (2004, Table 2). Reprinted by permission of Taylor & Francis.

of the cube and the center point. These partial data are given in Table 7.13. The full data are given in Table 10.13.

The Minitab output is shown in Display 7.2. It is easy to check that the constant term estimate is the average of all 12 observations, but the main effect and interactions are estimated only from the eight factorial observations; the center point observations do not enter into their calculation. For example,

$$\widehat{A} = \tfrac{1}{4}[(95 - 49) + (40 - 30) + (81 - 31) + (41 - 15)] = 33.$$

```
Term          Effect     Coef  SE Coef       T      P
Constant               47.750    1.966   24.29  0.000
A             33.000   16.500    1.966    8.39  0.004
B             32.500   16.250    1.966    8.27  0.004
C             11.500    5.750    1.966    2.92  0.061
A*B           15.000    7.500    1.966    3.82  0.032
A*C           -5.000   -2.500    1.966   -1.27  0.293
B*C            4.500    2.250    1.966    1.14  0.335
A*B*C          3.000    1.500    1.966    0.76  0.501
Ct Pt                  -5.000    3.405   -1.47  0.238
S = 5.56028    R-Sq = 98.24%    R-Sq(adj) = 93.55%
Analysis of Variance for Viscosity (coded units)
Source              DF   Seq SS    Adj SS    Adj MS      F      P
Main Effects         3  4555.00   4555.00   1518.33  49.11  0.005
2-Way Interactions   3   540.50    540.50    180.17   5.83  0.091
3-Way Interactions   1    18.00     18.00     18.00   0.58  0.501
  Curvature          1    66.67     66.67     66.67   2.16  0.238
Residual Error       3    92.75     92.75     30.92
  Pure Error         3    92.75     92.75     30.92
Total               11  5272.92
```

Display 7.2 Minitab output for effect estimates and ANOVA of mayonnaise viscosity data in Table 7.13.

The coefficient of the center point is the negative of the estimate of $\sum_k \beta_{kk}$, which equals $\overline{y}_c - \overline{y}_f = 42.75 - 47.75 = -5$.

The pure-error estimate of σ^2 is the sample variance of the four observations at the center point and equals 30.92 with three d.f.; hence $s = \sqrt{30.92} = 5.560$. The t-statistic for the test of curvature equals

$$|t| = \frac{47.75 - 42.75}{5.560\sqrt{1/8 + 1/4}} = 1.469$$

with three d.f. which has a two-sided p-value $= 0.238$. Therefore no significant curvature is indicated. ■

7.4 2^P FACTORIAL DESIGNS IN INCOMPLETE BLOCKS: CONFOUNDING OF EFFECTS

7.4.1 Construction of Designs

When the number of treatment combinations is large, all the runs cannot be made under identical conditions if sufficient experimental material of uniform quality or time for experimentation is not available. Then the 2^p treatment combinations can be divided among two or more incomplete blocks and a BIB design (see Section 5.2) can be used. As a result, some of the treatment effects will be confounded with the block effects. In this section we study schemes for assigning the treatment combinations to the blocks in such a way that the treatment effects of less importance are confounded with the block effects. According to the hierarchical ordering principle, these are usually taken to be the highest order interactions. We assume that there are no treatment–block interactions, that is, we assume the model

$$\mathrm{E}(y_{ij}) = \mu_i + \beta_j \qquad [i = (1), a, b, ab,\ j = 1, 2].$$

First consider a 2^2 design in two blocks. Between two blocks there is one d.f. corresponding to the difference $\beta_2 - \beta_1$. To confound the interaction $AB = \frac{1}{2}[\mu_{ab} - \mu_b - \mu_a + \mu_{(1)}]$ with $\beta_2 - \beta_1$, we must assign the treatment combinations having the same sign in the AB contrast (see Table 7.2) to the same block. Thus, the blocking arrangement should be as shown in Display 7.3.

With this arrangement, both $\beta_2 - \beta_1$ and AB are estimated by $\frac{1}{2}[y_{(ab)} + y_{(1)} - y_a - y_b]$ and its expected value is

$$\frac{1}{2}[\mu_{(ab)} + \beta_2 + \mu_{(1)} + \beta_2 - \mu_a - \beta_1 - \mu_b - \beta_1]$$
$$= \frac{1}{2}[\mu_{(ab)} - \mu_a - \mu_b + \mu_{(1)}] + (\beta_2 - \beta_1) = AB + (\beta_2 - \beta_1).$$

Thus the two effects are confounded. On the other hand, the main effects $\widehat{A} = \frac{1}{2}[y_{ab} - y_b + y_a - y_{(1)}]$ and $\widehat{B} = \frac{1}{2}[y_{ab} - y_a + y_b - y_{(1)}]$ are free of the block effects because in these two contrasts a treatment combination with a plus sign

Block 1	Block 2
a	(1)
b	ab

Display 7.3 Blocking scheme for 2^2 design to confound AB interaction between two blocks.

is matched with another one with a minus sign within the same block, and so the block effects cancel out. For example,

$$\begin{aligned} \mathrm{E}(\widehat{A}) &= \tfrac{1}{2}[\mu_{(ab)} + \beta_2 - \mu_b - \beta_1 + \mu_a + \beta_1 - \mu_{(1)} - \beta_2] \\ &= \tfrac{1}{2}[\mu_{(ab)} - \mu_b + \mu_a - \mu_{(1)}] = A. \end{aligned}$$

The above example suggests a scheme for confounding the highest order interaction between two blocks: From the contrast for that interaction, assign the treatment combinations having the same sign to the same block. For a 2^3 design the blocking scheme shown in Display 7.4 confounds the interaction ABC between the two blocks, but the main effects and two-factor interactions can be estimated free of the block effects.

More generally, we will consider blocking a 2^p design in 2^q blocks for $q < p$. Between 2^q blocks there are $2^q - 1$ d.f. As a result, $2^q - 1$ treatment effects will be confounded between the blocks. Of these, any q effects can be chosen independently. Additional $2^q - q - 1$ effects, called **generalized interactions**, obtained by taking all possible products (with letter powers mod 2) of the original q effects taken two at a time, three at a time, and so on, will be confounded automatically. Let us illustrate this by a few examples.

For $q = 1$, we have two blocks between which there is one independent contrast: $B_1 = \beta_2 - \beta_1$. We equate B_1 to the highest order interaction; for example, for $p = 3$ we equate $B_1 = ABC$. The interpretation of this equation is that the treatment combinations with a plus sign in the ABC contrast are assigned to one block and those with a minus sign are assigned to the other block.

For $q = 2$, we have four blocks between which there are three independent contrasts: B_1, B_2, and $B_{12} = B_1B_2$. We can choose two treatment contrasts or interactions to be confounded between the blocks; the third contrast (a generalized interaction) obtained by taking the product of the first two contrasts will be confounded automatically. Suppose $p = 4$ and we choose the interactions ABC and ABD to be confounded with B_1 and B_2, respectively, that is, $B_1 = ABC$

Block 1	Block 2
(1)	a
ab	b
ac	c
bc	abc

Display 7.4 Blocking scheme for 2^3 design to confound ABC interaction between two blocks.

and $B_2 = ABD$. Then $B_{12} = (ABC)(ABD) = A^2B^2CD = CD$ will be confounded automatically. The corresponding blocking scheme is found as follows: The treatment combinations that have the same pair of signs on the contrasts ABC and ABD are assigned to the same block. There are four pairs of signs: $(-, -)$, $(-, +)$, $(+, -)$, and $(+,+)$. We can find these treatment combinations from the contrast matrix shown in Table 7.11. By assigning them to block 1 through block 4, respectively, we obtain the blocking scheme shown in Display 7.5. It may be noted that the signs for the contrast CD are obtained by taking the products of the signs for the contrasts ABC and ABD. Thus the treatment combinations in block 1 and block 4 have a plus sign on the contrast CD, and the treatment combinations in block 2 and block 3 have a minus sign.

The above method extends in a natural way to 2^q blocks. After choosing the q basic interactions to be confounded between the blocks, we find the treatment combinations that have the same signs on the contrasts for these interactions. There are 2^q possible combinations of signs which divide the treatment combinations into 2^q blocks.

Instead of writing out and inspecting the entire contrast matrix to find the treatment combinations that have the same signs on the q contrasts,we can use the following simpler method: For each treatment combination we compute the value (mod 2) of the so-called **defining contrast** for each basic interaction chosen to be confounded between the blocks. For the ith basic interaction the defining contrast is given by

$$L_i = (a_{i1}\ell_1 + a_{i2}\ell_2 + \cdots + a_{ip}\ell_p) \bmod 2 \qquad (1 \le i \le q),$$

where $a_{ij} = 1$ if the jth factor appears in the ith interaction and $a_{ij} = 0$ otherwise $(1 \le j \le p)$ and $\ell_j = 0, 1$ depending on whether the jth factor is at low level or high level in the given treatment combination. The values of the vector $(L_1, L_2, \ldots, L_q)$ are computed for all treatment combinations and those having the same vector values are assigned to the same block. Note that this vector has 2^q distinct values because each $L_i = 0, 1$. The following examples illustrate this method.

Example 7.8 (2^3 Design in 2 Blocks)

Block 1	Block 2	Block 3	Block 4
(1)	*c*	*ac*	*a*
ab	*abc*	*bc*	*b*
acd	*ad*	*d*	*cd*
bcd	*bd*	*abd*	*abcd*

Display 7.5 Blocking scheme for 2^4 design to confound ABC, ABD, and CD interactions between four blocks.

We choose to confound the highest order interaction ABC. Since $a_{11} = a_{12} = a_{13} = 1$, the defining contrast for this interaction is

$$L_1 = (\ell_1 + \ell_2 + \ell_3) \bmod 2.$$

The values of this defining contrast for the eight treatment combinations are as follows.

$$
\begin{aligned}
(1):& \quad L_1 = (0 + 0 + 0) \bmod 2 = 0 \\
a:& \quad L_1 = (1 + 0 + 0) \bmod 2 = 1 \\
b:& \quad L_1 = (0 + 1 + 0) \bmod 2 = 1 \\
ab:& \quad L_1 = (1 + 1 + 0) \bmod 2 = 0 \\
c:& \quad L_1 = (0 + 0 + 1) \bmod 2 = 1 \\
ac:& \quad L_1 = (1 + 0 + 1) \bmod 2 = 0 \\
bc:& \quad L_1 = (1 + 0 + 1) \bmod 2 = 0 \\
abc:& \quad L_1 = (1 + 1 + 1) \bmod 2 = 1
\end{aligned}
$$

The treatment combinations $(1), ab, ac$, and bc have $L_1 = 0$ and are assigned to one block. The treatment combinations a, b, c, and abc have $L_1 = 1$ and are assigned to the other block. Thus we obtain the blocking scheme shown in Display 7.4. ■

Example 7.9 (2^4 Design in 4 Blocks)

We choose to confound the interactions ABC and ABD. The defining contrasts for these two interactions are

$$L_1 = (\ell_1 + \ell_2 + \ell_3) \bmod 2 \qquad \text{and} \qquad L_2 = (\ell_1 + \ell_2 + \ell_4) \bmod 2.$$

The values of these defining contrasts for the 16 treatment combinations are as follows:

$$
\begin{aligned}
(1):& \quad L_1 = (0 + 0 + 0) \bmod 2 = 0 \quad L_2 = (0 + 0 + 0) \bmod 2 = 0 \\
a:& \quad L_1 = (1 + 0 + 0) \bmod 2 = 1 \quad L_2 = (1 + 0 + 0) \bmod 2 = 1 \\
b:& \quad L_1 = (0 + 1 + 0) \bmod 2 = 1 \quad L_2 = (0 + 1 + 0) \bmod 2 = 1 \\
ab:& \quad L_1 = (1 + 1 + 0) \bmod 2 = 0 \quad L_2 = (1 + 1 + 0) \bmod 2 = 0 \\
c:& \quad L_1 = (0 + 0 + 1) \bmod 2 = 1 \quad L_2 = (0 + 0 + 0) \bmod 2 = 0 \\
ac:& \quad L_1 = (1 + 0 + 1) \bmod 2 = 0 \quad L_2 = (1 + 0 + 0) \bmod 2 = 1 \\
bc:& \quad L_1 = (1 + 0 + 1) \bmod 2 = 0 \quad L_2 = (0 + 1 + 0) \bmod 2 = 1 \\
abc:& \quad L_1 = (1 + 1 + 1) \bmod 2 = 1 \quad L_2 = (1 + 1 + 0) \bmod 2 = 0
\end{aligned}
$$

$$
\begin{array}{rll}
d: & L_1 = (0+0+0) \bmod 2 = 0 & L_2 = (0+0+0) \bmod 2 = 1 \\
ad: & L_1 = (1+0+0) \bmod 2 = 1 & L_2 = (1+0+1) \bmod 2 = 0 \\
bd: & L_1 = (0+1+0) \bmod 2 = 1 & L_2 = (0+1+1) \bmod 2 = 0 \\
abd: & L_1 = (1+1+0) \bmod 2 = 0 & L_2 = (1+1+1) \bmod 2 = 1 \\
cd: & L_1 = (0+0+1) \bmod 2 = 1 & L_2 = (0+0+1) \bmod 2 = 1 \\
acd: & L_1 = (1+0+1) \bmod 2 = 0 & L_2 = (1+0+1) \bmod 2 = 0 \\
bcd: & L_1 = (1+0+1) \bmod 2 = 0 & L_2 = (0+1+1) \bmod 2 = 0 \\
abcd: & L_1 = (1+1+1) \bmod 2 = 1 & L_2 = (1+1+1) \bmod 2 = 1
\end{array}
$$

By grouping the treatment combinations into four blocks with $(L_1, L_2) = (0, 0)$ for block 1, $(1, 0)$ for block 2, $(0, 1)$ for block 3, and $(1, 1)$ for block 4, the confounding scheme shown in Display 7.5 is obtained.

The q basic interactions to be confounded between the blocks must be chosen carefully. For instance, in the above example, had we chosen the interactions ABC and $ABCD$, their generalized interaction, $(ABC)(ABCD) = A^2B^2C^2D = D$, a main effect, would have been confounded automatically. For $q > 2$ there are many more generalized interactions to consider. For example, for $q = 3$ one can choose three basic interactions, which determine four generalized interactions for a total of seven confounded interactions; for $q = 4$ one can choose four basic interactions, which determine 11 generalized interactions for a total of 15 confounded interactions; and so on. The problem of how to choose the basic interactions so that none of the generalized interactions result in a main effect or some low-order interaction to be confounded is closely related to the choice of a fractional factorial design [specifically (2^{-q})th fraction of a 2^p design] discussed in the next chapter. ■

7.4.2 Statistical Analysis

Statistical analysis of a blocked 2^p factorial design is straightforward. The estimators of the unconfounded effects are the same as those from a full experiment; their sums of squares are also computed in the same way. The confounded effects cannot be estimated separately from the block effects; their pooled sum of squares equals the block sum of squares. The following example illustrates these calculations.

Example 7.10 (Iron Concentration Data: Analysis of a Blocked Experiment)

Refer to the iron concentration data from Example 7.5, but suppose that the experiment was carried out on four separate occasions with four runs made on

Occasion 1	Occasion 2	Occasion 3	Occasion 4
$(1) = 1.3553$	$c = 1.1782$	$ac = 0.5738$	$a = 1.3469$
$ab = 2.0817$	$abc = 0.4106$	$bc = 0.8728$	$b = 1.3090$
$acd = 1.7117$	$ad = 1.8326$	$d = 1.6072$	$cd = 0.8941$
$bcd = 0.7620$	$bd = 1.5318$	$abd = 2.2923$	$abcd = 1.6741$
$\bar{y}_{\cdot 1} = 1.4777$	$\bar{y}_{\cdot 2} = 1.2383$	$\bar{y}_{\cdot 3} = 1.3365$	$\bar{y}_{\cdot 4} = 1.3060$

Display 7.6 Iron concentration data assuming experiment is blocked in four occasions.

each occasion in a random order using the blocking scheme shown in Display 7.6, which also gives the data.

As we saw earlier, in this design the interactions ABC, ABD, and CD are confounded. The estimates of the remaining effects are the same as those obtained in Example 7.5. Using $\bar{\bar{y}} = 1.3396$, the unadjusted block sum of squares equals

$$4[(1.4777 - 1.3396)^2 + (1.2383 - 1.3396)^2 + (1.3365 - 1.3396)^2 + (1.3060 - 1.3396)^2] = 0.1218.$$

Using formula (7.16), the pooled sum of squares for the confounded interactions $\widehat{ABC} = -0.1349$, $\widehat{ABD} = -0.0367$, and $\widehat{CD} = 0.1044$ can be checked to be the same as the block sum of squares:

$$SS_{ABC} + SS_{ABD} + SS_{CD} = 4[(-0.1349)^2 + (-0.0367)^2 + (0.1044)^2] = 0.1218.$$

The sums of squares calculations are shown in Table 7.14. The conclusions are unchanged from Example 7.5, namely that the effects A, C, D, AD, and ACD are the important ones. ■

7.5 CHAPTER NOTES

7.5.1 Yates Algorithm

Yates (1937) proposed an algorithm which exploits the systematic and orthogonal nature of the effect contrasts. The algorithm is best explained through an example.

Example 7.11 (Bicycle Data: Yates Algorithm)

(From Tamhane and Dunlop 2000, Example 13.9. Reprinted by permission of Pearson Education, Inc.) To apply the Yates algorithm, first one must list the $2^p = 8$ treatment combinations and their sample means in the standard order. These are shown in the first two columns of Table 7.15, which also gives the calculations that follow. The $2^{p-1} = 4$ successive pairs of means are first added and then subtracted (subtracting the first mean from the second mean in each

Table 7.14 Sums of Squares for Iron Concentration Experiment Blocked in Four Blocks

Effect	Estimate	d.f.	Sum of Squares	Percent
A	0.3017	1	0.3641	8.494
B	0.0543	1	0.0118	0.275
C	−0.6599	1	1.7419	40.634
D	0.3972	1	0.6311	14.722
AB	0.1941	1	0.1507	3.515
AC	−0.1359	1	0.0739	1.724
AD	0.3772	1	0.5691	13.276
BC	−0.2139	1	0.1830	4.269
BD	−0.0007	1	0.0000	0.000
ACD	0.3218	1	0.4142	9.662
BCD	0.0754	1	0.0227	0.530
$ABCD$	0.0248	1	0.0025	0.058
Blocks (ABC, ABD, CD)	—	3	0.1218	2.841
Total	—	15	4.2868	100

pair) and the results are entered in the same order in the column labeled I. The first four and the last four entries in column I are obtained as follows:

$$52.5 + 42.0 = 94.5, \quad 57.0 + 43.5 = 100.5, \quad 49.0 + 39.0 = 88.0,$$

$$52.0 + 42.5 = 94.5,$$

$$-52.5 + 42.0 = -10.5, \quad -57.0 + 43.5 = -13.5, \quad -49.0 + 39.0 = -10.0,$$

$$-52.0 + 42.5 = -9.5.$$

This operation of pairwise addition and subtraction is repeated on column I and the results are entered in column II. For example, the first and the fifth entries in column II are $94.5 + 100.5 = 195.0$ and $-94.5 + 100.5 = 6.0$, respectively. This is continued on each successive column for a total of $p = 3$ times, resulting in columns I, II, and III. The first entry of column III is simply the total of all means; the remaining entries are the contrasts in treatment means of the effects in the standard order. We divide the first entry in column III by $2^p = 8$ to obtain the grand mean; the remaining entries are divided by $2^{p-1} = 4$ to obtain the effect estimates. The results are shown in the column labeled "$\widehat{\text{Effect}}$." Finally, the last column gives the sums of squares for the effects obtained using (7.16), which are the same as those given in Table 7.8. ■

7.5.2 Partial Confounding

If experimental resources are available to replicate a confounded 2^p factorial design, then it would be a good idea to confound a different set of interactions in each replicate so that estimates of all interactions can be computed. Obviously,

Table 7.15 Calculations for Yates Algorithm for Bicycle Data

Treatment Combination	Treatment Mean	I	II	III	$\widehat{\text{Effect}}$	SS for Effect
(1)	52.5	94.5	195.0	377.5	$\widehat{\beta}_0 = 47.1875$	—
a	42.0	100.5	182.5	−43.5	$\widehat{A} = -10.875$	473.0625
b	57.0	88.0	−24.0	12.5	$\widehat{B} = 3.125$	39.0625
ab	43.5	94.5	−19.5	−2.5	$\widehat{AB} = -0.625$	1.5625
c	49.0	−10.5	6.0	−12.5	$\widehat{C} = -3.125$	39.0625
ac	39.0	−13.5	6.5	4.5	$\widehat{AC} = 1.125$	5.0625
bc	52.0	−10.0	−3.0	0.5	$\widehat{BC} = 0.125$	0.0625
abc	42.5	−9.5	0.5	3.5	$\widehat{ABC} = 0.875$	3.0625

the interactions that are not confounded in any replicate would be estimated with a higher precision than those that are confounded in some of the replicates. The resulting design is called a **partially confounded design**.

Example 7.12 (Partially Confounded 2^3 Design)

Suppose that each block is of size four so that one replicate of the 2^3 full factorial requires two blocks between which one interaction will be confounded. With eight available blocks we will have four replicates and can confound a different interaction in each replicate. Choosing the interactions to be confounded as AB, AC, BC, and ABC, the partially confounded design is as shown in Table 7.16. With this design each interaction is estimable from three replicates, while the main effects are estimable from all four replicates. ■

7.6 CHAPTER SUMMARY

(a) A balanced 2^p design is an orthogonal design in which all main effects and interactions are orthogonal contrasts. Hence they can be estimated

Table 7.16 Partially Confounded 2^3 Design in Eight Blocks

Replicate	I		II		III		IV	
Confounded interaction	AB		AC		BC		ABC	
Treatment combinations	(1)	a	(1)	a	(1)	b	(1)	a
	ab	b	b	ab	a	ab	ab	b
	c	ac	ac	c	bc	c	ac	c
	abc	bc	abc	bc	abc	ac	bc	abc

independently of each other. Dropping or adding of any terms from the model does not affect the estimates of the remaining terms. Contributions to the total variability from individual terms are additive.

(b) If a single-replicate observation is made at each factorial combination, then the full model that includes all main effects and interactions is a saturated model. In this model there are no d.f. available for estimating the error variance. A normal or a half-normal plot of the estimated effects can be used to graphically identify large effects. Lenth's method is another alternative which is based on obtaining a pseudoestimate of the error variance. A common practice is to drop the effects identified as nonsignificant using these tests and pool their sums of squares to estimate the error variance. Sometimes an experimenter may choose to a priori ignore certain higher order interactions and fit a reduced model. Then the sums of squares for the ignored interactions can be used to estimate the error variance. In either case, it should be noted that the estimate so obtained is not a pure-error estimate and is positively biased.

(c) Adding a few observations at the center point to a 2^p design is always a good idea because (i) a pure-error estimate can be computed from these observations, (ii) a test for curvature can be performed by comparing the average of the observations at the vertex points with that of the observations at the center point, and (iii) these observations do not violate the orthogonality property of the design and the estimates of the main effects and interactions are unaffected.

(d) When a 2^p design must be blocked, there may be a physical limitation on the block size so that all treatment combinations cannot be accommodated in a single block. Therefore incomplete blocks have to be used. As a result, some effects get confounded between the blocks. For a BIB design with 2^q blocks, each of size 2^{p-q}, out of a total of $2^p - 1$ main and interaction effects, $2^q - 1$ effects are confounded between the blocks. Of these, q effects can be specified and the remaining $2^q - q - 1$ so-called generalized interactions are automatically confounded. Schemes are available for choosing the BIB design in such a way that only the higher order interactions are confounded.

EXERCISES

Section 7.1 (Estimation of Main Effects and Interactions)

Theoretical Exercises

7.1 Write the model matrix $\boldsymbol{X}$ for the regression model (7.4). Check that it is orthogonal.

7.2 Consider a 2^2 experiment conducted in an RB design with $n \geq 2$ complete blocks of size 4. Modify the regression model (7.1) to include the block effects. Can you estimate the interactions of the factors with the blocks? How will you estimate the error variance for this design and how many degrees of freedom will it have?

Applied Exercises

7.3 Two students (Steven Black and Blake Kirschner) from my Spring 2003 Statistics 351 class performed a stain removal experiment to study the effects of laundry detergent (A), stain stick (B), and water temperature (C) on removing ketchup stains. The levels of these factors are shown in Table 7.17. Sixteen pieces were cut from the same cotton cloth, each a 3×6-in. rectangle. A small ketchup stain was applied to each piece. After complete drying, each piece was scanned and the level of grayscale of the image was recorded on a scale of 0–255. The higher numbers represent a lighter color. The pieces were then washed for 3 min each in 3 gallons of water. After the piece was dried, it was scanned again and the grayscale of the image was measured. A 2^3 design was used to set the factor levels; the treatment combinations are shown in Table 7.17.

The experiment was replicated twice. The resulting data are shown in Table 7.18. Estimate the main effects and interactions of the three treatment factors.

7.4 Two students (Leah Marshall and Todd vanGoethem) from my Spring 2004 Statistics 351 class performed a bottle-rocket experiment. The goal was to study the effects of baking soda (A), vinegar (B), and bottle size (C) on the flight time of the rubber cork when baking soda and vinegar are mixed in the bottle, which is then corked. Buildup of pressure due to production of CO_2 gas as a result of chemical reaction causes the cork to be airborne. Each factor was studied at two levels and the whole experiment was replicated three times. The levels of the factors are given in Table 7.19.

For each run, both students made independent measurements of the flight time of the cork. All 24 runs were randomized for each student. The data are shown in Table 7.20. Note that the data are grouped into replicates merely for convenience of presentation, that is, replicate is not a blocking factor.

Table 7.17 Factors and Their Levels for Stain Removal Experiment

Factor	Name	Low (−)	High (+)
A	Laundry detergent	None	1 tsp/1 gallon water
B	Stain stick	No	Yes
C	Water temperature	Cold	Hot

Table 7.18 Grayscale Measurements

Treatment Combination	Replicate I		Replicate II	
	Before	After	Before	After
(1)	190.71	194.64	161.65	176.06
a	166.09	197.99	166.76	203.16
b	164.29	208.72	170.57	196.64
ab	166.32	200.24	163.35	204.09
c	162.25	183.30	168.64	178.63
ac	186.48	190.88	161.59	201.55
bc	168.79	197.03	175.24	206.91
abc	175.74	216.90	164.90	219.70

Table 7.19 Levels of Factors for Rocket Experiment

Factor	Low	High
Baking soda (A)	$\frac{1}{4}$ tsp	$\frac{1}{2}$ tsp
Vinegar (B)	2 oz	4 oz
Bottle volume (C)	12 oz	750 mL

Table 7.20 Airborne Times (sec) of Bottle Corks

Factor			Replicate I		Replicate II		Replicate III	
A	B	C	Student 1	Student 2	Student 1	Student 2	Student 1	Student 2
−1	−1	−1	1.81	1.78	1.57	1.94	2.18	2.06
+1	−1	−1	1.99	1.87	2.47	2.44	2.64	2.76
−1	+1	−1	1.56	1.46	1.57	1.40	1.20	1.29
+1	+1	−1	2.65	2.78	2.40	2.60	3.78	3.44
−1	−1	+1	0.99	0.78	1.04	1.00	0.80	0.60
+1	−1	+1	2.27	2.16	2.51	2.44	1.85	1.97
−1	+1	+1	0.71	0.96	0.66	0.76	1.03	1.03
+1	+1	+1	1.40	1.43	1.59	1.46	2.66	2.44

(a) Estimate the main effects and interactions of the three treatment factors. Provide physical explanations for the signs of the effects where possible.

(b) Estimate the student effect, that is, the mean difference between the two students.

Section 7.2 (Statistical Analysis)

Theoretical Exercises

7.5 Use the representation (7.9) for $\widehat{\beta}_j$ to derive the formula $\mathrm{Var}(\widehat{\beta}_j) = \sigma^2/N$.

Applied Exercises

7.6 Refer to Exercise 7.3.

(a) Make a detailed ANOVA table that includes all effects. Which effects are significant at $\alpha = 0.10$? Interpret these effects.
(b) Fit a reduced model by dropping nonsignificant effects but obeying the hierarchical principle.
(c) Make residuals plots and check if the normality and homoscedasticity assumptions are satisfied.

7.7 Refer to Exercise 7.4.

(a) Make a detailed ANOVA table that includes all effects (the three treatment factor main effects and their interactions and student effects). Which effects are significant at $\alpha = 0.05$?
(b) What information is lost if the responses are taken as $\overline{y} = (y_1 + y_2)/2$ for each run. Will the estimates of the treatment factor effects remain the same?

7.8 Sanders, Leitnaker, and McLean (2001–02) reported an experiment to study how the gloss on paper is affected when large rolls of paper are run through a machine, called a supercalendar. The three settings on the machine were temperature (factor A), pressure (factor B), and speed (factor C). A 2^3 experiment was run over two days. On each day, the experiment was replicated three times. The data are given in Table 7.21.

(a) Analyze the data from day 1 and identify the significant effects using $\alpha = 0.05$.

Table 7.21 Gloss Measurements

Day 1				Day 2			
Treatment	Gloss Measurement			Treatment	Gloss Measurement		
Combination	1	2	3	Combination	1	2	3
(1)	36	35	37	(1)	35	35	37
a	38	38	39	a	36	38	38
b	38	40	37	b	39	38	39
ab	41	39	40	ab	39	41	38
c	35	33	35	c	34	36	34
ac	35	34	35	ac	35	38	36
bc	35	36	36	bc	34	37	35
abc	35	37	35	abc	38	37	39

Source: Sanders, Leitnaker, and McLean (2001–02). Reprinted by permission of Pearson Education, Inc.

(b) Repeat part (a) for the data from day 2. Note that the significance of the AC effect changes from day 1 to day 2.

(c) Compute the ANOVA table for combined data by treating day as a blocking factor. Check that the interactions of the day with C and AC are significant.

(d) How will you use the information learned from this analysis to make the process more consistent?

Section 7.3 (Single-Replicate Case)

Theoretical Exercises

7.9 Consider a single replicate of a full 2^p factorial augmented by n_c replicate observations at the center point. Show that $E(\overline{y}_f) = \beta_0 + \sum_{k=1}^{p} \beta_{kk}$ and $E(\overline{y}_c) = \beta_0$ where $\overline{y}_f$ and $\overline{y}_c$ are the sample means of the observations at the factorial and center points, respectively. Hence $\overline{y}_f - \overline{y}_c$ is an unbiased estimator of $\sum_{k=1}^{p} \beta_{kk}$.

7.10 Consider a 2^3 experiment in which two factors are quantitative and one factor is qualitative. The design consists of a single replicate of the 2^3 factorial experiment augmented by two replicates each at the center point of the two quantitative factors at the low and high levels of the qualitative factor.

(a) Write the design matrix. Show that this is an orthogonal design.

(b) Show that the estimates of the main effects and interactions involving the quantitative factors are based only on the 8 factorial observations, while the main-effect estimate of the qualitative factor is based on all 12 observations.

(c) How many pure error d.f. does this design have?

7.11 Repeat the previous exercise but now assume that only one factor is quantitative and two factors are qualitative. Assume that a single replicate of the 2^3 factorial design is augmented by two replicates at the center point of the quantitative factor at each of the four combinations of the low and high levels of the two qualitative factors.

Applied Exercises

7.12 In direct marketing, the response rates to mail offers are quite low (1–2%) and so it is critically important to study factors that help improve the response rates. Hansotia (1990) provided the following example: A company developing a new offer for a continuation product (such as a monthly offer of a music CD or a book) conducted a 2^4 experiment to evaluate the effects of the following four factors on the response rates.

A: Payment terms (monthly/annual)
B: Premium (present/absent)
C: Sweeps (present/absent)
D: Trial membership card (paper/plastic)

Here premium refers to any special deal offered (e.g., freebies) and sweeps refers to entry into sweepstakes. The response rates under all factorial combinations (based on 10,000 mailings in each case) are given in Table 7.22.

(a) Observe that the response rate is higher for the high level of each factor compared to the low level of the same factor, keeping the levels of the other factors fixed. What does this say about the signs of the main effects of the factors?

(b) It is common to take a logistic transform of response rates before analysis. Thus make the transformation $y = \ln[p/(1-p)]$, where p is the response rate (expressed as a fraction) and use y as the response variable for the remainder of the analysis.

(c) Estimate all main effects and two-factor interactions. Do the signs of the main effects agree with those determined in part (a)? Make a half-normal plot of the estimated effects and identify the significant ones.

(d) Write the prediction model in terms of the significant effects. Predict the response rate when all factors are set at high levels. Compare this predicted rate with the observed rate of 1.44% at this setting.

Table 7.22 Response Rates in Direct Marketing Experiment

Run	A	B	C	D	Response Rate (%)	Run	A	B	C	D	Response Rate (%)
1	−	−	−	−	1.00	9	−	−	−	+	1.16
2	+	−	−	−	1.05	10	+	−	−	+	1.20
3	−	+	−	−	1.21	11	−	+	−	+	1.30
4	+	+	−	−	1.22	12	+	+	−	+	1.35
5	−	−	+	−	1.11	13	−	−	+	+	1.30
6	+	−	+	−	1.19	14	+	−	+	+	1.31
7	−	+	+	−	1.26	15	−	+	+	+	1.40
8	+	+	+	−	1.32	16	+	+	+	+	1.44

Source: Hansotia (1990). Reprinted by permission of John Wiley & Sons, Inc.

7.13 Refer to the previous exercise.

(a) Apply Lenth's method to the logistic-transformed data to identify significant effects. Are the results the same as those obtained from the half-normal plot?

(b) Compute the estimate of the common standard error of the effect estimates by pooling the sums of squares corresponding to the effects deemed to be nonsignificant using the half-normal plot method. Compare this estimate with that obtained by using Lenth's method.

7.14 A researcher wanted to study the effects of three factors on a response variable y. He augmented a 2^3 design with four additional runs by replicating observations at $(-, -, -)$, $(-, +, +)$, $(+, +, -)$, $(+, -, +)$. The data are shown in Table 7.23.

(a) Suppose we want to fit the model

$$y = \beta_0 + \beta_1 x_1 + \beta_2 x_2 + \beta_3 x_3 + e$$

to these data, where $e \sim N(0, \sigma^2)$ and $x_j = \pm 1$ depending on the level of factor j. Show that the model matrix $\boldsymbol{X}$ is orthogonal.

(b) Show how to calculate the LS estimates of the β_j's and carry out the calculations of $\widehat{\beta}_0$ and $\widehat{\beta}_1$.

(c) Show that all $\widehat{\beta}_j$'s have the same variance for this design. Give the formula for $\text{Var}(\widehat{\beta}_j)$.

(d) Calculate the pure-error estimate of σ^2 from the four pairs of replicate observations at $(-, -, -)$, $(-, +, +)$, $(+, +, -)$, $(+, -, +)$. How many degrees of freedom does this estimate have?

Table 7.23 Twelve-Run Design

	Factor			
Run	A	B	C	y
1	−	−	−	11
2	+	−	−	3
3	−	+	−	15
4	+	+	−	7
5	−	−	+	9
6	+	−	+	1
7	−	+	+	13
8	+	+	+	5
9	−	−	−	9
10	−	+	+	15
11	+	+	−	5
12	+	−	+	3

Table 7.24 Factors and Their Levels for Paper Helicopter Experiment

Factor	−1	0	1
A: Wing area (in.^{2})	11.20	11.80	12.40
B: Wing length–width ratio	2.25	2.52	2.78
C: Body width (in.)	1.00	1.25	1.50
D: Body length (in.)	1.50	2.00	2.50

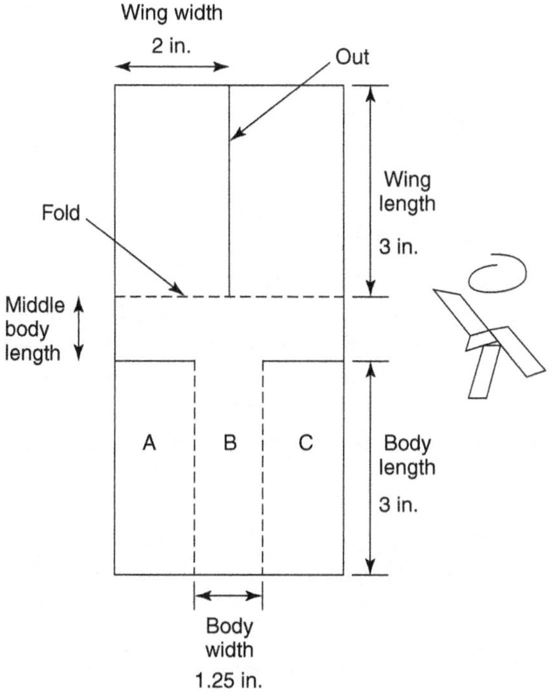

Figure 7.4 Paper helicopter. From Box and Liu (1999, Figure 7.1). Reprinted by permission of the American Society for Quality.

(e) Test for the significance of the main effect A, that is, test $H_0 : \beta_1 = 0$ versus $H_1 : \beta_1 \neq 0$ at $\alpha = 0.05$.

7.15 Box and Liu (1999) describe a 2^4 experiment conducted to study the effects of four factors shown in Table 7.24 on the flight times of paper helicopters. Figure 7.4 shows a diagram of the helicopter. The helicopter is dropped from a height of about 12 feet and its flight time (in seconds) is clocked until it hits the ground. To estimate the pure error and test for curvature, six

Table 7.25 Flight Times (centiseconds) from Paper Helicopter Experiment

Run	Treatment Combination	Flight Time	Run	Treatment Combination	Flight Time
1	(1)	367	12	abd	363
2	a	369	13	cd	344
3	b	374	14	acd	355
4	ab	370	15	bcd	370
5	c	372	16	$abcd$	362
6	ac	355	17	0	377
7	bc	397	18	0	375
8	abc	377	19	0	370
9	d	350	20	0	368
10	ad	373	21	0	369
11	bd	358	22	0	366

Source: Data adapted from Box and Liu (1999). Reprinted by permission of the American Society for Quality.

center points (denoted by 0) were added to the 2^4 design. (In the original experiment two center-point runs were made in one block and four center point runs were made in another block. Here we have combined them for simplicity.) The data in Table 7.25 are the averages (in centiseconds) for four independent drops. The objective of the experiment was to find the factors that significantly affect the flight time and to find the best combination of their levels to maximize the flight time.

(a) Estimate and test for significance all the main effects and interactions using the pure-error estimate from the center-point runs. Compare the results with a half-normal plot of effects. Which effects are significant at $\alpha = 0.05$? Write a predictive model for the flight time based on the significant effects.

(b) Test for curvature in the response surface using $\alpha = 0.05$. What is your conclusion?

Section 7.4 (2^p Factorial Designs in Incomplete Blocks: Confounding of Effects)

Theoretical Exercises

7.16 By using the defining contrasts method, find a scheme to block a 2^3 design in four blocks so that no main effect is confounded between the blocks.

7.17 By using the defining contrasts method, find a scheme to block a 2^4 design in two blocks by confounding the interaction $ABCD$.

Lot 1	Lot 2	Lot 3	Lot 4
(1) = 155	c = 156	c = 161	(1) = 164
bc = 152	b = 168	b = 175	bc = 162
ac = 150	a = 162	a = 171	ac = 153
ab = 157	abc = 161	abc = 173	ab = 171

Display 7.7 Design and data for Exercise 7.19. (Davies, 1963, Table 7.2).

7.18 By using the defining contrasts method, find a scheme to block a 2^5 design in four blocks by confounding the interactions ADE and BCE. Which additional interaction is confounded as a result?

Applied Exercises

7.19 The following experiment is described in Davies (1963). In an organic chemical manufacturing process it was of interest to assess the effects of the quality (A: coarse vs. finely ground) and the quantity (B: normal vs. +10%) of ammonium chloride on the yield of the chemical. Two identical units (C: unit 1 vs. unit 2) were used in the study. Base material used in the manufacture of the chemical was made by a batch process. Two batches of the base material were blended to form a lot, which was sufficient to produce four batches of the organic chemical product under four different treatment conditions. The full 2^3 factorial experiment was replicated twice using four lots of the base material. The design and the data (yield of the organic chemical in pounds) are shown in Display 7.7.

(a) Which interaction is confounded between the lots in each replication?

(b) Make an ANOVA table. Which effects are significant at $\alpha = 0.05$?

(c) Prior to the experiment it was anticipated that the yield would increase by changing from coarse to finely ground ammonium chloride and also by increasing the amount by 10%. The two units were thought to be essentially identical and were not expected to differ significantly from each other, and any interaction involving them was not expected to be significant. How much do the experimental results agree with these a priori expectations?

CHAPTER 8

Two-Level Fractional Factorial Experiments

In the last chapter we studied 2^p experiments, which are the smallest possible full-factorial experiments (since they use only two levels of each factor) for studying the effects of p factors on a response variable. However, even 2^p experiments become prohibitively large very quickly since the number of treatment combinations grows exponentially with p. For example, if $p = 10$, then the number of treatment combinations is $2^{10} = 1024$. Of the total 1023 effects, there are only 10 main effects and 45 two-factor interactions; the remaining 968 effects are used to estimate the third- and higher order interactions. Therefore it is wasteful to devote the majority of the data to estimate the effects that are less likely to be important than the main effects and two-factor interactions according to the **hierarchical ordering principle** mentioned in Chapter 7. The size of the experiment can be reduced dramatically by judiciously choosing the treatment combinations to run while still retaining the ability to estimate the main effects and lower order interactions. In this chapter we study fractional factorial experiments for this purpose. An important subclass of these designs are 2^{-q} fractions of 2^p factorial designs ($1 \leq q < p$), called 2^{p-q} **fractional factorial designs**. For example, a 2^{10-4} experiment (which is a 1/16th fraction of a 2^{10} factorial experiment) with 64 runs can be used to estimate all 10 main effects and 45 two-factor interactions with eight d.f. left over to estimate the experimental error if the higher order interactions can be neglected.

The chapter is organized as follows. Section 8.1 discusses 2^{p-q} fractional factorial designs in detail, including their construction and analysis. Section 8.2 studies a class of **saturated designs** called **Plackett–Burman designs** in which the number of runs is only one more than the number of factors, so that only their main effects can be estimated with no d.f. available for estimating the error. Section 8.3 introduces a generalization of Plackett–Burman designs, called **Hadamard designs**. These are saturated designs constructed using Hadamard

Statistical Analysis of Designed Experiments: Theory and Applications By Ajit C. Tamhane

matrices studied in combinatorial mathematics. Section 8.4 discusses **supersaturated designs**. As the name implies, these are highly economical designs in which the number of runs is less than the number of factors studied. These highly fractionated two-level factorial designs are especially useful in screening experiments, where the goal is to screen a large number of factors to identify a few active ones, and the runs are at premium. Section 8.5 introduces **orthogonal arrays (OAs)** as a generalization of 2^{p-q} and Hadamard designs. Section 8.6 discusses sequential experiments, where small fractional factorial experiments are carried out in stages by learning from previous experiments, and their results are assembled together. A summary of the chapter is given in Section 8.7.

8.1 2^{p-q} Fractional Factorial Designs

8.1.1 2^{p-1} Fractional Factorial Design

Let us begin by considering a half fraction of a 2^3 experiment. The question is: Which four of the eight treatment combinations, $(1), a, b, ab, c, ac, bc, abc$, should one run? We see from the contrast coefficients in Table 7.5, which is reproduced here as Table 8.1, that if we run the first four treatment combinations, then the main effect C cannot be estimated at all since the factor C is not varied, being kept at the low level. The same would be true if we run the last four treatment combinations in which C is kept at the high level. Clearly, any subset of treatment combinations in which some factor is not varied is not a good design.

By matching the signs of the contrast coefficients, we see that the above-mentioned two sets of four treatment combinations correspond to the equations $I = -C$ and $I = C$, respectively. These are called the **defining relations** of the respective designs. Since C has all plus or all minus signs in each design, it is not an estimable effect.

Let us try the defining relation $I = AB$, in which case the AB interaction is not estimable. From Table 8.1 we see that the corresponding design is

Table 8.1 Contrast Coefficients for Effects in 2^3 Experiment

Treatment Combination	Effect							
	I	A	B	AB	C	AC	BC	ABC
(1)	+	−	−	+	−	+	+	−
a	+	+	−	−	−	−	+	+
b	+	−	+	−	−	+	−	+
ab	+	+	+	+	−	−	−	−
c	+	−	−	+	+	−	−	+
ac	+	+	−	−	+	+	−	−
bc	+	−	+	−	+	−	+	−
abc	+	+	+	+	+	+	+	+

$\{(1), ab, c, abc\}$. Is this a good design? No, because the contrast coefficient signs are the same for the main effects A and B since the factors A and B are varied together, and hence their effects cannot be separated. In other words, the estimators of A and B are identical, or $A = B$. These two effects are said to be **aliases** of each other. If two effects are not aliases, then they are said to be **clear** of each other. Obviously, a design that confounds the main effects is not a good design. The same problem occurs if we choose the complementary design, $\{a, b, ac, bc\}$, which has the defining relation $I = -AB$. In this design the contrast coefficients of the main effects A and B have opposite signs, that is, $A = -B$. Thus, A and B are again aliases of each other since their estimators are identical but opposite in signs.

Lastly, let us try the defining relation $I = ABC$, in which case the ABC interaction is not estimable. However, being a three-factor interaction, it is less likely to be important according to the hierarchical ordering principle. This corresponds to the design $\{a, b, c, abc\}$. By inspection from Table 8.1 we see that the main effects A, B, and C are estimable contrasts, but A and BC, B and AC, and C and AB are aliases. We summarize these **alias relations** as $A = BC$, $B = AC$, and $C = AB$. Thus, if we are willing to assume that the two-factor interactions are negligible, the main effects can be estimated unbiasedly. This is the price paid for taking observations on only half the treatment combinations. If all effects must be clear of each other, then a full factorial experiment is necessary.

The three alias relations given above can be derived from the single defining relation $I = ABC$ as follows. First recall from Section 7.1.1 that the product of two effects, say A and B, equals the componentwise product of their contrast coefficients. By multiplying both sides of the defining relation $I = ABC$ by A, we get $A = A^2BC$. But $A^2 = I$ since squaring any column gives a column with all plus signs, which is the I column. So we get $A = IBC = BC$. Similarly the alias relations $B = AC$ and $C = AB$ follow.

The letter combination equated to I in a defining relation is called a **word** or a **generator**. It is called a generator because the associated design can be generated from it. For example, to derive the design associated with $I = ABC$ or equivalently $C = AB$, we first write a 2^2 full factorial design in A and B (called the **basic design**) and then equate the signs of C to the signs of the interaction AB, as illustrated in Table 8.2.

The defining relation $I = -ABC$ yields the complementary design $\{(1), ab, ac, bc\}$. For this design the alias relations are $A = -BC$, $B = -AC$, and $C = -AB$. The designs corresponding to the defining relations $I = ABC$ and $I = -ABC$ are called the **principal fraction** and the **alternate fraction**, respectively. These two fractions are shown as two sets of vertices of a cube in Figure 8.1. The principal-fraction vertices are indicated by open circles and alternate-fraction vertices are indicated by solid circles.

As can be checked from (7.6), the estimators for the aliased effects are identical (except possibly for a change of sign if a negative of an effect is aliased with another effect) and so they cannot be distinguished. Thus, for example, $\widehat{A} = \widehat{BC}$ for the principal fraction and $\widehat{A} = -\widehat{BC}$ for the alternate fraction. Furthermore,

Table 8.2 Generating 2^{3-1} Design from Defining Relation $I = ABC$

A	B	$C = AB$	Treatment Combination
$-$	$-$	$+$	c
$+$	$-$	$-$	a
$-$	$+$	$-$	b
$+$	$+$	$+$	abc

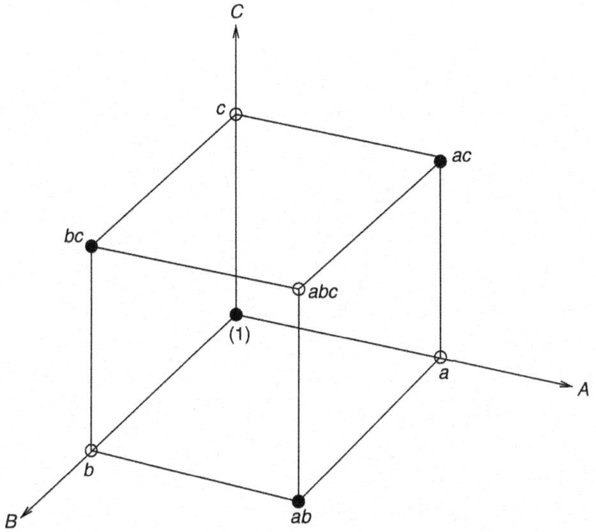

Figure 8.1 Principal and alternate 2^{3-1} fractional factorial designs (open circles represent principal fraction; solid circles represent alternate fraction).

it is easy to show that

$$\begin{aligned} E(\widehat{A}) &= E(\widehat{BC}) = A + BC \quad \text{if } A = BC, \\ E(\widehat{A}) &= -E(\widehat{BC}) = A - BC \quad \text{if } A = -BC. \end{aligned} \tag{8.1}$$

Thus, with some abuse of notation, we can denote $\widehat{A} = \widehat{BC}$ for the principal fraction by $\widehat{A} + \widehat{BC}$ since it estimates $A + BC$ and $\widehat{A} = -\widehat{BC}$ for the alternate fraction by $\widehat{A} - \widehat{BC}$ since it estimates $A - BC$. Similar relations hold for other aliased effects. This notation will be useful in the sequel for pooling the estimates from sequential assemblies of fractional factorial experiments; see Example 8.9.

The principal and alternate half fractions are equivalent in terms of their ability to estimate certain effects clear of others. This ability of a design is expressed by its **resolution**. A design is said to have resolution R if any two effects of order m and n are aliased with each other, then $m + n \geq R$. In other words, no mth-order effect is aliased with an nth-order effect if $m < R - n$. Here the order of an effect is the number of letters in that effect, for example, a main effect has order 1, a two-factor interaction has order 2, and so on. The resolution of a design is denoted by a roman numeral.

A high-resolution design does not confound low-order effects with other low-order effects, so a design with high resolution is desirable. In a resolution III design main effects are aliased with second- and higher order interactions but not with each other. In a resolution IV design main effects are aliased with third- and higher order interactions, but second-order interactions may be aliased with each other. In a resolution V design even two-factor interactions are clear of each other but may be aliased with third- and higher order interactions. These are the most commonly used fractional factorial designs in practice.

If we regard I as the zero-order effect, then it is readily checked that the two designs corresponding to the defining relations $I = C$ and $I = AB$ are of resolutions I and II, respectively. The defining relations $I = ABC$ and $I = -ABC$ give the highest possible resolution, III, when $p = 3$ and $q = 1$. We denote these designs by 2_{III}^{3-1}. Resolution III designs are also referred to as **main-effect plans** since they can estimate main effects clear of each other. Plackett–Burman designs discussed in Section 8.2 provide an example of main-effect plans.

A simple rule to determine the resolution of a half fraction of a full factorial (for any number of factors) can be deduced from this discussion, namely, the resolution equals the number of letters in the word (called the **word length**). This rule will need to be generalized for higher order fractions, as we shall see in Section 8.1.2.

The 2_{III}^{3-1} designs have a nice **projection property**, which we now explain. A three-factor design can be represented as the vertices of a cube as shown in Figure 8.2. The principal half-fraction vertices are indicated with solid circles. If we project this group of vertices (or the vertices for the alternate fraction) into one of the three planes representing a subset of two factors (i.e., A and B or A and C or B and C), we get a full factorial (a square with all four vertices) in those two factors. As can be readily checked, this is not true if we use a resolution I or resolution II design.

What is the significance of the projection property? In practice, after analyzing the data, it is common to find that not all factors are active. In such a case, the experimenter may wish to project the design on the active factors (or, equivalently, collapse the design on the inactive factors) with the hope that the resulting design will be a full factorial so that all main effects and interactions among the active factors can be estimated clear of each other. Of course, it is not a priori known which factors may turn out to be active. In general, a fractional-factorial design of resolution R has the property that its projection in *every* subset of $P = R - 1$ factors is a full factorial, where P is known as the **projectivity** of the design. Thus, no matter which $R - 1$ factors turn out to be active, the design projected in those factors will be a full factorial. If less than $R - 1$ factors turn

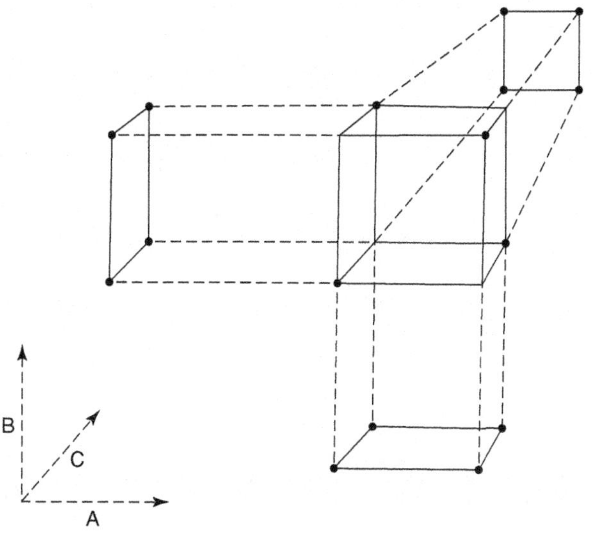

Figure 8.2 Projections of 2^{3-1}_{III} design into 2^2 factorials (From Box, Hunter, and Hunter, 2005, Figure 6.3). Reprinted by permission of John Wiley & Sons, Inc.

out to be active, then the design projected on those factors will be a full factorial with replicates.

Finally, note that these 2^{3-1}_{III} designs are orthogonal. This can be verified by checking that the columns for A, B, and C in the design given in Table 8.2 are mutually orthogonal. This in turn implies that the estimated main effects are uncorrelated.

Example 8.1 (Improving the Yield of a Spring Manufacturing Process: 2^{3-1} Experiment)

Refer to Example 6.1 on the percentage yield of crack-free springs as a function of three factors. Figure 8.3 shows the data. Suppose that only the principal half fraction, $\{a, b, c, abc\}$, is observed. Then the estimates of the three main effects, each of which is aliased with its complementary two-factor interaction, are computed as follows using the notational convention defined following (8.1):

$$\widehat{A} + \widehat{BC} = \tfrac{1}{2}(y_{abc} + y_a - y_b - y_c) = \tfrac{1}{2}(87 + 61 - 59 - 79) = +5.0,$$

$$\widehat{B} + \widehat{AC} = \tfrac{1}{2}(y_{abc} - y_a + y_b - y_c) = \tfrac{1}{2}(87 - 61 + 59 - 79) = +3.0,$$

$$\widehat{C} + \widehat{AB} = \tfrac{1}{2}(y_{abc} - y_a - y_b + y_c) = \tfrac{1}{2}(87 - 61 - 59 + 79) = +23.0.$$

Next suppose that only the alternate half fraction, $\{(1), ab, ac, bc\}$, was observed. Then the estimates of the three main effects, each of which is aliased with the negative of its complementary two-factor interaction, are computed as follows

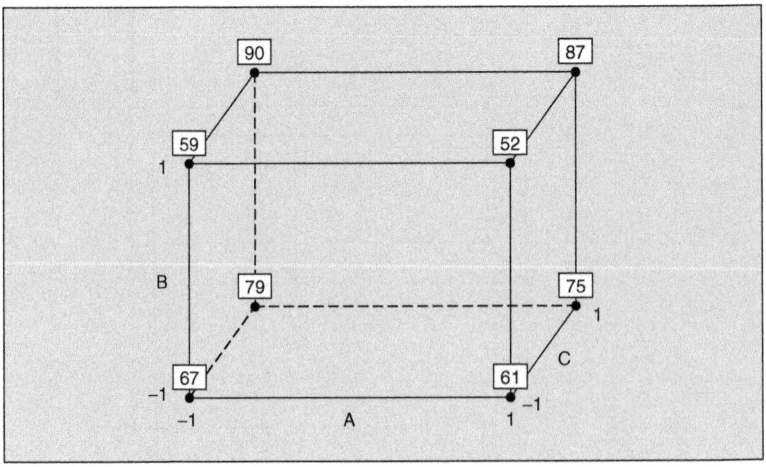

Figure 8.3 Percent yields of springs at different factor settings.

[again using the notational convention defined following (8.1)]:

$$\widehat{A} - \widehat{BC} = \tfrac{1}{2}(y_{ab} + y_{ac} - y_{bc} - y_{(1)}) = \tfrac{1}{2}(52 + 75 - 90 - 67) = -15.0,$$

$$\widehat{B} - \widehat{AC} = \tfrac{1}{2}(y_{ab} - y_{ac} + y_{bc} - y_{(1)}) = \tfrac{1}{2}(52 - 75 + 90 - 67) = 0.0,$$

$$\widehat{C} - \widehat{AB} = \tfrac{1}{2}(-y_{ab} + y_{ac} + y_{bc} - y_{(1)}) = \tfrac{1}{2}(-52 + 75 + 90 - 67) = 23.0.$$

The concepts introduced above extend naturally to general 2^{p-1} designs. In particular, the highest possible resolution designs are obtained by using the defining relation $I = AB \cdots P$ (principal fraction) or $I = -AB \cdots P$ (alternate fraction); the resolution of both designs is p. In these designs every mth-order effect is aliased with its complementary $(p - m)$th-order effect. The designs can be generated by first writing a full factorial in the first $p - 1$ factors and then equating the levels of the pth factor to the componentwise product of the columns of the design matrix of the first $p - 1$ factors for the principal fraction and the negative of the product for the alternate fraction.

As an example, consider the principal fraction of the 2^{4-1} design with the defining relation $I = ABCD$. Then the design can be constructed by first writing a full factorial in three factors, A, B, and C, and then setting the levels of D equal to the contrast coefficients of the interaction ABC. The alternate fraction with the defining relation $I = -ABCD$ is obtained by setting the levels of D equal to the negative of the contrast coefficients of the interaction ABC.

Example 8.2 (Alias Structure of a $2^{3-1} \times 2^1$ Crossed Array)

Consider a 2^{3-1} design with defining relation $I = ABC$ which is duplicated for the two levels of D giving a total of eight runs. We refer to this design (shown

Table 8.3 Design Matrix of $2^{3-1} \times 2^1$ Crossed Array

Factor				Treatment
A	B	C	D	Combination
−	−	+	−	c
+	−	−	−	a
−	+	−	−	b
+	+	+	−	abc
−	−	+	+	cd
+	−	−	+	ad
−	+	−	+	bd
+	+	+	+	$abcd$

in Table 8.3) as a $2^{3-1} \times 2^1$ **crossed array** or a **product array**. Such crossed arrays are used in Taguchi designs studied in Chapter 10. The 2^{3-1} design in factors A, B, C is called an **inner array** and the 2^1 design in factor D is called an **outer array**.

All aliases of the effects involving A, B, C are the same as those for the inner array and were derived earlier. The alias of any effect involving D can be obtained from the defining relation $I = ABC$ for the inner array. For example, $AD = BCD$, $BD = ACD$, and so on. ■

8.1.2 General 2^{p-q} Fractional Factorial Designs

8.1.2.1 Defining Relations and Alias Structure

Let us begin by considering a 2^{p-2} design, that is, a quarter fraction of a 2^p factorial experiment. Such a fraction can be constructed by first writing a full factorial in $p-2$ factors. Two additional factors are then added by equating their levels to two selected interactions among the first $p-2$ factors. Thus we require two defining relations or generators for a 2^{p-2} fractional factorial.

As an example, consider a 2^{5-2} design, and suppose that the generators are specified as $I = ABD$ and $I = ACE$ or $D = AB$ and $E = AC$. We first write the full factorial in three factors, A, B, and C, and then add two more factors, D and E, by equating $D = AB$ and $E = AC$. Thus we forgo information on interactions AB and AC by aliasing them with the main effects of D and E; in addition, interactions ABD and ACE are not estimable at all. The resulting design is shown in Display 8.1. There are in fact four possible fractions that can be obtained by using different plus/minus sign combinations in the defining relations $I = \pm ABD = \pm ACE$. The above design is the principal fraction obtained by using all plus signs; the other three quarter fractions are alternate fractions.

By inspecting the design in Display 8.1 we see that, in addition to $D = AB$ and $E = AC$, we have a third defining relation, $E = BCD$ or $I = BCDE$. This defining relation follows by taking the product of $I = ABD$ and $I = ACE$ (with

A	B	C	$D = AB$	$E = AC$	Treatment Combination
−	−	−	+	+	*de*
+	−	−	−	−	*a*
−	+	−	−	+	*be*
+	+	−	+	−	*abd*
−	−	+	+	−	*cd*
+	−	+	−	+	*ace*
−	+	+	−	−	*bc*
+	+	+	+	+	*abcde*

Display 8.1 Design matrix of a 2^{5-2}_{III} fractional factorial with defining relations $D = AB$ and $E = AC$.

I	=	ABD	=	ACE	=	$BCDE$
A	=	BD	=	CE	=	$ABCDE$
B	=	AD	=	$ABCE$	=	CDE
C	=	$ABCD$	=	AE	=	BDE
D	=	AB	=	$ACDE$	=	BCE
E	=	$ABDE$	=	AC	=	BCE
BC	=	ACD	=	ABE	=	DE
CD	=	ABC	=	ADE	=	BE

Display 8.2 Complete alias structure of a 2^{5-2}_{III} fractional factorial with defining relations $I = ABD$ and $I = ACE$.

letter powers mod 2), that is, $I = (ABD)(ACE) = A^2BCDE = BCDE$. Thus, in a 2^{p-2} experiment, one must specify two generators and the third one is their so-called **generalized interaction**. The **complete alias structure** can be derived from these three defining relations and is shown in Display 8.2. The use of the alias structure is explained in Section 8.1.2.2.

From this alias structure we see that the resolution of the design is III since the main effects are aliased with two-factor interactions but not with each other. The resolution of the design is given by the *minimum* word length among the three defining relations, $I = ABD = ACE = BCDE$. In this example, the minimum word length is 3, so this is a resolution III design. It turns out that this is the highest possible resolution for a 2^{5-2} design.

More generally, a 2^{p-q} fractional-factorial design requires specification of q independent defining relations.[1] The design is obtained by writing the full factorial in $p - q$ factors as the basic design and then equating q additional

[1]By independent we mean that no generator can be obtained by taking a product of any subset of the remaining generators. For example, ABC, ABD, and CD are not independent generators because the product of any two yields the third one.

factors to the specified interactions among the first $p - q$ factors which are the generators.

Additional $2^q - q - 1$ generators are the generalized interactions obtained by taking products of all pairs, triples, quadruples, and so on, of the specified independent q generators. Thus there are $2^q - 1$ generators in total. These together equated to I constitute a **complete defining relation**. In the 2^{5-2} design example above, the complete defining relation is $I = ABD = ACE = BCDE$.

The alias structure can be derived from the $2^q - 1$ generators by the usual multiplication process. Treating I as the zero-order effect, the totality of 2^p effects is divided into 2^{p-q} groups, each consisting of 2^q effects, which are aliased with each other. Note that I is aliased with $2^q - 1$ generators which gives the complete defining relation. For example, from Display 8.2 we see that for a 2^{5-2} design, there are $2^3 = 8$ groups of aliased effects; each group consists of $2^2 = 4$ effects.

8.1.2.2 Use of the Alias Structure to Accommodate Designated Interactions

What is the use of the alias structure? First, it tells us which effects cannot be estimated clear of each other. For example, the alias structure in Display 8.2 tells us that the main effect A and the two-factor interaction BD are aliases, so the letters A, B, and D should not be assigned to those factors whose corresponding effects may be important. Second, it tells us what follow-up experiment to run in order to separate the estimates of aliased effects of interest by pooling the data from the two experiments. This is discussed in Section 8.6.

The following is a common practical problem: Having chosen a particular 2^{p-q} fractional-factorial design, how to assign the experimental factors to the columns of the design matrix (i.e., how to assign the experimental factors to the letters $A, B, C, \cdots$) so that all main effects and at least certain two-factor interactions of interest are clear of each other. This is called **accommodating** these main effects and interactions. We limit to only two-factor interactions. A resolution V design leaves all main effects and two-factor interactions clear and so completely solves the accommodation problem. However, in practice often we must use resolution III or IV designs. **Linear graph** methods for this purpose were proposed by Taguchi and Wu (1985), Kacker and Tsui (1990), and Wu and Chen (1992). A simpler method was proposed by Bisgaard and Fuller (1994). It is illustrated in the following example.

Example 8.3 (Assigning Factors to Letters)

Consider the 2^{5-2}_{III} design given above. Suppose that this design is to be used to study how the surface finish in a milling operation is affected by the following factors: A', cutting speed; B', feed rate; C', depth of cut; D', type of coolant; and E', type of cutting tool. (Primes are put on the experimental factor labels to distinguish them from the unprimed generic factor labels.) Further suppose that in addition to all the main effects the experimenter wishes to estimate the interactions $A'B'$ and $A'C'$ clear of each other and of the main effects (all other interactions are assumed to be negligible). The problem is how to assign these five factors to the columns of the design matrix in Display 8.1.

We first write the alias structure up to two-factor interactions from the complete alias structure given in Display 8.2:

$$A = BD = CE, \quad B = AD, \quad C = AE, \quad D = AB, \quad E = AC,$$

$$BC = DE, \quad CD = BE.$$

We see that we cannot simply substitute $A \to A'$, $B \to B'$, and so on, since then the interactions $A'B'$ and $A'C'$ will be aliased with the main effects D' and E', respectively. Bisgaard and Fuller (1994) suggest typing the alias structure using a word processor and then use the "replace" command to find the proper assignment of primed letters to the unprimed ones by trial and error. By inspection of the alias structure up to two-factor interactions we see that all interactions involving A are aliased with the main effects; none are aliased with two-factor interactions among B, C, D, E. Since interactions involving D' or E' are not important, they may be aliased with some main effects. So we will assign, say, A to D'. We start by making the assignments $A \to D'$, $B \to A'$, $C \to B'$, $D \to C'$, and $E \to E'$. Then the above alias structure becomes

$$D' = A'C' = B'E', \quad A' = C'D', \quad B' = D'E', \quad C' = A'D', \quad E' = B'D',$$

$$A'B' = C'E', \quad B'C' = A'E'.$$

This is not a satisfactory assignment because $A'C'$ is still aliased with the main effect D'. By switching the last two assignments to $D \to E'$ and $E \to C'$, we get the alias structure

$$D' = A'E' = B'C', \quad A' = D'E', \quad B' = C'D', \quad E' = A'D', \quad C' = B'D',$$

$$A'B' = C'E', \quad B'E' = A'C'.$$

Now all main effects and interactions $A'B'$ and $A'C'$ are clear of each other, so our accommodation problem is solved. ■

This trial-and-error method does not always guarantee a solution. There may be more than one solution or a solution may not exist (a complete accommodation of all the desired effects may not be possible). In the latter case, one may prioritize the interactions that should be accommodated. A solution may be regarded as acceptable if the interactions that ranked high in this list are accommodated.

8.1.2.3 *Resolution*

As we saw before, the resolution of a design equals the minimum word length among the $2^q - 1$ generators. The q basic generators must be chosen carefully to maximize the resolution. For example, for the 2^{5-2} design considered in Section 8.1.2 we used the generators ABD and ACE resulting in the third generator $BCDE$ and resolution III. Suppose that, hoping to maximize the resolution, had we chosen the two generators to be the two highest order interactions, say $ABCD$ and $ABCDE$, we would have actually ended up with a resolution I design since the third generator would have been $(ABCD)(ABCDE) = E$. Many statistical packages (e.g., JMP and Minitab) give the highest possible resolution design for any specified p and q; they also give the complete alias structure up to any specified order of effects. Therefore we have not included a table of these designs. But for the sake of illustration, we give the defining relations for three designs with the highest resolutions for $p = 8$ with 16, 32, and 64 runs (i.e., $q = 4, 3, 2$):

$$2_{\text{IV}}^{8-4}: \quad I = \pm BCDE = \pm ACDF = \pm ABDH,$$

$$2_{\text{IV}}^{8-3}: \quad I = \pm ABCF = \pm ABDG = \pm BCDEH,$$

$$2_{\text{V}}^{8-2}: \quad I = \pm ABCDG = \pm ABEFH.$$

8.1.2.4 Projection Property

A 2^{p-q} design of resolution R projects into a full factorial (possibly with replicates) in a subset of any $R \le p - q$ factors if those R factors do not form a word in the complete defining relation. If those R factors form a word then the projected design is a fractional factorial (possibly with replicates).

Example 8.4 (Projecting 2_{III}^{5-2} Design on Three Factors)

Consider the 2_{III}^{5-2} design discussed in Section 8.1.2. The complete defining relation of this design is

$$I = ABD = ACE = BCDE.$$

Display 8.3 shows the design.

It can be checked that we get a full factorial if we project it into any three factors that do not form a word, for example, A, B, C or A, C, D or B, C, D. However, if we project it into A, B, D or A, C, E, we get two replicates of the respective half fractions. If we project it into B, C, D, E, we get a single half fraction.

Suppose we are doing an experiment with five factors, but a priori expect only up to three *specified* factors to be active. Then we should not assign the labels A, B, D or A, C, E to them, so that those factors do not form a generator. If, in fact, the other two factors turn out to be inactive, then we can project the data on the active factors and get a full factorial in them, thus being able to estimate all their main effects and interactions. ■

A	B	C	D	E
−	−	−	+	+
+	−	−	−	−
−	+	−	−	+
+	+	−	+	−
−	−	+	+	−
+	−	+	−	+
−	+	+	−	−
+	+	+	+	+

Display 8.3 A 2^{5-2} design.

8.1.3 Statistical Analysis

The formulas (7.9) and (7.11) for the coefficient estimators and their standard errors, respectively, for 2^p experiments apply to 2^{p-q} experiments as well (and more generally to two-level orthogonal arrays such as the Plackett–Burman designs discussed in Section 8.2). Thus we get

$$\widehat{\beta}_0 = \overline{\overline{y}} \quad \text{and} \quad \widehat{\beta}_j = \frac{\sum_i c_{ij}\overline{y}_i}{2^{p-q}}, \tag{8.2}$$

where i indexes the treatment combinations used in the experiment. The SEs of any coefficient estimator $\widehat{\beta}_j$ and the corresponding estimated effect equal

$$\text{SE}(\widehat{\beta}_j) = \frac{s}{\sqrt{n2^{p-q}}} = \frac{s}{\sqrt{N}} \quad \text{and} \quad \text{SE}(\widehat{\text{Effect}_j}) = \frac{s}{\sqrt{N/4}}. \tag{8.3}$$

The formula (7.16) for SS_{Effect} applies directly:

$$SS_{\text{Effect}} = \tfrac{1}{4}N(\widehat{\text{Effect}})^2. \tag{8.4}$$

From these formulas the usual t- and F-tests for the significance of the estimated effects follow.

Example 8.5 (Factors Affecting Yield of Peanut Oil)

Kligo (1988) reported a 2^{5-1} experiment to determine the effects of the factors listed in Table 8.4 on the yield (y) of peanut oil per batch. The data are given in Table 8.5. Based on these data we want to determine the significant factor effects and interactions and fit the corresponding predictive model. By inspection we readily see that the defining relation of this design is $I = -ABCDE$. Thus the design is of resolution V and so, if we ignore the third- and higher order interactions, then all the main effects and two-factor interactions are clear.

Table 8.4 Factors and Factor Levels Used in Peanut Oil Yield Experiment

	Level	
Factor	−	+
A: CO_2 pressure (bars)	415	550
B: Temperature (°C)	25	95
C: Moisture (% by weight)	5	15
D: Flow (L/min)	40	60
E: Particle Size (mm)	1.28	4.05

Source: Kligo (1988, Table 8.1). Reprinted by permission of Taylor & Francis.

Table 8.5 Yield of Peanut Oil

Treatment Combination	Yield (%)	Treatment Combination	Yield (%)
(1)	63	*de*	23
ae	21	*ad*	74
be	36	*bd*	80
ab	99	*abde*	33
ce	24	*cd*	63
ac	66	*acde*	21
bc	71	*bcde*	44
abce	54	*abcd*	96

Source: Kligo (1988, Table 8.2). Reprinted by permission of Taylor & Francis.

Residual plots for models fitted using the percentage yield (y) as the response variable show that the data are heteroscedastic. Therefore we decided to use $\log(y)$ as the response variable and fitted a second-order model that includes all main effects and two-factor interactions. The half-normal plot of the effects in Figure 8.4 shows that only the effects B and E are significant using $\alpha = 0.10$. We refitted the model with only those two effects. The normal and fitted-values plots of the residuals from this model are shown in Figure 8.5. The normal plot is satisfactory, but the fitted-value plot shows distinct curvature indicating a need to include an interaction term in B and E. The revised normal and fitted-values plots of the residuals are shown in Figure 8.6. Both plots are now quite satisfactory. The Minitab output for the final prediction model,

$$\widehat{\log(y)} = 1.6785 + 0.0943x_B - 0.1989x_E + 0.0387x_Bx_E, \tag{8.5}$$

is shown in Display 8.4. ■

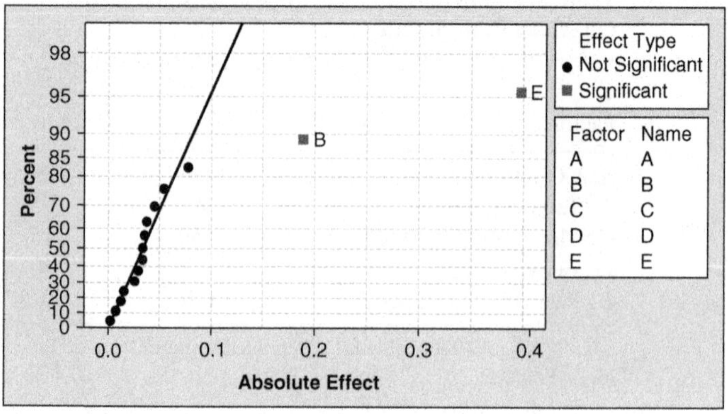

Figure 8.4 Half-normal plot of effects for peanut oil data.

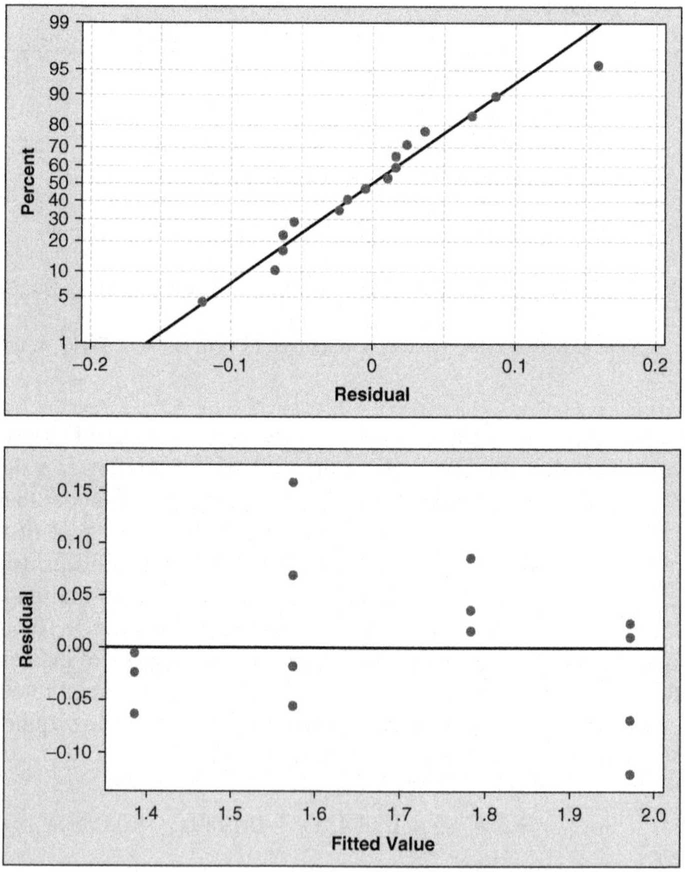

Figure 8.5 Normal (top) and fitted-values (bottom) plots of residuals from model with main effects B and E for peanut oil data.

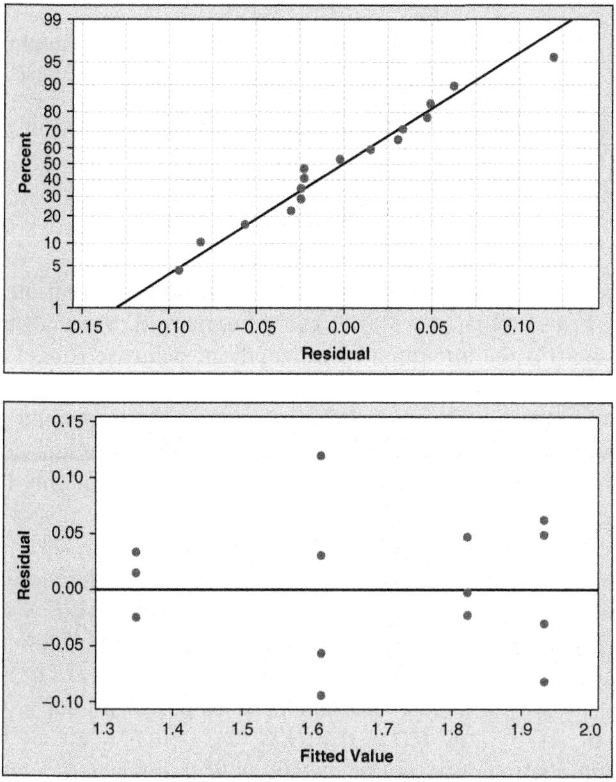

Figure 8.6 Normal (top) and fitted-values (bottom) plots of residuals from model with main effects B and E and interaction BE for peanut oil data.

Estimated Effects and Coefficients for log(y) (coded units)

Term	Effect	Coef	SE Coef	T	P
Constant		1.6785	0.01566	107.21	0.000
B	0.1886	0.0943	0.01566	6.02	0.000
E	-0.3978	-0.1989	0.01566	-12.71	0.000
B*E	0.0774	0.0387	0.01566	2.47	0.029

S = 0.0626247 R-Sq = 94.44

Analysis of Variance for log(y) (coded units)

Source	DF	Seq SS	Adj SS	Adj MS	F	P
Main Effects	2	0.77542	0.77542	0.387710	98.86	0.000
2-Way Interactions	1	0.02399	0.02399	0.023985	6.12	0.029
Residual Error	12	0.04706	0.04706	0.003922		
Pure Error	12	0.04706	0.04706	0.003922		
Total	15	0.84647				

Display 8.4 Minitab output for final fitted model (8.5) for peanut oil data.

8.1.4 Minimum Aberration Designs

In many cases there is more than one design with maximum resolution. For example, the following three 2^{7-2} designs all have resolution IV:

$$\begin{aligned}
&\text{Design 1:} \quad I = ABCDF = ABDEG = CEFG, \\
&\text{Design 2:} \quad I = ABCF = ADEG = BCDEFG, \\
&\text{Design 3:} \quad I = ABCF = BCDG = ADFG.
\end{aligned}$$

To choose among the three designs, the **minimum aberration criterion** introduced by Fries and Hunter (1980) can be employed. Since all three designs are of resolution IV, the minimum word length in each case is 4. However, design 1 has only one word of length 4, design 2 has two words of length 4, and design 3 has all three words of length 4. Design 1, having the least number of words with the shortest length (R, the resolution), is called the minimum-aberration design.

Why is this a desirable property for a design? Note that only four main effects are aliased with three-factor interactions in design 1; the others are aliased with higher order interactions. On the other hand, all seven main effects are aliased with three-factor interactions in designs 2 and 3. Furthermore, there are only three pairs of two-factor interactions aliased in design 1 ($CE = FG, CF = EG$, $CG = EF$), six pairs of two-factor interactions aliased in design 2 ($AB = CF$, $AC = BF$, $AD = EG$, $AE = DG$, $AF = BC$, $AG = DE$), and six pairs and one triple of two-factor interactions aliased in design 3 ($AB = CF$, $AC = BF$, $AD = FG$, $AG = DF$, $BD = CG$, $BG = CD$, $AF = BC = DG$). In general, a minimum aberration design of resolution R minimizes the number of k-factor effects that are aliased with $(R - k)$-factor effects for $1 \le k \le R$, and therefore is preferable.

A formal method for choosing a minimum aberration design is as follows. Let n_k denote the number of words of length k in a generator; for example, in design 3 with generators $ABCF$, $BCDG$, and $ADFG$, we have $n_4 = 3$ and $n_k = 0$ for $k \neq 4$. Let $(n_1, n_2, \ldots, n_p)$ denote the vector of number of words of length $1, 2, \ldots, p$ of all $2^q - 1$ generators in a 2^{p-q} design. Note that if the design has resolution R, then $n_k = 0$ for $k < R$. This vector is called the **word length pattern**.

Consider two 2^{p-q} designs, say D and D', of resolution R and with word length patterns $(n_1, n_2, \ldots, n_p)$ and $(n'_1, n'_2, \ldots, n'_p)$. Suppose that $s \ge R$ is the smallest word length for which $n_s \neq n'_s$. If $n_s < n'_s$, then D is said to have less aberration than D' and vice versa. If there is no other 2^{p-q} design with less aberration than D, then D is the minimum aberration design.

Example 8.6

For the three 2^{7-2}_{IV} designs given above, the word length patterns are as follows: design 1, $(0, 0, 0, 1, 2, 0, 0)$; design 2, $(0, 0, 0, 2, 0, 1, 0)$; design 3, $(0, 0, 0, 3, 0, 0, 0)$. Here $n_1 = n_2 = n_3 = 0$ for all three designs since they are of resolution IV. Design 1 minimizes n_4, so it is the minimum aberration design. ■

8.2 PLACKETT–BURMAN DESIGNS

Plackett and Burman (1946) proposed a class of resolution III orthogonal designs for studying p factors in $N = p + 1$ runs. These PB designs (considered here only for factors at two levels) are a special case of a more general class of designs, called Hadamard designs, which are discussed in Section 8.3. Being resolution III designs, they allow estimation of all main effects clear of each other. Table 8.6 shows a PB design for $N = 8$, $p = 7$.

PB designs are saturated because all the $N - 1 = p$ degrees of freedom are used up to estimate the main effects of the factors and none is left to estimate the error. However, one need not use a PB design in a saturated fashion. If less than p factors are studied using a PB design with $N = p + 1$ runs, then the contrasts corresponding to the unassigned columns can be used to estimate the error. For example, the PB design for $p = 7$ in Table 8.6 may be used to study only five factors. If the interactions corresponding to the two unassigned columns can be ignored, then the sums of squares of the respective contrasts can be pooled to obtain an error estimate. If the last two columns of that design are unassigned, then they can be used to estimate the error if the interactions BC and ABC (and their aliases) can be ignored.

PB designs exist only when N is a multiple of 4. If N is a power of 2, then they are simply 2^{p-q} designs. For example, if $N = 4$, $p = 3$, then a PB design is a 2^{3-1}_{III} design; if $N = 8$, $p = 7$, then a PB design is a 2^{7-4}_{III} design; if $N = 16$, $p = 15$, then a PB design is a 2^{15-11}_{III} design; and so on. They are called **geometric designs** because they can be represented as a subset of vertices of a hypercube, as seen in Figure 8.2.

PB designs are not so simple when N is not a power of 2. They are called **nongeometric designs** because they cannot be represented as vertices of a

Table 8.6 Plackett–Burman Design for $N = 8$, $p = 7$

	Factor						
Run	A	B	C	$D = AB$	$E = AC$	$F = BC$	$G = ABC$
1	−	−	−	+	+	+	−
2	+	−	−	−	−	+	+
3	−	+	−	−	+	−	+
4	+	+	−	+	−	−	−
5	−	−	+	+	−	−	+
6	+	−	+	−	+	−	−
7	−	+	+	−	−	+	−
8	+	+	+	+	+	+	+

hypercube. Since these designs are not 2^{p-q} fractional factorials, they cannot be defined in terms of generators. As a result, their alias patterns are not simple.

We now explain why geometric PB designs have simple alias patterns while nongeometric designs do not. From the preceding discussion and examples it is clear that when two effects are aliases of each other, their contrast vectors are identical except possibly for a change of sign; that is, they are perfectly correlated with a correlation of +1 or −1. On the other hand, two effects are not aliases of each other if their contrast vectors are orthogonal to each other; that is, they are uncorrelated. These are the only two cases that arise in 2^{p-q} fractional factorial designs of which the geometric PB designs are a special case. But for nongeometric PB designs (as well as nonorthogonal designs such as the supersaturated designs discussed in Section 8.4) contrast vectors with correlations strictly between −1 and +1 arise, which leads to **partial aliasing** of effects.

Refer to Exercise 14.25. Suppose that $\boldsymbol{X}_1$ and $\boldsymbol{X}_2$ are the model matrices of two sets of effects and $\boldsymbol{\beta}_1$ and $\boldsymbol{\beta}_2$ are the respective parameter vectors. Then part (a) of that exercise asks you to show that

$$\mathrm{E}(\widehat{\boldsymbol{\beta}}_1) = \boldsymbol{\beta}_1 + (\boldsymbol{X}_1'\boldsymbol{X}_1)^{-1}\boldsymbol{X}_1'\boldsymbol{X}_2\boldsymbol{\beta}_2. \tag{8.6}$$

In our context, the columns of $\boldsymbol{X}_1$ and $\boldsymbol{X}_2$ are the contrast vectors of those two sets of effects. If the columns are mutually orthogonal, then obviously the bias is zero, and the effects in the first set are not aliased with those in the second set. More generally, the bias of $\widehat{\boldsymbol{\beta}}_1$ is a weighted sum of the elements of $\boldsymbol{\beta}_2$ with the weighting coefficients, called the **alias coefficients**, given by the elements of the so-called **alias matrix**: $(\boldsymbol{X}_1'\boldsymbol{X}_1)^{-1}\boldsymbol{X}_1'\boldsymbol{X}_2$.

Consider the special case of a PB design with N rows and for the sake of simplicity focus only on the main effects and two-factor interactions. Let $\boldsymbol{X}_1$ and $\boldsymbol{X}_2$ correspond to the model matrices of the main effects and two-factor interactions, respectively. Then $(\boldsymbol{X}_1'\boldsymbol{X}_1)^{-1} = (1/N)\boldsymbol{I}$. If $\boldsymbol{c}_i$ denotes the contrast vector of the ith main effect, β_{1i}, and $\boldsymbol{c}_j$ that of the jth two-factor interaction, β_{2j}, then their alias coefficient is given by

$$r_{ij} = \frac{\boldsymbol{c}_i'\boldsymbol{c}_j}{N}, \tag{8.7}$$

which is also the correlation coefficient between the vectors $\boldsymbol{c}_i$ and $\boldsymbol{c}_j$.

In a geometric PB design, it is easy to check that, given any main-effect contrast $\boldsymbol{c}_i$ and a two-factor interaction contrast $\boldsymbol{c}_j$, either $\boldsymbol{c}_i$ is orthogonal to $\boldsymbol{c}_j$ or $\boldsymbol{c}_i = \pm\boldsymbol{c}_j$. In fact, most of the $r_{ij} = 0$ (the two effects are not aliased) and only a small number of $r_{ij} = \pm 1$ (the two effects are fully aliased). Therefore a geometric design is said to have a **simple aliasing pattern**. On the other hand, in a nongeometric design many $r_{ij} \neq 0$ and are strictly between −1 and +1. For example, in the PB design for $N = 12$, $p = 11$ shown in Table 8.8, the alias coefficient between every main effect and every two-factor interaction not involving that main effect is $\pm\frac{1}{3}$. As a check, consider the main effect A and the interaction BC. The correlation coefficient between the two effects can be computed to be $\frac{1}{3}$ from their contrast vectors,

$$\boldsymbol{c}_A = (+-+---+++-+-) \text{ and } \boldsymbol{c}_{BC} = (-+---++-++-+).$$

In this case we say that the two effects are **partially aliased** and the design has a **complex aliasing pattern**. In general, from (8.6) it follows that the expected value of the LS estimator of a main effect is given by

$$E(\widehat{\beta_{1i}}) = \beta_{1i} + \sum_{j \neq i} r_{ij}\beta_{2j}. \tag{8.8}$$

Because of the biases introduced by a large number of two-factor interactions, the analysis of experiments with complex aliasing patterns must be conducted with care, as the following example illustrates.

Example 8.7 (Weld-Repaired Cast Fatigue Experiment)

The following data and method of analysis are taken from Hamada and Wu (1992) who analyzed an experiment reported by Hunter, Hodi, and Eager (1982) to study the effects of seven factors on the fatigue life of weld-repaired castings using a 12-run PB design. The factors and their levels are listed in Table 8.7. The PB design and the data are shown in Table 8.8. Note that this is an unsaturated design since the columns e_1–e_4 are unassigned; they can be used to estimate the error.

The effect estimates for the main-effects model are shown in Display 8.5 and the half-normal plot of the effects is shown in Figure 8.7. We see that only the effect F is significant. The effect D is the second largest in magnitude. Applying the effect heredity principle, Hamada and Wu (1992) suggested examining all

Table 8.7 Factors and Levels for Cast Fatigue Experiment

	Level	
Factor	−	+
A: Initial structure	As received	β Treat
B: Bead size	Small	Large
C: Pressure treat	None	HIP
D: Heat treat	Anneal	Solution treat/age
E: Cooling rate	Slow	Rapid
F: Polish	Chemical	Mechanical
G: Final treat	None	Peen

Source: Hamada and Wu (1992, Table 8.1). Reprinted by permission of American Society for Quality.

Table 8.8 Design and Data for Cast Fatigue Experiment

	Factor							Unassigned				Logged
Run	A	B	C	D	E	F	G	e_1	e_2	e_3	e_4	Data
1	+	+	−	+	+	+	−	−	−	+	−	6.058
2	−	+	+	−	+	+	+	−	−	−	+	5.863
3	+	−	+	+	−	+	+	+	−	−	−	5.917
4	−	+	−	+	+	−	+	+	+	−	−	5.818
5	−	−	+	−	+	+	−	+	+	+	−	6.607
6	−	−	−	+	−	+	+	−	+	+	+	5.682
7	+	−	−	−	+	−	+	+	−	+	+	5.752
8	+	+	−	−	−	+	−	+	+	−	+	7.000
9	+	+	+	−	−	−	+	−	+	+	−	5.899
10	−	+	+	+	−	−	−	+	−	+	+	4.625
11	+	−	+	+	+	−	−	−	+	−	+	4.733
12	−	−	−	−	−	−	−	−	−	−	−	4.809

Source: Hamada and Wu (1992, Table 8.2). Reprinted by permission of the American Society for Quality.

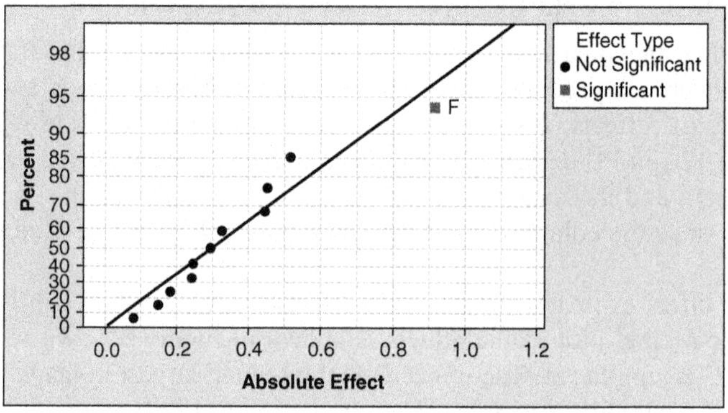

Figure 8.7 Half-normal plot of effects for cast fatigue experiment data.

two-factor interactions involving the factor F. Clearly, all these interactions are orthogonal to F, but only FG turns out to be significant with the resulting prediction model

$$\hat{y} = 5.730 + 0.458x_F - 0.459x_F x_G. \tag{8.9}$$

Both the coefficients in the model are highly significant ($p < 0.001$) and $R^2 =$ 89.25%.

The finding of the significant interaction $\widehat{FG}$ helps in interpreting the main-effect estimates of the other factors. First, if the main effect D is added to the above prediction model, then its R^2 only increases to 91.90%, indicating that D makes a nonsignificant contribution ($p = 0.145$). The relatively large value of $\widehat{D}$ in Display 8.5 is explained by the fact that the aliasing coefficient

```
Estimated Effects and Coefficients for log(y) (coded units)

Term          Effect      Coef
Constant                 5.7303
A             0.3258     0.1629
B             0.2938     0.1469
C            -0.2458    -0.1229
D            -0.5162    -0.2581
E             0.1498     0.0749
F             0.9152     0.4576
G             0.1832     0.0916
e1            0.4458     0.2229
e2            0.4525     0.2262
e3            0.0805     0.0403
e4           -0.2422    -0.1211
```

Display 8.5 Minitab output for effect estimates for cast fatigue experiment data.

between D and FG is $\frac{1}{3}$. Therefore $\widehat{D}$ actually estimates $D + \frac{1}{3}FG$. Since $\widehat{FG} = 2(-0.459) = -0.918$, the corrected estimate of D equals

$$\widehat{D} = -0.5162 - \tfrac{1}{3}\widehat{FG} = -0.5162 - \tfrac{1}{3}(-0.918) = -0.2102,$$

which is much smaller. In fact, the same correction can be made to all other main-effect estimates (except $\widehat{F}$ and $\widehat{G}$, which are uncorrelated with $\widehat{FG}$). The main effects A, B, and E have correlation $-\frac{1}{3}$ and C has correlation $\frac{1}{3}$ with FG. Hence we get the following corrected estimates:

$$\widehat{A} = 0.3258 + \tfrac{1}{3}(-0.918) = 0.0198,$$

$$\widehat{B} = 0.2938 + \tfrac{1}{3}(-0.918) = -0.0122,$$

$$\widehat{C} = -0.2458 - \tfrac{1}{3}(-0.918) = 0.0602,$$

$$\widehat{E} = 0.1498 + \tfrac{1}{3}(-0.918) = -0.1562.$$

We see that the corrected estimates $\widehat{A}$, $\widehat{B}$, and $\widehat{C}$ are much smaller.

The final prediction model (8.9) indicates that to maximize the response, one should use $F = +$ (mechanical polish) and $G = -$ (no peening), which gives $\widehat{y} = 5.730 + 0.458 + 0.459 = 6.647$. If the interaction FG was ignored and G was set at the $+$ level (peening), then $\widehat{y} = 5.729$. Thus setting $G = -$ (no peening) has increased the predicted life by 16% on the log scale. ■

The first 11 rows of the PB design for $N = 12$, $p = 11$ in Table 8.8 can be generated by cyclically permuting the first row, that is, by shifting all entries of every row, beginning with the first one, to the right by one column and placing the

Table 8.9 Generator Rows for Some Nongeometric Plackett–Burman Designs

p	N	Generator Row
11	12	+ + − + + + − − − + −
19	20	+ + − − + + + + − + − + − − − − + + −
23	24	+ + + + + − + − + + − − + + − − + − + − − − −
35	36	− + − + + + − − − + + + + + − + + + − − + − − − − + − + − + + − − + −

$$\boldsymbol{X} = \begin{bmatrix} + & - & + & + & + & + & - & - & - \\ + & + & - & + & + & + & - & - & - \\ - & + & + & + & + & + & - & - & - \\ - & - & - & + & - & + & + & + & + \\ - & - & - & + & + & - & + & + & + \\ - & - & - & - & + & + & + & + & + \\ + & + & + & - & - & - & + & - & + \\ + & + & + & - & - & - & + & + & - \\ + & + & + & - & - & - & - & + & + \end{bmatrix}$$

$$\boldsymbol{Y} = \begin{bmatrix} - & + & - & - & - & + & - & - & + \\ - & - & + & + & - & - & + & - & - \\ + & - & - & - & + & - & - & + & - \\ - & - & + & - & + & - & - & - & + \\ + & - & - & - & - & + & + & - & - \\ - & + & - & + & - & - & - & + & - \\ - & - & + & - & - & + & - & + & - \\ + & - & - & + & - & - & - & - & + \\ - & + & - & - & + & - & + & - & - \end{bmatrix}$$

$$\boldsymbol{Z} = \begin{bmatrix} + & + & - & + & - & + & + & - & + \\ - & + & + & + & + & - & + & + & - \\ + & - & + & - & + & + & - & + & + \\ + & - & + & + & + & - & + & - & + \\ + & + & - & - & + & + & + & + & - \\ - & + & + & + & - & + & - & + & + \\ + & - & + & + & - & + & + & + & - \\ + & + & - & + & + & - & - & + & + \\ - & + & + & - & + & + & + & - & + \end{bmatrix}.$$

Display 8.6 Matrices $\boldsymbol{X}, \boldsymbol{Y}, \boldsymbol{Z}$ for constructing Plackett–Burman design for $N = 28$, $p = 27$.

last entry in the first column, which yields the next row. To these 11 rows is added the last row of all minus signs. Check that this is an orthogonal design. We call the first row the **generator row**. Table 8.9 gives the generator rows of PB designs for $N = 12, 20, 24, 36$. However, not all nongeometric designs can be generated in this manner from a single generator row and more complicated constructions are needed. For example, the PB design for $N = 28$, $p = 27$ has the form

$$\begin{bmatrix} X & Y & Z \\ Z & X & Y \\ Y & Z & X \end{bmatrix},$$

with a last row of minus signs added where the matrices $\boldsymbol{X}$, $\boldsymbol{Y}$, $\boldsymbol{Z}$ are as shown in Display 8.6. For additional nongeometric PB designs, see Appendix 7A of Wu and Hamada (2000).

8.3 HADAMARD DESIGNS

A **Hadamard matrix $\boldsymbol{H}$** of order N is an $N \times N$ matrix with all entries ± 1 and which satisfies

$$\boldsymbol{H}'\boldsymbol{H} = \boldsymbol{H}\boldsymbol{H}' = n\boldsymbol{I},$$

where $\boldsymbol{I}$ is an identity matrix of order N. Two Hadamard matrices are said to be equivalent if one can be obtained from the other by permuting some rows and/or columns and/or by switching signs in some rows and/or columns. A Hadamard matrix with all entries in the first row *and* column equal to $+1$ is said to be **normalized**; if only the entries in the first row *or* column equal $+1$, then it is said to be **seminormalized**. Except for the trivial case of $N = 1$ when $\boldsymbol{H} = \pm 1$ and $N = 2$ when

$$\boldsymbol{H} = \begin{bmatrix} 1 & 1 \\ 1 & -1 \end{bmatrix},$$

it can be shown that, for a Hadamard matrix to exist, N must be a multiple of 4. For $N = 4$, the following is an example of a Hadamard matrix:

$$\boldsymbol{H} = \begin{bmatrix} 1 & -1 & -1 & 1 \\ 1 & 1 & -1 & -1 \\ 1 & -1 & 1 & -1 \\ 1 & 1 & 1 & 1 \end{bmatrix}.$$

This is just the main-effects model matrix of a PB design with $N = 4$, $p = 3$ (2^{3-1} fractional factorial); the first column corresponds to the intercept term of the model. If the first column of this matrix is deleted, then we get that PB design. In general, an $N \times N$ Hadamard matrix in which the first column has all entries equal to $+1$ is the main-effects model matrix for a PB design (also called a Hadamard design).

Next we address the question of construction of Hadamard matrices. Lower order Hadamard matrices can be combined to build higher order ones. To state this method, we first define the **Kronecker product** of two matrices: Let $\boldsymbol{A} = \{a_{ij}\} : m \times m$ and $\boldsymbol{B} = \{b_{ij}\} : n \times n$ be two matrices. Then their Kronecker product is an $mn \times mn$ matrix defined as

$$\boldsymbol{A} \otimes \boldsymbol{B} = \begin{bmatrix} a_{11}\boldsymbol{B} & a_{12}\boldsymbol{B} & \cdots & a_{1m}\boldsymbol{B} \\ a_{21}\boldsymbol{B} & a_{22}\boldsymbol{B} & \cdots & a_{2m}\boldsymbol{B} \\ \vdots & \vdots & \ddots & \vdots \\ a_{m1}\boldsymbol{B} & a_{m2}\boldsymbol{B} & \cdots & a_{mm}\boldsymbol{B} \end{bmatrix}. \tag{8.10}$$

If $\boldsymbol{G}$ and $\boldsymbol{H}$ are $m \times m$ and $n \times n$ Hadamard matrices, then their Kronecker product $\boldsymbol{G} \otimes \boldsymbol{H}$ is also a Hadamard matrix. The following is an example of this method:

$$\begin{bmatrix} 1 & 1 \\ 1 & -1 \end{bmatrix} \otimes \begin{bmatrix} 1 & 1 \\ 1 & -1 \end{bmatrix} = \begin{bmatrix} 1 & 1 & 1 & 1 \\ 1 & -1 & 1 & -1 \\ 1 & 1 & -1 & -1 \\ 1 & -1 & -1 & 1 \end{bmatrix},$$

which can be seen to be equivalent to the Hadamard matrix for $N = 4$ given above. A good reference on Hadamard matrices is the review article by Hedayat and Wallis (1978).

The PB designs are a special case of Hadamard designs in that they are obtained from seminormalized Hadamard matrices by deleting the extra column of all $+1$'s or -1's. For $N = 4, 8, 12$ all Hadamard matrices are equivalent (see Hedayat and Wallis, 1978), so the corresponding PB designs are the same as Hadamard designs. For $N = 16$, there are five equivalence classes of Hadamard matrices which result in nonequivalent Hadamard designs only one of which is the 2^{15-11} PB design. This design created using Minitab is shown in Display 8.7.

Table 8.10 gives an alternative Hadamard design for $N = 16$, $p = 15$. This design is fundamentally different from the PB design shown in Display 8.7. As seen in that display, each main effect in the PB design is aliased with seven two-factor interactions not involving that factor. On the other hand, in the Hadamard design shown in Table 8.10, only the A main effect is aliased with seven two-factor interactions, namely,

$$A = -BC = -DE = -FG = -HJ = -KL = -MN = -OP.$$

None of the other main effects is fully aliased with any two-factor interactions, but some are partially aliased. If the A column is not assigned to any factor when using this design with less than 15 factors, the main effects of the factors assigned to other columns can be estimated with less bias caused by two-factor interactions than the PB design shown in Display 8.7.

The methods of analysis for Hadamard designs are similar to those for PB designs and hence are not discussed separately.

Table 8.10 Hadamard Design for $N = 16, p = 15$

Run	A	B	C	D	E	F	G	H	J	K	L	M	N	O	P
1	−	−	−	−	−	−	−	+	+	+	+	+	+	+	+
2	−	−	−	+	+	+	+	−	−	−	−	+	+	+	+
3	−	−	−	+	+	+	+	+	+	+	+	−	−	−	−
4	−	+	+	−	−	+	+	−	−	+	+	−	−	+	+
5	−	+	+	−	−	+	+	+	+	−	−	+	+	−	−
6	−	+	+	+	+	−	−	−	−	+	+	+	+	−	−
7	−	+	+	+	+	−	−	+	+	−	−	−	−	+	+
8	+	−	+	−	+	−	+	−	+	−	+	−	+	−	+
9	+	−	+	−	+	+	−	−	+	+	−	+	−	+	−
10	+	−	+	+	−	−	+	+	−	+	−	−	+	+	−
11	+	−	+	+	−	+	−	+	−	−	+	+	−	−	+
12	+	+	−	−	+	−	+	+	−	+	−	+	−	−	+
13	+	+	−	−	+	+	−	+	−	−	+	−	+	+	−
14	+	+	−	+	−	−	+	−	+	−	+	+	−	+	−
15	+	+	−	+	−	+	−	−	+	+	−	−	+	−	+
16	−	−	−	−	−	−	−	−	−	−	−	−	−	−	−

8.4 SUPERSATURATED DESIGNS

In some exploratory screening experiments the number of factors to be studied far exceeds the number of runs that can be made due to budget or other resource constraints. According to the effect sparsity principle, only a small fraction of the factors are likely to be active. The goal of the experiment is to identify the active factors with large main effects. Obviously, the number of active factors that can be identified must be less than the number of runs. Designs used for this purpose are called supersaturated designs.

8.4.1 Construction of Supersaturated Designs

Two-level supersaturated designs are arrays of plus and minus signs with p columns of N-dimensional contrast vectors with $N < p$. Obviously, the columns cannot be mutually orthogonal or even linearly independent since their number exceeds the dimension of the contrast space, namely, $N - 1$. Therefore we look for designs that minimize correlations among the column vectors. In particular, no two column vectors must be perfectly correlated; otherwise the main effects of the factors assigned to that pair of columns will be completely aliased with each other.

Table 8.11 gives an example of a small supersaturated design with 6 rows and 10 columns. This design is obtained from the PB design for $N = 12, p = 11$ given in Table 8.8 by picking the 6 rows corresponding to the plus sign in column G and then dropping that column, which results in 10 columns. Clearly, we could have picked the other 6 rows, which would have also given a supersaturated

```
Design Generators:
E = ABC, F = ABD, G = ACD, H = BCD, J = ABCD, K = AB,
L = AC, M = AD, N = BC, O = BD, P = CD
```

Design Table

Run	A	B	C	D	E	F	G	H	J	K	L	M	N	O	P
1	−	−	−	−	−	−	−	−	+	+	+	+	+	+	+
2	+	−	−	−	+	+	+	−	−	−	−	−	+	+	+
3	−	+	−	−	+	+	−	+	−	−	+	+	−	−	+
4	+	+	−	−	−	−	+	+	+	+	−	−	−	−	+
5	−	−	+	−	+	−	+	+	−	+	−	+	−	+	−
6	+	−	+	−	−	+	−	+	+	−	+	−	−	+	−
7	−	+	+	−	−	+	+	−	+	−	−	+	+	−	−
8	+	+	+	−	+	−	−	−	−	+	+	−	+	−	−
9	−	−	−	+	−	+	+	+	−	+	+	−	+	−	−
10	+	−	−	+	+	−	−	+	+	−	−	+	+	−	−
11	−	+	−	+	+	−	+	−	+	−	+	−	−	+	−
12	+	+	−	+	−	+	−	−	−	+	−	+	−	+	−
13	−	−	+	+	+	+	−	−	+	+	−	−	−	−	+
14	+	−	+	+	−	−	+	−	−	−	+	+	−	−	+
15	−	+	+	+	−	−	−	+	−	−	−	−	+	+	+
16	+	+	+	+	+	+	+	+	+	+	+	+	+	+	+

```
Alias Structure (up to order 2)

A = BK = CL = DM = EN = FO = GP = HJ
B = AK = CN = DO = EL = FM = GJ = HP
C = AL = BN = DP = EK = FJ = GM = HO
D = AM = BO = CP = EJ = FK = GL = HN
E = AN = BL = CK = DJ = FP = GO = HM
F = AO = BM = CJ = DK = EP = GN = HL
G = AP = BJ = CM = DL = EO = FN = HK
H = AJ = BP = CO = DN = EM = FL = GK
J = AH = BG = CF = DE = KP = LO = MN
K = AB = CE = DF = GH = JP = LN = MO
L = AC = BE = DG = FH = JO = KN = MP
M = AD = BF = CG = EH = JN = KO = LP
N = AE = BC = DH = FG = JM = KL = OP
O = AF = BD = CH = EG = JL = KM = NP
P = AG = BH = CD = EF = JK = LM = NO
```

Display 8.7 Minitab output for Plackett–Burman design for $N = 16$, $p = 15$ (2^{15-11} fractional factorial).

design. It is easy to check that, in this design, the pairwise correlation between any pair of column contrast vectors is $\pm\frac{1}{3}$.

This example suggests a general procedure due to Lin (1993) for constructing a supersaturated design: Start with a PB design having $4k$ rows and $4k - 1$ columns. (Recall that, for a PB design to exist, the number of rows must be a multiple of 4.) Pick any column (called the **branching column**) and form two half fractions of the PB design corresponding to either the plus or minus signs in that column and

Table 8.11 Supersaturated Design for $N = 6, p = 10$

	Factor									
Run	A	B	C	D	E	F	G	H	I	J
1	+	+	−	+	+	+	−	−	−	+
2	+	−	+	+	−	+	+	+	−	−
3	−	+	−	+	+	−	+	+	+	−
4	−	−	+	−	+	+	−	+	+	+
5	+	+	+	−	−	−	+	−	+	+
6	−	−	−	−	−	−	−	−	−	−

then drop that column. This yields two supersaturated designs, each with $N = 2k$ rows and $p = 4k - 2 = 2(N - 1)$ columns. However, the resulting design may not necessarily have a good correlation structure among its column vectors. Lin (1995) proposed an algorithm that finds the maximum number of factors for a given number of runs when the degree of nonorthogonality is specified.

Nguyen (1996) gave an algorithm that attempts to minimize the correlations among the column vectors. Table 8.12 gives the designs constructed by this algorithm for $N = 6(2)26$ and $p = 10(4)50$ [which correspond to PB designs with $N = 4k$ for $k = 3(1)13$]. Two generator vectors of dimension $N - 1$ are given for each design. The design is created as follows: Write the first vector as a column vector followed by $N - 2$ additional column vectors formed by cyclically permuting its entries, that is, move the top entry to the bottom and move all other entries one place up, repeating this until the cycle is completed. Next, do the same with the second vector. Finally add the last row of all plus signs. This results in an array with N rows and $2N - 2$ columns. This construction is illustrated for $N = 8$ and $p = 14$ using the second design in Table 8.12, which results in the supersaturated design shown in Table 8.13. Obviously, the **foldover** of this design obtained by switching the signs on all factors will be an equivalent supersaturated design.

8.4.2 Statistical Analysis

We next discuss the methods of statistical analysis for supersaturated designs. We assume the main-effects model:

$$\mathrm{E}(y) = \beta_0 + \beta_1 x_1 + \beta_2 x_2 + \cdots + \beta_p x_p.$$

Because $N < p$, this model cannot be fitted using the least squares method. A naive method is to compute the main-effect estimates,

$$\widehat{\beta}_j = \frac{1}{N} \sum_{i=1}^{N} c_{ij} y_i \qquad (1 \le j \le p), \tag{8.11}$$

Table 8.12 Generator Vectors for Supersaturated Designs

N	p	Generator Column Vectors
6	10	(+ − − − +), (− + − + −)
8	14	(− + + − − − +), (− + + − + − −)
10	18	(+ + − + − − − + −), (− + + + − − + − −)
12	22	(− + + + − − + − − − +), (+ − − − − + + − + − +)
14	26	(− + + + − + + − − + − − −),
		(+ + − + − − + − + − − − +)
16	30	(+ − + + + + − − + − − − − + −),
		(− + − − + + − + − − − + + + −)
18	34	(− − − + − + + − + − + − + + + − −),
		(+ + − − + + − − − + − − + − − + +)
20	38	(− + + − + + + + − − + − − − + − − − +),
		(− + − + + − − − + − + + + − − − − + −+)
22	42	(+ + + + − + − − + − − − − + + − − − + − +),
		(− + + − − + + − − + − + + + − − − − + + + + − − + + +)
24	46	(+ + − − + + − + − + − − − − − + + + + − − + ++),
		(− + − + − − − − + − − + − − + + + + − − + + +)
26	50	(− + + − + + − − + − + + + + − − + + − + − − − − −),
		(− + − − − + + + + − + − + + − − − + + − + − + − −)

Source: Nguyen (1996, Table 8.1).

Table 8.13 Supersaturated Design for $N = 8$, $p = 14$ Constructed from Generator Vectors in Table 8.12

	Factor													
Run	A	B	C	D	E	F	G	H	I	J	K	L	M	N
1	−	+	−	−	−	+	+	−	−	−	+	−	+	+
2	+	−	+	−	−	−	+	+	−	−	−	+	−	+
3	+	+	−	+	−	−	−	+	+	−	−	−	+	−
4	−	+	+	−	+	−	−	−	+	+	−	−	−	+
5	−	−	+	+	−	+	−	+	−	+	+	−	−	−
6	−	−	−	+	+	−	+	−	−	+	−	+	+	−
7	+	−	−	−	+	+	−	−	−	+	−	+	+	−
8	+	+	+	+	+	+	+	+	+	+	+	+	+	+

and plot them on a normal or a half-normal plot. Here the c_{ij} are the contrast coefficients in the jth column. Unfortunately, this approach ignores the nonorthogonality of the design. These effect estimates are biased due to correlations r_{ij} between the factors. In particular, from (8.8) we have $E(\widehat{\beta}_j) = \beta_j + \sum_{i \neq j} r_{ij}\beta_i$. The larger the r_{ij}, the larger the bias caused by other main effects. Therefore this method for identifying significant effects is not recommended.

A better approach is to treat this as a regression problem with correlated predictors and use stepwise or best-subsets regression (see, e.g., Draper and Smith, 1998, Chapter 15) to find the "best"-fitting model with as few predictors as possible. Whichever method is employed to identify the most important factors, caution must be exercised in interpreting the results. The following conditions must be satisfied for the results to be meaningful: (i) the effect sparsity principle must hold and (ii) the correlations between the column vectors of the design must be small.

Example 8.8 (Factors Affecting Rubber-Making Process)

The data for this example are adapted from Lin (1993) and Abraham, Chipman, and Vijayan (1999). The experiment studied 24 factors in a PB design with 28 runs. Using one of the three unused columns of the PB design as a branching column, Lin (1993) created two half fractions. The data for one fraction are given in Table 8.14 and those for the other fraction are given in Table 8.26, which you are asked to analyze in Exercise 8.18.

Notice that in this design the column vectors for factors M and P are identical (the same is true for the complementary design shown in Table 8.26 in Exercise 8.18). Therefore these two factors are fully aliased with each other. This does not cause a problem in stepwise regression. Using α-to-enter and α-to-remove equal to 0.10 results in a model with 12 predictors, which is probably too many. The results for the first 5 predictors that entered the model are shown in Display 8.8. We see that the five most important factors according to this analysis are O, L, T, D, and J. Their $\widehat{\beta}_j$ values are $-71.3, -26.8, -28.0, 20.7$, and -9.4, respectively.

It is interesting to compare these estimates with the five largest effect estimates computed using (8.11). These correspond to the factors O, R, C, X, and G and have values $-53.2, -37.4, 23.2, 22.5$, and -20.2, respectively. Only the factor O is common in both cases, which demonstrates how wrong conclusions can result if we ignore the nonorthogonality of the design. ■

8.5 ORTHOGONAL ARRAYS

We have seen that 2^{p-q} fractional factorials and PB designs consist of matrices or arrays of ± 1's with mutually orthogonal columns. Such arrays are called orthogonal arrays. More generally, an orthogonal array, denoted by OA(N, p, s, t), is an $N \times p$ matrix whose elements consist of s distinct symbols such that in any subset of t $(2 \leq t \leq p)$ columns all s^t combinations of symbols occur as row vectors the same number of times, this number being $\lambda = N/s^t$. We call t the **strength** of the array and λ its **index**. In the literature on Taguchi methods (see Section 10.3) this OA is denoted simply as an $L_N(s^p)$ array.

In the context of a factorial design, an OA(N, p, s, t) gives a design matrix with N runs and p factors, each at s levels. Thus it corresponds to an s^p design.

Table 8.14 Rubber Data for Example 8.8

Run	Factor																								Response
	A	*B*	*C*	*D*	*E*	*F*	*G*	*H*	*I*	*J*	*K*	*L*	*M*	*N*	*O*	*P*	*Q*	*R*	*S*	*T*	*U*	*V*	*W*	*X*	*y*
1	+	+	+	−	−	−	+	+	+	+	+	−	+	−	−	+	+	−	−	+	−	−	−	+	133
2	+	−	−	−	−	−	+	+	+	−	−	−	+	+	+	+	−	+	−	−	+	+	−	−	62
3	+	+	−	+	+	−	−	−	−	+	−	+	+	+	+	+	+	−	−	−	−	+	+	−	45
4	+	+	−	+	−	+	−	−	−	+	+	−	+	−	+	+	+	−	−	+	−	+	+	+	52
5	−	−	+	+	+	+	−	+	+	−	−	−	+	−	+	+	+	−	−	+	−	+	+	+	56
6	−	−	+	+	+	+	+	−	+	+	+	−	−	+	+	−	+	+	+	+	+	+	−	−	47
7	−	−	−	−	+	−	−	+	−	+	−	+	+	+	−	+	+	+	+	+	+	−	−	+	88
8	−	+	+	−	−	+	−	+	−	+	−	−	−	−	−	−	−	−	+	−	+	+	+	−	193
9	−	−	−	−	−	+	+	−	−	−	+	+	−	−	+	−	+	+	−	−	−	−	+	+	32
10	+	+	+	+	−	+	+	+	−	−	−	+	−	+	+	−	+	−	+	−	+	−	−	+	53
11	−	+	−	+	+	−	−	+	+	−	+	−	−	+	−	−	−	+	+	−	−	−	+	+	276
12	+	−	−	−	+	+	+	−	+	+	+	+	+	−	−	+	−	−	+	−	+	+	+	+	145
13	+	+	+	+	+	−	+	−	+	−	−	+	−	−	−	−	−	+	−	+	+	−	+	−	130
14	−	−	+	−	−	−	−	−	−	−	+	+	−	+	−	−	−	−	−	+	−	+	−	−	127

Source: Lin (1993, Table 8.3).

```
Stepwise Regression: y versus A, B, ...

  Alpha-to-Enter: 0.1  Alpha-to-Remove: 0.1

Response is y on 24 predictors, with N = 14
Step              1       2       3        4        5
Constant      102.8   102.8   102.8    102.8    102.8

O             -53.2   -56.4   -60.5    -70.5    -71.3
T-Value       -4.54   -5.42   -7.75   -12.96   -15.96
P-Value       0.001   0.000   0.000    0.000    0.000

L                     -22.3   -26.4    -25.3    -26.8
T-Value               -2.14   -3.38    -5.19    -6.63
P-Value               0.055   0.007    0.001    0.000

T                             -24.8    -29.2    -28.0
T-Value                       -3.17    -5.86    -6.80
P-Value                       0.010    0.000    0.000

D                                       22.1     20.7
T-Value                                 4.09     4.64
P-Value                                0.003    0.002

J                                                -9.4
T-Value                                         -2.33
P-Value                                         0.048

S              43.9    38.5    28.5     17.8     14.5
R-Sq          63.17   74.01   87.05    95.48    97.30
R-Sq(adj)     60.11   69.29   83.17    93.47    95.62
```

Display 8.8 Minitab stepwise regression results for rubber data in Table 8.14.

Hence we use the notation OA(N, s^p, t) instead of OA(N, p, s, t). We shall see from the examples below that the resolution of the design equals $t + 1$. In Chapter 9 we shall generalize the definition of orthogonal arrays to factors with different numbers of levels. They are referred to as **asymmetrical (or mixed-level) orthogonal arrays**, while the ones in which all factors have the same number of levels are referred to as **symmetrical orthogonal arrays**.

First consider the PB design for $N = 8$, $p = 7$ (2^{7-4} design) in Table 8.6. Its array is shown in Display 8.9. In this design, we have $N = 8$, $p = 7$, $s = 2$, and $t = 2$. This is an array of strength 2, which can be seen from the fact that in any pair of columns the $2^2 = 4$ pairs of signs, $(-, -)$, $(-, +)$, $(+, -)$, $(+, +)$, occur $\lambda = \frac{8}{4} = 2$ times. On the other hand, for $t = 3$ all $2^3 = 8$ combinations of $(\pm, \pm, \pm)$ do not occur the same number of times. For example, in the first three columns each combination occurs exactly once, but in columns 1, 2, and 4 the combinations $(+, -, -)$, $(-, +, -)$, $(-, -, +)$, $(+, +, +)$ occur twice, and

$$\begin{bmatrix} - & - & - & + & + & + & - \\ + & - & - & - & - & + & + \\ - & + & - & - & + & - & + \\ + & + & - & + & - & - & - \\ - & - & + & + & - & - & + \\ + & - & + & - & + & - & - \\ - & + & + & - & - & + & - \\ + & + & + & + & + & + & + \end{bmatrix}$$

Display 8.9 Orthogonal array $OA(8, 2^7, 2)$.

the other four combinations do not occur at all. We denote the above array as $OA(8, 2^7, 2)$. Note that, when used as a 2^{7-4} design, this has resolution III, which is one more than its strength.

As a second example, consider a 2^{6-2} fractional factorial design with defining relations $I = ABCE = BCDF$. Its array is shown in Display 8.10. We can check that all $2^3 = 8$ combinations of the signs $(\pm, \pm, \pm)$ occur $\lambda = 2$ times in

$$\begin{bmatrix} - & - & - & - & - & - \\ + & - & - & - & + & - \\ - & + & - & - & + & + \\ + & + & - & - & - & + \\ - & - & + & - & + & + \\ + & - & + & - & - & + \\ - & + & + & - & - & - \\ + & + & + & - & + & - \\ - & - & - & + & - & + \\ + & - & - & + & + & + \\ - & + & - & + & + & - \\ + & + & - & + & - & - \\ - & - & + & + & + & - \\ + & - & + & + & - & - \\ - & + & + & + & - & + \\ + & + & + & + & + & + \end{bmatrix}.$$

Display 8.10 Orthogonal array $OA(16, 2^5, 3)$.

$$\begin{bmatrix} 0 & 0 & 0 & 0 \\ 0 & 1 & 1 & 1 \\ 0 & 2 & 2 & 2 \\ 1 & 0 & 1 & 2 \\ 1 & 1 & 2 & 0 \\ 1 & 2 & 0 & 1 \\ 2 & 0 & 2 & 1 \\ 2 & 1 & 0 & 2 \\ 2 & 2 & 1 & 0 \end{bmatrix}.$$

Display 8.11 Orthogonal array $OA(9, 3^4, 2)$.

every triplet of columns. Therefore this is an OA of strength $t = 3$ with $N = 16$ rows and $p = 6$ columns or an OA$(16, 2^6, 3)$. The corresponding 2^{6-2} fractional factorial design has resolution IV. Thus the formula $R = t + 1$ holds.

For another example, consider the OA$(9, 3^4, 2)$ shown in Display 8.11. Note that this is a 3^{4-2} design, that is, a 1/9th fraction of a 3^4 factorial. Three-level full and fractional factorials are discussed in Chapter 9.

Further details about OAs, including how to construct mixed-level OAs, are given in Chapter 9.

8.6 SEQUENTIAL ASSEMBLIES OF FRACTIONAL FACTORIALS

In Example 8.1 we saw that the estimators of the effects obtained from the two 2_{III}^{3-1} half fractions, $\{(1), ab, ac, bc\}$ and $\{a, b, c, abc\}$, are aliased. They can be dealiased by running both fractions and pooling the results. The two half fractions may be run sequentially because the resources (e.g., money, equipment, operators, time) may not be available to run a full factorial experiment at one time. When there are many factors, it is often a good strategy to not use the available resources to run one large experiment but to run smaller fractional factorials sequentially. In this section we study some systematic methods for this purpose.

One consequence of running the fractions sequentially on separate occasions, which are the blocks, is that certain interactions get confounded with the occasions, as discussed in Section 7.4.1. By using half fractions with the highest possible resolutions, only the higher order interactions can be compromised in this way. To illustrate this point, suppose that the half fractions $\{a, b, c, abc\}$ and $\{(1), ab, ac, bc\}$ are run sequentially on two separate occasions. The design is shown in Display 8.12. We see that in this design the three-factor interaction ABC is confounded between the occasions; other effects are not confounded.

Example 8.9 (Improving Yield of Spring Manufacturing Process: Pooling Estimates from Two Half Fractions)

Refer to the estimates calculated in Example 8.1. Separate estimates of the main effects and interactions can be computed as follows:

$$\begin{aligned}
\widehat{A} &= \tfrac{1}{2}\big[(\widehat{A} + \widehat{BC}) + (\widehat{A} - \widehat{BC})\big] = \tfrac{1}{2}[(5.0) + (-15.0)] = -5.0,\\
\widehat{B} &= \tfrac{1}{2}\big[(\widehat{B} + \widehat{AC}) + (\widehat{B} - \widehat{AC})\big] = \tfrac{1}{2}[(3.0) + (0.0)] = 1.5,\\
\widehat{C} &= \tfrac{1}{2}\big[(\widehat{C} + \widehat{AB}) + (\widehat{C} - \widehat{AB})\big] = \tfrac{1}{2}[(23.0) + (23.0)] = 23.0,\\
\widehat{BC} &= \tfrac{1}{2}\big[(\widehat{A} + \widehat{BC}) - (\widehat{A} - \widehat{BC})\big] = \tfrac{1}{2}[(5.0) - (-15.0)] = 10.0,
\end{aligned}$$

Occasion 1	Occasion 2
(1)	a
ab	b
ac	c
bc	abc

Display 8.12 2^3 Design in two blocks.

$$\widehat{AC} = \tfrac{1}{2}\left[(\widehat{B} + \widehat{AC}) - (\widehat{B} - \widehat{AC})\right] = \tfrac{1}{2}[(3.0) - (0.0)] = 1.5,$$
$$\widehat{AB} = \tfrac{1}{2}\left[(\widehat{C} + \widehat{AB}) - (\widehat{C} - \widehat{AB})\right] = \tfrac{1}{2}[(23.0) - (23.0)] = 0.00.$$

The same estimates would be obtained if all eight runs were considered together as a full 2^3 experiment. However, as noted above, the ABC interaction is confounded between the occasions and cannot be estimated. ■

8.6.1 Foldover of Resolution III Designs

In a resolution III design the main effects are aliased with the two-factor interactions. (We ignore the higher order interactions with which the main effects may also be aliased.) Suppose that in light of the results of an experiment conducted using this design a particular factor appears influential. Then we may want to run another experiment that would enable us to separate the main effect and all two-factor interactions involving that factor clear of all other main effects and two-factor interactions. The design needed for this purpose is obtained from the first design by switching the signs on the factor of interest. This process is called folding over that factor and the resulting design is called a **partial foldover**.

The basic defining relation of a foldover design is obtained by changing the sign on the factor on which the signs are switched. The following example illustrates how a foldover design enables separation of the desired effects.

Table 8.15 A 2_{III}^{6-3} Design Obtained by Dropping Factor G in Design from Table 8.6

	Factor					
Run	A	B	C	$D = AB$	$E = AC$	$F = BC$
1	−	−	−	+	+	+
2	+	−	−	−	−	+
3	−	+	−	−	+	−
4	+	+	−	+	−	−
5	−	−	+	+	−	−
6	+	−	+	−	+	−
7	−	+	+	−	−	+
8	+	+	+	+	+	+

Example 8.10 (A Foldover of a 2_{III}^{6-3} Design)

Consider the 2_{III}^{6-3} design shown in Table 8.15, which is obtained by dropping the column G of the PB design in Table 8.6. The basic generators of the design are then $I = ABD = ACE = BCF$. The alias structure up to second-order effects can be easily derived to be

$$A = BD = CE, \qquad B = AD = CF, \qquad C = AE = BF,$$
$$D = AB = EF, \qquad E = AC = DF, \qquad F = BC = DE.$$

Now suppose that, after running this experiment, factor A is thought to be influential. Therefore we run a second experiment by folding over A. The basic generators of this design are obtained by switching the signs on A, which yields $I = -ABD = -ACE = BCF$. The alias structure up to second-order effects of this design is

$$A = -BD = -CE, \qquad B = -AD = CF, \qquad C = -AE = BF,$$
$$D = -AB = EF, \qquad E = -AC = DF, \qquad F = BC = DE.$$

Denote the contrasts for estimating the effects for the first design by ℓ_A, ℓ_B, $\cdots$ and those for the foldover design by ℓ'_A, ℓ'_B, $\cdots$. For example, for the first design, $\ell_A = \frac{1}{4}(-y_1 + y_2 - y_3 + y_4 - y_5 + y_6 - y_7 + y_8)$ and for the second design $\ell'_A = \frac{1}{4}(y'_1 - y'_2 + y'_3 - y'_4 + y'_5 - y'_6 + y'_7 - y'_8)$, where y_i and y'_i are the observations at run i in the respective designs. The alias relations up to second-order effects yield the following relations:

$$\begin{aligned}
\text{E}(\ell_A) &= A + BD + CE, & \text{E}(\ell'_A) &= A - BD - CE,\\
\text{E}(\ell_B) &= B + AD + CF, & \text{E}(\ell'_B) &= B - AD + CF,\\
\text{E}(\ell_C) &= C + AE + BF, & \text{E}(\ell'_C) &= C - AE + BF,\\
\text{E}(\ell_D) &= D + AB + EF, & \text{E}(\ell'_D) &= D - AB + EF,\\
\text{E}(\ell_E) &= E + AC + DF, & \text{E}(\ell'_E) &= E - AC + DF,\\
\text{E}(\ell_F) &= F + BC + DE, & \text{E}(\ell'_F) &= F + BC + DE.
\end{aligned}$$

By combining the estimators from the two experiments we obtain the following unbiased estimators:

$$\begin{aligned}
\widehat{A} &= \tfrac{1}{2}(\ell_A + \ell'_A) & \widehat{BD} + \widehat{CE} &= \tfrac{1}{2}(\ell_A - \ell'_A)\\
\widehat{B} + \widehat{CF} &= \tfrac{1}{2}(\ell_B + \ell'_B) & \widehat{AD} &= \tfrac{1}{2}(\ell_B - \ell'_B)\\
\widehat{C} + \widehat{BF} &= \tfrac{1}{2}(\ell_C + \ell'_C) & \widehat{AE} &= \tfrac{1}{2}(\ell_C - \ell'_C)\\
\widehat{D} + \widehat{EF} &= \tfrac{1}{2}(\ell_D + \ell'_D) & \widehat{AB} &= \tfrac{1}{2}(\ell_D - \ell'_D)\\
\widehat{E} + \widehat{DF} &= \tfrac{1}{2}(\ell_E + \ell'_E) & \widehat{AC} &= \tfrac{1}{2}(\ell_E - \ell'_E)\\
\widehat{F} + \widehat{BC} + \widehat{DE} &= \tfrac{1}{2}(\ell_F + \ell'_F) & &
\end{aligned}$$

Note that the A main effect and all two-factor interactions involving A are clear of each other. Also note that ℓ_F and ℓ'_F provide two i.i.d. estimators of $F + BC + DE$; each estimator has variance $\sigma^2/2$. So the square of their difference is an unbiased estimator of σ^2. ■

If we fold over more than one factor, then, in general, the main effects and two-factor interactions involving those factors from their aliased effects cannot be separated. For example, if we fold over A and B in the design in Table 8.15, then the basic generators of the resulting design are $I = ABD = -ACE = -BCF$. The alias structure up to two-factor interactions is obtained by switching signs on both A and B; thus

$$A = BD = -CE, \qquad B = AD = -CF, \qquad C = -AE = -BF,$$

$$D = AB = EF, \qquad E = -AC = DF, \qquad F = -BC = DE.$$

Therefore A and BD, B and AD, and D and AB cannot be separated. The same is true of several two-factor interactions involving A or B. However, if a design is folded over *all* factors, then the signs of all three-letter words are reversed. This is called a **full foldover** or a **reflection** of the original design. In a combined design of a resolution III design with its full foldover, all main effects can be separated from all two-factor interactions; therefore we get a resolution IV design. The generators of the combined design are obtained by taking all pairwise products of the generators of the foldover or the original design. This is illustrated in the next example.

Example 8.11 (Foldover of 2^{6-3}_{III} Design)

Table 8.16 shows the foldover of the 2^{6-3}_{III} design given in Table 8.15. The basic defining relation of the foldover is

$$I = -ABD = -ACE = -BCF.$$

Table 8.16 A Full Foldover of 2^{6-3}_{III} Design in Table 8.15

	Factor					
Run	A	B	C	$D = -AB$	$E = -AC$	$F = -BC$
1	+	+	+	−	−	−
2	−	+	+	+	+	−
3	+	−	+	+	−	+
4	−	−	+	−	+	+
5	+	+	−	−	+	+
6	−	+	−	+	−	+
7	+	−	−	+	+	−
8	−	−	−	−	−	−

Hence the alias structure up to two-factor interactions is

$$A = -BD = -CE, \qquad B = -AD = -CF, \qquad C = -AE = -BF,$$
$$D = -AB = -EF, \qquad E = -AC = -DF, \qquad F = -BC = -DE.$$

Therefore all the main effects and two-factor interactions can be separated from each other in the combined design. The basic defining relation of the combined design, obtained by taking all pairwise products of ABD, ACE, and BCF, is

$$I = BCDE = ACDF = ABEF,$$

which shows that the design is of resolution IV. ■

8.6.2 Foldover of Resolution IV Designs

Foldovers of resolution IV designs are more complicated and must be done with care. We saw in the previous section that a full foldover (reflection) of a resolution III design combined with the original design gives a resolution IV design in which all main effects are clear of all two-factor interactions. Now consider a 2_{IV}^{4-1} design shown in Display 8.13. If the signs on all four factors are flipped, then we do not get a resolution V design, but the same design with the order of runs reversed. This is true for any resolution IV design. On the other hand, it is easy to check that the full foldover of a 2_{III}^{3-1} design is its complement, so that the combined design is a full 2^3 factorial.

If the design is folded on only one factor, then the combined design clears all two-factor interactions involving that factor from other two-factor interactions. In the above 2_{IV}^{4-1} design, if we fold over, say, A, then the combined design is a full 2^4 factorial and thus has no aliases. For a more interesting example, consider a 2_{IV}^{6-2} design shown in Display 8.14 with generators $E = ABC$ and $F = BCD$. Once again, you can check that a full foldover gives the same design. If we fold over only factor A, then the resulting design has generators $E = -ABC$ and $F = BCD$. The combined design has only one generator: $F = BCD$. Thus the resulting design is of resolution IV; furthermore, all two-factor interactions involving A as well as E are clear of other two-factor interactions.

Run	A	B	C	D
1	−	−	−	−
2	+	−	−	+
3	−	+	−	+
4	+	+	−	−
5	−	−	+	+
6	+	−	+	−
7	−	+	+	−
8	+	+	+	+

Display 8.13 A 2_{IV}^{4-1} design.

Run	A	B	C	D	E	F
1	−	−	−	−	−	−
2	+	−	−	−	+	−
3	−	+	−	−	+	+
4	+	+	−	−	−	+
5	−	−	+	−	+	+
6	+	−	+	−	−	+
7	−	+	+	−	−	−
8	+	+	+	−	+	−
9	−	−	−	+	−	+
10	+	−	−	+	+	+
11	−	+	−	+	+	−
12	+	+	−	+	−	−
13	−	−	+	+	+	−
14	+	−	+	+	−	−
15	−	+	+	+	−	+
16	+	+	+	+	+	+

Display 8.14 A 2_{IV}^{6-2} design.

A more efficient design that clears the same two-factor interactions from other two-factor interactions can be obtained by using the idea of semifolding (Mee and Peralta, 2000), that is, choosing only half the runs of the foldover on factor A (runs with either plus or minus signs); the resulting combined design requires only 24 runs instead of 32. Montgomery and Runger (1996) have discussed additional aspects of folding resolution IV designs.

8.7 CHAPTER SUMMARY

(a) A 2^{p-q} fractional factorial design is useful in screening experiments when one is interested in identifying factors with large main effects and possibly low-order interactions (assuming that principles of sparsity and hierarchy of effects hold).

(b) A consequence of not running all factorial combinations is that every effect is aliased with some others. One can choose the particular factorial combinations to run in such a way that low-order effects are aliased with high-order effects. Resolution is a measure of the ability of a design to estimate the low-order effects clear of other low-order effects.

(c) A fractional factorial design can be constructed from its generators. The relation obtained by equating the generators to I is called the defining relation of the design which can be used to derive its alias structure. The alias structure is useful for checking if an assignment of the treatments to the columns of the design matrix makes the effects of interest clear of each other.

(d) Two-level Plackett–Burman (PB) designs are resolution III orthogonal designs for studying p factors in $N = p + 1$ runs. These designs exist only when N is a multiple of 4. If N is a power of 2, then they are 2^{p-q} fractional factorial designs, in which case they are called geometric designs. If N is not a power of 2, then they are called nongeometric designs, for which special constructions are used. Nongeometric designs have complex aliasing patterns in which the main effects are partially aliased with two-factor interactions.

(e) Hadamard designs are resolution III designs obtained from Hadamard matrices. PB designs are a special case of Hadamard designs. Whereas the alias structure of a geometric PB design is simple and can be easily derived from its generators, that for a general Hadamard design is not so simple with the main effects being partially aliased with several two-factor interactions. The analysis methods are the same for both designs.

(f) Supersaturated designs have less runs than the number of factors. Therefore the main effects are partially aliased with each other, and not all are estimable. A simple way to construct a supersaturated design is to split a PB design in two equal halves based on the $\pm$ signs of a branching column. Stepwise or best subsets regression is the recommended method of analysis. The assumption of effect sparsity is essential in the analysis.

(g) An $N \times p$ orthogonal array (OA) of strength $t \leq p$ whose entries are taken from a set of $s \geq 2$ distinct symbols has the property that all s^t combinations of symbols occur as row vectors the same number of times in every t-tuple of columns. Such an OA can be used to construct a fractional factorial of an s^p design (p factors each at s levels) and therefore it is denoted by $\text{OA}(N, s^p, t)$. The resolution of this design is related to the strength of the OA by the equation $R = t + 1$.

(h) Folding over a single factor of a resolution III design enables us to estimate the main effect of that factor and all two-factor interactions involving that factor clear of other main effects and two-factor interactions. Folding over all factors yields a resolution IV design in which all main effects are clear of all two-factor interactions. Such a design is called a full foldover.

EXERCISES

Section 8.1 (2^{p-q} Fractional-Factorial Designs)

Theoretical Exercises

8.1 Consider a 2^{6-2} fractional factorial with the defining relations $I = ABCE = BCDF$. What is the resolution of this design? Write the alias structure up to two-factor interactions of this design.

8.2 Consider a 2^{6-2} fractional factorial with the defining relations $I = ABE = ACDF$. What is the resolution of this design? Write the alias structure up to two-factor interactions of this design.

8.3 Refer to Exercise 8.1. Project the design on factors A, B, C, and E. Do you get a full factorial? Now project on A, B, D, and E. Do you get a full factorial? Comment on the differences in results.

8.4 Refer to Exercise 8.2. Project the design on factors A, B, E and A, B, D. Which designs are obtained in the two cases? How do they differ and why?

8.5 Consider a crossed array design in which the inner array is the 2^{5-2} design given in Section 8.1.2.1 with the defining relations $I = ABD$ and $I = ACE$. Suppose that the outer array is a 2^{3-1} design in factors F, G, H with the defining relation $I = FGH$.

(a) Explain the structure of this design. How many runs does it have?

(b) Give the aliases of A, F, and AF up to three-factor interactions.

8.6 Consider the eight-run design in four factors shown in Display 8.15

(a) Give the generator of this design. What is the resolution of this design?

(b) Give a design that has a higher resolution. What is its generator?

Applied Exercises

8.7 Refer to Exercise 7.15 for the details of a paper helicopter experiment. Box (1992) reports another experiment in which eight factors (shown in Table 8.17) were studied in 16 runs using a 2^{8-4} design. The flight times, which are averages from four independent drops, are shown in Table 8.18.

(a) Check that $ABCE$, $BCDF$, $ACDG$, and $ABDH$ are the generators of this design.

(b) What is the resolution of this design? On how many maximum number of factors can this design be projected to get a full factorial (with possible replicates)?

A	B	C	D
−	−	−	+
+	−	−	−
−	+	−	−
+	+	−	+
−	−	+	+
+	−	+	−
−	+	+	−
+	+	+	+

Display 8.15 An 8-run design in four factors for Exercise 8.6

Table 8.17 Factors and Factor Levels for Paper Helicopter Experiment

	Level	
Factor	Low (−)	High (+)
A: Paper type	Regular	Bond
B: Wing length, in.	3.00	4.75
C: Body length, in.	3.00	4.75
D: Body width, in.	1.25	2.00
E: Paper clip	No	Yes
F: Fold	No	Yes
G: Taped body	No	Yes
H: Taped wing	No	Yes

Source: Box (1992, Figure 8.2). Reprinted by permission of Taylor & Francis.

Table 8.18 Flight Times (secs) from Paper Helicopter Experiment

	Factor								
Run	*A*	*B*	*C*	*D*	*E*	*F*	*G*	*H*	Flight Time (secs)
1	−	−	−	−	−	−	−	−	2.5
2	+	−	−	−	+	−	+	+	2.9
3	−	+	−	−	+	+	−	+	3.5
4	+	+	−	−	−	+	+	−	2.7
5	−	−	+	−	+	+	+	−	2.0
6	+	−	+	−	−	+	−	+	2.3
7	−	+	+	−	−	−	+	+	2.9
8	+	+	+	−	+	−	−	−	3.0
9	−	−	−	+	−	+	+	+	2.4
10	+	−	−	+	+	+	−	−	2.6
11	−	+	−	+	+	−	+	−	3.2
12	+	+	−	+	−	−	−	+	3.7
13	−	−	+	+	+	−	−	+	1.9
14	+	−	+	+	−	−	+	−	2.2
15	−	+	+	+	−	+	−	−	3.0
16	+	+	+	+	+	+	+	+	3.0

Source: Box (1992, Figure 8.2). Reprinted by permission of Taylor & Francis.

(c) Give the alias structure of this design up to two-factor interactions.

(d) Calculate the 15 effect estimates and make a half-normal plot. Which effect estimates are significant at $\alpha = 0.10$?

(e) Project the design on the important factors. Which factor-level combinations maximize the flight time?

Table 8.19 Factors and Factor Levels for Viscosity Experiment

Factor	Level	
	Low (−)	High (+)
A: Silica sand type	♯5	♯6
B: Reaction time (hr)	4	5
C: Agitation	100	150
D: Redaction temperature (°C)	190 ± 3	180 ± 3
E: Rate of temperature increase (°C/min)	30–35	15–20
F: Relative excess of silica sand (%)	20	10
G: Active carbon type	T_3	3A
H: Quantity of active carbon added (%)	0.13	0.06
J: Evacuated water temperature (°C)	80 ± 5	40 ± 5

Source: Bisgaard (1993, Table 8.1). Reprinted by permission of Taylor & Francis.

Table 8.20 Data from Viscosity Experiment

Run	Factor									Viscosity
	A	*B*	*C*	*D*	*E*	*F*	*G*	*H*	*J*	
1	−	−	−	−	−	−	−	−	−	17
2	+	−	−	−	−	+	+	+	+	21
3	−	+	−	−	+	−	+	+	+	41
4	+	+	−	−	+	+	−	−	−	20
5	−	−	+	−	+	+	−	−	+	10
6	+	−	+	−	+	−	+	+	−	42
7	−	+	+	−	−	+	+	+	−	14
8	+	+	+	−	−	−	−	−	+	58
9	−	−	−	+	+	+	−	+	−	8
10	+	−	−	+	+	−	+	−	+	18
11	−	+	−	+	−	+	+	−	+	7
12	+	+	−	+	−	−	−	+	−	15
13	−	−	+	+	−	−	−	+	+	8
14	+	−	+	+	−	+	+	−	−	10
15	−	+	+	+	+	−	+	−	−	12
16	+	+	+	+	+	+	−	+	+	10

Source: Bisgaard (1993, Table 8.2). Reprinted by permission of Taylor & Francis.

8.8 This exercise is based on the analysis by Bisgaard (1993) of a data set from Taguchi (1987). The experiment used a 2^{9-5} design to study the effects of the nine factors shown in Table 8.19 on the viscosity (measured in centipoise) of a sodium silicate liquid with the objective of finding the settings of the factors that maximize the viscosity. The data are shown in Table 8.20.

Table 8.21 Factors for Injection Molding Experiment

Factor	Factor Name	Factor	Factor Name
A	Mold temperature	E	Booster pressure
B	Moisture content	F	Cycle time
C	Holding pressure	G	Gate size
D	Cavity thickness	H	Screw speed

(a) The basic generators of this design can be checked to be $BCDE, ACDF, -ABG, ABDH$, and $ABCJ$. Find the alias structure up to second-order effects.

(b) Identify important effects using a normal plot. When choosing between the aliased effects, use the hierarchical principle that a two-factor interaction is unlikely to be important unless both main effects are important.

(c) Fit a model consisting of the effects identified as important in part (b) and plot the residuals versus normal scores and fitted values. Are there any outliers? Are any of the model assumptions violated?

(d) Apply the reciprocal transformation (motivated by the physical consideration that the viscosity is inversely proportional to the velocity of a steel ball dropped in the liquid) and redo the analyses in parts (b) and (c). Which effects are now indicated to be important? Are there any anomalies in the residual plots? Comment on the usefulness of the transformation.

8.9 Box, Hunter, and Hunter (1978) give data on the shrinkage of a mold produced by an injection molding process. Eight factors listed in Table 8.21 were studied in a 2^{8-4} experiment. The data are given in Table 8.22.

(a) This is a resolution IV design with the main effects clear of two-factor interactions, which are aliased with each other, as shown in Display 8.16. Obtain a basic defining relation of the design.

(b) Estimate the main effects and aliased two-factor interactions and make their normal plot. Show that the main effects C and E and the aliased two-factor interactions $AE = BF = CH = DG$ are significant at $\alpha = 0.10$. Which one of these interactions do you conjecture might be contributing to significance?

8.10 Refer to Exercise 8.1. Suppose six experimental factors labeled A', B', C', D', E', F' are to be studied using this design and all main effects and two-factor interactions $A'B'$, $A'E'$, $B'C'$, and $C'E'$ are to be accommodated. Find a suitable assignment of the labels $A, B, \ldots, F$ to the experimental factors.

8.11 Refer to Exercise 8.2. Suppose six experimental factors labeled A', B', C', D', E', F' are to be studied using this design and all main

Table 8.22 Design and Data for Injection Molding Experiment

Run	A	B	C	D	E	F	G	H	Shrinkage
1	−	−	−	+	+	+	−	+	14.0
2	+	−	−	−	−	+	+	+	16.8
3	−	+	−	−	+	−	+	+	15.0
4	+	+	−	+	−	−	−	+	15.4
5	−	−	+	+	−	−	+	+	27.6
6	+	−	+	−	+	−	−	+	24.0
7	−	+	+	−	−	+	−	+	27.4
8	+	+	+	+	+	+	+	+	22.6
9	+	+	+	−	−	−	+	−	22.3
10	−	+	+	+	+	−	−	−	17.1
11	+	−	+	+	−	+	−	−	21.5
12	−	−	+	−	+	+	+	−	15.5
13	+	+	−	−	+	+	−	−	15.9
14	−	+	−	+	−	+	+	−	21.9
15	+	−	−	+	+	−	+	−	16.7
16	−	−	−	−	−	−	−	−	20.3

Source: Box, Hunter, and Hunter (1978, Table 1). Reprinted by permission of John Wiley & Sons, Inc.

effects and two-factor interactions $A'C'$, $A'D'$, $A'E'$, $B'C'$, and $B'D'$ are to be accommodated. Find a suitable assignment of the labels $A, B, \ldots, F$ to the experimental factors.

Section 8.2 (Plackett–Burman Designs)

Theoretical Exercise

8.12 Consider the PB design for $N = 12$, $p = 11$ shown in Table 8.8. For simplicity assume that only the factors A, B, C and their two-factor interactions are active.

$$AB = CG = DH = EF$$

$$AC = BG = DF = EH$$

$$AD = BH = CF = EG$$

$$AE = BF = CH = DG$$

$$AF = BE = CD = GH$$

$$AG = BC = DE = FH$$

$$AH = BD = CE = FG$$

Display 8.16 Alias relations for the design in Table 8.22.

Table 8.23 Factors and Factor Levels for Wheel Cover Experiment

	Level	
Factor	Low (−)	High (+)
A: Mold temperature	80	110
B: Close time	16	21
C: Booster time	1.88	1.70
D: Plunger time	4	8
E: Pack pressure	1300	1425
F: Hold pressure	1100	700
G: Barrel	490	505

Source: Chiao and Hamada (2001, Table 8.1). Reprinted by permission of the American Society for Quality.

(a) Write the model matrices $\boldsymbol{X}_1$ and $\boldsymbol{X}_2$ for the three active main effects and their two-factor interactions, respectively.

(b) Calculate the alias matrix between the main effects and their two-factor interactions.

(c) Write the formula for the bias of the LS estimator of the main effect A.

Applied Exercises

8.13 Chiao and Hamada (2001) report on an experiment by Harper, Kosbe, and Peyton (1987) to find the optimum combination of seven factors of an injection molding process to minimize the imbalance of a plastic wheel cover component used in Ford Taurus car. The factors and their levels are listed in Table 8.23.

An eight-run PB design was used with five replicates (wheel covers) tested at each level. The response variable was the weight (in grams) of the wheel cover component. The data are shown in Table 8.24. Analyze the data and draw conclusions.

8.14 The data shown in Table 8.25 are taken from Box and Bisgaard (1993), who analyzed an experiment reported by Adam (1987). The experiment used a PB design for $N = 12$ runs to study how the surface defects on automobile instrument panels are affected by 10 factors. The response variable y is the square root of the defect counts multiplied by 10. There is only one unassigned column available to estimate the error. Calculate the main effects and make their half-normal plot. Which effects are shown to be significant?

Table 8.24 Wheel Cover Component Weights (grams)

Run	Factor							Replicate				
No.	*A*	*B*	*C*	*D*	*E*	*F*	*G*	1	2	3	4	5
1	−	−	−	−	−	−	−	711.9	713.4	712.3	712.4	711.9
2	−	−	−	+	+	+	+	725.0	720.1	711.8	723.9	720.9
3	−	+	+	−	−	+	+	711.6	711.7	711.3	712.1	711.7
4	−	+	+	+	+	−	−	733.7	724.1	732.0	732.7	733.3
5	+	−	+	−	+	−	+	725.4	721.6	722.6	723.1	721.1
6	+	−	+	+	−	+	−	728.7	721.1	722.9	723.0	719.7
7	+	+	−	−	+	+	−	726.6	731.4	731.4	729.6	731.3
8	+	+	−	+	−	−	+	714.3	714.4	713.6	716.3	714.6

Source: Chiao and Hamada (2001, Table 8.2). Reprinted by permission of the American Society for Quality.

Table 8.25 Data from Car Panel Surface Defects Experiment

Run	Factor											No. of
	A	*B*	*C*	*D*	*E*	*F*	*G*	*H*	*J*	*K*	Error	Defects
1	+	−	+	−	+	+	+	−	−	−	+	26
2	+	+	−	+	−	+	−	−	−	+	+	43
3	−	+	+	+	−	+	+	−	+	−	−	20
4	+	−	+	+	−	−	+	+	−	+	−	19
5	+	+	−	−	+	−	+	−	+	+	−	5
6	+	+	+	−	−	−	−	+	+	−	+	13
7	−	+	+	−	+	+	−	+	−	+	−	38
8	−	−	+	+	+	−	−	−	+	+	+	13
9	−	−	−	−	−	+	+	+	+	+	+	27
10	+	−	−	+	+	+	+	+	+	−	−	27
11	−	+	−	+	+	−	+	+	−	−	+	16
12	−	−	−	−	−	−	−	−	−	−	−	26

Source: Box and Bisgaard (1993, Table 8.1). Reprinted by permission of Taylor & Francis.

Section 8.3 (Hadamard Designs)

Theoretical Exercises

8.15 Show that if $\boldsymbol{H}$ is an $n \times n$ Hadamard matrix, then

$$\begin{bmatrix} 1 & 1 \\ 1 & -1 \end{bmatrix} \otimes \boldsymbol{H} = \begin{bmatrix} \boldsymbol{H} & \boldsymbol{H} \\ \boldsymbol{H} & -\boldsymbol{H} \end{bmatrix}$$

is a $2n \times 2n$ Hadamard matrix.

Table 8.26 Data for Exercise 8.18

	Factor																								Response
Run	*A*	*B*	*C*	*D*	*E*	*F*	*G*	*H*	*I*	*J*	*K*	*L*	*M*	*N*	*O*	*P*	*Q*	*R*	*S*	*T*	*U*	*V*	*W*	*X*	*y*
1	−	+	−	−	−	−	+	+	+	+	−	+	−	−	+	−	+	+	+	+	−	+	+	−	49
2	+	+	−	−	+	+	−	−	−	−	−	−	−	−	−	−	+	+	−	+	+	+	−	+	88
3	−	−	+	+	+	+	+	+	−	+	−	+	+	−	−	+	−	+	−	−	−	−	−	−	300
4	+	−	+	−	−	+	−	−	+	−	−	−	+	+	−	+	+	+	+	−	−	−	+	−	116
5	−	+	+	+	−	−	+	−	−	+	+	−	+	+	−	+	+	+	−	−	+	+	+	+	83
6	−	−	−	+	−	−	−	−	+	+	−	−	−	−	+	−	−	−	−	−	+	−	−	+	230
7	+	−	+	−	+	−	+	−	−	+	−	−	−	+	+	−	−	−	+	+	−	−	+	+	51
8	−	+	+	−	+	−	−	−	+	−	+	+	+	−	+	+	+	−	+	−	+	−	−	−	82
9	+	−	+	+	−	−	−	+	−	−	+	+	+	−	+	+	−	+	+	+	+	+	+	+	58
10	+	−	−	+	+	−	+	+	−	−	+	−	−	−	−	−	+	−	+	−	−	+	−	−	201
11	+	+	+	−	+	+	−	+	+	+	+	+	−	+	+	−	−	+	−	−	−	+	−	+	56
12	−	+	−	+	−	+	+	−	+	−	−	+	+	+	−	+	−	−	+	+	−	+	−	+	97
13	−	+	−	−	+	+	+	+	−	−	+	−	+	+	+	+	−	−	−	+	+	−	+	−	55
14	+	−	−	+	−	+	−	+	+	+	+	+	−	+	−	−	+	−	−	+	+	−	+	−	160

Source: Abraham, Chipman, and Vijayan (1999, Table 8.1).

8.16 As a generalization of the previous exercise, show that if $\boldsymbol{G}$ and $\boldsymbol{H}$ are $m \times m$ and $n \times n$ Hadamard matrices, then $\boldsymbol{G} \otimes \boldsymbol{H}$ is an $mn \times mn$ Hadamard matrix.

Section 8.4 (Supersaturated Designs)

Theoretical Exercise

8.17 Use the method given in Section 8.4 to construct a supersaturated design with $N = 6$ runs and $p = 10$ factors from the two generator columns in Table 8.12. Check that pairwise correlations between all column vectors are $\pm\frac{1}{3}$.

Applied Exercise

8.18 Refer to Example 8.8. Table 8.26 gives rubber data for the complementary half fraction of the PB design (note that in this half fraction also factors M and P are fully aliased with each other). Perform stepwise regression on these data with α-to-enter and α-to-remove equal to 0.10 to identify the most important factors. Compare your results with those from Example 8.8. Also compute the main effects of all factors. Which are the five largest (in absolute value) effects? How do they compare with the factors identified by stepwise regression?

Section 8.5 (Orthogonal Arrays)

Theoretical Exercise

8.19 Represent the PB design for $N = 12$, $p = 11$ shown in Table 8.6 in the orthogonal array notation. What is its strength?

Section 8.6 (Sequential Assemblies of Fractional Factorials)

Theoretical Exercises

8.20 Consider the Plackett–Burman design in Table 8.6. Its basic defining relation is $I = ABD = ACE = BCF = AFG$. Suppose that an experiment using this design shows that the main effects A, C, and E are significant. Since $A = CE$, $C = AE$, and $E = AC$, it is not evident whether the three factors are active through their main effects or whether only two of them are active through their main effects and interaction.

(a) What are the other two-factor interactions aliased with the main effects A, C, and E and what principle permits ruling them out as possibly significant?

AE	BF	DG	CH
+	−	−	−
+	+	−	−
+	−	+	−
+	+	+	+

Display 8.17 Matrix of signs for the interactions, AE, BF, DG, CH.

(b) Suppose we fold over only the three factors in question, namely, A, C, and E. Does the combined design untangle the main effects of the three factors from their two-factor interactions? Does the resulting design have resolution IV so that all main effects are clear of two-factor interactions?

(c) Suppose that in the foldover design the estimated main effects $\widehat{A}$, $\widehat{C}$, and $\widehat{E}$ continue to be significant but $\widehat{E}$ has switched the sign. Explain why this means that the interaction $\widehat{AC}$ is likely to be significant.

(d) If the estimated main effects $\widehat{A}$, $\widehat{B}$, and $\widehat{C}$ had turned out to be significant, can we fold over just those three factors to untangle their main effects from their two-factor interactions?

8.21 Consider the 2_{IV}^{8-3} design with the basic generators $ABCF$, $ABDG$, and $BCDEH$.

(a) Give the complete defining relation of this design and its alias structure up to three-factor interactions.

(b) Which factors should be folded over in order to reverse the signs of the basic generators with an even number of letters?

(c) What is the complete defining relation of the combined design and its resolution?

Applied Exercise

8.22 In Exercise 8.9 the main effects $\widehat{C}$, $\widehat{E}$, $\widehat{H}$ and the interaction $\widehat{AE} = \widehat{BF} = \widehat{DG} = \widehat{CH}$ were found to be significant. To resolve the ambiguity caused by the aliasing of the interactions, AE, BF, DG, and CH, Box, Hunter, and Hunter (1978, Table 12B.1) suggest making the four additional runs shown in Table 8.27; the observed shrinkage data are also shown in the table.

(a) Show that the matrix of signs for the interactions, AE, BF, DG, and CH, is as shown in Display 8.17.

(b) This matrix of signs can be viewed as the model matrix of a 2^2 design where AE is not a contrast but is fully aliased with the overall mean

Table 8.27 Four Additional Runs for Injection Molding Experiment

Run	A	B	C	D	E	F	G	H	Shrinkage
17	−	+	+	+	−	−	−	+	29.4
18	−	+	−	−	−	+	+	+	19.7
19	+	+	−	−	+	−	−	+	13.6
20	+	+	+	+	+	+	+	+	24.7

Source: Box, Hunter, and Hunter (1978, Table 12B.1). Reprinted by permission of John Wiley & Sons, Inc.

effect μ. But before this matrix can be used for estimating the three interactions BF, DG, and CH, the observed data in Table 8.27 must be adjusted for the effects $\widehat{C} = 5.5$ and $\widehat{E} = -3.8$ found significant in Exercise 8.9. Keeping in mind that the model coefficients equal $\frac{1}{2}$ times the respective effects, show that the estimates of BF, DG, and CH can be obtained by solving the following four equations:

$$\mu + \tfrac{1}{2}AE - \tfrac{1}{2}BF - \tfrac{1}{2}DG + \tfrac{1}{2}CH = 24.75$$

$$\mu + \tfrac{1}{2}AE + \tfrac{1}{2}BF - \tfrac{1}{2}DG - \tfrac{1}{2}CH = 20.55$$

$$\mu + \tfrac{1}{2}AE - \tfrac{1}{2}BF + \tfrac{1}{2}DG - \tfrac{1}{2}CH = 18.25$$

$$\mu + \tfrac{1}{2}AE + \tfrac{1}{2}BF + \tfrac{1}{2}DG + \tfrac{1}{2}CH = 23.85.$$

(c) Explain why the overall mean effect estimate of $\widehat{\mu} = 19.75$ from Exercise 8.9 should not be used as an estimate of μ for these four additional runs.

(d) From Exercise 8.9 we also obtain that the estimate of the aliased interaction effect $AE = BF = DG = CH$ is 4.6. Add the resulting fifth equation to the four equations from part (b) and solve the resulting system of equations for μ, AE, BF, DG, and CH. Which interaction is the largest? Is it the same as the one you conjectured in part (b) of Exercise 8.9?

CHAPTER 9

Three-Level and Mixed-Level Factorial Experiments

In the previous two chapters we studied two-level full and fractional factorial designs. We also studied saturated designs such as Plackett–Burman designs, Hadamard designs, and supersaturated designs. These are the most commonly used designs in the screening phase of experimentation. Subsequent phases of experimentation are generally confirmatory or modeling in nature. To model nonlinear effects, more than two levels of quantitative factors must be used. In this chapter we focus on three-level factorial designs. Full factorials are discussed in Section 9.1 and fractional factorials in Section 9.2. In Section 9.3 we discuss mixed-level (two- and three- and two- and four-level) factorial designs. Chapter notes in Section 9.4 are followed by a chapter summary in Section 9.5.

9.1 THREE-LEVEL FULL FACTORIAL DESIGNS

Three-level factorial designs are of interest when the factors are quantitative and their quadratic relationships with the response variable are to be estimated or when the factors are qualitative and come naturally in three levels, for example, three settings of a machine or three suppliers. As noted in Section 7.1.3, quadratic effects of the quantitative factors cannot be estimated using 2^p designs or their fractions. Augmenting a two-level factorial design with center points (see Section 7.3.3) allows testing for curvature but does not allow for estimation of the individual quadratic effects of the factors.

Be aware, however, that 3^p designs get large very quickly as p increases. For $p = 3$, one already needs 27 runs and for $p = 4$ one needs 81 runs, which is often beyond the resources commonly available for experimentation.

Statistical Analysis of Designed Experiments: Theory and Applications By Ajit C. Tamhane

Hence 3^p designs are not efficient for fitting quadratic models for $p \geq 4$. Second-order response surface designs discussed in Chapter 10 (see Section 10.1.3), for example, central composite designs, are much better suited for this purpose.

We begin by considering a 3^2 balanced design with factors A and B. Although in the previous chapters we used the ± 1 notation to denote the levels of factors in 2^p designs, here it is more convenient to use 0, 1, and 2 as the notation since we shall be using the mod 3 arithmetic later to determine the levels of the three-level factors in fractional factorial designs. (For the sake of consistency, we shall also use 0, 1 notation for two-level factors in this chapter.) If the factors are quantitative, then these levels would naturally correspond to the low, medium, and high levels of the factors respectively.

Let y_{ijk} be the kth replicate observation at the (i, j)th combination of A and B $(i, j = 0, 1, 2, k = 1, 2, \ldots, n)$. The y_{ijk} are assumed to be independent $N(\mu_{ij}, \sigma^2)$. The μ_{ij} are estimated by the sample means $\overline{y}_{ij}$ and σ^2 is estimated by

$$\text{MS}_e = \frac{\sum_{i=0}^{2} \sum_{j=0}^{3} \sum_{k=1}^{n} (y_{ijk} - \overline{y}_{ij})^2}{9(n-1)}$$

with $9(n-1)$ d.f. If $n = 1$, then a pure-error estimate is not available and the methods of Section 7.3 for singly replicated factorial designs must be employed.

The 3^2 design can be analyzed by using the methods for a two-way layout from Chapter 6. Let $\overline{y}_{i\cdot}$ and $\overline{y}_{\cdot j}$ denote the means of $\overline{y}_{ij}$ for fixed levels i of A and j of B, respectively, and let $\overline{y}_{\cdot\cdot}$ denote the overall mean. The main effects A and B each have two d.f. and their sums of squares are given by

$$\text{SS}_A = 3n \sum_{i=0}^{2} (\overline{y}_{i\cdot} - \overline{y}_{\cdot\cdot})^2 \quad \text{and} \quad \text{SS}_B = 3n \sum_{j=0}^{2} (\overline{y}_{\cdot j} - \overline{y}_{\cdot\cdot})^2. \tag{9.1}$$

The interaction AB has four d.f. and its sum of squares is given by

$$\text{SS}_{AB} = n \sum_{i=0}^{2} \sum_{j=0}^{2} (\overline{y}_{ij} - \overline{y}_{i\cdot} - \overline{y}_{\cdot j} + \overline{y}_{\cdot\cdot})^2. \tag{9.2}$$

If the factors are quantitative and equispaced, then a finer analysis can be done by partitioning the main effects and interaction of the factors into their orthogonal components by extending the orthogonal polynomial model for a single quantitative factor from Section 3.7. The resulting decomposition is called the **linear–quadratic system**, which is discussed in Section 9.1.1. This system gives more interpretable results than another system, called the **orthogonal component system**, discussed in Section 9.1.2. This latter system is primarily useful for constructing three-level fractional factorial designs and obtaining their alias structures.

9.1.1 Linear–Quadratic System

Recall from Section 3.7 that for a quantitative factor with $a \geq 2$ equispaced levels we can compute $a - 1$ orthogonal contrasts among the sample means $\bar{y}_i$. For $a = 3$ the two orthogonal contrast vectors that measure the linear and quadratic effects, respectively, are given by (see Table 3.7)

$$\boldsymbol{\ell} = \frac{1}{\sqrt{2}}(-1, 0, 1)' \quad \text{and} \quad \boldsymbol{q} = \frac{1}{\sqrt{6}}(1, -2, 1)'; \tag{9.3}$$

here $\sqrt{2}$ and $\sqrt{6}$ are normalizing constants, which make $\boldsymbol{\ell}$ and $\boldsymbol{q}$ of unit lengths. For a one-way layout with a sample mean vector $\bar{\boldsymbol{y}} = (\bar{y}_0, \bar{y}_1, \bar{y}_2)'$, the linear and quadratic effects equal

$$\boldsymbol{\ell}'\bar{\boldsymbol{y}} = \frac{1}{\sqrt{2}}(-\bar{y}_0 + \bar{y}_2) \quad \text{and} \quad \boldsymbol{q}'\bar{\boldsymbol{y}} = \frac{1}{\sqrt{6}}(\bar{y}_0 - 2\bar{y}_1 + \bar{y}_2).$$

For a two-way layout we can partition each main effect into its linear and quadratic components and the interaction AB into four components, namely, **linear-by-linear (LL), linear-by-quadratic (LQ), quadratic-by-linear (QL)**, and **quadratic-by-quadratic (QQ)**. As we shall see below, all of these contrasts are mutually orthogonal and have one d.f. each.

Let z_A and z_B be the coded variables for factors A and B such that the low and high values of the factors are coded as -1 and 1, respectively. (Note that this coding is different from the 0, 1, 2 coding used to denote the low, medium, and high levels of the factors. In the following linear model, both notations are used simultaneously.) From Section 3.7 we see that the linear and quadratic orthogonal polynomials for three-level factors are

$$\xi_1(z) = z \quad \text{and} \quad \xi_2(z) = 3(z^2 - \tfrac{2}{3}). \tag{9.4}$$

We want to fit the model

$$\begin{aligned} \mathrm{E}(y) = {} & \beta_{00} + \beta_{10}\xi_1(z_A) + \beta_{20}\xi_2(z_A) + \beta_{01}\xi_1(z_B) + \beta_{02}\xi_2(z_B) + \beta_{11}\xi_1(z_A)\xi_1(z_B) \\ & + \beta_{21}\xi_2(z_A)\xi_1(z_B) + \beta_{12}\xi_1(z_A)\xi_2(z_B) + \beta_{22}\xi_2(z_A)\xi_2(z_B). \end{aligned} \tag{9.5}$$

This model can be written in matrix form [see Eq. (2.27)], $\mathrm{E}(\bar{\boldsymbol{y}}) = \boldsymbol{X\beta}$, as follows. Here, as in Chapter 7, it is convenient to work with the vector $\bar{\boldsymbol{y}}$ of the sample means $\bar{y}_{ij}$ instead of the vector $\boldsymbol{y}$ of the y_{ijk}. Keeping in mind that $\xi_1(z) = -1, 0, 1$ and $\xi_2(z) = 1, -2, 1$ for $z = -1, 0, 1$, respectively, we have

$$\bar{\boldsymbol{y}} = \begin{bmatrix} \bar{y}_{00} \\ \bar{y}_{10} \\ \bar{y}_{20} \\ \bar{y}_{01} \\ \bar{y}_{11} \\ \bar{y}_{21} \\ \bar{y}_{02} \\ \bar{y}_{12} \\ \bar{y}_{22} \end{bmatrix},$$

$$\boldsymbol{X} = \begin{array}{c} \begin{array}{ccccccccc} \beta_{00} & \beta_{10} & \beta_{20} & \beta_{01} & \beta_{02} & \beta_{11} & \beta_{12} & \beta_{21} & \beta_{22} \end{array} \\ \begin{bmatrix} 1 & -1 & 1 & -1 & 1 & 1 & -1 & -1 & 1 \\ 1 & 0 & -2 & -1 & 1 & 0 & 0 & 2 & -2 \\ 1 & 1 & 1 & -1 & 1 & -1 & 1 & -1 & 1 \\ 1 & -1 & 1 & 0 & -2 & 0 & 2 & 0 & -2 \\ 1 & 0 & -2 & 0 & -2 & 0 & 0 & 0 & 4 \\ 1 & 1 & 1 & 0 & -2 & 0 & -2 & 0 & -2 \\ 1 & -1 & 1 & 1 & 1 & -1 & -1 & 1 & 1 \\ 1 & 0 & -2 & 1 & 1 & 0 & 0 & -2 & -2 \\ 1 & 1 & 1 & 1 & 1 & 1 & 1 & 1 & 1 \end{bmatrix} \end{array}, \qquad \boldsymbol{\beta} = \begin{bmatrix} \beta_{00} \\ \beta_{10} \\ \beta_{20} \\ \beta_{01} \\ \beta_{02} \\ \beta_{11} \\ \beta_{12} \\ \beta_{21} \\ \beta_{22} \end{bmatrix}.$$

Observe that the first column of $\boldsymbol{X}$ corresponds to the constant term β_{00} and the remaining columns are the contrast vectors corresponding to the linear and quadratic main effects of A followed by those of B, next followed by the four interaction terms obtained by taking the componentwise products of the respective main-effect contrast vectors. All columns are mutually orthogonal and

$$\boldsymbol{X}'\boldsymbol{X} = \text{diag}\{9, 6, 18, 6, 18, 4, 12, 12, 36\}.$$

Next, $\boldsymbol{X}'\bar{\boldsymbol{y}}$ gives the sum of the $\bar{y}_{ij}$ as the first term (corresponding to the first column of $\boldsymbol{X}$) and the contrasts (defined by the other columns of $\boldsymbol{X}$) among the $\bar{y}_{ij}$ as the remaining terms. Finally, applying the formula $\widehat{\boldsymbol{\beta}} = (\boldsymbol{X}'\boldsymbol{X})^{-1}\boldsymbol{X}'\bar{\boldsymbol{y}}$ from Eq. (2.29) gives the LS estimators. As in the case of 2^p factorial designs, it is common to use the estimated effects instead of the estimated **regression coefficients**, $\widehat{\beta}$'s, in the fitted regression model (9.5). For example, the estimated linear effect of A, denoted by $\widehat{A}_L$, may be used instead of $\widehat{\beta}_{10}$. The expressions for the LS estimators of the β's and of the corresponding effects are as follows:

$$\widehat{\beta}_{00} = \bar{y}_{..},$$

$$\widehat{\beta}_{10} = \frac{(\bar{y}_{20} - \bar{y}_{00}) + (\bar{y}_{21} - \bar{y}_{01}) + (\bar{y}_{22} - \bar{y}_{02})}{6} = \frac{\widehat{A}_L}{\sqrt{6}},$$

$$\widehat{\beta}_{20} = \frac{(\bar{y}_{20} - 2\bar{y}_{10} + \bar{y}_{00}) + (\bar{y}_{21} - 2\bar{y}_{11} + \bar{y}_{01}) + (\bar{y}_{22} - 2\bar{y}_{12} + \bar{y}_{02})}{18} = \frac{\widehat{A}_Q}{3\sqrt{2}},$$

$$
\begin{aligned}
\widehat{\beta}_{01} &= \frac{(\bar{y}_{02}-\bar{y}_{00})+(\bar{y}_{12}-\bar{y}_{10})+(\bar{y}_{22}-\bar{y}_{20})}{6} = \frac{\widehat{B}_L}{\sqrt{6}},\\
\widehat{\beta}_{02} &= \frac{(\bar{y}_{02}-2\bar{y}_{01}+\bar{y}_{00})+(\bar{y}_{12}-2\bar{y}_{11}+\bar{y}_{10})+(\bar{y}_{22}-2\bar{y}_{21}+\bar{y}_{20})}{18} = \frac{\widehat{B}_Q}{3\sqrt{2}},\\
\widehat{\beta}_{11} &= \frac{(\bar{y}_{22}-\bar{y}_{02})-(\bar{y}_{20}-\bar{y}_{00})}{4} = \frac{\widehat{AB}_{LL}}{2},\\
\widehat{\beta}_{12} &= \frac{(\bar{y}_{22}-2\bar{y}_{21}+\bar{y}_{20})-(\bar{y}_{02}-2\bar{y}_{01}+\bar{y}_{00})}{12} = \frac{\widehat{AB}_{LQ}}{2\sqrt{3}},\\
\widehat{\beta}_{21} &= \frac{(\bar{y}_{22}-2\bar{y}_{12}+\bar{y}_{02})-(\bar{y}_{20}-2\bar{y}_{10}+\bar{y}_{00})}{12} = \frac{\widehat{AB}_{QL}}{2\sqrt{3}},\\
\widehat{\beta}_{22} &= \frac{(\bar{y}_{22}-2\bar{y}_{21}+\bar{y}_{20})-2(\bar{y}_{12}-2\bar{y}_{11}+\bar{y}_{10})+(\bar{y}_{02}-2\bar{y}_{01}+\bar{y}_{00})}{36} = \frac{\widehat{AB}_{QQ}}{6}.
\end{aligned}
\tag{9.6}
$$

An alternate derivation of these expressions is given in Section 9.4 which may be more insightful to some.

Under the null hypothesis that the true component effect is zero, its estimator is normal with mean 0 and variance $= \sigma^2/n$. Therefore the sum of squares for each of the component effects under the corresponding null hypothesis is

$$
SS_{\text{Effect}} = n(\widehat{\text{Effect}})^2 \sim \sigma^2\chi_1^2. \tag{9.7}
$$

Since each contrast has one d.f., $SS_{\text{Effect}} = MS_{\text{Effect}}$ and its significance can be tested using the F-statistic:

$$
F_{\text{Effect}} = \frac{MS_{\text{Effect}}}{MS_e} \sim F_{1,9(n-1)}.
$$

Furthermore, since the contrasts are mutually orthogonal, their sums of squares add up to the total sum of squares for the corresponding main or interaction effect. Specifically,

$$
SS_A = n[\widehat{A}_L^2 + \widehat{A}_Q^2], \qquad SS_B = n[\widehat{B}_L^2 + \widehat{B}_Q^2], \qquad SS_{AB} = n[\widehat{AB}_{LL}^2 + \widehat{AB}_{LQ}^2 + \widehat{AB}_{QL}^2 + \widehat{AB}_{QQ}^2].
$$

Example 9.1 (Growing Stem Cells for Bone Implants: Linear–Quadratic Decomposition)

Section B.2 describes a detailed case study of an investigation to evaluate the effects of three factors on the growth of stem cells into mature bone cells. For the purpose of this example we focus on two of those factors with three levels each: growth factor (with levels 0, 15, and 30 ng/ml) and incubation period (with levels 3, 6, and 9 days). We only use the data for one level of the media factor (namely, osteogenic media). We also ignore the data on the covariate GAPDH.

The response variable of interest is elevated alkaline phosphatase (ALP) gene expression. To remove the heteroscedasticity in this variable, an inverse square-root transformation is applied (see the case study for details), which leads to the variable $y = 1/\sqrt{\text{ALP}}$ analyzed here. The resulting data are shown in Table 9.1 and the cell means are shown in Table 9.2.

From the ANOVA table in Display 9.1 we see that both the main effects and their interaction are highly significant. We now analyze them in finer detail. To

Table 9.1 Inverse Square-Root Transformed APL Data Classified by Growth Factor and Incubation Period for Osteogenic Media

	Incubation Period (B)		
Growth Factor (A)	3 Days	6 Days	9 Days
0 ng/ml	6.878	4.299	3.094
	6.718	3.167	4.299
	7.041	2.951	3.479
15 ng/ml	7.919	6.562	3.822
	7.735	7.556	3.647
	7.919	7.556	3.562
30 ng/ml	6.562	5.311	3.479
	4.835	4.299	3.094
	5.973	3.822	2.175

Table 9.2 Cell Means for Stem Cell Growth Data in Table 9.1 for Osteogenic Media

	Incubation Period (B)		
Growth Factor (A)	3 Days	6 Days	9 Days
0 ng/ml	$\bar{y}_{00} = 6.879$	$\bar{y}_{01} = 3.472$	$\bar{y}_{02} = 3.624$
15 ng/ml	$\bar{y}_{10} = 7.858$	$\bar{y}_{11} = 7.225$	$\bar{y}_{12} = 3.677$
30 ng/ml	$\bar{y}_{20} = 5.790$	$\bar{y}_{21} = 4.477$	$\bar{y}_{22} = 2.916$

```
Source                              DF      SS       MS      F      P
Growth Factor                        2  18.202    9.101  26.55  0.000
Incubation Period                    2  53.171   26.585  77.56  0.000
Growth Factor*Incubation Period      4  11.938    2.985   8.71  0.000
Error                               18   6.170    0.343
Total                               26  89.481

S = 0.585473   R-Sq = 93.10%   R-Sq(adj) = 90.04%
```

Display 9.1 ANOVA for stem cell growth data.

compute the various single d.f. effects, it is useful to first calculate $\widehat{A}_L$ and $\widehat{A}_Q$ conditional on B and $\widehat{B}_L$ and $\widehat{B}_Q$ conditional on A using the $\overline{y}_{ij}$ values from Table 9.2. These calculations are shown in Table 9.3. The calculations of the single d.f. orthogonal components of the main effects and interactions are shown in Display 9.2.

The sums of squares of these effects, each with one d.f., can be computed using formula (9.7) with $n = 3$. The resulting detailed ANOVA is shown in Table 9.4. From this table we see that the significant effects are $\widehat{A}_Q, \widehat{B}_L, \widehat{AB}_{LQ}, \widehat{AB}_{QL}$, and $\widehat{AB}_{QQ}$. The main-effects and interaction plots shown in Figure 9.1 give a better understanding of these effects. For example, the main-effect plot of growth factor (A) is essentially quadratic, while that of incubation period (B) is essentially linear. The interaction plot shows that there is significant interaction between the two factors. At $A = 30$, the mean response decreases linearly with B, while at $A = 0$ and 15, the mean response varies quadratically but in opposite ways (convex for $A = 0$ and concave for $A = 15$).

Table 9.3 $\widehat{A}_L, \widehat{A}_Q$ **Conditional on** B **and** $\widehat{B}_L, \widehat{B}_Q$ **Conditional on** A

	B					A			
Effect	0	1	2	Total	Effect	0	1	2	Total
$\widehat{A}_L$	−1.089	1.005	−0.708	−0.792	$\widehat{B}_L$	−3.255	−4.181	−2.874	−10.310
$\widehat{A}_Q$	−3.047	−6.501	−0.814	−10.362	$\widehat{B}_Q$	3.559	−2.915	−0.248	0.396

$$\widehat{A}_L = \frac{-0.792}{\sqrt{6}} = -0.323, \qquad \widehat{\beta}_{10} = \frac{-0.792}{6} = 0.132$$

$$\widehat{A}_Q = \frac{-10.362}{3\sqrt{2}} = -2.442, \qquad \widehat{\beta}_{20} = \frac{-10.362}{18} = -0.576$$

$$\widehat{B}_L = \frac{-10.310}{\sqrt{6}} = -4.209, \qquad \widehat{\beta}_{01} = \frac{-10.310}{6} = -1.718$$

$$\widehat{B}_Q = \frac{0.396}{3\sqrt{2}} = 0.093, \qquad \widehat{\beta}_{02} = \frac{0.396}{18} = 0.022$$

$$\widehat{AB}_{LL} = \frac{0.381}{2} = 0.190, \qquad \widehat{\beta}_{11} = \frac{0.381}{4} = 0.095$$

$$\widehat{AB}_{LQ} = \frac{-3.807}{2\sqrt{3}} = -1.099, \qquad \widehat{\beta}_{12} = \frac{-3.807}{12} = -0.317$$

$$\widehat{AB}_{QL} = \frac{2.233}{2\sqrt{3}} = 0.645, \qquad \widehat{\beta}_{21} = \frac{2.233}{12} = 0.186$$

$$\widehat{AB}_{QQ} = \frac{9.141}{6} = 1.524, \qquad \widehat{\beta}_{22} = \frac{9.141}{36} = 0.254$$

Display 9.2 Calculation of linear and quadratic effects and $\widehat{\beta}$ coefficients.

Table 9.4 Detailed ANOVA for Stem Cell Growth Data

Source	SS	d.f.	MS	F	p
Growth factor (A)	18.202	2	9.101	26.55	0.000
A linear	0.313	1	0.313	0.913	0.352
A quadratic	17.890	1	17.890	52.157	0.000
Incubation period (B)	53.171	2	26.585	77.560	0.000
B linear	53.147	1	53.147	154.95	0.000
B quadratic	0.026	1	0.026	0.076	0.786
Growth factor × incubation period	11.938	4	2.985	8.71	0.000
AB linear–linear	0.108	1	0.108	0.315	0.734
AB linear–quadratic	3.623	1	3.623	10.563	0.001
AB quadratic–linear	1.248	1	1.248	3.638	0.047
AB quadratic–quadratic	6.963	1	6.963	20.300	0.000
Error	6.170	18	0.343		
Total	89.481	26			

The single d.f. effects can be interpreted by plotting $\widehat{A}_L$ and $\widehat{A}_Q$ versus B and $\widehat{B}_L$ and $\widehat{B}_Q$ versus A. These plots are shown in Figure 9.2. The linear and quadratic main effects of A and B are proportional to the averages of the corresponding plots. Next, one can see that $\widehat{A}_L$ varies quadratically with B; the same is true of $\widehat{B}_L$ as a function of A. This explains the nonsignificance of the $\widehat{AB}_{LL}$ interaction and significance of the $\widehat{AB}_{LQ}$ and $\widehat{AB}_{QL}$ interactions. On the other hand, the $\widehat{A}_Q$ plot versus B has a significant curvature (and similarly the $\widehat{B}_Q$ plot vs. A). As a result, the $\widehat{AB}_{QQ}$ interaction is significant.

To fit an orthogonal polynomial, we define coded variables:

$$z_A = \tfrac{1}{15}(x_A - 15) \qquad \text{and} \qquad z_B = \tfrac{1}{3}(x_B - 6),$$

where x_A and x_B are the actual levels of growth factor (A) and incubation period (B). The β's are calculated in Display 9.2. Retaining only the significant coefficients (at the 0.05 level), the resulting equation is

$$\begin{aligned}\hat{y} = 5.102 &- 0.576\xi_2(z_A) - 1.718\xi_1(z_B) - 0.317\xi_1(z_A)\xi_2(z_B) \\ &+ 0.186\xi_2(z_A)\xi_1(z_B) + 0.254\xi_2(z_A)\xi_2(z_B).\end{aligned}$$

This response surface is plotted in Figure 9.3 as a function of growth factor (A) and incubation period (B) (both on coded scale). Suppose we want to predict ALP when the settings are $x_A = 10$ and $x_B = 8$. Then $z_A = \frac{1}{15}(10 - 15) = -\frac{1}{3}$ and $z_B = \frac{1}{3}(8 - 6) = \frac{2}{3}$. From (9.4) we obtain

$$\xi_1(z_A) = -\tfrac{1}{3}, \qquad \xi_2(z_A) = -\tfrac{5}{3}, \qquad \xi_1(z_B) = \tfrac{2}{3}, \qquad \xi_2(z_B) = -\tfrac{2}{3}.$$

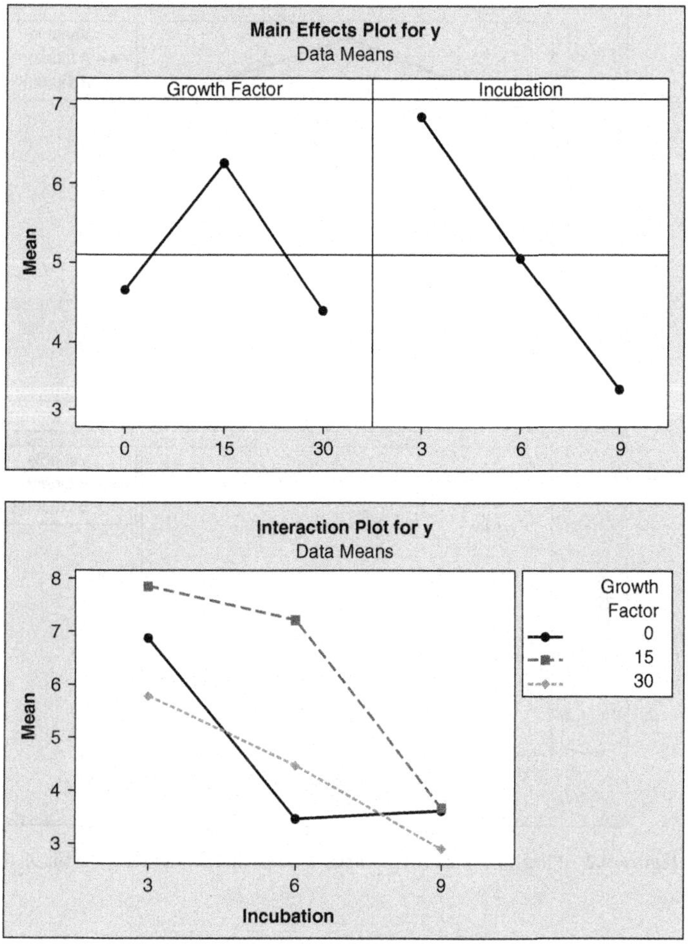

Figure 9.1 Main effects (top) and interaction (bottom) plots for stem cell data.

Substituting these values in the fitted response surface equation, we obtain $\widehat{y} = 4.922$ and hence $\widehat{\text{ALP}} = 1/(4.922)^2 = 0.0413$. The minimum value $\widehat{y}$ over the experimental range of A and B can be shown to be attained at their lower limits, that is, $z_A = z_B = -1$, which yields $\min \widehat{y} = 2.931$ and hence $\max \widehat{\text{ALP}} = 1/(2.931)^2 = 0.116$. ■

The linear–quadratic decomposition can be extended to a 3^3 design in a straightforward way. In this case there are three main effects, A, B, and C, each with two d.f., three two-factor interactions, AB, AC, and BC, each with four d.f., and one three-factor interaction, ABC, with eight d.f. Each main effect can be partitioned into its linear and quadratic components and each two-factor interaction into its four components, LL, LQ, QL, and QQ. When computing any

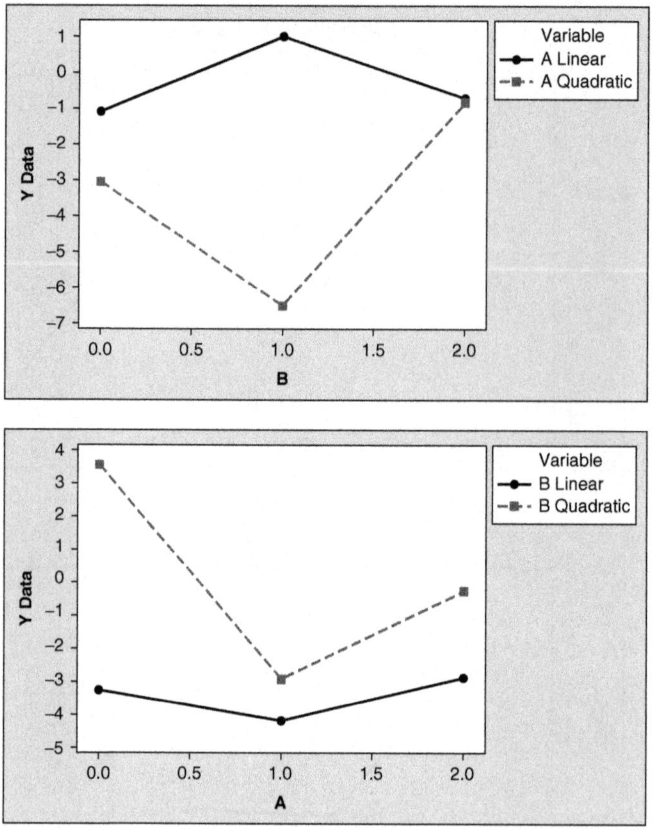

Figure 9.2 Plots of $\widehat{A}_L$ and $\widehat{A}_Q$ versus B (top) and $\widehat{B}_L$ and $\widehat{B}_Q$ versus A (bottom).

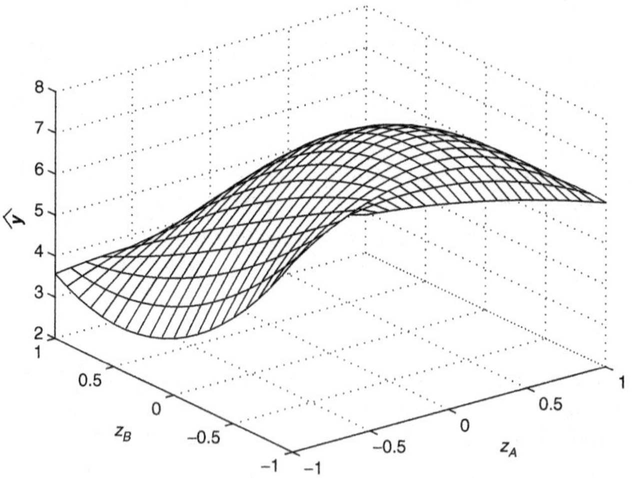

Figure 9.3 Fitted response surface $\widehat{y}$ as function of growth factor (A) and incubation period (B) (both variables in coded scale).

two-factor interaction the data can be averaged over the third factor. Decomposition of the ABC interaction into its eight single d.f. components is usually not of interest since these components are not very meaningful. If $n > 1$ replicates are observed at each factor-level combination, then there are $27(n-1)$ d.f. for estimating the error.

9.1.2 Orthogonal Component System

Consider a 3^2 design with two factors, A and B. Let $x_i = 0, 1, 2$ for the level of the ith factor ($i = 1, 2$). We add two columns to the 3^2 design matrix by defining their entries as

$$x_3 = (x_1 + x_2) \bmod 3 \qquad \text{and} \qquad x_4 = (x_1 + 2x_2) \bmod 3. \tag{9.8}$$

[Addition with respect to mod 3 equals the remainder left after dividing the sum by 3, e.g., $(1+1) \bmod 3 = 2$ but $(1+2) \bmod 3 = 0$.] The resulting matrix is shown in Display 9.3. Note that this is an orthogonal array of strength 2.

The two added columns are labeled AB and AB^2 using the coefficients of x_1 and x_2 in their definitions given by Eq. (9.8) as the exponents of A and B, respectively. Note that the exponent of any letter cannot be greater than 2 because of the mod 3 arithmetic. Further note that the exponent of one of the letters can be limited to 1; by convention we choose that to be the first letter. The reason for this is as follows: Consider the effect A^2B which corresponds to $(2x_1 + x_2) \bmod 3 = 0, 1, 2$. Multiplying this equation by 2 yields the equation $(x_1 + 2x_2) \bmod 3 = 0, 1, 2$ [since $(4x_1) \bmod 3 = x_1$], which is the same as the second equation in (9.8) and corresponds to $(A^2B)^2 = A^4B^2 = AB^2$. The mapping between A^2B and AB^2 is as follows: 0 is mapped to 0, 1 is mapped to 2, and 2 is mapped to 1. Although the labels 1 and 2 are swapped in this mapping, the same triples of $\bar{y}_{ij}$ are grouped with the three levels of A^2B and AB^2, and hence their sums of squares are equal. Thus, the effects A^2B and AB^2 are equivalent and only one of them needs to be considered. The common convention is to

A	B	AB	AB^2	$\bar{y}$
0	0	0	0	$\bar{y}_{00}$
0	1	1	2	$\bar{y}_{01}$
0	2	2	1	$\bar{y}_{02}$
1	0	1	1	$\bar{y}_{10}$
1	1	2	0	$\bar{y}_{11}$
1	2	0	2	$\bar{y}_{12}$
2	0	2	2	$\bar{y}_{20}$
2	1	0	1	$\bar{y}_{21}$
2	2	1	0	$\bar{y}_{22}$

Display 9.3 Main effects and interaction components for a 3^2 design.

choose that effect to be AB^2. Similarly, the effects A^2B^2 and AB are equivalent and we only consider the effect AB. In general, the square of any effect is equivalent to the same effect; therefore we do not consider the effect B^2, which is equivalent to B.

AB and AB^2 columns represent two orthogonal interactions. The previously defined AB interaction is different from the AB interaction defined here. To distinguish between the two, we denote the former as an $A \times B$ interaction. In fact, the sum of squares for the $A \times B$ interaction, which has four d.f., equals the total of the sums of squares for the AB and AB^2 interactions, each with two d.f. However, note that AB and AB^2 interactions do not have simple interpretations. In particular, they do not represent linear-by-linear and linear-by-quadratic interactions, respectively.

The $\overline{y}_{ij}$ values based on n replicates observed at each combination of A and B are shown in Display 9.3. The sums of squares for the main effects are the treatment sum of squares obtained by regarding the design as a two-way layout in factors A and B and are given by (9.1). Similarly, the sums of squares for AB and AB^2 are computed by regarding them as two separate pseudofactors with their levels as shown in the respective columns. These sums of squares have two d.f. each. Because the columns are mutually orthogonal, the sums of squares are mutually independent and additive.

Denote the groupings of the $\overline{y}_{ij}$ for the 0, 1, 2 levels of AB by α, β, γ, respectively. Similarly, denote the groupings for the 0, 1, 2 levels of AB^2 by a, b, c, respectively. For example, α represents the three observations $\{\overline{y}_{00}, \overline{y}_{12}, \overline{y}_{21}\}$ for the 0 level of AB and a represents the three observations $\{\overline{y}_{00}, \overline{y}_{11}, \overline{y}_{22}\}$ for the 0 level of AB^2. These groupings can be represented in the form of a 3×3 Graeco–Latin square (see Section 5.4.5) shown in Table 9.5.

Treating the three observations corresponding to α, β, γ as three replicates at the three levels of AB and similarly for AB^2, we can calculate their sums of squares by first calculating

$$\overline{y}_\alpha = \tfrac{1}{3}(\overline{y}_{00} + \overline{y}_{12} + \overline{y}_{21}),$$
$$\overline{y}_\beta = \tfrac{1}{3}(\overline{y}_{01} + \overline{y}_{10} + \overline{y}_{22}), \qquad \overline{y}_\gamma = \tfrac{1}{3}(\overline{y}_{02} + \overline{y}_{11} + \overline{y}_{20}) \tag{9.9}$$

Table 9.5 Graeco–Latin Square Representation of *AB* and *AB*^{2} Interactions

	B		
A	0	1	2
0	$a\alpha$ ($\overline{y}_{00}$)	$c\beta$ ($\overline{y}_{01}$)	$b\gamma$ ($\overline{y}_{02}$)
1	$b\beta$ ($\overline{y}_{10}$)	$a\gamma$ ($\overline{y}_{11}$)	$c\alpha$ ($\overline{y}_{12}$)
2	$c\gamma$ ($\overline{y}_{20}$)	$b\alpha$ ($\overline{y}_{21}$)	$a\beta$ ($\overline{y}_{22}$)

and

$$\overline{y}_a = \tfrac{1}{3}(\overline{y}_{00} + \overline{y}_{11} + \overline{y}_{22}),$$
$$\overline{y}_b = \tfrac{1}{3}(\overline{y}_{02} + \overline{y}_{10} + \overline{y}_{21}), \qquad \overline{y}_c = \tfrac{1}{3}(\overline{y}_{01} + \overline{y}_{12} + \overline{y}_{20}). \tag{9.10}$$

Then

$$SS_{AB} = 3n[(\overline{y}_\alpha - \overline{y}_{..})^2 + (\overline{y}_\beta - \overline{y}_{..})^2 + (\overline{y}_\gamma - \overline{y}_{..})^2],$$
$$SS_{AB^2} = 3n[(\overline{y}_a - \overline{y}_{..})^2 + (\overline{y}_b - \overline{y}_{..})^2 + (\overline{y}_c - \overline{y}_{..})^2]. \tag{9.11}$$

Furthermore, because of the mutual orthogonality of AB and AB^2, we have $SS_{A\times B} = SS_{AB} + SS_{AB^2}$.

Example 9.2 (Growing Stem Cells for Bone Implants: Orthogonal Component Decomposition)

Refer to Example 9.1 for details about this experiment and data. We now show calculation of the orthogonal components SS_{AB} and SS_{AB^2} of the interaction sum of squares $SS_{A\times B}$. By substituting the values of $\overline{y}_{ij}$ from Table 9.2 in Eqs. (9.9) and (9.10), we calculate

$$\overline{y}_\alpha = \tfrac{1}{3}(6.879 + 3.677 + 4.477) = 5.011,$$
$$\overline{y}_\beta = \tfrac{1}{3}(4.749 + 7.858 + 2.916) = 4.749, \overline{y}_\gamma = \tfrac{1}{3}(3.624 + 7.225 + 5.790) = 5.546$$

and

$$\overline{y}_a = \tfrac{1}{3}(6.879 + 7.225 + 2.916) = 5.673,$$
$$\overline{y}_b = \tfrac{1}{3}(3.624 + 7.858 + 4.477) = 5.320, \overline{y}_c = \tfrac{1}{3}(3.472 + 3.677 + 5.790) = 4.313.$$

The grand mean is $\overline{y}_{..} = 5.102$. Therefore,

$$SS_{AB} = 9[(5.011 - 5.102)^2 + (4.749 - 5.102)^2 + (5.546 - 5.102)^2] = 2.970$$

and

$$SS_{AB^2} = 9[(5.673 - 5.102)^2 + (5.320 - 5.102)^2 + (4.313 - 5.102)^2] = 8.964.$$

Note that $SS_{AB} + SS_{AB^2} = 2.970 + 8.964 = 11.934$, which equals $SS_{A\times B}$ from ANOVA in Display 9.1 except for a round-off error.

The F-statistics for AB and AB^2 equal

$$F_{AB} = \frac{2.970/2}{0.343} = 4.329 \qquad \text{and} \qquad F_{AB^2} = \frac{8.964/2}{0.343} = 13.067.$$

Comparing them with $f_{2,18,.05} = 3.55$, we find that both are significant at the 0.05 level. However, usually it is not of interest to test the significance of AB and AB^2. ■

For a 3^3 design, each two-factor interaction can be decomposed into two orthogonal components, as was done for the 3^2 design. The three-factor interaction (denoted by $A \times B \times C$ here) can be decomposed into four orthogonal components, ABC, ABC^2, AB^2C, and AB^2C^2. Denoting the level of the ith factor by $x_i = 0, 1, 2$, these four components are defined by the following equations, respectively:

$$x_4 = (x_1 + x_2 + x_3) \bmod 3, \qquad x_5 = (x_1 + x_2 + 2x_3) \bmod 3,$$
$$x_6 = (x_1 + 2x_2 + x_3) \bmod 3, \qquad x_7 = (x_1 + 2x_2 + 2x_3) \bmod 3.$$

The resulting total of 13 main-effects and interaction components are shown in Table 9.6. In general, there are $(3^p - 1)/2$ effects[1]. In the next section we shall use this table to construct three-level fractional factorial designs and to obtain their alias structures.

9.2 THREE-LEVEL FRACTIONAL FACTORIAL DESIGNS

As mentioned before, 3^p designs get large very quickly to be of practical use unless the runs are cheap. Therefore fractional factorials are almost a necessity for $p \geq 4$. Just as we restricted to fractions that are powers of 2 for two-level factorial designs, here we restrict to fractions that are powers of 3. We denote a $(1/3^q)$th fraction of a 3^p factorial as a 3^{p-q} fractional factorial.

To construct a 3^{3-1} design we begin with a 3^2 design with two factors, A and B. Letting $x_i = 0, 1, 2$ for the level of the ith factor ($i = 1, 2$), we add a third factor by defining its levels to be $x_3 = (x_1 + x_2) \bmod 3$ or $C = AB$. The resulting 3^{3-1} design is given by relabeling the AB column as factor C in Display 9.3.

It is easy to check that this design is an orthogonal array of strength 2. If we add $2x_3$ to both sides of the equation $x_3 = (x_1 + x_2) \bmod 3$ and remember that any integer multiple of 3 equals zero in the mod 3 arithmetic, we get $(x_1 + x_2 + 2x_3) \bmod 3 = 0$ as its equivalent representation. We write the first relation as $C = AB$ and the second relation as $I = ABC^2$, which are of course equivalent. The identity I here is the column of 0's. We refer to $I = ABC^2$ as the **defining relation** of this design.

[1]For p factors $A, B, \ldots, P$, denote an arbitrary effect by $A^{e_1} B^{e_2} \cdots P^{e_p}$, where the exponents $e_i = 0, 1, 2$. Hence there are 3^p possible combinations of which the one corresponding to all $e_i = 0$ is not an effect. Of the remaining $3^p - 1$ effects we can exclude exactly half since they have the first letter with exponent 2; e.g., B^2C can be excluded since it is equivalent to BC^2. This gives a total of $(3^p - 1)/2$ effects.

Table 9.6 Main-Effects and Interaction Components for 3^3 Design

Run	A	B	C	AB	AB^2	AC	AC^2	BC	BC^2	ABC	ABC^2	AB^2C	AB^2C^2
1	0	0	0	0	0	0	0	0	0	0	0	0	0
2	0	0	1	0	0	1	2	1	2	1	2	1	2
3	0	0	2	0	0	2	1	2	1	2	1	2	1
4	0	1	0	1	2	0	0	1	1	1	1	2	2
5	0	1	1	1	2	1	2	2	0	2	0	0	1
6	0	1	2	1	2	2	1	0	2	0	2	1	0
7	0	2	0	2	1	0	0	2	2	2	2	1	1
8	0	2	1	2	1	1	2	0	1	0	1	2	0
9	0	2	2	2	1	2	1	1	0	1	0	0	2
10	1	0	0	1	1	1	1	0	0	1	1	1	1
11	1	0	1	1	1	2	0	1	2	2	0	2	0
12	1	0	2	1	1	0	2	2	1	0	2	0	2
13	1	1	0	2	0	1	1	1	1	2	2	0	0
14	1	1	1	2	0	2	0	2	0	0	1	1	2
15	1	1	2	2	0	0	2	0	2	1	0	2	1
16	1	2	0	0	2	1	1	2	2	0	0	2	2
17	1	2	1	0	2	2	0	0	1	1	2	0	1
18	1	2	2	0	2	0	2	1	0	2	1	1	0
19	2	0	0	2	2	2	2	0	0	2	2	2	2
20	2	0	1	2	2	0	1	1	2	0	1	0	1
21	2	0	2	2	2	1	0	2	1	1	0	1	0
22	2	1	0	0	1	2	2	1	1	0	0	1	1
23	2	1	1	0	1	0	1	2	0	1	2	2	0
24	2	1	2	0	1	1	0	0	2	2	1	0	2
25	2	2	0	1	0	2	2	2	2	1	1	0	0
26	2	2	1	1	0	0	1	0	1	2	0	1	2
27	2	2	2	1	0	1	0	1	0	0	2	2	1

To find the alias structure of this design, we pick out from Table 9.6 the runs satisfying $C = AB$ or equivalently runs with level 0 for ABC^2. These runs are 1, 5, 9, 11, 15, 16, 21, 22, and 26. The submatrix of the main and interaction effects corresponding to these runs is shown in Table 9.7. By comparing the columns of this table, we see that the following triples of columns are identical to each other except possibly for an interchange of labels 1 and 2: (A, BC^2, AB^2C), (B, AC^2, AB^2C^2), (C, AB, ABC), and (AB^2, AC, BC). As a result, in each subset of three columns the same triples of $\overline{y}_{ij}$ are grouped with the three levels of those columns, and hence their sums of squares are equal, for example, $\text{SS}_A = \text{SS}_{BC^2} = \text{SS}_{AB^2C}$. Those columns are said to be aliases of each other. These alias relationships are summarized as

$$A = BC^2 = AB^2C, \qquad B = AC^2 = AB^2C^2,$$

$$C = AB = ABC, \qquad AB^2 = AC = BC.$$

Table 9.7 Main-Effects and Interaction Components for 3^{3-1} Design

Run[a]	A	B	C	AB	AB^2	AC	AC^2	BC	BC^2	ABC	ABC^2	AB^2C	AB^2C^2
1	0	0	0	0	0	0	0	0	0	0	0	0	0
5	0	1	1	1	2	1	2	2	0	2	0	0	1
9	0	2	2	2	1	2	1	1	0	1	0	0	2
11	1	0	1	1	1	2	0	1	2	2	0	2	0
15	1	1	2	2	0	0	2	0	2	1	0	2	1
16	1	2	0	0	2	1	1	2	2	0	0	2	2
21	2	0	2	2	2	1	0	2	1	1	0	1	0
22	2	1	0	0	1	2	2	1	1	0	0	1	1
26	2	2	1	1	0	0	1	0	1	2	0	1	2

[a]The run numbers refer to those in Table 9.6.

A mechanical method to derive the above relations is as follows. The basic defining relation is $I = ABC^2$. To obtain the two aliases of A, we multiply both sides of this defining relation by A and A^2, which yields

$$A = A^2BC^2 = (A^2BC^2)^2 = A^4B^2C^4 = AB^2C \qquad \text{and} \qquad A^2 = A^3BC^2 = BC^2.$$

But recall that A^2 is the same as A and hence the second relation is equivalent to $A = BC^2$. The other three relations can be derived in the same way.

A 3^{4-2} design can be obtained by labeling the column AB^2 in Display 9.3 as factor D, that is, $D = AB^2$ or $I = AB^2D^2$. This means that the levels of D are defined to be $x_4 = (x_1 + 2x_2) \bmod 3$ or, equivalently, $(x_1 + 2x_2 + 2x_4) \bmod 3 = 0$. Thus this design has two basic generators, ABC^2 and AB^2D^2. There are two additional ones obtained by taking the following products:

$$I = (ABC^2)(AB^2D^2) = A^2C^2D^2 = ACD \qquad \text{and}$$
$$I = (ABC^2)(AB^2D^2)^2 = BCD^2.$$

In the above, the mod 3 arithmetic is used for the exponents as before. Note that, in contrast to 2^{p-2} designs, where two basic generators give rise to one additional one, namely, their generalized interaction, here we get two interactions. The first is the product of the two basic generators, and the second is the product of one with the square of the other. Instead of taking the second interaction to be $(ABC^2)(AB^2D^2)^2$, we could have taken it to be $(ABC^2)^2(AB^2D^2)$. However, that would have given the same result since $(ABC^2)^2(AB^2D^2) = B^4C^4D^2 = BCD^2$. So the complete defining relation is

$$I = ABC^2 = AB^2D^2 = ACD = BCD^2. \tag{9.12}$$

The alias structure of this design is obtained from these four generators by multiplying each one by A, B, C, D and A^2, B^2, C^2, D^2, respectively, resulting in the following four sets, each consisting of nine aliased effects:

$$A = AB^2C = BC^2 = ABD = BD = AC^2D^2 = CD = ABCD^2 = AB^2C^2D,$$

$$B = AB^2C^2 = AC^2 = AD^2 = ABD^2 = ABCD = AB^2CD = BC^2D = CD^2,$$

$$C = AB = ABC = AB^2CD^2 = AB^2C^2D^2 = AC^2D = AD = BC^2D^2 = BD^2,$$

$$D = ABC^2D = ABC^2D^2 = AB^2 = AB^2D = ACD^2 = AC = BC = BCD.$$

These 36 effects together with the four generators from (9.12) account for the total of $(3^4 - 1)/2 = 40$ effects.

How do we know that these four generators give a complete defining relation? This question can be answered by writing out the matrix of main effects and interactions for the 3^4 design analogous to the matrix shown for the 3^3 design in Table 9.6. There are a total of 81 runs and 40 main effects and interactions as listed above. Then one can check that if we pick the 9 runs corresponding to the 3^{4-2} design in Display 9.3, then those runs have 0's for the columns ABC^2, AB^2D^2, ACD, and BCD^2; therefore they are the four generators of the design.

Example 9.3 (Chemical Process Yield: 3^{4-1} Experiment)

Moen, Nolan, and Provost (1999) give data on the yield of a chemical process as a function of four factors, each at three levels. The factors and their levels are given in Table 9.8. A 3^{4-1} experiment consisting of 27 runs was conducted. The design and the data are shown in Table 9.9.

It was anticipated that the main effects of all four factors would be significant; in addition, all three interactions between temperature, pressure, and inhibitor level would be significant, whereas catalyst concentration will act independently of other factors. Therefore a model that includes these effects was posited.

Table 9.8 Factors and Their Levels for Chemical Process Yield Experiment

Factor	Low (0)	Medium (1)	High (2)
A: Temperature (°F)	190	200	210
B: Pressure (psi)	300	350	400
C: Inhibitor Level (ppm)	40	50	60
D: Catalyst Concentration (lbs.)	8	10	12

Source: Moen, Nolan, and Provost (1999, Figure 8.11). Reprinted by permission of The McGraw Hill Companies.

Table 9.9 Design and Data for Chemical Process Yield (%) Experiment

Run	*A*	*B*	*C*	*D*	Yield	Run	*A*	*B*	*C*	*D*	Yield	Run	*A*	*B*	*C*	*D*	Yield
1	0	0	0	0	93.1	10	1	0	2	2	93.6	19	2	0	1	1	97.0
2	0	1	2	2	94.3	11	1	1	1	1	96.4	20	2	1	0	0	96.8
3	0	2	1	1	96.8	12	1	2	0	0	97.1	21	2	2	2	2	95.5
4	0	0	2	1	91.9	13	1	0	1	0	92.7	22	2	0	0	2	98.4
5	0	1	1	0	93.6	14	1	1	0	2	97.8	23	2	1	2	1	96.1
6	0	2	0	2	98.5	15	1	2	2	1	96.0	24	2	2	1	0	95.0
7	0	0	1	2	92.2	16	1	0	0	1	97.4	25	2	0	2	0	93.7
8	0	1	0	1	97.7	17	1	1	2	0	93.5	26	2	1	1	2	96.7
9	0	2	2	0	94.3	18	1	2	1	2	96.6	27	2	2	0	1	98.8

Source: Moen, Nolan, and Provost (1999, Figure 8.11). Reprinted by permission of The McGraw Hill Companies.

It can be checked that the defining relation of this design is $D = ABC^2$ or $I = ABC^2D^2$. Hence the aliases of the four main effects are given by

$$A = AB^2CD = BC^2D^2, \qquad B = AB^2C^2D^2 = AC^2D^2,$$
$$C = ABD^2 = ABCD^2, \qquad D = ABC^2 = ABC^2D.$$

Thus the main effects are clear of any two-factor interactions involving A, B, or C, for example, AB or AB^2.

The main-effect plots are shown in Figure 9.4 and the interaction plots are shown in Figure 9.5. These plots suggest that all four main effects are large as anticipated, but only the temperature–pressure interaction is large; inhibitor

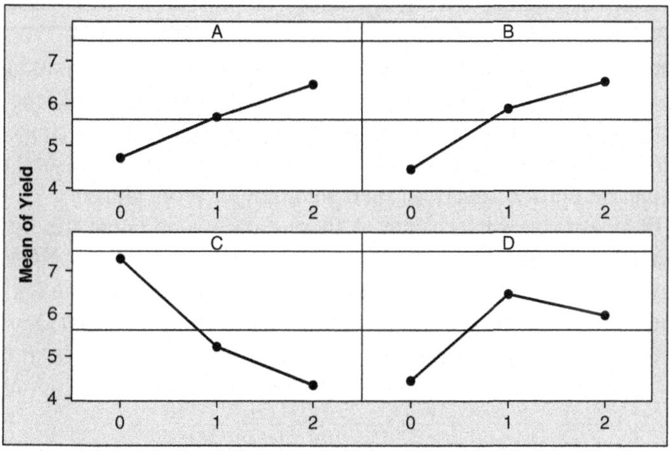

Figure 9.4 Main-effect plots for chemical process yield data.

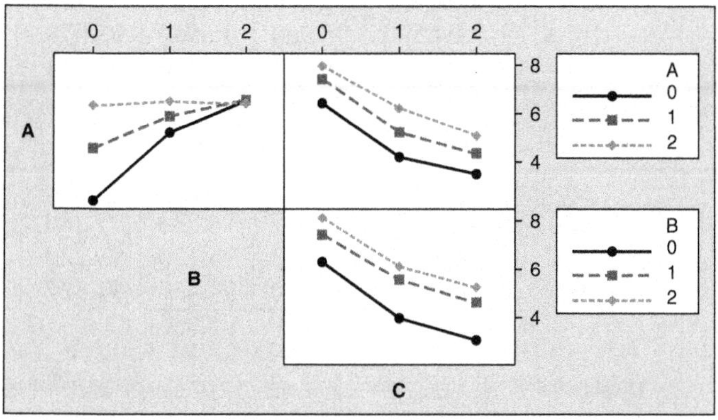

Figure 9.5 Interaction plots for chemical process yield example.

Source	DF	Seq SS	Adj SS	Adj MS	F	P
A	2	13.5800	13.5800	6.7900	305.55	0.000
B	2	20.1800	20.1800	10.0900	454.05	0.000
C	2	41.6467	41.6467	20.8233	937.05	0.000
D	2	20.2067	20.2067	10.1033	454.65	0.000
A*B	4	12.7867	12.7867	3.1967	143.85	0.000
A*C	4	0.2467	0.2467	0.0617	2.78	0.127
B*C	4	0.2267	0.2267	0.0567	2.55	0.147
Error	6	0.1333	0.1333	0.0222		
Total	26	109.0067				

S = 0.149071   R-Sq = 99.88%   R-Sq(adj) = 99.47%

Display 9.4 Minitab ANOVA output for chemical process yield data.

level does not interact with temperature or pressure. These visual impressions are confirmed by the Minitab ANOVA output in Display 9.4.

From the main-effect and interaction plots we see that, to maximize the yield, we should use the high levels of temperature and pressure, the low level of inhibitor, and the medium level of catalyst concentration. Since the last two factors do not interact with each other or with the other two factors, their optimum levels can be chosen based on their main-effect plots alone.

To fit a response surface plot to these data, we compute the linear–quadratic components of those effects found significant in the ANOVA shown in Display 9.4 using the means shown in Display 9.5. Denote by $\widehat{\beta}_{ijk\ell}$ the LS estimate of the coefficient $\beta_{ijk\ell}$ of the term $\xi_i(z_A)\xi_j(z_B)\xi_k(z_C)\xi_\ell(z_D)$ in the orthogonal polynomial representation of the response surface ($i, j, k, \ell = 0, 1, 2$), where recall that $\xi_0(z) = 1$. The results of these calculations are as follows:

		B			
		0	1	2	
A	0	92.400	95.200	96.530	94.711
	1	94.570	95.900	96.567	95.678
	2	96.370	96.533	96.430	96.444
		94.444	95.888	96.511	

		D			
		0	1	2	
C	0	95.670	97.967	98.233	97.289
	1	93.767	96.733	95.170	95.222
	2	93.833	94.670	94.467	94.322
		94.422	96.456	95.956	

Display 9.5 Cell means for $A \times B$ and $C \times D$ cross classifications.

$$\widehat{\beta}_{0000} = \overline{y}_{....} = 95.611$$

$$\begin{aligned}
\widehat{A}_L &= 2.1225, & \widehat{\beta}_{1000} &= 0.8665, \\
\widehat{A}_Q &= -0.1421, & \widehat{\beta}_{2000} &= -0.0335, \\
\widehat{B}_L &= 2.5315, & \widehat{\beta}_{0100} &= 1.0335, \\
\widehat{B}_Q &= -0.5664, & \widehat{\beta}_{0200} &= -0.1335, \\
\widehat{AB}_{LL} &= -2.0500, & \widehat{\beta}_{1100} &= -1.0250, \\
\widehat{AB}_{LQ} &= 0.3476, & \widehat{\beta}_{1200} &= 0.1003, \\
\widehat{AB}_{QL} &= 0.0566, & \widehat{\beta}_{2100} &= 0.0163, \\
\widehat{AB}_{QQ} &= -0.0683, & \widehat{\beta}_{2200} &= -0.0114, \\
\widehat{C}_L &= -3.6338, & \widehat{\beta}_{0010} &= -1.4835, \\
\widehat{C}_Q &= 0.8252, & \widehat{\beta}_{0020} &= 0.1945, \\
\widehat{D}_L &= 1.8788, & \widehat{\beta}_{0001} &= 0.7670, \\
\widehat{D}_Q &= -1.7918, & \widehat{\beta}_{0002} &= -0.0422.
\end{aligned}$$

A detailed ANOVA is shown in Table 9.10. In this table the sums of squares for the nonsignificant effects AC and BC are combined with the error term. The sums of squares for the linear–quadratic components do not exactly add up to the total sum of squares for the corresponding effects because of round-off errors. From this detailed ANOVA table we see that all effects except $\widehat{A}_Q$, $\widehat{AB}_{QL}$, and $\widehat{AB}_{QQ}$ are significant. Omitting these terms from the fitted model, the final prediction equation is given by

$$\begin{aligned}
\widehat{y} = {} & 95.611 + 0.8665\xi_1(z_A) + 1.0335\xi_1(z_B) - 0.1335\xi_2(z_B) - 1.4835\xi_1(z_C) \\
& + 0.1945\xi_2(z_C) + 0.7670\xi_1(z_D) - 0.0422\xi_2(z_D) \\
& - 1.0250\xi_1(z_A)\xi_1(z_B) + 0.1003\xi_1(z_A)\xi_2(z_B).
\end{aligned}$$

In this equation z_A, z_B, z_C, and z_D are coded variables defined by

$$z_A = \frac{x_A - 200}{10}, \qquad z_B = \frac{x_B - 350}{50}, \qquad z_C = \frac{x_C - 50}{10}, \qquad z_D = \frac{x_D - 10}{2},$$

where x_A, x_B, x_C, and x_D are the raw variables. ■

Table 9.10 Detailed ANOVA for Chemical Process Yield Data

Source	SS	d.f.	MS	F	p
A	13.5800	2	6.7900	156.81	0.000
A linear	13.5150	1	13.5150	312.12	0.000
A quadratic	0.0606	1	0.0606	1.400	0.256
B	20.1800	2	10.0900	233.03	0.000
B linear	19.2262	1	19.2262	444.02	0.000
B quadratic	0.9624	1	0.9624	22.23	0.000
C	41.6467	2	20.8233	480.90	0.000
C linear	39.6139	1	39.6139	914.88	0.000
C quadratic	2.0428	1	2.0428	47.18	0.000
D	20.2067	2	10.1033	233.33	0.000
D linear	10.5892	1	10.5892	244.55	0.000
D quadratic	9.6317	1	9.6317	222.44	0.000
AB	12.7867	4	3.1967	69.74	0.000
AB linear–linear	12.6075	1	12.6075	299.17	0.000
AB linear–quadratic	0.3624	1	0.3624	8.37	0.012
AB quadratic–linear	0.0096	1	0.0096	0.22	0.646
AB quadratic–quadratic	0.0140	1	0.0140	0.32	0.581
Error	0.6067	14	0.0433		
Total	109.0067	26			

9.3 MIXED-LEVEL FACTORIAL DESIGNS

Thus far we have studied factorial designs in which all factors have the same number of levels, either two or three. Many applications involve factors with different numbers of levels, usually two, three, or four. Such designs are called **mixed-level factorial** or **asymmetrical factorial** designs.

A simple way to construct such designs is by crossing full or fractional factorials for each level. However, such designs tend to be too large. Therefore it is of practical interest to accommodate mixed-level factors in smaller orthogonal arrays. In this section we review some methods for constructing such designs and methods of their analysis.

Before proceeding further it is useful to define **mixed-level orthogonal arrays**. For simplicity, we shall limit to arrays with factors (columns) having two distinct numbers of levels (symbols). An orthogonal array of N runs with p factors having r levels and q factors having s levels is an $N \times (p+q)$ array such that in any subset of $t = u + v \geq 2$ columns with $u \leq p$, $v \leq q$, all possible combinations of levels (namely, $r^u s^v$) occur the same number of times. We denote this orthogonal array by $\text{OA}(N, r^p s^q, t)$, where t is the **strength** of the array. Table 9.12 shows an $\text{OA}(8, 2^4 4^1, 2)$.

9.3.1 $2^p 4^q$ Designs

To construct an orthogonal array consisting of p two-level factors and q four-level factors, we begin with the saturated model matrix of a 2^r factorial design [or more generally an OA(2^r, 2^{r-1}, 2)]. As we have seen in Chapter 8, this model matrix has $2^r - 1$ columns of ±'s (which we replace with 0's and 1's in our new notation) corresponding to the $2^r - 1$ main effects and interactions. Then we proceed as follows:

(a) Take a pair of columns and replace the four combinations of their values, (0, 0), (1, 0), (0, 1), (1, 1), by the levels 0, 1, 2, 3 of a four-level factor using a one-to-one mapping, for example, $(0, 0) \to 0$, $(1, 0) \to 1$, $(0, 1) \to 2$, $(1, 1) \to 3$. The generalized interaction of the effects corresponding to the two columns will be confounded with the main effect of the four-level factor. Hence delete that column since it cannot be assigned to any other factor.

(b) Repeat this step a total of q times with mutually exclusive sets of triples of columns (consisting of a pair of columns with the third column being their generalized interaction) to accommodate q four-level factors. The remaining columns can be assigned to p two-level factors so that $p + 3q = 2^r - 1$.

This method of construction is called the **method of replacement**.

Example 9.4 (Construction of OA($8, 2^4 4^1, 2$))

Suppose we want to design an experiment with eight runs for four two-level factors, A, B, C, D, and one four-level factor, E. We can begin with any OA($8, 2^7, 2$), for example, the PB design in Table 8.6. Here we use Taguchi's $L_8(2^7)$ orthogonal array shown in Table 9.11. This choice and the factor assignments to the columns given below are made since the resulting design is used

Table 9.11 Orthogonal Array $L_8(2^7)$

A'	B'	C'	$-A'B'$	$-A'C'$	$-B'C'$	$A'B'C'$
0	0	0	0	0	0	0
1	0	0	1	1	0	1
0	1	0	1	0	1	1
1	1	0	0	1	1	0
0	0	1	0	1	1	1
1	0	1	1	0	1	0
0	1	1	1	1	0	0
1	1	1	0	0	0	1

later in Example 9.5. The columns in this array are labeled with pseudofactors A', B', C' and their interactions, so that when the actual factors A to E are assigned to these columns, their alias relationships can be easily derived. We assign the factors as follows:

$$A \to A', \qquad B \to -A'C', \qquad C \to -A'B', \qquad D \to A'B'C',$$

and

$$E \to (B', C') \quad \text{with } E = 0, 1, 2, 3 \Longleftrightarrow (B', C') = (0, 0), (1, 0), (0, 1), (1, 1).$$

The resulting array is shown in Table 9.12. In this array, for any pair of columns among A, B, C, D, each pair of symbols, (0, 0), (1, 0), (0, 1), (1, 1), occurs twice. Also for any pair of columns (A, E), (B, E), (C, E), (D, E), each pair of symbols, (0, 0), (0, 1), (0, 2), (0, 3), (1, 0), (1, 1), (1, 2), (1, 3), occurs once. Hence this OA has strength $t = 2$.

Finally note that this OA cannot accommodate a second four-level factor because there are no two disjoint triples of columns (S, T, ST) and (U, V, UV) in this array. For example, having assigned E to columns $(B', C', B'C')$, if we have another four-level factor, F, then there are no two other columns to which we can assign F such that their generalized interaction is not in the set $(B', C', B'C')$. For example, suppose we assign F to $(A', A'C')$; then their generalized interaction is C', which is in the set $(B', C', B'C')$. ■

If a four-level factor, say A, which has three d.f., is quantitative, then its sum of squares can be decomposed into mutually orthogonal linear, quadratic, and cubic components using the orthogonal contrasts from Table 3.7. However, a better system of orthogonal contrasts according to Wu and Hamada (2000, p. 265) is

$$A_1 = (-1, -1, 1, 1), \qquad A_2 = (1, -1, -1, 1), \qquad A_3 = (-1, 1, -1, 1). \tag{9.13}$$

Table 9.12 Orthogonal Array OA($8, 2^4 4^1, 2$)

A	B	C	D	E
0	0	0	0	0
1	1	1	1	0
0	0	1	1	1
1	1	0	0	1
0	1	0	1	2
1	0	1	0	2
0	1	1	0	3
1	0	0	1	3

This system has the following advantages:

(a) Contrasts A_1, A_2, and A_3 are highly correlated with the linear, quadratic, and cubic contrasts for $a = 4$ in Table 3.7 (of course, A_2 and the quadratic contrast are identical). Thus, they approximately represent the corresponding effects if A is a quantitative factor.

(b) If A is a qualitative factor, then these contrasts are more meaningful; for example, A_1 compares the average response of levels 3 and 4 with the average response of levels 1 and 2. On the other hand, the linear contrast is meaningless in this case.

(c) These contrasts correspond to the three columns of the two-level OA (one of which is the interaction of the other two columns) to which the four-level factor A is assigned. Therefore the contrasts and the sums of squares for A_1, A_2, and A_3 can be computed by treating each component as a two-level pseudofactor assigned to the respective column of the two-level OA. These three sums of squares, each with one d.f., add up to the sum of squares with three d.f. for factor A.

Example 9.5 (Clutch Rust Inhibition Study)

This example is taken from a study reported by Roy (2006) about a clutch plate production process. Many clutch plates delivered to customers were found to be stuck to each other requiring application of force to separate them; also some had rust spots on the surface. The study was undertaken to minimize these quality problems. Five factors were considered to be critical in the rust inhibition process, four of which had two levels and the fifth had four levels. The factors and their levels are listed in Table 9.13.

Eight test conditions were planned and the design shown in Table 9.12 was used. To derive the alias structure of this design we partition the four-level factor E into its three orthogonal components, E_1, E_2, E_3, defined according to Eq. (9.13). Then the factors are assigned as follows:

$$A \rightarrow A', \qquad B \rightarrow -A'C', \qquad C \rightarrow -A'B', \qquad D \rightarrow A'B'C',$$

Table 9.13 Factors and Their Levels for Clutch Rust Inhibition Study

Factor	Level 1	Level 2	Level 3	Level 4
A: Cure time in furnace	Short (−)	Delayed (+)		
B: Time from deburr to furnace	Short (−)	Standard (+)		
C: Rust inhibitor load rate	Slow (−)	Fast (+)		
D: Rust inhibitor load method	Spindle (−)	Head (+)		
E: Chemical concentration	50% (0)	75% (1)	100% (2)	125% (3)

Source: Roy (2006). Reprinted by permission of Dr. Ranjit Roy.

Table 9.14 Contrast Vectors for E_1, E_2, and E_3

E	E_1	E_2	E_3
0	0	1	0
0	0	1	0
1	0	0	1
1	0	0	1
2	1	0	0
2	1	0	0
3	1	1	1
3	1	1	1

and

$$E = (E_1, E_2, E_3) \to (C', B'C', B').$$

To see the last equivalence, write the three contrasts E_1, E_2, E_3 as they are applied to the four levels of E as shown in Table 9.14. For example, the contrast coefficient of E_1 is -1 when $E = 0$ or 1 and is $+1$ when $E = 2$ or 3. By comparing the columns E_1, E_2, E_3 in this table with the columns C', $-B'C'$, B' in Table 9.11 and recalling that 0 in that table corresponds to $-$ and 1 corresponds to $+$, we see that $E_1 \to C'$, $E_2 \to B'C'$, and $E_3 \to B'$.

From these assignments we can derive the following alias relations:

$$A = -BE_1 = DE_2 = -CE_3, \qquad B = -AE_1 = CE_2 = -DE_3,$$

$$C = -DE_1 = BE_2 = -AE_3, \qquad D = -CE_1 = AE_2 = -BE_3,$$

$$E_1 = -AB = -CD, \qquad E_2 = BC = AD, \qquad E_3 = -AC = -BD.$$

Thus, this is a resolution III design, and if we ignore two-factor interactions, then we can fit the main-effects model, decomposing the main effect of E into three single d.f. components, E_1, E_2, and E_3.

The data were collected as follows. For each test condition three batches of clutch plates were fabricated. Ten plates were randomly selected from each batch and their stickiness and rust properties were evaluated. The stickiness property was measured by the force necessary to separate two parts stuck together (maximum force necessary 2 lb). The amount of rust was subjectively judged on a scale of 0 (best) to 10 (worst) by comparing with a reference scale. The percentages of each property with respect to the respective maximums were weighted by 0.7 for stickiness and 0.3 for rust to arrive at an overall evaluation criterion (OEC). The average of the OEC readings for the 10 plates constituted a single reading. Three such samples were taken from three batches under each test condition resulting in three replicate readings. The design and data are shown in Table 9.15.

The details of the main-effects model fitted are shown in Display 9.6. A pure-error estimate with 16 d.f. is computed from the three replicates at each test condition. We see that all five main effects except E_2 are significant at the 5%

Table 9.15 Design and OEC Readings for Clutch Rust Inhibition Study

	Factor					Replicate		
Test Condition	*A*	*B*	*C*	*D*	*E*	1	2	3
1	0	0	0	0	0	44.6	43.4	43.2
2	1	1	1	1	0	57.4	53.0	56.0
3	0	0	1	1	1	38.5	47.1	30.6
4	1	1	0	0	1	47.4	41.3	47.1
5	0	1	0	1	2	30.2	35.6	39.8
6	1	0	1	0	2	51.0	52.6	45.8
7	0	1	1	0	3	44.4	38.6	38.4
8	1	0	0	1	3	24.4	31.8	23.2

Source: Roy (2006, Figure 3). Reprinted by permission of Dr. Ranjit Roy.

```
Term        Coef  SE Coef      T      P
Constant  41.8917   0.8975  46.68  0.000
A         -2.3583   0.8975  -2.63  0.018
B         -2.2083   0.8975  -2.46  0.026
C         -4.2250   0.8975  -4.71  0.000
D          2.9250   0.8975   3.26  0.005
E1         3.9083   0.8975   4.35  0.000
E2         0.3583   0.8975   0.40  0.695
E3         4.1583   0.8975   4.63  0.000

Analysis of Variance for OEC, using Adjusted SS for Tests

Source  DF   Seq SS   Adj SS   Adj MS      F      P
A        1   133.48   133.48   133.48   6.90  0.018
B        1   117.04   117.04   117.04   6.05  0.026
C        1   428.41   428.41   428.41  22.16  0.000
D        1   205.34   205.34   205.34  10.62  0.005
E1       1   366.60   366.60   366.60  18.96  0.000
E2       1     3.08     3.08     3.08   0.16  0.695
E3       1   415.00   415.00   415.00  21.47  0.000
Error   16   309.32   309.32    19.33
Total   23  1978.28

S = 4.39687   R-Sq = 84.36%   R-Sq(adj) = 77.52%
```

Display 9.6 Minitab output for effect estimates and ANOVA for clutch rust inhibition study data.

level. Of course, it is possible that the two-factor interactions aliased with the main effects could be causing significance. A full foldover must be run in order to resolve this ambiguity. Assuming that two-factor interactions are negligible, the prediction model is

$$\widehat{\text{OEC}} = 41.892 - 2.358A - 2.208B - 4.225C + 2.925D + 3.908E_1 + 4.158E_3.$$

Since the goal is to minimize OEC, the levels of the factors should be set as follows: A high, B high, C high, D low, and $E = 0$ (which corresponds to $(E_1, E_3) = (-1, -1)$ as seen from Table 9.14). The resulting minimum value of OEC is 22.110. ■

9.3.2 2^p3^q Designs

First we will consider the simplest type of a 2^p3^q design formed by crossing a fractional or a full factorial design in two-level factors with one in three-level factors. As a specific example, consider a crossed array formed by a 2^{3-1} design as the inner array and a 3^{3-1} design as the outer array with the two arrays as shown below:

$$\begin{array}{ccc} A & B & C \end{array} \\ \begin{bmatrix} 0 & 0 & 0 \\ 1 & 0 & 0 \\ 0 & 1 & 0 \\ 1 & 1 & 1 \end{bmatrix} \quad \text{and} \quad \begin{array}{ccc} D & E & F \end{array} \\ \begin{bmatrix} 0 & 0 & 1 \\ 1 & 0 & 2 \\ 2 & 0 & 0 \\ 0 & 1 & 2 \\ 1 & 1 & 0 \\ 2 & 1 & 1 \\ 0 & 2 & 0 \\ 1 & 2 & 1 \\ 2 & 2 & 2 \end{bmatrix}.$$

The resulting crossed-array design is shown in Table 9.16.

It is easy to check that this design is an orthogonal array of strength 2. For every pair of columns A, B, C, each of the four pairs of symbols occurs nine times. For every pair of columns D, E, F, each of the nine pairs of symbols occurs four times. Finally, for every pair of columns with one column from A, B, C and the other column from D, E, F, each of the six pairs of symbols occurs six times.

The 2^{3-1} design has the defining relation $I = ABC$ and the 3^{3-1} design has the defining relation $I = DEF^2$. From these two, we obtain the following two additional defining relations:

$$I = (ABC)(DEF^2) = ABCDEF^2 \qquad \text{and} \qquad I = (ABC)(DEF^2)^2 = ABCD^2E^2F,$$

whence it follows that the alias relations among the two- and three-level factors are unaffected by the crossing of the two arrays, for example, $A = BC$ and $D = EF^2$. Also, any two-factor interaction involving a two-level factor with a three-level factor such as AD is clear. However, any interactions within the sets of two- and three-level factors are aliased with the main effects.

Example 9.6 (Paint Thickness Experiment)

Hale-Bennett and Lin (1997) describe a designed experiment to improve the paint process for a charcoal grill. The current process resulted in paint thickness being above the nominal level and having too much within- and between-parts variation. A total of 12 paint guns were used to spray powder paint on two parts at a time, one hung on the top and the other on the bottom hook of a rack. The parts were then baked in an oven. The factors studied were positions of guns 1,

Table 9.16 Design Matrix and Side Paint Thickness Data for Top Part

	Gun 1	Gun 8	Gun 9	Pressure	Gun 7	Gun 11	Thickness (mils)	
Run	*A*	*B*	*C*	*D*	*E*	*F*	Measurement 1	Measurement 2
1	0	0	1	0	0	1	1.10	1.20
2	1	0	0	0	0	1	1.00	1.10
3	0	1	0	0	0	1	0.90	1.00
4	1	1	1	0	0	1	0.90	0.90
5	0	0	1	1	0	2	1.60	1.70
6	1	0	0	1	0	2	1.90	2.00
7	0	1	0	1	0	2	1.70	1.80
8	1	1	1	1	0	2	1.70	1.90
9	0	0	1	2	0	0	1.90	2.00
10	1	0	0	2	0	0	1.40	1.80
11	0	1	0	2	0	0	1.80	1.80
12	1	1	1	2	0	0	2.60	2.70
13	0	0	1	0	1	2	0.90	1.30
14	1	0	0	0	1	2	1.20	1.40
15	0	1	0	0	1	2	1.10	1.30
16	1	1	1	0	1	2	1.30	1.30
17	0	0	1	1	1	0	2.10	2.10
18	1	0	0	1	1	0	1.45	1.55
19	0	1	0	1	1	0	1.40	1.50
20	1	1	1	1	1	0	1.90	2.10
21	0	0	1	2	1	1	2.30	2.70
22	1	0	0	2	1	1	1.90	2.20
23	0	1	0	2	1	1	1.90	1.90
24	1	1	1	2	1	1	1.60	1.60
25	0	0	1	0	2	0	1.10	1.40
26	1	0	0	0	2	0	0.80	0.90
27	0	1	0	0	2	0	0.80	1.10
28	1	1	1	0	2	0	1.10	1.20
29	0	0	1	1	2	1	2.00	2.20
30	1	0	0	1	2	1	1.20	1.30
31	0	1	0	1	2	1	1.90	2.70
32	1	1	1	1	2	1	1.90	2.20
33	0	0	1	2	2	2	1.80	2.20
34	1	0	0	2	2	2	1.80	2.20
35	0	1	0	2	2	2	2.80	2.80
36	1	1	1	2	2	2	1.90	2.10

Source: Data adapted from Hale-Bennett and Lin (1997). Reprinted by permission of Taylor & Francis.

8, and 9 (labeled as factors A, B, and C, respectively) which were set at two levels each, powder pressure, and positions of guns 7 and 11 (labeled as factors D, E, and F, respectively) which were set at three levels each. A crossed-array $2^{3-1} \times 3^{3-1}$ design with 36 runs shown in Table 9.16 was used.

For the purpose of this example, we will focus on the paint thickness on the side (nominal value is 1.5 mils) of parts hung on the top hook (labeled as "top-side thickness"). Hale-Bennett and Lin (1997) report the mean and range of two measurements from which the individual measurements can be deduced. The design matrix and the data are shown in Table 9.16.

We will first fit a model with six main effects and nine pairwise interactions between the three two-level factors and three three-level factors. (Recall that interactions within each subset of two- and three-level factors are aliased with the main effects.) The Minitab ANOVA output for this fitted model is shown in Display 9.7. From this output we see that, at the 5% level, three main effects, A (gun 1 position), C (gun 9 position), and D (powder pressure), and four interactions, AE (gun 1 position $\times$ gun 7 position), AF (gun 1 position $\times$ gun 11 position), BE (gun 8 position $\times$ gun 7 position), and CF (gun 9 position $\times$ gun 11 position) are significant. Powder pressure does not interact with any other factor, but it has the biggest main effect by far.

We decompose the significant three-level factor main effects into their linear and quadratic components, for example, the D main effect into D_L and D_Q components. Similarly, we decompose the significant interactions between two- and three-level factors into their components, for example, the AE interaction into AE_L and AE_Q components. This is done by replacing the column for the levels

Source	DF	Seq SS	Adj SS	Adj MS	F	P
A	1	0.22222	0.22222	0.22222	4.52	0.039
B	1	0.09389	0.09389	0.09389	1.91	0.174
C	1	0.40500	0.40500	0.40500	8.23	0.006
D	2	12.18028	12.18028	6.09014	123.77	0.000
E	2	0.16444	0.16444	0.08222	1.67	0.200
F	2	0.19194	0.19194	0.09597	1.95	0.154
A*D	2	0.10528	0.10528	0.05264	1.07	0.352
A*E	2	0.70778	0.70778	0.35389	7.19	0.002
A*F	2	0.46528	0.46528	0.23264	4.73	0.014
B*D	2	0.11694	0.11694	0.05847	1.19	0.314
B*E	2	0.75111	0.75111	0.37556	7.63	0.001
B*F	2	0.17694	0.17694	0.08847	1.80	0.178
C*D	2	0.09083	0.09083	0.04542	0.92	0.405
C*E	2	0.04333	0.04333	0.02167	0.44	0.647
C*F	2	1.33583	1.33583	0.66792	13.57	0.000
Error	44	2.16500	2.16500	0.04920		
Total	71	19.21611				

S = 0.221821 R-Sq = 88.73% R-Sq(adj) = 81.82%

Display 9.7 Minitab ANOVA output for top side paint thickness data.

```
The regression equation is
Top_Side_Thickness = 1.66 - 0.0556 A + 0.0750 C + 0.483 DL - 0.0819 DQ
                     - 0.121 A*EL - 0.0069 A*EQ - 0.0208 A*FL
                     + 0.0556 A*FQ + 0.0583 B*EL + 0.0639 B*EQ
                     - 0.167 C*FL + 0.0042 C*FQ

Predictor       Coef   SE Coef      T      P
Constant     1.66111   0.02722  61.01  0.000
A           -0.05556   0.02722  -2.04  0.046
C            0.07500   0.02722   2.75  0.008
DL           0.48333   0.03334  14.50  0.000
DQ          -0.08194   0.01925  -4.26  0.000
A*EL        -0.12083   0.03334  -3.62  0.001
A*EQ        -0.00694   0.01925  -0.36  0.720
A*FL        -0.02083   0.03334  -0.62  0.535
A*FQ         0.05556   0.01925   2.89  0.005
B*EL         0.05833   0.03334   1.75  0.085
B*EQ         0.06389   0.01925   3.32  0.002
C*FL        -0.16667   0.03334  -5.00  0.000
C*FQ         0.00417   0.01925   0.22  0.829

S = 0.231011   R-Sq = 83.6%   R-Sq(adj) = 80.3%
```

Display 9.8 Minitab regression output for top side paint thickness data.

of each three-level factor by two columns—one corresponding to the contrast coefficients for the linear effect and the other corresponding to the contrast coefficients for the quadratic effect. The result of multiple regression using all these single d.f. terms is shown in Display 9.8. Keeping in mind that the terms in the regression model are mutually orthogonal, if we drop the nonsignificant terms (namely, AE_Q, AF_L, BE_L, and CF_Q), the coefficients of the remaining terms would remain unchanged. Therefore the final model is

$$\widehat{\text{Thickness}} = 1.66 - 0.0556A + 0.0750C + 0.483D_L - 0.0819D_Q$$
$$-0.121AE_L + 0.0556AF_Q + 0.0639BE_Q - 0.167CF_L.$$

This prediction equation can be used to find the settings of the significant factors in order to maintain the thickness at the nominal value of 1.5 mils. For this purpose the key factor is D (powder pressure). The main-effect plot for D is shown in Figure 9.6. From this plot we see that the coded level of D should be set approximately at -0.5. ■

The following example uses a mixed-level design with two- and three-level factors that is not a crossed array.

Example 9.7 (Radio Frequency Choke Experiment)

Xu, Cheng, and Wu (2004) present data from an experiment to study effects of one two-level factor (A) and seven three-level factors (B to H) on the performance of

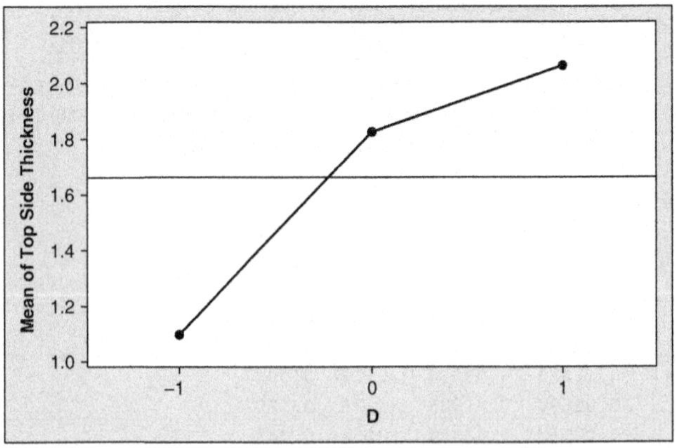

Figure 9.6 Main effect plot for factor D (powder pressure) for paint thickness data.

a radio frequency choke, which is a circuit element that presents high impedance to radio frequency energy while offering little resistance to direct current. An 18-run OA was used with two replicates per run. The design and data are shown in Table 9.17. In this design every level of the two-level factor A occurs with every level of the three-level factors, B to H, three times, whereas every level of any three-level factor occurs with every level of any other three-level factor twice.

According to the two-step analysis strategy suggested by Cheng and Wu (2001), we first fit a main-effects model to these data. The result is shown in Display 9.9. From this output we see that only the main effects B, E, G, and H are significant at the 5% level. In the second step of the analysis strategy, a second-order model is fitted in these four significant factors. The second-order model includes the linear and quadratic components of the four factors and linear-by-linear interactions among them. The regression output for this model is shown in Display 9.10. Note that this model has $R^2 = 97.5\%$.

We may ignore the nonsignificant terms without affecting the coefficients of the remaining terms because of the mutual orthogonality of the terms. Thus, the following equation may be used for prediction purposes:

$$\begin{aligned}\widehat{\text{Response}} = {} & 103 + 2.80B_L - 3.94E_L - 0.705E_Q - 7.81G_L + 3.24H_L \\ & + 2.01B_L \times E_L - 1.24B_L \times H_L - 3.34E_L \times G_L \\ & - 1.79E_L \times H_L + 2.67G_L \times H_L.\end{aligned}$$ ■

The third type of $2^p 3^q$ designs is a special case of orthogonal main-effect (OME) designs proposed by Addelman (1962). The idea of OME designs is based on the following theorem, which is proved in John (1998, pp. 69–70).

Theorem 9.1 *Consider two factors, A and B, with levels a and b, respectively. Denote by n_{ij} the number of observations at the ith level of A and the jth level of B ($1 \le i \le a, 1 \le j \le b$). Let $n_{i\cdot} = \sum_{j=1}^{b} n_{ij}, n_{\cdot j} = \sum_{i=1}^{a} n_{ij}$, and*

Table 9.17 Design Matrix and Response Data for Radio Frequency Choke Experiment

	Factor								Response	
Run	*A*	*B*	*C*	*D*	*E*	*F*	*G*	*H*	Measurement 1	Measurement 2
1	0	0	0	0	0	0	0	0	106.20	107.70
2	0	0	1	1	1	1	1	1	104.20	102.35
3	0	0	2	2	2	2	2	2	85.90	85.90
4	0	1	0	0	1	1	2	2	101.15	104.96
5	0	1	1	1	2	2	0	0	109.92	110.47
6	0	1	2	2	0	0	1	1	108.91	108.91
7	0	2	0	1	0	2	1	2	109.76	112.66
8	0	2	1	2	1	0	2	0	97.20	94.51
9	0	2	2	0	2	1	0	1	112.77	113.03
10	1	0	0	2	2	1	1	0	93.15	92.83
11	1	0	1	0	0	2	2	1	97.25	100.60
12	1	0	2	1	1	0	0	2	109.51	113.28
13	1	1	0	1	2	0	2	1	85.63	86.91
14	1	1	1	2	0	1	0	2	113.17	113.45
15	1	1	2	0	1	2	1	0	104.85	98.87
16	1	2	0	2	1	2	0	1	113.14	113.78
17	1	2	1	0	2	0	1	2	103.19	106.46
18	1	2	2	1	0	1	2	0	95.70	97.93

Source: Xu, Cheng, and Wu (2004, Table 1).

```
Source  DF    Seq SS    Adj SS   Adj MS      F      P
A        1     37.62     37.62    37.62   4.00  0.059
B        2    220.54    220.54   110.27  11.72  0.000
C        2     27.09     27.09    13.55   1.44  0.261
D        2     54.56     54.56    27.28   2.90  0.078
E        2    354.18    354.18   177.09  18.82  0.000
F        2     13.43     13.43     6.71   0.71  0.502
G        2   1720.18   1720.18   860.09  91.39  0.000
H        2    113.98    113.98    56.99   6.06  0.009
Error   20    188.23    188.23     9.41
Total   35   2729.82

S = 3.06784   R-Sq = 93.10%   R-Sq(adj) = 87.93%
```

Display 9.9 Minitab output for the main-effects model fit to the radio frequency choke data.

$N = \sum_{i=1}^{a} \sum_{j=1}^{b} n_{ij}$. *Then the necessary and sufficient condition for the main-effect estimators of A and B to be uncorrelated (orthogonal) with each other is that*

$$n_{ij} = \frac{n_{i\cdot} n_{\cdot j}}{N} \quad \text{for all } i, j. \tag{9.14}$$

```
The regression equation is
Response = 103 + 2.80 BL - 0.111 BQ - 3.94 EL - 0.705 EQ
            - 7.81 GL - 0.637 GQ + 3.24 HL - 0.068 HQ
           + 2.01 BL*EL + 1.27 BL*GL - 1.24 BL*HL - 3.34 EL*GL
           - 1.79 EL*HL + 2.67 GL*HL

Predictor      Coef   SE Coef        T      P
Constant    103.228     0.302   342.12  0.000
BL           2.8007    0.4814     5.82  0.000
BQ          -0.1112    0.3112    -0.36  0.725
EL          -3.9372    0.4243    -9.28  0.000
EQ          -0.7049    0.2684    -2.63  0.016
GL          -7.8103    0.4500   -17.36  0.000
GQ          -0.6366    0.3564    -1.79  0.088
HL           3.2390    0.4519     7.17  0.000
HQ          -0.0683    0.3047    -0.22  0.825
BL*EL        2.0072    0.8003     2.51  0.020
BL*GL        1.2683    0.7700     1.65  0.114
BL*HL        -1.238     1.130    -1.10  0.285
EL*GL       -3.3444    0.6634    -5.04  0.000
EL*HL       -1.7866    0.8557    -2.09  0.049
GL*HL        2.6705    0.8397     3.18  0.005

S = 1.81038   R-Sq = 97.5%   R-Sq(adj) = 95.8%

Analysis of Variance

Source           DF       SS       MS      F      P
Regression       14  2660.99   190.07  57.99  0.000
Residual Error   21    68.83     3.28
Total            35  2729.82
```

Display 9.10 Minitab regression output for the radio frequency choke data.

This is called the **proportional-balance condition**. An array that satisfies this condition is called an **orthogonal main-effects (OME) array**. OAs of strength 2 obviously satisfy this condition, but they satisfy a stronger condition that all n_{ij} are in fact equal for any given pair of columns (which implies that all $n_{i\cdot}$'s are equal and all $n_{\cdot j}$'s are equal). Thus, OMEs are a generalization of OAs, but with a weaker property that orthogonality holds only among the main-effect estimates.

Next we show how to construct OME designs using the **method of collapsing factors** of Addelman (1962) [referred to as the **dummy-level technique** by Phadke (1989, Section 7.8)]. Suppose we want to design an experiment of $N = 9$ runs with two-level factors A and B and three-level factors C and D. We begin with the OA(9, 3^4, 2) shown in Display 9.3.

First, C and D can be assigned to columns 3 and 4. To assign A and B to columns 1 and 2 while maintaining the proportional-balance condition (9.14),

$$\begin{array}{cccc} A & B & C & D \end{array}$$
$$\begin{bmatrix} 0 & 0 & 0 & 0 \\ 0 & 1 & 1 & 2 \\ 0 & 1 & 2 & 1 \\ 1 & 0 & 1 & 1 \\ 1 & 1 & 2 & 0 \\ 1 & 1 & 0 & 2 \\ 1 & 0 & 2 & 2 \\ 1 & 1 & 0 & 1 \\ 1 & 1 & 1 & 0 \end{bmatrix}.$$

Display 9.11 A $2^{2}3^{2}$ design.

the three levels of each column are collapsed to two levels by merging any two levels into one using a mapping such as $0 \to 0, 1 \to 1, 2 \to 1$.

The design that results by using the same mapping, $0 \to 0, 1 \to 1, 2 \to 1$ for factors A and B, is shown in Display 9.11. It is easy to check that this design satisfies the proportional-balance condition (9.14). For example, for factors A and B, we have $n_{00} = 1$, which satisfies $n_{00} = n_{0\cdot}n_{\cdot 0}/N$ since $n_{0\cdot} = n_{\cdot 0} = 3$. Similarly, for factors A and C, we have $n_{11} = 2$, which satisfies $n_{11} = n_{1\cdot}n_{\cdot 1}/N$ since $n_{1\cdot} = 6$ and $n_{\cdot 1} = 3$.

The following are some of the extensions, generalizations, and caveats of this method.

(a) Different mappings can be used for different columns. For instance, in the above example, we could use the mapping $0 \to 0, 1 \to 0, 2 \to 1$ for factor B.

(b) Any such mapping results in twice as many observations on one level of a factor compared to the other level. Which level to replicate more depends on the ease and cost of making observations and statistical requirements such as the desired precision of the estimated mean response at each level.

(c) Because of unequal numbers of the replications on the levels of two-level factors, there is some loss of efficiency in estimating their main effects. Suppose for a given factor level 0 is replicated n times and level 1 is replicated $2n$ times for a total of $N = 3n$ runs. Denote the estimated main effect of the factor by $\overline{y}_1 - \overline{y}_0$. Assuming the usual homoscedastic model with a common error variance σ^2, we have

$$\operatorname{Var}(\overline{y}_1 - \overline{y}_0) = \sigma^2\left(\frac{1}{2n} + \frac{1}{n}\right) = \frac{3\sigma^2}{2n}.$$

On the other hand, if an equal number of runs, $N/2 = 3n/2$ (assuming N is even, e.g., 18), were made on each level, then $\operatorname{Var}(\overline{y}_1 - \overline{y}_0) = 4\sigma^2/3n$, which is 11% smaller.

(d) The method of collapsing factors can be generalized to convert any s-level column to an r-level column for $r < s$. If the original array is orthogonal, then the resulting array maintains the proportional-balance property and hence the orthogonality of the main-effect estimators.

9.4 CHAPTER NOTES

9.4.1 Alternative Derivations of Estimators of Linear and Quadratic Effects

First consider the linear component of the A main effect, which is proportional to the sum of the three linear effects of A conditional on the levels of B, that is,

$$\begin{aligned}\widehat{A}_L &\propto (\widehat{A}_L|B=0) + (\widehat{A}_L|B=1) + (\widehat{A}_L|B=2)\\ &= (\bar{y}_{20} - \bar{y}_{00}) + (\bar{y}_{21} - \bar{y}_{01}) + (\bar{y}_{22} - \bar{y}_{02}).\end{aligned}$$

Then $\widehat{A}_L$ equals this contrast divided by its normalizing factor, $\sqrt{6}$, which makes the contrast vector to have unit length. Similarly, the quadratic component of the A main effect is given by

$$\begin{aligned}\widehat{A}_Q &\propto (\widehat{A}_Q|B=0) + (\widehat{A}_Q|B=1) + (\widehat{A}_Q|B=2)\\ &= (\bar{y}_{20} - 2\bar{y}_{10} + \bar{y}_{00}) + (\bar{y}_{21} - 2\bar{y}_{11} + \bar{y}_{01}) + (\bar{y}_{22} - 2\bar{y}_{12} + \bar{y}_{02})\end{aligned}$$

with the normalizing factor $\sqrt{18} = 3\sqrt{2}$. The linear and quadratic components of B are defined in an analogous manner:

$$\begin{aligned}\widehat{B}_L &\propto (\widehat{B}_L|A=0) + (\widehat{B}_L|A=1) + (\widehat{B}_L|A=2)\\ &= (\bar{y}_{02} - \bar{y}_{00}) + (\bar{y}_{12} - \bar{y}_{10}) + (\bar{y}_{22} - \bar{y}_{20})\end{aligned}$$

with the normalizing factor $\sqrt{6}$ and

$$\begin{aligned}\widehat{B}_Q &\propto (\widehat{B}_Q|A=0) + (\widehat{B}_Q|A=1) + (\widehat{B}_Q|A=2)\\ &= (\bar{y}_{02} - 2\bar{y}_{01} + \bar{y}_{00}) + (\bar{y}_{12} - 2\bar{y}_{11} + \bar{y}_{10}) + (\bar{y}_{22} - 2\bar{y}_{21} + \bar{y}_{20})\end{aligned}$$

with the normalizing factor $\sqrt{18} = 3\sqrt{2}$.

The orthogonal components of interactions are a bit harder to interpret. First consider the LL interaction effect, which is given by

$$\begin{aligned}\widehat{AB}_{LL} &\propto [(\widehat{A}_L|B=2) - (\widehat{A}_L|B=1)] + [(\widehat{A}_L|B=1) - (\widehat{A}_L|B=0)]\\ &= (\widehat{A}_L|B=2) - (\widehat{A}_L|B=0)\\ &= (\bar{y}_{22} - \bar{y}_{02}) - (\bar{y}_{20} - \bar{y}_{00})\end{aligned}$$

with the normalizing factor $\sqrt{4} = 2$. The $\widehat{AB}_{LL}$ interaction effect measures the *trend* in the linear effect of A with respect to the changes in the levels of B. By symmetry in A and B, we also have

$$\begin{aligned}\widehat{AB}_{LL} &\propto (\widehat{B}_L|A = 2) - (\widehat{B}_L|A = 0)\\ &= (\overline{y}_{22} - \overline{y}_{20}) - (\overline{y}_{02} - \overline{y}_{00})\end{aligned}$$

with the normalizing factor $\sqrt{4} = 2$.
The LQ interaction effect is given by

$$\begin{aligned}\widehat{AB}_{LQ} &\propto (\widehat{A}_L|B = 2) - 2(\widehat{A}_L|B = 1) + (\widehat{A}_L|B = 0)\\ &= (\overline{y}_{22} - \overline{y}_{02}) - 2(\overline{y}_{21} - \overline{y}_{01}) + (\overline{y}_{20} - \overline{y}_{00})\\ &= (\overline{y}_{22} - 2\overline{y}_{21} + \overline{y}_{20}) - (\overline{y}_{02} - 2\overline{y}_{01} + \overline{y}_{00})\\ &= (\widehat{B}_Q|A = 2) - (\widehat{B}_Q|A = 0)\end{aligned}$$

with the normalizing factor $\sqrt{12} = 2\sqrt{3}$. From the above equation we see that $\widehat{AB}_{LQ}$ measures the *curvature* in the linear effect of A with respect to the changes in the levels of B or equivalently the *trend* in the quadratic effect of B with respect to the changes in the levels of A. By analogy, the QL interaction effect is given by

$$\widehat{AB}_{QL} \propto (\overline{y}_{22} - 2\overline{y}_{12} + \overline{y}_{02}) - (\overline{y}_{20} - 2\overline{y}_{10} + \overline{y}_{00})$$

with the normalizing factor $\sqrt{12} = 2\sqrt{3}$.

Finally, the QQ interaction effect is given by

$$\begin{aligned}\widehat{AB}_{QQ} &\propto (\widehat{A}_Q|B = 2) - 2(\widehat{A}_Q|B = 1) + (\widehat{A}_Q|B = 0)\\ &= (\overline{y}_{22} - 2\overline{y}_{12} + \overline{y}_{02}) - 2(\overline{y}_{21} - 2\overline{y}_{11} + \overline{y}_{01}) + (\overline{y}_{20} - 2\overline{y}_{10} + \overline{y}_{00})\\ &= (\overline{y}_{22} - 2\overline{y}_{21} + \overline{y}_{20}) - 2(\overline{y}_{12} - 2\overline{y}_{11} + \overline{y}_{10}) + (\overline{y}_{02} - 2\overline{y}_{01} + \overline{y}_{00})\\ &= (\widehat{B}_Q|A = 2) - 2(\widehat{B}_Q|A = 1) + (\widehat{B}_Q|A = 0)\end{aligned}$$

with the normalizing factor $\sqrt{36} = 6$. This is a quadratic contrast among the quadratic effects of one factor with respect to the levels of the other factor.

The above contrasts are of the form $\sum_{i=0}^{2}\sum_{j=0}^{2} c_{ij}\overline{y}_{ij}$, where the contrast coefficients c_{ij} satisfy $\sum_{i=0}^{2} c_{ij} = 0$ for all j and $\sum_{j=0}^{2} c_{ij} = 0$ for all i. It is convenient to represent the contrast coefficients in the form of a 3×3 contrast coefficient matrix $\boldsymbol{C} = \{c_{ij}\}$, which can be obtained in a straightforward manner by taking products of the vectors $\boldsymbol{\ell}, \boldsymbol{q}$ given by (9.3) and the normalized unit vector $\boldsymbol{u} = (1/\sqrt{3})(1, 1, 1)'$ as given below:

A linear main effect: $$C = \boldsymbol{\ell u'} = \frac{1}{\sqrt{6}} \begin{bmatrix} -1 & -1 & -1 \\ 0 & 0 & 0 \\ 1 & 1 & 1 \end{bmatrix},$$

B linear main effect: $$C = \boldsymbol{u\ell'} = \frac{1}{\sqrt{6}} \begin{bmatrix} -1 & 0 & 1 \\ -1 & 0 & 1 \\ -1 & 0 & 1 \end{bmatrix},$$

A quadratic main effect: $$C = \boldsymbol{qu'} = \frac{1}{3\sqrt{2}} \begin{bmatrix} 1 & 1 & 1 \\ -2 & -2 & -2 \\ 1 & 1 & 1 \end{bmatrix},$$

B quadratic main effect: $$C = \boldsymbol{uq'} = \frac{1}{3\sqrt{2}} \begin{bmatrix} 1 & -2 & 1 \\ 1 & -2 & 1 \\ 1 & -2 & 1 \end{bmatrix},$$

AB linear-by-linear (LL) interaction: $$C = \boldsymbol{\ell\ell'} = \frac{1}{2} \begin{bmatrix} 1 & 0 & -1 \\ 0 & 0 & 0 \\ -1 & 0 & 1 \end{bmatrix},$$

AB linear-by-quadratic (LQ) interaction: $$C = \boldsymbol{\ell q'} = \frac{1}{2\sqrt{3}} \begin{bmatrix} -1 & 2 & -1 \\ 0 & 0 & 0 \\ 1 & -2 & 1 \end{bmatrix},$$

AB quadratic-by-linear (QL) interaction: $$C = \boldsymbol{q\ell'} = \frac{1}{2\sqrt{3}} \begin{bmatrix} -1 & 0 & 1 \\ 2 & 0 & -2 \\ -1 & 0 & 1 \end{bmatrix},$$

AB quadratic-by-quadratic (QQ) interaction: $$C = \boldsymbol{qq'} = \frac{1}{6} \begin{bmatrix} 1 & -2 & 1 \\ -2 & 4 & 2 \\ 1 & -2 & 1 \end{bmatrix}.$$

9.5 CHAPTER SUMMARY

(a) In a 3^2 experiment where both factors are quantitative, each main effect with two d.f. can be partitioned into linear and quadratic components, each with one d.f. The two-factor interaction with four d.f. can be partitioned into four effects: linear–linear, linear–quadratic, quadratic–linear, and quadratic–quadratic, with one d.f. each. These effects are mutually orthogonal. This is called the linear–quadratic system. These ideas can be generalized to more than two factors. There is another system, called the orthogonal components system, of partitioning the sums of squares for the main effects and interactions. But the components in this system do not have simple interpretations.

(b) A 3^{p-1} design requires specification of one generator. A single generator results in two aliases for each effect which can be obtained by multiplying the generator with the effect of interest and its square. A 3^{p-2} design requires specification of two generators. The two generators, say X and Y, result in two additional ones, which are the generalized interactions XY and XY^2. The complete alias structure can be derived by taking the products of the various effects and their squares with the four generators. Thus each effect is aliased with eight other effects. There are $(3^{p-2} - 1)/2$ such sets each consisting of nine aliased effects.

(c) A $2^p 4^q$ design can be constructed from an $\text{OA}(2^r, 2^{r-1}, 2)$ by the method of replacement. A pair of columns with levels $(0, 0)$, $(1, 0)$, $(0, 1)$, $(1, 1)$ is replaced by a single column for a four-level factor with levels 0, 1, 2, 3, respectively. The interaction of the two replaced columns is confounded with the main effect of the four-level factor as a consequence and cannot be assigned to another factor. To assign q four-level factors, one requires q disjoint triples of columns such that each triple consists of two columns and their generalized interaction. The remaining $p = 2^{r-1} - 3q$ columns can be assigned to two-level factors.

(d) A $2^p 3^q$ design can be constructed from an $\text{OA}(N, 3^{p+q}, 2)$ by the method of collapsing factors in which two levels in p columns are merged into a single level to make them into two-level columns. Although this causes the allocation of the runs on the two levels to be unbalanced, the proportional-frequencies condition is still satisfied, which makes the main-effect estimators mutually orthogonal. Such designs are a generalization of orthogonal arrays and are called orthogonal main-effects (OME) arrays.

EXERCISES

Section 9.1 (Three-Level Full-Factorial Designs)

Theoretical Exercise

9.1 For a 4^2 design there are three d.f. corresponding to the linear (L), quadratic (Q), and cubic (C) contrasts for each of the two main effects and nine d.f. for the two-factor interaction.

(a) Give the normalized contrast vectors, denoted by $\boldsymbol{\ell}$, $\boldsymbol{q}$, and $\boldsymbol{c}$, for the linear, quadratic, and cubic contrasts, respectively.

(b) Derive the contrast matrices for LL, QQ, CC, LQ, LC, and QC interaction effects.

Applied Exercise

9.2 In Example 9.1 we used the stem cell growth data by restricting the media factor to osteogenic media. In this exercise you will analyze the data for the other level of the media factor, namely regular media. The data on the variable $y = 1/\sqrt{\text{ALP}}$ are shown in Table 9.18 and the cell means are shown in Table 9.19.

(a) Give the ANOVA table treating the design as a 3×3 layout. Which main effects and interactions are significant at the 0.05 level?

(b) Give a more detailed ANOVA table in which the main effects and the interaction are decomposed into their single d.f. linear–quadratic effects. Identify the single d.f. effects significant at the 0.05 level.

(c) Give the fitted orthogonal polynomial equation retaining only the terms found significant in the previous part.

(d) Summarize the conclusions and compare them with those found in Example 9.1.

Table 9.18 Inverse Square-Root Transformed ALP Data Classified by Growth Factor and Incubation Period for Regular Media

	Incubation Period (B)		
Growth Factor (A)	3 Days	6 Days	9 Days
0 ng/ml	10.256	8.907	8.301
	10.500	10.749	7.736
	10.500	10.018	9.119
15 ng/ml	8.498	7.380	6.115
	10.749	6.562	6.878
	9.119	7.209	7.041
30 ng/ml	7.736	6.261	8.108
	7.041	5.973	6.878
	6.718	6.115	7.736

Table 9.19 Cell Means for Stem Cell Growth Data in Table 9.18 for Regular Media

	Incubation Period (B)		
Growth Factor (A)	3 Days	6 Days	9 Days
0 ng/ml	$\bar{y}_{00} = 10.419$	$\bar{y}_{01} = 9.891$	$\bar{y}_{02} = 8.385$
15 ng/ml	$\bar{y}_{10} = 9.455$	$\bar{y}_{11} = 7.050$	$\bar{y}_{12} = 6.678$
30 ng/ml	$\bar{y}_{20} = 7.165$	$\bar{y}_{21} = 6.116$	$\bar{y}_{22} = 7.574$

Section 9.2 (Three-Level Fractional-Factorial Designs)

Theoretical Exercises

9.3 (a) Write the design matrix of a 3^{4-1} design whose defining relation is $x_4 = (x_1 + x_2 + x_3) \bmod 3$.

(b) Obtain the alias structure of this design.

9.4 (a) How many distinct orthogonal component effects are there in a 3^5 design?

(b) How many aliases does each effect have in a 3^{5-2} design? How many groups of aliased effects are there?

9.5 Consider a 3^{5-2} design with basic generators ABD^2 and AB^2E^2.

(a) Write the complete defining relation of this design and give the design.

(b) Give the aliases of the five main effects.

Applied Exercise

9.6 Roy (1990) gave data from an experiment for optimizing the idle performance of a car engine. Three factors, each at three levels, were studied and performance was measured in terms of the deviation from the nominal idle speed. The factors and their levels are shown in Table 9.20. A 3^{3-1} design [OA(9, 3^3, 2)] was employed. Three replicate measurements were made at each test condition. This being a Taguchi experiment (see Section 10.3), in addition to minimizing the mean deviation from the nominal, it was desired to simultaneously minimize the variability of the deviation by maximizing the signal-to-noise (SN) ratio defined in Equation (10.16). Table 9.21 gives the design used and the data on SN ratios. To avoid negative responses, all SN ratios are multiplied by -1. Thus the response y should be minimized. Raw data from which the SN ratios are computed are not given.

(a) Check that the defining relation of this design is $C = AB^2$.

(b) Calculate the ANOVA table assuming the main-effects model.

Table 9.20 Factors and Their Levels for Engine Stability Experiment

	Level		
Factor	0	1	2
Indexing (A)	-5°	0°	$+5^\circ$
Overlap area (B)	0%	30%	60%
Spark advance (C)	25°	30°	35°

Source: Roy (1990, Table 9-4-1 (a)).

Table 9.21 Design and Data for Engine Stability Experiment

Run	A	B	C	$y = -$(SN) Ratio
1	0	0	0	27.54
2	0	1	1	30.19
3	0	2	2	31.10
4	1	0	2	25.96
5	1	1	0	31.81
6	1	2	1	28.87
7	2	0	1	33.06
8	2	1	2	35.60
9	2	2	0	33.12

Source: Roy (1990, Table 9-4-2(a)).

(c) Decompose each main effect into its orthogonal linear and quadratic components. Is any component significant at $\alpha = 0.10$? Verify the result by making the main-effects plot.

(d) Find the values of the three factor settings that minimize y, that is, maximize the SN ratio.

Section 9.3 (Mixed-Level Factorial Designs)

Theoretical Exercise

9.7 Consider the OA$(16, 2^{15}, 2)$ given by the contrast matrix in Table 7.11 for a 2^4 design without the I column.

(a) What is the maximum number of mutually exclusive triples of columns such that each triple contains two columns and their interaction? List these triples.

(b) Assign three triples to three four-level factors, X, Y, Z and the other six columns to four two-level factors, T, U, V, W, leaving the remaining two columns to estimate the interactions TU and VW. Show the resulting design.

Applied Exercises

9.8 Roy (1990, Example 9-6) discussed an experiment to determine the worst combination of vehicle body style and options to be used as a test specimen in crashworthiness tests. A $2^4 4^1$ experiment was carried out using the same design as in Example 9.5 but with a single replicate observation for each test condition. The factors studied and their levels are shown in Table 9.22. The response variable was a predefined occupant injury index

Table 9.22 Factors and Their Levels for Crashworthiness Study

Factor	Level 1	Level 2	Level 3	Level 4
A: Type of vehicle	Style 1	Style 2		
B: Power train	Light duty	Heavy duty		
C: Roof structure	Hard top	Sun-roof		
D: Seat structure	Standard	Reinforced		
E: Test type	0°F	30°R	30°L	NCAP

Source: Roy (1990, Table 9-6-1(a)).

(OII). The observed values of OII under the eight test conditions are given in Table 9.23.

(a) Compute the effect estimates. Decompose the E main effect into its linear, quadratic, and cubic components.

(b) Why is it not possible to test the significance of the main effects calculated in part (a)? Use a graphical method to determine the most important main effects.

9.9 Table 9.24 gives data on the bottom-part side paint thickness. Analyze these data in the same manner as the top-part side paint thickness data are analyzed in Example 9.6.

9.10 Table 9.24 also gives data on the weight of the paint applied to each part (measured by weighing the part before and after paint application and taking the difference). Identify the factors with significant effects (main effects or two-factor interactions) on the paint weight. Use $\alpha = 0.05$. Which factor settings minimize the paint weight?

Table 9.23 Design and Data for Crashworthiness Study

	Factor					Observed
Test Condition	A	B	C	D	E	OII
1	0	0	0	0	0	45
2	1	1	1	1	0	65
3	0	0	1	1	1	38
4	1	1	0	0	1	48
5	0	1	0	1	2	59
6	1	0	1	0	2	32
7	0	1	1	0	3	36
8	1	0	0	1	3	38

Source: Roy (1990, Table 9-6-2(a)).

Table 9.24 Design Matrix and Side Paint Thickness and Paint Weight Data for Bottom Part

	Gun 1	Gun 8	Gun 9	Pressure	Gun 7	Gun 11	Thickness (mils)		
Run	A	B	C	D	E	F	Measurement 1	Measurement 2	Weight
1	0	0	1	0	0	1	0.900	1.000	0.118
2	1	0	0	0	0	1	0.900	1.000	0.110
3	0	1	0	0	0	1	1.300	1.500	0.108
4	1	1	1	0	0	1	0.900	1.000	0.102
5	0	0	1	1	0	2	1.800	2.000	0.166
6	1	0	0	1	0	2	1.800	1.900	0.145
7	0	1	0	1	0	2	1.600	1.700	0.182
8	1	1	1	1	0	2	1.700	1.900	0.166
9	0	0	1	2	0	0	2.300	2.900	0.242
10	1	0	0	2	0	0	1.400	1.500	0.205
11	0	1	0	2	0	0	2.400	2.500	0.209
12	1	1	1	2	0	0	2.300	2.400	0.220
13	0	0	1	0	1	2	1.000	1.200	0.126
14	1	0	0	0	1	2	1.300	1.400	0.112
15	0	1	0	0	1	2	1.200	1.300	0.124
16	1	1	1	0	1	2	1.000	1.400	0.122
17	0	0	1	1	1	0	1.700	1.800	0.166
18	1	0	0	1	1	0	1.300	1.300	0.176
19	0	1	0	1	1	0	1.500	1.600	0.186
20	1	1	1	1	1	0	2.000	2.100	0.150
21	0	0	1	2	1	1	2.825	2.875	0.236
22	1	0	0	2	1	1	2.200	2.200	0.230
23	0	1	0	2	1	1	2.200	2.200	0.224
24	1	1	1	2	1	1	1.800	2.100	0.220
25	0	0	1	0	2	0	1.000	1.200	0.110
26	1	0	0	0	2	0	1.100	1.100	0.104
27	0	1	0	0	2	0	1.100	1.100	0.105
28	1	1	1	0	2	0	1.000	1.200	0.120
29	0	0	1	1	2	1	1.800	2.100	0.184
30	1	0	0	1	2	1	1.400	1.400	0.184
31	0	1	0	1	2	1	2.325	2.375	0.180
32	1	1	1	1	2	1	2.100	2.200	0.176
33	0	0	1	2	2	2	2.300	2.400	0.220
34	1	0	0	2	2	2	2.100	2.200	0.224
35	0	1	0	2	2	2	2.100	2.300	0.210
36	1	1	1	2	2	2	2.200	2.500	0.226

Source: Hale-Bennett and Lin (1997, Appendix: Data). Reprinted by permission of Taylor & Francis.

CHAPTER 10

Experiments for Response Optimization

The subject of this chapter is design and analysis of experiments for optimizing response. An engineer is often faced with the problem of optimizing a process output or a product property that depends upon several input variables or factors. For example, a chemical engineer may want to maximize the yield of a chemical process by varying the process parameters such as the reaction temperature, pressure, and concentration of the catalyst or minimize the viscosity of a polymer product by varying the composition of its ingredients. In this chapter we study three types of problems in this area. **Response surface methodology (RSM)** was proposed by Box and Wilson (1951) to optimize response when there are no constraints on the input variables and they can be varied independently of each other. Comprehensive reference books on this topic are by Box and Draper (2007), Khuri and Cornell (1996), and Myers and Montgomery (2002). RSM is discussed in Section 10.1.

Section 10.2 deals with **mixture experiments**, which are a special type of response surface experiments in which the factors are the proportions of components or ingredients of a product; the product property is a function of these proportions. Since the proportions must add up to 1, the factor levels cannot be varied independently of each other. Mixture experiments are useful in designing product formulations, for example, gasoline blending. A comprehensive reference book on this topic is by Cornell (1990a).

Section 10.3 deals with **Taguchi methodology** for quality improvement in which the goal is to minimize the variation of the response around a nominal target value. Note the emphasis here on the variation rather than the mean of the response. This approach to improving quality through designed experiments was proposed by a Japanese engineer, Genichi Taguchi. Phadke (1989) is a very readable reference on the Taguchi approach to quality improvement.

A summary of the chapter is given in Section 10.4.

Statistical Analysis of Designed Experiments: Theory and Applications By Ajit C. Tamhane

10.1 RESPONSE SURFACE METHODOLOGY

10.1.1 Outline of Response Surface Methodology

Denote the response variable by y and the input variables by $x_1, x_2, \ldots, x_p$. We assume that all variables are continuous. Furthermore, we assume that the input variables are coded so that each one is unitless with the range $[-1, 1]$. If the raw input variables are denoted by X_i, then the coded variables are

$$x_i = \frac{X_i - (H_i + L_i)/2}{(H_i - L_i)/2} \qquad (1 \le i \le p),$$

where $[L_i, H_i]$ is the range of X_i. The functional relationship between the response and input variables may be represented as

$$y = f(x_1, x_2, \ldots, x_p) + e, \tag{10.1}$$

where $f(\cdot)$ is an unknown function and e is a $\mathrm{N}(0, \sigma^2)$ random error. The plot of $\mathrm{E}(y)$ as a function of $x_1, x_2, \ldots, x_p$ is called the **response surface**. The goal is to determine the optimum (maximum or minimum) of the response surface. Since $f(\cdot)$ is unknown, optimization must be performed experimentally. Without loss of generality we will assume that the goal is to find the maximum.

In RSM one proceeds from a starting point to the maximum by conducting a series of experiments. This is done in two phases. In the first phase, generally one is far from the maximum and, locally, the first-order model

$$\mathrm{E}(y) = \beta_0 + \sum_{i=1}^{p} \beta_i x_i \tag{10.2}$$

provides a good fit to the response surface. Here the β_i $(1 \le i \le p)$ represent the linear (main) effects of the factors. After fitting this model the direction of the greatest improvement is determined and a step is taken along that direction to a new experimental region. This is called the **steepest-ascent** method.

In the second phase one is in the neighborhood of the maximum and the response surface is likely to have a significant curvature. Locally it can be adequately approximated by fitting a **second-order model**:

$$\mathrm{E}(y) = \beta_0 + \sum_{i=1}^{p} \beta_i x_i + \sum_{i=1}^{p-1} \sum_{j=i+1}^{p} \beta_{ij} x_i x_j + \sum_{i=1}^{p} \beta_{ii} x_i^2. \tag{10.3}$$

Here the β_{ii} represent the quadratic effects and the β_{ij} represent the linear-by-linear interactions of the factors. The stationary point of the fitted surface is determined and the region around it is explored to determine its nature. In the following sections we describe each experimental phase in detail and explain how the path toward the maximum is followed.

10.1.2 First-Order Experimentation Phase

10.1.2.1 First-Order Designs

To fit the model (10.2), any orthogonal design that can estimate the main effects clear of each other can be used. These designs have resolution III or higher. They commonly consist of fractional factorials, for example, 2^{p-q} fractional factorials or Plackett–Burman designs, augmented by some replicate runs at the center point. Replicates are used to provide a test for curvature (see Section 7.3.3) and an estimate of the pure error.

Let N denote the total number of runs and let x_{ij} denote the value of the jth variable in the ith run ($1 \le i \le N, 1 \le j \le p$). The design matrix is

$$\boldsymbol{D} = \begin{array}{c} \begin{matrix} x_1 & x_2 & \cdots & x_p \end{matrix} \\ \begin{bmatrix} x_{11} & x_{12} & \cdots & x_{1p} \\ x_{21} & x_{22} & \cdots & x_{2p} \\ \vdots & \vdots & \ddots & \vdots \\ x_{N1} & x_{N2} & \cdots & x_{Np} \end{bmatrix} \end{array}.$$

and the model matrix for the first-order model (10.2) is

$$\boldsymbol{X} = \begin{array}{c} \begin{matrix} 1 & x_1 & x_2 & \cdots & x_p \end{matrix} \\ \begin{bmatrix} 1 & x_{11} & x_{12} & \cdots & x_{1p} \\ 1 & x_{21} & x_{22} & \cdots & x_{2p} \\ \vdots & \vdots & \vdots & \ddots & \vdots \\ 1 & x_{N1} & x_{N2} & \cdots & x_{Np} \end{bmatrix} \end{array} = \{\mathbf{1}|\boldsymbol{D}\},$$

where $\mathbf{1}$ is an N-vector of all 1's. The condition for orthogonality of the design is that $\boldsymbol{X}'\boldsymbol{X}$ be a diagonal matrix, which translates into the condition

$$\sum_{i=1}^{N} x_{ij} = 0 \qquad (1 \le j \le p) \qquad \text{and} \qquad \sum_{i=1}^{N} x_{ij}x_{ik} = 0 \qquad (1 \le j \ne k \le p). \tag{10.4}$$

As noted in Section 7.3.3, adding center points does not violate these conditions. If the goal is to obtain an estimate of the pure error, then adding center points is more economical than replicating the whole design.

The **simplex design** is another design used in the first phase. Geometrically, a $(p+1)$-simplex is a regular p-dimensional figure with $p+1$ vertices (representing the design points) such that any two adjacent vertices subtend the same angle, $\cos^{-1}(-1/p)$, with the origin. For $p = 2$ and $p = 3$, a simplex is an equilateral triangle and a regular tetrahedron, respectively, as shown in Figure 10.1. A simplex is a saturated design since there are $N = p + 1$ design points for estimating $p + 1$ parameters in the first-order model (10.2). Therefore there are no degrees of freedom available for estimating the error unless either the whole design is replicated or some center-point runs are added.

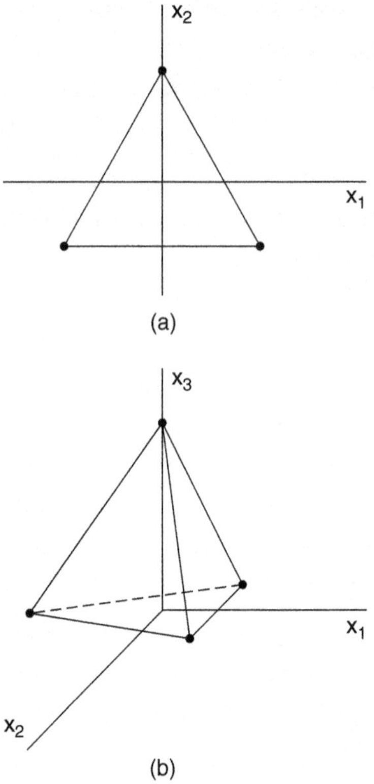

Figure 10.1 Simplex Designs for $p = 2$ and $p = 3$ (Montgomery, 2005, Figure 11-19). Reprinted by permission of John Wiley & Sons, Inc.

For $p = 2$, a simplex is an equilateral triangle as shown in Figure 10.1. The coordinates of its vertices form the design matrix (scaled such that the distance between any two vertices is 2) given by

$$\boldsymbol{D} = \begin{array}{c} \\ \end{array}\begin{bmatrix} \overset{x_1}{-1} & \overset{x_2}{-\frac{1}{\sqrt{3}}} \\ 1 & -\frac{1}{\sqrt{3}} \\ 0 & \frac{2}{\sqrt{3}} \end{bmatrix}.$$

It is easy to check that this design is orthogonal and satisfies the conditions in (10.4). Any orthogonal rotation of this simplex is still a simplex. For any such design the model matrix satisfies

$$\boldsymbol{X}'\boldsymbol{X} = 3\boldsymbol{I}.$$

This result can be generalized to obtain a simplex design for any $p \geq 2$. Let $\boldsymbol{\Gamma}$ be an orthogonal matrix of dimension $p + 1$ such that its first column has all

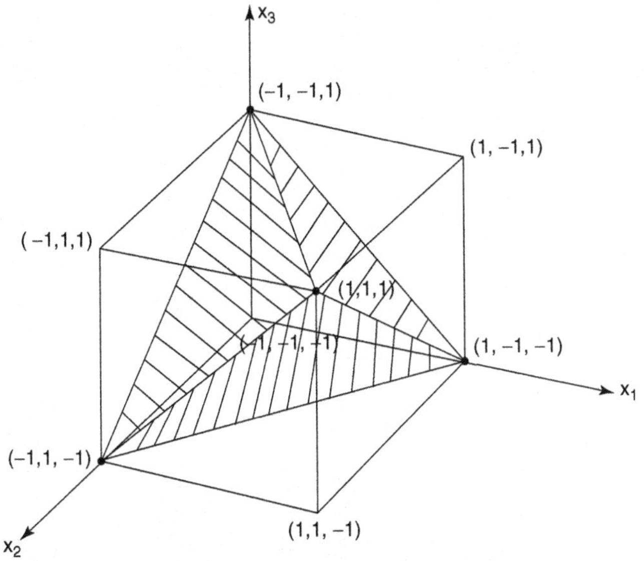

Figure 10.2 A 2^{3-1} Design as a tetrahedron (4-simplex).

entries equal to $1/\sqrt{p+1}$. Let the model matrix be

$$X = (p+1)^{1/2}\boldsymbol{\Gamma} = \{\mathbf{1}|\boldsymbol{D}\}.$$

Then the required design matrix of the simplex is $\boldsymbol{D}$.

A simple application of this rule shows that for $p = 3$ the 2_{III}^{3-1} design with the defining relation $I = ABC$ (or its alternate fraction) is a simplex design with the design matrix

$$\boldsymbol{D} = \begin{array}{c} \begin{matrix} x_1 & x_2 & x_3 \end{matrix} \\ \begin{bmatrix} -1 & -1 & 1 \\ 1 & -1 & -1 \\ -1 & 1 & -1 \\ 1 & 1 & 1 \end{bmatrix} \end{array}.$$

This design is shown in Figure 10.2. It can be seen that it is the same regular tetrahedron (4-simplex) shown in Figure 10.1.

10.1.2.2 Steepest-Ascent Method

Denote the LS fitted first-order model by

$$\widehat{y} = \widehat{\beta}_0 + \sum_{i=1}^{p} \widehat{\beta}_{ix_i}.$$

Note that the slope estimators $\widehat{\beta}_i$ $(1 \le i \le p)$ are unaffected by the center-point runs; only the intercept estimator $\widehat{\beta}_0$ depends on the center-point runs and equals the grand average $\bar{y}$ of all the runs. The gradient of the fitted surface at the origin (the current center point of the design) equals (see Exercise 14.3)

$$\frac{d\widehat{y}}{d\boldsymbol{x}} = \left(\frac{\partial \widehat{y}}{\partial x_1}, \ldots, \frac{\partial \widehat{y}}{\partial x_p}\right)' = (\widehat{\beta}_1, \ldots, \widehat{\beta}_p)' = \widehat{\boldsymbol{\beta}}.$$

In Exercise 10.2 you are asked to show that this is the direction of the steepest ascent. If we change x_i by an amount Δx_i along this direction, then x_j changes by an amount Δx_j, where the ratio of the two changes equals

$$\frac{\Delta x_i}{\Delta x_j} = \frac{\widehat{\beta}_i}{\widehat{\beta}_j} \quad (1 \le i \ne j \le p);$$

that is, the change in any variable x_i is proportional to its estimated slope coefficient $\widehat{\beta}_i$.

The next step is to determine the **step size**. Clearly, the largest change will be in the variable with the largest value of $|\widehat{\beta}_i|$, so the step size may be determined by how much change in this variable is feasible. However, it must be remembered that these changes are in the coded variables x_i; the changes in the raw variables X_i may be quite different depending on their scales. The choice of the step size depends on the feasible ranges of the X_i's and the experimenter's judgment about which changes are too small to achieve a significant improvement and which changes are too large to possibly overshoot the location of the desired maximum. As mentioned before, when the search reaches the neighborhood of the maximum, the response surface has a significant curvature (as when approaching the top of the hill). So the first-order model no longer gives a satisfactory fit. A test for curvature, described in Section 7.3.3, is performed in order to decide whether to switch to second-order experiments.

Another indication for switching to second-order experiments is that the first-order fitted model has significant interaction terms but nonsignificant main effects (thus violating the effect heredity principle). This suggests the presence of curvature for which a second-order model is required.

Example 10.1 (Chemical Process Yield Maximization: First-Order Experiment)

The percent yield (y) of a chemical process depends on the reaction temperature (X_1) and reaction time (X_2). It was desired to determine the optimum settings of these two factors to maximize the yield. In the first phase of response surface analysis, a 2^2 experiment was replicated twice with readings at $X_1 = 80° \pm 10°$C and $X_2 = 60 \pm 30$ sec. There were no replicates at the center point, which was the current setting. The data are shown in Table 10.1.

The first-order fitted model is

$$\widehat{y} = 61.688 + 3.437x_1 + 9.812x_2;$$

Table 10.1 Chemical Process Yield Data from First-Order Experiment

		Percent Yield (y)	
x_1	x_2	Replicate 1	Replicate 2
−1	−1	49.8	48.1
1	−1	57.3	52.3
−1	1	65.7	69.4
1	1	73.1	77.8

```
Source             DF    Seq SS    Adj SS    Adj MS      F      P
Main Effects        2   864.812   864.812   432.406  63.71  0.000
Residual Error      5    33.936    33.936     6.787
  Lack of Fit       1     2.101     2.101     2.101   0.26  0.634
  Pure Error        4    31.835    31.835     7.959
Total               7   898.749

Estimated Effects and Coefficients for Yield (coded units)

Term       Effect     Coef  SE Coef      T      P
Constant            61.688   0.9211  66.97  0.000
x_1         6.875    3.437   0.9211   3.73  0.014
x_2        19.625    9.812   0.9211  10.65  0.000
```

Display 10.1 Minitab output for linear fit to first-order experiment data from Example 10.1.

the corresponding ANOVA table computed using Minitab is shown in Display 10.1. Note that the pure-error estimate with four d.f. is obtained from the replicates at the four factorial design points. The lack-of-fit test is the test of significance of the interaction term, x_1x_2, which is omitted from the first-order model. We see that this term is nonsignificant, so it is safe to omit it and pool it with the pure error to obtain an overall error estimate of 6.787 with five d.f. The tests of significance of the two slope coefficients using this estimate are shown in the output. Both slope coefficients are significant.

The direction of the steepest ascent is given by

$$\frac{\Delta x_1}{\Delta x_2} = \frac{3.437}{9.812} = 0.350.$$

Therefore one unit of change in x_2 is accompanied by 0.35 unit of change in x_1. This direction is shown in Figure 10.3.

Suppose that a step size of 45 secs in X_2 is thought suitable. This corresponds to a change of $45/30 = 1.5$ units in the coded variable x_2, which gives

$$\Delta x_1 = (0.35)(1.5) = 0.525 \quad \Rightarrow \Delta X_1 = (0.525)(10) = 5.25.$$

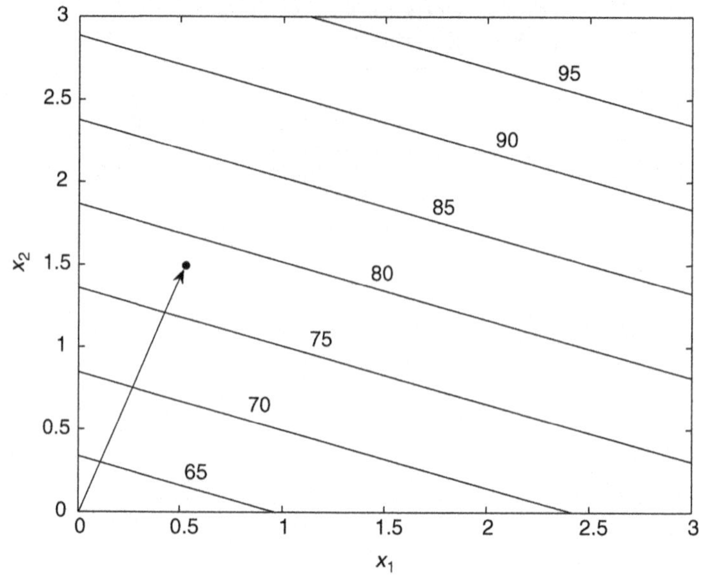

Figure 10.3 Contours of first-order model, direction of steepest ascent, and center of new experimental region.

Hence the center of the next experiment will be at $X_1 = 80 + 5.25 = 85.25°C$, $X_2 = 60 + 45 = 105$ sec. The predicted percent yield at the new center equals $\widehat{y} = 61.688 + (3.437)(0.525) + (9.812)(1.5) = 78.210$. ■

10.1.3 Second-Order Experimentation Phase

The second-order surface (10.3) has $m = (p + 1)(p + 2)/2$ unknown parameters; for example, $m = 10$ for $p = 3$ and $m = 15$ for $p = 4$. Therefore any second-order design must have at least that many distinct design points. Also, there must be at least three levels of each factor in order to estimate the second-order effects. The 3^p full factorials or 3^{p-q} fractional factorials are often too large, for example, a 3^3 experiment has 27 design points and a 3^4 experiment has 81 design points. Two classes of smaller and more efficient designs are **central composite (CC) designs** and **Box–Behnken (BB) designs**, which are discussed in the following two sections.

10.1.3.1 Central Composite Designs

A central composite (CC) design has three parts:

(a) Single replicates at each of the n_f factorial points forming an orthogonal design. This is called the **factorial portion** of the design.
(b) n_c replicates at the center point.
(c) Single replicates at each of the $2p$ **axial** or **star points**, which have coordinates $x_i = \pm\alpha$ and $x_j = 0$ for $j \neq i$ $(1 \leq i \leq p)$ for some $\alpha > 0$.

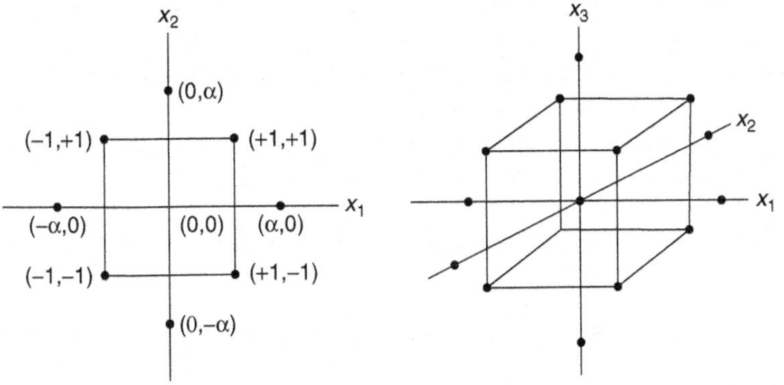

Figure 10.4 CC designs for $p = 2$ and $p = 3$ (Montgomery, 2005, Figure 11-20). Reprinted by permission of John Wiley & Sons, Inc.

$$D = \begin{bmatrix} -1 & -1 \\ 1 & -1 \\ -1 & 1 \\ 1 & 1 \\ -\alpha & 0 \\ \alpha & 0 \\ 0 & -\alpha \\ 0 & \alpha \\ 0 & 0 \\ 0 & 0 \end{bmatrix}$$

(columns: x_1, x_2)

Display 10.2 Design matrix of CC design for $p = 2$ with two center points.

If a 2^{p-q} design is used for the factorial portion, then it must have resolution at least V so that all two-factor interactions are clear of each other. Figure 10.4 shows CC designs for $p = 2$ and $p = 3$. The design matrix of a CC design for $p = 2$ with $n_c = 2$ is shown in Display 10.2.

A CC design can be viewed as a first-order design augmented by axial points. It is a natural extension of a first-order design to which axial points are added to model the second-order effects. The choice of α is up to the experimenter. If $\alpha > 1$ (< 1), then the axial points are outside (inside) the hypercube and there are five levels for each factor; if $\alpha = 1$, then the axial points are on the face of the hypercube and there are three levels for each factor. This is called a **face-centered cube (FCC)**.

We now discuss how to choose the three parameters of a CC design, namely, n_f, n_c, and α. We assume that each factorial design point is singly replicated. The condition on the number of distinct design points requires that

$$n_f + 2p + 1 \geq \tfrac{1}{2}(p+1)(p+2).$$

If a 2^{p-q} design is used for the factorial portion of a CC design, then $n_f = 2^{p-q}$.

Usually α is chosen between 1 and $\sqrt{p}$. For $\alpha = 1$, an FCC is a useful design when the experimental region is cuboidal, which occurs when there are independent restrictions of the type $a_i \le X_i \le b_i$ on all factors. On the other hand, $\alpha = \sqrt{p}$ places all factorial and axial points at an equal radial distance from the center point. The resulting design is called a **spherical design**, which is useful when the experimental region is spherical, a less common occurrence in practice.

Between these two extremes, α may be chosen to satisfy the **rotatability** criterion defined as follows:

$$\mathrm{Var}(\widehat{y}(\boldsymbol{x})) = \text{ const. for all } \boldsymbol{x} \text{ s.t. } \boldsymbol{x}'\boldsymbol{x} = \sum_{i=1}^{p} x_i^2 = \text{ const.} \tag{10.5}$$

In other words, the variance of the predicted response $\widehat{y}(\boldsymbol{x})$ is the same for all equiradial distance points $\boldsymbol{x}$ from the center point. Thus the contours of constant variance are concentric circles (or hyperspheres in higher dimensions). Let $\boldsymbol{X}$ be the model matrix of the second-order model (10.3). For example, for the design matrix in Display 10.2, the model matrix is shown in Display 10.3. From (2.39), we see that

$$\mathrm{Var}(\widehat{y}(\mathbf{x})) = \sigma^2(1, \boldsymbol{x}')(\boldsymbol{X}'\boldsymbol{X})^{-1}\begin{pmatrix} 1 \\ \boldsymbol{x} \end{pmatrix}.$$

It can be shown that $\alpha = (n_f)^{1/4}$ makes the CC design rotatable. For example, for a 2^3 design $\alpha = (8)^{1/4} = 1.682$. Furthermore, for a CC design to be orthogonal, it must satisfy (see Khuri and Cornell, 1996, p. 122)

$$(n_f + 2\alpha^2)^2 = n_f N,$$

where $N = n_f + n_c + 2p$ is the total number of runs. If we choose $\alpha = (n_f)^{1/4}$ to make the design rotatable, then by choosing

$$n_c \approx 4\sqrt{n_f} + 4 - 2p,$$

the design can also be made nearly orthogonal.

$$\begin{array}{c} \\ \boldsymbol{X} = \end{array}
\begin{array}{c}
\begin{array}{cccccc} 1 & x_1 & x_2 & x_1x_2 & x_1^2 & x_2^2 \end{array} \\
\left[\begin{array}{cccccc}
1 & -1 & -1 & 1 & 1 & 1 \\
1 & 1 & -1 & -1 & 1 & 1 \\
1 & -1 & 1 & -1 & 1 & 1 \\
1 & 1 & 1 & 1 & 1 & 1 \\
1 & -\alpha & 0 & 0 & \alpha^2 & 0 \\
1 & \alpha & 0 & 0 & \alpha^2 & 0 \\
1 & 0 & -\alpha & 0 & 0 & \alpha^2 \\
1 & 0 & \alpha & 0 & 0 & \alpha^2 \\
1 & 0 & 0 & 0 & 0 & 0 \\
1 & 0 & 0 & 0 & 0 & 0
\end{array}\right]
\end{array}$$

Display 10.3 Quadratic model matrix for CC design with $p = 2$.

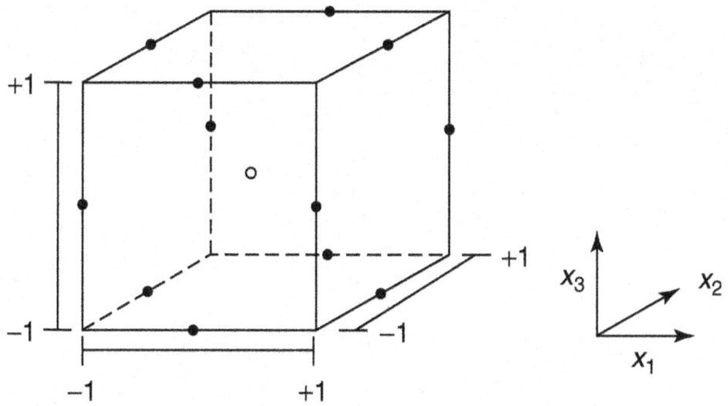

Figure 10.5 Box–Behnken design for $p = 3$ (Montgomery, 2005, Figure 11-22). Reprinted by permission of John Wiley & Sons, Inc.

It should be noted that the rotatability criterion is in terms of the coded variables x_i, not in terms of the raw variables X_i. Rotatability is not invariant to scale changes in the variables and therefore should not be the primary criterion. Smaller designs which are nearly rotatable are generally preferred.

Example 10.2

For $p = 5$, a 2^{5-1} design with resolution V exists with $n_f = 16$. Hence $\alpha = (16)^{1/4} = 2$ gives a rotatable design and $n_c = 4\sqrt{16} + 4 - (2)(5) = 10$ gives an orthogonal design. Often, n_c is chosen between 2 and 5. ■

10.1.3.2 Box-Behnken Designs

A Box-Behnken (BB) design (Box and Behnken 1960) consists of all edge centers of the $[-1, 1]^p$ hypercube augmented by n_c center points. Figure 10.5 shows a BB design for $p = 3$.

Every edge center has two $x_i = \pm 1$ and all other $x_i = 0$. Thus a BB design has $4\binom{p}{2} = 2p(p-1)$ edge centers or a total of $N = 2p(p-1) + n_c$ runs. For example, $N = 12 + n_c$ for $p = 3$, $N = 24 + n_c$ for $p = 4$, and $N = 40 + n_c$ for $p = 5$. The design matrix of a BB design can be obtained by a composition of a BIB design for p treatments with block size of 2 and number of blocks equal to $\binom{p}{2}$ and a 2^2 design; n_c rows of 0's are added corresponding to the center-point runs. This construction is illustrated in the following example.

Example 10.3 (Design Matrix of BB Design for $p = 3$)

Let

$$\boldsymbol{D}_1 = \begin{bmatrix} X & X & - \\ X & - & X \\ - & X & X \end{bmatrix}$$

be the design matrix of a BIB design for three treatments (columns) in three blocks (rows) of size 2; here X indicates the presence of the treatment corresponding to the column. Next let

$$\boldsymbol{D}_2 = \begin{bmatrix} -1 & -1 \\ 1 & -1 \\ -1 & 1 \\ 1 & 1 \end{bmatrix}$$

be the design matrix of a 2^2 design. Then composition of $\boldsymbol{D}_1$ with $\boldsymbol{D}_2$ (denoted by $\boldsymbol{D}_1 \otimes \boldsymbol{D}_2$) means that for each block of $\boldsymbol{D}_1$ we have design $\boldsymbol{D}_2$ in terms of the variables (treatments) present in that block (indicated by X) and the variable not present in the block is set at zero (the midpoint of its range). The resulting design matrix is

$$\boldsymbol{D} = \boldsymbol{D}_1 \otimes \boldsymbol{D}_2 = \begin{array}{c} \begin{matrix} x_1 & x_2 & x_3 \end{matrix} \\ \begin{bmatrix} -1 & -1 & 0 \\ 1 & -1 & 0 \\ -1 & 1 & 0 \\ 1 & 1 & 0 \\ -1 & 0 & -1 \\ 1 & 0 & -1 \\ -1 & 0 & 1 \\ 1 & 0 & 1 \\ 0 & -1 & -1 \\ 0 & 1 & -1 \\ 0 & -1 & 1 \\ 0 & 1 & 1 \end{bmatrix} \end{array}.$$

This design can be supplemented by center-point runs (usually between two and five). ■

The FCC and BB designs are both three-level designs, but they differ in the experimental regions covered by them. As mentioned before, the FCC design is a cuboidal design while the BB design is a spherical design since all edge centers are equidistant from the center point. For example, in the BB design for $p = 3$ shown above, all design points are at a distance $\sqrt{2}$ from the center. On the other hand, in the FCC design for p factors, the vertex (factorial) points are at a distance $\sqrt{p}$ from the center. The radius of the spherical region of a BB design is generally much smaller than $\sqrt{p}$. Therefore a BB design should not be used if prediction of the response is desired near the vertex points. Data from a real experiment using a BB design are analyzed in Example 10.5.

10.1.3.3 Analysis of Second-Order Experiments

It is convenient to write the second-order model (10.3) in matrix notation. Denote $\boldsymbol{\beta} = (\beta_1, \ldots, \beta_p)'$ and

$$\boldsymbol{B} = \begin{bmatrix} \beta_{11} & \frac{1}{2}\beta_{12} & \cdots & \frac{1}{2}\beta_{1p} \\ \frac{1}{2}\beta_{12} & \beta_{22} & \cdots & \frac{1}{2}\beta_{2p} \\ \vdots & \vdots & \vdots & \vdots \\ \frac{1}{2}\beta_{1p} & \frac{1}{2}\beta_{2p} & \cdots & \beta_{pp} \end{bmatrix}.$$

Then the second-order model (10.3) can be written as

$$\mathrm{E}(y) = \beta_0 + \boldsymbol{x}'\boldsymbol{\beta} + \boldsymbol{x}'\boldsymbol{B}\boldsymbol{x}.$$

The LS fitted model is given by

$$\widehat{y} = \widehat{\beta}_0 + \boldsymbol{x}'\widehat{\boldsymbol{\beta}} + \boldsymbol{x}'\widehat{\boldsymbol{B}}\boldsymbol{x}, \qquad (10.6)$$

where $\widehat{\boldsymbol{\beta}}$ and $\widehat{\boldsymbol{B}}$ are the LS estimators of $\boldsymbol{\beta}$ and $\boldsymbol{B}$ (obtained by replacing the elements of $\boldsymbol{\beta}$ and $\boldsymbol{B}$ by their LS estimators), respectively. Using the results from Exercise 14.3, we obtain the stationary point of the fitted response surface by setting

$$\frac{d\widehat{y}}{d\boldsymbol{x}} = \widehat{\boldsymbol{\beta}} + 2\widehat{\boldsymbol{B}}\boldsymbol{x} = \boldsymbol{0},$$

where $\boldsymbol{0}$ is the null vector. The stationary point is given by (assuming $\widehat{\boldsymbol{B}}^{-1}$ exists)

$$\boldsymbol{x}_0 = -\tfrac{1}{2}\widehat{\boldsymbol{B}}^{-1}\widehat{\boldsymbol{\beta}}$$

and the corresponding fitted value of the response equals

$$\widehat{y}_0 = \widehat{\beta}_0 + \boldsymbol{x}_0'\widehat{\boldsymbol{\beta}} + \boldsymbol{x}_0'\widehat{\boldsymbol{B}}\boldsymbol{x}_0 = \widehat{\beta}_0 + \boldsymbol{x}_0'\widehat{\boldsymbol{\beta}} + \boldsymbol{x}_0'\widehat{\boldsymbol{B}}\left(-\tfrac{1}{2}\widehat{\boldsymbol{B}}^{-1}\widehat{\boldsymbol{\beta}}\right) = \widehat{\beta}_0 + \tfrac{1}{2}\boldsymbol{x}_0'\widehat{\boldsymbol{\beta}}. \qquad (10.7)$$

To determine whether the stationary point is a maximum, minimum, or saddle point, we perform a **canonical analysis** of the response surface. First we translate the coordinate system by subtracting $\boldsymbol{x}_0$ from $\boldsymbol{x}$ so that the origin is at the stationary point. Next we rotate the coordinate axes orthogonally so that they are aligned with the axes of the contours of the quadratic fitted response surface. These contours are conics. The columns of the desired rotation matrix $\boldsymbol{P}$ are the normalized eigenvectors of $\widehat{\boldsymbol{B}}$. If $\boldsymbol{\Lambda} = \mathrm{diag}(\lambda_1, \ldots, \lambda_p)$ denotes a diagonal matrix whose entries λ_i $(1 \le i \le p)$ are the eigenvalues of $\widehat{\boldsymbol{B}}$, then we have

$$\boldsymbol{P}'\widehat{\boldsymbol{B}}\boldsymbol{P} = \boldsymbol{\Lambda} \qquad \text{and} \qquad \boldsymbol{P}'\boldsymbol{P} = \boldsymbol{P}\boldsymbol{P}' = \boldsymbol{I}.$$

The desired transformation is

$$z = \boldsymbol{P}'(\boldsymbol{x} - \boldsymbol{x}_0).$$

Substituting $\boldsymbol{x} = \boldsymbol{P}z + \boldsymbol{x}_0$ in Eq. (10.6) leads to

$$\begin{aligned}
\widehat{y} &= \widehat{\beta}_0 + (\boldsymbol{P}z + \boldsymbol{x}_0)'\widehat{\boldsymbol{\beta}} + (\boldsymbol{P}z + \boldsymbol{x}_0)'\widehat{\boldsymbol{B}}(\boldsymbol{P}z + \boldsymbol{x}_0) \\
&= (\widehat{\beta}_0 + \boldsymbol{x}_0'\widehat{\boldsymbol{\beta}} + \boldsymbol{x}_0'\widehat{\boldsymbol{B}}\boldsymbol{x}_0) + z'\boldsymbol{P}'\widehat{\boldsymbol{\beta}} + 2z'\boldsymbol{P}'\widehat{\boldsymbol{B}}\boldsymbol{x}_0 + z'\boldsymbol{P}'\widehat{\boldsymbol{B}}\boldsymbol{P}z \\
&= \widehat{y}_0 + z'\boldsymbol{P}'\widehat{\boldsymbol{\beta}} + 2z'\boldsymbol{P}'\widehat{\boldsymbol{B}}\left(-\tfrac{1}{2}\widehat{\boldsymbol{B}}^{-1}\widehat{\boldsymbol{\beta}}\right) + z'\boldsymbol{\Lambda}z \\
&= \widehat{y}_0 + \sum_{i=1}^{p} \lambda_i z_i^2. \qquad (10.8)
\end{aligned}$$

From this canonical representation the following conclusions follow immediately:

(a) If all the λ_i are nonpositive (nonnegative), then $\widehat{y}_0$ is a local maximum (minimum) since moving in any direction z_i from the origin (i.e., the stationary point $\boldsymbol{x}_0$) decreases (increases) $\widehat{y}$ from $\widehat{y}_0$. In this case the contours of the response surface are elliptical, as shown in Figure 10.6*a* and we have an **elliptic system**.

(b) If the λ_i have mixed signs, then the stationary point $\boldsymbol{x}_0$ is a **saddle point** since moving in the direction z_i for which λ_i is negative (positive) decreases (increases) $\widehat{y}$ from $\widehat{y}_0$. In this case the contours of the response surface are hyperbolic, as shown in Figure 10.6*b*, and we have a **hyperbolic system**.

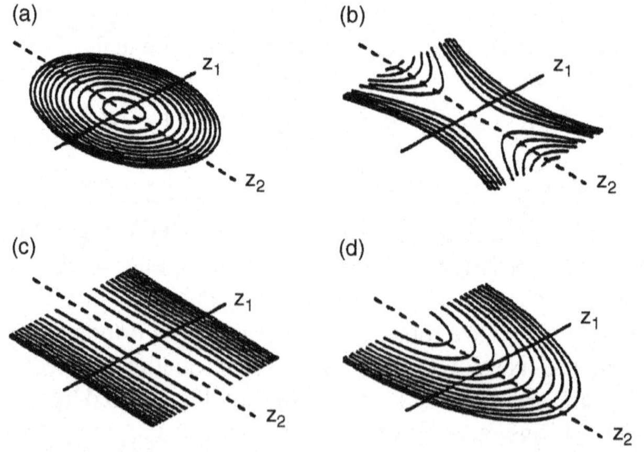

Figure 10.6 Contours of different second-order response surfaces: (*a*) elliptic system (simple maximum); (*b*) hyperbolic system (saddle point); (*c*) stationary-ridge system; (d) rising-ridge system (Wu and Hamada, 2000, Figure 9.7). Reprinted by permission of John Wiley & Sons, Inc.

(c) If $\lambda_i = 0$ for some i, then there is no change in the response surface along the z_i direction and the contours along that direction are parallel lines. If the stationary point is within the experimental region, then we have a **stationary-ridge system**, shown in Figure 10.6*c*. If all other λ_i are negative (positive), then the maximum (minimum) response $\widehat{y}_0$ is attained along the entire z_i line passing through the origin. This affords a flexible choice of the factor levels to attain the desired optimum.

(d) If $\lambda_i \approx 0$ for some i but the stationary point is outside the experimental region, then we have a **rising-ridge system** (if the stationary point is a maximum) or a **falling-ridge system** (if the stationary point is a minimum), shown in Figure 10.6*d*. In this case the response rises (falls) slowly in the z_i direction. The existence of such a system tells the user to move outside the current experimental region.

Example 10.4 (Chemical Process Yield Maximization: Second-Order Experiment Using a CC Design)

Refer to Example 10.1. After following a steepest-ascent path through a series of experiments, suppose a second phase was entered and an experiment was conducted using a CC design at the center point: $X_1 = 135.9°\text{C}$, $X_2 = 195$ sec. The design consisted of a 2^2 factorial portion with ranges $\pm 10°\text{C}$ and ± 23.1 sec, four axial points with $\alpha = \sqrt{2}$, and two center-point runs. The data are shown in Table 10.2.

The second-order fitted model is

$$\widehat{y} = 96.6 + 0.0302x_1 - 0.3112x_2 + 0.5750x_1x_2 - 1.9813x_1^2 - 1.8312x_2^2.$$

A detailed Minitab output is shown in Display 10.4. We see that both $\widehat{\beta}_1$ and $\widehat{\beta}_2$ are not significant at $\alpha = 0.05$; however, we will keep those terms in the model

Table 10.2 Chemical Process Yield Data from CC Experiment using CC Design

x_1	x_2	Percent Yield
-1	-1	93.6
1	-1	92.5
-1	1	91.7
1	1	92.9
$-\sqrt{2}$	0	92.7
$\sqrt{2}$	0	92.8
0	$-\sqrt{2}$	93.4
0	$\sqrt{2}$	92.7
0	0	96.2
0	0	97.0

```
Estimated Regression Coefficients for Yield

Term          Coef   SE Coef        T      P
Constant   96.6000    0.2382  405.511  0.000
x1          0.0302    0.1191    0.253  0.812
x2         -0.3112    0.1191   -2.613  0.059
x1*x1     -1.9813   0.1576  -12.574  0.000
x2*x2      -1.8312    0.1576  -11.622  0.000
x1*x2       0.5750    0.1684    3.414  0.027

S = 0.3369   R-Sq = 98.2%   R-Sq(adj) = 96.1%

Analysis of Variance for Yield

Source           DF   Seq SS   Adj SS   Adj MS       F      P
Regression        5  25.4510  25.4510   5.0902   44.85  0.001
  Linear          2   0.7823   0.7823   0.3911    3.45  0.135
  Square          2  23.3463  23.3463  11.6731  102.85  0.000
  Interaction     1   1.3225   1.3225   1.3225   11.65  0.027
Residual Error    4   0.4540   0.4540   0.1135
  Lack-of-Fit     3   0.1340   0.1340   0.0447    0.14  0.925
  Pure Error      1   0.3200   0.3200   0.3200
Total             9  25.9050
```

Display 10.4 Minitab output for second-order model fit to yield data in Table 10.2.

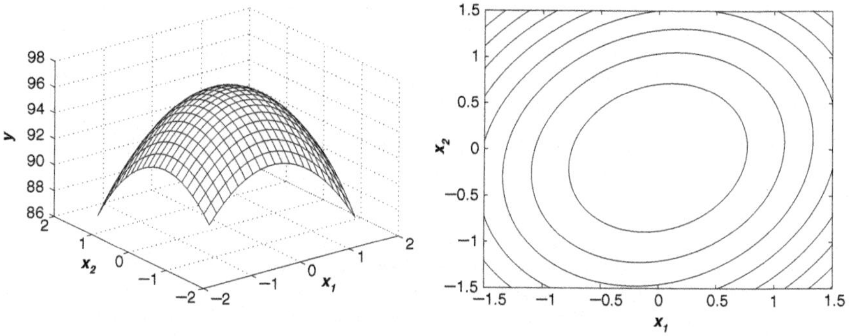

Figure 10.7 Response surface and contour plots for second-order model fit to yield data in Table 10.2.

respecting the hierarchical modeling principle. The response surface and contour plots for this model are shown in Figure 10.7.

To find the stationary point of the response surface, let

$$\widehat{\boldsymbol{\beta}} = (\widehat{\beta}_1, \widehat{\beta}_2)' = (0.0302, -0.3112)' \qquad \text{and}$$

$$\widehat{\boldsymbol{B}} = \begin{bmatrix} \widehat{\beta}_{11} & \frac{1}{2}\widehat{\beta}_{12} \\ \frac{1}{2}\widehat{\beta}_{12} & \widehat{\beta}_{22} \end{bmatrix} = \begin{bmatrix} -1.9813 & 0.2875 \\ 0.2875 & -1.8312 \end{bmatrix}.$$

Then the stationary point is

$$x_0 = -\frac{1}{2}\widehat{B}^{-1}\widehat{\beta} = -\frac{1}{2}\begin{bmatrix} -0.5165 & -0.0811 \\ -0.0811 & -0.5588 \end{bmatrix}\begin{bmatrix} 0.0302 \\ -0.3112 \end{bmatrix} = \begin{bmatrix} 0.0048 \\ 0.0857 \end{bmatrix}.$$

In terms of the raw variables, the stationary point is

$$X_1 = 135.9 + (0.0048)(10) = 135.95^\circ\text{C} \qquad \text{and}$$

$$X_2 = 195 + (0.0857)(23.1) = 196.98 \text{ sec}.$$

Notice that the stationary point is very close to the current center point.

From the response surface and contour plots it is evident that the stationary point of the surface corresponds to a maximum. We shall check this by canonical analysis. (Commonly, one uses a software package, e.g., JMP, Minitab, or Design-Expert, to do this canonical analysis.) The eigenvalues are the roots of the determinantal equation

$$|\widehat{B} - \lambda I| = \begin{vmatrix} -(\lambda + 1.9813) & 0.2875 \\ 0.2875 & -(\lambda + 1.8312) \end{vmatrix} = \lambda^2 + 3.8125\lambda + 3.5455 = 0.$$

The roots are $\lambda_1 = -2.2034$ and $\lambda_2 = -1.6091$. Both λ_1 and λ_2 are negative, which shows that the stationary point is a maximum and the corresponding predicted maximum yield is nearly the same as that at the center point, which equals $\widehat{y}_0 = 96.6\%$. ■

Example 10.5 (Robust VLSI Circuit Design Using a BB Design)

Tarim, Kuntman, and Ismail (1998) present results of a computer simulation experiment for robust design of a low-voltage complementary metal–oxide–semiconductor (CMOS) analog very large scale integrated (VLSI) circuit. The areas of three transistors in the circuit were the three factors A, B, C and the outcome variable y was the standard deviation of the relative current mismatch from the desired nominal value which was to be minimized. The low and high settings of the three factors are shown in Table 10.3. A BB design with three center-point runs was used. The center-point runs give a pure-error

Table 10.3 Factor-Level Settings for Robust VLSI Design Experiment

	Area of Transistor (μm^2)		
Level	A	B	C
Low (−1)	4	200	30
High (+1)	40	2000	300

Source: Tarim et al. (1998, Table 1).

Table 10.4 Data from Robust VLSI Design Experiment Using Box–Behnken Design

A	B	C	y
−1	−1	0	7.7295
1	−1	0	3.4893
−1	1	0	5.2752
1	1	0	4.0788
−1	0	−1	5.8381
1	0	−1	3.7844
−1	0	1	6.0744
1	0	1	3.7401
0	−1	−1	4.1782
0	1	−1	4.1700
0	−1	1	3.9802
0	1	1	4.2750
0	0	0	4.1032
0	0	0	4.2125
0	0	0	4.2070

Source: Tarim et al. (1998, Table 2).

estimate with two d.f. The simulated values of y obtained at the design points are given in Table 10.4. We will fit a second-order model to these data.

The Minitab output for a second-order fit is shown in Display 10.5. We see that factor C is inactive; factor B is active only through the interaction AB, while factor A is active through its linear and quadratic effects as well. We drop C from the model and fit a second-order model in A and B; the results are shown in Display 10.6. The contour plot shown in Figure 10.8 indicates that the stationary point of the quadratic fit is a saddle point, and the surface is ridgelike with the minimum achieved well outside the experimental region. This observation is confirmed by the eigenvalues 1.0161 and −0.0235 of $\widehat{\boldsymbol{B}}$ which have opposite signs and one eigenvalue is very close to zero, indicating the presence of a ridge. ■

10.2 MIXTURE EXPERIMENTS

Let $x_1, x_2, \ldots, x_p$ be the proportions of $p \geq 2$ components of a mixture which are subject to the constraints

$$0 \leq x_i \leq 1 \quad (1 \leq i \leq p) \quad \text{and} \quad \sum_{i=1}^{p} x_i = 1.$$

These constraints define the experimental region, which is a p-simplex. Simplexes for $p = 3$ and $p = 4$ are shown in Figure 10.1. The vertices of a simplex satisfy $x_i = 1, x_j = 0$ for $j \neq i$ and are called **pure blends** since only one

Estimated Regression Coefficients for y

Term	Coef	SE Coef	T	P
Constant	4.17423	0.2212	18.873	0.000
A	-1.22808	0.1354	-9.067	0.000
B	-0.19728	0.1354	-1.457	0.205
C	0.01237	0.1354	0.091	0.931
A*A	0.83868	0.1994	4.207	0.008
B*B	0.13028	0.1994	0.654	0.542
C*C	-0.15367	0.1994	-0.771	0.476
A*B	0.76095	0.1915	3.973	0.011
A*C	-0.07015	0.1915	-0.366	0.729
B*C	0.07575	0.1915	0.395	0.709

S = 0.383079 PRESS = 11.6357
R-Sq = 95.98% R-Sq(pred) = 36.31% R-Sq(adj) = 88.75%

Analysis of Variance for y

Source	DF	Seq SS	Adj SS	Adj MS	F	P
Regression	9	17.5354	17.5354	1.94838	13.28	0.005
Linear	3	12.3779	12.3779	4.12597	28.12	0.001
Square	3	2.7987	2.7987	0.93288	6.36	0.037
Interaction	3	2.3588	2.3588	0.78627	5.36	0.051
Residual Error	5	0.7337	0.7337	0.14675		
Lack-of-Fit	3	0.7262	0.7262	0.24205	63.84	0.015
Pure Error	2	0.0076	0.0076	0.00379		
Total	14	18.2691				

Display 10.5 Minitab output for full second-order fit to robust VLSI design experiment data.

Estimated Regression Coefficients for y

Term	Coef	SE Coef	T	P
Constant	4.0797	0.1489	27.397	0.000
A	-1.2281	0.1096	-11.206	0.000
B	-0.1973	0.1096	-1.800	0.105
A*A	0.8505	0.1608	5.288	0.001
B*B	0.1421	0.1608	0.884	0.400
A*B	0.7609	0.1550	4.910	0.001

S = 0.309982 PRESS = 5.02248
R-Sq = 95.27% R-Sq(pred) = 72.51% R-Sq(adj) = 92.64%

Display 10.6 Minitab output for partial second-order fit to robust VLSI design experiment data.

component is present. The edges joining the vertices satisfy $x_i + x_j = 1, x_k = 0$ for $k \neq i, j$ and are called **binary blends** since only two components are present. The response of interest, denoted by y, is some unknown function of $x_1, x_2, \ldots, x_p$ according to the general model (10.1). The goal is the same as in RSM, namely, to find the combination of the component proportions x_i's to optimize the expected response.

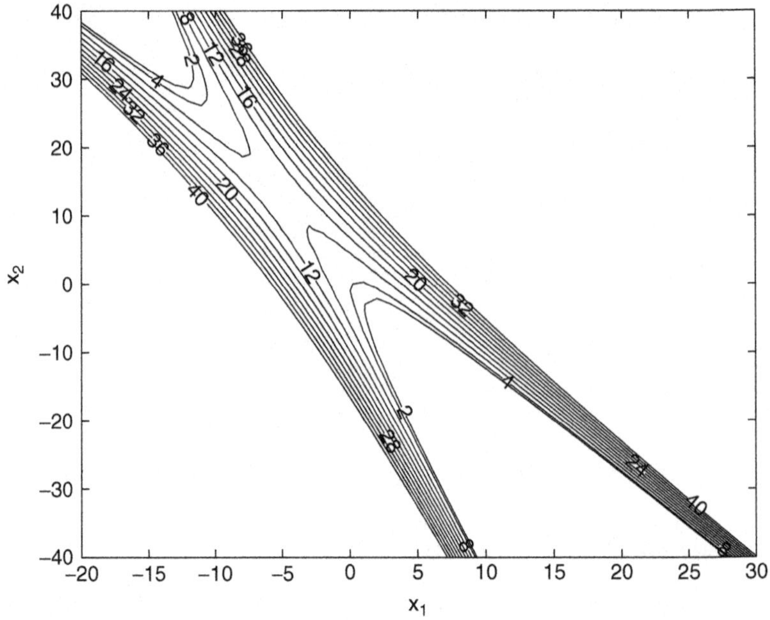

Figure 10.8 Contour plot for robust VLSI design experiment data.

10.2.1 Designs for Mixture Experiments

Designs for mixture experiments are modifications of the simplex designs discussed in Section 10.1.2.1. One such modification is called a (p, q)-**simplex lattice design**. This design adds $q - 1$ equally spaced points along each edge of a p-dimensional simplex. Along the edge joining the vertices i and j (i.e., pure blends of components i and j), x_i takes values $0, 1/q, 2/q, \ldots, (q-1)/q, 1$ and $x_j = 1 - x_i$. The number of design points in a (p, q)-simplex lattice design equals

$$N = \frac{(p+q-1)!}{q!(p-1)!}. \tag{10.9}$$

For $(p = 3, q = 2), (p = 3, q = 3), (p = 4, q = 2),$ and $(p = 4, q = 3)$ these designs are shown in Figure 10.9.

One problem with simplex lattice designs is that all runs are made on the edges of the simplex, which means that only the pure and binary blends are studied. **Simplex centroid designs** correct this problem to some extent by placing points also at the centers of all q-dimensional faces $(2 \le q \le p - 1)$. The vertices have coordinates $(0, \ldots, 1, \ldots, 0)$ and the centers of the edges have coordinates $(0, \ldots, \frac{1}{2}, 0, \ldots, \frac{1}{2}, \ldots, 0)$. In general, the coordinates of the center of a q-dimensional face have $q + 1$ elements equal to $1/(q + 1)$ and the other elements equal to zero. For $q = p - 1$ there is a single point with coordinates

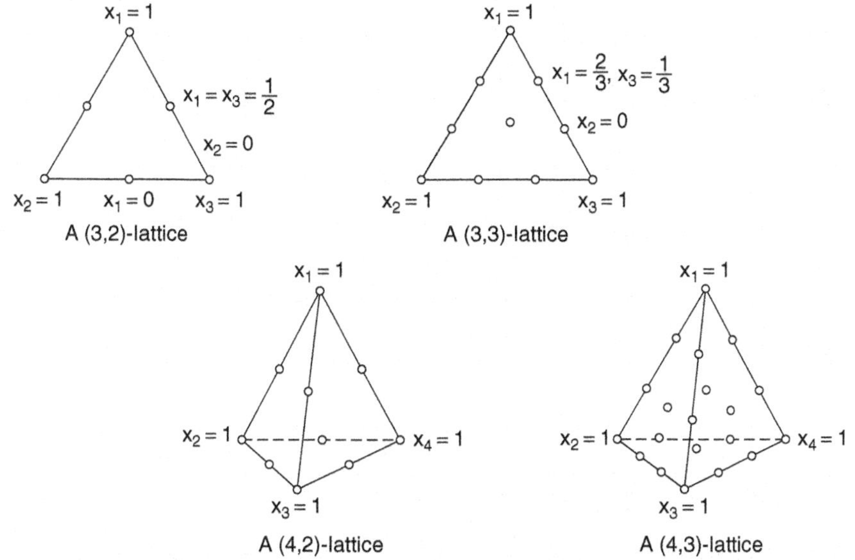

Figure 10.9 Examples of simplex lattice designs for $p = 3$ and $p = 4$ (Montgomery, 2005, Figure 11-34). Reprinted by permission of John Wiley & Sons, Inc.

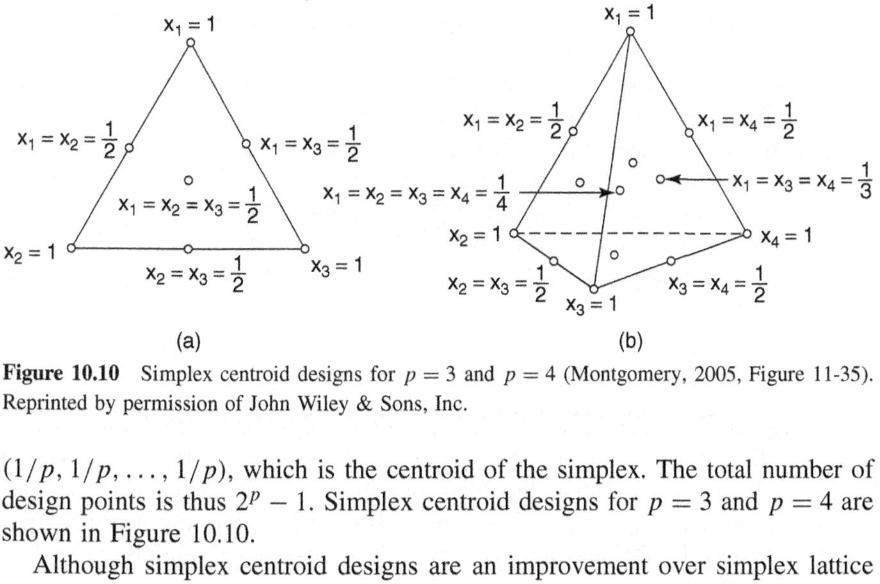

Figure 10.10 Simplex centroid designs for $p = 3$ and $p = 4$ (Montgomery, 2005, Figure 11-35). Reprinted by permission of John Wiley & Sons, Inc.

$(1/p, 1/p, \ldots, 1/p)$, which is the centroid of the simplex. The total number of design points is thus $2^p - 1$. Simplex centroid designs for $p = 3$ and $p = 4$ are shown in Figure 10.10.

Although simplex centroid designs are an improvement over simplex lattice designs in that they study blends of order higher than just the pure and binary blends, they still have only one run for a complete blend, namely the run at the centroid of the simplex; all other runs are on the boundaries of the simplex. To obtain better information on the response surface in the interior of the simplex, it is recommended that this design be augmented with **axial points** (note that these are different from the axial points in central composite designs). An **axis**

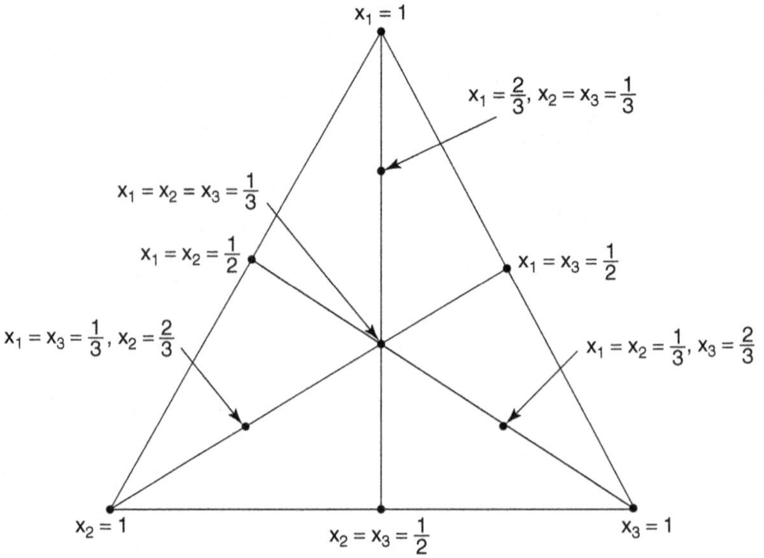

Figure 10.11 Simplex centroid design augmented by axial points for $p = 3$ (Montgomery, 2005, Figure 11-37). Reprinted by permission of John Wiley & Sons, Inc.

of component i is the perpendicular line from vertex i on the opposite boundary. On the axis, we have $x_1 = \cdots = x_{i-1} = x_{i+1} = \cdots = x_p$ with $x_i = 1, x_j = 0$ for $j \neq i$ at the vertex and $x_i = 0, x_j = 1/(p-1)$ for $j \neq i$ at the opposite boundary. The intersection of all axes is the centroid, $x_1 = \cdots = x_p = 1/p$. A recommended choice is to have axial points located midway between each vertex and the centroid. Since the distance between the centroid and any vertex is $(p-1)/p$, the coordinates of the ith axial point are $x_i = (p-1)/2p$ and $x_j = (p+1)/2p(p-1)$ for $j \neq i$. The total number of design points is thus $2^p + p - 1$. A simplex centroid design augmented by axial points for $p = 3$ is shown in Figure 10.11.

10.2.2 Analysis of Mixture Experiments

Linear, quadratic, and cubic models are commonly used to fit mixture experiment data. The general forms of these models can be simplified by using the constraint $\sum_{i=1}^{p} x_i = 1$. For example, the intercept term β_0 can be excluded from all models by writing $\beta_0 = \beta_0 \left(\sum_{i=1}^{p} x_i\right)$ and thus absorbing β_0 into the coefficients β_i of the linear terms. These simplified forms, called **canonical models** (Scheffé 1958, 1963), are given below.

Linear Model:

$$\mathrm{E}(y) = \sum_{i=1}^{p} \beta_i x_i. \tag{10.10}$$

This model has p parameters and is associated with the $(p, 1)$-simplex lattice design which has the same number of design points; see (10.9).

Quadratic Model:

$$E(y) = \sum_{i=1}^{p} \beta_i x_i + \sum_{j=i+1}^{p} \sum_{i=1}^{p-1} \beta_{ij} x_i x_j. \tag{10.11}$$

This model has $p(p+1)/2$ parameters and is associated with the $(p, 2)$-simplex lattice design which has the same number of design points; see (10.9).

Full Cubic Model:

$$E(y) = \sum_{i=1}^{p} \beta_i x_i + \sum_{j=i+1}^{p} \sum_{i=1}^{p-1} \beta_{ij} x_i x_j + \sum_{j=i+1}^{p} \sum_{i=1}^{p-1} \gamma_{ij} x_i x_j (x_i - x_j) +$$
$$\sum_{k=j+1}^{p} \sum_{j=i+1}^{p-1} \sum_{i=1}^{p-2} \beta_{ijk} x_i x_j x_k. \tag{10.12}$$

This model has $p(p+1)(p+2)/6$ parameters and is associated with the $(p, 3)$-simplex lattice design which has the same number of design points; see (10.9).

Special Cubic Model:

$$E(y) = \sum_{i=1}^{p} \beta_i x_i + \sum_{j=i+1}^{p} \sum_{i=1}^{p-1} \beta_{ij} x_i x_j + \sum_{k=j+1}^{p} \sum_{j=i+1}^{p-1} \sum_{i=1}^{p-2} \beta_{ijk} x_i x_j x_k. \tag{10.13}$$

This model has $p(p^2+5)/6$ parameters and is obtained by dropping the terms with the γ_{ij} coefficients from the full cubic model.

The coefficients in these models have relatively simple interpretations. For example, for the pure blend i, we see that $E(y) = \beta_i$ for all models. The LS estimators have particularly simple forms if these models are used in conjunction with their associated designs. For example, for the $(p, 2)$-simplex lattice design, with n_i replicates on the pure blend, $x_i = 1$, and n_{ij} replicates on the binary blend, $x_i = x_j = \frac{1}{2}$, the LS estimators for the canonical quadratic model are

$$\widehat{\beta}_i = \overline{y}_i \quad (1 \le i \le p) \quad \text{and} \quad \widehat{\beta}_{ij} = 4\overline{y}_{ij} - 2(\overline{y}_i + \overline{y}_j) \quad (1 \le i < j \le p), \tag{10.14}$$

where $\overline{y}_i$ and $\overline{y}_{ij}$ are the sample means of the replicates for the corresponding pure blends and binary blends, respectively.

Example 10.6 (Effect of Plasticizers on Vinyl Thickness)

The effects of three plasticizers on the thickness of vinyl used in automobile seat covers were studied using a (3,2)-simplex lattice design (Cornell, 1990b). Higher thickness is desirable. The plasticizers constituted 42% of the formulation with the remaining 58% coming from other ingredients. However, the plasticizer percentages can be scaled by 42% so that they add up to 100%. The remaining ingredients may be ignored as long as their percentages are fixed and they do not interact with plasticizers. Three replicates were made for each design point. The data on vinyl thickness (scaled) is shown in Table 10.5.

The special cubic model (10.13) was fitted to the data using Minitab. The output is shown in Display 10.7. The fitted model is

$$\widehat{y} = 11.667x_1 + 5.333x_2 + 8.667x_3 + 23.333x_1x_2 - 4.667x_1x_3 + 12.000x_2x_3 + 1.000x_1x_2x_3.$$

The coefficients of the terms x_1x_3 and $x_1x_2x_3$ are seen to be nonsignificant. The Minitab output for the refitted model obtained by dropping these two terms is shown in Display 10.8. The final refitted model is

$$\widehat{y} = 11.257x_1 + 5.364x_2 + 8.257x_3 + 23.597x_1x_2 + 12.264x_2x_3.$$

The contour plot, also called the **ternary plot**, of this model is shown in Figure 10.12. We see that the estimated thicknesses using pure blends are 11.257 with A, 5.364 with B, and 8.257 with C. Both interaction coefficients are positive with the coefficient of the AB interaction being much larger than all other coefficients. Therefore we can expect that the optimum blend for maximizing the vinyl thickness would contain only these two plasticizers. This is confirmed by the results of response optimization, also shown in Display 10.8. The optimum

Table 10.5 Vinyl Thickness Data Using (3,2)-Simplex Lattice Design

Plasticizer Proportion			Scaled Vinyl
A (x_1)	B (x_2)	C (x_3)	Thickness (y)
1	0	0	10, 12, 13
0	1	0	5, 4, 7
0	0	1	8, 8, 10
1/2	1/2	0	12, 16, 15
1/2	0	1/2	10, 8, 9
0	1/2	1/2	9, 9, 12
1/3	1/3	1/3	13, 9, 14

Source: Cornell (1990b, Table 2). Reprinted by permission of Taylor & Francis.

```
Estimated Regression Coefficients for Thickness (component proportions)

Term       Coef  SE Coef      T      P     VIF
A        11.667    1.008      *      *   1.611
B         5.333    1.008      *      *   1.611
C         8.667    1.008      *      *   1.611
A*B      23.333    4.938   4.73  0.000   1.796
A*C      -4.667    4.938  -0.95  0.361   1.796
B*C      12.000    4.938   2.43  0.029   1.796
A*B*C     1.000   34.740   0.03  0.977   1.630

S = 1.74574       PRESS = 96
R-Sq = 77.84%     R-Sq(pred) = 50.15%    R-Sq(adj) = 68.35%

Analysis of Variance for Thickness (component proportions)

Source            DF   Seq SS    Adj SS    Adj MS     F      P
Regression         6  149.905  149.9048   24.9841  8.20  0.001
  Linear           2   48.933   60.2222   30.1111  9.88  0.002
  Quadratic        3  100.969   88.9000   29.6333  9.72  0.001
  Special Cubic    1    0.003    0.0025    0.0025  0.00  0.977
Residual Error    14   42.667   42.6667    3.0476
Total             20  192.571
```

Display 10.7 Minitab output for full special cubic model fit to vinyl thickness data.

blend has $x_1 = 0.6255, x_2 = 0.3745, x_3 = 0$ with the predicted maximum $\widehat{y} = 14.578$. Thus the optimum blend does not use the third plasticizer; in practice, a small percentage of it may be required to satisfy other considerations. The standard error of $\widehat{y}$ calculated using Minitab is 0.858; hence a 95% CI for the average vinyl thickness obtained with this optimum blend is (using $t_{16,0.025} = 2.120$) $[14.578 \pm (2.120)(0.858)] = [12.760, 16.396]$. ■

In many mixture experiments there are constraints on the proportions of the ingredients or there are other practical constraints which reduce the experimental region from the original simplex to a possibly irregular shaped subset of it. Designs for such situations must be found using a computer. These **computer-aided designs** are discussed in Cornell (1990a), Nachtsheim (1987), and Snee (1985b).

10.3 TAGUCHI METHOD OF QUALITY IMPROVEMENT

Traditionally, quality control has meant stipulating specification limits (abbreviated as specs) on key quality (functional) characteristics of a product and using end-of-process inspection to eliminate nonconforming items. Sampling inspection is generally employed for this purpose. However, any form of inspection to screen out defectives does not get at the root of the problem, leading to continued production of scrap and waste. Although inspection is often necessary to guarantee outgoing quality, it does not improve it.

```
Estimated Regression Coefficients for y (component proportions)

Term     Coef  SE Coef     T      P    VIF
A      11.257   0.8888     *      *  1.336
B       5.364   0.9717     *      *  1.597
C       8.257   0.8888     *      *  1.336
A*B    23.597   4.4643  5.29  0.000  1.566
B*C    12.264   4.4643  2.75  0.014  1.566

S = 1.69037       PRESS = 77.6178
R-Sq = 76.26%     R-Sq(pred) = 59.69%    R-Sq(adj) = 70.32%

Analysis of Variance for y (component proportions)

Source          DF   Seq SS   Adj SS  Adj MS      F      P
Regression       4  146.854  146.854  36.713  12.85  0.000
   Linear        2   48.933   57.722  28.861  10.10  0.001
   Quadratic     2   97.920   97.920  48.960  17.13  0.000
Residual Error  16   45.718   45.718   2.857
   Lack-of-Fit   2    3.051    3.051   1.526   0.50  0.617
   Pure Error   14   42.667   42.667   3.048
Total           20  192.571

Response Optimization

Global Solution

Components

A   =   0.625528
B   =   0.374472
C   =          0

Predicted Responses

y   =   14.5779
```

Display 10.8 Minitab output for reduced special cubic model fit to vinyl thickness data along with response optimization.

Statistical process control (SPC) is a more proactive approach which aims at preventing production of defectives by alerting the operator when the process goes out of control, that is, when the observed variation is in excess of the expected variation for a stable process (a process subject only to **common causes** of variation). Such an alert, usually triggered by a point falling outside the control limits on a control chart, indicates a possible presence of **special causes** which need to be identified and removed. Although SPC is a step forward from sampling inspection, it still does not improve the process itself. It only helps to detect problems that throw a process out of control.

The Taguchi method is fundamentally different from the previous approaches in that it aims to design quality into the product rather than inspecting into it. As we shall see below, the Taguchi method does not achieve this by tightening

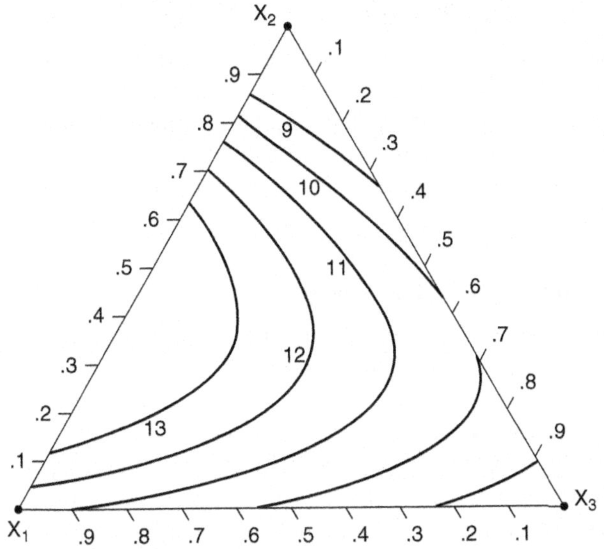

Figure 10.12 Ternary plot for special cubic model for vinyl thickness data.

the part tolerances or using more high quality parts, which are costly alternatives to improve quality. Instead, the Taguchi method aims to design the products and processes so that their performance is not affected (and hence their quality is not degraded) by variations in factors or conditions that are difficult to control.

Taguchi deserves the credit for popularizing the use of designed experiments to make the products and processes robust to environmental and other noise factor variations; however, this problem has a long history. Box (1999) and Box and Jones (1992a, b, 2000) have pointed out that it was Michaels (1964), who first showed that this problem is best dealt with by using a split-plot design (see Chapter 12). As will be discussed in Section 10.3.3, Taguchi's crossed-array designs for this problem are often not the most efficient designs for this problem.

The following is an example from Taguchi and Wu (1979, p. 50) and discussed in Phadke (1989, p. 5) which conveys the essence of the Taguchi method.

Example 10.7 (The Ina Tile Company)

The Ina Tile Company in Japan had a quality problem because of high variability in the dimensions of its tiles. The problem was traced to nonuniform temperature distribution inside the kiln in which the tiles were baked. The temperature in the center of the kiln was lower than that at the periphery. A redesign of the kiln to make the temperature uniform would have cost half a million dollars. Instead, experiments were carried out by varying the composition of the tile and it was found that increasing the lime content of the tile from 1 to 5% made the tile dimensions insensitive to temperature variations. Since lime was the least expensive ingredient of the tiles, this solution was also cost efficient. ■

10.3.1 Philosophy Underlying Taguchi Method

The basic premise of the Taguchi method is that it is not sufficient to meet the specs. Rather one must attempt to minimize the variation of the quality characteristic of the product (the response variable) around its nominal target value. The following example from Phadke (1989, p. 15), based on a customer preference study done by the Japanese newspaper *The Asahi* in 1979, illustrates this point.

Example 10.8 (TV Color Density)

Figure 10.13 shows the distributions of color densities of the TV sets made by Sony-USA and Sony-Japan. In the figure, T is the target color density and $T \pm 5$ are the specs. Grades A, B, C, and D (nonconforming) are progressively worse grades of TV sets as their color density deviates from the target. The Sony-USA distribution is more or less uniform over the specs with no nonconforming sets. Such a distribution typically arises when screening inspection is used to eliminate defectives. The standard deviation of this distribution is $5/\sqrt{3}$. On the other hand, the Sony-Japan distribution is normal centered at the target with a standard deviation of $\frac{5}{3}$, which results in 0.3% nonconforming sets. Despite this, customers preferred Sony-Japan sets to Sony-USA sets because a much larger proportion of the former were A grade. The moral here is that tighter variation around the target is more critical than conforming to specs. ■

Taguchi introduced the idea of **societal loss** resulting from lack of quality, defined as deviation of the functional characteristic of a product from its target. He

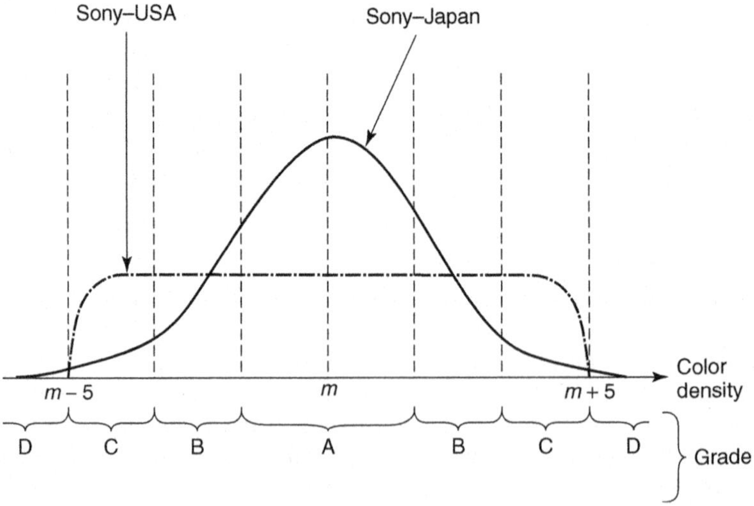

Figure 10.13 Distribution of color density in Sony-USA and Sony-Japan TV sets (Phadke, 1989, Figure 2.1). Courtesy of AT&T Archives and History Center.

suggested the use of the **quadratic loss function** (or more generally a convex loss function), which has the form $L(y) = K(y - T)^2$, where K is a positive constant. This is in contrast to the traditional focus on conformance to specification limits, which implies a so-called **goal post loss function**. This is a step function that equals zero for y within specs and a positive constant K for y outside specs. For the quadratic loss function, the expected loss is proportional to the variance of the distribution if the process is on target on the average. In the above example, the expected loss for Sony-USA sets is three times that for Sony-Japan sets.

As in Chapter 1, we will classify the factors affecting the response into design factors (also referred to as controllable factors), denoted by $\boldsymbol{x} = (x_1, x_2, \ldots, x_p)'$, and noise factors, denoted by $\boldsymbol{z} = (z_1, z_2, \ldots, z_q)'$. The levels of the design factors can be easily set by the investigator. On the other hand, the noise factors are not easy to control or are too expensive to control. In the Ina Tile example, the lime content is a design factor while the kiln temperature is a noise factor. Other examples of noise factors include variations in raw materials, process conditions, environmental conditions (e.g., room temperature and humidity) during production, deviations in the field use of the product from its recommended use, and degradation of the product over time.

To minimize the variation transmitted in the product functional characteristic by the noise factors, one can either adopt a costly solution such as tightening the tolerances on the raw materials or controlling the environmental conditions by using a clean room. Sometimes such costly solutions are unavoidable, for example, in the production of semiconductor chips. Taguchi suggested investigating a more economical solution first, which is to set the levels of the design factors so that the transmitted variation due to noise factors is minimized. This is called **robust parameter design** or simply **parameter design**. If further reduction in variation is desired, then the costlier approach of tightening the tolerances may be employed, which is called **tolerance design**. In the Ina Tile example, by setting the lime content level at 5%, the variation in tile dimensions caused by the kiln temperature variation was minimized, which is an application of parameter design.

Parameter design reduces variation by exploiting (i) interactions between the design factors and noise factors and (ii) nonlinearity in the response function. To see how interactions can be exploited to reduce variation, see Figure 10.14. In the left panel, the design factor x and the noise factor z do not interact. Therefore, as z varies over its range, the variation in y is the same whether x is set equal to -1 or $+1$. Here x affects only the mean of y, not its variance. Such a design factor is called a **signal factor** or an **adjustment factor**. In the right panel, the design factor x and the noise factor z do interact with a flat mean profile for $x = -1$ and a steep mean profile for $x = +1$. As shown in the figure, the variation in z is attenuated in its transmission to y by choosing $x = -1$. Here x affects the variance of y as well as its mean. Such a design factor is called a **control factor**. Thus design factors may be classified into three categories:

(a) *Signal factors*: Influence only the mean of the response but not its variance.

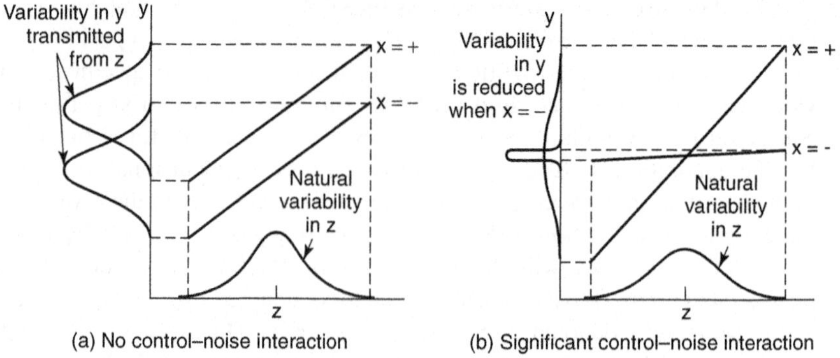

Figure 10.14 Interaction plots between two control factors and a noise factor (Montgomery, 2005, Figure 12-1). Reprinted by permission of John Wiley & Sons, Inc.

(b) *Control factors*: Influence the variance of the response and may or may not influence the mean.

(c) *Inactive factors*: Influence neither the mean nor the variance of the response.

The primary goal of the Taguchi method is to use designed experiments to classify the design factors into these categories and choose their best levels.

We next discuss how nonlinearity in the response function can be exploited. Consider the following example from Phadke (1989, p. 28).

Example 10.9 (Output Voltage of a Power Supply Circuit)

An electrical power supply circuit is to be designed with a target output of 110 volts. The output is a nonlinear increasing function of the transistor gain and a linear decreasing function of the resistance of the dividing resistor, as shown in Figure 10.15. The target output voltage can be attained if the transistor gain is set at A_1, but the slope of the curve is very steep at this point, which means that the transmitted variation in the voltage due to variation in the transistor gain (the noise factor) will be large. On the other hand, if the transistor gain is set at A_2 in a flat portion of the curve, the transmitted variation will be much smaller. However, the output voltage at this setting equals 125 volts. To bring it down to the target level, we can set the resistance of the dividing resistor at B_2. Note that because the relationship between the output voltage and the resistance is linear, the transmitted variation due to variation in the resistance is the same at all settings. In other words, the transistor gain has an effect on both the mean and variance of the output voltage, whereas the resistance has an effect only on the mean. The former is a control factor while the latter is a signal factor. ■

In summary, the three basic tenets underlying the Taguchi method are as follows:

(a) Focus on minimizing the variation around the target instead of conforming to specs.

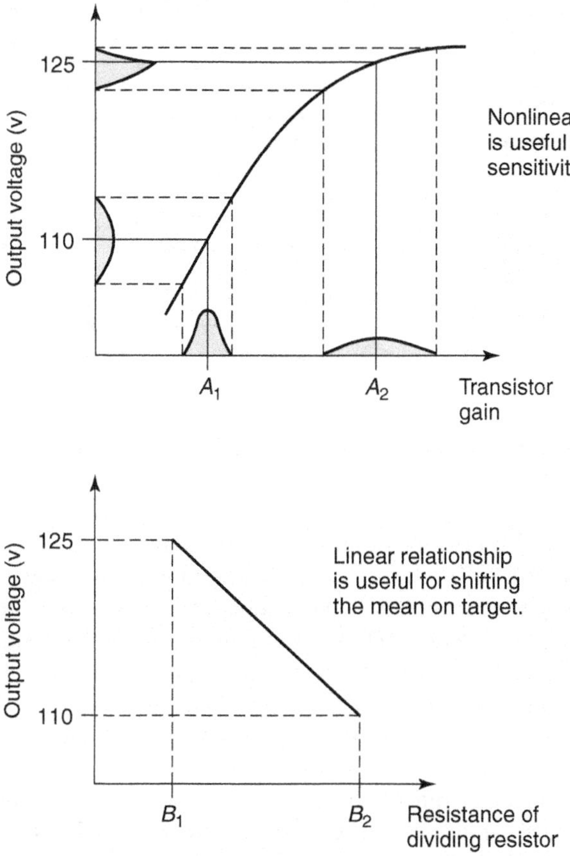

Figure 10.15 Output voltage as nonlinear function of transistor gain and linear function of resistance of dividing resistor (Phadke, 1989, Figure 2.6). Courtesy of AT&T Archives and History Center.

(b) Design the products and processes so that they are insensitive (robust) to noise factor variations.

(c) Use designed experiments to choose the levels of design factors to make the products and processes robust.

10.3.2 Implementation of Taguchi Method

The Taguchi method is implemented by carrying out a factorial experiment in design factors and noise factors. Note that, although the noise factors are infeasible to control in practice, they are controlled in the experiment. Taguchi proposed a **crossed-array design** (see Example 8.2), which consists of an orthogonal array in design factors, called an **inner array**, and an orthogonal array in noise factors, called an **outer array**, which is crossed with each run of the inner array. This design is depicted in Figure 10.16. Suppose there are m runs in the inner array and n runs in the outer array. Then there are mn runs in the crossed array.

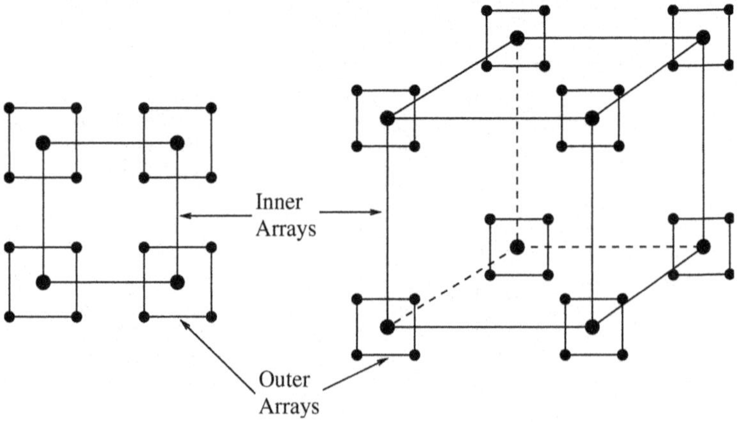

Figure 10.16 Crossed-array designs used in Taguchi method.

Let y_{ij} denote the observation at the ith run of the inner array and the jth run of the outer array and let $\overline{y}_i$ and s_i be the sample mean and standard deviation of the y_{ij}. These may be regarded as the sample estimates of the corresponding population mean and population SD for the ith run setting of the design factors as noise factors are allowed to vary. (Note that the noise factors are not allowed to vary randomly in the experiment but are varied systematically over their natural ranges of variation.) As an inverse measure of variation, Taguchi proposed the use of SN ratios. A particular SN ratio is chosen depending on the problem and is computed for each run of the inner array. To minimize variation, one must maximize the SN ratio. The three most commonly used SN ratios are the following:

Nominal Is Best: This measure is used when a particular nominal value is the target as in Example 10.8 and is given by

$$(\mathrm{SN})_i = 20 \log_{10}\left(\frac{\overline{y}_i}{s_i}\right). \tag{10.15}$$

Smaller Is Better: This measure is used when the functional characteristic is to be minimized, for example, surface defects. The SN ratio is given by

$$(\mathrm{SN})_i = -10 \log_{10}\left(\frac{1}{n}\sum_{j=1}^{n} y_{ij}^2\right). \tag{10.16}$$

Larger Is Better: This measure is used when the functional characteristic is to be maximized, for example, the shear strength of a steel shaft. The SN ratio is given by

$$(\mathrm{SN})_i = -10 \log_{10}\left(\frac{1}{n}\sum_{j=1}^{n}\frac{1}{y_{ij}^2}\right). \tag{10.17}$$

For specific problems different SN ratios may be devised depending on the particulars of the problem. The following example illustrates a crossed-array experiment and calculation of the nominal-is-best SN ratio.

Example 10.10 (Free Height of Leaf Springs: Data and SN Ratios)

Pignatiello and Ramberg (1985) analyzed an experiment to study a process for inducing camber in a leaf spring used in a truck assembly. The description of the process given by the authors is as follows: "The leaf spring assembly is transported by a conveyor system through a high temperature furnace. Upon exiting the furnace, the part is transferred to a forming machine where the camber (curvature) is induced by holding the spring in a high pressure press for a short length of time. Next, the spring is submerged into an oil quench and then removed from the processing area." The effects of the five factors shown in Table 10.6 on the free height of the spring were investigated.

The factors A, B, C, D were design factors and E was a noise factor (but was controlled at its low and high settings for the experiment). The overall goal of the experiment was to find the settings of these factors so as to attain the target of 8 in. free height for the spring with minimum variation around this target. The main effects of all four design factors and three two-factor interactions, AB, AC, and BC, were deemed important.

A 2^{4-1} design in A, B, C, D with the defining relation $I = ABCD$ and resolution IV was used as the inner array and a 2^1 design in E was used as the outer array. The alias structure for this design can be found from Example 8.2. Interactions CD, BD, and AD, which were aliased with AB, AC, and BC, respectively, were assumed to be negligible. The L_8 orthogonal array used for this design is shown in Display 10.9.

Table 10.6 Factors and Factor Levels for Leaf Spring Experiment

	Level	
Factor	Low (−)	High (+)
A: Furnace temperature (°F)	1840	1880
B: Heating time (sec)	25	23
C: Transfer time (sec)	12	10
D: Hold-down time (sec)	2	3
E: Quench oil temperature (°F)	130–150	150–170

Source: Pignatiello and Ramberg (1985, Table 1). Reprinted by permission of the American Society for Quality.

A	B	C	D	AB	AC	BC
−	−	−	−	+	+	+
+	−	−	+	−	−	+
−	+	−	+	−	−	+
−	+	−	+	−	+	−
−	−	+	+	+	−	−
+	−	+	−	−	+	−
−	+	+	−	−	−	+
+	+	+	+	+	+	+

Display 10.9 L_8 Orthogonal array used as inner array in truck leaf spring experiment.

Three replicates were observed for each combination of the factors. The data for the experiment are shown in Table 10.7. What is curious about these data is that, of the 48 observations, only 5 are 8 in. or higher, suggesting that the experimental region was not appropriately chosen. In this table, the sample means $\overline{y}_i$ and the sample standard deviations s_i are calculated for each run. Thus $\overline{y}_1 = 7.540$ and $s_1 = 0.300$ are the sample mean and sample standard deviation for the 6 observations (three replicates each at $E = -$ and $E = +$). The nominal-is-best SN ratio defined in (10.15) is used. We will analyze these data in subsequent examples. ■

The next task is to identify the control and signal factors and choose their appropriate levels to minimize variation and set the mean on target. Either a formal ANOVA or informal graphical plots (with the SN ratio and the sample means used as response variables) are used toward this end. Taguchi recommends the following two-step procedure:

Step 1: Choose the levels of the control factors to minimize variation by maximizing SN ratios.

Step 2: Choose the levels of the signal factors to bring the mean on target.

The levels of the factors that are neither control nor signal factors can be chosen based on side considerations such as cost and convenience. We will now apply this two-step process to the truck leaf spring free height data.

Example 10.11 (Free Height of Leaf Springs: Identifying Control and Signal Factors)

Refer to Example 10.10 and the summary statistics calculated in Table 10.7. Treating the SN ratio as a response, the estimated effects are shown in Display 10.10. Since there is only a single replicate observation for SN at each

Table 10.7 Truck Leaf Spring Free Height Data (in.)

	Inner Array				Outer Array						Summary Statistics		
					$E = -$			$E = +$					
Run	A	B	C	D	Rep. I	Rep. II	Rep. III	Rep. I	Rep. II	Rep. III	$\overline{y}_i$	s_i	$(\text{SN})_i$
1	−	−	−	−	7.78	7.78	7.81	7.50	7.25	7.12	7.540	0.300	28.005
2	+	−	−	+	8.15	8.18	7.88	7.88	7.88	7.44	7.902	0.266	29.457
3	−	+	−	+	7.50	7.56	7.50	7.50	7.56	7.50	7.520	0.031	47.697
4	+	+	−	−	7.59	7.56	7.75	7.63	7.75	7.56	7.640	0.089	38.674
5	−	−	+	+	7.94	8.00	7.88	7.32	7.44	7.44	7.670	0.301	28.125
6	+	−	+	−	7.69	8.09	8.06	7.56	7.69	7.62	7.785	0.230	30.591
7	−	+	+	−	7.56	7.62	7.44	7.18	7.18	7.25	7.372	0.195	31.551
8	+	+	+	+	7.56	7.81	7.69	7.81	7.50	7.59	7.660	0.131	35.339

Source: Pignatiello and Ramberg (1985, Table 2). Reprinted by permission of the American Society for Quality.

```
Estimated Effects and Coefficients for SN (coded units)

Term        Effect     Coef
Constant              33.680
A           -0.329   -0.165
B            9.271    4.635
C           -4.557   -2.278
D            2.949    1.475
A*B         -2.288   -1.144
A*C          3.456    1.728
B*C         -5.184   -2.592
```

Display 10.10 Minitab output for effect estimates with $(\mathrm{SN})_i$ as response.

inner array run, there are no error d.f. for testing the significance of the effects. Pignatiello and Ramberg (1985) created an SS_e term by pooling the SS of those effects whose cumulative contribution to SS_{tot} was less than 10%. These are the two smallest effects, $\widehat{A}$ and $\widehat{AB}$. Using Eq. (7.16) for SS_{Effect}, we get

$$SS_e = \tfrac{1}{4}N[\widehat{A}^2 + (\widehat{AB})^2] = \tfrac{8}{4}[(-0.329)^2 + (-2.288)^2] = 10.686 \qquad \text{and}$$

$$MS_e = \tfrac{1}{2}(10.686) = 5.343.$$

Using this MS_e with two d.f. they carried out F-tests on the remaining effects and found that the two largest effects, $\widehat{B}$ and $\widehat{BC}$, were significant at $\alpha = 0.10$. Therefore, although the $\widehat{C}$ main effect barely fails to be significant at the 0.10 level, choice of its level is important. Thus, B and C are *control factors*. The BC interaction plot in Figure 10.17 shows that, to maximize the SN ratio, B should be set at high level and C should be set at low level (which choice happens to be in accord with the signs of their main effects).

Next we do a similar analysis using the mean as the response. The estimated effects are shown in Display 10.11. Here all three estimated interactions and the

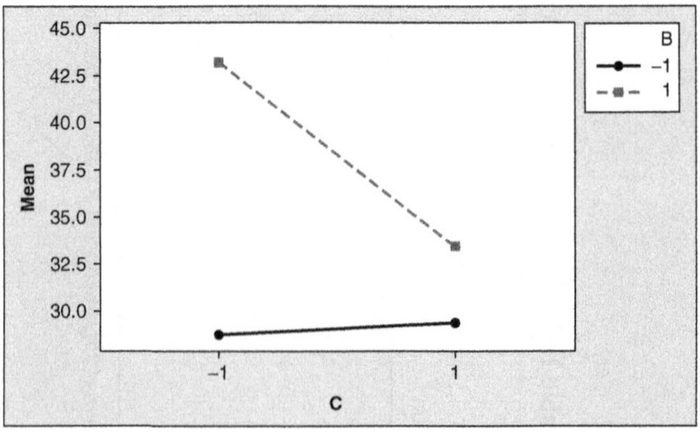

Figure 10.17 Plot of BC interaction with SN ratio as response.

```
Estimated Effects and Coefficients for Mean (coded units)

Term          Effect       Coef
Constant                  7.63613
A            0.22125     0.11063
B           -0.17625    -0.08813
C           -0.02875    -0.01437
D            0.10375     0.05188
A*B         -0.01725    -0.00862
A*C         -0.01975    -0.00987
B*C         -0.03525    -0.01762
```

Display 10.11 Minitab output for effect estimates with $\overline{y}_i$ as response.

main effect $\widehat{C}$ are nonsignificant in that they contribute less than 10% to SS_{tot}. Pooling their SS, we obtain

$$\begin{aligned}\text{SS}_e &= \tfrac{1}{4}N[\widehat{C}^2 + (\widehat{AB})^2 + (\widehat{AC})^2 + (\widehat{BC})^2] \\ &= \tfrac{8}{4}[(-0.02875)^2 + (-0.01725)^2 + (-0.01975)^2 + (-0.03525)^2] \\ &= 0.0055\end{aligned}$$

and

$$\text{MS}_e = \tfrac{1}{4}(0.0055) = 0.0014.$$

Now

$$\text{MS}_A = 2(0.22125)^2 = 0.0979, \qquad \text{MS}_B = 2(-0.17625)^2 = 0.0621,$$

$$\text{MS}_D = 2(0.10375)^2 = 0.0215.$$

Therefore

$$F_A = \frac{0.0979}{0.0014} = 69.93, \qquad F_B = \frac{0.0622}{0.0014} = 44.43, \qquad F_D = \frac{0.0215}{0.0014} = 15.36,$$

which are all highly significant ($p = 0.001, 0.003$, 0.017, respectively). So A and D are *signal factors* and their effects are positive. Since all of the mean free heights in Table 10.7 are below the target value of 8 in., we should set both A and D at high levels. Thus the final choice of factor levels is

$$A\,\text{high},\ B\,\text{high},\ C\,\text{low, and}\ D\,\text{high}.$$

A confirmatory experiment should be conducted to verify that the design objectives are met. The estimated mean free height at this factor-level combination (ignoring the nonsignificant effects) is

$$\widehat{y} = 7.63613 + 0.11063 - 0.08813 + 0.01437 + 0.05188 = 7.72488.$$

This is still well below the target value of 8 in. One way to increase the mean height would be to explore still higher values of A and D. ■

10.3.3 Critique of Taguchi Method

There is a general agreement that Taguchi's philosophy about quality improvement is right on mark, but the design and analysis methods that he proposed are not necessarily the best available. In particular, the three basic tenets summarized at the end of Section 10.3.1 provide an ideal working framework, but the statistical methods have come under much criticism. In this section we will review these criticisms.

First consider the designs used in Taguchi experiments. The inner–outer array construction in crossed-array designs is aimed at estimating design and noise factor interactions, which are necessary to identify control factors. However, these designs are generally too large for the resolution that they offer. The orthogonal arrays that Taguchi recommends for the inner array often have resolution III (to keep their size small), which means that interactions between the design factors are aliased. Instead, a **combined array** in both design and noise factors offers a more economical approach with a higher resolution. For example, the leaf spring experiment uses a $2^{4-1} \times 2^1$ crossed array with 16 runs and resolution IV. On the other hand, a combined array between all five factors consisting of a 2^{5-1} experiment with defining relation $I = ABCDE$ and the same number of runs has resolution V.

Taguchi does not emphasize the importance of randomization in the design and analysis of the experiment. As has been mentioned in earlier chapters, lack of randomization could introduce biases due to uncontrolled noise factors. Often, in crossed-array experiments, it is more practical to carry out two separate randomizations, first with respect to the design factors in the inner array and then with respect to the noise factors in the outer array. Such designs are called **split-plot designs**, which are discussed in Chapter 12. The analysis of these designs involves two separate errors, the so-called **whole-plot error** for testing the main effects and interactions among the design factors (which are referred to as whole plot factors) and the **subplot error** for testing all other effects involving noise factors (which are referred to as subplot factors), in particular, interactions among design and noise factors. The subplot error is generally smaller than the whole-plot error, thus rendering the tests on the effects involving noise factors more powerful. This aspect of analysis is ignored in Taguchi's approach and the designs are analyzed as completely randomized designs with a single error which can lead to wrong conclusions, as Example 12.6 shows.

Taguchi focuses on estimating the curvature of the response function at the expense of estimating the interactions among the design factors. Therefore, for inner arrays, he recommends three-level fractional factorial designs which have complicated alias structures. An example of such a design is given in Exercise 10.17. That design uses 72 runs and is yet unable to estimate two-factor interactions between design factors. Montgomery (1997, p. 597) has given a

2^{7-2} combined array with only 32 runs and resolution IV in which not only all main effects are clear of all two-factor interactions but also all two-factor interactions between design and noise factors are clear of each other.

Next consider the statistical analysis strategies recommended by Taguchi. Here the main critique is that Taguchi recommends choosing the levels of design factors based on their main effects ignoring interactions. This could lead to the wrong choice of levels of design factors. Second, maximizing an SN ratio does not necessarily translate into minimizing variation. For instance, the nominal-is-best SN ratio confounds the effects of the mean and variance, as has been shown by Box (1988). It is much more direct to minimize the variance itself. Example 10.12 gives an alternative analysis of the leaf spring experiment data using this approach.

Example 10.12 (Free Height of Leaf Springs: Alternative Analysis)

Pignatiello and Ramberg (1985) observed that if the noise factor E (the oil quench temperature) is treated as a design factor, then quite different results emerge. We will now carry out this alternative analysis. Table 10.8 gives the corresponding summary statistics. Note that, as opposed to the standard deviations calculated in Table 10.7, here they are calculated within each level of E from three replicates and thus represent "pure error" independent of the variation caused by E.

Display 10.12 shows the Minitab output for the effect estimates when $z = -10 \log_{10} s^2$ is taken as the response and E is treated as a design factor. Note that, to minimize variation, this response must be maximized. Since we have

Table 10.8 Summary Statistics for Truck Leaf Spring Free Height Data Treating E as Design Factor

Run	A	B	C	D	E	$\bar{y}_i$	s_i	$-10 \log_{10} s_i^2$
1	−	−	−	−	−	7.790	0.017	35.299
2	+	−	−	+	−	8.070	0.165	15.638
3	−	+	−	+	−	7.520	0.035	29.208
4	+	+	−	−	−	7.633	0.102	19.816
5	−	−	+	+	−	7.940	0.060	24.437
6	+	−	+	−	−	7.947	0.223	13.042
7	−	+	+	−	−	7.540	0.092	20.757
8	+	+	+	+	−	7.687	0.125	18.060
9	−	−	−	−	+	7.290	0.193	14.283
10	+	−	−	+	+	7.733	0.254	11.902
11	−	+	−	+	+	7.520	0.035	29.208
12	+	+	−	−	+	7.647	0.096	20.346
13	−	−	+	+	+	7.400	0.069	23.188
14	+	−	+	−	+	7.623	0.065	23.733
15	−	+	+	−	+	7.203	0.040	27.689
16	+	+	+	+	+	7.633	0.159	15.946

```
Estimated Effects and Coefficients for z (coded units)

Term          Effect      Coef
Constant                21.427
A             -8.213    -4.106
B              2.482     1.241
C             -1.061    -0.531
D             -0.954    -0.477
E             -1.199    -0.599
A*B           -0.017    -0.009
A*C            1.836     0.918
A*D           -2.907    -1.453
A*E            2.550     1.275
B*E            2.603     1.301
C*E            4.838     2.419
D*E           -0.576    -0.288
A*B*E         -4.746    -2.373
A*C*E         -1.886    -0.943
A*D*E         -3.700    -1.850
```

Display 10.12 Minitab output for effect estimates with $z_i = -10 \log_{10} s_i^2$ as response and treating E as design factor.

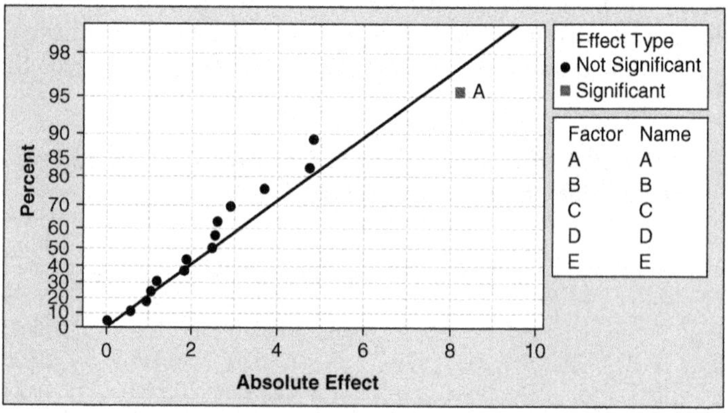

Figure 10.18 Half-normal plot of effect estimates with $z_i = -10 \log_{10} s_i^2$ as response and treating E as design factor.

only a single observation on z at each run, we determine significant effects using a half-normal plot shown in Figure 10.18. We see that only the main effect A is indicated to be significant. Since this effect is negative, A must be set at low level to maximize z.

Next, to identify the signal factors, we analyze the raw free height data. Since we have three replicates at each run of the experiment, a pure-error estimate with 32 d.f. and hence the t- or F-tests for the significance of the effects can be computed as shown in Display 10.13. The half-normal plot of the effects is shown in Figure 10.19. We see that the main effects A, B, D, E and the interactions AE

```
Estimated Effects and Coefficients for Free (coded units)

Term          Effect      Coef     SE Coef       T      P
Constant                7.6360     0.01857  411.18  0.000
A             0.2212    0.1106     0.01857    5.96  0.000
B            -0.1763   -0.0881     0.01857   -4.75  0.000
C            -0.0287   -0.0144     0.01857   -0.77  0.445
D             0.1038    0.0519     0.01857    2.79  0.009
E            -0.2596   -0.1298     0.01857   -6.99  0.000
A*B          -0.0171   -0.0085     0.01857   -0.46  0.649
A*C          -0.0196   -0.0098     0.01857   -0.53  0.602
A*D          -0.0354   -0.0177     0.01857   -0.95  0.347
A*E           0.0846    0.0423     0.01857    2.28  0.030
B*E           0.1654    0.0827     0.01857    4.45  0.000
C*E          -0.0538   -0.0269     0.01857   -1.45  0.158
D*E           0.0271    0.0135     0.01857    0.73  0.471
A*B*E        -0.0104   -0.0052     0.01857   -0.28  0.781
A*C*E         0.0404    0.0202     0.01857    1.09  0.285
A*D*E        -0.0471   -0.0235     0.01857   -1.27  0.214

Analysis of Variance for Free (coded units)

Source                 DF     Seq SS     Adj SS     Adj MS      F      P
Main Effects            5    1.90788    1.90788    0.38158  23.05  0.000
2-Way Interactions      7    0.48083    0.48083    0.06869   4.15  0.002
3-Way Interactions      3    0.04751    0.04751    0.01584   0.96  0.425
Residual Error         32    0.52973    0.52973    0.01655
  Pure Error           32    0.52973    0.52973    0.01655
Total                  47    2.96595
```

Display 10.13 Minitab output for effect estimates with free height measurement (y_{ij}) as response and treating E as design factor.

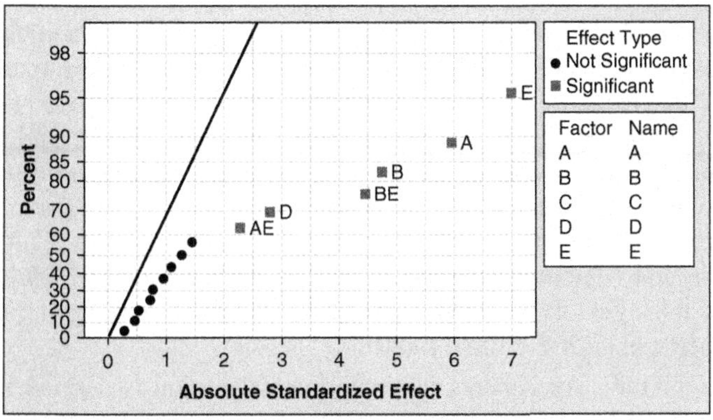

Figure 10.19 Half-normal plot of effect estimates with free height as response and treating E as design factor.

and BE are significant. The factor C is inactive. Interestingly, E, which was a noise factor in the original experiment, has the largest effect size. Therefore, if the cost of controlling E is not prohibitively high, it makes practical sense to treat E as a design factor and control it at a proper level.

The target free height is 8 in., but all except one factor setting give the average height less than 8 in. Therefore we need to maximize the free height. Factor C can be set at either level depending on side considerations. For the remaining four factors, the combination $\{A$ high, B low, D high, E low$\}$ gives the highest mean free height (8.0568 in.). However, we must choose A low to minimize variation as discussed before. Therefore the final recommended combination is $\{A$ low, B low, D high, E low$\}$, for which the mean free height is

$$\begin{aligned}\widehat{y} &= 7.6361 - 0.1106 + 0.0881 + 0.0519 + 0.1298 + 0.0423 + 0.0827 \\ &= 7.9203.\end{aligned}$$

This value is slightly less than the target height of 8 in., but the variation at this setting is smaller. ■

10.4 CHAPTER SUMMARY

(a) In response surface optimization, there are two phases of experimentation. In the first phase, generally one is far from the maximum and locally a first-order model provides a good fit to the response surface. This fit is used to determine the direction of the steepest slope and a step is taken in that direction (assuming that the goal is to maximize the response); this is called the steepest-ascent method. The first-order designs used in this phase include resolution III fractional factorials with center points and simplex designs.

(b) In the second phase, experimentation is done in the neighborhood of the maximum. In this region it is important to be able to estimate the curvature of the response surface. Therefore second-order models are fitted using second-order designs, which include central composite and Box–Behnken designs. A key part of the analysis is the characterization of the stationary point of the response surface. This point may be a local maximum/minimum or a saddle point or a ridge system (stationary or rising). Canonical analysis is used to determine the nature of the stationary point which is characterized by the signs of the eigenvalues of the matrix formed from the second-order coefficients of the fitted model.

(c) In mixture experiments the experimental region is defined by a constraint on the values of the design variables that they must sum to a fixed constant (typically the design variables are proportions which sum to 1). Simplex lattice and simplex centroid designs are commonly used. Four types of models are fitted: linear, quadratic, full cubic, and special cubic.

(d) The basic idea behind the Taguchi method is to design products or processes so that their performance is insensitive to deviations from ideal operating conditions; in other words, minimize the variation caused by noise factors around the target level of performance. This is done through a designed experiment in which an inner array in the design factors and an outer array in the noise factors are used in a crossed-array design. The data analysis consists of two steps. In the first step, SN ratios (which are measures of variation caused by noise factors at design factor settings) are analyzed to identify the design factors that affect variability (called control factors); their levels are then chosen to minimize variability. In the second step, the performance measurements are analyzed to identify the design factors that affect only the mean performance (called signal factors); their levels are then chosen to bring the mean performance on target. This is called the robust parameter design.

(e) Taguchi's philosophy of robust design through designed experiments that incorporate systematic changes in not only the design factors but also the noise factors is widely accepted. However, his statistical methods, in particular, use of crossed arrays and two-step analysis of SN ratios and mean performance measures, are criticized by many statisticians. Better experimental designs and methods of data analysis are available.

EXERCISES

Section 10.1 (Response Surface Methodology)

Theoretical Exercises

10.1 Show that the following design matrix corresponds to the tetrahedron 2^{3-1} design shown in Figure 10.2:

$$\boldsymbol{D} = \begin{array}{c} \\ \end{array}\begin{bmatrix} -1/\sqrt{2} & -1/\sqrt{6} & -1/2\sqrt{3} \\ 1/\sqrt{2} & -1/\sqrt{6} & -1/2\sqrt{3} \\ 0 & \sqrt{2/3} & -1/2\sqrt{3} \\ 0 & 0 & \sqrt{3}/2 \end{bmatrix}.$$

(columns: x_1, x_2, x_3)

10.2 Show that the gradient of the planar surface $\widehat{y} = \widehat{\beta}_0 + \sum_{i=1}^{p} \widehat{\beta}_i x_i$, namely $(\partial\widehat{y}/\partial x_1, \ldots, \partial\widehat{y}/\partial x_p)' = (\widehat{\beta}_1, \ldots, \widehat{\beta}_p)'$ is the direction of the steepest ascent. (*Hint*: Set up the Lagrangian to maximize $\widehat{y}$ subject to taking a unit step in any direction from the origin, i.e., subject to $\sum x_i^2 = 1$.)

10.3 Show that a 2^{p-q} fractional factorial design for fitting a first-order model is rotatable.

10.4 Write the design matrix of a Box–Behnken design for $p = 4$ with two center-point runs using the construction method of Example 10.3.

Applied Exercises

10.5 The following experimental study was reported in Yang, Lin, and Wen (1995). The background of a study is as follows: An aqueous polyurethane (PU) dispersion consists of PU particles dispersed in a continuous water phase. It is useful for coatings and adhesives applications. PU particle size is a key variable for the effectiveness of the dispersion with small average particle sizes ($<$ 200 nm) preferred because such dispersions are storage stable and possess high energy, resulting in a strong driving force for film formation. Yang et al. conducted a second-order experiment using a CC design with two center points and the following three factors: acetone/PU ratio (A), phase inversion temperature in degrees Celsius (B), and water addition rate in milliliters per minute (C). The coded and actual levels of the three factors are shown in Table 10.9. Experimental data are given in Table 10.10. Because of the wide range of data values, it is recommended to log-transform the data and then do the analysis.

(a) Fit a second-order response surface model.

(b) How should the factor levels be changed to minimize the particle size?

10.6 Wu and Hamada (2000) report data from an electrophoresis process experiment by Morris et al. (1997) to separate ranitidine hydrochloride (an active ingredient of Zantac, a commonly used drug for stomach ulcer). The experiment used a CC design augmented by six center points to study three factors: pH of the buffer solution (A) (levels: $2, 3.42, 5.5, 7.58, 9$), the voltage (kV) used in electrophoresis (B) (levels: $9.9, 14, 20, 26, 30.1$), and the concentration (mM) of α-CD (C), a component of the buffer solution (levels: $0, 2, 5, 8, 10$). The efficacy of

Table 10.9 Factors and Factor Levels for Polyurethane Experiment

	Actual Factor Level		
Coded Level	Acetone/PU Ratio	Phase Inversion Temperature (°C)	Water Addition Rate (mL/min)
−1.68	2.22	23.2	1.3
−1	2.80	30.0	2.0
0	3.65	40.0	3.0
+1	4.50	50.0	4.0
+1.68	5.08	56.8	4.7

Source: Yang, Lin, and Wen (1995, Table 5).

Table 10.10 Data from Polyurethane Experiment

	Factor			Average Particle
Run	A	B	C	Size (nm)
1	−1	−1	−1	70
2	+1	−1	−1	192
3	−1	+1	−1	120
4	+1	+1	−1	50
5	−1	−1	+1	152
6	+1	−1	+1	257
7	−1	+1	+1	86
8	+1	+1	+1	212
9	−1.68	0	0	173
10	+1.68	0	0	385
11	0	−1.68	0	1485
12	0	+1.68	0	1365
13	0	0	−1.68	90
14	0	0	+1.68	235

Source: Data adapted from Yang, Lin, and Wen (1995, Table 6).

separation was measured by the chromatographic exponential function (CEF), which is a quality measure of separation achieved and time of final separation. The natural log of CEF is used as the response variable. The goal is to minimize ln(CEF). The data using coded levels of the factors are given in Table 10.11. Wu and Hamada (2000) mention that data at run 7 was faulty; hence carry out the analyses by omitting this run.

(a) Fit a second-order model. Which factors appear to be active?

(b) Fit a reduced second-order model in active factors and find its stationary point.

(c) Make a contour plot and characterize the nature of the stationary point. Check using eigenanalysis.

10.7 Wu and Hamada (2000) give data on a follow-up experiment which was run as a CC design on two factors, pH of the buffer solution (A) and voltage (kV) used in electrophoresis (B) using narrower ranges of the factor levels (levels of A: 4.19, 4.50, 5.25, 6.00, 6.31; levels of B: 11.5, 14, 20, 26, 28.5). In terms of the coded levels, these correspond to $\pm\alpha$, ± 1, and 0 with $\alpha = 1.41$. In addition, there were five center-point runs. The data are shown in Table 10.12.

(a) Fit a second-order model.

(b) Make a contour plot and check that the stationary point gives a minimum. Show this using eigenanalysis.

Table 10.11 Ranitidine Separation Data

	Factor			
Run	*A*	*B*	*C*	ln(CEF)
1	−1	−1	−1	2.850
2	+1	−1	−1	3.817
3	−1	+1	−1	2.333
4	+1	+1	−1	9.372
5	−1	−1	+1	2.830
6	+1	−1	+1	3.235
7	−1	+1	+1	10.364
8	+1	+1	+1	9.396
9	−1.68	0	0	9.714
10	+1.68	0	0	10.179
11	0	−1.68	0	2.411
12	0	+1.68	0	1.897
13	0	0	−1.68	2.011
14	0	0	+1.68	1.842
15	0	0	0	2.288
16	0	0	0	2.262
17	0	0	0	2.182
18	0	0	0	2.173
19	0	0	0	2.081
20	0	0	0	2.087

Source: Wu and Hamada (2000, Table 9.2). Reprinted by permission of John Wiley & Sons, Inc.

Table 10.12 Ranitidine Separation Data

	Factor		
Run	*A*	*B*	ln(CEF)
1	−1	−1	2.390
2	1	−1	6.248
3	−1	1	2.065
4	1	1	3.252
5	−1.41	0	2.100
6	1.41	0	9.445
7	0	−1.41	6.943
8	0	1.41	1.781
9	0	0	2.034
10	0	0	2.009
11	0	0	2.022
12	0	0	1.925
13	0	0	2.113

Source: Wu and Hamada (2000, Table 9.12). Reprinted by permission of John Wiley & Sons, Inc.

Table 10.13 Mayonnaise Viscosity Data

A	*B*	*C*	Viscosity
−1	−1	−1	15
1	−1	−1	41
−1	1	−1	31
1	1	−1	81
−1	−1	1	30
1	−1	1	40
−1	1	1	49
1	1	1	95
0	0	0	35
0	0	0	48
0	0	0	43
0	0	0	45
−1	0	0	27
1	0	0	64
0	−1	0	22
0	1	0	56
0	0	−1	50
0	0	1	45

Source: Bjerke et al. (2004, Table 2). Reprinted by permission of Taylor & Francis.

(c) Find the factor levels corresponding to this minimum point and the predicted minimum value of the response.

10.8 Refer to Example 7.7 for the description of an experiment to study the effects of three production factors on the viscosity of low-fat mayonnaise. The full data (including the axial points with $\alpha = 1$, i.e., face centers) are given in Table 10.13. Fit a second-order response surface model and characterize the nature of the stationary point.

10.9 Refer to Exercise 7.15 for details of a paper helicopter experiment. The analysis of the data in that exercise shows that a second-order model needs to be fitted. Therefore a CC design with axial points at ± 2 was run with the factor levels shown in Table 10.14 and the results shown in Table 10.15.

(a) Combine the results of this design with those of the design from Exercise 7.15 to fit a second-order model to the data.

(b) Characterize the nature of the stationary point of the fitted second-order model.

10.10 Bae and Shoda (2005) used a BB design to optimize the culture conditions for bacterial cellulose (BC) production with respect to four factors:

Table 10.14 Factors and Their Levels for Paper Helicopter Experiment

Factor	−2	0	2
A: Wing area (in.^{2})	10.60	11.80	13.00
B: Wing length/width ratio	1.98	2.52	3.04
C: Body width (in.)	0.75	1.25	1.75
D: Body length (in.)	1.00	2.00	3.00

Source: Box and Liu (1999, Table 7). Reprinted by permission of the American Society for Quality.

Table 10.15 Helicopter Flight Times (centiseconds) Using Central Composite Design

Run	*A*	*B*	*C*	*D*	Flight Time
1	−2	0	0	0	361
2	2	0	0	0	364
3	0	−2	0	0	355
4	0	2	0	0	373
5	0	0	−2	0	361
6	0	0	2	0	360
7	0	0	0	−2	380
8	0	0	0	2	360

Source: Box and Liu (1999, Table 8). Reprinted by permission of the American Society for Quality.

A, fructose; *B*, corn steep liquor (CSP); *C*, dissolved oxygen (DO); and *D*, agar. The factor levels are given in Table 10.16.

The BC production data are given in Table 10.17.

(a) Fit a second-order model to the data. Observe that all two-factor interactions are highly nonsignificant, so fit a reduced model by dropping these interactions.

Table 10.16 Factors and Their Levels for BC Production Experiment

Factor	Factor Levels		
	−1	0	1
A: Fructose concentration (%)	3	4	5
B: CSP concentration (%)	2	3	4
C: DO concentration (%)	25	30	35
D: Agar concentration (%)	0.3	0.4	0.5

Source: Bae and Shoda (2005, Table I).

Table 10.17 BC Production Experiment Data

	Factors				BC Concentration
Run	*A*	*B*	*C*	*D*	(g/L)
1	−1	−1	0	0	8.80
2	1	−1	0	0	13.40
3	−1	1	0	0	7.67
4	1	1	0	0	12.87
5	0	0	−1	−1	12.08
6	0	0	1	−1	11.50
7	0	0	−1	1	11.88
8	0	0	1	1	11.73
9	−1	0	0	−1	8.20
10	1	0	0	−1	13.70
11	−1	0	0	1	8.90
12	1	0	0	1	13.55
13	0	−1	−1	0	10.10
14	0	1	−1	0	9.25
15	0	−1	1	0	9.76
16	0	1	1	0	9.60
17	−1	0	−1	0	7.32
18	1	0	−1	0	13.15
19	−1	0	1	0	9.18
20	1	0	1	0	13.26
21	0	−1	0	−1	10.52
22	0	1	0	−1	10.40
23	0	−1	0	1	9.90
24	0	1	0	1	10.40
25	0	0	0	0	12.98
26	0	0	0	0	12.98
27	0	0	0	0	12.98

Source: Bae and Shoda (2005, Table II).

(b) Characterize the nature of the stationary point of the reduced model. Suggest which factor-level settings should be chosen to maximize BC production or whether the maximum does not fall within the experimental region and further exploration is necessary.

Section 10.2 (Mixture Experiments)

Theoretical Exercises

10.11 Show that the usual quadratic model

$$\mathrm{E}(y) = \beta_0 + \sum_{i=1}^{p} \beta_i x_i + \sum_{j=i+1}^{p} \sum_{i=1}^{p-1} \beta_{ij} x_i x_j + \sum_{i=1}^{p} \beta_{ii} x_i^2$$

reduces to the canonical form (10.11). How are the parameters in the canonical model related to those in the above model? [*Hint*: Use $x_i^2 = x_i(1 - \sum_{j \neq i} x_j)$.]

10.12 Consider a $(p, 2)$-simplex lattice design with a single replicate observation at each design point.

(a) Derive the LS estimators given by (10.14). (Hint: Use the result of Exercise 14.6.)

(b) Find $\text{Var}(\widehat{\beta}_i)$ and $\text{Var}(\widehat{\beta}_{ij})$.

Applied Exercises

10.13 Snee (1973) describes a gasoline blending experiment to model the dependence of research octane (y) on four ingredients: light FCC (x_1), alkylate (x_2), butane (x_3), and reformate (x_4). The data for the experiment are given in Table 10.18.

(a) Fit the first-order model (10.10) to these data. (It would be helpful to subtract 100 from all responses to obtain a numerically more stable regression.) Notice that the coefficients of x_1, x_2, x_3 are similar, suggesting similar effects of the first three ingredients on the octane value in the experimental region.

(b) Fit the reduced model in which the regression coefficients for the first three ingredients are set equal. What are the advantages of this reduced model?

(c) Can you conclude that the pure blends of the first three ingredients give the same or similar octane values because their coefficients may be assumed to be equal? Why or why not?

Table 10.18 Gasoline Blending Data

x_1	x_2	x_3	x_4	y
0.250	0.400	0.100	0.250	102.6
0.250	0.400	0.030	0.320	102.0
0.250	0.200	0.100	0.450	101.2
0.150	0.400	0.030	0.420	101.7
0.250	0.200	0.030	0.520	100.7
0.150	0.200	0.100	0.550	100.7
0.150	0.400	0.100	0.350	101.8
0.150	0.200	0.030	0.620	100.2
0.200	0.300	0.065	0.435	101.2

Source: Snee (1973, Table 1).

Table 10.19 Elongation Data from Yarn Experiment

Component Proportions			Elongation Values
x_1	x_2	x_3	
1	0	0	11.0, 12.4
$\frac{1}{2}$	$\frac{1}{2}$	0	15.0, 14.8, 16.1
0	1	0	8.8, 10.0
0	$\frac{1}{2}$	$\frac{1}{2}$	10.0, 9.7, 11.8
0	0	1	16.8, 16.0
$\frac{1}{2}$	0	$\frac{1}{2}$	17.7, 16.4, 16.6

Source: Cornell (1990a, Table 2.4). Reprinted by permission of John Wiley & Sons, Inc.

10.14 Cornell (1990a) gave data from an experiment in which the effects on the elongation of yarn for draperies made from three constituents, polyethylene (x_1), polystyrene (x_2), and polypropylene (x_3), were studied. A (3,2)-simplex lattice design was used. Yarn elongations measured in kilograms of force applied are given in Table 10.19. Higher yarn elongations are desirable.

(a) Fit a canonical quadratic model to the data. Are all coefficients significant at the 5% level?

(b) Make a contour plot. Which combination would you use to maximize yarn elongation? What is the maximum estimated elongation?

(c) Find a 95% confidence interval to accompany this estimate.

10.15 Batra and Parsad (date unknown) give the data shown in Table 10.20 on the counts of mites on plants spread with three pesticides in different proportions. Two replicates were observed at each of the design points of a three-dimensional simplex centroid design supplemented by three axial points as shown in Figure 10.11.

(a) Fit the special cubic model to the data.

(b) Find the combination that minimizes the number of mites. What is the predicted minimum number of mites?

Section 10.3 (Taguchi Method of Quality Improvement)

Theoretical Exercise

10.16 Refer to Example 10.8. Suppose that the specification range $T \pm 5$ is divided into three equal regions for classifying the TV sets into grades A,

Table 10.20 Mites Data

Pesticide Proportions			Mites ($\times 10^{-2}$)	
x_1	x_2	x_3	y_1	y_2
1	0	0	3.8	3.0
0	1	0	4.0	4.7
0	0	1	5.1	4.6
$\frac{1}{2}$	$\frac{1}{2}$	0	1.8	2.2
$\frac{1}{2}$	0	$\frac{1}{2}$	2.3	3.0
0	$\frac{1}{2}$	$\frac{1}{2}$	3.5	2.9
$\frac{1}{3}$	$\frac{1}{3}$	$\frac{1}{3}$	3.6	4.5
$\frac{2}{3}$	$\frac{1}{6}$	$\frac{1}{6}$	2.9	3.5
$\frac{1}{6}$	$\frac{2}{3}$	$\frac{1}{6}$	4.6	4.0
$\frac{1}{6}$	$\frac{1}{6}$	$\frac{2}{3}$	3.5	2.8

Source: Batra and Parsad (date unknown).

B, and C. A grade sets satisfy $|y - T| \leq 1.67$, B grade sets satisfy $1.67 < |y - T| \leq 3.34$, and C grade sets satisfy $3.34 < |y - T| \leq 5$. Find the proportion of Sony-USA and Sony-Japan sets falling in each grade.

Applied Exercise

10.17 Byrne and Taguchi (1987) report an experiment conducted to find a method to assemble an elastomeric connector to a nylon tube for use in automotive engine components. One of the objectives was to maximize the pull-off force. Researchers studied four design factors, each at three levels, and three noise factors, each at two levels. The factors and their levels are given in Table 10.21. A 3^{4-2} fractional factorial was used for the inner array and a 2^3 full factorial was used for the outer array. The data are shown in Table 10.22.

(a) Calculate the sample mean and the larger-is-better SN ratio for each run of the inner array.

(b) Plot the average SN ratios for the three levels of each design factor. Similarly plot the average $\overline{y}$ values. Based on these plots identify control factors and signal factors. Which levels of these factors would you choose to minimize variation and maximize the pull-off force? Is this graphical analysis valid if there are interactions between the design factors?

(c) Suppose that after examining all pairwise interactions between design factors and noise factors the AG and DE interactions were

Table 10.21 Design and Noise Factors and Their Levels for Elastomeric Connector Experiment

Design factors	
A: Interference	Low, medium, high
B: Connector wall thickness	Thin, medium, thick
C: Insertion depth	Shallow, medium, deep
D: Percent adhesive in connector preip	Low, medium, high
Noise factors	
E: Conditioning time	24 h, 120 h
F: Conditioning temperature	72°F, 150°F
G: Conditioning relative humidity	25%, 75%

Source: Byrne and Taguchi (1987, Table 1). Reprinted by permission of the American Society for Quality.

Table 10.22 Pull-Off Force Data

	Inner Array					Outer Array							
					E	1	1	1	1	2	2	2	2
					F	1	1	2	2	1	1	2	2
Run	*A*	*B*	*C*	*D*	*G*	1	2	1	2	1	2	1	2
1	1	1	1	1		15.6	9.5	16.9	19.9	19.6	19.6	20.0	19.1
2	1	2	2	2		15.0	16.2	19.4	19.2	19.7	19.8	24.2	21.9
3	1	3	3	3		16.3	16.7	19.1	15.6	22.6	18.2	23.3	20.4
4	2	1	2	3		18.3	17.4	18.9	18.6	21.0	18.9	23.2	24.7
5	2	2	3	1		19.7	18.6	19.4	25.1	25.6	21.4	27.5	25.3
6	2	3	1	2		16.2	16.3	20.0	19.8	14.7	19.6	22.5	24.7
7	3	1	3	2		16.4	19.1	18.4	23.6	16.8	18.6	24.3	21.6
8	3	2	1	3		14.2	15.6	15.1	16.8	17.8	19.6	23.2	24.2
9	3	3	2	1		16.1	19.9	19.3	17.3	23.1	22.7	22.6	28.6

Source: Byrne and Taguchi (1987, Table 4). Reprinted by permission of the American Society for Quality.

found to be the most revealing. Make these two interaction plots. Do these plots change the choice of levels selected in part (b)? Taking into account the fact that a thin wall is less costly than medium or thick, what is your final choice of the levels of design factors?

CHAPTER 11

Random and Mixed Crossed-Factors Experiments

Thus far in this book we have focused on fixed factors. In this chapter we study designs with random or mixed factors. The main point of departure in random- and mixed-effects models is the focus on estimating multiple sources of variation (called variance components), rather than just a single source of variation due to random or experimental error. This problem is of importance in many applications where it is of interest to analyze total variability in the data and identify the major sources of variation that could then be addressed to reduce their contributions. In quality control applications it is critical to minimize the product variability. Different sources of variability may include workers, machines, raw materials, operating conditions, and so on. In customer service applications, it is necessary to minimize not only the mean of the service time distribution but also its variance. Here the sources of variability may include the service provider, customer needs, time of the day, and so on. This chapter develops methods for estimation and testing of variance components

We first discuss a single random factor (one-way layout) design in Section 11.1. Next we consider a two-way crossed-factors layout in Section 11.2. We discuss cases where both factors are random in Section 11.2.1 or one factor is fixed and the other is random (the mixed-effects model) in Section 11.2.2. Section 11.3 extends these considerations to three-way crossed-factors layouts where one or two or all three factors may be random. A new wrinkle arises in these designs, namely that there are no exact F-tests available for certain effects and approximate F-tests must be used. Mathematical derivations of the E(MS) expressions used to obtain F-ratios for ANOVA hypothesis tests and for variance component estimates are given in Section 11.4. Section 11.5 gives a summary of the chapter.

Statistical Analysis of Designed Experiments: Theory and Applications By Ajit C. Tamhane

Throughout we confine ourselves to balanced layouts. Analyses of random-effects unbalanced designs are rather involved and are not discussed here; the reader is referred to Searle, Casella, and McCulloch (1992).

11.1 ONE-WAY LAYOUTS

As an example of a one-way random-effects design, consider an experiment to compare different batches of raw material in terms of some physical property. The experimental batches may be thought of as a random sample from the population of all batches if the batches are independently drawn.[1] As such, it would be meaningless to compare the means of specific batches that *happened* to be selected for the experiment. Rather, it would be of interest to estimate the variability among the population of all batches. Thus the focus is on the population variance among the batches. There is also the variance within the samples taken from each batch. These two variance components contribute to the total variance of a random sample taken from a random batch. This is the crux of the difference between the random- and fixed-effects models.

11.1.1 Random-Effects Model

Let A denote the treatment factor of interest. We assume that A has many possible levels (generically referred to as treatments), which may be taken to be infinite in number.[2] We further assume that the population of the treatment means can be modeled by a $\mathrm{N}(\mu, \sigma_A^2)$ distribution, where μ represents the overall mean and σ_A^2 represents the variance of this population. We select $a \geq 2$ treatments at random from this infinite population. Thus the means of the selected treatments, $\mu_1, \mu_2, \ldots, \mu_a$, form a random sample from a $\mathrm{N}(\mu, \sigma_A^2)$ distribution. Note that the μ_i are not fixed unknown parameters as in the case of the fixed-effects model studied in Chapter 3.

Having selected a random sample of the treatments with means μ_i, we next take a random sample of observations y_{ij} $(1 \leq j \leq n_i)$ from treatment i $(1 \leq i \leq a)$. This can be viewed as a two-stage sampling process as depicted in Figure 11.1. We assume that, conditional on the μ_i, the y_{ij} are independently distributed as $\mathrm{N}(\mu_i, \sigma_e^2)$ r.v.'s, where we have added a subscript "e" to σ^2 to emphasize that it is the **error variance** as distinct from the factor A variance σ_A^2.

We assume a balanced one-way layout with $n_i \equiv n$. The data satisfy the following model:

$$y_{ij} = \mu_i + e_{ij} = \mu + \alpha_i + e_{ij} \qquad (1 \leq i \leq a, 1 \leq j \leq n), \tag{11.1}$$

[1] In practice, this is not always exactly possible since the batches are made sequentially. Even if there are no changes in the manufacturing process, the batches may need to be spaced sufficiently apart to minimize any serial correlations.

[2] The case of finite treatment populations is covered in Dunn and Clark (1974); for most practical applications the additional complication caused by finiteness of populations is unnecessary.

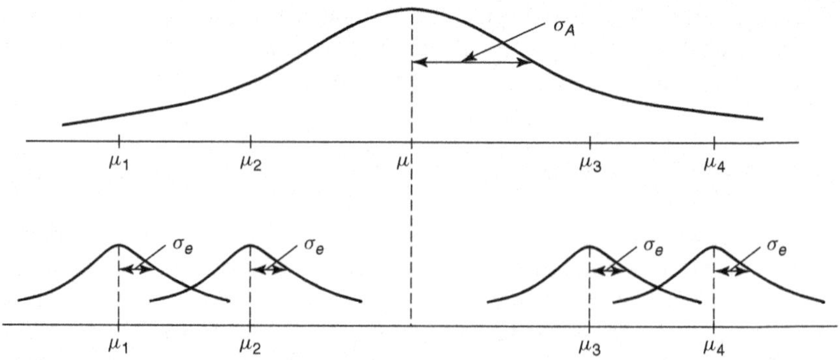

Figure 11.1 Two-stage sampling process underlying one-way random-effects model.

where the e_{ij} are i.i.d. $N(0, \sigma_e^2)$ random errors independent of the μ_i and the $\alpha_i = \mu_i - \mu$ are i.i.d. $N(0, \sigma_A^2)$ random **treatment effects**. We refer to σ_A^2 and σ_e^2 as the **variance components** because the variance of y_{ij} is their sum, that is,

$$\sigma_y^2 = \mathrm{Var}(\mu_i) + \mathrm{Var}(e_{ij}) = \sigma_A^2 + \sigma_e^2.$$

Note that the observations from each treatment are correlated since

$$\mathrm{Cov}(y_{ij}, y_{ik}) = \mathrm{Cov}(\mu_i + e_{ij}, \mu_i + e_{ik}) = \mathrm{Cov}(\mu_i, \mu_i) = \mathrm{Var}(\mu_i) = \sigma_A^2 > 0,$$

but the observations from different treatments are independent. The correlation coefficient between the observations from a given treatment is

$$\rho = \mathrm{Corr}(y_{ij}, y_{ik}) = \frac{\mathrm{Cov}(y_{ij}, y_{ik})}{\sqrt{\mathrm{Var}(y_{ij})\mathrm{Var}(y_{ik})}} = \frac{\sigma_A^2}{\sigma_A^2 + \sigma_e^2}, \tag{11.2}$$

which is known as the **intraclass correlation coefficient**. This is often used as a measure of similarity among the sampled items from each treatment.

11.1.2 Analysis of Variance

The total variation can be decomposed into two components, just as in the case of the fixed-effects model, resulting in the **ANOVA identity**:

$$\mathrm{SS}_{\mathrm{tot}} = \mathrm{SS}_A + \mathrm{SS}_e, \tag{11.3}$$

where the three sums of squares are defined in the same way as in (3.8):

$$\mathrm{SS}_{\mathrm{tot}} = \sum_{i=1}^{a}\sum_{j=1}^{n}(y_{ij} - \bar{\bar{y}})^2, \qquad \mathrm{SS}_A = n\sum_{i=1}^{a}(\bar{y}_i - \bar{\bar{y}})^2, \qquad \mathrm{SS}_e = \sum_{i=1}^{a}\sum_{j=1}^{n}(y_{ij} - \bar{y}_i)^2.$$

In Section 11.4.2 we show that SS_A and SS_e are distributed independently as

$$SS_A \sim (n\sigma_A^2 + \sigma_e^2)\chi_{a-1}^2 \quad \text{and} \quad SS_e \sim \sigma_e^2 \chi_\nu^2, \tag{11.4}$$

where $\nu = a(n-1)$ is the error d.f. From this result it follows that

$$E(MS_A) = n\sigma_A^2 + \sigma_e^2 \quad \text{and} \quad E(MS_e) = \sigma_e^2. \tag{11.5}$$

Consider testing

$$H_0 : \sigma_A^2 = 0 \quad \text{vs.} \quad H_1 : \sigma_A^2 > 0.$$

This testing problem is the random-effects analog of the fixed-effects problem (3.6) of testing $H_0 : \mu_1 = \mu_2 = \cdots = \mu_a$ versus $H_1 : \mu_i \neq \mu_j$ for some $i \neq j$. Under H_0, $E(MS_A) = E(MS_e) = \sigma_e^2$. Therefore $MS_A/MS_e \sim F_{a-1,\nu}$ and an α-level test of H_0 rejects if

$$F_A = \frac{MS_A}{MS_e} > f_{a-1,\nu,\alpha}, \tag{11.6}$$

where $f_{a-1,\nu,\alpha}$ is the upper α critical point of the F-distribution with $a-1$ and ν d.f. Note that this test is identical to the fixed-effects one-way ANOVA F-test; see (3.9). These calculations are summarized in Table 11.1.

Example 11.1 (Compressive Modulus Data: Analysis of Variance) In a quality control project at an electronics manufacturing firm, silicone rubber used in high-voltage transformers was tested for various properties. Here we consider data on compressive modulus from five batches of rubber with 10 sample sheets from each batch. The data are given in Table 11.2. It is of interest to determine if the between-batch variation is statistically significant.

First we make a side-by-side box plot of the data in Figure 11.2. We see that the within-batch variation is roughly stable across batches as can also be checked from the batch standard deviations with $s_{\max}/s_{\min} = 56.2/33.2 = 1.693$, but there is a great deal of batch-to-batch variation.

The batch is a random factor if we consider the five batches as a random sample from the population of all batches. The resulting ANOVA is given in

Table 11.1 ANOVA Table for Random-Effects One-Way Layout Model

Source	SS	d.f.	MS	F	E(MS)
Factor A	$SS_A = n\sum(\bar{y}_i - \bar{\bar{y}})^2$	$a-1$	MS_A	MS_A/MS_e	$\sigma_e^2 + n\sigma_A^2$
Error	$SS_e = \sum\sum(y_{ij} - \bar{y}_i)^2$	$a(n-1)$	MS_e		σ_e^2
Total	$SS_{tot} = \sum\sum(y_{ij} - \bar{\bar{y}})^2$	$an-1$			

Table 11.2 Compressive Modulus (lb/in.^{2}) of Silicone Rubber Samples

	Batch 1	Batch 2	Batch 3	Batch 4	Batch 5
	997.5	870.1	1018.7	973.0	1018.6
	972.4	1064.1	993.6	1048.7	1061.3
	1064.2	925.9	939.6	1058.3	1066.5
	972.0	906.2	1006.3	961.9	969.0
	994.4	969.8	1027.0	1002.2	1063.9
	1044.5	982.4	877.1	973.3	955.7
	982.6	936.0	926.4	967.6	967.9
	956.3	930.0	980.2	992.3	1018.1
	991.3	927.2	966.6	956.1	925.4
	1001.5	924.5	925.0	973.5	1088.4
$\bar{y}_i$	997.7	943.6	966.0	990.7	1013.5
s_i	33.2	52.4	48.2	35.8	56.2

Display 11.1. We see that the variability between the batches is statistically significant ($p = 0.013$). ■

11.1.3 Estimation of Variance Components

The primary focus in a random-effects model is the estimation of the variance components σ_A^2 and σ_e^2. From (11.5) it follows that unbiased estimators of the two variance components are

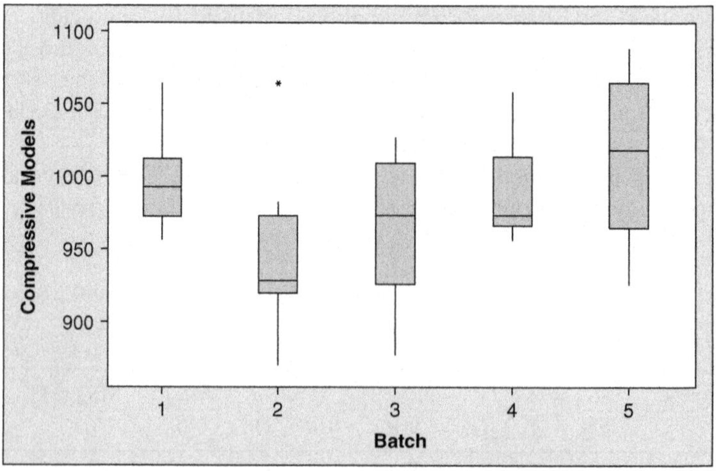

Figure 11.2 Side-by-side box plot of compressive modulus data from five batches.

```
Source  DF      SS    MS     F      P
Batch    4   30390  7598  3.58  0.013
Error   45   95554  2123
Total   49  125944
S = 46.08   R-Sq = 24.13%   R-Sq(adj) = 17.39%
```

Display 11.1 Minitab ANOVA output for compressive modulus of rubber data.

$$\widehat{\sigma}_A^2 = \frac{\mathrm{MS}_A - \mathrm{MS}_e}{n} \quad \text{and} \quad \widehat{\sigma}_e^2 = \mathrm{MS}_e, \tag{11.7}$$

which are known as the **ANOVA estimators**. For unbalanced designs, n in the formula for $\widehat{\sigma}_A^2$ is replaced by

$$\tilde{n} = \frac{1}{a-1}\left[N - \frac{\sum_{i=1}^{a} n_i^2}{N}\right]. \tag{11.8}$$

Note that $\widehat{\sigma}_A^2$ will be negative if $\mathrm{MS}_A < \mathrm{MS}_e$, that is, if $F_A = \mathrm{MS}_A/\mathrm{MS}_e < 1$. In practice, it is common to replace a negative estimate with a zero; however, this violates the unbiasedness property. An alternative to the ANOVA method of estimation is the ML method or the **restricted maximum likelihood (REML)** method. In both these methods certain likelihood functions (see Tamhane and Dunlop, 2000, Chapter 15) are maximized with respect to μ, σ_A^2, and σ_e^2 subject to a nonnegativity constraint on the variance components. These methods are discussed in Section 11.4.1.

Exact CIs and tests can be constructed for $\theta = \sigma_A/\sigma_e$ and σ_e, but not for σ_A. A $100(1-\alpha)\%$ CI for σ_e follows from elementary statistics:

$$\widehat{\sigma}_e\sqrt{\frac{a(n-1)}{\chi^2_{a(n-1),\alpha/2}}} \le \sigma_e \le \widehat{\sigma}_e\sqrt{\frac{a(n-1)}{\chi^2_{a(n-1),1-\alpha/2}}}, \tag{11.9}$$

where $\chi^2_{a(n-1),\alpha/2}$ and $\chi^2_{a(n-1),1-\alpha/2}$ are the upper and lower $\alpha/2$ critical points, respectively, of the chi-square distribution with $a(n-1)$ d.f. To obtain a CI for θ, we use the distributional result (11.4) to conclude that

$$\left(\frac{\sigma_e^2}{n\sigma_A^2 + \sigma_e^2}\right)\frac{\mathrm{MS}_A}{\mathrm{MS}_e} = \left(\frac{1}{n\theta^2 + 1}\right) F_A \sim F_{a-1,\nu}. \tag{11.10}$$

Hence we can write the probability statement

$$\mathrm{P}\left\{ f_{a-1,\nu,1-\alpha/2} \le \left(\frac{1}{n\theta^2+1}\right) F_A \le f_{a-1,\nu,\alpha/2}\right\} = 1-\alpha.$$

This probability statement can be readily manipulated to obtain the following $100(1-\alpha)\%$ CI for θ:

$$\left[\frac{1}{n}\left(\frac{F_A}{f_{a-1,\nu,\alpha/2}}-1\right)\right]^{1/2} \leq \theta \leq \left[\frac{1}{n}\left(\frac{F_A}{f_{a-1,\nu,1-\alpha/2}}-1\right)\right]^{1/2}. \quad (11.11)$$

An approximate CI and a test can be constructed for σ_y using the Welch–Satterthwaite approximation. We do not discuss this method here, but it is explained in Section 11.3.3 and the formula for the CI is derived in Exercise 11.5.

Example 11.2 (Compressive Modulus Data: Estimation of Variance Components)

The variance component estimates are as follows:

$$\widehat{\sigma}_e^2 = \text{MS}_e = 2123.0 \quad \text{and} \quad \widehat{\sigma}_{\text{batch}}^2 = \frac{\text{MS}_A - \text{MS}_e}{n} = \frac{7598.0 - 2123.0}{10} = 547.5.$$

These can be checked to be the same as in the Minitab output shown in Display 11.2. This output also gives the E(MS) expressions in coded notation—not in the usual algebraic notation. In this coded notation, (1) refers to the first entry, namely Batch, and (2) refers to the second entry, namely Error, in the ANOVA table. Thus, the E(MS) expression (2) + 10.0000(1) for Batch denotes $\sigma_e^2 + 10\sigma_{\text{batch}}^2$ and the E(MS) expression (2) for Error denotes σ_e^2. This notational convention is followed in all the Minitab outputs given in this and the following chapters.

Adding $\widehat{\sigma}_A^2$ and $\widehat{\sigma}_e^2$, we get $\widehat{\sigma}_y^2 = 2123.0 + 547.5 = 2670.5$. Note that the batch-to-batch variation accounts for about 20% of the total variation. Thus, although the batch-to-batch variation is statistically significant, the within-batch variation is the major cause of variation and needs corrective action to be reduced.

```
Expected Mean Squares, using Adjusted SS
             Expected Mean Square
   Source    for Each Term
1  Batch     (2) + 10.0000 (1)
2  Error     (2)
Error Terms for Tests, using Adjusted SS
                                       Synthesis
   Source    Error DF   Error MS    of Error MS
1  Batch        45.00       2123    (2)
Variance Components, using Adjusted SS
          Estimated
Source       Value
Batch        547.5
Error       2123.4
```

Display 11.2 Minitab variance components output for compressive modulus of rubber data.

Note that part of this variation may be due to the measurement error; however, since there are no repeat measurements, this component of variance cannot be estimated. The estimate $\widehat{\sigma}_y = \sqrt{2670.5} = 51.677$ can be used to provide a prediction interval for future y-values.

The relative contributions of different causes to total variation are useful for diagnostic and corrective action purposes, but ultimately, it is the total variation that must be minimized to improve the consistency of the product. A point estimate $\widehat{\sigma}_y$ alone is not sufficient to monitor changes in σ_y since it does not take into account the sampling error. Unfortunately, an exact CI for σ_y does not exist, but an approximate 95% CI can be calculated using the χ^2-approximation to the distribution of $\widehat{\sigma}_y^2$ derived in Exercise 11.5. Substituting $\text{MS}_A = 7598, \text{MS}_e = 2123, \widehat{\sigma}_y = 51.677, a = 5$, and $n = 10$ in the formula for the d.f., $\widehat{q}$, of χ^2, we get $\widehat{q} = 31.63$. Truncating $\widehat{q}$ to 31 d.f., the critical constants for the 95% CI are $\chi^2_{31,0.975} = 17.538$ and $\chi^2_{31,0.025} = 48.231$ from Table C.3. Hence the desired CI for σ_y is

$$\left[51.677\sqrt{\frac{31}{48.231}}, 51.677\sqrt{\frac{31}{17.538}}\right] = [41.430, 68.705].$$

If the rubber-making process is changed, then one should collect new data and compare the new CI for σ_y with this one to see if σ_y has decreased.

Next we compute a 95% CI for $\theta = \sigma_A/\sigma_e$ using (11.11). The critical constants needed are $f_{4,45,0.975} = 0.119$ and $f_{4,45,0.025} = 3.086$. The lower and upper confidence limits for θ are

$$L = \left[\frac{1}{n}\left\{\frac{F_A}{f_{a-1,\nu,\alpha/2}} - 1\right\}\right]^{1/2} = \left[\frac{1}{10}\left\{\frac{3.58}{3.086} - 1\right\}\right]^{1/2} = 0.400,$$

$$U = \left[\frac{1}{n}\left\{\frac{F_A}{f_{a-1,\nu,1-\alpha/2}} - 1\right\}\right]^{1/2} = \left[\frac{1}{10}\left\{\frac{3.58}{0.119} - 1\right\}\right]^{1/2} = 1.705.$$

This CI is rather wide. It shows that σ_A could be as small as $0.400\sigma_e$ and as large as $1.705\sigma_e$ with 95% confidence. To narrow this CI down, we need more batches.

In conclusion, the within-batch variation contributes 80% to the total variation and should be reduced first. The batch-to-batch variation is also significant and needs to be addressed next. ■

11.2 TWO-WAY LAYOUTS

11.2.1 Random-Effects Model

In this design both factors are random and crossed with each other. An example of this design is a **gage reproducibility and repeatability (R&R) study** conducted in quality control applications to assess the capability of a measuring instrument (gage). In this experiment a number of randomly chosen parts are measured using the same gage by several inspectors. If the inspectors are also randomly chosen, then the parts and the inspectors are crossed random factors; see Example 11.3 for data from such a study.

In general, denote the two random factors by A and B. Let $a \geq 2$ and $b \geq 2$ be the numbers of levels of A and B, respectively. We assume a balanced two-way layout in which $n \geq 2$ observations are taken at each treatment combination. Let y_{ijk} denote the kth observation on the (i, j)th treatment combination. We assume the following model for y_{ijk}:

$$y_{ijk} = \mu + \alpha_i + \beta_j + (\alpha\beta)_{ij} + e_{ijk} \qquad (1 \leq i \leq a, 1 \leq j \leq b, 1 \leq k \leq n), \tag{11.12}$$

where μ is a fixed constant and the $\alpha_i \sim \mathrm{N}(0, \sigma_A^2)$, $\beta_j \sim \mathrm{N}(0, \sigma_B^2)$, $(\alpha\beta)_{ij} \sim \mathrm{N}(0, \sigma_{AB}^2)$, and $e_{ijk} \sim \mathrm{N}(0, \sigma_e^2)$ are mutually independent. Here μ is the overall mean effect, the α_i are the factor A main effects, the β_j are the factor B main effects, the $(\alpha\beta)_{ij}$ are the AB interactions, and the e_{ijk} are random errors. We refer to $\sigma_A^2, \sigma_B^2, \sigma_{AB}^2$, and σ_e^2 as the corresponding variance components. Because of the mutual independence between these effects and the random error, we have

$$\mathrm{Var}(y_{ijk}) = \sigma_y^2 = \sigma_A^2 + \sigma_B^2 + \sigma_{AB}^2 + \sigma_e^2.$$

The variance components have the following interpretations: σ_A^2 and σ_B^2 are the contributions to the total variance of the response variable due to the differences in the levels of factor A and factor B, respectively, and σ_{AB}^2 is the additional contribution due to the joint effect of the variation between the levels of the two factors.

Note that the above model can also be stated in terms of the following correlation structure:

$$\mathrm{Corr}(y_{ijk}, y_{i'j'k'}) = \begin{cases} \rho_1 = 1 & \text{if } i = i', j = j', k = k' \\ \rho_2 = \dfrac{\sigma_A^2 + \sigma_B^2 + \sigma_{AB}^2}{\sigma_y^2} & \text{if } i = i', j = j', k \neq k' \\ \rho_3 = \dfrac{\sigma_A^2}{\sigma_y^2} & \text{if } i = i', j \neq j' \\ \rho_4 = \dfrac{\sigma_B^2}{\sigma_y^2} & \text{if } i \neq i', j = j' \\ \rho_5 = 0 & \text{if } i \neq i', j \neq j. \end{cases}$$

The ANOVA decomposition of the total sum of squares and the degrees of freedom are the same as that for the two-way fixed-effects layout; see (6.8) in Chapter 6. Analogous to that case, it can be shown that SS_A, SS_B, SS_{AB}, and SS_e

are mutually independent. The hypotheses tested using these sums of squares are as follows:

$$H_{0A} : \sigma_A^2 = 0, \qquad H_{0B} : \sigma_B^2 = 0, \qquad H_{0AB} : \sigma_{AB}^2 = 0,$$

which are the hypotheses of no main effects and no interactions, respectively. For the fixed-effects model, to test the null hypotheses for the main effects and interactions, we use the F-ratios composed of the respective mean squares in the numerator and MS_e in the denominator. For the random-effects model, as we shall see, it is not always the case. We need to examine the expected mean squares E(MS) and choose the numerator and the denominator of the F-ratios so that their expected values are equal under the null hypothesis to be tested. The E(MS) expressions are shown in Table 11.3. By examining these expressions we see that the respective F-ratios are as follows:

$$F_A = \frac{\text{MS}_A}{\text{MS}_{AB}}, \qquad F_B = \frac{\text{MS}_B}{\text{MS}_{AB}}, \qquad F_{AB} = \frac{\text{MS}_{AB}}{\text{MS}_e}.$$

For example, $\text{E}(\text{MS}_A) = \sigma_e^2 + n\sigma_{AB}^2 + bn\sigma_A^2$. Therefore, under $H_{0A} : \sigma_A^2 = 0$, we have $\text{E}(\text{MS}_A) = \text{E}(\text{MS}_{AB}) = \sigma_e^2 + n\sigma_{AB}^2$ and hence the ratio $\text{MS}_A/\text{MS}_{AB}$ provides the desired F-statistic. Thus the proper denominator term (often referred to as the **error term**) is MS_{AB}. Note that $\text{E}(\text{MS}_A) \neq \text{E}(\text{MS}_e)$ under H_{0A}. Therefore MS_e cannot be used as the error term in the F-statistic. On the other hand, to test H_{0AB} the proper error term is MS_e. The ANOVA estimators of the variance components can be derived directly from the expected mean square expressions, and they are as follows:

$$\widehat{\sigma}_e^2 = \text{MS}_e, \qquad \widehat{\sigma}_{AB}^2 = \frac{\text{MS}_{AB} - \text{MS}_e}{n},$$

$$\widehat{\sigma}_B^2 = \frac{\text{MS}_B - \text{MS}_{AB}}{an}, \qquad \widehat{\sigma}_A^2 = \frac{\text{MS}_A - \text{MS}_{AB}}{bn}. \tag{11.13}$$

Example 11.3 (Gage R&R Study for Semiconductor Power Modules)

A gage refers to a measurement system that includes the measuring instrument and the inspector. A gage R&R study is conducted in quality control applications

Table 11.3 ANOVA Table for Balanced Two-Way Layout with Random Effects

Source	SS	d.f.	MS	E(MS)	F
A main effect	SS_A	$a-1$	MS_A	$\sigma_e^2 + n\sigma_{AB}^2 + bn\sigma_A^2$	$\text{MS}_A/\text{MS}_{AB}$
B main effect	SS_B	$b-1$	MS_B	$\sigma_e^2 + n\sigma_{AB}^2 + an\sigma_B^2$	$\text{MS}_B/\text{MS}_{AB}$
AB interaction	SS_{AB}	$(a-1)(b-1)$	MS_{AB}	$\sigma_e^2 + n\sigma_{AB}^2$	$\text{MS}_{AB}/\text{MS}_e$
Error	SS_e	$ab(n-1)$	MS_e	σ_e^2	
Total	SS_{tot}	$N-1$			

to compare the measurement variation with the variation between the parts being measured (or the specified tolerance on the parts) to determine if the measurement system is precise and capable enough to distinguish between good parts and bad parts.

Houf and Berman (1988) report a gage R&R study in which three inspectors measured the thermal resistance (in degrees Celsius per watt) of 10 semiconductor power modules. Each inspector made three measurements on each module. According to the article (p. 519), "The order in which the samples were measured was randomized to assure that any drift or external influence will spread randomly throughout the study. After recording the ten measurements for each trial, the data were sequestered so that the previous readings would not affect the next trial measurements. Also, the inspectors were not allowed to exchange information during the study. Each inspector had been trained prior to the study and had experience using the equipment." The data are given in Table 11.4. We will assume that the three inspectors are a random sample from a large population of inspectors.

Figure 11.3 shows a labeled scatterplot of the measurements versus the module with inspectors as labels. We see that there is a large variation between modules with modules 3, 5, 7, and 9 having systematically lower measurements than the other modules (which might raise the question of whether alternate modules vary systematically in an up-and-down fashion). Inspector A often has the lowest measurements than the other two inspectors, but more to the point here, he also has the least variation. Thus there appears to be a large inspector effect as well.

The Minitab output of ANOVA and variance components is shown in Display 11.3. From this output we compute the following quantities useful for gage R&R analysis:

Table 11.4 Data from Gage R&R Study for Semiconductor Power Modules

Module	Inspector A Measurements			Inspector B Measurements			Inspector C Measurements		
No.	1	2	3	1	2	3	1	2	3
1	0.37	0.38	0.37	0.41	0.41	0.40	0.41	0.42	0.41
2	0.42	0.41	0.43	0.42	0.42	0.42	0.43	0.42	0.43
3	0.30	0.31	0.31	0.31	0.31	0.31	0.29	0.30	0.28
4	0.42	0.43	0.42	0.43	0.43	0.43	0.42	0.42	0.42
5	0.28	0.30	0.29	0.29	0.30	0.29	0.31	0.29	0.29
6	0.42	0.42	0.43	0.45	0.45	0.45	0.44	0.46	0.45
7	0.25	0.26	0.27	0.28	0.28	0.30	0.29	0.27	0.27
8	0.40	0.40	0.40	0.43	0.42	0.42	0.43	0.43	0.41
9	0.25	0.25	0.25	0.27	0.29	0.28	0.26	0.26	0.26
10	0.35	0.34	0.34	0.35	0.35	0.34	0.35	0.34	0.35

Source: Houf and Berman (1988, Figure 4).

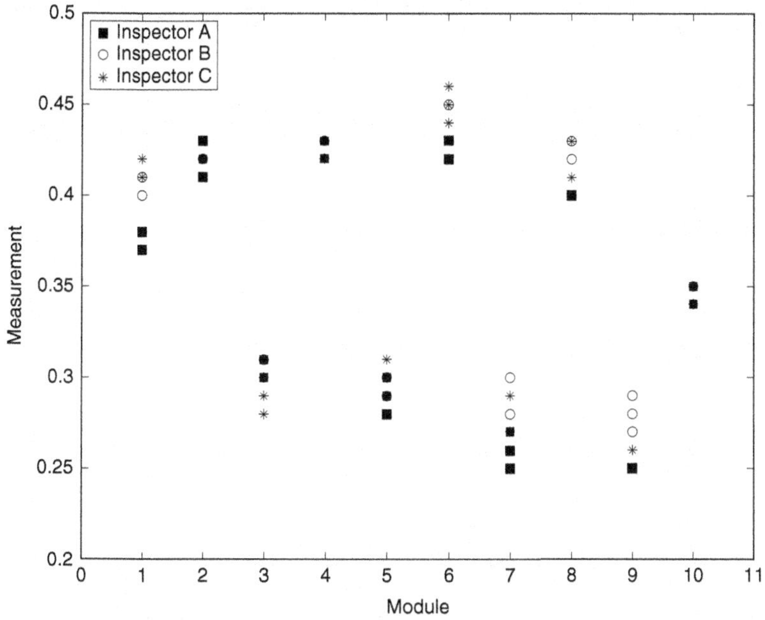

Figure 11.3 Labeled scatterplot of gage R&R data from Table 11.4.

$$\widehat{\sigma}^2_{\text{Repeatability}} = \widehat{\sigma}^2_e = 0.00005,$$

$$\widehat{\sigma}^2_{\text{Reproducibility}} = \widehat{\sigma}^2_{\text{Inspector}} + \widehat{\sigma}^2_{\text{Module} \times \text{Inspector}} = 0.00013,$$

$$\widehat{\sigma}^2_{\text{Gage}} = \widehat{\sigma}^2_{\text{Repeatability}} + \widehat{\sigma}^2_{\text{Reproducibility}} = 0.00018.$$

To assess the gage capability, we compare $\widehat{\sigma}_{\text{Gage}} = \sqrt{0.00018} = 0.0134$ (called the **gage error**) with $\widehat{\sigma}_{\text{Module}} = \sqrt{0.00483} = 0.0695$. The ratio of the two is 0.193, which is nearly twice the maximum of 10% that is generally regarded as acceptable in practice. Thus the thermal resistance measurement system is not capable. ■

11.2.2 Mixed-Effects Model

In some two-factor experiments one of the factors is fixed while the other is random, which results in a two-way mixed model. As an example, suppose we want to compare different methods of treating raw rubber. Samples from several batches are randomly assigned to the different treatment methods and the properties of interest are measured. Here the treatment methods are fixed (factor A), the batches are random (factor B), and the two factors are crossed.

We will assume a balanced two-way layout with $n \geq 2$ observations per factor-level combination. There are two models commonly employed for this

```
Source              DF    Seq SS     Adj SS     Adj MS        F      P
Module               9  0.393596   0.393596   0.043733   162.27  0.000
Inspector            2  0.003927   0.003927   0.001963     7.28  0.005
Module*Inspector    18  0.004851   0.004851   0.000270     5.27  0.000
Error               60  0.003067   0.003067   0.000051
Total               89  0.405440
S = 0.00714920   R-Sq = 99.24%   R-Sq(adj) = 98.88%
Expected Mean Squares, using Adjusted SS
   Source             Expected Mean Square for Each Term
1  Module             (4) + 3.0000 (3) + 9.0000 (1)
2  Inspector          (4) + 3.0000 (3) + 30.0000 (2)
3  Module*Inspector   (4) + 3.0000 (3)
4  Error              (4)
Variance Components, using Adjusted SS
                     Estimated
Source                   Value
Module                 0.00483
Inspector              0.00006
Module*Inspector       0.00007
Error                  0.00005
```

Display 11.3 Minitab output of ANOVA and variance components for gage R&R data for semiconductor power modules.

design: the unrestricted model discussed in Section 11.2.2.1 and the restricted model discussed in Section 11.2.2.2. In the unrestricted model the random effects are assumed to be mutually independent with no constraints, while the restricted model imposes a certain constraint on them. In fact, the two models are equivalent ways of defining the parameters, as shown in Section 11.4.3. For pedagogical reasons we present the unrestricted model first, as it is simpler.

11.2.2.1 Unrestricted Model

The unrestricted model has the same form as (11.12), except that the α_i are now fixed effects subject to the constraint $\sum_{i=1}^{a} \alpha_i = 0$. The β_j, $(\alpha\beta)_{ij}$ and the e_{ij} are mutually independent r.v.'s and have the same distributions as before, namely, $N(0, \sigma_B^2)$, $N(0, \sigma_{AB}^2)$, and $N(0, \sigma_e^2)$, respectively. Therefore

$$\text{Var}(y_{ijk}) = \sigma_y^2 = \sigma_B^2 + \sigma_{AB}^2 + \sigma_e^2. \tag{11.14}$$

Note that A, being a fixed factor, does not contribute to σ_y^2.

The ANOVA for this model is given in Table 11.5. The SS expressions are the same as those for the two-way fixed-effects model and are given in Table 6.4. The table also gives the expressions for the expected mean squares E(MS). In these expressions,

$$Q_A = \frac{\sum \alpha_i^2}{(a-1)},$$

Table 11.5 ANOVA Table for Balanced Two-Way Layout with Mixed Effects (Unrestricted Model)

Source	SS	d.f.	MS	E(MS)	F
A main effect	SS_A	$a-1$	MS_A	$\sigma_e^2 + n\sigma_{AB}^2 + bnQ_A$	MS_A/MS_{AB}
B main effect	SS_B	$b-1$	MS_B	$\sigma_e^2 + n\sigma_{AB}^2 + an\sigma_B^2$	MS_B/MS_{AB}
AB interaction	SS_{AB}	$(a-1)(b-1)$	MS_{AB}	$\sigma_e^2 + n\sigma_{AB}^2$	MS_{AB}/MS_e
Error	SS_e	$N-ab$	MS_e	σ_e^2	
Total	SS_{tot}	$N-1$			

which is a finite-population analog of the variance component of the α_i's since $\overline{\alpha} = 0$. With this analogy, the expression for $E(MS_A)$ is identical to the corresponding expression for the two-way random-effects model. The expressions for $E(MS_B)$, $E(MS_{AB})$, and $E(MS_e)$ are also identical. Therefore the F-statistics for the tests of the hypotheses are the same for the two models. (Note that the null hypothesis for the A main effects is $H_{0A} : \alpha_1 = \cdots = \alpha_a = 0$, which is equivalent to $Q_A = 0$.) Section 11.3 gives rules for writing the E(MS) expressions.

The ANOVA estimators of the variance components, $\sigma_B^2, \sigma_{AB}^2$, and σ_e^2, are the same as those given in (11.13). The LS estimators of the fixed effects are given by

$$\widehat{\mu} = \overline{y}_{...} \qquad \text{and} \qquad \widehat{\alpha}_i = \overline{y}_{i..} - \overline{y}_{...} \qquad (1 \le i \le a). \tag{11.15}$$

11.2.2.2 Restricted Model

The restricted model has the same general form as the unrestricted model, but the interpretations of the random-effect parameters are now different because of the constraints imposed on them. To distinguish the parameters of the unrestricted model from those of the restricted model, we put primes on the latter. Thus $\mu' = \mu$ and the $\alpha_i' = \alpha_i$ are the fixed effects subject to the constraint $\sum_{i=1}^a \alpha_i' = 0$. The β_j' are i.i.d. $N(0, \sigma_B'^2)$ and are independent of the $(\alpha\beta)_{ij}'$, which are subject to the constraint that $\sum_{i=1}^a (\alpha\beta)_{ij}' = 0$ for each j, which is the usual type of constraint associated with a fixed factor (e.g., $\sum_{i=1}^a \alpha_i' = 0$). This makes $(\alpha\beta)_{ij}'$ correlated; in particular, we get

$$(\alpha\beta)_{ij}' \sim N(0, \{(a-1)/a\}\sigma_{AB}'^2) \qquad \text{and}$$

$$\text{Corr}((\alpha\beta)_{ij}', (\alpha\beta)_{ij'}') = -1/(a-1) \quad \text{for } j \ne j'.$$

Here the variance of the $(\alpha\beta)_{ij}'$ is specified as $\{(a-1)/a\}\sigma_{AB}'^2$ instead of the usual $\sigma_{AB}'^2$ in order to obtain simpler expressions for the expected mean squares given in the ANOVA given in Table 11.6. Finally, the e_{ijk} are independent of

Table 11.6 ANOVA Table for Balanced Two-Way Layout with Mixed Effects (Restricted Model)

Source	SS	d.f.	MS	E(MS)	F
A main effect	SS_A	$a-1$	MS_A	$\sigma_e'^2 + n\sigma_{AB}'^2 + bnQ_A$	MS_A/MS_{AB}
B main effect	SS_B	$b-1$	MS_B	$\sigma_e'^2 + an\sigma_B'^2$	MS_B/MS_e
AB interaction	SS_{AB}	$(a-1)(b-1)$	MS_{AB}	$\sigma_e'^2 + n\sigma_{AB}'^2$	MS_{AB}/MS_e
Error	SS_e	$N-ab$	MS_e	$\sigma_e'^2$	
Total	SS_{tot}	$N-1$			

the β_j' and the $(\alpha\beta)_{ij}'$ and are i.i.d. $N(0, \sigma_e'^2)$ with $\sigma_e'^2 = \sigma_e^2$. Thus the variance component model is

$$\text{Var}(y_{ijk}) = \sigma_y^2 = \sigma_B'^2 + \left(\frac{a-1}{a}\right)\sigma_{AB}'^2 + \sigma_e^2. \qquad (11.16)$$

In (11.20) we show that $\sigma_B'^2 = \sigma_B^2 + (1/a)\sigma_{AB}^2$ and in (11.21) we show that $\sigma_{AB}'^2 = \sigma_{AB}^2$. Then it follows that the models (11.14) and (11.16) are equivalent. The relationship between the two models is discussed in Section 11.4.3.

By comparing the E(MS) expressions for the restricted and unrestricted models, we see that only E(MS_B) is different for the two models, and therefore the test of $H_{0B} : \sigma_B^2 = 0$ is also different. In particular, for the unrestricted model the statistic used to test H_{0B} is $F_B = MS_B/MS_{AB}$, while for the restricted model the statistic is $F_B = MS_B/MS_e$. Generally, the test of H_{0B} for the restricted model is less conservative because MS_e tends to be smaller than MS_{AB} [although the denominator d.f. are different for the two F-statistics, namely, $(a-1)(b-1)$ and $ab(n-1)$, respectively]. The ANOVA estimators of the variance components are as follows:

$$\widehat{\sigma_e'^2} = MS_e, \qquad \widehat{\sigma_{AB}'^2} = \frac{MS_{AB} - MS_e}{n}, \qquad \widehat{\sigma_B'^2} = \frac{MS_B - MS_e}{an}.$$

Note that the estimator of $\sigma_B'^2$ is different in this case. The LS estimators of μ and the α_i are the same as in (11.15).

Example 11.4 (Tear Strength of Silicone Rubber: ANOVA and Variance Component Estimates)

An electronics manufacturing company conducted a study to evaluate the consistency of testing silicone rubber samples at three different test sites. One of the rubber properties of interest was the tear strength. Twenty samples were tested at each of the three sites from three different batches of rubber. The summary data (mean tear strength and the sample SD for each sample of 20) are shown in Table 11.7.

Table 11.7 Means and Standard Deviations of Tear Strengths of Rubber Samples from Three Batches Tested at Three Sites

Test site	Batch 1	Batch 2	Batch 3	Row mean
1	26.73 (1.424)	28.24 (1.297)	29.45 (1.386)	28.140
2	26.43 (1.178)	20.86 (1.286)	29.11 (1.924)	25.467
3	29.66 (1.434)	32.65 (1.137)	33.48 (1.508)	31.930
Column mean	27.607	27.250	30.680	28.512

The sums of squares can be calculated in the usual manner as follows:

$$\begin{aligned}
SS_A &= bn\sum_i \overline{y}_{i\cdot\cdot}^2 - abn\overline{y}_{\cdots}^2 \\
&= (60)(28.140^2 + 25.457^2 + 31.930^2) - (180)(28.512)^2 = 1238.45, \\
SS_B &= an\sum_j \overline{y}_{\cdot j\cdot}^2 - abn\overline{y}_{\cdots}^2 \\
&= (60)(27.607^2 + 27.250^2 + 30.680^2) - (180)(28.512)^2 = 430.13, \\
SS_{AB} &= n\sum_{i,j} \overline{y}_{ij\cdot}^2 - SS_A - SS_B - abn\overline{y}_{\cdots}^2 \\
&= 148,540.36 - 1238.45 - 430.13 - 146,328.15 = 543.64, \\
SS_e &= (n-1)\sum_{i,j} s_{ij}^2 = (19)(1.424^2 + \cdots + 1.508^2) = 341.95.
\end{aligned}$$

Table 11.8 shows the ANOVA for these data. We see that the most significant effect is the site–batch interaction. The site main effect is somewhat significant, but the batch-to-batch variation is not significant.

The high site–batch interaction can be explained as follows. For each site compute the variance between the batch means. They are 1.860 for site 1, 17.711 for site 2, and 4.037 for site 3. Thus the batch-to-batch variances differ quite a bit across the sites, and hence the high site–batch interaction. The high variance for site 2 is caused by the low mean for batch 2. The ANOVA estimates of the

Table 11.8 Analysis of Variance for Rubber Tear Strength Data

Source	SS	d.f.	MS	E(MS)	F	p
Sites (A)	1238.45	2	619.23	$\sigma_e^2 + 20\sigma_{AB}^2 + 60Q_A$	4.56	0.093
Batches (B)	430.13	2	215.07	$\sigma_e^2 + 20\sigma_{AB}^2 + 60\sigma_B^2$	1.58	0.312
Sites$\times$batches (AB)	543.64	4	135.91	$\sigma_e^2 + 20\sigma_{AB}^2$	67.96	0.000
Error	341.95	171	2.00	σ_e^2		
Total	2554.17	179				

variance components are as follows:

$$\widehat{\sigma}_e^2 = \text{MS}_e = 2.000,$$

$$\widehat{\sigma}_{AB}^2 = \frac{\text{MS}_{AB} - \text{MS}_e}{20} = \frac{135.91 - 2.00}{20} = 6.696,$$

$$\widehat{\sigma}_B^2 = \frac{215.07 - 135.91}{60} = 1.319.$$

Hence

$$\widehat{\sigma}_y^2 = \widehat{\sigma}_B^2 + \widehat{\sigma}_{AB}^2 + \widehat{\sigma}_e^2 = 10.015.$$

We see that two-thirds of the variation comes from the site–batch interaction. ■

11.3 THREE-WAY LAYOUTS

11.3.1 Random- and Mixed-Effects Models

Consider three factors, A, B, C, having levels a, b, c, respectively. All three factors are assumed to be crossed, resulting in a total of abc treatment combinations. Assume n replicate observations at each factor-level combination with a total of $N = abcn$ observations. We shall consider a random-effects model in which all three factors are random and two mixed-effects models in which one or two factors are fixed and the others are random. As an example, consider a gage R&R study in which the three factors are inspectors (A), measuring instruments (B), and parts (C). It is assumed that each inspector measures each part with each instrument n times independently. Generally, parts are always randomly selected from production lots, whereas inspectors and instruments may be random or fixed depending on the situation.

Denote the data by $\{y_{ijk\ell}\ (1 \le i \le a, 1 \le j \le b, 1 \le k \le c, 1 \le \ell \le n)\}$. The linear model for $y_{ijk\ell}$ is the same as in (6.13):

$$y_{ijk\ell} = \mu + \alpha_i + \beta_j + \gamma_k + (\alpha\beta)_{ij} + (\alpha\gamma)_{ik} + (\beta\gamma)_{jk} + (\alpha\beta\gamma)_{ijk} + e_{ijk\ell}. \tag{11.17}$$

The assumptions that we make on different parameters in this model depend upon the nature of the factors (fixed or random). The assumptions for the three models are summarized below.

All Three Factors Random (Random-Effects Model) In this case, all parameters (except the overall mean μ) are mutually independent with the following distributions:

$$\alpha_i \sim \text{N}(0, \sigma_A^2), \quad \beta_j \sim \text{N}(0, \sigma_B^2), \quad \gamma_k \sim \text{N}(0, \sigma_C^2),$$
$$(\alpha\beta)_{ij} \sim \text{N}(0, \sigma_{AB}^2), \quad (\alpha\gamma)_{ik} \sim \text{N}(0, \sigma_{AC}^2), \quad (\beta\gamma)_{jk} \sim \text{N}(0, \sigma_{BC}^2),$$
$$(\alpha\beta\gamma)_{ijk} \sim \text{N}(0, \sigma_{ABC}^2),$$

and

$$e_{ijk} \sim \mathrm{N}(0, \sigma_e^2).$$

As before, the various σ^2's are the variance components with σ_e^2 being the error variance. The null hypothesis concerning any effect is that the variance component of that effect is zero, for example, the null hypothesis H_{0A} concerning the A main effect is $\sigma_A^2 = 0$.

A Fixed, B and C Random (Mixed-Effects Model 1) Assuming a straightforward extension of the unrestricted model of Section 11.2.2.1, the only change from the random-effects model given above is that the α_i are fixed unknown quantities subject to the constraint that $\sum_{i=1}^{a} \alpha_i = 0$. [Recall that any parameters involving random factors are random, e.g., $(\alpha\beta)_{ij}$ and $(\alpha\beta\gamma)_{ijk}$.] In this case, there is no variance component σ_A^2, but its finite-population analog, namely

$$Q_A = \frac{\sum_{i=1}^{a} \alpha_i^2}{a-1},$$

occurs in the $\mathrm{E}(\mathrm{MS}_A)$ expression and H_{0A} is $Q_A = 0$.

A and B Fixed, C Random (Mixed-Effects Model 2) In this case, we extend the unrestricted model for the previous case. Thus, in addition to the α_i, the β_j and the $(\alpha\beta)_{ij}$ are also fixed subject to the constraints

$$\sum_{j=1}^{b} \beta_j = 0, \qquad \sum_{i=1}^{a} (\alpha\beta)_{ij} = 0 \quad \text{for all } j, \quad \sum_{j=1}^{b} (\alpha\beta)_{ij} = 0 \quad \text{for all } i.$$

In this case, instead of σ_A^2, σ_B^2, and σ_{AB}^2, we have the following finite-population analogs:

$$Q_A = \frac{\sum_{i=1}^{a} \alpha_i^2}{a-1}, \qquad Q_B = \frac{\sum_{j=1}^{b} \beta_j^2}{b-1}, \qquad Q_{AB} = \frac{\sum_{i=1}^{a} \sum_{j=1}^{b} (\alpha\beta)_{ij}^2}{(a-1)(b-1)}.$$

The corresponding null hypotheses are

$$H_{0A} : Q_A = 0, \qquad H_{0B} : Q_B = 0, \qquad H_{0AB} : Q_{AB} = 0.$$

11.3.2 Analysis of Variance

The sums of squares and the degrees of freedom (and hence the mean squares) in the ANOVA table are the same regardless of the nature of the factors and are as shown in Table 6.12. The expected mean squares E(MS) are different for the three models and are summarized in Table 11.19. The table also gives for each main and interaction effect the error term for the F-statistic for testing the null hypothesis that the corresponding variance component (or in the case of

Table 11.9 Expected Mean Squares and Error Terms for Random- and Two Mixed-Effects Models for Three-Way Layout

	All Factors Random		A Fixed, B, C Random		A, B Fixed, C Random	
Source	E(MS)	Error Term	E(MS)	Error Term	E(MS)	Error Term
A	$\sigma_e^2 + n\sigma_{ABC}^2 + cn\sigma_{AB}^2 + bn\sigma_{AC}^2 + bcn\sigma_A^2$	—[a]	$\sigma_e^2 + n\sigma_{ABC}^2 + cn\sigma_{AB}^2 + bn\sigma_{AC}^2 + bcnQ_A$	—[a]	$\sigma_e^2 + bn\sigma_{AC}^2 + bcnQ_A$	MS_{AC}
B	$\sigma_e^2 + n\sigma_{ABC}^2 + cn\sigma_{AB}^2 + an\sigma_{BC}^2 + acn\sigma_B^2$	—[a]	$\sigma_e^2 + an\sigma_{BC}^2 + acn\sigma_B^2$	MS_{BC}	$\sigma_e^2 + an\sigma_{BC}^2 + acnQ_B$	MS_{BC}
C	$\sigma_e^2 + n\sigma_{ABC}^2 + bn\sigma_{AC}^2 + an\sigma_{BC}^2 + abn\sigma_C^2$	—[a]	$\sigma_e^2 + an\sigma_{BC}^2 + abn\sigma_C^2$	MS_{BC}	$\sigma_e^2 + abn\sigma_C^2$	MS_e
AB	$\sigma_e^2 + n\sigma_{ABC}^2 + cn\sigma_{AB}^2$	MS_{ABC}	$\sigma_e^2 + n\sigma_{ABC}^2 + cn\sigma_{AB}^2$	MS_{ABC}	$\sigma_e^2 + n\sigma_{ABC}^2 + cnQ_{AB}$	MS_{ABC}
AC	$\sigma_e^2 + n\sigma_{ABC}^2 + bn\sigma_{AC}^2$	MS_{ABC}	$\sigma_e^2 + n\sigma_{ABC}^2 + bn\sigma_{AC}^2$	MS_{ABC}	$\sigma_e^2 + bn\sigma_{AC}^2$	MS_e
BC	$\sigma_e^2 + n\sigma_{ABC}^2 + an\sigma_{BC}^2$	MS_{ABC}	$\sigma_e^2 + an\sigma_{BC}^2$	MS_e	$\sigma_e^2 + an\sigma_{BC}^2$	MS_e
ABC	$\sigma_e^2 + n\sigma_{ABC}^2$	MS_e	$\sigma_e^2 + n\sigma_{ABC}^2$	MS_e	$\sigma_e^2 + n\sigma_{ABC}^2$	MS_e
Error	σ_e^2		σ_e^2		σ_e^2	

Note: The error term refers to the denominator of the F-statistic for testing the null hypothesis of the given effect. The numerator of the F-statistic is the mean square for that effect.

[a]Exact F-test does not exist.

a fixed effect, the corresponding Q term) equals zero. The numerator is the mean square for the effect being tested. The effects for which an exact F-test does not exist are marked. Approximate F-tests for these hypotheses are discussed in Section 11.3.3.

We have omitted the derivations of the E(MS) expressions, but they follow the same general outline of the derivations for the two-way layout given in Section 11.4. Common patterns can be seen in all these E(MS) expressions for balanced crossed layouts. These patterns are summarized by a set of rules in Section 11.3.2.1.

Exact F-tests are derived by the usual method of finding a numerator and a denominator mean square such that their expected mean squares are equal under the given null hypothesis. The ratio is F-distributed because of the independence of the mean squares which follows from the balanced nature of the design. Thus, for example, for the random-effects model, under $H_{0AB}: \sigma^2_{AB} = 0$, we have

$$E(MS_{AB}) = E(MS_{ABC}) = \sigma^2_e + n\sigma^2_{ABC}.$$

Therefore the ratio $F_{AB} = MS_{AB}/MS_{ABC}$ is F-distributed with $(a-1)(b-1)$ and $(a-1)(b-1)(c-1)$ d.f. As before, note that the error terms for the F-statistics are not necessarily MS_e.

For null hypotheses of certain effects, there do not exist two mean squares whose expected values match. In this case a possible solution is an approximate F-test. This topic is discussed in the following section.

Although exact F-tests do not exist for certain variance components, the ANOVA estimates of all variance components can be computed by the usual back-substitution method. Here are the formulas for the random-effects model:

$$\widehat{\sigma}^2_e = MS_e, \qquad \widehat{\sigma}^2_{ABC} = \frac{MS_{ABC} - MS_e}{n},$$

$$\widehat{\sigma}^2_{AB} = \frac{MS_{AB} - MS_{ABC}}{cn} \quad \text{with similar expressions for } \widehat{\sigma}^2_{AC} \text{ and } \widehat{\sigma}^2_{BC},$$

$$\widehat{\sigma}^2_A = \frac{MS_A - MS_e - n\widehat{\sigma}^2_{ABC} - cn\widehat{\sigma}^2_{AB} - bn\widehat{\sigma}^2_{AC}}{bcn} \quad \text{with similar expressions}$$

for $\widehat{\sigma}^2_B$ and $\widehat{\sigma}^2_C$.

11.3.2.1 Rules for Expected Mean Squares

The rule for E(MS) of any effect depends on the combination of the factors constituting that effect (e.g., AB) and whether the factors are fixed or random:

Rule 1 Every E(MS) includes σ^2_e and $E(MS_e) = \sigma^2_e$.

Rule 2 If all factors are random, then E(MS) for any effect includes all variance components of the sets of factors of which the factors in the given effect are a subset. For example, $E(MS_A)$ includes $\sigma^2_A, \sigma^2_{AB}, \sigma^2_{AC}$, and σ^2_{ABC}, while $E(MS_{AB})$ includes only σ^2_{AB} and σ^2_{ABC}.

Rule 3 If any factors are fixed, then three cases must be considered to obtain E(MS) for any effect. Begin with the E(MS) expression for that effect assuming all factors are random.

Rule 3A If the effect includes only random factors, then delete all terms consisting of variance components of effects involving fixed factors. For example, if A is fixed and B, C are random, then delete the terms σ^2_{ABC} and σ^2_{AB} from the $\text{E}(\text{MS}_B)$ expression for the random-effects model. Similarly, delete the term σ^2_{ABC} from the $\text{E}(\text{MS}_{BC})$ expression for the random-effects model.

Rule 3B If the effect includes fixed as well as random factors, then delete only those variance component terms that involve fixed factors not included in the effect. For example, if A and B are fixed and C is random, then from the $\text{E}(\text{MS}_{AC})$ expression for the random-effects model delete the σ^2_{ABC} term since it involves B, which is not included in AC, but do not delete any term involving A (e.g., σ^2_{AC}).

Rule 3C If the effect includes only fixed factors then delete any variance component terms that do not include those fixed factors but keep any terms that include random factors. Furthermore, replace σ^2 for the effect by its finite-population analog, Q. For example, if A and B are fixed and C is random, then from the $\text{E}(\text{MS}_A)$ expression for the random-effects model delete the variance components involving B (e.g., σ^2_{AB} and σ^2_{ABC}), but do not delete any term involving C (e.g., σ^2_{AC}). In addition, replace σ^2_A by Q_A.

Rule 4 The coefficient of any variance component term equals the total number of observations at each factor-level combination for those factors included in the effect summed over the remaining factors. For example, the coefficient of the σ^2_{ABC} term is n because there are n observations at each combination (i, j, k) of A, B, C. Similarly the coefficient of the σ^2_{AB} (or Q_{AB}) term is cn because there are cn observations at each combination (i, j) of A, B summed over the levels of C.

By applying these rules, the E(MS) expressions for any balanced crossed layout with a combination of fixed and random factors can be obtained.

11.3.3 Approximate F-Tests

Suppose all three factors are random. Then under $H_{0A} : \sigma^2_A = 0$, we have

$$\text{E}(\text{MS}_A) = \sigma^2_e + n\sigma^2_{ABC} + cn\sigma^2_{AB} + bn\sigma^2_{AC}.$$

But there is no other mean square which has the same expectation. If we make the assumption that $\sigma^2_{AC} = 0$, then $\text{E}(\text{MS}_A) = \text{E}(\text{MS}_{AB})$ and we can use $F_A = \text{MS}_A/\text{MS}_{AB}$ as an F-statistic with $a - 1$ and $(a - 1)(b - 1)$ d.f. But there must be a valid basis for making such an assumption. One possibility is to test this assumption using the F-statistic $F_{AC} = \text{MS}_{AC}/\text{MS}_{ABC}$ and adopt the assumption only if the test is nonsignificant, but there is a type II error risk that a nonzero σ^2_{AC} may not be detected because of insufficient d.f. for the F-statistic.

Another possibility is to find a linear combination of two or more mean squares whose expected value is equal to that of the mean square for the effect being tested under the given null hypothesis. Their ratio can then be used as an approximate F-statistic. The approximation comes from the fact that such a linear combination is generally not chi-square distributed, but it may be approximated by a scaled chi-square using the moment-matching method.

We shall illustrate this method for testing $H_{0A} : \sigma^2_A = 0$. Observe that under this null hypothesis

$$\mathrm{E}(\mathrm{MS}_A) = \mathrm{E}(\mathrm{MS}_{AB} + \mathrm{MS}_{AC} - \mathrm{MS}_{ABC}) = \sigma^2_e + n\sigma^2_{ABC} + cn\sigma^2_{AB} + bn\sigma^2_{AC}.$$

Hence we can attempt to use

$$F_A = \frac{\mathrm{MS}_A}{\mathrm{MS}_{AB} + \mathrm{MS}_{AC} - \mathrm{MS}_{ABC}}$$

as an F-statistic. Now, $\mathrm{MS}_A \sim \mathrm{E}(\mathrm{MS}_A)\chi^2_{a-1}/(a-1)$ and is independent of $\mathrm{MS}_{AB} + \mathrm{MS}_{AC} - \mathrm{MS}_{ABC}$, but the latter does not have a scaled chi-square distribution. Nonetheless, we approximate its distribution by $p\chi^2_q/q$, where the constants p and q are determined by equating the first two moments (mean and variance) of $\mathrm{MS}_{AB} + \mathrm{MS}_{AC} - \mathrm{MS}_{ABC}$ to those of $p\chi^2_q/q$. This is known as the **Welch–Satterthwaite approximation**, which was used in Section 2.2 for the t-test to compare two means under heteroscedasticity. Recalling from Section A.4 that $\mathrm{E}(\chi^2_m) = m$ and $\mathrm{Var}(\chi^2_m) = 2m$, we get

$$\mathrm{E}\left(\frac{p\chi^2_q}{q}\right) = p = \mathrm{E}(\mathrm{MS}_{AB} + \mathrm{MS}_{AC} - \mathrm{MS}_{ABC}) = \sigma^2_e + n\sigma^2_{ABC} + cn\sigma^2_{AB} + bn\sigma^2_{AC}$$

and

$$\mathrm{Var}\left(\frac{p\chi^2_q}{q}\right) = \frac{2p^2}{q} = \mathrm{Var}(\mathrm{MS}_{AB}) + \mathrm{Var}(\mathrm{MS}_{AC}) + \mathrm{Var}(\mathrm{MS}_{ABC})$$

$$= \frac{2[\mathrm{E}(\mathrm{MS}_{AB})]^2}{(a-1)(b-1)} + \frac{2[\mathrm{E}(\mathrm{MS}_{AC})]^2}{(a-1)(c-1)} + \frac{2[\mathrm{E}(\mathrm{MS}_{ABC})]^2}{(a-1)(b-1)(c-1)}$$

since $\mathrm{MS}_{AB}, \mathrm{MS}_{AC}$, and MS_{ABC} are independently distributed as $\mathrm{E}(\mathrm{MS}_{AB})$ $\chi^2_{(a-1)(b-1)}/(a-1)(b-1)$, $\mathrm{E}(\mathrm{MS}_{AC})\chi^2_{(a-1)(c-1)}/(a-1)(c-1)$, and $\mathrm{E}(\mathrm{MS}_{ABC})$ $\chi^2_{(a-1)(b-1)(c-1)}/(a-1)(b-1)(c-1)$, respectively.

Substituting the value of p from the first equation into the second, we obtain the following formula for the d.f. q:

$$q = \frac{[\mathrm{E}(\mathrm{MS}_{AB} + \mathrm{MS}_{AC} - \mathrm{MS}_{ABC})]^2}{\dfrac{[\mathrm{E}(\mathrm{MS}_{AB})]^2}{(a-1)(b-1)} + \dfrac{[\mathrm{E}(\mathrm{MS}_{AC})]^2}{(a-1)(c-1)} + \dfrac{[\mathrm{E}(\mathrm{MS}_{ABC})]^2}{(a-1)(b-1)(c-1)}}.$$

The expected mean squares in this expression involve unknown E(MS) terms, but they can be replaced by the corresponding sample mean squares (which are their unbiased estimators), thus obtaining an estimator, $\widehat{q}$, of q. We use $\widehat{q}$ as the estimated d.f. of the chi-square and thereby approximate the distribution of F_A by $F_{a-1,\widehat{q}}$. Note that $\widehat{q}$ will in general be fractional, and so a suitable interpolation must be employed or $\widehat{q}$ must be truncated down to the next lower integer to find a conservative critical constant for F_A. The same method can be employed for testing H_{0B} and H_{0C}.

There are some problems with this approximation. One problem is that the linear combination of mean squares may turn out to be negative or zero, in which case it cannot be used. Another problem is that there is not a unique way to form an F-ratio. One can find different linear combinations of mean squares for the numerator and denominator such that their expected values are equal under the given null hypothesis. Exercise 11.14 gives an alternative F-statistic for the testing problem considered here; it involves only the sums of mean squares both in the numerator and denominator and so cannot be negative.

Example 11.5 (Gage R&R Study for Tape Head Testers)

Adamec and Burdick (2003) describe a gage R&R study for the measurement of "forward resolution" of magnetic tape heads used for computerized data storage. Three automated test stations (factor A) were used to evaluate the quality of nine tape heads (factor B). Each test station tested each head on three tapes, and each measurement was replicated three times. Thus we have $a = 3, b = 9, c = 3$, and $n = 3$ for a total of $N = 243$ observations. All three factors are assumed to be random in the following analysis. The authors do not give raw data but only give a summary ANOVA table (d.f. and means squares). The final ANOVA computed using this summary data is shown in Table 11.10. We see that, except for the main effect of test stations, all other effects are significant ($p < 0.05$).

The calculations for approximate F-tests for testing significance of the main effects of A, B, and C are as follows. First we calculate the approximate

Table 11.10 ANOVA for Gage R&R Study of Tape Head Testers

Source	SS	d.f.	MS	Error Term	F	Error d.f.	p
A (test stations)	26.161	2	13.081	3.588[a]	3.645[a]	5.782[a]	0.095[a]
B (tape heads)	822.773	8	102.847	2.540[a]	40.490[a]	21.549[a]	0.000[a]
C (tapes)	58.462	2	29.231	4.938[a]	5.920[a]	10.261[a]	0.020[a]
AB	17.191	16	1.074	0.418	2.569	32	0.011
AC	11.728	4	2.932	0.418	7.014	32	0.000
BC	30.137	16	1.884	0.418	4.507	32	0.000
ABC	13.385	32	0.418	0.139	3.007	81	0.000
Error	11.289	81	0.139				
Total	991.126	161					

[a]Entries are for approximate F-tests.

F-statistics:

$$F_A = \frac{\text{MS}_A}{\text{MS}_{AB} + \text{MS}_{AC} - \text{MS}_{ABC}} = \frac{13.081}{1.0744 + 2.932 - 0.418} = 3.645,$$

$$F_B = \frac{\text{MS}_B}{\text{MS}_{AB} + \text{MS}_{BC} - \text{MS}_{ABC}} = \frac{102.847}{1.0744 + 1.844 - 0.418} = 40.490,$$

$$F_C = \frac{\text{MS}_C}{\text{MS}_{AC} + \text{MS}_{BC} - \text{MS}_{ABC}} = \frac{29.231}{2.932 + 1.844 - 0.418} = 5.920.$$

The denominators of the F-statistics are the error terms shown in the ANOVA table. The error d.f. are calculated as follows:

$$\widehat{q}_A = \frac{(\text{MS}_{AB} + \text{MS}_{AC} - \text{MS}_{ABC})^2}{\dfrac{(\text{MS}_{AB})^2}{(a-1)(b-1)} + \dfrac{(\text{MS}_{AC})^2}{(a-1)(c-1)} + \dfrac{(\text{MS}_{ABC})^2}{(a-1)(b-1)(c-1)}}$$

$$= \frac{3.588^2}{\dfrac{1.074^2}{16} + \dfrac{2.932^2}{4} + \dfrac{0.418^2}{32}} = 5.782,$$

$$\widehat{q}_B = \frac{(\text{MS}_{AB} + \text{MS}_{BC} - \text{MS}_{ABC})^2}{\dfrac{(\text{MS}_{AB})^2}{(a-1)(b-1)} + \dfrac{(\text{MS}_{BC})^2}{(b-1)(c-1)} + \dfrac{(\text{MS}_{ABC})^2}{(a-1)(b-1)(c-1)}}$$

$$= \frac{2.540^2}{\dfrac{1.074^2}{16} + \dfrac{1.884^2}{16} + \dfrac{0.418^2}{32}} = 21.549,$$

$$\widehat{q}_C = \frac{(\text{MS}_{AC} + \text{MS}_{BC} - \text{MS}_{ABC})^2}{\dfrac{(\text{MS}_{AC})^2}{(a-1)(c-1)} + \dfrac{(\text{MS}_{BC})^2}{(b-1)(c-1)} + \dfrac{(\text{MS}_{ABC})^2}{(a-1)(b-1)(c-1)}}$$

$$= \frac{4.938^2}{\dfrac{2.932^2}{4} + \dfrac{1.884^2}{16} + \dfrac{0.418^2}{32}} = 10.261.$$

The variance component estimates are calculated as follows:

$$\widehat{\sigma}_e^2 = 0.139,$$

$$\widehat{\sigma}_{ABC}^2 = \frac{\text{MS}_{ABC} - \text{MS}_e}{n} = \frac{0.418 - 0.139}{3} = 0.093,$$

$$\widehat{\sigma}_{AB}^2 = \frac{\text{MS}_{AB} - \text{MS}_{ABC}}{cn} = \frac{1.074 - 0.418}{9} = 0.073,$$

$$\widehat{\sigma}^2_{AC} = \frac{\text{MS}_{AC} - \text{MS}_{ABC}}{bn} = \frac{2.932 - 0.418}{27} = 0.093,$$

$$\widehat{\sigma}^2_{BC} = \frac{\text{MS}_{BC} - \text{MS}_{ABC}}{an} = \frac{1.884 - 0.418}{9} = 0.163,$$

$$\widehat{\sigma}^2_{A} = \frac{\text{MS}_{A} - \text{MS}_{e} - n\widehat{\sigma}^2_{ABC} - cn\widehat{\sigma}^2_{AB} - bn\widehat{\sigma}^2_{AC}}{bcn}$$

$$= \frac{13.081 - 0.139 - 0.0279 - 0.656 - 2.514}{81} = 0.117,$$

$$\widehat{\sigma}^2_{B} = \frac{\text{MS}_{B} - \text{MS}_{e} - n\widehat{\sigma}^2_{ABC} - cn\widehat{\sigma}^2_{AB} - an\widehat{\sigma}^2_{BC}}{acn}$$

$$= \frac{102.847 - 0.139 - 0.0279 - 0.656 - 1.466}{27} = 3.715,$$

$$\widehat{\sigma}^2_{C} = \frac{\text{MS}_{C} - \text{MS}_{e} - n\widehat{\sigma}^2_{ABC} - bn\widehat{\sigma}^2_{AC} - an\widehat{\sigma}^2_{BC}}{abn}$$

$$= \frac{29.231 - 0.139 - 0.0279 - 2.514 - 1.466}{81} = 0.307.$$

The gage capability calculations can be done along the same lines as in Example 11.3. Thus we have

$$\widehat{\sigma}^2_{\text{Repeatability}} = \widehat{\sigma}^2_e = 0.139,$$

$$\widehat{\sigma}^2_{\text{Reproducibility}} = \widehat{\sigma}^2_{\text{Test stations}} + \widehat{\sigma}^2_{\text{Test stations}\times\text{Tape heads}} = 0.117 + 0.073 = 0.190,$$

$$\widehat{\sigma}^2_{\text{Gage}} = \widehat{\sigma}^2_{\text{Repeatability}} + \widehat{\sigma}^2_{\text{Reproducibility}} = 0.329.$$

Hence $\widehat{\sigma}_{\text{Gage}} = \sqrt{0.329} = 0.574$. The ratio of $\widehat{\sigma}_{\text{Gage}}$ to $\widehat{\sigma}_{\text{Tape heads}}$ is $0.574/\sqrt{3.715} = 0.298$, which is nearly three times the minimum of 10% that is generally regarded as acceptable. Therefore the measurement system is not capable. ■

11.4 CHAPTER NOTES

11.4.1 Maximum Likelihood and Restricted Maximum Likelihood (REML) Estimation of Variance Components

This exposition follows along the lines of Searle, Casella, and McCulloch (1992, Chapter 3). For simplicity, we will restrict to the balanced one-way random-effects model. Denote the vector of observations from the ith treatment by $\boldsymbol{y}_i = (y_{i1}, y_{i2}, \dots, y_{in})'$ and the complete data vector by $\boldsymbol{y} = (\boldsymbol{y}'_1, \boldsymbol{y}'_2, \dots, \boldsymbol{y}'_a)'$. Then from the model (11.1) it follows that the y_{ij} are identically distributed as $\text{N}(\mu, \sigma^2_y = \sigma^2_A + \sigma^2_e)$. Within each treatment they are equicorrelated with common correlation ρ given by (11.2), while between treatments they are independent. Thus the $\boldsymbol{y}_i$ are i.i.d. n-variate MVN$(\mu\mathbf{1}, \sigma_y^2\boldsymbol{R})$, where $\mathbf{1}$ is an n-vector of all 1's and $\boldsymbol{R}$ is an $n \times n$ correlation matrix with all correlations equal to ρ. Therefore the likelihood function is given by

$$\begin{aligned} L(\mu, \sigma_A^2, \sigma_e^2|\boldsymbol{y}) &= \prod_{i=1}^{a} L_i(\mu, \sigma_A^2, \sigma_e^2|\boldsymbol{y}_i) \\ &= \prod_{i=1}^{a} \frac{1}{(2\pi)^{n/2}(\sigma_y^n)|\boldsymbol{R}|^{1/2}} \\ &\quad \times \exp\left\{-\frac{1}{2\sigma_y^2}(\boldsymbol{y}_i - \mu\mathbf{1})'\boldsymbol{R}^{-1}(\boldsymbol{y}_i - \mu\mathbf{1})\right\}. \end{aligned} \tag{11.18}$$

Now $\boldsymbol{R}$ is a **compound symmetric matrix** of the form $a\boldsymbol{I} + b\boldsymbol{J}$, where $\boldsymbol{I}$ is an identity matrix and $\boldsymbol{J}$ is a matrix with all entries equal to 1; for $\boldsymbol{R}$ we have $a = 1 - \rho$ and $b = \rho$. It is well known that

$$\boldsymbol{R}^{-1} = \frac{1}{a}\left[\boldsymbol{I} - \frac{b}{a + bn}\boldsymbol{J}\right] = \frac{1}{1-\rho}\left[\boldsymbol{I} - \frac{\rho}{1 + (n-1)\rho}\boldsymbol{J}\right]$$

and

$$|\boldsymbol{R}| = (a - b)^{n-1}(a + bn) = (1 - \rho)^{n-1}[1 + (n-1)\rho].$$

Substituting these expressions in the likelihood function and after some simplification, we obtain

$$\begin{aligned} L_i(\mu, \sigma_A^2, \sigma_e^2|\boldsymbol{y}_i) &= \frac{1}{(2\pi)^{n/2}(\sigma_y^n)(1-\rho)^{(n-1)/2}[1 + (n-1)\rho]^{1/2}} \\ &\quad \times \exp\left\{-\frac{1}{2\sigma_y^2(1-\rho)}\left[\sum_j (y_{ij} - \mu)^2 - \frac{n^2\rho}{1 + (n-1)\rho}(\bar{y}_i - \mu)^2\right]\right\}. \end{aligned}$$

Substituting $\rho = \sigma_A^2/\sigma_y^2$ from (11.12) and noting that $\sigma_y^2(1 - \rho) = \sigma_e^2$ and $\sigma_y^2[1 + (n-1)\rho] = \sigma_e^2 + n\sigma_A^2$, more algebra leads to

$$\begin{aligned} L_i(\mu, \sigma_A^2, \sigma_e^2|\boldsymbol{y}_i) &= \frac{1}{(2\pi)^{n/2}(\sigma_e^n)(\sigma_e^2 + n\sigma_A^2)^{1/2}} \\ &\quad \times \exp\left\{-\frac{1}{2\sigma_e^2}\left[\sum_j (y_{ij} - \mu)^2 - \frac{n^2\sigma_A^2}{\sigma_e^2 + n\sigma_A^2}(\bar{y}_i - \mu)^2\right]\right\}. \end{aligned}$$

Therefore the overall likelihood from (11.18) equals

$$L(\mu, \sigma_A^2, \sigma_e^2 | \mathbf{y}) = \frac{1}{(2\pi)^{an/2}(\sigma_e^{an})(\sigma_e^2 + n\sigma_A^2)^{a/2}} \times \exp\left\{-\frac{1}{2\sigma_e^2}\left[\sum_{i,j}(y_{ij} - \mu)^2 - \frac{n^2\sigma_A^2}{\sigma_e^2 + n\sigma_A^2}\sum_i(\bar{y}_i - \mu)^2\right]\right\}. \tag{11.19}$$

The expression inside the exponential equals

$$\begin{aligned}
&-\frac{1}{2\sigma_e^2}\left[\sum_{i,j}(y_{ij} - \bar{y}_i)^2 + n\sum_i(\bar{y}_i - \mu)^2 - \frac{n^2\sigma_A^2}{\sigma_e^2 + n\sigma_A^2}\sum_i(\bar{y}_i - \mu)^2\right] \\
&= -\frac{1}{2\sigma_e^2}\left[\mathrm{SS}_e + n\left(1 - \frac{n\sigma_A^2}{\sigma_e^2 + n\sigma_A^2}\right)\sum_i(\bar{y}_i - \bar{\bar{y}} + \bar{\bar{y}} - \mu)^2\right] \\
&= -\frac{1}{2\sigma_e^2}\left[\mathrm{SS}_e + \frac{\sigma_e^2}{\sigma_e^2 + n\sigma_A^2}\left\{n\sum_i(\bar{y}_i - \bar{\bar{y}})^2 + an(\bar{\bar{y}} - \mu)^2\right\}\right] \\
&= -\frac{1}{2\sigma_e^2}\left[\mathrm{SS}_e + \frac{\sigma_e^2}{\sigma_e^2 + n\sigma_A^2}\left\{\mathrm{SS}_A + N(\bar{\bar{y}} - \mu)^2\right\}\right].
\end{aligned}$$

Substituting this expression in (11.19) and putting $\tau^2 = \sigma_e^2 + n\sigma_A^2$, we see that the full likelihood factors into two marginal likelihoods as follows:

$$L(\mu, \sigma_A^2, \sigma_e^2 | \mathbf{y}) = L(\mu | \bar{\bar{y}}) L(\sigma_A^2, \sigma_e^2 | \mathrm{SS}_A, \mathrm{SS}_e),$$

where

$$L(\mu | \bar{\bar{y}}) = \frac{1}{(2\pi)^{1/2}(\tau^2/N)^{1/2}} \exp\left\{-\frac{N}{2\tau^2}(\bar{\bar{y}} - \mu)^2\right\}$$

and

$$L(\sigma_A^2, \sigma_e^2 | \mathrm{SS}_A, \mathrm{SS}_e) = \frac{1}{(2\pi)^{(an-1)/2}(\sigma_e^{a(n-1)})\tau^{a-1}N^{1/2}} \exp\left\{-\frac{1}{2}\left(\frac{\mathrm{SS}_e}{\sigma_e^2} + \frac{\mathrm{SS}_A}{\tau^2}\right)\right\}.$$

We see that the marginal likelihood of μ depends only on $\bar{\bar{y}}$, while that of σ_A^2 and σ_e^2 depends only on SS_e and SS_A. The ML estimators are found by maximizing the full likelihood (11.19). On the other hand, the REML estimators of the variance components are obtained by maximizing only that part of the full likelihood that does not depend on μ, that is, the marginal likelihood of σ_A^2 and σ_e^2. Another way to think about REML estimators is that they maximize the likelihood based on the joint distribution of the residuals obtained after fitting the location parameter (μ) part of the model. For ML as well as for REML maximization is done subject to $\sigma_e^2, \sigma_A^2 \geq 0$.

The estimators in each case can be derived by taking the partial derivatives of the appropriate log-likelihood functions and setting them equal to zeros and taking account of the nonnegativity constraints on σ_e^2 and σ_A^2. The resulting formulas are as follows:

ML Estimators:

$$\widehat{\sigma}_A^2 = \frac{1}{n}\left[\left(1 - \frac{1}{a}\right)\text{MS}_A - \text{MS}_e\right] \text{ and } \widehat{\sigma}_e^2 = \text{MS}_e \quad \text{if } \left(1 - \frac{1}{a}\right) F_A \geq 1,$$

$$\widehat{\sigma}_A^2 = 0 \quad \text{and} \quad \widehat{\sigma}_e^2 = \frac{\text{SS}_{\text{tot}}}{N} \quad \text{if } \left(1 - \frac{1}{a}\right) F_A < 1.$$

REML Estimators:

$$\widehat{\sigma}_A^2 = \frac{1}{n}[\text{MS}_A - \text{MS}_e] \qquad \text{and} \qquad \widehat{\sigma}_e^2 = \text{MS}_e \quad \text{if } F_A \geq 1,$$

$$\widehat{\sigma}_A^2 = 0 \qquad \text{and} \qquad \widehat{\sigma}_e^2 = \frac{\text{SS}_{\text{tot}}}{N - 1} \quad \text{if } F_A < 1.$$

Note that ML uses $(1 - 1/a)\,\text{MS}_A = \text{SS}_A/a$ whereas REML uses $\text{MS}_A = \text{SS}_A/(a-1)$; similarly ML uses N as the divisor for SS_{tot} whereas REML uses $N - 1$. Thus REML adjusts for the loss of one d.f. due to the estimation of μ, which ML does not. Also note that REML estimators are identical to the ANOVA estimators. This is true for balanced designs but not for unbalanced designs. In general, REML estimators do not have closed formulas and must be computed using specialized software such as SAS, which can handle complex models, including mixed effects, nested effects, and so on.

11.4.2 Derivations of Results for One- and Two-Way Random-Effects Designs

The following two lemmas are useful to derive the distributions of the sums of squares in the one- and two-way random effects ANOVAs. They follow directly from Lemmas A.2 and A.6, respectively.

Lemma 11.1 *Let $x_1, x_2, \ldots, x_n$ be independent r.v.'s with $x_i \sim \text{N}(\mu_i, \sigma^2)$ for $i = 1, 2, \ldots, n$. Then*

$$\sum_{i=1}^{n}(x_i - \overline{x})^2 \sim \sigma^2 \chi^2_{n-1}(\lambda^2),$$

where $\chi^2_{n-1}(\lambda^2)$ *is a noncentral chi-square r.v. (see Section A.4) with* $n - 1$ *d.f. and n.c.p.* $\lambda^2 = \sum_{i=1}^{n}(\mu_i - \overline{\mu})^2/\sigma^2$. *Hence*

$$\mathrm{E}\left[\sum_{i=1}^{n}(x_i - \overline{x})^2\right] = \sigma^2[(n-1) + \lambda^2].$$

If all μ_i*'s are equal, then* $\lambda^2 = 0$ *and the distribution is central* χ^2_{n-1}. ■

Lemma 11.2 *Let* x_{ij} $(1 \le i \le a, 1 \le j \le b)$ *be an array of independent r.v.'s with* $x_{ij} \sim \mathrm{N}(\mu_{ij}, \sigma^2)$. *Then*

$$\sum_{i=1}^{a}\sum_{j=1}^{b}(x_{ij} - \overline{x}_{i\cdot} - \overline{x}_{\cdot j} + \overline{x}_{\cdot\cdot})^2 \sim \sigma^2 \chi^2_{(a-1)(b-1)}(\lambda^2)$$

where the n.c.p.

$$\lambda^2 = \frac{1}{\sigma^2}\sum_{i=1}^{a}\sum_{j=1}^{b}(\mu_{ij} - \overline{\mu}_{i\cdot} - \overline{\mu}_{\cdot j} + \overline{\mu}_{\cdot\cdot})^2.$$

Hence

$$\mathrm{E}\left[\sum_{i=1}^{a}\sum_{j=1}^{b}(x_{ij} - \overline{x}_{i\cdot} - \overline{x}_{\cdot j} + \overline{x}_{\cdot\cdot})^2\right] = \sigma^2[(a-1)(b-1) + \lambda^2].$$

If all μ_{ij} *are equal, then* $\lambda^2 = 0$ *and the distribution is central* $\chi^2_{(a-1)(b-1)}$. ■

First we shall derive (11.4) by using Lemma 11.1. Substitute the model (11.1) in the expressions for SS_A and SS_e to obtain

$$\mathrm{SS}_A = n\sum_{i=1}^{a}(\alpha_i - \overline{\alpha} + \overline{e}_{i\cdot} - \overline{e}_{\cdot\cdot})^2 \quad \text{and} \quad \mathrm{SS}_e = \sum_{i=1}^{a}\sum_{j=1}^{n}(e_{ij} - \overline{e}_{i\cdot})^2.$$

The independence between SS_A and SS_e follows from the facts that the e_{ij} are independent of the α_i and the $e_{ij} - \overline{e}_{i\cdot}$ are independent of the $\overline{e}_{i\cdot} - \overline{e}_{\cdot\cdot}$. The latter follows from Corollary 2 to Lemma A.6.

To derive the distribution of SS_e, note that the summands $\sum_{j=1}^{n}(e_{ij} - \overline{e}_{i\cdot})^2$ for $i = 1, 2, \ldots, a$ are distributed independently as $\sigma_e^2\chi^2_{n-1}$ using Lemma 11.1, and hence SS_e, which is their sum, is distributed as $\sigma_e^2\chi_\nu^2$ with $\nu = a(n-1)$. Moving on to the distribution of SS_A, put $x_i = \alpha_i - \overline{e}_{i\cdot\cdot}$. Then the x_i are i.i.d. $N(0, \sigma_A^2 + \sigma_e^2/n)$ r.v.'s and $SS_A = n\sum_{i=1}^a (x_i - \overline{x})^2$. Hence, by Lemma 11.1, it follows that $SS_A \sim n(\sigma_A^2 + \sigma_e^2/n)\chi_{a-1}^2 = (n\sigma_A^2 + \sigma_e^2)\chi_{a-1}^2$, which completes the proof of (11.4). The expected values given in (11.5) follow immediately.

Next consider the two-way random-effects design. By substituting the model (11.12) into the sums-of-squares expressions, we obtain

$$SS_e = \sum_{i=1}^{a}\sum_{j=1}^{b}\sum_{k=1}^{n}(e_{ijk} - \overline{e}_{ij\cdot})^2,$$

$$SS_{AB} = n\sum_{i=1}^{a}\sum_{j=1}^{b}[((\alpha\beta)_{ij} - \overline{(\alpha\beta)}_{i\cdot} - \overline{(\alpha\beta)}_{\cdot j} + \overline{(\alpha\beta)}_{\cdot\cdot} + \overline{e}_{ij\cdot} - \overline{e}_{i\cdot\cdot} - \overline{e}_{\cdot j\cdot} + \overline{e}_{\cdot\cdot\cdot}]^2,$$

$$SS_B = an\sum_{j=1}^{b}(\beta_j - \overline{\beta} + \overline{(\alpha\beta)}_{\cdot j} - \overline{(\alpha\beta)}_{\cdot\cdot} + \overline{e}_{\cdot j\cdot} - \overline{e}_{\cdot\cdot\cdot})^2,$$

$$SS_A = bn\sum_{i=1}^{a}(\alpha_i - \overline{\alpha} + \overline{(\alpha\beta)}_{i\cdot} - \overline{(\alpha\beta)}_{\cdot\cdot} + \overline{e}_{i\cdot\cdot} - \overline{e}_{\cdot\cdot\cdot})^2.$$

Since the e_{ijk} are i.i.d. $N(0, \sigma_e^2)$, it follows that

$$SS_e \sim \sigma_e^2\chi_{ab(n-1)}^2$$

and hence

$$E(MS_e) = \sigma_e^2.$$

Next, in the expression for SS_{AB} substitute $x_{ij} = (\alpha\beta)_{ij} + \overline{e}_{ij\cdot} \sim N(0, \sigma_{AB}^2 + (1/n)\sigma_e^2)$. Then using Lemma 11.2, it follows that

$$SS_{AB} = n\sum_{i=1}^{a}\sum_{j=1}^{b}(x_{ij} - \overline{x}_{i\cdot} - \overline{x}_{\cdot j} + \overline{x}_{\cdot\cdot})^2 \sim n\left(\sigma_{AB}^2 + \frac{1}{n}\sigma_e^2\right)\chi_{(a-1)(b-1)}^2$$

and hence

$$E(MS_{AB}) = n\sigma_{AB}^2 + \sigma_e^2.$$

Next, in the expression for SS_B substitute $x_j = \beta_j + \overline{(\alpha\beta)}_{\cdot j} + \overline{e}_{\cdot j\cdot} \sim N(0, \sigma_B^2 + (1/a)\sigma_{AB}^2 + (1/an)\sigma_e^2)$. Then using Lemma 11.1, it follows that

$$SS_B = an\sum_{j=1}^{b}(x_j - \overline{x})^2 \sim an\left(\sigma_B^2 + \frac{1}{a}\sigma_{AB}^2 + \frac{1}{an}\sigma_e^2\right)\chi_{b-1}^2$$

and hence

$$E(MS_B) = an\sigma_B^2 + n\sigma_{AB}^2 + \sigma_e^2.$$

The proof of the expression for $E(MS_A)$ is analogous.

11.4.3 Relationship between Unrestricted and Restricted Models

The relationship between the two models can be derived by noting that the restricted model can be obtained from the unrestricted model by defining

$$\beta_j' = \beta_j + \overline{(\alpha\beta)}_{\cdot j} \qquad \text{and} \qquad (\alpha\beta)_{ij}' = (\alpha\beta)_{ij} - \overline{(\alpha\beta)}_{\cdot j}.$$

Then it is immediately clear that the two models are equivalent since $\beta_j' + (\alpha\beta)_{ij}' = \beta_j + (\alpha\beta)_{ij}$ (the μ, α_i, and the e_{ij} terms are the same in both models). Furthermore,

$$\sum_{i=1}^{a}(\alpha\beta)_{ij}' = \sum_{i=1}^{a}[(\alpha\beta)_{ij} - \overline{(\alpha\beta)}_{\cdot j}] = 0,$$

which is the specified constraint on the $(\alpha\beta)_{ij}'$. Now using the independence of the β_j and the $(\alpha\beta)_{ij}$ we can write

$$\begin{aligned}\sigma_B'^2 &= \text{Var}(\beta_j') \\ &= \text{Var}(\beta_j) + \text{Var}(\overline{(\alpha\beta)}_{\cdot j}) \\ &= \sigma_B^2 + \frac{1}{a}\sigma_{AB}^2.\end{aligned} \tag{11.20}$$

Next,

$$\begin{aligned}\left(\frac{a-1}{a}\right)\sigma_{AB}'^2 &= \text{Var}((\alpha\beta)_{ij}') \\ &= \text{Var}((\alpha\beta)_{ij} - \overline{(\alpha\beta)}_{\cdot j}) \\ &= \text{Var}((\alpha\beta)_{ij}) + \text{Var}(\overline{(\alpha\beta)}_{\cdot j}) - 2\,\text{Cov}((\alpha\beta)_{ij}, \overline{(\alpha\beta)}_{\cdot j}) \\ &= \sigma_{AB}^2 + \frac{1}{a}\sigma_{AB}^2 - \frac{2}{a}\sigma_{AB}^2 \\ &= \left(\frac{a-1}{a}\right)\sigma_{AB}^2,\end{aligned} \tag{11.21}$$

and so $\sigma_{AB}'^2 = \sigma_{AB}^2$. These relations can also be obtained by equating the expected mean squares in the two cases. In particular,

$$E(MS_{AB}) = n\sigma_{AB}^2 + \sigma_e^2 = n\sigma_{AB}'^2 + \sigma_e'^2$$

yields $\sigma_{AB}^2 = \sigma_{AB}'^2$ since $\sigma_e^2 = \sigma_e'^2$ and

$$E(MS_B) = an\sigma_B^2 + n\sigma_{AB}^2 + \sigma_e^2 = an\sigma_B'^2 + \sigma_e'^2$$

yields $\sigma_B'^2 = \sigma_B^2 + (1/a)\sigma_{AB}^2$. The ANOVA estimators of the variance components also satisfy these relations.

From these relations one can see that since $\sigma_B'^2 > \sigma_B^2$ (and hence $\widehat{\sigma}_B'^2 > \widehat{\sigma}_B^2$), the chance of finding the B main effect significant is greater using the restricted model. A question then arises as to which model should one use since the conclusions drawn from the two models could be different. However, it should be understood that different parameters are under test when H_{0B} is tested for the two models, and so the interpretation of conclusions is different in the two cases. If the parameters in the two models are interpreted correctly, then the same conclusions follow from the two models. We follow the unrestricted model because of its simplicity of interpretation.

11.5 CHAPTER SUMMARY

(a) Random effects are assumed to follow normal distributions with zero means and variances specific to the effects. These variances are known as the variance components. Under the assumption of independence between the effects, the total variance of a random observation is the sum of the variance components due to all random effects including the random error. The main focus in random-effects models is estimation of the variance components.

(b) The one-way random-effects ANOVA is formally similar to that for the fixed-effects case. The ANOVA F-statistic tests the hypothesis that the variance component σ_A^2 due to the random factor is zero. An unbiased estimator of σ_A^2 is given by $(MS_A - MS_e)/n$, which can be negative. Exact CIs can be computed for σ_e and $\theta = \sigma_A/\sigma_e$.

(c) The two-way random-effects ANOVA has the same entries as those for the fixed-effects case, but the F-tests for both main effects use MS_{AB}, the mean square for interaction, as the error term. This model is used to analyze gage R&R studies. The gage variability is the sum of the variance components due to repeatability, which is the error variance, and due to reproducibility, which is the sum of the variance components due to inspectors and parts-by-inspectors interaction. The gage variability must be small compared to the parts variability in order for the gage to be regarded as capable of distinguishing between "good" and "bad" parts.

(d) In a two-way layout, if one factor is fixed and the other is random, then we have a mixed-effects design. Two models are used to analyze these designs:

unrestricted and restricted. Mathematically the two models are equivalent, but the tests of hypotheses and the variance component estimators are different, which can be confusing. We recommend using the unrestricted model because of its simplicity of interpretation.

(e) In a three-way layout with random factors, exact F-tests do not exist for tests of null hypotheses of certain effects because there are no two mean squares whose expectations are equal under the given null hypotheses. Approximate F-tests use ratios of linear combinations of mutually exclusive sets of mean squares whose expectations are equal. Generally, the numerator is the mean square for the effect under test and hence has an exact scaled chi-square distribution. The distribution of the denominator is approximated by a scaled chi-square by matching the first two moments. This gives a formula for estimating the error d.f. for F (the scaling factors from the numerator and denominator cancel).

EXERCISES

Section 11.1 (One-Way Layouts)

Theoretical Exercises

11.1 (a) Derive the CI formula (11.11) from the probability statement preceding it.

(b) From the CI on θ, obtain a CI on the intraclass correlation coefficient ρ.

(c) Use the results of Example 11.2 to calculate a point estimate and a 95% CI on ρ.

11.2 Show that the power of the test (11.6) can be expressed as a function of $\theta^2 = \sigma_A^2/\sigma_e^2$ in terms of the central F-distribution.

11.3 Use the distributional result (11.4) and generalize the test in (11.6) for testing $H_0 : \theta = \theta_0$ versus $H_1 : \theta > \theta_0$ for any $\theta_0 \geq 0$.

11.4 Show that for an unbalanced one-way layout $\mathrm{E}(\mathrm{MS}_A) = \sigma_e^2 + \widetilde{n}\sigma_A^2$, where $\widetilde{n}$ is given by (11.8). Hence $(\mathrm{MS}_A - \mathrm{MS}_e)/\widetilde{n}$ is an unbiased estimator of σ_A^2. [*Hint*: Write $\mathrm{SS}_A = \sum_{i=1}^{a} n_i(\overline{y}_i - \overline{\overline{y}})^2$ as a quadratic form in $\overline{\boldsymbol{y}} = (\overline{y}_1, \overline{y}_2, \ldots, \overline{y}_a)'$ by noting that $\mathrm{SS}_A = (\overline{\boldsymbol{y}} - \overline{\overline{\boldsymbol{y}}})'\boldsymbol{N}(\overline{\boldsymbol{y}} - \overline{\overline{\boldsymbol{y}}})$ and $\overline{\overline{\boldsymbol{y}}} = (1/N)\boldsymbol{J}\boldsymbol{N}\overline{\boldsymbol{y}}$, where $\overline{\overline{\boldsymbol{y}}}$ is an $a \times 1$ vector all of whose entries are $\overline{\overline{y}}$, $\boldsymbol{N} = \mathrm{diag}(n_1, n_2, \ldots, n_a)$, and $\boldsymbol{J}$ is an $a \times a$ matrix of all 1's. Then compute $\mathrm{E}(\mathrm{SS}_A)$ by using Lemma A.1.]

11.5 There is no exact CI for σ_y, but an approximate CI can be constructed as follows:

(a) Show that $\widehat{\sigma}_y^2 = \widehat{\sigma}_A^2 + \widehat{\sigma}_e^2$ is distributed as

$$\frac{E(MS_A)}{n(a-1)}\chi^2_{a-1} + \left(\frac{n-1}{n}\right)\frac{E(MS_e)}{a(n-1)}\chi^2_{a(n-1)},$$

where $E(MS_A) = \sigma_e^2 + n\sigma_A^2$ and $E(MS_e) = \sigma_e^2$.

(b) Use the Welch–Satterthwaite approximation given in Section 11.3.3 to show that this distribution can be approximated by that of $p\chi^2_{\widehat{q}}/\widehat{q}$, where

$$p = \sigma_A^2 + \sigma_e^2 = \sigma_y^2 \quad \text{and} \quad \widehat{q} = \frac{\widehat{\sigma}_y^4}{\frac{MS_A^2}{n^2(a-1)} + \left(\frac{n-1}{n}\right)^2 \frac{MS_e^2}{a(n-1)}}.$$

(c) Hence show that an approximate $100(1-\alpha)\%$ CI for σ_y is given by

$$\widehat{\sigma}_y\sqrt{\frac{\widehat{q}}{\chi^2_{\widehat{q},\alpha/2}}} \le \sigma_y \le \widehat{\sigma}_y\sqrt{\frac{\widehat{q}}{\chi^2_{\widehat{q},1-\alpha/2}}}.$$

Applied Exercises

11.6 Uniformly strong cables are needed to transmit high-voltage electricity. In one design, each cable was composed of 12 wires. Wires in a sample of nine cables were tested for tensile strength (measured in kilograms). The results are given in Table 11.11. Regard the nine cables used in the experiment as a random sample of all cables.

(a) Perform an ANOVA on these data and test if the variability between cables is significantly greater than zero. Use $\alpha = 0.05$.

(b) Estimate the variance components for the cables and for the wires within the cables.

11.7 Table 11.12 gives percentages of sulfur in core samples taken from five coal seams in Texas. Assume that these coal seams are chosen as a random sample from many coal seams.

(a) Calculate the ANOVA table and test whether there are significant differences between the seams.

(b) Estimate the variance component for the variation between the seams. What is the average sample size $\widetilde{n}$ used to compute this estimate?

Table 11.11 Tensile Strengths of Cable Wires

Cable 1	Cable 2	Cable 3	Cable 4	Cable 5	Cable 6	Cable 7	Cable 8	Cable 9
329	340	345	328	341	339	347	339	342
327	330	327	344	340	340	341	340	346
332	325	335	342	335	342	345	347	347
348	328	338	350	336	341	340	345	348
337	338	330	335	339	336	350	350	355
328	332	334	332	340	342	346	348	351
328	335	335	328	342	347	345	341	333
330	340	340	340	345	345	342	342	347
345	336	337	335	341	341	340	337	350
334	339	342	337	338	340	339	346	347
328	335	333	337	346	336	330	340	348
330	329	335	340	347	342	338	345	341

Source: Hald (1952, Table 16.9).

Table 11.12 Sulfur Content of Coal

		Seam		
A	*B*	*C*	*D*	*E*
1.51	1.69	1.56	1.30	0.73
1.92	0.64	1.22	0.75	0.80
1.08	0.90	1.32	1.26	0.90
2.04	1.41	1.39	0.69	1.24
2.14	1.01	1.33	0.62	0.82
1.76	0.84	1.54	0.90	0.72
1.17	1.28	1.04	1.20	0.57
	1.59	2.25	0.32	1.18
		1.49		0.54
				1.30

Source: Elmore, Hettmansperger, and Xuan (2004, Table 2). Reprinted by permission of the Institute of Mathematical Statistics.

Section 11.2 (Two-Way Layouts)

Theoretical Exercises

11.8 Derive the expected mean square expressions shown in Table 11.5 for the balanced two-way mixed-effects unrestricted model.

11.9 Derive the expected mean square expressions shown in Table 11.6 for the balanced two-way mixed-effects restricted model.

Applied Exercises

11.10 Hancock et al. (1997) did a gage capability study on a flexible optical coordinate measurement (OCM) system used for making robotic measurements at different locations on a car door. Variation was thought to result mainly from two different sources: part to part and setup to setup (every time the same OCM system is set up to measure a new type of part, there are changes that occur which affect the measurements). So a two-factor experiment was run in which three randomly selected parts were measured three times each with one setup. That setup was taken down and reassembled, and the same three parts were measured again. This procedure was repeated a third time. Thus the parts and setups may be regarded as crossed random factors. The measurements giving the deviations from the target location are given in Table 11.3. Compute the ANOVA table and the variance components. Which is the major contributor to measurement variability?

11.11 Breaking strength data were collected on roof shingles. A force (measured in grams) was applied to a piece of shingle in a testing machine until the piece broke. Since shingles are brittle, the measured breaking strengths are highly variable. Two operators made five measurements each on 20 shingles. The data are shown in Table 11.14. Perform gage R&R analysis and determine if the measurement system is capable.

11.12 Box (1993b) gave the data shown in Table 11.15 on the coded measurements of output from five heads of a machine taken during six successive periods of eight-hour shifts (i.e., three shifts each day over two successive days). The order of sampling of the heads within each shift was random.

(a) The study was conducted because variability was suspected among the machine heads. Perform an ANOVA using heads as a fixed factor and periods as random factor. Are there significant differences between the heads using $\alpha = 0.05$?

Table 11.13 OCM Gage Capability Study Data (mm)

Part	Setup 1			Setup 2			Setup 3		
1	2.53,	2.53,	2.53	2.44,	2.43,	2.43	2.49,	2.50,	2.50
2	2.34,	2.36,	2.36	2.50,	2.51,	2.51	2.71,	2.71,	2.70
3	2.33,	2.33,	2.33	2.41,	2.42,	2.42	2.44,	2.47,	2.46

Note: All entries are multiplied by −1.

Source: Hancock et al. (1997, Table 6). Reprinted by permission of Taylor & Francis.

Table 11.14 Roof Shingle Breaking Strength Data

Shingle	Operator A	Operator B
1	1408, 1738, 1617, 1830, 1711	1610, 1117, 1139, 1485, 1754
2	1691, 1057, 1027, 1198, 1587	1220, 1072, 1117, 1415, 1450
3	1027, 1499, 1394, 1206, 1443	0958, 1386, 1242, 1401, 981.2
4	1154, 0773, 1415, 0981, 1102	1034, 0996, 0928, 0742, 2033
5	0989, 1344, 1457, 2136, 1534,	1365, 1206, 1784, 0843, 1213
6	1464, 1882, 1471, 2344, 1596,	1843, 1308, 0913, 1228, 1711
7	2242, 2271, 1778, 1901, 2288	2051, 1970, 1862, 2692, 2189
8	1977, 1731, 2142, 2367, 2470,	2569, 2817, 1804, 3016, 1830
9	2027, 2454, 1520, 1478, 1678,	1534, 1322, 1351, 1336, 1499
10	2070, 1989, 1751, 1562, 1698,	1457, 1264, 1169, 1415, 1862
11	1764, 2311, 0966, 1933, 2166,	1139, 2509, 1933, 1582, 1958
12	1882, 1610, 1300, 1920, 1983,	2136, 1169, 1778, 1758, 2027
13	2142, 1849, 1791, 1534, 1778	1804, 2634, 1751, 2002, 1555
14	1907, 1637, 1843, 1836, 1888,	1379, 1738, 2350, 1882, 1286
15	2367, 2614, 2316, 2184, 1513,	1882, 2178, 2282, 2265, 1810
16	2189, 2288, 2522, 2299, 2448,	3267, 1983, 2322, 2378, 1452
17	2389, 2130, 2522, 1372, 2219	2574, 2063, 2688, 1758, 2219
18	2629, 2288, 2908, 2668, 2277,	2195, 2654, 1751, 1945, 1576
19	2130, 2438, 2316, 2166, 2475,	2184, 2744, 2254, 2361, 2242
20	2502, 2400, 2378, 2629, 2564,	2411, 2378, 2438, 1386, 1623

Table 11.15 Coded Data from Five Heads of Machine Over Six Periods

	Heads				
Period	1	2	3	4	5
Day 1, shift 1	20	14	17	12	22
Day 1, shift 2	16	19	16	17	21
Day 1, shift 3	25	32	31	24	24
Day 2, shift 1	18	22	9	15	17
Day 2, shift 2	21	22	16	20	17
Day 2, shift 3	19	28	22	29	29

Source: Box (1993b). Reprinted by permission of Taylor & Francis.

(b) Next do a more detailed ANOVA by taking into account the two-way factorial structure of the period factor with day (random) and shift (fixed) as two separate factors. Which factor causes most variability? Identify the particular level of that factor that is causing variability.

11.13 (From Tamhane and Dunlop, 2005, Exercise 13.28. Reprinted by permission of Pearson Education, Inc.) The housing for a pregnancy test kit contains a disposable strip with a hole and a reagent that is jetted

Table 11.16 Deviations (mm) from True Distance as Measured by Four Technicians Using Three Measuring Instruments

Instrument	Technician 1	Technician 2	Technician 3	Technician 4
Calipers	0.20, 0.14, 0.15	0.24, 0.35, 0.21	0.50, 0.38, 0.54	0.33, 0.25, 0.29
Optical comparator	0.26, 0.25, 0.30	0.20, 0.19, 0.28	0.14, 0.15, 0.17	0.13, 0.15, 0.16
CMM	0.14, 0.19, 0.17	0.21, 0.23, 0.24	0.17, 0.18, 0.21	0.25, 0.35, 0.27

Source: Tamhane and Dunlop (2005, Exercise 13.28). Reprinted by permission of Pearson Education, Inc.

onto the strip. When the disposable strip is manufactured, the distance between the alignment hole and a mark on the strip is measured in order to assure visibility of test results when the strip is assembled in the housing. An experiment was performed to determine which of three measuring instruments, calipers, an optical comparator, or a coordinate measuring machine (CMM), should be used to measure the distance. The experiment was run by having four randomly selected technicians make three readings with each of the three measuring instruments in a random order. The values of the absolute difference between the distance reading and the actual distance are given in Table 11.16.

(a) Regard the instruments as fixed and technicians as random. Calculate the ANOVA table and perform tests of significance of the effects at $\alpha = 0.05$.

(b) Estimate the appropriate variance components.

Section 11.3 (Three-Way Layouts)

Theoretical Exercises

11.14 Assume the three-way random-effects model of Section 11.3 and consider the problem of testing $H_{0A} : \sigma_A^2 = 0$. In Section 11.3.3 we saw that an exact F-test does not exist for this problem and an approximate F-test can be based on $F_A = \text{MS}_A/(\text{MS}_{AB} + \text{MS}_{AC} - \text{MS}_{ABC})$. Consider an alternative "F"-statistic, $F_A = (\text{MS}_A + \text{MS}_{ABC})/(\text{MS}_{AB} + \text{MS}_{AC})$.

(a) Show that $\text{E}(\text{MS}_A + \text{MS}_{ABC}) = \text{E}(\text{MS}_{AB} + \text{MS}_{AC})$ under $H_{0A} : \sigma_A^2 = 0$.

(b) To approximate the distributions of $\text{MS}_A + \text{MS}_{ABC}$ and $\text{MS}_{AB} + \text{MS}_{AC}$ by $p_1\chi^2_{q_1}/q_1$ and $p_2\chi^2_{q_2}/q_2$, respectively, find p_1, q_1, p_2, q_2 by the moment-matching method. [*Note:* $p_1 = p_2 =$ common expected value derived in part (a), so you only need to derive formulas for q_1 and q_2.]

Table 11.17 Spring Lengths (in.)

Day	Spring 1	Spring 2	Spring 3	Spring 4	Spring 5
Operator 1					
1	72.750	73.000	73.125	73.500	73.875
2	72.750	73.031	73.250	73.438	73.750
3	72.813	72.938	73.000	73.531	73.750
Operator 2					
1	73.000	73.000	73.156	73.438	73.813
2	72.844	73.000	73.250	73.438	73.750
3	72.750	73.125	73.125	73.500	73.500
Operator 3					
1	72.813	73.063	73.125	73.500	73.750
2	72.813	73.125	73.063	73.500	73.813
3	72.750	72.938	73.125	73.500	73.813

Source: Mason, Gunst, and Hess (2003, Figure 12.2). Reprinted by permission of John Wiley & Sons, Inc.

(c) Show how to use the F-approximation with the estimated d.f. for the numerator and denominator.

11.15 Consider a balanced four-way crossed layout with four factors A, B, C, D having levels a, b, c, d, respectively, and with n observations per cell (factor-level combination). Assume that A, B are fixed and C, D are random. Apply the expected mean-squares rules to obtain the expressions for $\mathrm{E}(\mathrm{MS}_A)$, $\mathrm{E}(\mathrm{MS}_{AB})$, $\mathrm{E}(\mathrm{MS}_{AC})$, $\mathrm{E}(\mathrm{MS}_{ABC})$, and $\mathrm{E}(\mathrm{MS}_{CD})$.

Applied Exercise

11.16 Mason, Gunst, and Hess (2003) describe a gage R&R study for measurements of the lengths of mattress innerspings. Three operators (A) measured the same five springs (B) in a random order over three consecutive days (C). The springs were sampled at random from several days of production and days can also be regarded as a random sample of all days. The data are shown in Table 11.17. Assess the gage capability.

CHAPTER 12

Nested, Crossed–Nested, and Split-Plot Experiments

As explained in Chapter 1, two or more factors are crossed if the same levels of the factors are used in combination with each other. If different levels of one factor are used in combination with each level of another factor, then the former factor is said to be nested in the latter factor. Practical considerations on how the levels of the factors can be chosen determine whether two factors will be crossed or nested. For example, if the two factors are sites and operators then the investigator generally must choose different operators from different sites; thus operators will be nested in sites. So far in this book we have only studied designs with crossed factors. In this chapter we study designs with nested or a combination of nested and crossed factors.

As an example of a nested design, consider an experiment to study the variation in thickness of silicon wafers in semiconductor manufacturing (Jensen 2002). Integrated circuitry is built on wafers by a photolithography process in lots of a fixed number of wafers. A sample of wafers is taken from each lot and thickness is measured at several randomly selected locations on the sampled wafers. Here wafers are nested in lots and locations are nested in wafers. Figure 12.1 gives a schematic picture of this so-called three-stage nested design.

Another important class of designs studied in this chapter is split-plot designs. They are used for factorial experiments in which different-sized experimental units are given independent applications of the treatments corresponding to different factors. For example, in a two-factor experiment, factor A treatments are applied to larger experimental units, called whole plots, which are then "split" into smaller experimental units, called subplots, to which factor B treatments are applied. This terminology comes from agricultural experiments, where the whole plots refer to large plots of land to which treatments such as different amounts of watering are applied and subplots are smaller parcels of whole plots to which

Statistical Analysis of Designed Experiments: Theory and Applications By Ajit C. Tamhane

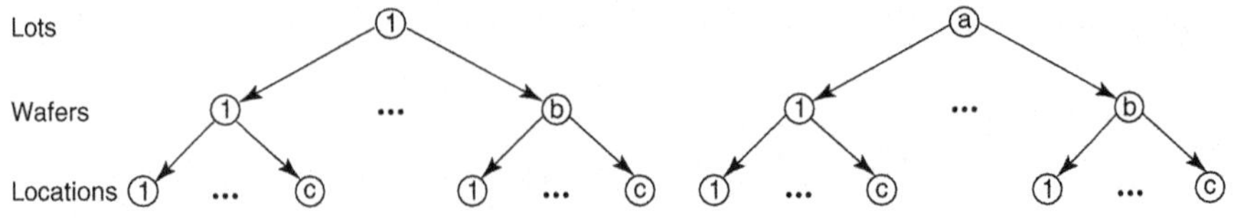

Figure 12.1 Three-stage nested design used in semiconductor manufacturing industry.

another set of treatments such as varieties of crop are applied. Because two separate randomizations are performed to assign treatments to the different-size plots, the error structure is two-tiered. Split-plot designs are also used when all factor-level combinations cannot be randomized simultaneously because some factors are harder to change than others, so once they are set at certain levels, it is convenient to keep those levels fixed and randomize the levels of the other factors. Naturally, the harder-to-change factor is assigned to whole plots and the easier-to-change factor to subplots.

The chapter is organized as follows. Section 12.1 introduces two-stage nested designs. Section 12.2 extends the model and the analysis methods to three-stage nested designs. Section 12.3 considers designs in which some factors are crossed and others are nested. Section 12.4 covers split-plot designs. Mathematical derivations of the E(MS) expressions used to obtain the F-ratios for ANOVA hypothesis tests and for variance component estimators are given in Section 12.5. A summary of the chapter is given in Section 12.6.

12.1 TWO-STAGE NESTED DESIGNS

Let us first consider two factors, A and B, with a and b levels, respectively, with B nested in A [denoted by $B(A)$]. This is a **two-stage nested design**. Suppose we make n independent replicate measurements, y_{ijk} $(1 \le k \le n)$, on each (i, j) combination of levels of A and B $(1 \le i \le a, 1 \le j \le b)$. Because the same number of levels of B are nested under each level of A and the same number of replicate observations are made at each (i, j) combination, this design is **balanced**. Throughout this chapter we will restrict to balanced designs.

12.1.1 Model

To write the model for this design, note that the effect of the jth level of B is different for each level i of A since the levels of B are different for different levels of A. Denote this effect by $\beta_{j(i)}$. Furthermore, there can be no interaction between A and B because they are not crossed. Thus the model is

$$y_{ijk} = \mu + \alpha_i + \beta_{j(i)} + e_{ijk} \qquad (1 \le i \le a, 1 \le j \le b, 1 \le k \le n). \qquad (12.1)$$

The random errors e_{ijk} are i.i.d. $N(0, \sigma_e^2)$ r.v.'s. The α_i and the $\beta_{j(i)}$ are fixed or random depending on whether A and B are fixed or random, respectively. For example, if both factors are fixed, then the α_i and the $\beta_{j(i)}$ are fixed unknown parameters subject to the constraints $\sum_{i=1}^{a} \alpha_i = 0$ and $\sum_{j=1}^{b} \beta_{j(i)} = 0$ for all i. If both factors are random, then the α_i are i.i.d. $N(0, \sigma_A^2)$ and the $\beta_{j(i)}$ are i.i.d. $N(0, \sigma_{B(A)}^2)$. Here $\sigma_A^2, \sigma_{B(A)}^2$, and σ_e^2 are the **variance components** of the total variance of y, which equals $\sigma_y^2 = \sigma_A^2 + \sigma_{B(A)}^2 + \sigma_e^2$. If A is fixed and B is random, then there is no variance component for A and $\sigma_y^2 = \sigma_{B(A)}^2 + \sigma_e^2$.

12.1.2 Analysis of Variance

Corresponding to the model (12.1), we have the following orthogonal decomposition:

$$y_{ijk} = \bar{y}_{...} + (\bar{y}_{i..} - \bar{y}_{...}) + (\bar{y}_{ij.} - \bar{y}_{i..}) + (y_{ijk} - \bar{y}_{ij.}).$$

From Lemma A.6 it follows that each of the terms on the right-hand side, when considered as an N-dimensional vector ($N = abn$), is orthogonal to every other term. Hence sums of their cross products (i.e., the dot products of the corresponding vectors) are zero, resulting in the following decomposition of the total sum of squares:

$$\underbrace{\sum_{i=1}^{a}\sum_{j=1}^{b}\sum_{k=1}^{n}(y_{ijk} - \bar{y}_{...})^2}_{\text{SS}_{\text{tot}}} = \underbrace{bn\sum_{i=1}^{a}(\bar{y}_{i..} - \bar{y}_{...})^2}_{\text{SS}_A} + \underbrace{n\sum_{i=1}^{a}\sum_{j=1}^{b}(\bar{y}_{ij.} - \bar{y}_{i..})^2}_{\text{SS}_{B(A)}}$$

$$+ \underbrace{\sum_{i=1}^{a}\sum_{j=1}^{b}\sum_{k=1}^{n}(y_{ijk} - \bar{y}_{ij.})^2}_{\text{SS}_e}, \qquad (12.2)$$

where we have labeled the three sums of squares on the right-hand side as that due to factor A (SS_A), to factor B nested in factor A ($\text{SS}_{B(A)}$), and to error (SS_e), respectively. The corresponding decomposition of the d.f. is

$$\underbrace{N - 1}_{\text{total d.f.}} = \underbrace{a - 1}_{A \text{ main-effects d.f.}} + \underbrace{a(b - 1)}_{B(A) \text{ main-effects d.f.}} + \underbrace{N - ab}_{\text{error d.f.}}.$$

As we saw in the previous chapter, in the case of two-way random- and mixed-effects models, the statistics for testing the significance of the effects A and $B(A)$ depend on whether A and B are fixed or random. In the following we will assume that both A and B are random. [The case where both A and B are fixed is covered in Exercise 12.1 and the case where A is fixed and B is random (mixed effects) is covered in Exercise 12.2.] The hypotheses to be tested are

$$H_{0A} : \sigma_A^2 = 0 \quad \text{and} \quad H_{0B(A)} : \sigma_{B(A)}^2 = 0.$$

Table 12.1 ANOVA Table for Two-Stage Nested Design with Both Factors Random

Source	SS	d.f.	MS	E(MS)	F
A main effect	SS_A	$a-1$	MS_A	$\sigma_e^2 + n\sigma_{B(A)}^2 + bn\sigma_A^2$	$\text{MS}_A/\text{MS}_{B(A)}$
$B(A)$ main effect	$\text{SS}_{B(A)}$	$a(b-1)$	$\text{MS}_{B(A)}$	$\sigma_e^2 + n\sigma_{B(A)}^2$	$\text{MS}_{B(A)}/\text{MS}_e$
Error	SS_e	$N-ab$	MS_e	σ_e^2	
Total	SS_{tot}	$N-1$			

The ANOVA in Table 12.1 gives the F-statistics for testing the above hypotheses. These F-statistics are obtained by matching the appropriate E(MS) expressions (also shown in the table) under H_{0A} and $H_{0B(A)}$. The derivations of the E(MS) expressions are given in Section 12.5.1.

The E(MS) expressions can be used to obtain the following unbiased estimators of the variance components:

$$\widehat{\sigma}_e^2 = \text{MS}_e, \widehat{\sigma}_{B(A)}^2 = \frac{\text{MS}_{B(A)} - \text{MS}_e}{n} \quad \text{and} \quad \widehat{\sigma}_A^2 = \frac{\text{MS}_A - \text{MS}_{B(A)}}{bn}.$$

The overall variance of the response variable is estimated as $\widehat{\sigma}_y^2 = \widehat{\sigma}_A^2 + \widehat{\sigma}_{B(A)}^2 + \widehat{\sigma}_e^2$. The percentage contributions to the total variance from the three sources can be estimated from this expression.

Example 12.1 (Analyzing Variation in a Measurement System)

Moen, Nolan, and Provost (1999) gave an example of an experiment conducted to analyze the variation in a measurement system. Six randomly selected parts were measured using three setups (different for each part). Three measurements were made for each part–setup combination. Here setup and part are both random factors with setup nested within part. The data are shown in Table 12.2. The ANOVA table computed using Minitab is shown in Display 12.1.

We see that both the part and setup(part) effects are statistically significant, but $\widehat{\sigma}_{\text{part}}^2 = 0.31674$ accounts for almost three-fourths of the total variance, $\widehat{\sigma}_y^2 = 0.31674 + 0.07684 + 0.03426 = 0.42784$. Therefore part-to-part variation is the major contributor to the total variance and should be minimized. ■

12.2 THREE-STAGE NESTED DESIGNS

Now consider three factors, A, B, and C, with a, b, and c levels, respectively. Factor B is nested in A [denoted by $B(A)$], and C is nested in both A and B [denoted by $C(AB)$]. Thus there are b different levels of B for each level of A and c different levels of C for each combination of levels of A and B. The semiconductor manufacturing example given at the beginning of this chapter illustrates a three-stage nested design with the lot being factor A, the wafer being factor B, and the location being factor C.

Table 12.2 Measurements (mm)

Part	Setup	Measurement
1	1	3.40, 3.45, 3.70
1	2	3.95, 3.80, 4.00
1	3	3.95, 4.00, 4.15
2	1	4.55, 4.70, 4.70
2	2	4.30, 4.20, 4.50
2	3	4.75, 5.15, 5.20
3	1	4.30, 4.70, 4.70
3	2	4.50, 4.55, 4.65
3	3	4.15, 4.45, 4.50
4	1	2.60, 2.65, 2.50
4	2	3.00, 3.15, 3.15
4	3	3.40, 3.55, 3.40
5	1	3.90, 4.85, 4.60
5	2	4.35, 4.90, 4.35
5	3	4.10, 4.00, 3.95
6	1	3.95, 4.10, 4.00
6	2	3.90, 3.90, 4.00
6	3	4.40, 4.30, 4.40

Source: Moen, Nolan, and Provost (1999, Figure 7.8). Reprinted by permission of The McGraw Hill Companies.

```
Source         DF   Seq SS   Adj SS   Adj MS      F       P
Part            5  15.5770  15.5770   3.1154  11.77   0.000
Setup(Part)    12   3.1772   3.1772   0.2648   7.73   0.000
Error          36   1.2333   1.2333   0.0343
Total          53  19.9876

S = 0.185093   R-Sq = 93.83%   R-Sq(adj) = 90.92%

Source          Variance Error Expected Mean Square for Each Term
               component  term (using unrestricted model)
1 Part            0.31674   2    (3) + 3(2) + 9(1)
2 Setup(Part)     0.07684   3    (3) + 3(2)
3 Error           0.03426        (3)
```

Display 12.1 Minitab output of ANOVA and variance components for measurement data.

12.2.1 Model

Consider a balanced design with n independent replicate measurements, $y_{ijk\ell}$ $(1 \le \ell \le n)$, on each (i, j, k) combination of levels of A, B, and C $(1 \le i \le a, 1 \le j \le b, 1 \le k \le c)$. The basic model is

$$y_{ijk\ell} = \mu + \alpha_i + \beta_{j(i)} + \gamma_{k(ij)} + e_{ijk\ell}$$
$$(1 \le i \le a, 1 \le j \le b, 1 \le k \le c, 1 \le \ell \le n). \quad (12.3)$$

We will assume that all three factors are random. (The case where all three factors are fixed is covered in Exercise 12.6 and the case where A is fixed and B and C are random or A and B are fixed and C is random is covered in Exercise 12.7.) Then the α_i are i.i.d. $N(0, \sigma_A^2)$, the $\beta_{j(i)}$are i.i.d. $N(0, \sigma_{B(A)}^2)$, and the $\gamma_{k(ij)}$ are i.i.d. $N(0, \sigma_{C(AB)}^2)$ r.v.'s. The total variance of the response variable equals

$$\sigma_y^2 = \sigma_A^2 + \sigma_{B(A)}^2 + \sigma_{C(AB)}^2 + \sigma_e^2.$$

12.2.2 Analysis of Variance

xtending the results for two-stage nested designs, we get the following orthogonal decomposition corresponding to the model (12.3):

$$y_{ijk\ell} = \bar{y}_{....} + (\bar{y}_{i...} - \bar{y}_{....}) + (\bar{y}_{ij..} - \bar{y}_{i...}) + (\bar{y}_{ijk.} - \bar{y}_{ij..}) + (y_{ijk\ell} - \bar{y}_{ijk.}).$$

This results in the following decomposition of the total sum of squares:

$$\underbrace{\sum_{i=1}^{a}\sum_{j=1}^{b}\sum_{k=1}^{c}\sum_{\ell=1}^{n}(y_{ijk\ell} - \bar{y}_{....})^2}_{\text{SS}_{\text{tot}}} = bcn \underbrace{\sum_{i=1}^{a}(\bar{y}_{i...} - \bar{y}_{....})^2}_{\text{SS}_A} + cn \underbrace{\sum_{i=1}^{a}\sum_{j=1}^{b}(\bar{y}_{ij..} - \bar{y}_{i...})^2}_{\text{SS}_{B(A)}}$$

$$+ \underbrace{n\sum_{i=1}^{a}\sum_{j=1}^{b}\sum_{k=1}^{c}(\bar{y}_{ijk.} - \bar{y}_{ij..})^2}_{\text{SS}_{C(AB)}}$$

$$+ \underbrace{\sum_{i=1}^{a}\sum_{j=1}^{b}\sum_{k=1}^{c}\sum_{\ell=1}^{n}(y_{ijk\ell} - \bar{y}_{ijk.})^2}_{\text{SS}_e}, \quad (12.4)$$

where we have labeled the four sums of squares on the right-hand side as that due to factor A (SS_A), factor B nested in factor A ($\text{SS}_{B(A)}$), factor C nested in factors A and B ($\text{SS}_{C(AB)}$), and error (SS_e). The corresponding decomposition of the d.f. is

$$\underbrace{N-1}_{\text{total d.f.}} = \underbrace{a-1}_{A\text{ main-effects d.f.}} + \underbrace{a(b-1)}_{B(A)\text{ main-effects d.f.}} + \underbrace{ab(c-1)}_{C(AB)\text{ main-effects d.f.}} + \underbrace{N-abc}_{\text{error d.f.}}.$$

The hypotheses to be tested are

$$H_{0A} : \sigma_A^2 = 0, \qquad H_{0B(A)} : \sigma_{B(A)}^2 = 0, \qquad H_{0C(AB)} : \sigma_{C(AB)}^2 = 0.$$

Table 12.3 ANOVA Table for Three-Stage Nested Design with All Factors Random

Source	SS	d.f.	MS	E(MS)	F
A main effect	SS_A	$a-1$	MS_A	$\sigma_e^2 + n\sigma_{C(AB)}^2 + cn\sigma_{B(A)}^2 + bcn\sigma_A^2$	$MS_A/MS_{B(A)}$
$B(A)$ main effect	$SS_{B(A)}$	$a(b-1)$	$MS_{B(A)}$	$\sigma_e^2 + n\sigma_{C(AB)}^2 + cn\sigma_{B(A)}^2$	$MS_{B(A)}/MS_{C(AB)}$
$C(AB)$ main effect	$SS_{C(AB)}$	$ab(c-1)$	$MS_{C(AB)}$	$\sigma_e^2 + n\sigma_{C(AB)}^2$	$MS_{C(AB)}/MS_e$
Error	SS_e	$N-abc$	MS_e	σ_e^2	
Total	SS_{tot}	$N-1$			

The ANOVA table, including the E(MS) expressions and the F-statistics for testing the above hypotheses, is given in Table 12.3. The derivations of the E(MS) expressions are analogous to those for the two-stage nested model and hence are omitted.

The E(MS) expressions can be used to obtain the following unbiased estimators of the variance components:

$$\widehat{\sigma}_e^2 = MS_e, \qquad \widehat{\sigma}_{C(AB)}^2 = \frac{MS_{C(AB)} - MS_e}{n},$$

$$\widehat{\sigma}_{B(A)} = \frac{MS_{B(A)} - MS_{C(AB)}}{cn}, \qquad \widehat{\sigma}_A^2 = \frac{MS_A - MS_{B(A)}}{bcn}.$$

The overall variance of the response variable is estimated as $\widehat{\sigma}_y^2 = \widehat{\sigma}_A^2 + \widehat{\sigma}_{B(A)}^2 + \widehat{\sigma}_{C(AB)}^2 + \widehat{\sigma}_e^2$. The percentage contributions to the total variance from each of the sources can be estimated from this expression.

Example 12.2 (Variability in Assay Values of Raw Materials)

Gonzalez-de la Parra and Rodriguez-Loaiza (2003) described a case study in which the components of variability in assay values of a raw material used in drug manufacture were investigated. Raw material samples were obtained as follows. First, two suppliers were randomly selected from a list of approved suppliers. Then three lots of raw material were selected from each supplier and four containers were selected from each lot. Three determinations of assay values were made on each container by using an analytical method based on high-performance liquid chromatography. The measurements were expressed as percent values with respect to a primary reference standard. Here factor A is supplier with $a = 2$ levels, factor B is lot with $b = 3$ levels, and factor C is container with $c = 4$ levels. The number of replicates is $n = 3$. The data are shown in Table 12.4. The Minitab ANOVA and variance component calculations are shown in Display 12.2.

Table 12.4 Assay Data

Supplier	Lot	Container	Assay Values (%) −100		
1	1	1	0.3031,	0.3839,	−0.1716
1	1	2	−0.0908,	−0.0908,	0.0506
1	1	3	0.0405,	−0.2120,	0.1617
1	1	4	−0.2423,	−0.2524,	0.0203
1	2	1	0.4445,	0.0304,	0.3637
1	2	2	0.3637,	0.2324,	0.2829
1	2	3	0.4344,	0.5960,	0.4445
1	2	4	0.2223,	0.0001,	0.2223
1	3	1	0.1718,	0.3738,	0.3031
1	3	2	0.2021,	0.4546,	0.3738
1	3	3	0.2930,	0.4647,	0.1819
1	3	4	0.3233,	0.1516,	0.4748
2	1	1	1.7979,	1.4141,	1.3636
2	1	2	1.8080,	1.0808,	1.3333
2	1	3	0.9697,	0.2021,	0.7677
2	1	4	0.8283,	0.9091,	0.3738
2	2	1	−0.0807,	0.8990,	−0.6261
2	2	2	0.8788,	0.8485,	0.9091
2	2	3	0.7475,	−0.3029,	−0.0100
2	2	4	−0.0706,	−0.5453,	−0.3635
2	3	1	0.2223,	−0.7574,	−0.4645
2	3	2	−0.0807,	−0.6564,	−0.6362
2	3	3	2.0504,	2.4342,	1.1010
2	3	4	0.0102,	0.3132,	1.1515

Source: Gonzalez-de la Parra and Rodriguez-Loaiza (2003, Table 1). Reprinted by permission of Taylor & Francis.

```
Source                    DF       SS         MS       F      P
Supplier                   1   2.1750     2.1750    1.47  0.292
Lot(Supplier)              4   5.9148     1.4787    1.80  0.173
Containe(Supplier Lot)    18  14.7907     0.8217    6.87  0.000
Error                     48   5.7432     0.1196
Total                     71  28.6237


Source                    Variance Error Expected Mean Square for
Each Term
                         component term (using unre-
stricted model)
 1 Supplier               0.01934 2 (4) + 3(3) + 12(2) + 36(1)
 2 Lot(Supplier)          0.05475 3 (4) + 3(3) + 12(2)
 3 Containe(Supplier Lot) 0.23402 4 (4) + 3(3)
 4 Error                  0.11965    (4)
```

Display 12.2 Minitab output of ANOVA and variance components for assay data.

As a numerical check, we calculate the variance component estimates from the mean squares given in the ANOVA table:

$$\widehat{\sigma}_e^2 = \text{MS}_e = 0.1196,$$

$$\widehat{\sigma}_{C(AB)}^2 = \frac{0.8217 - 0.1196}{3} = 0.2340,$$

$$\widehat{\sigma}_{B(A)}^2 = \frac{1.4787 - 0.8217}{12} = 0.0548,$$

$$\widehat{\sigma}_A^2 = \frac{2.1750 - 1.4787}{36} = 0.0193.$$

The total variance is estimated as $\widehat{\sigma}_y^2 = 0.0193 + 0.0548 + 0.2340 + 0.1196 = 0.4277$. We see that the suppliers and lots do not have significant effects on the variability; only containers do. In fact, containers account for $0.2340/0.4277 = 54.7\%$ of the total variability. Thus efforts should be directed toward minimizing the variability between containers. ■

12.3 CROSSED AND NESTED DESIGNS

Many designs used in practice involve a combination of crossed and nested factors. For instance, in the semiconductor manufacturing example described at the beginning of this chapter, if the locations on the wafers are fixed, then wafers and locations would be crossed but the wafers would be nested within lots.

12.3.1 Model

Consider two crossed fixed factors, A and B, and a random factor C which is nested in A. Let the numbers of levels of A, B, and C be a, b, and c, respectively, and suppose that $n \geq 1$ replicate measurements are made on each treatment combination. The model for this design is

$$y_{ijk\ell} = \mu + \alpha_i + \beta_j + \gamma_{k(i)} + (\alpha\beta)_{ij} + (\beta\gamma)_{jk(i)} + e_{ijk\ell} \qquad (1 \leq i \leq a, 1 \leq j \leq b, 1 \leq k \leq c, 1 \leq \ell \leq n). \qquad (12.5)$$

Notice that the model allows interactions between the crossed factors but not between the nested factors; thus there are interactions $(\alpha\beta)_{ij}$ and $(\beta\gamma)_{jk(i)}$, but there are no interactions $(\alpha\gamma)_{ik(i)}$ and $(\alpha\beta\gamma)_{ijk(i)}$. Here the α_i, β_j and the $(\alpha\beta)_{ij}$ are fixed parameters subject to

$$\sum_{i=1}^a \alpha_i = \sum_{j=1}^b \beta_j = 0, \qquad \sum_{i=1}^a (\alpha\beta)_{ij} = 0 \quad \text{for all } j,$$

$$\sum_{j=1}^b (\alpha\beta)_{ij} = 0 \quad \text{for all } i.$$

We will assume the unrestricted model (see Section 11.2.2.1). Thus the $\gamma_{k(i)}$ are i.i.d. $N(0, \sigma^2_{C(A)})$ and the $(\beta\gamma)_{jk(i)}$ are i.i.d. $N(0, \sigma^2_{BC(A)})$.

12.3.2 Analysis of Variance

The following is the orthogonal decomposition corresponding to the above linear model:

$$\begin{aligned} y_{ijk\ell} &= \overline{y}_{....} + (\overline{y}_{i...} - \overline{y}_{....}) + (\overline{y}_{.j..} - \overline{y}_{....}) + (\overline{y}_{i\cdot k\cdot} - \overline{y}_{i...}) \\ &\quad + (\overline{y}_{ij..} - \overline{y}_{i...} - \overline{y}_{.j..} + \overline{y}_{....}) + (\overline{y}_{ijk\cdot} - \overline{y}_{ij..} - \overline{y}_{i\cdot k\cdot} + \overline{y}_{i...}) \\ &\quad + (y_{ijk\ell} - \overline{y}_{ijk\cdot}). \end{aligned}$$

Note that the term

$$(\overline{y}_{i\cdot k\cdot} - \overline{y}_{i...}) = (\overline{y}_{..k\cdot} - \overline{y}_{....}) + (\overline{y}_{i\cdot k\cdot} - \overline{y}_{i...} - \overline{y}_{..k\cdot} + \overline{y}_{....})$$

corresponding to $\gamma_{k(i)}$ estimates the effect of level k of factor C within level i of factor A and equals the sum of what would be the main effect of level k of factor C and the interaction between levels i and k of factors A and C, respectively, if A and C were crossed. Similarly, the term

$$\begin{aligned} (\overline{y}_{ijk\cdot} - \overline{y}_{ij..} - \overline{y}_{i\cdot k\cdot} + \overline{y}_{i...}) &= (\overline{y}_{i\cdot k\cdot} - \overline{y}_{i...} - \overline{y}_{..k\cdot} + \overline{y}_{....}) + (\overline{y}_{ijk\cdot} - \overline{y}_{ij..} - \overline{y}_{i\cdot k\cdot} \\ &\quad - \overline{y}_{\cdot jk\cdot} + \overline{y}_{i...} + \overline{y}_{.j..} + \overline{y}_{..k\cdot} - \overline{y}_{....}) \end{aligned} \tag{12.6}$$

corresponding to $(\beta\gamma)_{jk(i)}$ estimates the interaction between the levels j and k of factors B and C within level i of factor A and equals the sum of what would be the two-way interaction between those levels of factors B and C and the three-way interaction between the levels i, j, k of factors A, B, C, respectively, if these three factors were crossed.

This results in the following decomposition of the total sum of squares:

$$\begin{aligned} \underbrace{\sum_{i=1}^{a}\sum_{j=1}^{b}\sum_{k=1}^{c}\sum_{\ell=1}^{n}(y_{ijk\ell} - \overline{y}_{....})^2}_{SS_{tot}} &= \underbrace{bcn\sum_{i=1}^{a}(\overline{y}_{i...} - \overline{y}_{....})^2}_{SS_A} + \underbrace{acn\sum_{j=1}^{b}(\overline{y}_{.j..} - \overline{y}_{....})^2}_{SS_B} \\ &\quad + \underbrace{bn\sum_{i=1}^{a}\sum_{k=1}^{c}(\overline{y}_{i\cdot k\cdot} - \overline{y}_{i...})^2}_{SS_{C(A)}} \\ &\quad + \underbrace{cn\sum_{i=1}^{a}\sum_{j=1}^{b}(\overline{y}_{ij..} - \overline{y}_{i...} - \overline{y}_{.j..} + \overline{y}_{....})^2}_{SS_{AB}} \end{aligned}$$

$$+ \underbrace{n \sum_{i=1}^{a} \sum_{j=1}^{b} \sum_{k=1}^{c} (\overline{y}_{ijk\cdot} - \overline{y}_{ij\cdot\cdot} - \overline{y}_{i\cdot k\cdot} + \overline{y}_{i\cdots})^2}_{SS_{BC(A)}}$$

$$+ \underbrace{\sum_{i=1}^{a} \sum_{j=1}^{b} \sum_{k=1}^{c} \sum_{\ell=1}^{n} (y_{ijk\ell} - \overline{y}_{ijk\cdot})^2}_{SS_e}.$$

The sums of squares have the usual meanings. The corresponding decomposition of the total d.f. is

$$\underbrace{N-1}_{\text{total d.f.}} = \underbrace{a-1}_{A \text{ main-effects d.f.}} + \underbrace{b-1}_{B \text{ main-effects d.f.}} + \underbrace{a(c-1)}_{C(A) \text{ main-effects d.f.}}$$

$$+ \underbrace{(a-1)(b-1)}_{AB \text{ interaction d.f.}} + \underbrace{a(b-1)(c-1)}_{BC(A) \text{ interaction d.f.}} + \underbrace{N-abc}_{\text{error d.f.}}.$$

The hypotheses to be tested are

$$\begin{aligned} H_{0A} : \alpha_i &= 0 \quad \text{for all } i, \\ H_{0B} : \beta_j &= 0 \quad \text{for all } j, \\ H_{0AB} : (\alpha\beta)_{ij} &= 0 \quad \text{for all } i, j, \\ H_{0C(A)} : \sigma^2_{C(A)} &= 0, \\ H_{0BC(A)} : \sigma^2_{BC(A)} &= 0. \end{aligned}$$

The ANOVA, including the E(MS) expressions and the F-statistics for testing these hypotheses, is given in Table 12.5. The following finite-population analogs of variance components are used in the E(MS) expressions for the fixed effects:

$$Q_A = \frac{\sum_{i=1}^{a} \alpha_i^2}{a-1}, \qquad Q_B = \frac{\sum_{j=1}^{b} \beta_j^2}{b-1}, \qquad Q_{AB} = \frac{\sum_{i=1}^{a} \sum_{j=1}^{b} (\alpha\beta)_{ij}^2}{(a-1)(b-1)}. \tag{12.7}$$

The derivations of the E(MS) expressions are given in Section 12.5.2.

Example 12.3 (Mouse Two-Cell Embryo Assay: Analysis of Variance)

Gorrill et al. (1991) conducted a study to compare three assays (hamster sperm motility assay, mouse one-cell embryo assay, and mouse two-cell embryo assay)

Table 12.5 ANOVA Table for Three-Way Layout with Two Crossed Fixed Factors and a Nested Random Factor

Source	SS	d.f.	MS	E(MS)	F
A main effect	SS_A	$a-1$	MS_A	$\sigma_e^2 + n\sigma_{BC(A)}^2 + bn\sigma_{C(A)}^2 + bcnQ_A$	$\text{MS}_A/\text{MS}_{C(A)}$
B main effect	SS_B	$b-1$	MS_B	$\sigma_e^2 + n\sigma_{BC(A)}^2 + acnQ_B$	$\text{MS}_B/\text{MS}_{BC(A)}$
$C(A)$ main effect	$\text{SS}_{C(A)}$	$a(c-1)$	$\text{MS}_{C(A)}$	$\sigma_e^2 + n\sigma_{BC(A)}^2 + bn\sigma_{C(A)}^2$	$\text{MS}_{C(A)}/\text{MS}_{BC(A)}$
AB interaction	SS_{AB}	$(a-1)(b-1)$	MS_{AB}	$\sigma_e^2 + n\sigma_{BC(A)}^2 + cnQ_{AB}$	$\text{MS}_{AB}/\text{MS}_{BC(A)}$
$BC(A)$ interaction	$\text{SS}_{BC(A)}$	$a(b-1)(c-1)$	$\text{MS}_{BC(A)}$	$\sigma_e^2 + n\sigma_{BC(A)}^2$	$\text{MS}_{BC(A)}/\text{MS}_e$
Error	SS_e	$N-abc$	MS_e	σ_e^2	
Total	SS_{tot}	$N-1$			

with regard to their ability to discriminate between four culture media, in particular, two types of solutions and two types of water (tap water or ultrapure water). The goal was to choose an assay for quality control use in an in vitro fertilization (IVF) laboratory. In this example we focus on the bioassay using mouse two-cell embryos. Four culture media were prepared by combining each one of two solutions (modified Ham's F-10 solution, labeled as solution 1, or modified Tyrode's solution, labeled as solution 2) with each one of two types of water. The percentage of total embryos reaching a certain development stage (expanded blastocyst or hatched blastocyst), which was the response variable, was measured by three analysts. A total of 16 mice were used in the experiment—4 per culture. The data are given in Table 12.6.

The medium (i.e., solution plus type of water) is obviously a fixed factor. Initially we ignore the two-factor structure of the medium factor and regard it as a nominal factor with four levels. For convenience, we assume that analyst is also a fixed factor, which is justified if these are the only analysts available for the experiment. (Assuming that analyst is a random factor results in a model in which exact F-tests do not exist for certain terms. Approximate F-tests can be derived along the lines of Section 11.3.3 but are not pursued here.) Mouse is clearly a random factor and is nested under medium. We assume the model (12.5) with the medium as factor A, analyst as factor B, and mouse as factor C. Here $n = 1$, so σ_e^2 cannot be estimated. An inspection of the E(MS) column in the ANOVA in Table 12.5 shows that, as a consequence, the $BC(A)$ interaction is not testable, but all other effects are testable.

Table 12.6 Percentage of Mouse Two-Cell Embryos Reaching Development Stage as Measured by Three Analysts Using Four Culture Media

Culture Medium	Mouse	Analyst 1	Analyst 2	Analyst 3
Solution 1 + Tap water	1	82.6	71.4	54.5
	2	70.9	68.6	54.5
	3	62.5	68.0	60.0
	4	22.7	22.7	23.8
Solution 1 + MilliQ water	1	50.0	73.3	62.5
	2	64.5	73.3	50.0
	3	69.2	61.5	46.4
	4	26.9	50.0	22.2
Solution 2 + Tap water	1	73.6	88.2	72.2
	2	77.4	77.4	82.1
	3	77.7	75.0	74.1
	4	82.6	85.7	85.0
Solution 2 + MilliQ water	1	77.7	81.5	67.8
	2	73.0	81.2	78.3
	3	80.6	87.5	79.4
	4	78.9	82.6	61.9

Source: Gorrill et al. (1991, Table 3).

```
Source                  DF    Seq SS    Adj SS   Adj MS      F      P
Medium                   3   6778.60   6778.60  2259.53   4.35  0.027
Analyst                  2    941.21    941.21   470.60  10.20  0.001
Medium*Analyst           6    404.65    404.65    67.44   1.46  0.233
Mouse(Medium)           12   6230.04   6230.04   519.17  11.25  0.000
Analyst*Mouse(Medium)   24   1107.68   1107.68    46.15     **
Error                    0         *         *        *
Total                   47  15462.18

** Denominator of F-test is zero.

S = *

Expected Mean Squares, using Adjusted SS

   Source                   Expected Mean Square for Each Term
1  Medium                   (6) + (5) + 3.0000 (4) + Q[1, 3]
2  Analyst                  (6) + (5) + Q[2, 3]
3  Medium*Analyst           (6) + (5) + Q[3]
4  Mouse(Medium)            (6) + (5) + 3.0000 (4)
5  Analyst*Mouse(Medium)    (6) + (5)
6  Error                    (6)

Error Terms for Tests, using Adjusted SS
                                                    Synthesis
                                                    of Error
   Source                   Error DF   Error MS     MS
1  Medium                      12.00     519.17     (4)
2  Analyst                     24.00      46.15     (5)
3  Medium*Analyst              24.00      46.15     (5)
4  Mouse(Medium)               24.00      46.15     (5)
5  Analyst*Mouse(Medium)           *          *     (6)

Variance Components, using Adjusted SS

                         Estimated
Source                       Value
Mouse(Medium)              157.672
Analyst*Mouse(Medium)       46.153
Error                        0.000
```

Display 12.3 Minitab output of ANOVA and variance components for mouse two-cell embryo bioassay data.

The Minitab output is shown in Display 12.3. We see that all three main effects, medium, analyst, and mouse, are significant; only the medium–analyst interaction is not significant. Both mouse and analyst–mouse variance components are positively biased because they include σ_e^2, which is assumed to be zero.

A more detailed ANOVA can be computed by considering the factorial structure of the medium factor. Thus the medium main effect and the medium–analyst interaction can each be partitioned into three independent

Table 12.7 Detailed ANOVA Table for Mouse Two-Cell Embryo Data

Source	SS	d.f.	MS	F	p
Medium	6778.60	3	2259.53	4.35	0.027
Solution	6754.5	1	6754.5	13.01	0.004
Water	22.7	1	22.7	0.043	0.849
Solution*Water	1.4	1	1.4	0.003	0.957
Analyst	941.21	2	470.60	10.20	0.001
Medium–analyst	404.65	6	67.44	1.46	0.233
Solution–analyst	129.2	2	64.6	1.40	0.266
Water–analyst	191.3	2	95.6	2.07	0.148
Solution–water–analyst	84.2	2	42.1	0.91	0.416
Mouse (medium)	6230.04	12	519.17	11.25	0.000
Analyst–mouse (medium)	1107.68	24	46.15		
Error	0	0			
Total	15462.18	47			

components, the main-effect components having one d.f. each and the interaction components having two d.f. each. The corresponding sums of squares can be computed by treating the design as a three-way layout in three fixed factors: solution, water, and analyst, regarding mice as replicates. However, note that the error terms for testing these effects come from the ANOVA table in Display 12.3, which takes into account the fact that mice are nested under the medium factor. The resulting detailed ANOVA is shown in Table 12.7.

The conclusion is that this assay is able to discriminate between the four media; the differences are primarily due to the solutions. The water quality makes no difference either through its main effect or through any of its interactions. These conclusions are in agreement with the main and interaction effect calculations in Example 7.1. In that example we saw that the solution effect is large positive, meaning that solution 2 (modified Tyrode's solution) produces significantly higher percentage of embryos that reach the development stage. Since this is the desired outcome, use of this solution is recommended. The analysts differ significantly from each other and there is also significant heterogeneity among mice. ■

12.4 SPLIT-PLOT DESIGNS

As mentioned in the introduction to this chapter, a split-plot design uses two different-sized experimental units—whole plots and subplots. The treatment factor applied to whole plots is called the **whole-plot factor** and that applied to subplots is called the **subplot factor**. The error structure is two-tiered: The **whole-plot error** associated with the first randomization is generally larger than the **subplot error** associated with the second randomization. We shall illustrate the key concepts of split-plot designs using the following example from Box (1996).

Example 12.4 (Corrosion Resistance of Coatings: Experimental Setup)
Four different coatings, C_1, C_2, C_3, and C_4, applied to steel bars were tested for corrosion resistance at three different furnace temperatures, 360, 370, and 380°C. In each heat, the furnace was set to the desired temperature and four steel bars treated with different coatings were randomly assigned to the positions in the furnace and baked for a prescribed time. They were then removed, cooled, and finally tested. This procedure was replicated twice. The order of application of temperatures (heats) was randomized in the two replications. Clearly, this is not a completely randomized factorial experiment since the $4 \times 3 = 12$ runs in each replicate were not made in random order. Once a particular furnace temperature was set, all four steel bars with different coatings were baked together. This is the practical way of doing the experiment as setting the furnace temperature 12 different times and baking one bar at a time is not feasible. Table 12.8 shows how the experiment was run along with the data. Temperature is the whole-plot factor; different temperatures are independently applied to whole plots of four steel bars. Coating is the subplot factor; different coatings are applied to subplots of single bars. Note that the whole plots are split into smaller subplots for application of subplot factor treatments (coatings). ■

Because the subplot error is generally smaller than the whole-plot error, the factor of primary interest should be allocated to the subplots, if possible, and the factor of secondary interest should be allocated to the whole plots. However, if the factor of primary interest is harder to change, then practical considerations dictate that it be allocated to whole plots.

The following example uses the experimental setup and data in Table 12.8 to elucidate the concepts and calculations of whole-plot and subplot errors.

Table 12.8 Corrosion Resistance Data from Split-Plot Experiment

Heats (Whole Plots) (°C)	Replicate 1 Positions (Subplots) 1	2	3	4	Heats (Whole Plots) (°C)	Replicate 2 Positions (Subplots) 1	2	3	4
360	C_2	C_3	C_1	C_4	380	C_4	C_3	C_2	C_1
	73	83	67	89		153	90	100	108
370	C_1	C_3	C_4	C_2	370	C_4	C_1	C_3	C_2
	65	87	86	91		150	140	121	142
380	C_3	C_1	C_2	C_4	360	C_1	C_4	C_2	C_3
	147	155	127	212		33	54	8	46

Source: Box (1996). Reprinted by permission of Taylor & Francis.

Table 12.9 Mean Corrosion Resistances for Different Temperatures (Whole Plots)

Temperature (°C)	Replicate 1	Replicate 2	Row Average
360	78.00	35.25	56.63
370	82.25	138.25	110.25
380	160.25	112.75	136.50

Example 12.5 (Corrosion Resistance of Coatings: Whole-Plot and Subplot Errors)

To simplify the exposition, suppose that there is no blocking effect of replicates. Then, we have a completely randomized design in terms of the temperature factor with two replicates per temperature. The data, which are the averages of the corrosion resistances of four coatings, are given in Table 12.9.

Since each data value in this table is an average of four observations, SS_e can be calculated as

$$\begin{aligned} SS_e &= 4[(78.00 - 56.63)^2 + (35.25 - 56.63)^2 + (82.25 - 110.25)^2 \\ &\quad + (138.25 - 110.25)^2 + (160.25 - 136.50)^2 + (112.75 - 136.50)^2] \\ &= 14440 \end{aligned}$$

with $3(2 - 1) = 3$ d.f.; hence $MS_e = 14440/3 = 4813.3$. In what follows, these quantities will be used to estimate the whole-plot error and will be denoted by SS_{e1} and MS_{e1}, respectively. The differences $78.00 - 56.63$, $35.25 - 56.63$, and so on, used to compute SS_{e1} are called the **whole-plot residuals**.

Turning attention to the subplot treatments (coatings), we see that for a given temperature there are two replicate observations for each coating. For example, for 360°C, the two observations for coating C_1 are 67 and 33. The deviations from their average are $67 - 50$ and $33 - 50$. To remove the effect of the whole-plot errors on these deviations, we subtract from them the respective whole-plot residuals. The resulting quantities, $(67.00 - 50.00) - (78.00 - 56.63)$ and $(33.00 - 50.00) - (35.25 - 56.63)$, are called **subplot residuals**. Squaring and adding these residuals give the sum of squares for the subplot error (denoted by SS_{e2}), which works out to be

$$\begin{aligned} SS_{e2} &= [(67.00 - 50.00) - (78.00 - 56.63)]^2 \\ &\quad + [(33.00 - 50.00) - (35.25 - 56.63)]^2 + \cdots \\ &\quad + [(212 - 182.5) - (160.25 - 136.50)]^2 \\ &\quad + [(153 - 182.5) - (112.75 - 136.50)]^2 \\ &= 1121. \end{aligned}$$

The subplot error d.f. equal 9. (There are 12 d.f. associated with 12 independent differences, $67 - 50, 33 - 50$, etc., and 3 d.f. are deducted corresponding to the whole-plot residuals that are subtracted.) Hence $\text{MS}_{e2} = 1121/9 = 124.5$.

Denote the whole-plot and subplot error variances by σ_{e1}^2 and σ_{e2}^2, respectively. It will be shown in the sequel that

$$\text{E}(\text{MS}_{e_1}) = \sigma_{e2}^2 + b\sigma_{e1}^2 \qquad \text{and} \qquad \text{E}(\text{MS}_{e2}) = \sigma_{e2}^2.$$

Hence the unbiased estimates of these error variances are

$$\widehat{\sigma}_{e2}^2 = \text{MS}_{e2} = 124.5 \qquad \text{and} \qquad \widehat{\sigma}_{e1}^2 = \frac{\text{MS}_{e1} - \text{MS}_{e2}}{b} = \frac{4813.3 - 124.5}{4} = 1172.2.$$

The corresponding estimated standard deviations are $\widehat{\sigma}_{e2} = \sqrt{124.5} = 11.158$ and $\widehat{\sigma}_{e1} = \sqrt{1172.2} = 34.237$. Note that the whole-plot standard deviation is more than three times larger than the subplot standard deviation. ■

Next we will present a general model for the above example that will also include the possible blocking effect of the replicates.

12.4.1 Model

Suppose that the whole-plot factor A has a levels and the subplot factor B has b levels. Further suppose that there are c replicates (blocks), which we will call factor C. The total number of observations is thus $N = abc$. We assume that the two treatment factors are fixed while the replicates are random. We use the following model to analyze this design:

$$\begin{aligned} y_{ijk} &= \mu + \alpha_i + \gamma_k + e_{ik}^{(1)} \\ &\quad + \beta_j + (\alpha\beta)_{ij} + e_{ijk}^{(2)} \qquad (1 \le i \le a, 1 \le j \le b, 1 \le k \le c). \end{aligned} \tag{12.8}$$

The first line of the model pertains to the whole-plot treatments with the $e_{ik}^{(1)}$ being i.i.d. $\text{N}(0, \sigma_{e1}^2)$ whole-plot errors. The second line pertains to the subplot treatments and their interactions with the whole-plot treatments with the $e_{ijk}^{(2)}$ being i.i.d. $\text{N}(0, \sigma_{e2}^2)$ subplot errors. The α_i, β_j, and $(\alpha\beta)_{ij}$ are fixed parameters satisfying $\sum_{i=1}^{a} \alpha_i = 0$, $\sum_{j=1}^{b} \beta_j = 0$, and $\sum_{i=1}^{a} (\alpha\beta)_{ij} = 0$ for all j and $\sum_{j=1}^{b} (\alpha\beta)_{ij} = 0$ for all i. The random parameters γ_k are i.i.d. $\text{N}(0, \sigma_C^2)$.

By comparing the split-plot design model (12.8) with the three-way layout model (6.13), we see that formally $e_{ik}^{(1)} = (\alpha\gamma)_{ik}$ and $e_{ijk}^{(2)} = (\beta\gamma)_{jk} + (\alpha\beta\gamma)_{ijk}$. Thus the whole-plot error can be estimated from SS_{AC} and the subplot error can be estimated from $\text{SS}_{BC} + \text{SS}_{ABC}$ with the corresponding (pooled) d.f. It may be noted that in terms of the whole-plot treatments the design is an RB design. As seen in Chapter 4, the treatment–block interaction $MS_{AC} = MS_{e1}$ will therefore be the estimate of the whole-plot error used to test the whole-plot treatment and block effects.

12.4.2 Analysis of Variance

The ANOVA decomposition corresponding to the model (12.8) is

$$\underbrace{\sum_{i,j,k}(y_{ijk}-\bar{y}_{\cdots})^2}_{SS_{tot}} = \underbrace{bc\sum_{i}(\bar{y}_{i\cdot\cdot}-\bar{y}_{\cdots})^2}_{SS_A} + \underbrace{ab\sum_{k}(\bar{y}_{\cdot\cdot k}-\bar{y}_{\cdots})^2}_{SS_C}$$

$$+ \underbrace{b\sum_{i,k}(\bar{y}_{i\cdot k}-\bar{y}_{i\cdot\cdot}-\bar{y}_{\cdot\cdot k}+\bar{y}_{\cdots})^2}_{SS_{e_1}}$$

$$+ \underbrace{ac\sum_{j}(\bar{y}_{\cdot j\cdot}-\bar{y}_{\cdots})^2}_{SS_B} + \underbrace{c\sum_{i,j}(\bar{y}_{ij\cdot}-\bar{y}_{i\cdot\cdot}-\bar{y}_{\cdot j\cdot}+\bar{y}_{\cdots})^2}_{SS_{AB}}$$

$$+ \underbrace{\sum_{i,j,k}(y_{ijk}-\bar{y}_{ij\cdot}-\bar{y}_{i\cdot k}+\bar{y}_{i\cdot\cdot})^2}_{SS_{e_2}}.$$

The corresponding decomposition of the degrees of freedom is

$$\underbrace{abc-1}_{\text{total d.f}} = \underbrace{a-1}_{\text{factor } A \text{ d.f.}} + \underbrace{c-1}_{\text{blocks d.f.}} + \underbrace{(a-1)(c-1)}_{\text{whole-plot error d.f.}}$$

$$+ \underbrace{b-1}_{\text{factor } B \text{ d.f.}} + \underbrace{(a-1)(b-1)}_{AB \text{ interaction d.f.}} + \underbrace{a(b-1)(c-1)}_{\text{subplot error d.f.}}.$$

The whole-plot residuals are $\bar{y}_{i\cdot k}-\bar{y}_{i\cdot\cdot}-\bar{y}_{\cdot\cdot k}+\bar{y}_{\cdots}$ and the subplot residuals are $y_{ijk}-\bar{y}_{ij\cdot}-\bar{y}_{i\cdot k}+\bar{y}_{i\cdot\cdot}$.

In this decomposition, SS_A, SS_B, SS_C, and SS_{AB} are the usual main-effect and interaction sums of squares and SS_{e1} and SS_{e2} are the whole-plot and subplot error sums of squares, respectively. We now explain the SS_{e1} and SS_{e2} terms. As noted before, if the design were analyzed as a singly replicated three-way crossed design in A, B, and C, then $SS_{e1} = SS_{AC}$ and its d.f. are $(a-1)(c-1)$. By comparison, when the blocking effect of factor C is ignored, as in Example 12.5, then SS_C is pooled with SS_{AC} to give the whole-plot error sums of squares as $SS_C + SS_{AC} = b\sum_{i,k}(\bar{y}_{i\cdot k}-\bar{y}_{i\cdot\cdot})^2 = SS_{C(A)}$, which is the sum of squares for C nested in A. Next, the subplot error sum of squares SS_{e2} equals $SS_{BC} + SS_{ABC} = SS_{BC(A)}$ whether the blocking effect of C is ignored or not; the d.f. for SS_{e2}

equal $a(b-1)(c-1)$. Thus, if the whole-plot treatment structure is completely randomized, then the sums of squares for the ANOVA can be computed by regarding C as nested in A.

The hypotheses tested are

$$H_{0A}: Q_A = 0, \qquad H_{0B}: Q_B = 0, \qquad H_{0AB}: Q_{AB} = 0, \qquad H_{0C}: \sigma_C^2 = 0, \tag{12.9}$$

where

$$Q_A = \frac{\sum_{i=1}^{a} \alpha_i^2}{a-1}, \qquad Q_B = \frac{\sum_{j=1}^{b} \beta_j^2}{b-1}, \qquad Q_{AB} = \frac{\sum_{i=1}^{a}\sum_{j=1}^{b} (\alpha\beta)_{ij}^2}{(a-1)(b-1)}. \tag{12.10}$$

The ANOVA is shown in Table 12.10. The E(MS) expressions given in the table are derived in Section 12.5.3.

Example 12.6 (Corrosion Resistance of Coatings: Analysis of Variance)

As a first step in analyzing the data, we compute mean responses shown in Table 12.11. From this table we see that coating C_4 is uniformly best (has the highest corrosion resistance) at all three temperatures. The corrosion resistance increases with temperature for every coating, but the disparity between C_4 and other coatings is the largest at the highest temperature (380°C). Figure 12.2 displays these effects graphically. There appear to be temperature and coating main effects as well as interactions. To determine whether they are statistically significant or not, we now carry out an ANOVA.

The ANOVA is shown in Table 12.12. We see that the furnace temperature effects and the replicate (block) effects are both nonsignificant. This is partly due

Table 12.10 ANOVA Table for Split-Plot Design

Source	SS	d.f.	MS	E(MS)	F
A (whole plot) main effect	SS_A	$a-1$	MS_A	$\sigma_{e2}^2 + b\sigma_{e1}^2 + bcQ_A$	MS_A/MS_{e1}
C (blocks) main effect	SS_C	$c-1$	MS_C	$\sigma_{e2}^2 + b\sigma_{e1}^2 + ab\sigma_C^2$	MS_C/MS_{e1}
Whole-plot error	SS_{e1}	$(a-1)(c-1)$	MS_{e1}	$\sigma_{e2}^2 + b\sigma_{e1}^2$	
B (subplot) main effect	SS_B	$b-1$	MS_B	$\sigma_{e2}^2 + acQ_B$	MS_B/MS_{e2}
AB interaction	SS_{AB}	$(a-1)(b-1)$	MS_{AB}	$\sigma_{e2}^2 + cQ_{AB}$	MS_{AB}/MS_{e2}
Subplot error	SS_{e2}	$a(b-1)(c-1)$	MS_{e2}	σ_{e2}^2	
Total	SS_{tot}	$abc-1$			

Table 12.11 Means for Coating Corrosion Resistance Data

	Coating				
Temperature (°C)	C_1	C_2	C_3	C_4	Row Mean
360	50.0	40.5	64.5	71.5	56.63
370	102.5	116.5	104.0	118.0	110.25
380	131.5	113.5	118.5	182.5	136.50
Column mean	94.67	90.17	95.67	124.0	101.13

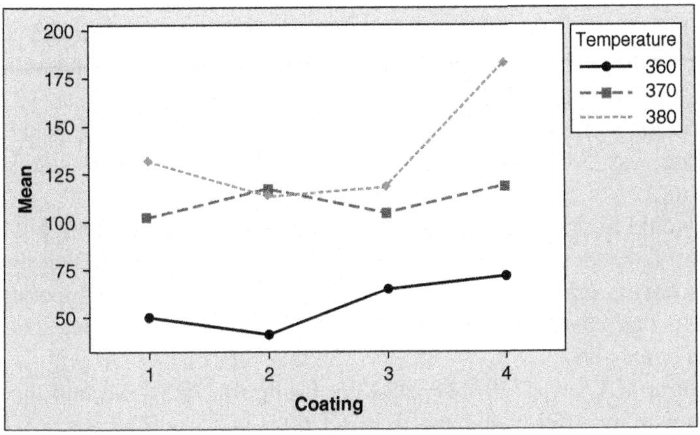

Figure 12.2 Interaction plot for mean corrosion resistance for four coatings baked at three temperatures.

Table 12.12 ANOVA Table for Coatings Data

Source	SS	d.f.	MS	F	p
Temperature (whole plots)	26,519	2	13,259.5	1.94	0.340
Replicate	782	1	782.0	0.14	0.744
Whole-plot error	13,658	2	6,829.0		
Coatings (subplots)	4,289	3	1429.7	11.44	0.002
Temperature × coatings	3,270	6	545.0	4.36	0.024
Subplot error	1,121	9	124.5		
Total	49,639	23			

to the large whole-plot error and the small numbers of d.f. On the other hand, the coating effects and temperature–coating interactions are highly significant. Coatings were of primary interest and the experiment has been effective in revealing the differences between them.

We can estimate the variance components as follows:

$$\widehat{\sigma}_{e2}^2 = \text{MS}_{e2} = 124.5, \qquad \widehat{\sigma}_{e1}^2 = \frac{\text{MS}_{e1} - \text{MS}_{e2}}{b} = \frac{6829.0 - 124.5}{4} = 1676.1,$$

and

$$\widehat{\sigma}_C^2 = \frac{\text{MS}_C - \text{MS}_{e1}}{ab} = \frac{782.0 - 6829.0}{12} = -503.9.$$

The negative estimate of σ_C^2 is in agreement with the high nonsignificance ($p = 0.744$) of the replicate effect. The whole-plot standard deviation estimate is $\sqrt{1676.1} = 40.940$, which is almost four times that of the subplot standard deviation estimate $\sqrt{124.5} = 11.158$. (Recall from Example 12.5 that when the blocking effect of replicates was ignored the whole-plot standard deviation estimate was 34.23, about three times that of the subplot standard deviation estimate.)

It would be interesting to see how the analysis would change if the split-plot nature of the design is ignored and it is regarded as an RB design with treatments having crossed factorial structure with two factors, temperature and coatings. In that case, the error term would be the treatment–replicate interaction, which equals $\text{SS}_e = \text{SS}_{e1} + \text{SS}_{e2} = 13658 + 1121 = 14779$ with $2 + 9 = 11$ d.f. Therefore $\text{MS}_e = 14779/11 = 1343.5$. Using this MS_e, we find that for the temperature main effect we have $F = 13260/1343.5 = 9.87$ $(p = 0.003)$, for the coatings main effect we have $F = 1430/1343.5 = 1.06$ $(p = 0.405)$, and for the temperature–coating interaction we have $F = 545/1343.5 = 0.41$ $(p = 0.858)$. These results are exactly opposite to those obtained above. This demonstrates the danger in analyzing data from a design without considering how treatments were randomized. The moral is that a data table does not define design. ■

Multiple comparisons between whole-plot treatments or between subplot treatments can be performed using the procedures described in Chapter 4. Appropriate error terms should be used for this purpose, that is, whole-plot error for comparing whole-plot treatments and subplot error for comparing subplot treatments. See Federer and McCulloch (1984) for additional details.

12.4.3 Extensions of Split-Plot Designs

12.4.3.1 Split-Plot Designs with Factorial Structures

So far we have considered a single whole-plot factor and a single subplot factor. More generally, there can be several whole and subplot factors arranged in full- or fractional-factorial designs. An example is Taguchi's crossed-array design in

Chapter 10 for robust parameter design. Its analysis there assumed that it is run as a completely randomized design. However, this is often infeasible, and such designs are more efficiently run as split-plot designs, as noted in Section 10.3.3. Depending on the practical constraints, either the design or the noise factors may be assigned to the whole plots and the other set of factors to the subplots. Any main effects or interactions involving only the whole-plot factors are tested against the whole-plot error, while any other effects that involve subplot factors are tested against the subplot error. The following example illustrates these ideas.

Example 12.7 (Choosing the Best Cake Mix Robust to Noise Factors)

Box and Jones (2000) discuss an experiment for choosing the best tasting cake mix that is robust to noise (environmental) factors of baking temperature and time. There were three factors that make up the mix: amounts of flour (A), shortening powder (B), and egg powder (C). These are design factors since they can be set by the manufacturer. Baking temperature (D) and time (E) were noise factors because they were not controlled by the manufacturer and may vary among the users. The manufacturer was interested in designing the mix so that it is good-tasting over chosen ranges of baking temperature and time.

Two levels (low and high) of each of the five factors were selected for the experiment, which was run as follows. Large batches of dough were made from the eight cake mixes according to the 2^3 design in the factors A, B, C which were divided into four pieces. These four pieces from each batch of dough were baked under four different oven settings according to a 2^2 design in the factors D, E; thus the oven was operated a total of 32 times. In this case, A, B, C are whole-plot factors while D, E are subplot factors. The data consist of average taste scores (on a scale of 1–7 with higher scores for better tasting cakes) assigned to the cakes by a panel of tasters and are shown in Table 12.13.

Table 12.13 Data for Cake Mix Experiment

	Design Factors			Noise Factors						
				D	−	+	−	+		
Cake Mix	A	B	C	E	−	−	+	+	Mean	Range
1	−	−	−		1.1	1.4	1.0	2.9	1.6	1.9
2	+	−	−		1.8	5.1	2.8	6.1	4.0	4.3
3	−	+	−		1.7	1.6	1.9	2.1	1.8	0.5
4	+	+	−		3.9	3.7	4.0	4.4	4.0	0.7
5	−	−	+		1.9	3.8	2.6	4.7	3.2	2.8
6	+	−	+		4.4	6.4	6.2	6.6	5.6	2.2
7	−	+	+		1.6	2.1	2.3	1.9	2.0	0.7
8	+	+	+		4.9	5.5	5.2	5.7	5.3	0.8

Source: Box and Jones (2000, Table 1). Reprinted by permission of Taylor & Francis.

A quick assessment of the data can be made by studying the mean and range columns (which give the means of the average taste scores and their ranges over four settings of the oven for the eight cake mixes). We see that cake mix 8 provides the best combination of high average (reflecting good-tasting cake) and low range (reflecting a robust cake mix whose taste scores do not vary much with changes in the oven settings). To determine which factors contribute most to the goodness of taste, we computed the estimates of all the main effects and interactions. To save space, the results for only the main effects and two-factor interactions are shown in Display 12.4.

Since the design is singly replicated (because only the averages of the taste panelists are reported and not their individual scores), we do not have a pure estimate of the error. In fact, since the experiment was run as a split-plot design, there are two different errors—the whole plot and subplot. Therefore we must make separate normal plots of the whole-plot main effects and interactions, which are compared to the whole-plot error and subplot main effects and whole plot–subplot interactions, which are compared to the subplot error. These two plots are shown in Figure 12.3. We see that the normal plot of the whole-plot effects is more spread out than that for the subplot effects, indicating that the whole-plot error is larger than the subplot error. Among the whole-plot factors, flour (A) and egg powder (C) have significant main effects (both positive), while among the subplot factors, the temperature (D) main effect and shortening–temperature (BD) interaction are significant (the former is positive and the latter is negative). Thus, if the goodness of taste is the only criterion, then the manufacturer should use the higher amounts of flour and egg powder and the lower amount of shortening powder (cake mix 6) and recommend the higher temperature in the baking instructions on the package.

```
Estimated Effects and Coefficients for score (coded units)

Term          Effect      Coef
Constant                3.4781
A             2.6313    1.3156
B            -0.3937   -0.1969
C             1.2688    0.6344
D             1.0437    0.5219
E             0.5938    0.2969
A*B           0.1312    0.0656
A*C           0.3688    0.1844
A*D           0.2438    0.1219
A*E           0.0687    0.0344
B*C          -0.5312   -0.2656
B*D          -0.8562   -0.4281
B*E          -0.2813   -0.1406
C*D          -0.0938   -0.0469
C*E          -0.0188   -0.0094
D*E           0.0063    0.0031
```

Display 12.4 Minitab output for main effect and two-factor interaction estimates for cake mix data.

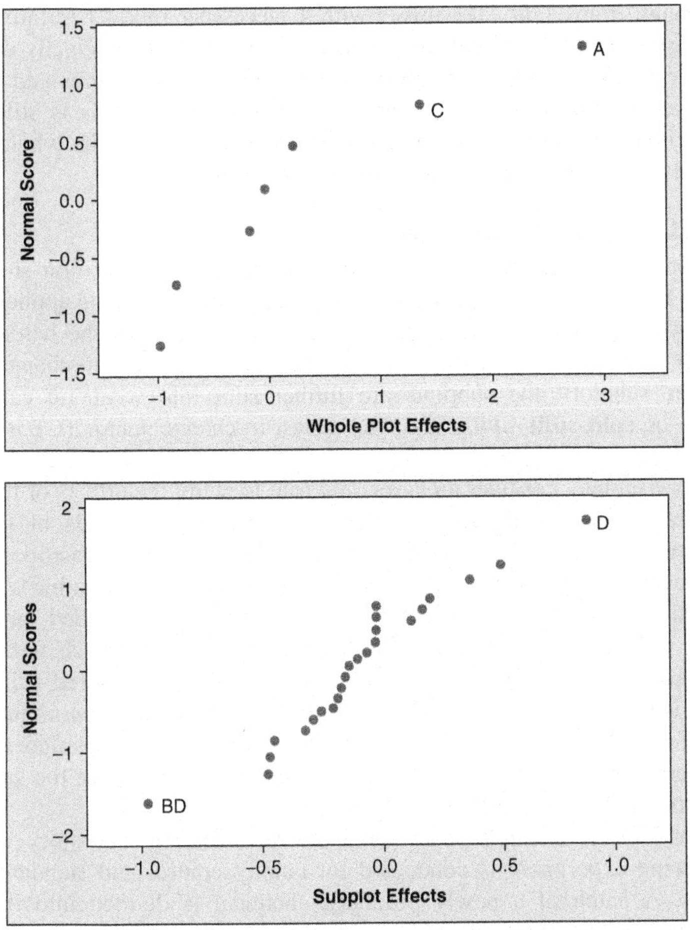

Figure 12.3 Normal plots of whole-plot (top) and subplot (bottom) effects for cake mix data.

However, as noted above, from the point of view of robustness, cake mix 8 is preferred. Box and Jones (2000) provided a nice explanation of why using the higher amount of shortening (B) improves robustness. Since the temperature is a noise factor with a significant main effect, they suggested canceling that effect by exploiting the significant BD interaction by adjusting the level of B. If x_B and x_D denote the coded levels of B and D and D is set at high level $(x_D = 1)$, then the net effect of D is modulated by B according to the equation $\widehat{D} + \widehat{BD} \times x_B = 1.0437 - 0.8562x_B$. Setting this equal to zero, we get $x_B = 1.219$, or approximately 1. This explains why the variation is low when B is at high level (cake mixes 3, 4, 7, and 8). ■

In the above experimental setup the oven is operated 32 times, which may be impractical. An alternative way to run the experiment is to prepare 32 pieces of dough from eight cake mixes with 4 pieces per mix. These are divided into four groups of 8 pieces of each, one from each cake mix. Finally each group of 8 pieces is baked together using one of the four temperatures and time settings of oven. Thus the oven is operated only four times. This is still a split-plot experiment with oven settings (temperature and time) as the whole-plot factors and the cake mix ingredients as the subplot factors.

12.4.3.2 Split-Split-Plot Designs

In some applications it is necessary to perform randomization successively in more than two stages. The idea of the split-plot design can be applied recursively in such applications. After performing randomization of the hardest-to-change factor (A) on the whole plots and the second hardest-to-change factor (B) on the subplots, the subplots are further split into what are called **sub-sub plots** or **split-split plots**, and the easiest-to-change factor (C) is randomized over them.

For example, consider an agronomy trial to study the effects of three different fertilizers and two different watering regimes on the yields of four different varieties of a crop. One might think of this as a $3 \times 2 \times 4$ factorial experiment. However, a split-split-plot approach is the only practically feasible way in modern agriculture. In this approach, the experimental field is divided into three large areas (whole plots) which are fertilized using airplanes. Each fertilized area is divided into two smaller areas (split plots) which are watered using sprinklers programmed according to different regimes. Each smaller area is further divided into four split-split plots which are planted with the varieties under study. Here the fertilizers are the whole-plot factor, watering regimes are the subplot factor, and the varieties are the sub-subplot factor.

For another example, consider a chemical testing laboratory in which the following experiment is conducted for quality control and standardization purposes. A batch of a newly produced chemical is divided into three samples and assigned to three technicians. Each technician divides the chemical sample into two subsamples; one subsample is tested immediately and the other subsample is tested after storing it for one week. Finally, each of these subsamples is further subdivided into two sub-subsamples, which are tested using two different procedures. Here the technicians are the whole-plot factor, duration of storage is the subplot factor, and the testing procedures are the sub-subplot factor.

Denote the levels of the factors A, B, C by a, b, c, respectively. Assume that these factors are fixed and replicates (factor D with d levels) are random. Then the following generalization of the split-plot design model (12.8) may be used:

$$\begin{aligned} y_{ijk\ell} = {} & \mu + \alpha_i + \delta_\ell + e_{i\ell}^{(1)} \\ & + \beta_j + (\alpha\beta)_{ij} + e_{ij\ell}^{(2)} + \gamma_k + (\alpha\gamma)_{ik} + (\beta\gamma)_{jk} \\ & + (\alpha\beta\gamma)_{ijk} + e_{ijk\ell}^{(3)} \qquad (1 \le i \le a, 1 \le j \le b, 1 \le k \le c, 1 \le \ell \le d), \end{aligned}$$

where the $\alpha_i, \beta_j, \gamma_k, (\alpha\beta)_{ij}, (\alpha\gamma)_{ik}, (\beta\gamma)_{jk}, (\alpha\beta\gamma)_{ijk}$ are fixed parameters subject to the usual side constraints, the $\delta_\ell \sim \text{N}(0, \sigma_D^2)$ are the random replicate effects, and the $e_{i\ell}^{(1)} \sim \text{N}(0, \sigma_{e1}^2)$, $e_{ij\ell}^{(2)} \sim \text{N}(0, \sigma_{e2}^2)$, and $e_{ijk\ell}^{(3)} \sim \text{N}(0, \sigma_{e3}^2)$ are whole-plot, subplot, and sub-subplot errors, respectively. All random quantities are assumed to be mutually independent. The ANOVA for this design is given in Table 12.14.

12.4.3.3 Strip Block Designs

The ideas from the two experimental setups described in the previous section can be combined to run an even more efficient experiment: Prepare large batches of dough using the eight cake mixes and divide each batch into 4 pieces as in the first setup. Then divide the resulting 32 pieces into four groups and bake each group of 8 pieces together as in the second setup. This design is shown schematically in Table 12.15 as an 8×4 array. The batches are called **strips** and the design is called a **strip block design** or **split-block design**. This terminology comes from agriculture where a rectangular plot of land is divided into a grid pattern with strips running perpendicular to each other. Strips in one direction are treated with the levels of one factor, say A (e.g., fertilizer). Strips in the other direction are treated with the levels of another factor, say B (e.g., amount of watering).

Note that in the agricultural example both factors are whole-plot factors. Strips to which different fertilizers are applied are one set of whole plots and strips to which different amounts of watering are applied are another set of whole plots. These two randomizations are exchangeable, that is, their order of application can be reversed. On the other hand, in the cake mix experiment the two randomizations are sequential (first make dough batches using different cake mixes and then bake them at different oven settings). Nonetheless, both A and B can be thought of as whole-plot factors but with different errors. Note that there are no subplots in this design. If the strip block design is replicated c times, then treating replicates as a random factor C and A and B as fixed factors leads to the following linear model:

$$
\begin{aligned}
y_{ijk} &= \mu + \alpha_i + \gamma_k + e_{ik}^{(1)} \\
&\quad + \beta_j + e_{jk}^{(2)} + (\alpha\beta)_{ij} + e_{ijk}^{(3)} \qquad (1 \le i \le a, 1 \le j \le b, 1 \le k \le c),
\end{aligned}
$$

where the α_i, β_j, and $(\alpha\beta)_{ij}$ are fixed parameters subject to the usual side constraints, $\gamma_k \sim \text{N}(0, \sigma_C^2)$ are random replicate effects, and $e_{ik}^{(1)} \sim \text{N}(0, \sigma_{e1}^2)$, $e_{jk}^{(2)} \sim \text{N}(0, \sigma_{e2}^2)$, and $e_{ijk}^{(3)} \sim \text{N}(0, \sigma_{e3}^2)$ are whole-plot random errors for A, B, and AB.

An abbreviated ANOVA for this linear model is given in Table 12.16. The sums of squares can be computed by treating the design as a three-way crossed array with a single observation per cell. In the E(MS) expressions Q_A, Q_B, and Q_{AB} are as defined in (12.10).

Table 12.14 ANOVA Table for Split-Split-Plot Design

Source	SS	d.f.	E(MS)	F
A	SS_A	$a-1$	$\sigma_{e3}^2 + c\sigma_{e2}^2 + bc\sigma_{e1}^2 + bcdQ_A$	$F_A = MS_A/MS_{e1}$
D	SS_D	$d-1$	$\sigma_{e3}^2 + c\sigma_{e2}^2 + abc\sigma_D^2$	$F_D = MS_D/MS_{e1}$
Error$_1$	$SS_{e1} = SS_{AD}$	$(a-1)(d-1)$	$\sigma_{e3}^2 + c\sigma_{e2}^2 + bc\sigma_{e1}^2$	
B	SS_B	$b-1$	$\sigma_{e3}^2 + c\sigma_{e2}^2 + acdQ_B$	$F_B = MS_B/MS_{e2}$
AB	SS_{AB}	$(a-1)(b-1)$	$\sigma_{e3}^2 + c\sigma_{e2}^2 + cdQ_{AB}$	$F_{AB} = MS_{AB}/MS_{e2}$
Error$_2$	$SS_{e2} = SS_{BD(A)}$	$a(b-1)(d-1)$	$\sigma_{e3}^2 + c\sigma_{e2}^2$	
C	SS_C	$c-1$	$\sigma_{e3}^2 + abdQ_C$	$F_C = MS_C/MS_{e3}$
AC	SS_{AC}	$(a-1)(c-1)$	$\sigma_{e3}^2 + bdQ_{AC}$	$F_{AC} = MS_{AC}/MS_{e3}$
BC	SS_{BC}	$(b-1)(c-1)$	$\sigma_{e3}^2 + adQ_{BC}$	$F_{BC} = MS_{BC}/MS_{e3}$
ABC	SS_{ABC}	$(a-1)(b-1)(c-1)$	$\sigma_{e3}^2 + dQ_{ABC}$	$F_{ABC} = MS_{ABC}/MS_{e3}$
Error$_3$	$SS_{e3} = SS_{CD(AB)}$	$ab(c-1)(d-1)$	σ_{e3}^2	

Table 12.15 Strip Block Design for Baking Cake Mixes

	Oven Setting (B)			
Cake Mix (A)	B_1	B_2	B_3	B_4
A_1	A_1B_1	A_1B_2	A_1B_3	A_1B_4
A_2	A_2B_1	A_2B_2	A_2B_3	A_2B_4
A_3	A_3B_1	A_3B_2	A_3B_3	A_3B_4
A_4	A_4B_1	A_4B_2	A_4B_3	A_4B_4
A_5	A_5B_1	A_5B_2	A_5B_3	A_5B_4
A_6	A_6B_1	A_6B_2	A_6B_3	A_6B_4
A_7	A_7B_1	A_7B_2	A_7B_3	A_7B_4
A_8	A_8B_1	A_8B_2	A_8B_3	A_8B_4

Table 12.16 ANOVA Table for Strip Block Design

Source	SS	d.f.	E(MS)	F
A	SS_A	$a-1$	$\sigma_{e3}^2 + b\sigma_{e1}^2 + bcQ_A$	$F_A = MS_A/MS_{e1}$
C	SS_C	$c-1$	$\sigma_{e3}^2 + b\sigma_{e1}^2 + ab\sigma_C^2$	$F_C = MS_C/MS_{e1}$
Error$_1$	$SS_{e1} = SS_{AC}$	$(a-1)(c-1)$	$\sigma_{e3}^2 + b\sigma_{e1}^2$	
B	SS_B	$b-1$	$\sigma_{e3}^2 + a\sigma_{e2}^2 + acQ_B$	$F_B = MS_B/MS_{e2}$
Error$_2$	$SS_{e2} = SS_{BC}$	$(b-1)(c-1)$	$\sigma_{e3}^2 + a\sigma_{e2}^2$	
AB	SS_{AB}	$(a-1)(b-1)$	$\sigma_{e3}^2 + cQ_{AB}$	$F_{AB} = MS_{AB}/MS_{e3}$
Error$_3$	$SS_{e3} = SS_{ABC}$	$(a-1)(b-1)(c-1)$	σ_{e3}^2	
Total	SS_{tot}	$abc-1$		

12.5 CHAPTER NOTES

12.5.1 Derivations of E(MS) Expressions for Two-Stage Nested Design of Section 12.1 with Both Factors Random

In this and the following section we make repeated uses of Lemmas 11.1 and 11.2 without specifically mentioning them. By substituting the model (12.1) into the sums-of-squares expressions (12.2), we obtain

$$\begin{aligned} SS_e &= \sum_{i=1}^{a}\sum_{j=1}^{b}\sum_{k=1}^{n}(y_{ijk} - \overline{y}_{ij\cdot})^2 \\ &= \sum_{i=1}^{a}\sum_{j=1}^{b}\sum_{k=1}^{n}(e_{ijk} - \overline{e}_{ij\cdot})^2. \end{aligned}$$

Since the e_{ijk} are i.i.d. $N(0, \sigma_e^2)$, it follows that

$$SS_e \sim \sigma_e^2 \chi^2_{ab(n-1)}$$

and hence

$$E(MS_e) = \sigma_e^2.$$

Next,

$$\begin{aligned} SS_{B(A)} &= n \sum_{i=1}^{a} \sum_{j=1}^{b} (\bar{y}_{ij\cdot} - \bar{y}_{i\cdot\cdot})^2 \\ &= n \sum_{i=1}^{a} \sum_{j=1}^{b} (\beta_{j(i)} - \bar{\beta}_{\cdot(i)} + \bar{e}_{ij\cdot} - \bar{e}_{i\cdot\cdot})^2. \end{aligned}$$

Put $x_{ij} = \beta_{j(i)} + \bar{e}_{ij\cdot} \sim N(0, \sigma^2_{B(A)} + (1/n)\sigma_e^2)$. Then it follows that

$$SS_{B(A)} = n \sum_{i=1}^{a} \sum_{j=1}^{b} (x_{ij} - \bar{x}_{i\cdot})^2 \sim n \left(\sigma^2_{B(A)} + \frac{1}{n} \sigma_e^2 \right) \chi^2_{a(b-1)}$$

and hence

$$E(MS_{B(A)}) = \sigma_e^2 + n\sigma^2_{B(A)}.$$

Finally,

$$\begin{aligned} SS_A &= bn \sum_{i=1}^{a} (\bar{y}_{i\cdot\cdot} - \bar{y}_{\cdot\cdot\cdot})^2 \\ &= bn \sum_{i=1}^{a} (\alpha_i + \bar{\beta}_{\cdot(i)} + \bar{e}_{i\cdot\cdot} - \bar{\alpha} - \bar{\beta}_{\cdot(\cdot)} - \bar{e}_{\cdot\cdot\cdot})^2. \end{aligned}$$

Put $x_i = \alpha_i + \bar{\beta}_{\cdot(i)} + \bar{e}_{i\cdot\cdot} \sim N(0, \sigma_A^2 + (1/b)\sigma^2_{B(A)} + (1/bn)\sigma_e^2)$. Then

$$SS_A = bn \sum_{i=1}^{a} (x_i - \bar{x})^2 \sim (bn\sigma_A^2 + n\sigma^2_{B(A)} + \sigma_e^2)\chi^2_{a-1}$$

and hence

$$E(MS_A) = \sigma_e^2 + n\sigma_{B(A)}^2 + bn\sigma_A^2.$$

12.5.2 Derivations of E(MS) Expressions for Design of Section 12.3 with Crossed and Nested Factors

By substituting the model (12.8) into the sums-of-squares expressions, we obtain

$$\begin{aligned} SS_e &= \sum_{i=1}^{a}\sum_{j=1}^{b}\sum_{k=1}^{c}(y_{ijk\ell} - \overline{y}_{ijk\cdot})^2 \\ &= \sum_{i=1}^{a}\sum_{j=1}^{b}\sum_{k=1}^{c}(e_{ijk\ell} - \overline{e}_{ijk\cdot})^2. \end{aligned}$$

Since the $e_{ijk\ell}$ are i.i.d. $N(0, \sigma_e^2)$, it follows that

$$SS_e \sim \sigma_e^2 \chi_{abc(n-1)}^2$$

and hence

$$E(MS_e) = \sigma_e^2.$$

Next,

$$\begin{aligned} SS_{BC(A)} &= n\sum_{i=1}^{a}\sum_{j=1}^{b}\sum_{k=1}^{c}(\overline{y}_{ijk\cdot} - \overline{y}_{ij\cdot\cdot} - \overline{y}_{i\cdot k\cdot} + \overline{y}_{i\cdot\cdot\cdot})^2 \\ &= n\sum_{i=1}^{a}\sum_{j=1}^{b}\sum_{k=1}^{c}\Big[(\beta\gamma)_{jk(i)} + \overline{e}_{ijk\cdot} - \overline{(\beta\gamma)}_{j\cdot(i)} - \overline{e}_{ij\cdot\cdot} \\ &\quad - \overline{(\beta\gamma)}_{\cdot k(i)} - \overline{e}_{i\cdot k\cdot} + \overline{(\beta\gamma)}_{\cdot\cdot(i)} + \overline{e}_{i\cdot\cdot\cdot}\Big]^2. \end{aligned}$$

Put $x_{ijk} = (\beta\gamma)_{jk(i)} + \overline{e}_{ijk\cdot} \sim N\left(0, \sigma_{BC(A)}^2 + (1/n)\sigma_e^2\right)$. Then

$$SS_{BC(A)} = n\sum_{i=1}^{a}\sum_{j=1}^{b}\sum_{k=1}^{c}(x_{ijk} - \overline{x}_{ij\cdot} - \overline{x}_{i\cdot k} + \overline{x}_{i\cdot\cdot})^2 \sim (n\sigma_{BC(A)}^2 + \sigma_e^2)\chi_{a(b-1)(c-1)}^2,$$

and hence

$$E(MS_{BC(A)}) = \sigma_e^2 + n\sigma_{BC(A)}^2.$$

Next,

$$\begin{aligned}
SS_{C(A)} &= bn \sum_{i=1}^{a} \sum_{k=1}^{c} (\overline{y}_{i\cdot k\cdot} - \overline{y}_{i\ldots})^2 \\
&= bn \sum_{i=1}^{a} \sum_{k=1}^{n} \left[\gamma_{k(i)} + \overline{(\beta\gamma)}_{\cdot k(i)} + \overline{e}_{i\cdot k\cdot} - \overline{\gamma}_{\cdot(i)} - \overline{(\beta\gamma)}_{\cdot\cdot(i)} - \overline{e}_{i\ldots} \right]^2.
\end{aligned}$$

Put

$$x_{ik} = \gamma_{k(i)} + \overline{(\beta\gamma)}_{\cdot k(i)} + \overline{e}_{i\cdot k\cdot} \sim \mathrm{N}\left(0, \sigma^2_{C(A)} + \frac{1}{b}\sigma^2_{BC(A)} + \frac{1}{bn}\sigma^2_e\right).$$

Then

$$SS_{C(A)} = bn \sum_{i=1}^{a} \sum_{k=1}^{n} (x_{ik} - \overline{x}_{i\cdot})^2 \sim (bn\sigma^2_{C(A)} + n\sigma^2_{BC(A)} + \sigma^2_e)\chi^2_{a(c-1)},$$

and hence

$$\mathrm{E}(MS_{C(A)}) = \sigma^2_e + n\sigma^2_{BC(A)} + bn\sigma^2_{C(A)}.$$

Next,

$$\begin{aligned}
SS_{AB} &= cn \sum_{i=1}^{a} \sum_{j=1}^{b} (\overline{y}_{ij\cdot\cdot} - \overline{y}_{i\ldots} - \overline{y}_{\cdot j\cdot\cdot} + \overline{y}_{\ldots\cdot})^2 \\
&= cn \sum_{i=1}^{a} \sum_{j=1}^{b} \Big[(\alpha\beta)_{ij} + \overline{(\beta\gamma)}_{j\cdot(i)} + \overline{e}_{ij\cdot\cdot} - \overline{(\beta\gamma)}_{\cdot\cdot(i)} - \overline{e}_{i\ldots} \\
&\qquad - \overline{(\beta\gamma)}_{j\cdot(\cdot)} - \overline{e}_{\cdot j\cdot\cdot} + \overline{(\beta\gamma)}_{\cdot\cdot(\cdot)} + \overline{e}_{\ldots\cdot} \Big]^2.
\end{aligned}$$

Put

$$x_{ij} = \overline{(\beta\gamma)}_{j\cdot(i)} + \overline{e}_{ij\cdot\cdot} \sim \mathrm{N}\left(0, \frac{1}{c}\sigma^2_{BC(A)} + \frac{1}{cn}\sigma^2_e\right).$$

Then

$$\begin{aligned}
SS_{AB} &= cn \sum_{i=1}^{a} \sum_{j=1}^{b} \left[(\alpha\beta)_{ij} + x_{ij} - \overline{x}_{i\cdot} - \overline{x}_{\cdot j} + \overline{x}_{\cdot\cdot} \right]^2 \\
&\sim (n\sigma^2_{BC(A)} + \sigma^2_e)\chi^2_{(a-1)(b-1)}(\lambda^2),
\end{aligned}$$

where

$$\lambda^2 = \frac{cn \sum_{i=1}^{a} \sum_{j=1}^{b} (\alpha\beta)_{ij}^2}{n\sigma_{BC(A)}^2 + \sigma_e^2}$$

is the n.c.p. Hence

$$\begin{aligned} \mathrm{E}(\mathrm{MS}_{AB}) &= \frac{1}{(a-1)(b-1)}(n\sigma_{BC(A)}^2 + \sigma_e^2) \\ &\quad \times \left[(a-1)(b-1) + \frac{cn \sum_{i=1}^{a} \sum_{j=1}^{b} (\alpha\beta)_{ij}^2}{n\sigma_{BC(A)}^2 + \sigma_e^2}\right] \\ &= \sigma_e^2 + n\sigma_{BC(A)}^2 + cnQ_{AB}, \end{aligned}$$

where Q_{AB} is given by (12.10).

Next,

$$\begin{aligned} \mathrm{SS}_A &= bcn \sum_{i=1}^{a} (\overline{y}_{i\ldots} - \overline{y}_{\ldots.})^2 \\ &= bcn \sum_{i=1}^{a} \left[\alpha_i + \overline{\gamma}_{\cdot(i)} + \overline{(\beta\gamma)}_{\cdot\cdot(i)} + \overline{e}_{i\ldots} - \overline{\gamma}_{\cdot(\cdot)} - \overline{(\beta\gamma)}_{\cdot\cdot(\cdot)} - \overline{e}_{\ldots.}\right]^2. \end{aligned}$$

Put

$$x_i = \overline{\gamma}_{\cdot(i)} + \overline{(\beta\gamma)}_{\cdot\cdot(i)} + \overline{e}_{i\ldots} \sim \mathrm{N}\left(0, \frac{1}{c}\sigma_{C(A)}^2 + \frac{1}{bc}\sigma_{BC(A)}^2 + \frac{1}{bcn}\sigma_e^2\right).$$

Then

$$\mathrm{SS}_A = bcn \sum_{i=1}^{a} (\alpha_i + x_i - \overline{x})^2 \sim (bn\sigma_{C(A)}^2 + n\sigma_{BC(A)}^2 + \sigma_e^2)\chi_{a-1}^2(\lambda^2),$$

where

$$\lambda^2 = \frac{bcn \sum_{i=1}^{a} \alpha_i^2}{bn\sigma_{C(A)}^2 + n\sigma_{BC(A)}^2 + \sigma_e^2}$$

is the n.c.p. Hence

$$\begin{aligned} \mathrm{E}(\mathrm{MS}_A) &= \frac{1}{a-1}(bn\sigma_{C(A)}^2 + n\sigma_{BC(A)}^2 + \sigma_e^2)\left[a - 1 + \frac{bcn \sum_{i=1}^{a} \alpha_i^2}{bn\sigma_{C(A)}^2 + n\sigma_{BC(A)}^2 + \sigma_e^2}\right] \\ &= \sigma_e^2 + n\sigma_{BC(A)}^2 + bn\sigma_{C(A)}^2 + bcnQ_A, \end{aligned}$$

where Q_A is given by (12.10).

Finally,

$$\begin{aligned}\mathrm{SS}_B &= acn\sum_{j=1}^{b}(\overline{y}_{\cdot j\cdot\cdot} - \overline{y}_{\cdot\cdot\cdot\cdot})^2 \\ &= acn\sum_{j=1}^{b}\left[\beta_j + \overline{(\beta\gamma)}_{j\cdot(\cdot)} + \overline{e}_{\cdot j\cdot\cdot} - \overline{(\beta\gamma)}_{\cdot\cdot(\cdot)} - \overline{e}_{\cdot\cdot\cdot\cdot}\right]^2.\end{aligned}$$

Put

$$x_j = \overline{(\beta\gamma)}_{j\cdot(\cdot)} + \overline{e}_{\cdot j\cdot\cdot} \sim \mathrm{N}\left(0, \frac{1}{ac}\sigma^2_{BC(A)} + \frac{1}{acn}\sigma^2_e\right).$$

Then

$$\mathrm{SS}_B = acn\sum_{j=1}^{b}(\beta_j + x_j - \overline{x})^2 \sim (n\sigma^2_{BC(A)} + \sigma^2_e)\chi^2_{b-1}(\lambda^2),$$

where

$$\lambda^2 = \frac{acn\sum_{j=1}^{b}\beta_j^2}{n\sigma^2_{BC(A)} + \sigma^2_e}$$

is the n.c.p. Hence

$$\begin{aligned}\mathrm{E}(\mathrm{MS}_B) &= \frac{1}{b-1}(n\sigma^2_{BC(A)} + \sigma^2_e)\left[b - 1 + \frac{acn\sum_{j=1}^{b}\beta_j^2}{n\sigma^2_{BC(A)} + \sigma^2_e}\right] \\ &= \sigma^2_e + n\sigma^2_{BC(A)} + acnQ_B,\end{aligned}$$

where Q_B is given by (12.10).

12.5.3 Derivations of E(MS) Expressions for Split-Plot Design

By substituting the model (12.8) in the sums-of-squares expressions, we obtain

$$\begin{aligned}\mathrm{SS}_{e2} &= \sum_{i=1}^{a}\sum_{j=1}^{b}\sum_{k=1}^{c}(y_{ijk} - \overline{y}_{ij\cdot} - \overline{y}_{i\cdot k} + \overline{y}_{i\cdot\cdot})^2 \\ &= \sum_{i=1}^{a}\sum_{j=1}^{b}\sum_{k=1}^{c}(e^{(2)}_{ijk} - \overline{e}^{(2)}_{ij\cdot} - \overline{e}^{(2)}_{i\cdot k} + \overline{e}^{(2)}_{i\cdot\cdot})^2.\end{aligned}$$

Using the fact that $e_{ijk}^{(2)} \sim \mathrm{N}(0, \sigma_{e2}^2)$, it follows that

$$\mathrm{SS}_{e2} \sim \sigma_{e2}^2 \chi_{a(b-1)(c-1)}^2$$

and hence

$$\mathrm{E}(\mathrm{MS}_{e2}) = \sigma_{e2}^2.$$

Next,

$$\begin{aligned}
\mathrm{SS}_{e1} &= b \sum_{i=1}^{a} \sum_{k=1}^{c} (\overline{y}_{i \cdot k} - \overline{y}_{i \cdot\cdot} - \overline{y}_{\cdot\cdot k} + \overline{y}_{\cdots})^2 \\
&= b \sum_{i=1}^{a} \sum_{k=1}^{c} (e_{ik}^{(1)} + \overline{e}_{i \cdot k}^{(2)} - \overline{e}_{i \cdot}^{(1)} - \overline{e}_{i \cdot\cdot}^{(2)} - \overline{e}_{\cdot k}^{(1)} - \overline{e}_{\cdot\cdot k}^{(2)} + \overline{e}_{\cdot\cdot}^{(1)} + \overline{e}_{\cdots}^{(2)})^2.
\end{aligned}$$

Put $x_{ik} = e_{ik}^{(1)} + \overline{e}_{i \cdot k}^{(2)} \sim \mathrm{N}(0, \sigma_{e1}^2 + (1/b)\sigma_{e2}^2)$. Then

$$\mathrm{SS}_{e1} = b \sum_{i=1}^{a} \sum_{k=1}^{c} (x_{ik} - \overline{x}_{i \cdot} - \overline{x}_{\cdot k} + \overline{x}_{\cdot\cdot})^2 \sim (b\sigma_{e1}^2 + \sigma_{e2}^2) \chi_{(a-1)(c-1)}^2.$$

Hence

$$\mathrm{E}(\mathrm{MS}_{e1}) = \sigma_{e2}^2 + b\sigma_{e1}^2.$$

Next,

$$\begin{aligned}
\mathrm{SS}_{AB} &= c \sum_{i=1}^{a} \sum_{j=1}^{b} (\overline{y}_{ij \cdot} - \overline{y}_{i \cdot\cdot} - \overline{y}_{\cdot j \cdot} + \overline{y}_{\cdots})^2 \\
&= c \sum_{i=1}^{a} \sum_{j=1}^{b} \left[(\alpha\beta)_{ij} + \overline{e}_{ij \cdot}^{(2)} - \overline{e}_{i \cdot\cdot}^{(2)} - \overline{e}_{\cdot j \cdot}^{(2)} + \overline{e}_{\cdots}^{(2)} \right]^2.
\end{aligned}$$

Put $x_{ij} = \overline{e}_{ij \cdot}^{(2)} \sim \mathrm{N}(0, (1/c)\sigma_{e2}^2)$. Then

$$\mathrm{SS}_{AB} = c \sum_{i=1}^{a} \sum_{j=1}^{b} \left[(\alpha\beta)_{ij} + x_{ij} - \overline{x}_{i \cdot} - \overline{x}_{\cdot j} + \overline{x}_{\cdot\cdot} \right]^2 \sim \sigma_{e2}^2 \chi_{(a-1)(b-1)}^2 (\lambda^2),$$

where

$$\lambda^2 = \frac{c \sum_{i=1}^{a} \sum_{j=1}^{b} (\alpha\beta)_{ij}^2}{\sigma_{e2}^2}$$

is the n.c.p. Hence

$$\begin{aligned}
\mathrm{E}(\mathrm{MS}_{AB}) &= \frac{\sigma_{e2}^2}{(a-1)(b-1)}\left[(a-1)(b-1) + \frac{c\sum_{i=1}^{a}\sum_{j=1}^{b}(\alpha\beta)_{ij}^2}{\sigma_{e2}^2}\right] \\
&= \sigma_{e2}^2 + cQ_{AB}.
\end{aligned}$$

Next,

$$\begin{aligned}
\mathrm{SS}_C &= ab\sum_{k=1}^{c}(\overline{y}_{..k} - \overline{y}_{...})^2 \\
&= ab\sum_{k=1}^{c}(\gamma_k - \overline{\gamma}_{.} + \overline{e}_{.k}^{(1)} + \overline{e}_{..k}^{(2)} - \overline{e}_{..}^{(1)} - \overline{e}_{...}^{(2)})^2.
\end{aligned}$$

Put

$$x_k = \gamma_k + \overline{e}_{.k}^{(1)} + \overline{e}_{..k}^{(2)} \sim \mathrm{N}\left(0, \sigma_C^2 + \frac{1}{a}\sigma_{e1}^2 + \frac{1}{ab}\sigma_{e2}^2\right).$$

Hence

$$\mathrm{SS}_C = ab\sum_{k=1}^{c}(x_k - \overline{x})^2 \sim (ab\sigma_C^2 + b\sigma_{e1}^2 + \sigma_{e2}^2)\chi_{c-1}^2.$$

Hence

$$\mathrm{E}(\mathrm{MS}_C) = ab\sigma_C^2 + b\sigma_{e1}^2 + \sigma_{e2}^2.$$

Next,

$$\begin{aligned}
\mathrm{SS}_B &= ac\sum_{j=1}^{b}(\overline{y}_{.j.} - \overline{y}_{...})^2 \\
&= ac\sum_{j=1}^{b}(\beta_j + \overline{e}_{.j.}^{(2)} - \overline{e}_{...}^{(2)})^2.
\end{aligned}$$

Put $x_j = \overline{e}_{.j.}^{(2)} \sim \mathrm{N}(0, (1/ac)\sigma_{e2}^2)$. Then

$$\mathrm{SS}_B \sim \sigma_{e2}^2\chi_{b-1}^2(\lambda^2),$$

where

$$\lambda^2 = \frac{ac\sum_{j=1}^{b}\beta_j^2}{\sigma_{e2}^2}.$$

Hence

$$\mathrm{E}(\mathrm{MS}_B) = \sigma_{e2}^2 + acQ_B.$$

Finally,

$$\begin{aligned} \mathrm{SS}_A &= bc\sum_{i=1}^{a}(\bar{y}_{i..}-\bar{y}_{...})^2 \\ &= bc\sum_{i=1}^{a}(\alpha_i+\bar{e}_{i.}^{(1)}+\bar{e}_{i..}^{(2)}-\bar{e}_{..}^{(1)}+\bar{e}_{...}^{(2)})^2. \end{aligned}$$

Put $x_i = \bar{e}_{i.}^{(1)} + \bar{e}_{i..}^{(2)} \sim \mathrm{N}(0, (1/c)\sigma_{e1}^2 + (1/bc)\sigma_{e2}^2)$. Then

$$\mathrm{SS}_A \sim bn\left(\frac{1}{c}\sigma_{e1}^2 + \frac{1}{bc}\sigma_{e2}^2\right)\chi_{b-1}^2(\lambda^2),$$

where

$$\lambda^2 = \frac{bc\sum_{i=1}^{a}\alpha_i^2}{\sigma_{e2}^2 + b\sigma_{e1}^2}.$$

Hence

$$\mathrm{E}(\mathrm{MS}_A) = \sigma_{e2}^2 + b\sigma_{e1}^2 + bcQ_A.$$

12.6 CHAPTER SUMMARY

(a) A factor is said to be nested in another factor if different levels of the first factor are selected for each level of the second factor. The main consequence of nesting is that an interaction between the two factors is not estimable or testable.

(b) In a multifactor experiment some factors may be nested while others may be crossed, and the factors may be fixed or random. It is important to specify the relationship between the factors and their nature in order to obtain the correct E(MS) expressions, the tests of hypotheses, and variance component estimates.

(c) In a split-plot design, first the whole-plot treatments are randomly assigned to experimental units (called whole plots). Then each whole plot is divided into subplots to which subplot treatments are randomly assigned; thus whole plots are bigger experimental units than subplots. This restricted randomization scheme is adopted when one factor that is more difficult to change is designated to be the whole-plot factor and the other factor is the subplot factor. Once a particular whole-plot treatment is chosen, observations are made on all subplot treatments. Even though the whole-plot and subplot factors are crossed, the design cannot be analyzed as a crossed two-way layout because all treatment combinations are not randomly assigned to the experimental units.

The split-plot design has a two-tiered error structure with the whole-plot error being generally larger than the subplot error. Therefore the treatment factor of primary interest should be assigned to subplots if practically feasible.

EXERCISES

Section 12.1 (Two-Stage Nested Designs)

Theoretical Exercises

12.1 Show that the E(MS) expressions for the two-stage nested design when both factors A and B are fixed are as follows: $\mathrm{E}(\mathrm{MS}_A) = \sigma_e^2 + bnQ_A$, $\mathrm{E}(\mathrm{MS}_{B(A)}) = \sigma_e^2 + nQ_{B(A)}$ and $\mathrm{E}(\mathrm{MS}_e) = \sigma_e^2$, where

$$Q_A = \frac{\sum_{i=1}^a \alpha_i^2}{a-1} \quad \text{and} \quad Q_{B(A)} = \frac{\sum_{i=1}^a \sum_{j=1}^b \beta_{j(i)}^2}{a(b-1)}.$$

Give the F-statistics for testing $H_{0A} : \alpha_i = 0$ for all i and $H_{0B(A)} : \beta_{j(i)} = 0$ for all i, j.

12.2 Show that the E(MS) expressions for the two-stage nested design when A is fixed and B is random are as follows: $\mathrm{E}(\mathrm{MS}_A) = \sigma_e^2 + n\sigma_{B(A)}^2 + bnQ_A$, $\mathrm{E}(\mathrm{MS}_{B(A)}) = \sigma_e^2 + n\sigma_{B(A)}^2$, and $\mathrm{E}(\mathrm{MS}_e) = \sigma_e^2$. Give the F-statistics for testing $H_{0A} : \alpha_i = 0$ for all i and $H_{0B(A)} : \sigma_{B(A)}^2 = 0$.

Applied Exercises

12.3 Give an example of a two-stage nested design for each of the following cases:

(a) A and B are both fixed.
(b) A and B are both random.
(c) A is fixed and B is random.

Can you have A random and B fixed?

12.4 (From Tamhane and Dunlop, 2000, Exercise 13.30. Reprinted by permission of Pearson Education, Inc.) A strong risk factor for skin cancer in adults is childhood sun exposure. An educational program using a CD-ROM presenting information on sun exposure and skin cancer was developed for children in elementary schools. Three interventions were

Table 12.17 Students' Knowledge of Exposure to Sun and Skin Cancer

Control		CD-ROM		Teacher Presentation	
Class 1	Class 2	Class 3	Class 4	Class 5	Class 6
43.75, 3.75	38.75, 8.75	18.75, 2.50	10.00, 3.75	72.50, 2.50	0.00, 32.50
67.50, 3.75	53.75, 1.25	35.00, 11.25	8.75, 5.00	81.25, 12.50	1.25, 37.50
56.25, 3.75	45.00, 7.50	46.25, 0.00	3.75, 3.75	80.00, 7.50	2.50, 55.50
78.75, 8.75	43.75, 7.50	47.50, 10.00	25.00, 5.00	93.75, 38.75	1.25, 67.50
71.25, 5.00	56.25, 18.75	51.25, 16.25	13.75, 1.25	81.25, 53.75	5.00, 70.00
81.25, 11.25	60.00, 16.25	57.50, 21.25	23.75, 3.75	75.00, 66.25	2.50, 75.00
72.50, 31.25	43.75, 16.25	26.25, 31.25	15.00, 7.50	82.50, 76.25	15.00, 77.50
70.00, 51.25	62.50, 33.75	36.25, 27.50	12.50, 8.75	91.25, 62.50	8.75, 77.50
58.75, 22.50	78.75, 41.25	38.75, 27.50	6.25, 20.00	88.75, 63.75	32.50, 75.00
68.75, 45.00	67.50, 61.25	57.50, 40.00	6.25, 7.50	80.00, 78.75	40.00, 81.25

Source: Tamhane and Dunlop (2000, Exercise 13.30). Reprinted by permission of Pearson Education, Inc.

tested: the CD-ROM intervention, regular teacher-led skin health presentations, and a control which received no skin health information. Six classes of grade school students, with 20 students per class, were randomly selected to test the interventions. Two classes were assigned to each group. The students' knowledge of skin cancer and the relationship of sun exposure to future skin cancer were tested before the intervention and three months after the intervention. The changes in their test scores are given in Table 12.17.

(a) Regard the intervention effect as fixed and the classroom effect as random. Calculate the ANOVA table. Show that there is not a significant difference in types of interventions.

(b) Estimate the appropriate variance components.

Section 12.2 (Three-Stage Nested Designs)

Theoretical Exercises

12.5 Derive the E(MS) expressions for a three-stage nested design given in Table 12.3 when all three factors are random. Indicate how these expressions would change if factor A is fixed and B and C are random.

12.6 Show that the E(MS) expressions for the three-stage nested design when all three factors A, B, and C are fixed are as follows: $\mathrm{E(MS}_A) = \sigma_e^2 + bcnQ_A$, $\mathrm{E(MS}_{B(A)}) = \sigma_e^2 + cnQ_{B(A)}$, $\mathrm{E(MS}_{C(AB)}) = \sigma_e^2 + nQ_{C(AB)}$, and $E(MS_e) = \sigma_e^2$, where

$$Q_{C(AB)} = \frac{\sum_{i=1}^{a} \sum_{j=1}^{b} \sum_{k=1}^{c} \gamma_{k(ij)}^2}{ab(c-1)}.$$

Give the F-statistics for testing $H_{0A}: \alpha_i = 0$ for all i and $H_{0B(A)}: \beta_{j(i)} = 0$ for all i, j and $H_{0C(AB)}: \gamma_{k(ij)} = 0$ for all i, j, k.

12.7 Show that the E(MS) expressions for the three-stage nested design when A and B are fixed and C is random are as follows: $E(MS_A) = \sigma_e^2 + n\sigma_{C(AB)}^2 + bcnQ_A$, $E(MS_{B(A)}) = \sigma_e^2 + n\sigma_{C(AB)}^2 + cnQ_{B(A)}$, $E(MS_{C(AB)}) = \sigma_e^2 + n\sigma_{C(AB)}^2$, and $E(MS_e) = \sigma_e^2$. Give the F-statistics for testing $H_{0A}: \alpha_i = 0$ for all i and $H_{0B(A)}: \beta_{j(i)} = 0$ for all i, j and $H_{0C(AB)}: \sigma_{C(AB)}^2 = 0$.

12.8 Consider a three-stage fully nested design with all three factors random and the same setup as in Section 12.2 but with $n = 1$. Naive estimators of $\sigma_{C(AB)}^2, \sigma_{B(A)}^2$, and σ_A^2 can be obtained as follows. First, to estimate $\sigma_{C(AB)}^2$, pool the sample variances s_{ij}^2 of the c observations in cell (i, j) corresponding to level i of A and level j of B. Thus,

$$\widehat{\sigma}_{C(AB)}^2 = \frac{1}{ab}\sum_{i=1}^{a}\sum_{j=1}^{b} s_{ij}^2 \quad \text{where } s_{ij}^2 = \frac{1}{c-1}\sum_{k=1}^{c}(y_{ijk} - \overline{y}_{ij\cdot})^2.$$

Similarly, let

$$\widehat{\sigma}_{B(A)}^2 = \frac{1}{a}\sum_{i=1}^{a} s_i^2 \quad \text{where } s_i^2 = \frac{1}{b-1}\sum_{j=1}^{b}(\overline{y}_{ij\cdot} - \overline{y}_{i\cdot\cdot})^2$$

and

$$\widehat{\sigma}_A^2 = \frac{1}{a-1}\sum_{i=1}^{a}(\overline{y}_{i\cdot\cdot} - \overline{y}_{\cdot\cdot\cdot})^2.$$

Show that these estimators are positively biased.

Applied Exercises

12.9 Bliss (1967) gave an example in which six labs were compared. A well-mixed can of dried whole eggs was packaged in two samples that were labeled G and H. Two analysts in each laboratory measured the fat content of each sample twice using the same chemical technique. Since the aliquot of the sample tested by each analyst is different, the samples are nested under analysts. The data are shown in Table 12.18. Since all the samples came from a well-mixed can, a significant finding on

Table 12.18 Percentage Fat Content of Dried Whole Eggs

		Sample	
Lab	Analyst	G	H
I	1	0.62, 0.55	0.34, 0.24
	2	0.80, 0.68	0.76, 0.65
II	1	0.30, 0.40	0.33, 0.43
	2	0.39, 0.40	0.29, 0.18
III	1	0.46, 0.38	0.27, 0.37
	2	0.37, 0.42	0.45, 0.54
IV	1	0.18, 0.47	0.53, 0.32
	2	0.40, 0.37	0.31, 0.43
V	1	0.35, 0.39	0.37, 0.33
	2	0.42, 0.36	0.20, 0.41
VI	1	0.37, 0.43	0.28, 0.36
	2	0.18, 0.20	0.26, 0.06

Note: Data give %fat −41.70.
Source: Bliss (1967). Reprinted by permission of The McGraw Hill Companies.

any null hypothesis indicates a lack of consistency in the results for the corresponding factor.

(a) Compute the ANOVA table and determine which factors show a lack of consistency. Use $\alpha = 0.05$.

(b) Estimate the variance components.

12.10 Jensen (2002) gave a case study in which variation in a semiconductor manufacturing process was analyzed. A sample of two wafers was taken from each of 20 lots. Measurements on a certain variable (not mentioned in the article for proprietary reasons) were made at nine locations on each wafer. The lot, wafer, and location are all considered random factors. The data are shown in Table 12.19. Compute the ANOVA table and the variance components. What are the percentage contributions to the total variability due to the three factors? Which factors should one focus on to minimize variability?

Section 12.3 (Crossed and Nested Designs)

Theoretical Exercises

12.11 Derive the E(MS) expressions for the ANOVA in Table 12.5 but assuming that factor B is random. Give the F-statistics for the tests of all effects.

Table 12.19 Wafer Data

		Site Measurement −100								
Lot	Wafer	1	2	3	4	5	6	7	8	9
1	1	81.247	81.280	85.021	80.144	92.570	78.741	84.153	84.353	83.117
1	2	75.267	79.844	81.146	77.338	86.057	74.399	77.372	76.336	81.514
2	1	67.718	69.956	69.990	71.560	73.597	70.791	74.165	77.706	73.530
2	2	70.758	71.292	74.332	72.261	82.649	69.322	73.764	73.029	75.668
3	1	69.054	80.211	75.635	73.230	89.497	69.856	76.436	83.451	82.449
3	2	68.988	67.017	71.493	69.388	84.821	69.489	75.134	72.061	73.397
4	1	63.152	68.664	68.263	67.962	80.989	62.585	68.864	69.165	73.407
4	2	63.486	69.566	71.937	69.265	82.960	65.758	71.987	70.835	67.428
5	1	71.760	67.284	68.186	70.691	82.015	70.424	72.628	64.949	65.480
5	2	66.048	67.150	63.710	67.718	76.036	66.449	66.616	69.422	66.148
6	1	80.077	78.507	74.800	75.601	86.457	81.013	76.938	75.869	76.236
6	2	69.121	67.919	65.580	68.052	74.399	69.155	69.722	65.914	66.649
7	1	81.156	82.025	78.751	85.465	84.396	83.962	87.736	83.929	81.791
7	2	84.764	84.864	84.396	89.139	91.678	85.966	92.947	87.870	86.434
8	1	82.893	80.020	79.820	81.858	90.375	86.701	87.636	79.887	78.785
8	2	76.213	76.446	77.682	73.841	86.233	74.643	80.020	76.747	78.678
9	1	76.537	75.334	72.962	75.000	75.668	72.829	78.841	73.564	68.052
9	2	74.833	77.105	75.902	71.259	81.948	70.825	71.526	76.102	71.493
10	1	75.334	75.100	76.436	75.201	87.794	76.938	76.670	78.741	78.374
10	2	83.384	80.445	78.374	79.042	83.017	81.781	80.445	84.487	77.539
11	1	66.493	74.810	68.397	77.081	71.503	71.904	81.423	82.158	69.532
11	2	60.380	65.257	64.355	68.931	68.697	63.820	72.204	67.294	64.522
12	1	77.706	77.071	76.770	77.539	88.996	79.109	83.117	77.839	78.808
12	2	73.864	74.466	78.474	71.626	90.700	66.048	71.359	76.503	78.641
13	1	61.148	62.050	65.090	60.948	70.467	57.240	62.084	66.893	66.125
13	2	55.002	58.510	58.075	58.343	63.253	56.305	59.011	57.574	61.382
14	1	72.538	76.179	77.048	75.110	80.822	71.670	69.866	73.039	71.670
14	2	70.033	72.271	75.945	72.505	83.962	68.397	71.603	75.645	76.680
15	1	84.253	84.219	87.727	77.873	95.476	77.505	75.167	79.543	87.660
15	2	87.526	83.217	89.430	82.783	100.520	83.451	83.184	83.952	89.965
16	1	66.382	70.858	69.121	72.061	79.810	68.119	72.795	70.858	70.658
16	2	68.486	68.954	69.722	71.626	77.572	72.395	73.998	72.428	71.660
17	1	81.046	81.514	82.750	77.639	93.973	81.414	75.134	78.942	85.088
17	2	73.707	77.071	80.545	75.167	95.243	73.096	72.328	78.775	81.447
18	1	63.309	70.123	67.885	71.626	73.664	68.820	71.560	62.374	66.215
18	2	65.347	64.612	64.612	68.019	73.430	67.585	73.297	59.701	61.006
19	1	73.707	74.175	70.568	79.753	80.154	75.712	83.594	77.415	69.766
19	2	65.357	68.998	66.726	73.975	72.705	67.962	73.507	71.670	67.896
20	1	71.770	69.399	66.359	61.716	76.413	72.004	70.701	68.330	65.825
20	2	62.084	64.154	58.142	61.783	60.013	60.981	62.384	63.720	61.716

Source: Jensen (2002). Reprinted by permission of Taylor & Francis.

Table 12.20 Percentage of Mouse One-Cell Embryos Reaching Development Stage as Measured by Three Analysts Using Four Culture Media

Culture Medium	Mouse	Analyst 1	Analyst 2	Analyst 3
Solution 1 + Tap water	1	46.6	40.6	27.2
	2	27.5	25.0	25.0
	3	48.1	52.3	29.2
	4	31.2	30.3	6.4
Solution 1 + MilliQ water	1	53.8	50.0	37.0
	2	33.3	34.4	20.6
	3	62.5	64.0	21.4
	4	37.5	48.1	16.7
Solution 2 + Tap water	1	34.8	34.6	19.2
	2	16.0	10.7	3.5
	3	37.5	47.8	25.9
	4	42.8	46.8	26.1
Solution 2 + MilliQ water	1	48.3	44.8	41.3
	2	6.1	6.1	7.1
	3	33.3	31.8	28.5
	4	42.8	48.6	32.5

Source: Gorrill et al. (1991, Table 3).

12.12 Derive the E(MS) expressions for the ANOVA in Table 12.5 but assuming that both factors A and B are random. Give the F-statistics for the tests of all effects.

Applied Exercise

12.13 Refer to Example 12.3. Here you will analyze the data for the mouse one-cell embryo experiment which are given in Table 12.20.

(a) Regarding the culture medium as a fixed factor and analyst as a random factor, make an ANOVA table and test the significance of the two factors and their interaction.

(b) Compute the variance components due to all random effects.

(c) In Example 7.1 the main effects of the solution and type of water and their interaction were calculated. Test their significance using appropriate error terms from the ANOVA table.

Section 12.4 (Split-Plot Designs)

Theoretical Exercises

12.14 Consider a split-plot design and a two-stage nested design. Note that both designs have two hierarchically ordered factors. Explain clearly the difference between the two designs.

12.15 A 2^2 experiment is conducted as a split-plot experiment with A as the whole-plot factor and B as the subplot factor. Let $e_i^{(1)} \sim \mathrm{N}(0, \sigma_{e1}^2)$ denote whole-plot errors and $e_{ij}^{(2)} \sim \mathrm{N}(0, \sigma_{e2}^2)$ denote subplot errors, where $i, j = 1, 2$ denote the low and high levels of the two factors, respectively. The model for this design is

$$y_{ij} = \mu + \alpha_i + e_i^{(1)} + \beta_j + (\alpha\beta)_{ij} + e_{ij}^{(2)} \qquad (i, j = 1, 2),$$

where the various quantities have the usual meanings.

(a) Show that for the effect estimates $\widehat{A}$, $\widehat{B}$, and $\widehat{AB}$ [where $\widehat{A} = \frac{1}{2}(y_{22} - y_{12} + y_{21} - y_{11})$, etc.],

$$\mathrm{Var}(\widehat{A}) = 2\sigma_{e1}^2 + \sigma_{e2}^2 \qquad \text{and} \qquad \mathrm{Var}(\widehat{B}) = \mathrm{Var}(\widehat{AB}) = \sigma_{e2}^2.$$

To simplify the derivation, you may assume that all fixed effects [μ, α_i, β_j, and $(\alpha\beta)_{ij}$] are zero.

(b) Show that the observations on each whole-plot treatment are correlated with

$$\mathrm{Corr}(y_{ij}, y_{ik}) = \frac{\sigma_{e1}^2}{\sigma_{e1}^2 + \sigma_{e2}^2}.$$

(c) If a completely randomized design was used in this case, what would be the common variance of all three estimated effects? How does it compare with the variances obtained in part (a)?

12.16 Generalize the results of the previous exercise to a 2^3 experiment in which A and B are whole-plot factors with four whole-plot treatments and C is a subplot factor with two subplot treatments. List the whole-plot effects whose estimates have a common larger variance and subplot effects whose estimates have a common smaller variance.

12.17 Consider the split-plot design of Section 12.4 but suppose that we have $n \geq 2$ i.i.d. observations $y_{ijk\ell}$ $(1 \leq \ell \leq n)$ at each of the abc factor-level combinations of the whole-plot factor A, subplot factor B, and blocking factor C, where A and B are fixed and C is random. In this design, besides the whole-plot error $e_{ik}^{(1)} \sim \mathrm{N}(0, \sigma_{e1}^2)$ and the subplot error $e_{ijk}^{(2)} \sim \mathrm{N}(0, \sigma_{e2}^2)$, we can also estimate the replication error $e_{ijk\ell} \sim \mathrm{N}(0, \sigma_e^2)$. Thus we postulate the following extension of the model (12.8):

$$y_{ijk\ell} = \mu + \alpha_i + \gamma_k + (\alpha\gamma)_{ik} + e_{ik}^{(1)} + \beta_j + (\alpha\beta)_{ij}$$
$$+ e_{ijk}^{(2)} + e_{ijk\ell} \qquad (1 \leq i \leq a, 1 \leq j \leq b, 1 \leq k \leq c, 1 \leq \ell \leq n),$$

Table 12.21 ANOVA Table for Split-Plot Design with Replicate Observations

Source	SS	d.f.	MS	E(MS)	F
A	$SS_A = bcn \sum_i (\overline{y}_{i\ldots} - \overline{y}_{\ldots.})^2$	$a-1$	MS_A	$\sigma_e^2 + n\sigma_{e2}^2 + bn\sigma_{e1}^2 + bcnQ_A$	MS_A/MS_{e1}
C	$SS_C = abn \sum_k (\overline{y}_{\cdot\cdot k\cdot} - \overline{y}_{\ldots.})^2$	$c-1$	MS_C	$\sigma_e^2 + n\sigma_{e2}^2 + bn\sigma_{e1}^2 + abn\sigma_C^2$	MS_C/MS_{e1}
Whole-plot error	$SS_{e1} = bn \sum_{i,k} (\overline{y}_{i\cdot k\cdot} - \overline{y}_{i\ldots} - \overline{y}_{\cdot\cdot k\cdot} + \overline{y}_{\ldots.})^2$	$(a-1)(c-1)$	MS_{e1}	$\sigma_e^2 + n\sigma_{e2}^2 + bn\sigma_{e1}^2$	
B	$SS_B = acn \sum_j (\overline{y}_{\cdot j\cdot\cdot} - \overline{y}_{\ldots.})^2$	$b-1$	MS_B	$\sigma_e^2 + n\sigma_{e2}^2 + acnQ_B$	MS_B/MS_{e2}
AB	$SS_{AB} = cn \sum_{i,j} (\overline{y}_{ij\cdot\cdot} - \overline{y}_{i\ldots} - \overline{y}_{\cdot j\cdot\cdot} + \overline{y}_{\ldots.})^2$	$(a-1)(b-1)$	MS_{AB}	$\sigma_e^2 + n\sigma_{e2}^2 + cnQ_{AB}$	MS_{AB}/MS_{e2}
Subplot error	$SS_{e2} = n \sum_{i,j,k} (\overline{y}_{ijk\cdot} - \overline{y}_{ij\cdot\cdot} - \overline{y}_{i\cdot k\cdot} + \overline{y}_{i\ldots})^2$	$a(b-1)(c-1)$	MS_{e2}	$\sigma_e^2 + n\sigma_{e2}^2$	
Replication error	$SS_e = \sum_{i,j,k,\ell} (y_{ijk\ell} - \overline{y}_{ijk\cdot})^2$	$abc(n-1)$	MS_e	σ_e^2	
Total	$SS_{tot} = \sum_{i,j,k,\ell} (y_{ijk\ell} - \overline{y}_{\ldots.})^2$	$abcn-1$			

where the α_i, β_j, and $(\alpha\beta)_{ij}$ are fixed parameters subject to the usual side constraints and the $\gamma_k \sim N(0, \sigma_C^2)$ and $(\alpha\gamma)_{ik} \sim N(0, \sigma_{AC}^2)$ are random parameters. All random quantities are assumed to be mutually independent. Derive the E(MS) expressions shown in Table 12.21. The F-ratios for the tests of the hypotheses in (12.9) follow immediately from the E(MS) expressions. The quantities Q_A, Q_B, and Q_{AB} are defined in (12.10).

Applied Exercises

12.18 Potcner and Kowalski (2004) describe the following experiment to study how the water resistance of wood depends on two types of wood pretreatment and four types of stain. To conduct this as a CR experiment for each replicate would require eight pieces of wood to be randomly assigned to eight treatment combinations. However, it is more practical to apply a given pretreatment to a single wooden board and then cut it into four pieces to which the four stains are applied at random. This split-plot experiment was conducted with six boards (three replicates per pretreatment). The data are shown in Table 12.22.

(a) Identify the whole-plot and subplot treatments.

(b) Ignoring the split-plot nature, analyze this as a two-factor CR experiment by computing the ANOVA table. Which effects are significant at the 0.05 level?

(c) Now analyze this as a split-plot experiment. Which effects are now significant at the 0.05 level? Explain any discrepancies between the results of the two analyses.

12.19 An experiment was conducted to determine the effect of the date of last cutting of alfalfa on the yield in the following year. Six blocks were used,

Table 12.22 Water Resistance of Wood

		Stain			
Board	Pretreatment	1	2	3	4
1	1	43.0	51.8	40.8	45.5
2	1	57.4	60.9	51.1	55.3
3	1	52.8	59.2	51.7	55.3
4	2	46.6	53.5	35.4	32.5
5	2	52.2	48.3	45.9	44.6
6	2	32.1	34.4	32.2	30.1

Source: Potcner and Kowalski (2004, Table 1). Reprinted by permission of the American Society for Quality.

Table 12.23 Alfalfa Yield Data

Variety	Block 1	Variety	Block 2	Variety	Block 3
A	D_4, 2.23 D_1, 2.17 D_3, 2.29 D_2, 1.58	C	D_3, 1.61 D_4, 1.72 D_2, 1.47 D_1, 1.95	B	D_4, 2.01 D_3, 1.81 D_1, 1.70 D_2, 1.85
C	D_2, 1.52 D_4, 1.56 D_3, 1.55 D_1, 1.75	B	D_2, 1.30 D_1, 2.01 D_4, 1.81 D_3, 1.70	C	D_1, 2.13 D_3, 1.82 D_4, 1.99 D_2, 1.80
B	D_1, 2.33 D_3, 1.86 D_4, 2.27 D_2, 1.38	A	D_1, 1.88 D_3, 1.60 D_2, 1.26 D_4, 2.01	A	D_1, 1.62 D_3, 1.67 D_4, 1.82 D_2, 1.22
Variety	Block 4	Variety	Block 5	Variety	Block 6
C	D_2, 1.37 D_1, 1.78 D_3, 1.56 D_4, 1.55	A	D_1, 1.58 D_4, 1.66 D_3, 1.39 D_2, 1.25	B	D_3, 0.88 D_4, 1.06 D_1, 1.35 D_2, 1.06
A	D_1, 2.34 D_3, 1.91 D_4, 2.10 D_2, 1.59	B	D_2, 1.13 D_3, 1.67 D_4, 1.31 D_1, 1.42	C	D_4, 1.33 D_1, 1.30 D_3, 1.13 D_2, 1.31
B	D_2, 1.09 D_3, 1.54 D_1, 1.78 D_4, 1.40	C	D_4, 1.51 D_1, 1.31 D_2, 1.01 D_3, 1.23	A	D_1, 1.66 D_2, 0.94 D_4, 1.10 D_3, 1.12

Source: http://www.york.ac.uk/depts/maths/teaching/gm/splot.doc.

each of which was divided into three large plots on which three varieties of alfalfa [Ladak (A), Cossack (B), and Ranger (C)] were planted. Each plot was cut on two different occasions on the same dates. For the third and final cut, the plots were divided into four subplots, each of which was assigned at random to one of four cutting dates: D_1, no third cut; D_2, September 1; D_3, September 20; D_4, October 7. The experimental layout and the data are given in Table 12.23. Analyze the data and determine if there are significant differences between the varieties and between the cutting dates on the yield. Use $\alpha = 0.05$.

12.20 Refer to the previous exercise.

(a) Determine using the Tukey procedure which cutting dates differ significantly from each other at $\alpha = 0.05$.

(b) Use the Dunnett procedure from Section 4.3.1 to compare the three cutting dates with no cutting. Is there any discernible pattern on the effect of earlier cutting on the yield in the following year? Use $\alpha = 0.05$.

12.21 Bisgaard et al. (1996) describe an experiment in which security paper was treated with plasma (a highly heated gas in which molecules start breaking up into unstable particles) to make it more susceptible to ink.

Table 12.24 Wettability Data

Process Factors				E	
A	B	C	D	−	+
−	−	−	−	48.6	57.0
+	−	−	−	41.2	38.2
−	+	−	−	55.8	62.9
+	+	−	−	53.5	51.3
−	−	+	−	37.6	43.5
+	−	+	−	47.2	44.8
−	+	+	−	47.2	54.6
+	+	+	−	48.7	44.4
−	−	−	+	5.0	18.1
+	−	−	+	56.8	56.2
−	+	−	+	25.6	33.0
+	+	−	+	41.8	37.8
−	−	+	+	13.3	23.7
+	−	+	+	47.5	43.2
−	+	+	+	11.3	23.9
+	+	+	+	49.5	48.2

Source: Bisgaard et al. (1996, Table 2). Reprinted by permission of Taylor & Francis.

There were four process factors each at two levels: pressure (A), low and high; power (B), low and high; gas flow rate (C), low and high; and type of gas (D), O_2 or $SiCl_4$. Scientists also wanted to test two types of paper (E). Because changing the process factor combinations involved pumping out the reactor to create vacuum, a time-consuming operation, before the next run could be made, both types of paper were placed in the reactor together (their position determined by a flip of a coin). The response variable was "wettability" measured by placing a water droplet on the treated paper and measuring the contact angle between the droplet and the paper surface. The data are shown in Table 12.24.

(a) Explain why A, B, C, D are whole-plot factors while E is a subplot factor.

(b) Calculate the estimates of all effects and divide them into whole-plot effects (those not involving E) and subplot effects (those involving E). Make two separate normal plots of the estimated effects and identify the significant effects in each plot. Which plot has a smaller slope and why?

(c) Make a combined normal plot of all effects. Which effects are now shown to be significant? Why are the results obtained from this plot not trustworthy? Comment on the shape of the plot.

CHAPTER 13

Repeated Measures Experiments

Many experiments are performed to study and compare the temporal effects of different treatments. For example, in a clinical trial to compare different drugs, it is often of interest to evaluate how the effects of the drugs vary with time, that is, when the effects manifest first, how long they last, and when they reach their peaks. Repeated measures (RM) designs, in which the experimental units are measured under different treatment conditions or at different times, are useful to address these study goals. These designs are distinguished by the fact that temporal observations on each experimental unit are autocorrelated. Under certain conditions on the correlation matrix of the observations the data can be analyzed using univariate techniques; otherwise multivariate techniques must be used.

The outline of the chapter is as follows. Statistical analysis of RM designs using the univariate approach is discussed in Section 13.1 and using the multivariate approach in Section 13.2. Some complementary topics and mathematical derivations are given in Section 13.3. Section 13.4 gives a chapter summary.

13.1 UNIVARIATE APPROACH

As described in the introduction to this chapter, in an RM design measurements are made on the same variable for each subject at several points in time or under different treatments (conditions). Because within-subject variability is generally smaller than between-subject variability, an RM design is a good choice for comparing within-subject trends under different treatments. If the order of application of the treatments is randomized for each subject, then, as discussed in Section 5.1.1, under certain conditions, these experiments may be analyzed as RB designs with subjects as blocks.

13.1.1 Model

In its simplest form, an RM design is a CR design in which subjects are whole plots that are allocated to the treatments at random. More generally, the treatments

Statistical Analysis of Designed Experiments: Theory and Applications By Ajit C. Tamhane

may have a factorial structure. The repeated measures are the subplots within each subject. Viewed this way, an RM design is a split-plot design. However, the similarity ends there since the repeated measures are not randomized within subjects.

Consider an RM design in which there are a treatments and n_i subjects are assigned at random to the ith treatment $(1 \leq i \leq a)$. Let $N = \sum_{i=1}^{a} n_i$ be the total number of subjects. Suppose that measurements are made on each subject at $m \geq 2$ fixed time periods or under $m \geq 2$ conditions.

The treatments are referred to as the **between-subjects factor** and the time periods are referred to as the **within-subjects factor**. Let y_{ijk} be the measurement at the kth time period on the jth subject given the ith treatment $(1 \leq i \leq a, 1 \leq j \leq n_i, 1 \leq k \leq m)$. Then the basic model is

$$y_{ijk} = \mu + \alpha_i + \beta_{j(i)} + \gamma_k + (\alpha\gamma)_{ik} + e_{ijk}. \tag{13.1}$$

Here the α_i, γ_k, and $(\alpha\gamma)_{ik}$ are fixed effects subject to $\sum_{i=1}^{a} \alpha_i = 0$, $\sum_{k=1}^{m} \gamma_k = 0$, and $\sum_{i=1}^{a}(\alpha\gamma)_{ik} = 0$ for all k and $\sum_{k=1}^{m}(\alpha\gamma)_{ik} = 0$ for all i. The random effects $\beta_{j(i)}$ are i.i.d. N$(0, \sigma^2_{B(A)})$. Note that the subjects are nested under treatments, so there is no treatment-by-subject interaction in the model. The repeated measurements are correlated within each subject but are independent between subjects. We assume that the observation vectors $\boldsymbol{y}_{ij} = (y_{ij1}, y_{ij2}, \ldots, y_{ijm})'$ are independently distributed as multivariate normal (MVN) with a common unknown covariance matrix $\boldsymbol{\Sigma} = \{\sigma_{k\ell}\}$.

In the univariate approach to RM designs the covariance matrix $\boldsymbol{\Sigma}$ is treated as a scalar matrix of the form $\sigma^2 \boldsymbol{I}$. For this analysis to be valid, the $\boldsymbol{\Sigma}$ matrix must have the so-called spherical form (13.2). On the other hand, the multivariate approach is valid without any condition on $\boldsymbol{\Sigma}$ (except the homoscedasticity condition that $\boldsymbol{\Sigma}$ is the same for all observation vectors). We discuss the univariate ANOVA approach in Section 13.1.2. The multivariate approach is discussed in Section 13.2.

13.1.2 Univariate Analysis of Variance for RM Designs

Denote the diagonal entries of $\boldsymbol{\Sigma}$ by $\sigma_k^2 = \text{Var}(y_{ijk})$ $(1 \leq k \leq m)$ and $\text{Corr}(y_{ijk}, y_{ij\ell}) = \rho_{k\ell} = \sigma_{k\ell}/\sigma_k\sigma_\ell$. Huynh and Feldt (1970) showed that the following is a necessary condition on $\boldsymbol{\Sigma}$ in order for the univariate ANOVA, in particular, the tests of hypotheses concerning the within-subjects effects (i.e., the time and treatment–time effects), to be valid:

Sphericity (Huynh–Feldt) Condition: In its simplest form this condition states that

$$\text{Var}(y_{ijk} - y_{ij\ell}) = \sigma_k^2 + \sigma_\ell^2 - 2\rho_{k\ell}\sigma_k\sigma_\ell = 2\lambda \quad \text{for all } k \neq \ell \tag{13.2}$$

for some $\lambda > 0$. This means that the variance of the difference between any pair of measurements on a given subject is the same. A special case of the sphericity condition is **compound symmetry**:

$$\sigma_k^2 = \sigma^2 \quad \text{for all } k,\ \rho_{k\ell} = \rho \quad \text{for all } k \neq \ell. \tag{13.3}$$

Thus the observations y_{ijk} on the same subject j in any treatment group i over time periods $k = 1, 2, \ldots, m$ have a common variance σ^2 and are equicorrelated with a common correlation coefficient ρ. Sphericity condition (13.2) can be checked by using a test due to Mauchly (1940) discussed in Section 13.1.2.1.

According to model (13.1), the total sum of squares can be decomposed as follows:

$$\begin{aligned}
\underbrace{\sum_{i=1}^{a}\sum_{j=1}^{n_i}\sum_{k=1}^{m}(y_{ijk} - \bar{y}_{\dots})^2}_{\text{SS}_{\text{tot}}} &= \underbrace{m\sum_{i=1}^{a} n_i(\bar{y}_{i\cdot\cdot} - \bar{y}_{\dots})^2}_{\text{SS}_A} \\
&\quad + \underbrace{m\sum_{i=1}^{a}\sum_{j=1}^{n_i}(\bar{y}_{ij\cdot} - \bar{y}_{i\dots})^2}_{\text{SS}_{B(A)}} + \underbrace{N\sum_{k=1}^{m}(\bar{y}_{\cdot\cdot k} - \bar{y}_{\dots})^2}_{\text{SS}_C} \\
&\quad + \underbrace{\sum_{i=1}^{a}\sum_{k=1}^{m} n_i(\bar{y}_{i\cdot k} - \bar{y}_{i\cdot\cdot} - \bar{y}_{\cdot\cdot k} + \bar{y}_{\dots})^2}_{\text{SS}_{AC}} \\
&\quad + \underbrace{\sum_{i=1}^{a}\sum_{j=1}^{n_i}\sum_{k=1}^{m}(y_{ijk} - \bar{y}_{i\cdot k} - \bar{y}_{\cdot jk} + \bar{y}_{\cdot\cdot k})^2}_{\text{SS}_e}.
\end{aligned}$$

The sums of squares have the usual meanings. The corresponding decomposition of the degrees of freedom is

$$\begin{aligned}
\underbrace{mN - 1}_{\text{total d.f.}} = \underbrace{a - 1}_{A \text{ main-effects d.f.}} + \underbrace{N - a}_{B(A) \text{ main-effects d.f.}} + \underbrace{m - 1}_{C \text{ main-effects d.f.}} \\
+ \underbrace{(a-1)(m-1)}_{AC \text{ interaction d.f.}} + \underbrace{(N-a)(m-1)}_{\text{error d.f.}}.
\end{aligned}$$

The hypotheses tested are

$$H_{0A} : Q_A = 0, \qquad H_{0C} : Q_C = 0, \qquad H_{0AC} : Q_{AC} = 0, \tag{13.4}$$

where

$$Q_A = \frac{\sum_{i=1}^{a} n_i \alpha_i^2}{a-1}, \quad Q_C = \frac{N\sum_{k=1}^{m}\gamma_k^2}{m-1}, \quad Q_{AC} = \frac{\sum_{i=1}^{a} n_i \sum_{k=1}^{m}(\alpha\gamma)_{ik}^2}{(a-1)(m-1)}. \tag{13.5}$$

The subject effects hypothesis $H_{0B(A)} : \sigma^2_{B(A)} = 0$ is not testable. Tests of H_{0A}, H_{0C}, and H_{0AC} are shown in the ANOVA in Table 13.1. These tests are based on the E(MS) expressions (also given in the table), which are derived under the compound symmetry assumption (13.3). The derivations are given in Section 13.3.1. It can be seen from the ANOVA table that there is no exact F-test for the subject effects since, under the null hypothesis $H_{0B(A)}$, its E(MS) (which equals $\sigma^2[1 + (m - 1)\rho]$) does not match any other E(MS).

A more detailed analysis can be done of the time and treatment–time effects by partitioning their SS into orthogonal polynomial (linear, quadratic, etc.) components. This can be done by replacing the original data for each subject by their contrast values and doing an appropriate ANOVA for each set of contrast values (linear, quadratic, etc.) separately. One can compute $m - 1$ orthogonal contrasts from m repeated measurements. If the treatments have a factorial or a linear ordered structure (e.g., if the treatments are doses of a drug), then their SS can also be partitioned into appropriate orthogonal components, for example, main effects and interactions. See Example 13.2 and the accompanying ANOVA in Table 13.4 for this detailed analysis.

Example 13.1 (Histamine Levels in Dogs: Univariate ANOVA)

Cole and Grizzle (1966) gave data from a study conducted to investigate the effects of the drugs morphine and trimethaphan on histamine release and hypotension in dogs. Sixteen dogs were divided into four treatment groups of four dogs each. The dogs in the first two groups received intravenous morphine sulfate, while the other dogs received intravenous trimethaphan. In addition, the dogs in the second and fourth groups were inoculated with treatment drugs so that their supplies of available histamine were depleted at the time of inoculation. Thus the treatment had a 2^2 factorial structure. Blood histamine levels were measured at baseline and at intervals of 1, 3, and 5 min after administration of morphine and trimethaphan. The baseline data are not considered in the analysis. A missing value on dog 2 in group II (morphine depleted) at 5 min was replaced by the average of the values at that time for the other three dogs in order to treat this as a balanced design. To account for this estimation of a missing value, one d.f. was deducted from the total and the error. Thus this is an RM design with $a = 4$ treatments, $n_i = 4$ dogs per treatment group, and $m = 3$ repeated measurements on each dog. The data are shown in Table 13.2.

Paralleling the analysis given in Cole and Grizzle (1967), we first log-transform (using natural logs) the blood histamine level data. The resulting values (multiplied by -1 to avoid too many negative data values) are shown in matrix form in Display 13.1. From this data matrix we can compute the matrix of sample means with entries $\overline{y}_{i \cdot k}$. This matrix is shown in Display 13.2.

The ANOVA for the data in Display 13.1 is shown in Table 13.4. We see that group and time main effects and group–time interaction are highly significant. According to Cole and Grizzle (1966), the group and time effects were a priori expected to be significant because the histamine levels follow a time response

Table 13.1 Univariate ANOVA for RM Design Assuming Compound Symmetry

Source	SS	d.f.	MS	E(MS)	F
Treatments (A)	SS_A	$a-1$	MS_A	$\sigma^2[1+(m-1)\rho]+m\sigma^2_{B(A)}+mQ_A$	$MS_A/MS_{B(A)}$
Subjects ($B(A)$)	$SS_{B(A)}$	$N-a$	$MS_{B(A)}$	$\sigma^2[1+(m-1)\rho]+m\sigma^2_{B(A)}$	
Time (C)	SS_C	$m-1$	MS_C	$\sigma^2(1-\rho)+Q_C$	MS_C/MS_e
Interaction (AC)	SS_{AC}	$(a-1)(m-1)$	MS_{AC}	$\sigma^2(1-\rho)+Q_{AC}$	MS_{AC}/MS_e
Error	SS_e	$(N-a)(m-1)$	MS_e	$\sigma^2(1-\rho)$	
Total	SS_{tot}	$mN-1$			

Table 13.2 Blood Histamine Levels (μg/ml) of Dogs

Group	Dog	Time: 1 min	3 min	5 min
	1	0.20	0.10	0.08
Group I	2	0.06	0.02	0.02
(morphine intact)	3	1.40	0.48	0.24
	4	0.57	0.35	0.24
	1	0.09	0.13	0.14
Group II	2	0.11	0.10	0.09[a]
(morphine depleted)	3	0.07	0.07	0.07
	4	0.07	0.06	0.07
	1	0.62	0.31	0.22
Group III	2	1.05	0.73	0.60
(trimethaphan intact)	3	0.83	1.07	0.80
	4	3.13	2.06	1.23
	1	0.09	0.09	0.08
Group IV	2	0.09	0.09	0.10
(trimethaphan depleted)	3	0.10	0.12	0.12
	4	0.05	0.05	0.05

[a] A missing value is replaced by an average.

Source: Cole and Grizzle (1966, Table 2). Reprinted by permission of the International Biometric Society.

$$
\boldsymbol{Y} = \begin{bmatrix}
1.6094 & 2.3026 & 2.5257 \\
2.8134 & 3.9120 & 3.9120 \\
-0.3365 & 0.7340 & 1.4271 \\
0.5621 & 1.0498 & 1.4271 \\
2.4080 & 2.0402 & 1.9661 \\
2.2073 & 2.3026 & 2.4080 \\
2.6593 & 2.6593 & 2.6593 \\
2.6593 & 2.8134 & 2.6593 \\
0.4780 & 1.1712 & 1.5141 \\
-0.0488 & 0.3147 & 0.5108 \\
0.1863 & -0.0677 & 0.2231 \\
-1.1410 & -0.7227 & -0.2070 \\
2.4080 & 2.4080 & 2.5257 \\
2.4080 & 2.4080 & 2.3026 \\
2.3026 & 2.1203 & 2.1203 \\
2.9957 & 2.9957 & 2.9957
\end{bmatrix}
$$

Display 13.1 Log-transformed dog data from Table 13.2 (all values are multiplied by -1).

$$\overline{Y} = \begin{bmatrix} 1.1621 & 1.9996 & 2.3230 \\ 2.4835 & 2.4539 & 2.4232 \\ -0.1314 & 0.1739 & 0.5103 \\ 2.5286 & 2.4830 & 2.4861 \end{bmatrix}$$

Display 13.2 Sample means of log-transformed dog data.

Table 13.3 Univariate ANOVA for Log-Transformed Histamine Data on Dogs

Source	SS	d.f.	MS	*F*	*p*
Group	42.1625	3	14.0542	7.67	0.004
Drug	7.6633	1	7.6633	4.18	0.064
Deplete	25.9340	1	25.9340	14.15	0.003
Drug–Deplete	8.5652	1	8.5652	4.67	0.052
Dog(Group)	21.9982	12	1.8332		
Time	1.4765	2	0.7381	19.65	0.000
Linear	1.4450	1	1.4450	24.79	0.000
Quadratic	0.0315	1	0.0315	1.72	0.217
Group–time	2.2318	6	0.3720	9.90	0.000
Linear	2.0850	3	0.6950	11.92	0.000
Quadratic	0.1468	3	0.0489	2.67	0.109
Error	0.9014	23	0.0392		
Linear	0.7000	12	0.0583		
Quadratic	0.2014	11	0.0183		
Total	68.7704	46			

[a]*Note:* The error term for the group F-statistic is the mean square for dog(group). The error term for the time and group–time F-statistics is the overall MS_e. The error terms for the linear and quadratic components of these effects are the corresponding components of the overall MS_e.

Table 13.4 Group Means of Blood Histamine Levels (μg/ml) of Dogs

	Deplete		
Drug	No	Yes	Row mean
Morphine	1.8282	2.4535	2.1409
Trimethapan	0.1843	2.4992	1.3418
Column mean	1.0063	2.4764	

curve that was known to rise sharply during the first minute and fall off in the next 4 min. Furthermore, it was known that the amount of rise and fall in the curve are much greater for dogs with undepleted reserves of available histamine than those with depleted reserves. What was not predictable was how the type of drug (morphine or trimethapan) would affect the histamine response relationship.

Since the group factor has a 2^2 factorial structure, its SS can be partitioned into three orthogonal components, the drug and deplete main effects and drug–deplete interaction, each with one d.f. (see Chapter 7 for details). We see that the drug main effect and drug–deplete interaction are significant at $\alpha = 0.10$ with p-values equal to 0.064 and 0.052, respectively. Thus the type of drug may be affecting the response curves differently for the depleted versus undepleted histamine groups of dogs.

The linear and quadratic components of the effects involving the time factor are obtained as follows. Compute these contrasts for each dog:

$$L_{ij} = \frac{-y_{ij1} + y_{ij3}}{\sqrt{2}} \quad \text{and} \quad Q_{ij} = \frac{y_{ij1} - 2y_{ij2} + y_{ij3}}{\sqrt{6}} \quad (1 \le i \le 4, 1 \le j \le 4),$$

where $\sqrt{2}$ and $\sqrt{6}$ are normalizing constants. The matrices of these contrasts (with rows as groups and columns as subjects) are as follows:

$$\boldsymbol{L} = \begin{bmatrix} 0.6479 & 0.7768 & 1.2470 & 0.6116 \\ -0.3124 & 0.1419 & 0.0000 & 0.0000 \\ 0.7326 & 0.3957 & 0.0260 & 0.6605 \\ 0.0833 & -0.0745 & -0.1289 & 0.0000 \end{bmatrix},$$

$$\boldsymbol{Q} = \begin{bmatrix} -0.1919 & -0.4485 & -0.1540 & -0.0451 \\ 0.1199 & 0.0041 & 0.0000 & -0.1259 \\ -0.1430 & -0.0683 & 0.2224 & 0.0397 \\ 0.0481 & -0.0430 & 0.0744 & 0.0000 \end{bmatrix}.$$

Using the L_{ij} and Q_{ij} separately as the response variables, run one-way ANOVAs with group (treatment) as the only factor. Then the entries in the two ANOVA tables for the constant term, the group effect, and the error term give the corresponding (linear or quadratic) components of the time effect, group–time interaction, and error, respectively. We deduct one d.f. from the quadratic error term because, if the missing value is not replaced from that dog, then we would have data only at two time points, so we would still be able to estimate the linear effect of time for that dog but not the quadratic effect.

The linear and quadratic components of the time effect can also be obtained directly from the above contrasts as follows. Compute the average linear and quadratic contrasts from L_{ij} and Q_{ij}, respectively, or from the matrix $\overline{\boldsymbol{Y}}$ of the sample means given in Display 13.2:

$$\overline{L} = \frac{\sum_{i=1}^{4}\sum_{j=1}^{4} L_{ij}}{16} = \frac{\sum_{i=1}^{4}(\overline{y}_{i\cdot 3} - \overline{y}_{i\cdot 1})}{4\sqrt{2}} = 0.3005,$$

$$\overline{Q} = \frac{\sum_{i=1}^{4}\sum_{j=1}^{4} Q_{ij}}{16} = \frac{\sum_{i=1}^{4}(\overline{y}_{i\cdot 1} - 2\overline{y}_{i\cdot 2} + \overline{y}_{i\cdot 2})}{4\sqrt{6}} = -0.0444.$$

Comparing these expressions with the corresponding expressions (3.24) and (3.26) for $\widehat{\beta}_k$ and SS_k, we see that the sums of squares for the linear and quadratic effects of the time factor are

$$SS_L = an\overline{L}^2 = 16(0.3005)^2 = 1.4450,$$

$$SS_Q = an\overline{Q}^2 = 16(-0.0444)^2 = 0.0315.$$

Each has one d.f. and their sum equals $SS_{\text{Time}} = 1.4765$ with two d.f.

The following conclusions can be drawn from the ANOVA shown in Table 13.3.

(a) The group effect is highly significant, primarily due to the significant effect of depleting the histamine reserves in treated dogs and to a lesser extent due to the drug–deplete interaction. The corresponding mean responses (means taken over the three observation times and four dogs per group) are as shown in Table 13.4. We see that the mean for the undepleted group is significantly lower, that is, the histamine levels for the undepleted group are significantly higher (because all log-transformed histamine levels are multiplied by -1). The drug–deplete interaction is the result of trimethapan raising the histamine levels much more than morphine for the undepleted group, but for the depleted group there is hardly any difference between the two drugs.

(b) The time effect is mainly due to linear decreases in the histamine levels of groups 1 and 3. The group–time interaction is the result of different linear changes in the histamine levels of the four groups (decreases in groups 1 and 3 and slight increases in groups 2 and 4) as seen in Figure 13.3 ■

13.1.2.1 A Test of the Sphericity Condition and Adjustments to Univariate ANOVA When the Condition Is Not Met

As seen in (13.2), the sphericity condition states that all pairwise differences between the measurements on the same subject, $y_{ijk} - y_{ij\ell}$, have the same variance. This assumption may hold when the within-subjects factor is different conditions, but when that factor is time, the measurements tend to be more positively correlated when they are closer together and less positively correlated when they are farther apart. As a result, the differences between the measurements that are closer tend to have smaller variances than the differences between the measurements that are farther apart. Therefore the sphericity condition does not hold. In that case the within-subjects F-statistics for the time effect and for the treatment–time interaction do not have exact F-distributions. The corresponding F-tests are liberal, that is, they reject a true hypothesis more often than indicated by the nominal α level. We first describe a test to check the sphericity condition and then give adjustments to the d.f. of the F-statistics.

Mauchly Test Let $\boldsymbol{C}$ be an $(m - 1) \times m$ matrix whose rows are normalized orthogonal contrasts. In other words, if $\boldsymbol{c}_i' = (c_{i1}, c_{i2}, \ldots, c_{im})$ is the ith row of

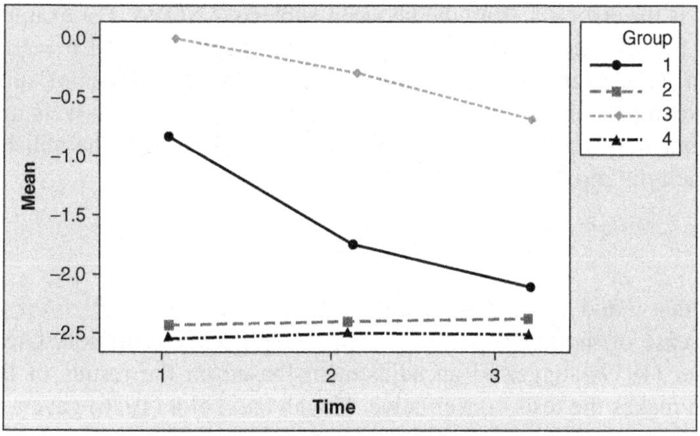

Figure 13.1 Time trends in mean histamine levels of dogs in four groups.

$\boldsymbol{C}$ $(1 \le i \le m-1)$, then $\sum_{k=1}^{m} c_{ik} = 0$, $\sum_{k=1}^{m} c_{ik}^2 = 1$, and $\sum_{k=1}^{m} c_{ik}c_{jk} = 0$ for all row pairs $i \neq j$. Let $\boldsymbol{C}\boldsymbol{y}$ be a vector of $m-1$ normalized orthogonal contrasts among the repeated measures $\boldsymbol{y} = (y_1, y_2, \ldots, y_m)'$. Then the sphericity condition is equivalent to

$$\mathrm{Cov}(\boldsymbol{C}\boldsymbol{y}) = \boldsymbol{C}\boldsymbol{\Sigma}\boldsymbol{C}' = \lambda \boldsymbol{I} \tag{13.6}$$

for some $\lambda\rangle 0$. In other words, the mutually orthogonal contrasts $\boldsymbol{C}\boldsymbol{y}$ are uncorrelated and have the same variance λ. Mauchly (1940) proposed the following test of the null hypothesis that this condition holds: Calculate an unbiased sample estimate of $\boldsymbol{\Sigma}$ provided by the $m \times m$ pooled (over all treatment groups) within-subjects sample covariance matrix $\boldsymbol{S}$ with entries

$$s_{k\ell} = \frac{1}{N-a} \sum_{i=1}^{a} \sum_{j=1}^{n_i} (y_{ijk} - \overline{y}_{i \cdot k})(y_{ij\ell} - \overline{y}_{i \cdot \ell}) \qquad (1 \le k, \ell \le m) \tag{13.7}$$

and with error d.f. $N - a$. Next calculate $\boldsymbol{T} = \boldsymbol{C}\boldsymbol{S}\boldsymbol{C}'$ and

$$W = \frac{(m-1)^{m-1}\det(\boldsymbol{T})}{[\mathrm{tr}(\boldsymbol{T})]^{m-1}}.$$

Note that the value of W is invariant under the choice of $\boldsymbol{C}$ since any other orthogonal matrix of normalized contrasts, say $\boldsymbol{D}$, equals $\boldsymbol{\Gamma}\ \boldsymbol{C}$, where $\boldsymbol{\Gamma}$ is an orthogonal matrix, and $\det(\boldsymbol{T})$ and $\mathrm{tr}(\boldsymbol{T})$ remain unchanged if $\boldsymbol{C}$ is replaced by $\boldsymbol{D}$. The final statistic is $-c \ln W$ where

$$c = \nu - \frac{2m^2 - 3m + 3}{6(m-1)}$$

and ν is the error d.f. from the between-subjects ANOVA. For example, $\nu = N - a$ for the CR design with N subjects and a treatments and $\nu = (a-1)(b-1)$ for an RB design with a treatments and b complete blocks of subjects. It can be shown that under the null hypothesis of sphericity $-c \ln W$ is approximately distributed as χ^2_f where $f = m(m-1)/2 - 1$. Therefore the null hypothesis of sphericity is rejected at level α if

$$-c \ln W = \chi^2 > \chi^2_{f,\alpha}.$$

Note that ν and f are two different d.f.

In case of modest departures from the sphericity condition, Greenhouse and Geisser (1959) suggested an adjustment [based on the results of Box (1954b)] which makes the tests conservative. Huynh and Feldt (1976) gave a modification which provides a less conservative approximation. Denote the elements of $\boldsymbol{T}$ by $\{t_{ij}\}$. Then the **Greenhouse–Geisser (GG) adjustment factor** is

$$\widehat{\varepsilon} = \frac{\left(\sum_{i=1}^{m-1} t_{ii}\right)^2}{(m-1)\sum_{i=1}^{m-1}\sum_{j=1}^{m-1} t_{ij}^2}.$$

Note that the true ε (obtained by putting $\boldsymbol{\Sigma}$ in place of $\boldsymbol{S}$ in the definition of $\boldsymbol{T}$) equals 1 if the sphericity condition is satisfied and is < 1 otherwise. The **Huynh–Feldt (HF) adjustment factor** is

$$\widetilde{\varepsilon} = \frac{(m-1)N\widehat{\varepsilon} - 2}{(m-1)\{\nu - (m-1)\widehat{\varepsilon}\}}.$$

If $\widetilde{\varepsilon} \geq 1$, then it is set equal to 1. The adjustment is used only for the F-tests of the within-subject effects, that is, the time effect and the treatment–time interaction, since the times are not randomly assigned. The adjustment is not required for the F-test on the treatment factor since the treatments are assumed to be randomly assigned to the subjects. Suppose that the F-statistics are based on ν_1 and ν_2 d.f. Then we use $\varepsilon\nu_1$ and $\varepsilon\nu_2$ d.f., where $\varepsilon = \widehat{\varepsilon}$ for the GG adjustment and $\varepsilon = \widetilde{\varepsilon}$ for the HF adjustment. The effect of these adjustments is to reduce ν_1 and ν_2 and thus make the tests less liberal. The use of the HF adjustment factor is recommended since it is less conservative.

Example 13.2 (Histamine Levels in Dogs: Mauchly's Test and Adjustments to Univariate ANOVA)

The pooled covariance and correlation matrices of the log-histamine measurements for $m = 3$ time periods (ignoring the baseline data) are as follows:

$$\boldsymbol{S} = \begin{bmatrix} 0.6228 & 0.6433 & 0.5249 \\ & 0.7417 & 0.6273 \\ & & 0.5438 \end{bmatrix}, \qquad \boldsymbol{R} = \begin{bmatrix} 1 & 0.946 & 0.908 \\ & 1 & 0.989 \\ & & 1 \end{bmatrix}.$$

We choose

$$C = \begin{bmatrix} 0 & -1/\sqrt{2} & 1/\sqrt{2} \\ 2/\sqrt{6} & -1/\sqrt{6} & -1/\sqrt{6} \end{bmatrix}.$$

Then

$$T = CSC' = \begin{bmatrix} 0.0155 & -0.0112 \\ -0.0112 & 0.0598 \end{bmatrix}.$$

Calculate

$$\det(T) = 8.0146 \times 10^{-4} \qquad \text{and} \qquad \text{tr}(T) = 0.0753.$$

Therefore

$$W = \frac{2^2(8.0146 \times 10^{-4})}{(0.0753)^2} = 0.5654.$$

Also, one can readily check that

$$\nu = 16 - 4 = 12, \qquad c = 12 - \frac{2(3)^2 - 3(3) + 3}{6(3-1)} = 11,$$

$$f = \frac{3(3-1)}{2} - 1 = 2.$$

Hence Mauchly's test statistic is

$$\chi^2 = -c \ln W = -11 \ln(0.5654) = 6.2724$$

with a p-value of 0.043 when referred to the χ_2^2-distribution. Therefore the sphericity condition is not met at the 0.05 level of significance.

The GG adjustment factor equals

$$\widehat{\varepsilon} = \frac{(0.0155 + 0.0598)^2}{2[(0.0155)^2 + 2(-0.0112)^2 + (0.0598)^2]} = 0.6971.$$

The HF adjustment factor equals (using $\nu = 12$)

$$\tilde{\varepsilon} = \frac{(2)(16)(0.6971) - 2}{2[12 - (2)(0.6971)]} = 0.9574.$$

The HF adjustment factor gives the following adjusted d.f.:

Time effect: $\nu_1 = (0.9574)(2) = 1.9148$,

$\nu_2 = (0.9574)(24) = 22.9776$,

Treatment–time interaction: $\nu_1 = (0.9574)(6) = 5.7444$,

$\nu_2 = (0.9574)(24) = 22.9776$.

The F-statistics are 19.65 and 9.90 with $p < 0.001$, so the conclusions of the study are unchanged. ■

13.2 MULTIVARIATE APPROACH

Multivariate analysis of variance (MANOVA) is a generalization of univariate ANOVA when the responses are multivariate and are assumed to follow MVN distributions with different mean vectors depending on the treatment groups but a common unknown covariance matrix (the homoscedasticity assumption). We begin by introducing basic MANOVA for one-way layouts as a background for its application to the RM design.

13.2.1 One-Way Multivariate Analysis of Variance

Consider a one-way layout consisting of $a \geq 2$ treatment groups with n_i i.i.d. observations $\boldsymbol{y}_{ij} = (y_{ij1}, \ldots, y_{ijm})' \sim \text{MVN}(\boldsymbol{\mu}_i, \boldsymbol{\Sigma})$ $(1 \leq j \leq n_i)$ on the ith treatment $(1 \leq i \leq a)$. The null hypothesis tested is a generalization of the univariate overall null hypothesis (3.6), namely, $H_0 : \boldsymbol{\mu}_1 = \boldsymbol{\mu}_2 = \cdots = \boldsymbol{\mu}_a$. Denote the sample mean vector for the ith treatment group by $\overline{\boldsymbol{y}}_i = (\overline{y}_{i\cdot 1}, \ldots, \overline{y}_{i\cdot m})'$ $(1 \leq i \leq a)$ and the sample covariance matrix by

$$\boldsymbol{S} = \frac{1}{N-a} \sum_{i=1}^{a} \sum_{j=1}^{n_i} (\boldsymbol{y}_{ij} - \overline{\boldsymbol{y}}_i)(\boldsymbol{y}_{ij} - \overline{\boldsymbol{y}}_i)'$$

with $\nu = N - a = \sum_{i=1}^{a} n_i - a$ d.f. Also denote the overall sample mean vector by $\overline{\boldsymbol{y}} = (\overline{y}_{\cdot\cdot 1}, \ldots, \overline{y}_{\cdot\cdot m})'$. For a two-group $(a = 2)$ comparison, **Hotelling's T^2-test** is used to test $H_0 : \boldsymbol{\mu}_1 = \boldsymbol{\mu}_2$ based on the statistic

$$T^2 = \frac{n_1 n_2}{n_1 + n_2} (\overline{\boldsymbol{y}}_1 - \overline{\boldsymbol{y}}_2)' \boldsymbol{S}^{-1} (\overline{\boldsymbol{y}}_1 - \overline{\boldsymbol{y}}_2).$$

From Lemma A.9, we have the result that under H_0

$$T^2 \sim \frac{(n_1 + n_2 - 2)m}{(n_1 + n_2 - m - 1)} F_{m, n_1 + n_2 - m - 1}.$$

Therefore an α-level test of H_0 rejects if

$$T^2 > \frac{(n_1 + n_2 - 2)m}{(n_1 + n_2 - m - 1)} f_{m, n_1 + n_2 - m - 1, \alpha}.$$

For $a > 2$, the test of H_0 is derived from two matrices,

$$\boldsymbol{H} = \sum_{i=1}^{a} n_i (\overline{\boldsymbol{y}}_i - \overline{\boldsymbol{y}})(\overline{\boldsymbol{y}}_i - \overline{\boldsymbol{y}})', \quad \boldsymbol{E} = (N - a)\boldsymbol{S} = \sum_{i=1}^{a} \sum_{j=1}^{n_i} (\boldsymbol{y}_{ij} - \overline{\boldsymbol{y}}_i)(\boldsymbol{y}_{ij} - \overline{\boldsymbol{y}}_i)',$$

called the **hypothesis matrix** and **error matrix**, respectively (which correspond to the treatment sum of squares, SS_{trt}, and error sum of squares, SS_e, for the univariate one-way ANOVA). The test statistics are functions of the solutions of the characteristic equation

$$|\boldsymbol{H} - \lambda \boldsymbol{E}| = 0,$$

that is, of the eigenvalues λ_i $(1 \le i \le s)$ of the matrix $\boldsymbol{H}\boldsymbol{E}^{-1}$, where $s = \min(a-1, m)$ is the number of nonzero eigenvalues. The four test statistics commonly used in MANOVA are as follows:

Wilks's Lambda (Determinant) Statistic:

$$\Lambda = \frac{\det(\boldsymbol{E})}{\det(\boldsymbol{H}+\boldsymbol{E})} = \prod_{i=1}^{s} \frac{1}{1+\lambda_i}. \tag{13.8}$$

Lawley–Hotelling's Trace Statistic:

$$\operatorname{tr}(\boldsymbol{H}\boldsymbol{E}^{-1}) = \sum_{i=1}^{s} \lambda_i. \tag{13.9}$$

Pillai's Trace Statistic:

$$\operatorname{tr}[\boldsymbol{H}(\boldsymbol{H}+\boldsymbol{E})^{-1}] = \sum_{i=1}^{s} \frac{\lambda_i}{1+\lambda_i}. \tag{13.10}$$

Roy's Largest Root Statistic[1]:

$$\lambda_{\max} = \text{max eigenvalue}[\boldsymbol{H}\boldsymbol{E}^{-1}]. \tag{13.11}$$

Small values of Wilks's Λ indicate rejection of H_0, while large values of the other three statistics do. The tables of these critical constants are not given here but can be found in many multivariate texts (e.g., Timm, 1975). For $a = 2$, we have $s = 1$; thus all four test statistics are functions of a single λ_1 and it can be shown that they all reduce to Hotelling's T^2-test.

13.2.2 Multivariate Analysis of Variance for RM Designs

The MANOVA model postulates that the observation vectors $\boldsymbol{y}_{ij} = (y_{ij1}, y_{ij2}, \ldots, y_{ijm})'$ are independently distributed as

$$\boldsymbol{y}_{ij} \sim \text{MVN}(\boldsymbol{\mu}_i, \boldsymbol{\Sigma}) \qquad (1 \le i \le a, 1 \le j \le n_i), \tag{13.12}$$

[1] In some multivariate texts, Roy's statistic is given as $\theta_{\max} = \lambda_{\max}/(1+\lambda_{\max})$, which is the maximum eigenvalue of $[H(H+E)^{-1}]$. Published tables are generally given for this statistic.

where $\boldsymbol{\mu}_i = (\mu_{i1}, \mu_{i2}, \ldots, \mu_{im})'$. In contrast to the one-way MANOVA model in which the μ_{ik} are completely unstructured, here we assume that they follow the following structural model:

$$\mu_{ik} = \mathrm{E}(y_{ijk}) = \mu + \alpha_i + \gamma_k + (\alpha\gamma)_{ik},$$

which is obtained by putting $\mathrm{E}(\beta_{j(i)}) = 0$ in (13.1). This model is the same as the univariate ANOVA model except that the common covariance matrix $\boldsymbol{\Sigma}$ is an arbitrary symmetric positive-definite matrix.

The following three hypotheses are tested in the MANOVA approach:

No Treatment Effect: The hypothesis tested is that the treatment means averaged over time periods (or conditions) are equal:

$$H_{0A} : \overline{\mu}_{1\cdot} = \overline{\mu}_{2\cdot} = \cdots = \overline{\mu}_{a\cdot}, \tag{13.13}$$

where $\overline{\mu}_{i\cdot} = (1/m)\sum_{k=1}^{m} \mu_{ik}$. Using (13.1), it can be readily checked that this hypothesis is equivalent to testing $\alpha_i = 0$ for all i in the univariate ANOVA. If we define $\overline{y}_{ij\cdot} = (1/m)\sum_{k=1}^{m} y_{ijk}$ as an average score for the jth subject in the ith treatment group, then the $\overline{y}_{ij\cdot}$ are independent $\mathrm{N}(\overline{\mu}_{i\cdot}, \sigma^2)$ r.v.'s, where σ^2 is a common unknown variance that is a function of the elements of $\boldsymbol{\Sigma}$. Thus the ANOVA F-test applied to the data $\{\overline{y}_{ij\cdot}\ (1 \le i \le a, 1 \le j \le n_i)\}$ can be used to test H_{0A}. This F-test is identical to the univariate ANOVA F-test for H_{0A} given in Table 13.1.

No Time Effect: The hypothesis tested is that the time (condition) means averaged over the treatments are equal:

$$H_{0C} : \overline{\mu}_{\cdot 1} = \overline{\mu}_{\cdot 2} = \cdots = \overline{\mu}_{\cdot m}, \tag{13.14}$$

where $\overline{\mu}_{\cdot k} = (1/a)\sum_{i=1}^{a} \mu_{ik}$. Using (13.1), it can be readily checked that this hypothesis is equivalent to testing $\gamma_k = 0$ for all k in the univariate ANOVA. The equality of the $\overline{\mu}_{\cdot k}$ is tested based on their LS estimators $\overline{y}_{\cdot\cdot k} = (1/N)\sum_{i=1}^{a}\sum_{j=1}^{n_i} y_{ijk}$. These estimators are correlated, so a multivariate test must be used. In fact, the random vector $\overline{\boldsymbol{y}} = (\overline{y}_{\cdot\cdot 1}, \overline{y}_{\cdot\cdot 2}, \ldots, \overline{y}_{\cdot\cdot m})'$ is MVN with mean vector $\overline{\boldsymbol{\mu}} = (\overline{\mu}_{\cdot 1}, \overline{\mu}_{\cdot 2}, \ldots, \overline{\mu}_{\cdot m})'$ and covariance matrix $(1/N)\boldsymbol{\Sigma}$. If we define a contrast matrix

$$\boldsymbol{D}_{m\times(m-1)} = \begin{bmatrix} 1 & 0 & \cdots & 0 \\ 0 & 1 & \cdots & 0 \\ \vdots & \vdots & \ddots & \vdots \\ 0 & 0 & \cdots & 1 \\ -1 & -1 & \cdots & -1 \end{bmatrix},$$

then H_{0C} is equivalent to $\boldsymbol{D}'\overline{\boldsymbol{\mu}} = \boldsymbol{0}$, and under this hypothesis,

$$\boldsymbol{D}'\overline{\boldsymbol{y}} \sim \text{MVN}\left(\boldsymbol{0}, \frac{1}{N}\boldsymbol{D}'\boldsymbol{\Sigma}\boldsymbol{D}\right).$$

An unbiased estimator of $\boldsymbol{\Sigma}$ is $\boldsymbol{S}$ with $N - a$ d.f. whose elements are given by (13.7). Setting $\nu \to N - a$ and $m \to m - 1$ in Lemma A.6, we see that

$$T^2 = N\overline{\boldsymbol{y}}'\boldsymbol{D}(\boldsymbol{D}'\boldsymbol{S}\boldsymbol{D})^{-1}\boldsymbol{D}'\overline{\boldsymbol{y}} \sim \frac{(N-a)(m-1)}{(N-a-m+2)}F_{m-1,N-a-m+2}. \quad (13.15)$$

This result is the basis of **Hotelling's T^2-test**, which rejects H_{0C} at level α if

$$F = \frac{(N-a-m+2)}{(N-a)(m-1)}T^2 > F_{m-1,N-a-m+2,\alpha}. \quad (13.16)$$

No Treatment–Time Interaction: The hypothesis tested is that the the mean profiles of the treatments are parallel, that is, $\mu_{ik} - \mu_{i'k} = \mu_{ik'} - \mu_{i'k'}$ for all $i \neq i'$ and $k \neq k'$. In vector notation,

$$H_{0AC}: \begin{bmatrix} \mu_{11} - \mu_{12} \\ \mu_{12} - \mu_{13} \\ \vdots \\ \mu_{1,m-1} - \mu_{1m} \end{bmatrix} = \begin{bmatrix} \mu_{21} - \mu_{22} \\ \mu_{22} - \mu_{23} \\ \vdots \\ \mu_{2,m-1} - \mu_{2m} \end{bmatrix}$$

$$= \cdots = \begin{bmatrix} \mu_{a1} - \mu_{a2} \\ \mu_{a2} - \mu_{a3} \\ \vdots \\ \mu_{a,m-1} - \mu_{am} \end{bmatrix}. \quad (13.17)$$

In terms of the univariate ANOVA model, this hypothesis is equivalent to testing $(\alpha\gamma)_{ik} = 0$ for all i, k.

Let

$$\boldsymbol{M}_{a\times m} = \begin{bmatrix} \boldsymbol{\mu}_1' \\ \boldsymbol{\mu}_2' \\ \vdots \\ \boldsymbol{\mu}_a' \end{bmatrix} = \begin{bmatrix} \mu_{11} & \mu_{12} & \cdots & \mu_{1m} \\ \mu_{21} & \mu_{22} & \cdots & \mu_{2m} \\ \vdots & \vdots & \vdots & \vdots \\ \mu_{a1} & \mu_{a2} & \cdots & \mu_{am} \end{bmatrix}.$$

Then all three hypotheses are of the general form

$$\boldsymbol{CMD} = \boldsymbol{O},$$

where the matrices $\boldsymbol{C}$ and $\boldsymbol{D}$ depend on the particular hypothesis and $\boldsymbol{O}$ is a null matrix of appropriate dimensions. The $\boldsymbol{C}$ and $\boldsymbol{D}$ matrices for the individual hypotheses are as follows:

Hypothesis H_{0A}:

$$\boldsymbol{C}_{(a-1)\times a} = \begin{bmatrix} 1 & 0 & \cdots & 0 & -1 \\ 0 & 1 & \cdots & 0 & -1 \\ \vdots & \vdots & \ddots & \vdots & \vdots \\ 0 & 0 & \cdots & 1 & -1 \end{bmatrix} \quad \text{and} \quad \boldsymbol{D}_{m\times 1} = \begin{bmatrix} 1 \\ 1 \\ \vdots \\ 1 \end{bmatrix}.$$

Hypothesis H_{0C}:

$$\boldsymbol{C}_{1\times a} = [1, 1, \ldots, 1] \quad \text{and} \quad \boldsymbol{D}_{m\times(m-1)} = \begin{bmatrix} 1 & 0 & \cdots & 0 \\ 0 & 1 & \cdots & 0 \\ \vdots & \vdots & \ddots & \vdots \\ 0 & 0 & \cdots & 1 \\ -1 & -1 & \cdots & -1 \end{bmatrix}.$$

Hypothesis H_{0AC}:

$$\boldsymbol{C}_{(a-1)\times a} = \begin{bmatrix} 1 & 0 & \cdots & 0 & -1 \\ 0 & 1 & \cdots & 0 & -1 \\ \vdots & \vdots & \ddots & \vdots & \vdots \\ 0 & 0 & \cdots & 1 & -1 \end{bmatrix} \quad \text{and}$$

$$\boldsymbol{D}_{m\times(m-1)} = \begin{bmatrix} 1 & 0 & \cdots & 0 \\ 0 & 1 & \cdots & 0 \\ \vdots & \vdots & \ddots & \vdots \\ 0 & 0 & \cdots & 1 \\ -1 & -1 & \cdots & -1 \end{bmatrix}.$$

Denote the data matrix by $\boldsymbol{Y}_{N\times m}$, whose rows are the data vectors $\boldsymbol{y}'_{ij}$ $(1 \le i \le a, 1 \le j \le n_i)$. Next denote the matrix of sample means by $\overline{\boldsymbol{Y}}_{a\times m}$, whose rows are the sample mean vectors $\overline{\boldsymbol{y}}'_{i\cdot} = (\overline{y}_{i\cdot 1}, \overline{y}_{i\cdot 2}, \ldots, \overline{y}_{i\cdot m})$ $(1 \le i \le a)$. Finally, let $\boldsymbol{X}$ be the model matrix of the design given by

$$\boldsymbol{X}_{N\times a} = \begin{bmatrix} 1 & 0 & \cdots & 0 \\ \vdots & \vdots & \ddots & \vdots \\ 1 & 0 & \cdots & 0 \\ 0 & 1 & \cdots & 0 \\ \vdots & \vdots & \ddots & \vdots \\ 0 & 1 & \cdots & 0 \\ \vdots & \vdots & \ddots & \vdots \\ 0 & 0 & \cdots & 1 \\ \vdots & \vdots & \ddots & \vdots \\ 0 & 0 & \cdots & 1 \end{bmatrix}. \tag{13.18}$$

Then analogous to the hypothesis and error sum of squares in univariate ANOVA, we have the hypothesis and error sum of squares and cross-product (SSCP) matrices given by

$$\boldsymbol{H} = \boldsymbol{D}'\overline{\boldsymbol{Y}}'\boldsymbol{C}'[\boldsymbol{C}(\boldsymbol{X}'\boldsymbol{X})^{-1}\boldsymbol{C}']^{-1}\boldsymbol{C}\overline{\boldsymbol{Y}}\boldsymbol{D} \quad \text{and} \quad \boldsymbol{E} = \boldsymbol{D}'\boldsymbol{Y}'[\boldsymbol{I} - \boldsymbol{X}(\boldsymbol{X}'\boldsymbol{X})^{-1}\boldsymbol{X}']\boldsymbol{Y}\boldsymbol{D}. \tag{13.19}$$

Note that $(\boldsymbol{X}'\boldsymbol{X})^{-1} = \text{diag}(1/n_1, \ldots, 1/n_a)$ and so the above expressions simplify somewhat.

For H_{0A}, all four tests (13.8)–(13.11) reduce to the univariate ANOVA F-test since $\boldsymbol{D}$ is a column vector of all 1's. For H_{0C}, all four tests reduce to the Hotelling T^2-test (13.16) since $\boldsymbol{C}$ is a row vector of all 1's. In this case, $\boldsymbol{HE}^{-1}$ has only one nonzero eigenvalue, that is, $s = 1$. Only in case of H_{0AC}, must one of these multivariate tests be used. We will focus on Wilks' Λ-test. This test rejects H_{0AC} if

$$\Lambda < c_\alpha(u, \nu_h, \nu_e),$$

where c_α is the lower α critical point of the null distribution of Λ which depends on $u = m - 1$, $\nu_h = a - 1$, and $\nu_e = N - a$. These critical points are tabulated in Table C.10.

Example 13.3 (Histamine Levels in Dogs: Multivariate ANOVA)

The test for the group effect is the same as the univariate ANOVA test and hence is not repeated here. Using SAS, Wilks's Λ is computed to be 0.3197 ($p = 0.002$) for the time effect and 0.1945 ($p = 0.003$) for the group–time interaction. As noted above, for the time effect, Wilks's Λ (as well as the other three MANOVA statistics) are equivalent to Hotelling's T^2-statistic. Although these calculations are always done using statistical software, it would be pedagogically instructive to go through these calculations in detail. All matrix calculations below are done using Matlab.

The data matrix $\boldsymbol{Y}$ and the matrix of sample means are given in Displays 13.1 and 13.2, respectively. The sample covariance matrix $\boldsymbol{S}$ was calculated in Example 13.2. The 16×4 $\boldsymbol{X}$ matrix is as shown in (13.18) with each column containing four 1's and the rest 0's. Thus $\boldsymbol{X}'\boldsymbol{X} = 4\boldsymbol{I}$.

First we consider the test for the time main effect. In this case

$$\boldsymbol{C} = [\,1 \quad 1 \quad 1 \quad 1\,] \quad \text{and} \quad \boldsymbol{D} = \begin{bmatrix} 1 & 0 \\ 0 & 1 \\ -1 & -1 \end{bmatrix}.$$

Also,

$$\overline{\boldsymbol{y}} = \begin{bmatrix} 1.5107 \\ 1.7776 \\ 1.9357 \end{bmatrix} \quad \text{and} \quad \boldsymbol{S} = \begin{bmatrix} 0.6228 & 0.6433 & 0.5249 \\ & 0.7417 & 0.6273 \\ & & 0.5438 \end{bmatrix}.$$

The T^2-statistic for this test can be computed from (13.15) to be 25.5273. Using (13.16) with $N = 16, m = 3$, and $a = 4$, the corresponding F-statistic equals

$$F = \frac{(N-a-m+2)}{(N-a)(m-1)}T^2 = \frac{11}{(12)(2)} \times 25.5273 = 11.700.$$

This is referred to the F-distribution with $m - 1 = 2$ and $N - a - m + 2 = 11$ d.f., which gives a p-value of 0.002.

Next we compute Wilks's Λ. Using (13.19) the hypothesis and error SSCP matrices are computed to be

$$\boldsymbol{H} = \begin{bmatrix} 2.8890 & 1.0744 \\ 1.0744 & 0.3996 \end{bmatrix} \quad \text{and} \quad \boldsymbol{E} = \begin{bmatrix} 1.4001 & 0.4176 \\ 0.4176 & 0.3697 \end{bmatrix}.$$

Hence

$$\boldsymbol{H}\boldsymbol{E}^{-1} = \begin{bmatrix} 1.8045 & 0.8679 \\ 0.6711 & 0.3228 \end{bmatrix}.$$

This matrix has only one nonzero eigenvalue (i.e., $s = 1$), namely, $\lambda_1 = 2.1273$. Hence Wilks's Λ equals

$$\Lambda = \frac{1}{1 + 2.1273} = 0.3197,$$

which can also be computed as

$$\Lambda = \frac{\det(\boldsymbol{E})}{\det(\boldsymbol{H}+\boldsymbol{E})} = \frac{0.3432}{1.0733} = 0.3197.$$

The other three test statistics can also be computed readily. Since all four test statistics are functions of a single λ_1, they are equivalent to each other and have the same p-value. In turn, all four test statistics are equivalent to the T^2-statistic by the relation

$$\lambda_1 = \frac{T^2}{N-a} = \frac{25.5273}{12} = 2.1273.$$

Hence, under H_{0C},

$$F = \left(\frac{N-a-m+2}{m-1}\right)\left(\frac{1-\Lambda}{\Lambda}\right) \sim F_{m-1, N-a-m+2}.$$

Next consider testing H_{0AC}. In this case,

$$\boldsymbol{C} = \begin{bmatrix} 1 & 0 & 0 & -1 \\ 0 & 1 & 0 & -1 \\ 0 & 0 & 1 & -1 \end{bmatrix} \quad \text{and} \quad \boldsymbol{D} = \begin{bmatrix} 1 & 0 \\ 0 & 1 \\ -1 & -1 \end{bmatrix}.$$

The hypothesis and error SSCP matrices are computed using (13.19) to be

$$\boldsymbol{H} = \begin{bmatrix} 4.1699 & 1.2974 \\ 1.2974 & 0.4752 \end{bmatrix} \quad \text{and} \quad \boldsymbol{E} = \begin{bmatrix} 1.4001 & 0.4176 \\ 0.4176 & 0.3697 \end{bmatrix}.$$

This gives

$$\boldsymbol{H}\boldsymbol{E}^{-1} = \begin{bmatrix} 2.9132 & 0.2185 \\ 0.8194 & 0.3596 \end{bmatrix}.$$

Here $\boldsymbol{H}\boldsymbol{E}^{-1}$ has two nonzero eigenvalues ($s = 2$), which are $\lambda_1 = 2.9815$ and $\lambda_2 = 0.2913$. Hence Wilks's Λ equals

$$\Lambda = \left(\frac{1}{1 + 2.9815}\right)\left(\frac{1}{1 + 0.2913}\right) = 0.1945,$$

which can also be computed as

$$\Lambda = \frac{\det(\boldsymbol{E})}{\det(\boldsymbol{H} + \boldsymbol{E})} = \frac{0.3432}{1.7645} = 0.1945.$$

The critical constant $c_{0.05}(2, 3, 12) = 0.348$ from Table C.10, and since $\Lambda = 0.1945 < 0.348$, we conclude that the group–time interaction is significant.

The Λ-statistic can be transformed into an F-statistic in this case as follows. From (13.17) we see that here H_{0AC} is testing the equality of two-dimensional mean vectors, $(\mu_{i1} - \mu_{i2}, \mu_{i2} - \mu_{i3})'$ $(1 \le i \le 4)$, which results in $s = 2$. From Johnson and Wichern (2002, Table 6.3, p. 300) we see that the following exact distributional result holds in this case:

$$F = \left(\frac{N - a - 1}{a - 1}\right)\left(\frac{1 - \sqrt{\Lambda}}{\sqrt{\Lambda}}\right) \sim F_{2(a-1),2(N-a-1)}.$$

Substituting $N = 16, a = 4$, and $\Lambda = 0.1945$ in the above, we get $F = 4.647$, which when referred to the F-distribution with 6 and 22 d.f. yields a p-value of 0.0034. In conclusion, both the time main effect and group–time interaction are shown to be highly significant using multivariate tests. Note that for the test of H_{0AC} there are two nonzero eigenvalues. Therefore the four test statistics are not equivalent and give different p-values when referred to their individual null distributions. ■

13.3 CHAPTER NOTES

13.3.1 Derivations of E(MS) Expressions for Repeated Measures Design Assuming Compound Symmetry

In case of compound symmetry we have $\text{Corr}(e_{ijk}, e_{ij\ell}) = \rho$ for all $k \ne \ell$ and for all i, j. We assume $\rho \ge 0$. In that case we can write

$$e_{ijk} = \sqrt{\rho}\eta_{ij0} + \sqrt{1 - \rho}\eta_{ijk},$$

where the η_{ij0} and η_{ijk} are i.i.d. $N(0, \sigma^2)$ r.v.'s. Substituting this representation and the model (13.1) in the sums-of-squares expressions, we get

$$\begin{aligned} SS_e &= \sum_{i=1}^{a} \sum_{j=1}^{n_i} \sum_{k=1}^{m} (e_{ijk} - \bar{e}_{ij\cdot} - \bar{e}_{i\cdot k} + \bar{e}_{i\cdot\cdot})^2 \\ &= (1-\rho) \sum_{i=1}^{a} \sum_{j=1}^{n_i} \sum_{k=1}^{m} (\eta_{ijk} - \bar{\eta}_{ij\cdot} - \bar{\eta}_{i\cdot k} + \bar{\eta}_{i\cdot\cdot})^2. \end{aligned}$$

It follows that

$$SS_e \sim (1-\rho)\sigma^2 \chi^2_{(N-a)(m-1)}$$

and hence

$$E(MS_e) = \sigma^2(1-\rho).$$

Next,

$$\begin{aligned} SS_{AC} &= \sum_{i=1}^{a} \sum_{k=1}^{m} n_i (\bar{y}_{i\cdot k} - \bar{y}_{i\cdot\cdot} - \bar{y}_{\cdot\cdot k} + \bar{y}_{\cdots})^2 \\ &= \sum_{i=1}^{a} \sum_{k=1}^{m} n_i [(\alpha\gamma)_{ik} + \sqrt{1-\rho}(\bar{\eta}_{i\cdot k} - \bar{\eta}_{i\cdot\cdot} - \bar{\eta}_{\cdot\cdot k} + \bar{\eta}_{\cdots})^2]. \end{aligned}$$

Put $x_{ik} = \bar{\eta}_{i\cdot k} \sim N\left(0, \sigma^2/n_i\right)$. Then

$$\begin{aligned} SS_{AC} &= \sum_{i=1}^{a} \sum_{k=1}^{m} n_i [(\alpha\gamma)_{ik} + \sqrt{1-\rho}(x_{ik} - \bar{x}_{i\cdot} - \bar{x}_{\cdot k} + \bar{x}_{\cdot\cdot})]^2 \\ &\sim \sigma^2(1-\rho)\chi^2_{(a-1)(m-1)}(\lambda^2), \end{aligned}$$

where

$$\lambda^2 = \frac{\sum_{i=1}^{a} \sum_{k=1}^{m} n_i (\alpha\gamma)^2_{ik}}{\sigma^2(1-\rho)}.$$

Hence

$$\begin{aligned} E(MS_{AC}) &= \frac{\sigma^2(1-\rho)}{(a-1)(m-1)} \left[(a-1)(m-1) + \frac{\sum_{i=1}^{a} \sum_{k=1}^{m} n_i (\alpha\gamma)^2_{ik}}{\sigma^2(1-\rho)} \right] \\ &= \sigma^2(1-\rho) + Q_{AC}, \end{aligned}$$

where Q_{AC} is defined in (13.5).

Next,

$$\begin{aligned} SS_C &= N \sum_{k=1}^{m} (\overline{y}_{..k} - \overline{y}_{...})^2 \\ &= N \sum_{k=1}^{m} (\gamma_k + \overline{e}_{..k} - \overline{e}_{...})^2 \\ &= N \sum_{k=1}^{m} [\gamma_k + \sqrt{1-\rho}(\overline{\eta}_{..k} - \overline{\eta}_{...})]^2. \end{aligned}$$

Using similar arguments as before, it follows that

$$SS_C \sim \sigma^2(1-\rho)\chi^2_{m-1}(\lambda^2),$$

where

$$\lambda^2 = \frac{N \sum_{k=1}^{m} \gamma_k^2}{\sigma^2(1-\rho)}.$$

Hence,

$$E(MS_C) = \sigma^2(1-\rho) + Q_C,$$

where Q_C is defined in (13.5).

Next,

$$\begin{aligned} SS_{B(A)} &= m \sum_{i=1}^{a} \sum_{j=1}^{n_i} (\overline{y}_{ij.} - \overline{y}_{i..})^2 \\ &= m \sum_{i=1}^{a} \sum_{j=1}^{n_i} (\beta_{j(i)} + \overline{e}_{ij.} - \overline{\beta}_{.(i)} - \overline{e}_{i..})^2 \\ &= m \sum_{i=1}^{a} \sum_{j=1}^{n_i} \Big(\beta_{j(i)} + \sqrt{\rho}\eta_{ij0} + \sqrt{1-\rho}\overline{\eta}_{ij.} - \overline{\beta}_{\cdot(i)} - \sqrt{\rho}\overline{\eta}_{i\cdot 0} \\ &\qquad + \sqrt{1-\rho}\overline{\eta}_{i..}\Big)^2. \end{aligned}$$

Put

$$x_{ij} = \beta_{j(i)} + \sqrt{\rho}\eta_{ij0} + \sqrt{1-\rho}\overline{\eta}_{ij.} \sim N\left(0, \sigma^2_{B(A)} + \rho\sigma^2 + \frac{(1-\rho)\sigma^2}{m}\right).$$

Then

$$SS_{B(A)} = m \sum_{i=1}^{a} \sum_{j=1}^{n_i} (x_{ij} - \overline{x}_{i.})^2 \sim [m\sigma^2_{B(A)} + m\rho\sigma^2 + (1-\rho)\sigma^2]\chi^2_{N-a}.$$

Hence,

$$\mathrm{E}(\mathrm{MS}_{B(A)}) = m\sigma^2_{B(A)} + [1 + (m-1)\rho]\sigma^2.$$

Finally,

$$\begin{aligned}\mathrm{SS}_A &= m\sum_{i=1}^{a} n_i(\overline{y}_{i\cdot\cdot} - \overline{y}_{\cdot\cdot\cdot})^2\\ &= m\sum_{i=1}^{a} n_i\left(\alpha_i + \overline{\beta}_{\cdot(i)} + \overline{e}_{i\cdot\cdot} - \overline{\beta}_{\cdot(\cdot)} + \overline{e}_{\cdot\cdot\cdot}\right)^2\\ &= m\sum_{i=1}^{a} n_i\Big(\alpha_i + \overline{\beta}_{\cdot(i)} + \sqrt{\rho}\overline{\eta}_{i\cdot 0} + \sqrt{1-\rho}\overline{\eta}_{i\cdot\cdot} - \overline{\beta}_{\cdot(\cdot)} + \sqrt{\rho}\overline{\eta}_{\cdot\cdot 0}\\ &\qquad + \sqrt{1-\rho}\overline{\eta}_{\cdot\cdot\cdot}\Big)^2.\end{aligned}$$

Put

$$x_i = \overline{\beta}_{\cdot(i)} + \sqrt{\rho}\overline{\eta}_{i\cdot 0} + \sqrt{1-\rho}\overline{\eta}_{i\cdot\cdot} \sim \mathrm{N}\left(0, \frac{1}{n_i}\left[\sigma^2_{B(A)} + \rho\sigma^2 + \frac{(1-\rho)\sigma^2}{m}\right]\right),$$

whence it follows that

$$\mathrm{SS}_A = m\sum_{i=1}^{a} n_i(\alpha_i + x_i - \overline{x})^2 \sim \{m\sigma^2_{B(A)} + [1+(m-1)\rho]\sigma^2\}\chi^2_{a-1}(\lambda^2),$$

where

$$\lambda^2 = \frac{m\sum_{i=1}^{a} n_i\alpha_i^2}{m\sigma^2_{B(A)} + [1+(m-1)\rho]\sigma^2}.$$

Hence,

$$\mathrm{E}(\mathrm{MS}_A) = m\sigma^2_{B(A)} + [1+(m-1)\rho]\sigma^2 + Q_A,$$

where Q_A is defined in (13.5).

13.4 CHAPTER SUMMARY

(a) In an RM design the treatments are randomly assigned to experimental units (say, subjects) and then the response of each subject is observed over time for a fixed number of periods. Therefore we have a vector of correlated measurements on each subject and subjects are nested in treatments. If the covariance matrix (assumed to be the same across all treatment groups according to the homoscedasticity assumption) satisfies the so-called sphericity condition, then a univariate ANOVA can be performed on the data by ignoring the correlations. Mauchly's test can be used to check the sphericity assumption. Huynh and Feldt's adjustment to the d.f. of the F-statistics for the within-subject effects approximately adjusts for small deviations from the sphericity condition.

(b) A multivariate ANOVA should be employed if the sphericity condition is not satisfied. The test statistics for no treatment effect, no time effect, and no treatment–time interaction can be expressed in a unified form in terms of hypothesis and error sums of squares and cross-product matrices $\boldsymbol{H}$ and $\boldsymbol{E}$. However, the test statistics simplify to the univariate F-statistic for the no treatment effect hypothesis and Hotelling's T^2-statistic for the no time effect hypothesis. Only for the no treatment–time interaction hypothesis is a MANOVA test required.

EXERCISES

Section 13.1 Univariate Approach

Theoretical Exercise

13.1 Suppose that the common covariance matrix $\boldsymbol{\Sigma}$ is compound symmetric.

(a) Explain why the significance of subject effects, that is, the null hypothesis $H_0 : \sigma^2_{B(A)} = 0$, cannot be tested.

(b) Suggest an estimator of ρ and explain how you can use it to obtain an estimator of $\sigma^2_{B(A)}$ from the expressions for $\text{E}(\text{MS}_e)$ and $\text{E}(\text{MS}_{B(A)})$.

Applied Exercises

13.2 Table 13.5 gives cholesterol data from Hirotsu (1991) on 23 subjects, of whom 12 were treated with a drug and 11 were treated with a placebo. Measurements were taken every 4 weeks over a 24-week period.

(a) Make mean profile plots for the two groups. Comment on the time trends in the two groups and the difference between their trends.

(b) Perform Mauchly's test for sphericity of the covariance matrix. Is the sphericity assumption acceptable at $\alpha = 0.10$?

(c) Perform univariate ANOVA of the data with the Huynh–Feldt adjustment for the F-tests of the within-subjects effects. Which effects are significant at $\alpha = 0.10$?

(d) Summarize your conclusions.

Table 13.5 Cholesterol Measurements

		Week					
Group	Subject	4	8	12	16	20	24
Drug	1	317	280	275	270	274	266
	2	186	189	190	135	197	205
	3	377	395	368	334	338	334
	4	229	258	282	272	264	265
	5	276	310	306	309	300	264
	6	272	250	250	255	228	250
	7	219	210	236	239	242	221
	8	260	245	264	268	317	314
	9	284	256	241	242	243	241
	10	365	304	294	287	311	302
	11	298	321	341	342	357	335
	12	274	245	262	263	235	246
Placebo	1	232	205	244	197	218	233
	2	367	354	358	333	338	355
	3	253	256	247	228	237	235
	4	230	218	245	215	230	207
	5	190	188	212	201	169	179
	6	290	263	291	312	299	279
	7	337	337	383	318	361	341
	8	283	279	277	264	269	271
	9	325	257	288	326	293	275
	10	266	258	253	284	245	263
	11	338	343	307	274	262	309

Source: Hirotsu (1991, Table 2). Reprinted by permission of the Oxford University Press

13.3 The following experiment is described in Kuehl (2000, p. 493). The objective of the experiment was to explore mechanisms for early detection of phlebitis during amiodarone therapy. Three treatments, (i) amiodarone with a vehicle solution, (ii) vehicle solution, and (iii) saline solution, were compared using 15 rabbits in a CR design with 5 rabbits per treatment. The treatments were administered to the rabbits by an interavenous needle inserted in the vein of one ear. An increase of temperature in the treated ear was considered a possible early indicator of phlebitis. The data in Table 13.6 give the difference of temperatures between treated and untreated ears taken every 30 min over 90 min starting at baseline.

(a) Make a profile plot of the mean temperature differences for three groups. Comment on the plot, for example, in which group the temperature differences seem to increase with time and what is the nature of the time effect (linear or nonlinear) in the three treatment groups.

Table 13.6 Difference in Temperatures (°C) Between Treated and Untreated Ears of Rabbits

		Time of Observation (min)			
Treatment	Rabbit	0	30	60	90
Amiodarone	1	−0.3	−0.2	1.2	3.1
	2	−0.5	2.2	3.3	3.7
	3	−1.1	2.4	2.2	2.7
	4	1.0	1.7	2.1	2.5
	5	−0.3	0.8	0.6	0.9
Vehicle	6	−1.1	−2.2	0.2	0.3
	7	−1.4	−0.2	−0.5	−0.1
	8	−0.1	−0.1	−0.5	−0.3
	9	−0.2	0.1	−0.2	0.4
	10	−0.1	−0.2	0.7	−0.3
Saline	11	−1.8	0.2	0.1	0.6
	12	−0.5	0.0	1.0	0.5
	13	−1.0	−0.3	−2.1	0.6
	14	0.4	0.4	−0.7	−0.3
	15	−0.5	0.9	−0.4	−0.3

Source: Kuehl (2000, Table 15.1)

(b) Perform Mauchly's test for sphericity of the covariance matrix and show that it is highly nonsignificant. Perform univariate analysis without the Hyunh–Feldt adjustment.

(c) Calculate the univariate ANOVA table. Are there significant treatment and time differences? What about treatment–time interaction?

(d) Is amiodarone therapy more effective than the vehicle and saline solutions (which may be treated as controls) for early detection of phlebitis?

13.4 Refer to the previous exercise. Give a detailed ANOVA table by partitioning the time effect into its orthogonal linear, quadratic, and cubic components. Apply the same partitioning to the treatment–time interaction. Summarize your conclusions.

13.5 The following experiment is described in Kuehl (2000, p. 503). The objective of the experiment was to study the effect of moisture levels on soil aeration as measured by CO_2 evolution due to microbial activity. Soils fertilized with nutrient sludge tend to have lower aeration, which is essential for active plant root growth and microbial activity. Four soil treatments were compared: control consisting of soil with no sludge added and moisture content of 0.24 kg water/kg soil and three soils to which sludge was added and having moisture contents of 0.24, 0.26, and 0.28 kg water/kg soil. The experiment was conducted as a CR design with 12

Table 13.7 Percent Increase in CO_2 Evolution in Soils Under Different Moisture Conditions

Moisture Content (kg water/kg soil)	Sample No.	Day 1	Day 2	Day 3	Day 4
Control	1	0.22	0.56	0.66	0.89
	2	0.68	0.91	1.06	0.80
	3	0.68	0.45	0.72	0.89
0.24	4	2.53	2.70	2.10	1.50
	5	2.59	1.43	1.35	0.74
	6	0.56	1.37	1.87	1.21
0.26	7	0.22	0.22	0.20	0.11
	8	0.45	0.28	1.24	0.86
	9	0.22	0.33	0.34	0.20
0.28	10	0.22	0.80	0.80	0.37
	11	0.22	0.62	0.89	0.95
	12	0.22	0.56	0.69	0.63

Source: Kuehl (2000, Table 15.6)

samples of soils randomly assigned to the four treatments. The samples were placed in airtight containers and incubated under conditions conducive to microbial activity. Measurements on CO_2 evolution were made on days 2, 4, 6, and 8. The data in Table 13.7 give the percent increase in CO_2 evolution above atmospheric levels.

(a) Make a mean profile plot for the four treated soils. What do you conclude about the time trends in microbial activity in them?

(b) Perform Mauchly's test for sphericity of the covariance matrix and show that it is significant. Perform univariate analysis but with the Hyunh–Feldt adjustment.

(c) Calculate the univariate ANOVA table. Are there significant differences between soils with different moisture levels and between days? What about soil–day interaction? Do the results change if the Hyunh–Feldt adjustment is applied?

(d) Summarize your conclusions.

13.6 Table 13.8 gives dental measurements data from Potthoff and Roy (1964) on 16 boys and 11 girls taken at ages 8, 10, 12, and 14 years.

(a) Make mean profile plots for the two groups. Comment on the time trends in the two groups and the difference between their trends.

(b) Perform Mauchly's test for sphericity on the common covariance matrix. Is the sphericity assumption acceptable at $\alpha = 0.10$?

Table 13.8 Dental Measurements (mm) on 16 Boys and 11 Girls

		Age (years)			
Group	No.	8	10	12	14
Boys	1	26.0	25.0	29.0	31.0
	2	21.5	22.5	23.0	26.5
	3	23.0	22.5	24.0	27.5
	4	25.5	27.5	26.5	27.0
	5	20.0	23.5	22.5	26.0
	6	24.5	25.5	27.0	28.5
	7	22.0	22.0	24.5	26.5
	8	24.0	21.5	24.5	26.5
	9	23.0	20.5	31.0	26.0
	10	27.5	28.0	31.0	31.5
	11	23.0	23.0	23.5	25.0
	12	21.5	23.5	24.0	28.0
	13	17.0	24.5	26.0	29.5
	14	22.5	25.5	25.5	26.0
	15	23.0	24.5	26.0	30.0
	16	22.0	21.5	23.5	25.0
Girls	1	21.0	20.0	21.5	23.0
	2	21.0	21.5	24.0	25.5
	3	20.5	24.0	24.5	26.0
	4	23.5	24.5	25.0	26.5
	5	21.5	23.0	22.5	23.5
	6	20.0	21.0	21.0	22.5
	7	21.5	22.5	23.0	25.0
	8	23.0	23.0	23.5	24.0
	9	20.0	21.0	22.0	21.5
	10	16.5	19.0	19.0	19.5
	11	24.5	25.0	28.0	28.0

Source: Potthoff and Roy (1964, Table 1). Reprinted by permission of Oxford University Press

(c) Perform univariate ANOVA of the data with the Huynh–Feldt adjustment for the F-tests of the within-subjects effects.

(d) Summarize your conclusions.

13.7 Crowder (1995) gave data on the logarithms of times (in seconds) to dissolve pills from four different storage conditions. The data are shown in Table 13.9.

(a) Make mean profile plots for the two groups. Comment on the time trends in the two groups and the difference between their trends.

(b) Perform Mauchly's test for sphericity on the common covariance matrix. Is the sphericity assumption acceptable at $\alpha = 0.10$?

Table 13.9 Logarithms of Times (sec) to Dissolve Pills from Four Storage Conditions

Storage		Fraction Dissolved					
Condition	Pill	0.10	0.30	0.50	0.70	0.75	0.90
1	1	2.56	2.77	2.94	3.14	3.18	3.33
	2	2.64	2.89	3.09	3.26	3.33	3.47
	3	2.94	3.18	3.33	3.50	3.50	3.66
	4	2.56	2.83	3.04	3.22	3.26	3.37
	5	2.64	2.77	2.94	3.14	3.18	3.26
	6	2.56	2.77	2.94	3.14	3.18	3.26
2	1	2.56	2.83	3.07	3.26	3.33	3.51
	2	2.44	2.74	3.02	3.20	3.28	3.44
	3	2.34	2.67	2.91	3.16	3.22	3.39
	4	2.41	2.71	2.94	3.16	3.21	3.36
3	1	2.46	2.83	3.09	3.32	3.37	3.54
	2	2.60	2.93	3.21	3.40	3.46	3.62
	3	2.48	2.84	3.12	3.35	3.41	3.58
	4	2.49	2.82	3.05	3.29	3.37	3.52
4	1	2.40	2.67	2.94	3.20	3.26	3.47
	2	2.64	2.94	3.18	3.40	3.45	3.66
	3	2.40	2.64	2.86	3.09	3.16	3.38

Source: Crowder (1995, Table 1)

(c) Perform univariate ANOVA of the data with the Huynh–Feldt adjustment for the F-tests of the within-subjects effects. Which effects are significant at $\alpha = 0.10$?

(d) Summarize your conclusions.

Section 13.2 Multivariate Approach

Theoretical Exercise

13.8 Simplify the $\boldsymbol{H}$ and $\boldsymbol{E}$ matrices defined in (13.19) for testing the hypotheses H_{0A} and show that the MANOVA test reduces to the univariate ANOVA F-test for H_{0A}.

Applied Exercises

13.9 Refer to Exercise 13.6 Perform multivariate ANOVA of the data and check if the results agree with those obtained from univariate ANOVA. Comment.

13.10 Refer to Exercise 13.3 Perform multivariate ANOVA of the data and check if the results agree with those obtained from univariate ANOVA. Comment.

13.11 Refer to Exercise 13.5 Perform multivariate ANOVA of the data and check if the results agree with those obtained from univariate ANOVA. Comment.

CHAPTER 14

Theory of Linear Models with Fixed Effects

In previous chapters we have used the ANOVA and multiple regression methods to analyze data from designed experiments. The statistical models that underlie these methods are linear models. Since they provide a common basis for analyses of different designs, it is of interest to study their theory. The multiple regression model was introduced in Chapter 2, but the basic results were stated without proofs. The purpose of this chapter is to give some basic theory of linear models. For those interested in a more detailed exposition of the subject, there are many excellent full-length texts which treat it in greater depth (Christensen, 1996; Graybill, 1961; Hocking, 1985; Searle, 1971; Scheffé, 1959; Weber and Skillings, 2000).

The outline of the chapter is as follows. Section 14.1 recaps the basic linear model from Section 2.3 and derives the LS estimators for this model. Section 14.2 discusses the sampling distributions of the LS estimators and statistical inference based on them, in particular, confidence intervals and hypothesis tests on linear functions of the parameters of the model. Section 14.3 studies the power of the ANOVA F-test and sample size determination. Section 14.4 gives some mathematical details and proofs. Section 14.5 gives a summary of the chapter.

14.1 BASIC LINEAR MODEL AND LEAST SQUARES ESTIMATION

We refer the reader to the background material on multiple regression in Section 2.3 and the terminology and notation defined there. We begin with the matrix form of the regression model [see (2.27)]:

$$\boldsymbol{y} = \boldsymbol{X}\boldsymbol{\beta} + \boldsymbol{e}, \tag{14.1}$$

Statistical Analysis of Designed Experiments: Theory and Applications By Ajit C. Tamhane

where the components e_i $(1 \le i \le N)$ of $\boldsymbol{e}$ are i.i.d. $\mathrm{N}(0, \sigma^2)$ r.v.'s or equivalently $\boldsymbol{e}$ has an MVN distribution (see Section A.3 for a discussion of the MVN distribution) with mean vector $\mathbf{0}$ (a vector of zeros, called a null vector) and covariance matrix $\sigma^2\boldsymbol{I}$; here $\boldsymbol{I}$ represents an $N \times N$ identity matrix. We denote this by $\boldsymbol{e} \sim \mathrm{MVN}(\mathbf{0}, \sigma^2\boldsymbol{I})$. It follows that $\boldsymbol{y} \sim \mathrm{MVN}(\boldsymbol{\mu}, \sigma^2\boldsymbol{I})$, where $\boldsymbol{\mu} = \boldsymbol{X\beta} = (\mu_1, \mu_2, \ldots, \mu_N)'$.

The **LS** estimator of $\boldsymbol{\beta}$ minimizes

$$Q = (\boldsymbol{y} - \boldsymbol{X\beta})'(\boldsymbol{y} - \boldsymbol{X\beta}) = \boldsymbol{y}'\boldsymbol{y} - 2\boldsymbol{\beta}'\boldsymbol{X}'\boldsymbol{y} + \boldsymbol{\beta}'\boldsymbol{X}'\boldsymbol{X\beta} = ||\boldsymbol{y} - \boldsymbol{X\beta}||^2, \quad (14.2)$$

which is the squared norm of the vector $\boldsymbol{y} - \boldsymbol{X\beta}$. To minimize Q, we take its vector derivative (see Exercise 14.3) with respect to $\boldsymbol{\beta}$ and set it equal to $\mathbf{0}$, resulting in the equation

$$\frac{dQ}{d\boldsymbol{\beta}} = \begin{bmatrix} \dfrac{\partial Q}{\partial \beta_0} \\ \dfrac{\partial Q}{\partial \beta_1} \\ \vdots \\ \dfrac{\partial Q}{\partial \beta_p} \end{bmatrix} = -2\boldsymbol{X}'\boldsymbol{y} + 2\boldsymbol{X}'\boldsymbol{X\beta} = \mathbf{0}.$$

This gives the so-called normal equations for an LS estimator of $\boldsymbol{\beta}$:

$$(\boldsymbol{X}'\boldsymbol{X})\boldsymbol{\beta} = \boldsymbol{X}'\boldsymbol{y}. \quad (14.3)$$

If $\boldsymbol{X}$ is a full column rank matrix, then $\boldsymbol{X}'\boldsymbol{X}$, which is a symmetric $(p+1)\times(p+1)$ matrix, is invertible and there exists a unique LS estimator $\widehat{\boldsymbol{\beta}}$ of $\boldsymbol{\beta}$ given by

$$\widehat{\boldsymbol{\beta}} = (\boldsymbol{X}'\boldsymbol{X})^{-1}\boldsymbol{X}'\boldsymbol{y}, \quad (14.4)$$

which is the formula (2.29). Two remarks are worth noting about this formula.

(a) $\widehat{\boldsymbol{\beta}}$ is a linear function of the random vector $\boldsymbol{y}$. Since $\boldsymbol{X}$ is a fixed matrix, this greatly simplifies the derivation of the sampling distribution of $\widehat{\boldsymbol{\beta}}$ under the normality assumption, as we shall see in Section 14.2.1.

(b) The LS estimation method can be applied to nonlinear models as well, but in general one does not obtain a closed formula for $\widehat{\boldsymbol{\beta}}$ and iterative methods must be used to compute it. Furthermore, the LS estimator is not a linear function of $\boldsymbol{y}$; as a result, the sampling distribution of $\widehat{\boldsymbol{\beta}}$ is not easy to derive.

For later use, define the fitted values as

$$\widehat{y}_i = \widehat{\mu}_i = \widehat{\beta}_0 + \widehat{\beta}_1 x_{i1} + \cdots + \widehat{\beta}_p x_{ip} = \boldsymbol{x}_i'\widehat{\boldsymbol{\beta}} \qquad (1 \le i \le N),$$

where $\boldsymbol{x}_i' = (1, x_{i1}, \ldots, x_{ip})$ is the ith row of $\boldsymbol{X}$. The vector of fitted values equals

$$\widehat{\boldsymbol{y}} = \widehat{\boldsymbol{\mu}} = \begin{bmatrix} \boldsymbol{x}'_1 \\ \boldsymbol{x}'_2 \\ \vdots \\ \boldsymbol{x}'_N \end{bmatrix} \widehat{\boldsymbol{\beta}} = \boldsymbol{X}\widehat{\boldsymbol{\beta}}. \tag{14.5}$$

Also define the residuals as

$$\widehat{e_i} = y_i - \widehat{y_i} \qquad (1 \leq i \leq N).$$

The vector of residuals equals

$$\widehat{\boldsymbol{e}} = \boldsymbol{y} - \widehat{\boldsymbol{y}}. \tag{14.6}$$

Finally, define

$$\mathrm{SS}_e = \widehat{\boldsymbol{e}}'\widehat{\boldsymbol{e}} = \sum_{i=1}^{N} \widehat{e}_i^2 \tag{14.7}$$

as the error sum of squares (SS_e). An unbiased estimator of σ^2 is obtained by dividing SS_e by the error d.f. $\nu = N - (p+1)$. This ratio is called the mean square error (MS_e); thus

$$\widehat{\sigma}^2 = s^2 = \mathrm{MS}_e = \frac{\mathrm{SS}_e}{N - (p+1)} \tag{14.8}$$

and

$$\mathrm{E}(s^2) = \mathrm{E}(\mathrm{MS}_e) = \sigma^2.$$

For a proof of this fact, see Section 14.2.2.

14.1.1 Geometric Interpretation of Least Squares Estimation

The model (14.1) can be expressed as

$$\mathrm{E}(\boldsymbol{y}) = \boldsymbol{\mu} = \beta_0\boldsymbol{\xi}_0 + \beta_1\boldsymbol{\xi}_1 + \beta_2\boldsymbol{\xi}_2 + \cdots + \beta_p\boldsymbol{\xi}_p, \tag{14.9}$$

where $\boldsymbol{\xi}_j$ is the jth column vector of $\boldsymbol{X}$ for $j = 0, 1, \ldots, p$; note that $\boldsymbol{\xi}_0 = \mathbf{1}$, the vector of all 1's. This model states that $\boldsymbol{\mu}$ is some unknown linear combination of the column vectors of $\boldsymbol{X}$. Thus $\mathrm{E}(\boldsymbol{y})$ lies in the column space of $\boldsymbol{X}$, denoted by $\mathcal{X}$ (the space spanned by the column vectors $\boldsymbol{\xi}_0, \boldsymbol{\xi}_1, \ldots, \boldsymbol{\xi}_p$ of $\boldsymbol{X}$). This is called the **model space** or the **estimation space**. The dimension of the model space $\mathcal{X}$ equals the column rank r of $\boldsymbol{X}$ (called the **model d.f.**). The observation vector $\boldsymbol{y}$ lies in the N-dimensional real space $\mathcal{R}^N$, called the **observation space**. The orthogonal complement of $\mathcal{X}$ in $\mathcal{R}^N$ is called the **error space**, denoted by $\mathcal{E}$. In other words, every vector in $\mathcal{E}$ is orthogonal to every vector in $\mathcal{X}$ and every vector in $\mathcal{R}^N$ can be expressed as the sum of its orthogonal projections on $\mathcal{X}$

and $\mathcal{E}$. The dimension of the error space $\mathcal{E}$ equals $N - r$ (called the **error d.f.**). For the most part, we restrict attention to the case where $r = p + 1$, that is, the columns of $\boldsymbol{X}$ are linearly independent or $\boldsymbol{X}$ is full column rank.

From (14.2) we see that the LS estimators of the model parameters are obtained by finding the vector in the model space that is closest to the observation vector $\boldsymbol{y}$ in terms of the minimum Euclidean distance. This closest vector is the fitted vector $\widehat{\boldsymbol{y}} = \boldsymbol{X}\widehat{\boldsymbol{\beta}}$ defined in (14.5), which is the **projection** of $\boldsymbol{y}$ onto the model space. The fitted vector $\widehat{\boldsymbol{y}}$ is unique because the model space is uniquely defined even in the singular case; see Section 14.1.2. The residual vector $\widehat{\boldsymbol{e}}$ defined in (14.6) is orthogonal to $\widehat{\boldsymbol{y}}$. These relationships are shown in Figure 14.1.

In what follows, projection matrices will be useful. Here is a definition and some of their properties. An $n \times n$ symmetric matrix $\boldsymbol{A}$ is called a **projection matrix** or an **idempotent matrix** if $\boldsymbol{A}$ is symmetric and $\boldsymbol{AA} = \boldsymbol{A}$. A projection matrix $\boldsymbol{A}$ of order n and rank $r \le n$ projects $\mathcal{R}^n$ onto an r-dimensional subspace of $\mathcal{R}^n$. The simplest example of a projection matrix is the identity matrix $\boldsymbol{I}$. The following properties of projection matrices are useful in what follows:

(a) If $\text{rank}(\boldsymbol{A}) = r$, then r of the eigenvalues of $\boldsymbol{A}$ equal 1 and the rest equal 0.

(b) It follows that $\text{rank}(\boldsymbol{A}) = \text{tr}(\boldsymbol{A}) = r$.

(c) Let $\boldsymbol{A}_1, \boldsymbol{A}_2, \ldots, \boldsymbol{A}_k$ be symmetric matrices of order n with ranks $r_i = \text{rank}(\boldsymbol{A}_i)$. If $\boldsymbol{I} = \boldsymbol{A}_1 + \boldsymbol{A}_2 + \cdots + \boldsymbol{A}_k$ and $\sum_{i=1}^{k} r_i = n$, then the following two properties hold:

   (i) $\boldsymbol{A}_1, \boldsymbol{A}_2, \ldots, \boldsymbol{A}_k$ are idempotent.

   (ii) $\boldsymbol{A}_i \boldsymbol{A}_j = \boldsymbol{O}$ for all $i \ne j$.

From (14.5) the fitted vector $\widehat{\boldsymbol{y}}$ can be written as

$$\widehat{\boldsymbol{y}} = \widehat{\boldsymbol{\mu}} = \boldsymbol{X}\widehat{\boldsymbol{\beta}} = \boldsymbol{X}(\boldsymbol{X}'\boldsymbol{X})^{-1}\boldsymbol{X}'\boldsymbol{y} = \boldsymbol{H}\boldsymbol{y}, \tag{14.10}$$

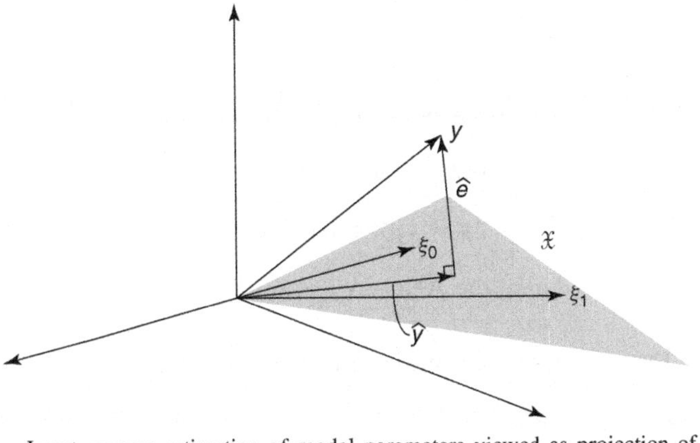

Figure 14.1 Least squares estimation of model parameters viewed as projection of observation vector $\boldsymbol{y}$ onto model space.

where $\boldsymbol{H} = \boldsymbol{X}(\boldsymbol{X}'\boldsymbol{X})^{-1}\boldsymbol{X}'$ is an $N \times N$ projection matrix (see Exercise 14.8) that projects $\boldsymbol{y}$ onto the model space of dimension $p+1$; $\boldsymbol{H}$ is called the **hat matrix**. Similarly, the residual vector $\widehat{\boldsymbol{e}}$ can be written as

$$\widehat{\boldsymbol{e}} = \boldsymbol{y} - \widehat{\boldsymbol{y}} = [\boldsymbol{I} - \boldsymbol{X}(\boldsymbol{X}'\boldsymbol{X})^{-1}\boldsymbol{X}']\boldsymbol{y} = (\boldsymbol{I} - \boldsymbol{H})\boldsymbol{y}, \tag{14.11}$$

where $\boldsymbol{I} - \boldsymbol{H}$ is also an $N \times N$ projection matrix (see Exercise 14.8) that projects $\boldsymbol{y}$ onto the error space of dimension $N - (p+1)$. Since the error space is orthogonal to the column space of $\boldsymbol{X}$, it follows that $\widehat{\boldsymbol{e}}$ is orthogonal to every column of $\boldsymbol{X}$ (see Exercise 14.9 for an algebraic proof of this result), which yields

$$\boldsymbol{\xi}_j'\widehat{\boldsymbol{e}} = \sum_{i=1}^{N} x_{ij}\widehat{e}_i = 0 \qquad (0 \le j \le p). \tag{14.12}$$

In particular,

$$\boldsymbol{\xi}_0'\widehat{\boldsymbol{e}} = \boldsymbol{1}'\widehat{\boldsymbol{e}} = \sum_{i=1}^{N} \widehat{e}_i = 0. \tag{14.13}$$

Thus the residuals sum to zero if the constant term is included in the model.

The projection $\widehat{\boldsymbol{y}}$ can be further decomposed into two mutually orthogonal projections $\overline{\boldsymbol{y}} = \overline{y}\boldsymbol{1}$ (which is an $N \times 1$ vector all of whose entries are $\overline{y}$) and $\widehat{\boldsymbol{y}} - \overline{\boldsymbol{y}}$. Let $\boldsymbol{J}$ be an $N \times N$ matrix all of whose entries are 1. Then it is readily seen that

$$\overline{\boldsymbol{y}} = \left(\frac{1}{N}\right)\boldsymbol{J}\boldsymbol{y} \qquad \text{and} \qquad \widehat{\boldsymbol{y}} - \overline{\boldsymbol{y}} = \left[\boldsymbol{H} - \left(\frac{1}{N}\right)\boldsymbol{J}\right]\boldsymbol{y}. \tag{14.14}$$

Furthermore, $(1/N)\boldsymbol{J}$, $\boldsymbol{H} - (1/N)\boldsymbol{J}$, and $\boldsymbol{I} - \boldsymbol{H}$ are projection matrices of ranks 1, p, and $N - (p+1)$. They are mutually orthogonal, which follows from property (b) above since they sum to $\boldsymbol{I}$ and their ranks sum to N. Thus the observation vector $\boldsymbol{y}$ can be projected onto three mutually orthogonal subspaces of dimensions 1, p, and $N - (p+1)$ as follows:

$$\boldsymbol{y} = \left(\frac{1}{N}\right)\boldsymbol{J}\boldsymbol{y} + \left[\boldsymbol{H} - \left(\frac{1}{N}\right)\boldsymbol{J}\right]\boldsymbol{y} + (\boldsymbol{I} - \boldsymbol{H})\boldsymbol{y} = \overline{\boldsymbol{y}} + (\widehat{\boldsymbol{y}} - \overline{\boldsymbol{y}}) + \widehat{\boldsymbol{e}}. \tag{14.15}$$

Hence, using the Pythagoras theorem, we get

$$||\boldsymbol{y} - \overline{\boldsymbol{y}}||^2 = ||\widehat{\boldsymbol{y}} - \overline{\boldsymbol{y}}||^2 + ||\widehat{\boldsymbol{e}}||^2, \tag{14.16}$$

which is used to derive the ANOVA identity (14.30).

14.1.2 Least Squares Estimation in Singular Case

Typically, in the ANOVA models $\boldsymbol{X}$ is *not* full column rank and hence $\boldsymbol{X}'\boldsymbol{X}$ is not invertible. As a result, the normal equations (14.3) do not have a unique solution. This is called the singular case. Various approaches are possible to deal with this case. One approach is to use a generalized inverse of $\boldsymbol{X}'\boldsymbol{X}$ to obtain a particular solution $\widehat{\boldsymbol{\beta}}$. As seen in the previous section, although $\widehat{\boldsymbol{\beta}}$ is nonunique, $\widehat{\boldsymbol{y}} = \boldsymbol{X}\widehat{\boldsymbol{\beta}}$ is unique.

A second approach is to restrict inferences to the so-called estimable functions: Let $\boldsymbol{c} = (c_0, c_1, \ldots, c_p)'$ be a vector of constants. Then a parametric function $\boldsymbol{c}'\boldsymbol{\beta} = c_0\beta_0 + c_1\beta_1 + \cdots + c_p\beta_p$ is called an **estimable function** if there exists a linear unbiased estimator of $\boldsymbol{c}'\boldsymbol{\beta}$; in other words, there exists a linear function $\sum_{i=1}^{N} \ell_i y_i = \boldsymbol{\ell}'\boldsymbol{y}$ such that $\mathrm{E}(\boldsymbol{\ell}'\boldsymbol{y}) = \boldsymbol{c}'\boldsymbol{\beta}$. For estimable functions, it can be shown that even though $\widehat{\boldsymbol{\beta}}$ is not unique the LS estimator $\boldsymbol{c}'\widehat{\boldsymbol{\beta}} = \boldsymbol{\ell}'\boldsymbol{y}$ is unique.

We will follow a third approach: Suppose that the column rank of $\boldsymbol{X}$ is $r < p+1$ (say). Then one can augment the normal equations $(\boldsymbol{X}'\boldsymbol{X})\boldsymbol{\beta} = \boldsymbol{X}'\boldsymbol{y}$ by a system of $q = (p+1) - r$ additional independent linear constraints on the β_j's, say $\boldsymbol{A}\boldsymbol{\beta} = \boldsymbol{b}$, where $\boldsymbol{A}$ is a $q \times (p+1)$ full column rank matrix and $\boldsymbol{b}$ is a $q \times 1$ vector of constants. This augmented system of equations has a unique solution $\widehat{\boldsymbol{\beta}}$. The constraints follow from the problem under consideration as seen for the one-way layout design in Chapter 3, as explained in the following example.

Example 14.1 (One-Way Layout: LS Estimation)

The basic one-way layout model (3.2) is $\mathrm{E}(y_{ij}) = \mu_i$ $(1 \le i \le a, 1 \le j \le n_i)$ from which it follows that the LS estimators of the μ_i that minimize $Q = \sum_{i=1}^{a}\sum_{j=1}^{n_i}(y_{ij} - \mu_i)^2$ are $\widehat{\mu}_i = \overline{y}_i$ $(1 \le i \le a)$. These estimators can be derived using the matrix formulation of this model by defining $\boldsymbol{X}$ as an $N \times a$ matrix (where $N = \sum n_i$ is the total sample size) in which the ith column consists of an ith block of n_i 1's and other entries equal to 0's. Similarly $\boldsymbol{y}$ is the column in which the ith block of n_i entries consists of observations y_{ij} $(1 \le j \le n_i)$. Finally, $\boldsymbol{\beta} = (\mu_1, \mu_2, \ldots, \mu_a)'$. Then $\boldsymbol{X}'\boldsymbol{X} = \mathrm{diag}\{n_1, n_2, \ldots, n_a\}$ and $\boldsymbol{X}'\boldsymbol{y} = (\sum y_{1j}, \sum y_{2j}, \ldots, \sum y_{aj})'$, which gives $\widehat{\boldsymbol{\beta}} = (\widehat{\mu}_1, \widehat{\mu}_2, \ldots, \widehat{\mu}_a)' = (\overline{y}_1, \overline{y}_2, \ldots, \overline{y}_a)'$.

The reparameterized model (3.3) is $\mathrm{E}(y_{ij}) = \mu + \alpha_i$ $(1 \le i \le a, 1 \le j \le n_i)$. The parameter vector for this model is $\boldsymbol{\beta} = (\mu, \alpha_1, \alpha_2, \ldots, \alpha_a)'$, which adds a column of 1's (due to the constant term, μ) to the $\boldsymbol{X}$ matrix given above, resulting in $a+1$ columns. Since the column of 1's equals the sum of the remaining a columns, they are linearly dependent. Therefore

$$
\boldsymbol{X}'\boldsymbol{X} = \begin{bmatrix} N & n_1 & n_2 & \cdots & n_a \\ n_1 & n_1 & \cdots & 0 & \\ n_2 & 0 & n_2 & \cdots & 0 \\ \vdots & \vdots & \vdots & \ddots & \vdots \\ n_a & 0 & 0 & \cdots & n_a \end{bmatrix}
$$

is noninvertible. To obtain unique estimators of μ and the α_i, we need to add a linear constraint on the $\boldsymbol{\beta}$ vector. If we restrict $\mu = 0$, then we get the basic model with $\alpha_i = \mu_i$. Another possible constraint is to set one of the $\alpha_i = 0$, say $\alpha_1 = 0$. This has the effect of deleting the column of the $\boldsymbol{X}$ matrix corresponding to α_1, making it a full column rank matrix. In this case, $\mu = \mu_1$ and $\alpha_i = \mu_i - \mu_1$ $(2 \le i \le a)$.

The constraint used in Chapter 3 was $\sum n_i \alpha_i = 0$. With this constraint, the normal equations become

$$\begin{array}{lllllllll}
N\mu & + & n_1\alpha_1 & + & n_2\alpha_2 & + \cdots + & n_a\alpha_a & = & \sum\sum y_{ij}, \\
n_1\mu & + & n_1\alpha_1 & & & & & = & \sum y_{1j}, \\
n_2\mu & & & + & n_2\alpha_2 & & & = & \sum y_{2j}, \\
\vdots & & & & & \ddots & & & \vdots \\
n_a\mu & & & & & + & n_a\alpha_a & = & \sum y_{aj}, \\
 & & n_1\alpha_1 & + & n_2\alpha_2 & + \cdots + & n_a\alpha_a & = & 0.
\end{array}$$

Substituting the last equation in the first one, we get $\widehat{\mu} = \sum\sum y_{ij}/N = \overline{\overline{y}}$. Substituting this estimate in the remaining equations, we get $\widehat{\alpha}_i = \overline{y}_i - \overline{\overline{y}}$. Still another possible constraint is $\sum \alpha_i = 0$. The advantage of this constraint is that it is independent of the sample sizes. Under this constraint it can be shown that the unique LS estimators are:

$$\widehat{\mu} = \frac{\sum \overline{y}_i}{N} \quad \text{and} \quad \widehat{\alpha}_i = \overline{y}_i - \frac{\sum \overline{y}_i}{N}.$$

Finally note that, regardless of the constraint used, the LS estimators of the basic parameters μ_i are unchanged, namely, $\widehat{\mu}_i = \widehat{\mu} + \widehat{\alpha}_i = \overline{y}_i$. ■

14.1.3 Least Squares Estimation in Orthogonal Case

Suppose that $\boldsymbol{X}$ is partitioned into $[\boldsymbol{X}_1 \vdots \boldsymbol{X}_2]$ where $\boldsymbol{X}_1$ is an $N \times p_1$ matrix and $\boldsymbol{X}_2$ is an $N \times p_2$ matrix such that $p_1 + p_2 = p + 1$. Further suppose that every column of $\boldsymbol{X}_1$ is orthogonal to every column of $\boldsymbol{X}_2$. Hence $\boldsymbol{X}_1'\boldsymbol{X}_2 = \boldsymbol{O}$, a $p_1 \times p_2$ null matrix.

Let the parameter vector $\boldsymbol{\beta}$ be correspondingly partitioned into $\boldsymbol{\beta}_1 : p_1 \times 1$ and $\boldsymbol{\beta}_2 : p_2 \times 1$. We will show that the LS estimators of $\boldsymbol{\beta}_1$ and $\boldsymbol{\beta}_2$ can be computed independently of each other. Therefore, if the predictor variables corresponding to $\boldsymbol{\beta}_2$ are dropped from the linear model, the LS estimator of $\boldsymbol{\beta}_1$ remains unchanged and vice versa.

Write

$$\begin{aligned}
\widehat{\boldsymbol{\beta}} = \begin{bmatrix} \widehat{\boldsymbol{\beta}}_1 \\ \widehat{\boldsymbol{\beta}}_2 \end{bmatrix} &= (\boldsymbol{X}'\boldsymbol{X})^{-1}\boldsymbol{X}'\boldsymbol{y} \\
&= \left(\begin{bmatrix} \boldsymbol{X}_1' \\ \boldsymbol{X}_2' \end{bmatrix} \begin{bmatrix} \boldsymbol{X}_1 \vdots \boldsymbol{X}_2 \end{bmatrix} \right)^{-1} \begin{bmatrix} \boldsymbol{X}_1' \\ \boldsymbol{X}_2' \end{bmatrix} \boldsymbol{y} \\
&= \begin{bmatrix} \boldsymbol{X}_1'\boldsymbol{X}_1 & \boldsymbol{X}_1'\boldsymbol{X}_2 \\ \boldsymbol{X}_2'\boldsymbol{X}_1 & \boldsymbol{X}_2'\boldsymbol{X}_2 \end{bmatrix}^{-1} \begin{bmatrix} \boldsymbol{X}_1'\boldsymbol{y} \\ \boldsymbol{X}_2'\boldsymbol{y} \end{bmatrix}
\end{aligned}$$

$$= \begin{bmatrix} X_1'X_1 & O \\ O' & X_2'X_2 \end{bmatrix}^{-1} \begin{bmatrix} X_1'y \\ X_2'y \end{bmatrix}$$

$$= \begin{bmatrix} (X_1'X_1)^{-1} & O \\ O' & (X_2'X_2)^{-1} \end{bmatrix} \begin{bmatrix} X_1'y \\ X_2'y \end{bmatrix}$$

$$= \begin{bmatrix} (X_1'X_1)^{-1}X_1'y \\ (X_2'X_2)^{-1}X_2'y \end{bmatrix},$$

which gives $\widehat{\beta}_1 = (X_1'X_1)^{-1}X_1'y$ and $\widehat{\beta}_2 = (X_2'X_2)^{-1}X_2'y$. Thus $\widehat{\beta}_1$ does not depend on X_2 and $\widehat{\beta}_2$ does not depend on X_1.

The above result generalizes naturally to the case where X can be partitioned into $m \geq 2$ mutually orthogonal matrices: $X = [X_1 \vdots X_2 \vdots \cdots \vdots X_m]$ such that $X_i'X_j = O$ for all $i \neq j$. In particular, if all columns of X are mutually orthogonal, then the LS estimators of the β_j are independent of each other (both in a numerical sense, as shown here, and in a statistical sense, as will be shown in Section 14.2). The $X'X$ matrix is diagonal in that case. Therefore computation of the LS estimators is very simple. The idea of orthogonal X_i matrices underlies orthogonal designs studied in earlier chapters.

14.2 CONFIDENCE INTERVALS AND HYPOTHESIS TESTS

Inferences on the β_j's are based on the sampling distributions of $\widehat{\beta}$ and s^2. In the following sections we derive these distributions.

14.2.1 Sampling Distribution of $\widehat{\beta}$

The sampling distribution of $\widehat{\beta}$ is MVN since $\widehat{\beta}$ is a linear transform of y, which is MVN distributed. Next we find its mean vector $\text{E}(\widehat{\beta})$ and the covariance matrix $\text{Cov}(\widehat{\beta})$. Using (A.4) we obtain

$$\begin{aligned} \text{E}(\widehat{\beta}) &= \text{E}[(X'X)^{-1}X'y] \\ &= (X'X)^{-1}X'\text{E}(y) \\ &= (X'X)^{-1}X'(X\beta) \\ &= \beta. \end{aligned} \tag{14.17}$$

Thus $\widehat{\beta}$ is an unbiased estimator of β. Next, using the sandwich formula (A.5), we obtain

$$\begin{aligned} \text{Cov}(\widehat{\beta}) &= \text{Cov}[(X'X)^{-1}X'y] \\ &= (X'X)^{-1}X'\text{Cov}(y)X(X'X)^{-1} \\ &= (X'X)^{-1}X'(\sigma^2 I)X(X'X)^{-1} \\ &= \sigma^2(X'X)^{-1}. \end{aligned} \tag{14.18}$$

Denote $\boldsymbol{V} = (\boldsymbol{X}'\boldsymbol{X})^{-1}$ and let v_{jk} be the (j, k)th element of $\boldsymbol{V}$. The LS estimators $\widehat{\beta}_j$ are distributed as $\mathrm{N}(\beta_j, \sigma^2 v_{jj})$ and $\mathrm{Cov}(\widehat{\beta}_j, \widehat{\beta}_k) = \sigma^2 v_{jk}$. In Section 14.1.3 we saw that if $\boldsymbol{X}$ is partitioned into mutually orthogonal matrices $\boldsymbol{X}_1$ and $\boldsymbol{X}_2$, then $\boldsymbol{X}'\boldsymbol{X}$ and hence $\boldsymbol{V} = (\boldsymbol{X}'\boldsymbol{X})^{-1}$ are block diagonal. This means that $\mathrm{Cov}(\widehat{\beta}_j, \widehat{\beta}_k) = 0$ for every pair of $\widehat{\beta}_j$ from $\widehat{\boldsymbol{\beta}}_1$ and $\widehat{\beta}_k$ from $\widehat{\boldsymbol{\beta}}_2$. Thus $\widehat{\boldsymbol{\beta}}_1$ and $\widehat{\boldsymbol{\beta}}_2$ are uncorrelated and under the normality assumption are independent.

The above results imply that if $\theta = \boldsymbol{c}'\boldsymbol{\beta}$ is any linear parametric function, then its LS estimator $\widehat{\theta} = \boldsymbol{c}'\widehat{\boldsymbol{\beta}}$ is normally distributed with

$$\mathrm{E}(\widehat{\theta}) = \boldsymbol{c}'\mathrm{E}(\widehat{\boldsymbol{\beta}}) = \boldsymbol{c}'\boldsymbol{\beta} = \theta \qquad \text{and} \qquad \mathrm{Var}(\widehat{\theta}) = \sigma^2 \boldsymbol{c}'\boldsymbol{V}\boldsymbol{c}. \tag{14.19}$$

One might ask if the LS estimators are optimal in some sense. An affirmative answer is provided by the **Gauss–Markov theorem** stated below.

Theorem 14.1 *Consider the linear model (14.1) and assume that the e_i are independent with $\mathrm{E}(e_i) = 0$ and $\mathrm{Var}(e_i) = \sigma^2$. (Note that the e_i are not assumed to be normally distributed.) Then, among all linear unbiased estimators (i.e., unbiased estimators that are linear functions of y_i's) of any linear parametric function $\theta = \boldsymbol{c}'\boldsymbol{\beta}$, the LS estimator $\widehat{\theta} = \boldsymbol{c}'\widehat{\boldsymbol{\beta}}$ has the smallest variance. Thus the LS estimator is the* ***best linear unbiased estimator (BLUE).***

Proof: See Section 14.4.1. ■

14.2.2 Sampling Distribution of s^2

From Lemma A.2 we have

$$\mathrm{SS}_e = \widehat{\boldsymbol{e}}'\widehat{\boldsymbol{e}} = \boldsymbol{y}'(\boldsymbol{I} - \boldsymbol{H})\boldsymbol{y} \sim \sigma^2 \chi^2_\nu(\lambda^2),$$

where $\chi^2_\nu(\lambda^2)$ is a noncentral chi-square r.v. with $\nu = \mathrm{rank}(\boldsymbol{I} - \boldsymbol{H}) = N - (p + 1)$ d.f. (see Exercise 14.10) and the n.c.p.

$$\lambda^2 = \frac{\boldsymbol{\mu}'(\boldsymbol{I} - \boldsymbol{H})\boldsymbol{\mu}}{\sigma^2} = \boldsymbol{\beta}'\boldsymbol{X}'(\boldsymbol{I} - \boldsymbol{X}(\boldsymbol{X}'\boldsymbol{X})^{-1}\boldsymbol{X}')\boldsymbol{X}\boldsymbol{\beta}\sigma^2 = 0.$$

Therefore SS_e/σ^2 has a central chi-square distribution with $\nu = N - (p + 1)$ d.f. Hence

$$\frac{\nu s^2}{\sigma^2} = \frac{\mathrm{SS}_e}{\sigma^2} \sim \chi^2_\nu. \tag{14.20}$$

Exercise 14.23 shows that $\widehat{\boldsymbol{\beta}}$ and $\widehat{\boldsymbol{e}}$ (and hence s^2) are independently distributed.

From (14.20) it follows that

$$\mathrm{E}(s^2) = \frac{\sigma^2 \mathrm{E}(\chi^2_\nu)}{\nu} = \sigma^2$$

since $E(\chi_\nu^2) = \nu$. In fact, s^2 can be shown to be an unbiased estimator of σ^2, even without using the above chi-square distribution result (which requires the normality assumption) as follows:

$$
\begin{aligned}
E(s^2) &= \frac{1}{N-(p+1)} E(\widehat{e'e}) \\
&= \frac{1}{N-(p+1)} E[\boldsymbol{y}'(\boldsymbol{I}-\boldsymbol{H})\boldsymbol{y}] \\
&= \frac{\sigma^2}{N-(p+1)} [\lambda^2 + \text{tr}(\boldsymbol{I}-\boldsymbol{H})] \quad \text{(using Lemma A.1)} \\
&= \frac{\sigma^2(N-(p+1))}{N-(p+1)} \quad [\text{since } \lambda^2 = 0 \text{ and } \text{tr}(\boldsymbol{I}-\boldsymbol{H}) = N-(p+1)] \\
&= \sigma^2.
\end{aligned}
$$

14.2.3 Inferences on Scalar Parameters

Using the results from the previous section it follows that

$$\frac{\widehat{\beta}_j - \beta_j}{s\sqrt{v_{jj}}}$$

has a Student's t-distribution with $\nu = N - (p+1)$ d.f. Let $t_{\nu,\alpha/2}$ denote the upper $\alpha/2$ critical point of this distribution. Then a $100(1-\alpha)\%$ CI on β_j is given by

$$\beta_j \in \left[\widehat{\beta}_j \pm t_{\nu,\alpha/2} s \sqrt{v_{jj}}\right]. \tag{14.21}$$

A $100(1-\alpha)\%$ CI on any linear parametric function θ is given by

$$\theta \in \left[\widehat{\theta} \pm t_{\nu,\alpha/2} s \sqrt{\boldsymbol{c}'\boldsymbol{V}\boldsymbol{c}}\right].$$

The t-statistic for testing the hypothesis $H_0 : \theta = \theta_0$, where θ_0 is a specified constant, is

$$t = \frac{\widehat{\theta} - \theta_0}{s\sqrt{\mathbf{c}'\mathbf{V}\mathbf{c}}}.$$

14.2.4 Inferences on Vector Parameters

Often it is of interest to make simultaneous inferences on $q \geq 2$ linear parametric functions:

$$\theta_1 = \boldsymbol{c}_1'\boldsymbol{\beta}, \theta_2 = \boldsymbol{c}_2'\boldsymbol{\beta}, \ldots, \theta_q = \boldsymbol{c}_q'\boldsymbol{\beta}.$$

For instance, the overall null hypothesis $H_0 : \mu_1 = \cdots = \mu_a$ in a one-way layout setting is equivalent to testing $a - 1$ hypotheses $H_{0i} : \theta_i = \mu_i - \mu_a = 0$ $(1 \le i \le a - 1)$ simultaneously.

Let $\boldsymbol{C}$ be a $q \times (p+1)$ matrix whose rows are $\boldsymbol{c}'_1, \boldsymbol{c}'_2, \ldots, \boldsymbol{c}'_q$ and $\boldsymbol{\theta} = (\theta_1, \ldots, \theta_q)' = \boldsymbol{C\beta}$. Assume that the rows of $\boldsymbol{C}$ are linearly independent (so the θ_j are linearly independent) and $q \le p + 1$. A confidence region for $\boldsymbol{\theta}$ is derived from the following distributional result [which generalizes the result in (14.19)]: $\widehat{\boldsymbol{\theta}} = \boldsymbol{C}\widehat{\boldsymbol{\beta}}$ has an MVN distribution with

$$\mathrm{E}(\widehat{\boldsymbol{\theta}}) = \boldsymbol{\theta} \qquad \text{and} \qquad \mathrm{Cov}(\widehat{\boldsymbol{\theta}}) = \sigma^2 \boldsymbol{CVC}'. \tag{14.22}$$

Since $\boldsymbol{C}$ is a full-row rank matrix, $\boldsymbol{CVC}'$ is invertible. Using Lemma A.5 it follows that

$$\frac{(\widehat{\boldsymbol{\theta}} - \boldsymbol{\theta})'(\boldsymbol{CVC}')^{-1}(\widehat{\boldsymbol{\theta}} - \boldsymbol{\theta})}{\sigma^2} \sim \chi^2_q.$$

Furthermore, this r.v. is independent of $s^2 \sim \sigma^2\chi^2_\nu/\nu$. Combining these two results, we obtain that

$$\begin{aligned} F &= \frac{\{(\widehat{\boldsymbol{\theta}} - \boldsymbol{\theta})'(\boldsymbol{CVC}')^{-1}(\widehat{\boldsymbol{\theta}} - \boldsymbol{\theta})\}/q}{s^2} \\ &= \frac{(\widehat{\boldsymbol{\theta}} - \boldsymbol{\theta})'(\boldsymbol{CVC}')^{-1}(\widehat{\boldsymbol{\theta}} - \boldsymbol{\theta})/q\sigma^2}{s^2/\sigma^2} \\ &\sim \frac{\chi^2_q/q}{\chi^2_\nu/\nu} \\ &\sim F_{q,\nu}. \end{aligned}$$

Hence a $100(1 - \alpha)\%$ simultaneous confidence region for $\boldsymbol{\theta}$ is given by

$$\left\{\boldsymbol{\theta} : (\widehat{\boldsymbol{\theta}} - \boldsymbol{\theta})'(\boldsymbol{CVC}')^{-1}(\widehat{\boldsymbol{\theta}} - \boldsymbol{\theta}) \le q s^2 f_{q,\nu,\alpha}\right\}. \tag{14.23}$$

This region is a q-dimensional ellipsoid; hence it is called a **confidence ellipsoid**.

Consider the following hypothesis testing problem:

$$H_0 : \boldsymbol{\theta} = \boldsymbol{\theta}_0 \quad \text{vs.} \quad H_1 : \boldsymbol{\theta} \ne \boldsymbol{\theta}_0, \tag{14.24}$$

where $\boldsymbol{\theta} = \boldsymbol{C\beta}$ as before and $\boldsymbol{\theta}_0$ is a $q \times 1$ vector of specified constants, θ_{i0}'s. The hypothesis H_0 is called a **general linear hypothesis**. An α-level test of H_0 : $\boldsymbol{\theta} = \boldsymbol{\theta}_0$ rejects if $\boldsymbol{\theta}_0$ lies outside the $100(1 - \alpha)\%$ confidence ellipsoid (14.23), which is equivalent to rejecting if

$$F = \frac{(\widehat{\boldsymbol{\theta}} - \boldsymbol{\theta}_0)'(\boldsymbol{CVC}')^{-1}(\widehat{\boldsymbol{\theta}} - \boldsymbol{\theta}_0)}{q s^2} > f_{q,\nu,\alpha}. \tag{14.25}$$

Example 14.2 (Confidence Ellipse for Two Pairwise Contrasts)

Consider a one-way layout with $a = 3$ groups having sample sizes $n_1 = 2$, $n_2 = 3, n_3 = 4$. Suppose that group 3 is a control group with which we want to compare groups 1 and 2. The contrasts of interest are

$$\theta_1 = \mu_1 - \mu_3 \qquad \text{and} \qquad \theta_2 = \mu_2 - \mu_3$$

and we want to construct a 95% confidence ellipse (in two dimensions) for θ_1 and θ_2.

Here we have

$$\boldsymbol{V} = \begin{bmatrix} \frac{1}{2} & 0 & 0 \\ 0 & \frac{1}{3} & 0 \\ 0 & 0 & \frac{1}{4} \end{bmatrix}, \boldsymbol{C} = \begin{bmatrix} 1 & 0 & -1 \\ 0 & 1 & -1 \end{bmatrix}.$$

Hence

$$\boldsymbol{CVC}' = \begin{bmatrix} \frac{3}{4} & \frac{1}{4} \\ \frac{1}{4} & \frac{7}{12} \end{bmatrix} \qquad \text{and} \qquad (\boldsymbol{CVC}')^{-1} = \begin{bmatrix} \frac{14}{9} & -\frac{2}{3} \\ -\frac{2}{3} & 2 \end{bmatrix}.$$

Substituting these values in (14.23) and noting that $q = 2$, $\nu = 6$, and so $f_{2,6,.05} = 5.14$, we obtain the confidence ellipse as

$$\begin{aligned} \tfrac{14}{9}[(\bar{y}_1 - \bar{y}_3) - \theta_1]^2 - \tfrac{4}{3}[(\bar{y}_1 - \bar{y}_3) - \theta_1][(\bar{y}_2 - \bar{y}_3) - \theta_2] + 2[(\bar{y}_2 - \bar{y}_3) - \theta_2]^2 \\ \le 2(5.14)s^2. \end{aligned}$$

Suppose $\bar{y}_1 = 4.0, \bar{y}_2 = 3.0, \bar{y}_3 = 2.0$, and $s^2 = 1.0$. The 95% confidence ellipse for these data is shown in Figure 14.2. The parameter point $\theta_1 = \theta_2 = 0$ falls inside the ellipse. So the null hypothesis $H_0 : \theta_1 = \theta_2 = 0$, which is the same as $H_0 : \mu_1 = \mu_2 = \mu_3$, is not rejected at $\alpha = 0.05$. Equivalently, the F-statistic equals the left-hand side of the above inequality divided by $2s^2$, which can be calculated as $F = 2.778 < f_{2,6,0.05} = 5.14$, and hence H_0 is not rejected at $\alpha = 0.05$. ■

14.2.5 Extra Sum of Squares Method

Another method for deriving the above F-test of the general linear hypothesis is by comparing the error sums of squares from two fits: SS_e from the **full-model** fit under (14.1) and SS_{e0} from the **reduced- (partial)-model** fit under $H_0 : \boldsymbol{\theta} = \boldsymbol{C\beta} = \boldsymbol{\theta}_0$. Clearly, $\text{SS}_{e0} \ge \text{SS}_e$. We refer to $\text{SS}_{e0} - \text{SS}_e$ as the **hypothesis sum of squares** (SS_{H_0}). The **hypothesis d.f.** equals q, the number of linearly independent restrictions that H_0 imposes on the model space. This is the same as the number of linearly independent parametric functions of $\boldsymbol{\beta}$ that are set equal to specified constants θ_{i0}.

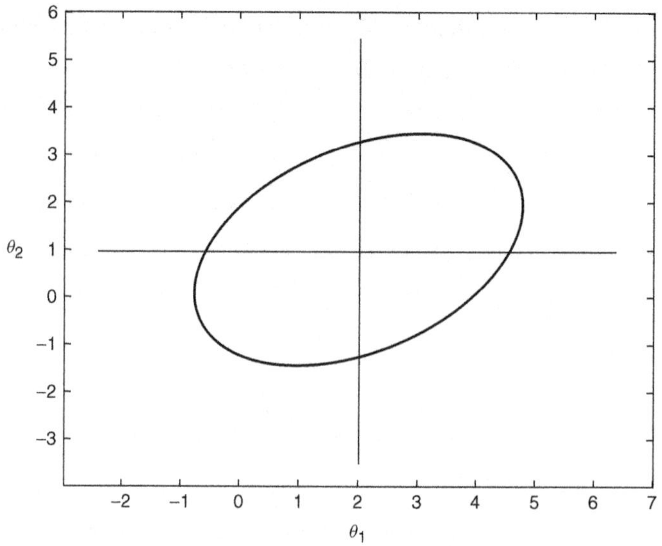

Figure 14.2 The 95% confidence ellipse for two pairwise contrasts.

A geometric interpretation of the extra sum of squares method provides useful insights. The null hypothesis $H_0 : \boldsymbol{\theta} = \boldsymbol{\theta}_0$ puts q linearly independent constraints on the β_j's. Therefore the vector $\mathrm{E}(\boldsymbol{y})$ is constrained to a $(p+1-q)$-dimensional subspace $\mathcal{X}_0 \subseteq \mathcal{X}$. The LS fit of the model under H_0 is obtained by projecting $\boldsymbol{y}$ onto $\mathcal{X}_0$; see Figure 14.3. Denote this projection by $\widehat{\boldsymbol{y}}_0$ and the corresponding residual vector by $\widehat{\boldsymbol{e}}_0 = \boldsymbol{y} - \widehat{\boldsymbol{y}}_0$. Then the vector

$$\widehat{\boldsymbol{e}}_0 - \widehat{\boldsymbol{e}} = (\boldsymbol{y} - \widehat{\boldsymbol{y}}_0) - (\boldsymbol{y} - \widehat{\boldsymbol{y}}) = \widehat{\boldsymbol{y}} - \widehat{\boldsymbol{y}}_0$$

is orthogonal to the vector $\widehat{\boldsymbol{e}}$ (since $\widehat{\boldsymbol{y}} - \widehat{\boldsymbol{y}}_0$ lies in the model space) and therefore they are independent. Additionally, by the Pythagoras theorem

$$\mathrm{SS}_{H_0} = \mathrm{SS}_{e0} - \mathrm{SS}_e = ||\widehat{\boldsymbol{e}}_0||^2 - ||\widehat{\boldsymbol{e}}||^2 = ||\widehat{\boldsymbol{e}}_0 - \widehat{\boldsymbol{e}}||^2 = ||\widehat{\boldsymbol{y}} - \widehat{\boldsymbol{y}}_0||^2. \quad (14.26)$$

Therefore $\mathrm{SS}_{H_0} = \mathrm{SS}_{e0} - \mathrm{SS}_e$ and SS_e are independently distributed, which fact is used in the derivation of the F-distribution result in (14.28). It can be shown (see Exercise 14.29) that

$$\mathrm{SS}_{H_0} = (\widehat{\boldsymbol{\theta}} - \boldsymbol{\theta}_0)'(\boldsymbol{C}\boldsymbol{V}\boldsymbol{C}')^{-1}(\widehat{\boldsymbol{\theta}} - \boldsymbol{\theta}_0). \quad (14.27)$$

Since $\mathrm{SS}_{H_0} \sim \sigma^2\chi^2_q$ under H_0, the ratio

$$F = \frac{\mathrm{SS}_{H_0}/q}{\mathrm{SS}_e/\nu} = \frac{\mathrm{MS}_{H_0}}{\mathrm{MS}_e} \sim \frac{\sigma^2\chi^2_q/q}{\sigma^2\chi^2_\nu/\nu} \sim F_{q,\nu}, \quad (14.28)$$

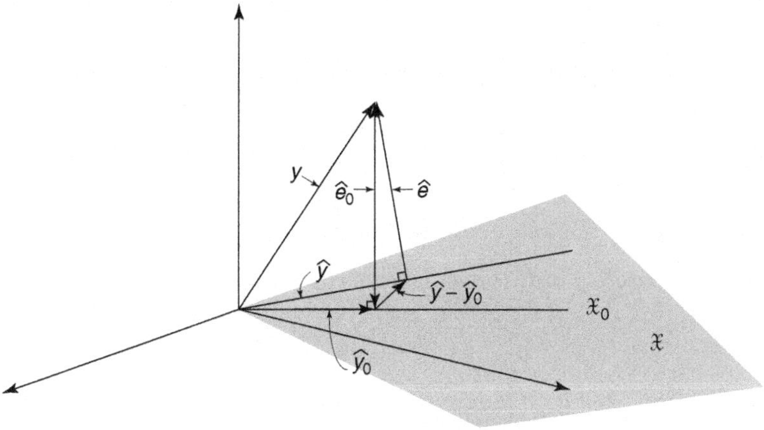

Figure 14.3 Least squares estimation of model parameters under general linear hypothesis.

where $MS_{H_0} = SS_{H_0}/q$ is called the **mean square for hypothesis**. Hence an α-level test of H_0 rejects if this F-ratio exceeds $f_{q,\nu,\alpha}$. So the extra sum of squares method leads to the same test as in (14.25). Generally, the extra sum of squares method provides an easier way to derive the F-test.

14.2.6 Analysis of Variance

A linear hypothesis that is often tested first when fitting a linear model is the **overall null hypothesis** $H_0 : \beta_1 = \beta_2 = \cdots = \beta_p = 0$. Rejection of this hypothesis implies that at least one of the β_j's is nonzero.

The partial model under H_0 is simply $E(y) = \beta_0$. The LS estimator of β_0 under the partial model is $\widehat{\beta}_0 = \overline{y}$, the overall mean of the y_i's. Therefore

$$SS_{e0} = \sum (y_i - \overline{y})^2 = ||\boldsymbol{y} - \overline{\boldsymbol{y}}||^2 = SS_{\text{tot}}, \qquad (14.29)$$

which is called the **total sum of squares**. Using (14.16) this equals $||\widehat{\boldsymbol{y}} - \overline{\boldsymbol{y}}||^2 + ||\widehat{\boldsymbol{e}}||^2$. We have defined $||\widehat{\boldsymbol{e}}||^2 = \widehat{\boldsymbol{e}}'\widehat{\boldsymbol{e}} = \sum \widehat{e}_i^2$ as the error sum of squares (SS_e) in (14.7). In the context of regression, $SS_{H_0} = ||\widehat{\boldsymbol{y}} - \overline{\boldsymbol{y}}||^2$ is referred to as the **regression sum of squares** (SS_{reg}), which is the part of the variation in y that is accounted for by regression of y on the predictor variables. Hence we get the **ANOVA identity**:

$$SS_{\text{tot}} = SS_{\text{reg}} + SS_e. \qquad (14.30)$$

The corresponding decomposition of the d.f. is

$$\underbrace{N-1}_{\text{total d.f.}} = \underbrace{p}_{\text{regression d.f.}} + \underbrace{N-(p+1)}_{\text{error d.f. } \nu}. \qquad (14.31)$$

These identities were used in Chapter 2.

In the following theorem we obtain an expression for $\mathrm{SS_{reg}}$ and its distribution, which form the basis of the ANOVA F-test.

Theorem 14.2 *The sum of squares for testing $H_0 : \beta_1 = \beta_2 = \cdots = \beta_p = 0$ is given by*

$$SS_{\mathrm{reg}} = \widehat{\boldsymbol{\beta}}'\boldsymbol{X}'\boldsymbol{y} - N\bar{y}^2 \tag{14.32}$$

Furthermore, SS_{reg} is distributed independently of SS_e as

$$SS_{\mathrm{reg}} \sim \sigma^2\chi_p^2(\lambda^2), \tag{14.33}$$

where $\chi_p^2(\lambda^2)$ denotes a noncentral chi-square r.v. (see Section A.4) with n.c.p.

$$\lambda^2 = \frac{\boldsymbol{\beta}'\boldsymbol{X}'[\boldsymbol{H} - (1/N)\boldsymbol{J}]\boldsymbol{X}\boldsymbol{\beta}}{\sigma^2}, \tag{14.34}$$

which equals zero under H_0. Hence SS_{reg} is distributed as σ^2 times a central chi-square r.v. with p d.f. under H_0.

Proof: See Section 14.4.2. ■

Since $\mathrm{SS_{reg}} \sim \sigma^2\chi_p^2$ under H_0 independently of $\mathrm{SS}_e \sim \sigma^2\chi^2_{N-(p+1)}$, the ratio

$$F = \frac{\mathrm{SS_{reg}}/p}{\mathrm{SS}_e/(N-(p+1))} = \frac{\mathrm{MS_{reg}}}{\mathrm{MS}_e} \sim F_{p,N-(p+1)},$$

where $\mathrm{MS_{reg}} = \mathrm{SS_{reg}}/p$ is the **regression mean square ($\mathbf{MS_{reg}}$)**. Therefore an α-level F-test of H_0 rejects if $F > f_{p,\nu,\alpha}$. These calculations are summarized in Table 14.1.

A measure of **goodness of fit** of the model is provided by the ratio of $\mathrm{SS_{reg}}$ to $\mathrm{SS_{tot}}$. This ratio, which represents the proportion of variation in the y_i's that is accounted for by the model, is denoted by R^2:

$$R^2 = \frac{\mathrm{SS_{reg}}}{\mathrm{SS_{tot}}} = 1 - \frac{\mathrm{SS}_e}{\mathrm{SS_{tot}}}.$$

Table 14.1 ANOVA Table for Multiple Regression

Source	SS	d.f.	MS	F
Regression	$\mathrm{SS_{reg}} = \widehat{\boldsymbol{\beta}}'\boldsymbol{X}'\boldsymbol{y} - N\bar{y}^2$	p	$\mathrm{MS_{reg}}$	$\mathrm{MS_{reg}}/\mathrm{MS}_e$
Error	$\mathrm{SS}_e = \widehat{\boldsymbol{e}}'\widehat{\boldsymbol{e}}$	$N-(p+1)$	MS_e	
Total	$\mathrm{SS_{tot}} = \boldsymbol{y}'\boldsymbol{y} - N\bar{y}^2$	$N-1$		

Example 14.3 (Simple Linear Regression: Analysis of Variance)

We now apply the results from Table 14.1 to the simple linear regression model (2.15). In this case, SS_{reg} is the sum of squares for testing $H_0 : \beta_1 = 0$ and is given by

$$\begin{aligned}
\text{SS}_{\text{reg}} &= (\widehat{\beta}_0, \widehat{\beta}_1)\boldsymbol{X}'\boldsymbol{y} - N\overline{y}^2 && \text{[using (14.32)]} \\
&= (\widehat{\beta}_0, \widehat{\beta}_1)\begin{bmatrix} \sum y_i \\ \sum x_i y_i \end{bmatrix} - N\overline{y}^2 && \text{(using the formula for } \boldsymbol{X}'\boldsymbol{y} \text{ from Example 2.13)} \\
&= \widehat{\beta}_0 \sum_{i=1}^N y_i + \widehat{\beta}_1 \sum_{i=1}^N x_i y_i - N\overline{y}^2 \\
&= (\overline{y} - \widehat{\beta}_1 \overline{x})N\overline{y} + \widehat{\beta}_1 \sum_{i=1}^N x_i y_i - N\overline{y}^2 \\
&= N\overline{y}^2 + \widehat{\beta}_1 \left(\sum_{i=1}^N x_i y_i - N\overline{x}\overline{y} \right) - N\overline{y}^2 \\
&= \widehat{\beta}_1 S_{xy} \\
&= \frac{S_{xy}^2}{S_{xx}},
\end{aligned}$$

where S_{xy}, S_{xx}, and S_{yy} are defined in Equation (2.17). Next, $\text{SS}_{\text{tot}} = \boldsymbol{y}'\boldsymbol{y} - N\overline{y}^2 = \sum y_i^2 - N\overline{y}^2 = S_{yy}$. Then SS_e is obtained by subtraction. The d.f. are obtained by setting $p = 1$. The resulting ANOVA is shown in Table 14.2. The α-level test of $H_0 : \beta_1 = 0$ rejects if

$$F = \frac{\text{SS}_{\text{reg}}/1}{\text{MS}_e} = \frac{S_{xy}^2}{s^2 S_{xx}} = \frac{\widehat{\beta}_1^2 S_{xx}}{s^2} > f_{1,N-2,\alpha},$$

which is the same test that was given in Equation (2.22). ■

Table 14.2 ANOVA Table for Simple Linear Regression

Source	SS	d.f.	MS	F
Regression	$\text{SS}_{\text{reg}} = \dfrac{S_{xy}^2}{S_{xx}}$	1	MS_{reg}	$\text{MS}_{\text{reg}}/\text{MS}_e$
Error	$\text{SS}_e = S_{yy} - \dfrac{S_{xy}^2}{S_{xx}}$	$N - 2$	MS_e	
Total	$\text{SS}_{\text{tot}} = S_{yy}$	$N - 1$		

Example 14.4 (One-Way ANOVA F-Test)

Consider the one-way layout setup of Chapter 3. We will derive the F-test of $H_0 : \mu_1 = \mu_2 = \cdots = \mu_a$ given by (3.9) using the extra sum of squares method. Under H_0, denote the common value of the μ_i's by μ. Then the LS estimator $\widehat{\mu}$, which minimizes

$$\sum_{i=1}^{a}\sum_{j=1}^{n_i}(y_{ij} - \mu)^2,$$

equals $\overline{\overline{y}}$, the grand mean of all the y_{ij} (note the change of notation from $\overline{y}$ to $\overline{\overline{y}}$ to denote the grand mean). Hence

$$\mathrm{SS}_{e0} = \sum_{i=1}^{a}\sum_{j=1}^{n_i}(y_{ij} - \overline{\overline{y}})^2 = ||\boldsymbol{y} - \overline{\overline{\boldsymbol{y}}}||^2 = \mathrm{SS}_{\mathrm{tot}}.$$

Let $\widehat{\boldsymbol{y}}$ denote the $N \times 1$ fitted vector which has the first n_1 components equal to $\overline{y}_1$, the next n_2 components equal to $\overline{y}_2$, and so on, and let $\widehat{\boldsymbol{e}}$ denote the $N \times 1$ residual vector whose (i, j)th element equals $\widehat{e}_{ij} = y_{ij} - \overline{y}_i$ $(1 \le i \le a, 1 \le j \le n_i)$. From (14.26) we see that

$$\mathrm{SS}_{H_0} = ||\widehat{\boldsymbol{y}} - \overline{\overline{\boldsymbol{y}}}||^2 = \sum_{i=1}^{a} n_i(\overline{y}_i - \overline{\overline{y}})^2.$$

Finally, the hypothesis d.f. are $q = a - 1$, the error d.f. are $\nu = N - a$, and the F-test rejects H_0 at level α if

$$F = \frac{\mathrm{MS}_{H_0}}{\mathrm{MS}_e} = \frac{\sum_{i=1}^{a} n_i(\overline{y}_i - \overline{\overline{y}})^2/(a-1)}{\sum_{i=1}^{a}\sum_{j=1}^{n_i}(y_{ij} - \overline{y}_i)^2/(N-a)} > f_{a-1,N-a,\alpha}.$$

Applying this test to the data from Example 14.2, we get $\overline{\overline{y}} = \frac{25}{9}$. Therefore,

$$\mathrm{SS}_{H_0} = 2\left(4 - \tfrac{25}{9}\right)^2 + 3\left(3 - \tfrac{25}{9}\right)^2 + 4\left(2 - \tfrac{25}{9}\right)^2 = \frac{50}{9}.$$

Also, $s^2 = 1$. Therefore the F-statistic equals

$$F = \frac{(50/9)/2}{1} = \frac{25}{9} = 2.778,$$

which is identical to the F-statistic calculated in that example using a different method. This is the consequence of the fact, proven in Exercise 14.28, that testing any nonsingular transformation of the hypothesis $\boldsymbol{\theta} = \boldsymbol{0}$ yields the same SS_{H_0} and hence the same F-statistic. Hence testing any $a - 1$ linearly independent contrasts among the μ_i equal zeros is equivalent to testing $H_0 : \mu_1 = \mu_2 = \cdots = \mu_a$. ■

14.3 POWER OF F-TEST

Consider testing the general linear hypothesis (14.24). To assess the sensitivity of an already conducted experiment or to determine the sample size for a future planned experiment, it is necessary to study the power of the F-test (14.25). An expression for the power for specified α and nonnull configuration $\boldsymbol{\theta} \neq \boldsymbol{\theta}_0$ is given by

$$\mathrm{P}_{\boldsymbol{\theta}}\left\{\frac{(\widehat{\boldsymbol{\theta}}-\boldsymbol{\theta}_0)'(\boldsymbol{C}\boldsymbol{V}\boldsymbol{C}')^{-1}(\widehat{\boldsymbol{\theta}}-\boldsymbol{\theta}_0)}{qs^2} > f_{q,\nu,\alpha}\right\}. \tag{14.35}$$

Using the results from Section A.4 it follows that

$$F = \frac{(\widehat{\boldsymbol{\theta}}-\boldsymbol{\theta}_0)'(\boldsymbol{C}\boldsymbol{V}\boldsymbol{C}')^{-1}(\widehat{\boldsymbol{\theta}}-\boldsymbol{\theta}_0)}{qs^2}$$

has a noncentral F-distribution under H_1 with q and $\nu = N-(p+1)$ d.f. and n.c.p. equal to

$$\lambda^2 = \frac{(\boldsymbol{\theta}-\boldsymbol{\theta}_0)'(\boldsymbol{C}\boldsymbol{V}\boldsymbol{C}')^{-1}(\boldsymbol{\theta}-\boldsymbol{\theta}_0)}{\sigma^2}. \tag{14.36}$$

The n.c.p. depends on the true parameter vector $\boldsymbol{\theta}$ as well as σ^2, which must be specified for power calculation.

The expected value of $\mathrm{MS}_{H_0} = \mathrm{SS}_{H_0}/q$ is directly related to the n.c.p. It can be shown that, under H_1, $\mathrm{SS}_{H_0}/\sigma^2$ has a noncentral chi-square distribution with q d.f. and n.c.p. given by (14.36). Using the formula (A.9) for the expected value of a noncentral chi-square r.v., we get

$$\begin{aligned}
\mathrm{E}(\mathrm{MS}_{H_0}) &= \mathrm{E}\left[\frac{\sigma^2}{q}\chi_q^2(\lambda^2)\right] \\
&= \frac{\sigma^2}{q}(q+\lambda^2) \\
&= \sigma^2 + \frac{(\boldsymbol{\theta}-\boldsymbol{\theta}_0)'(\boldsymbol{C}\boldsymbol{V}\boldsymbol{C}')^{-1}(\boldsymbol{\theta}-\boldsymbol{\theta}_0)}{q}.
\end{aligned} \tag{14.37}$$

Note that λ^2 equals $1/\sigma^2$ times $(\boldsymbol{\theta}-\boldsymbol{\theta}_0)'(\boldsymbol{C}\boldsymbol{V}\boldsymbol{C}')^{-1}(\boldsymbol{\theta}-\boldsymbol{\theta}_0)$, which is just the SS_{H_0} expression with $\widehat{\boldsymbol{\theta}}$ replaced by its expected value $\boldsymbol{\theta}$. Similarly, $\mathrm{E}(\mathrm{MS}_{H_0})$ equals σ^2 plus $(\boldsymbol{\theta}-\boldsymbol{\theta}_0)'(\boldsymbol{C}\boldsymbol{V}\boldsymbol{C}')^{-1}(\boldsymbol{\theta}-\boldsymbol{\theta}_0)/q$, which is just the MS_{H_0} expression with $\widehat{\boldsymbol{\theta}}$ replaced by its expected value $\boldsymbol{\theta}$. From these observations the following simple rule can be deduced:

Rule for Obtaining n.c.p. of F-statistic for Test of Linear Hypothesis The n.c.p. λ^2 equals $1/\sigma^2$ times the SS_{H_0} expression with the LS estimators of all linear parametric functions (such as $\widehat{\theta}$) replaced by their expected (true) values, and $E(MS_{H_0}) = \sigma^2(1 + \lambda^2/q)$, where q is the hypothesis d.f.

From (14.37) we see that $E(MS_{H_0})$ exceeds σ^2 by a quantity that is proportional to λ^2 and $E(MS_{H_0}) = \sigma^2$ under H_0. On the other hand, $E(MS_e) = \sigma^2$ regardless of whether H_0 is true or not. Therefore $F = MS_{H_0}/MS_e$ is a ratio of two independent unbiased scaled chi-square distributed estimators of σ^2 under H_0, and this ratio has a central F-distribution. Under H_1, $E(MS_{H_0}) > E(MS_e)$ and the F-statistic has a noncentral F-distribution. Therefore we reject H_0 for large values of F.

A special chart (called the **Pearson–Hartley chart**) or a computer program is needed to evaluate (14.35), which gives the power [or, equivalently, the **operating characteristic (OC) function**, which is the complement of the power function] as the tail probability under a noncentral F-distribution. Figure 14.4 shows this chart for one- and two-sample problems discussed in Sections 2.1 and 2.2 (when the numerator d.f. of the noncentral F-distribution is $\nu_1 = 1$) and $\alpha = 0.01, 0.05$. The power function is plotted against

$$\phi = \frac{\lambda}{\sqrt{\nu_1 + 1}}.$$

Note that different ϕ-scales are used for $\alpha = 0.01, 0.05$. Also note that, for each α, a set of power curves are given for selected values of ν_2. The following example illustrates the use of this chart.

Example 14.5 (Cloud-Seeding Experiment: Power of Independent Samples *t*-Test)

Refer to the independent samples t-test from Section 2.2.1.1 for testing $H_0 : \mu_1 - \mu_2 = 0$ versus $H_1 : \mu_1 - \mu_2 \neq 0$. For $n_1 = n_2 = n$, the α-level test rejects H_0 if

$$|t| = \frac{|\bar{y}_1 - \bar{y}_2|}{s}\sqrt{\frac{n}{2}} > t_{\nu,\alpha/2} \iff F = t^2 = \frac{n(\bar{y}_1 - \bar{y}_2)^2}{2s^2} > f_{1,\nu,\alpha},$$

where $\nu = 2n - 2$. Using the rule given above, the n.c.p. of F is

$$\lambda^2 = \frac{n\delta^2}{2\sigma^2},$$

where $\delta = \mu_1 - \mu_2 \neq 0$ (obtained by replacing $\bar{y}_1$ and $\bar{y}_2$ in the F-statistic by their expected values).

In Example 2.6 we considered the sample size problem for the cloud-seeding experiment for detecting a difference in the mean rainfall amounts of $\delta = \ln 3 = 1.099$ on the log-scale. For $\alpha = 0.05$ and $1 - \beta = 0.80$ and assuming a *known* $\sigma = 2$, we computed the sample size per group to be 52 for the two-sided z-test.

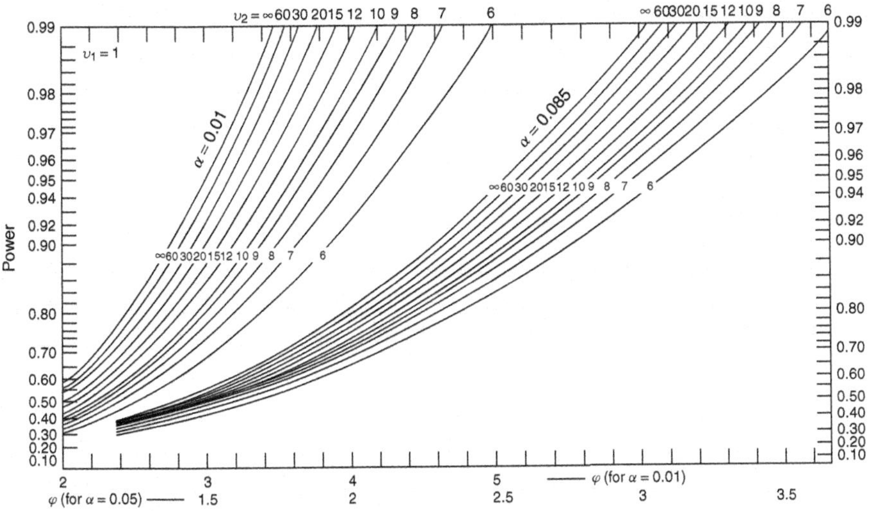

Figure 14.4 Pearson–Hartley chart for power function of F-Test for $\nu_1 = 1$ (Pearson and Hartley, 1951). Reprinted by permission of Oxford University Press.

In fact, one would use the t-test in this case since σ is estimated from the data. A two-sided t-test is equivalent to an F-test. To find the power of this F-test, we first calculate

$$\lambda = \frac{\delta}{\sigma}\sqrt{\frac{n}{2}} = \frac{1.099}{2}\sqrt{\frac{52}{2}} = 2.802.$$

Now enter the chart in Figure 14.4 with

$$\phi = \frac{\lambda}{\sqrt{\nu_1 + 1}} = \frac{2.802}{\sqrt{2}} = 1.981$$

and $\nu_2 = 2n - 2 = 102$. By visually interpolating between $\phi = 1.5$ and 2, and $\nu_2 = 60$ and ∞, we see that the power is about 0.79. (An exact calculation of power using Minitab yields a value of 0.7926.) Thus, as remarked in Example 2.6, when σ is estimated and the t-test is used instead of the z-test, the power is slightly less than the specified value of 0.80. Here the shortfall is rather small since ν_2 is quite large. In fact, using Minitab, we can readily calculate that $n = 53$ is the minimum sample size required to guarantee 80% power (the exact power is 0.8003). ■

Example 14.6 (Comparison of Corn Yields: Sample Size Determination)

Refer to Example 2.11, where we calculated a sample size of 43 per group to detect a difference of 30 lb/acre between the mean yields of regular and kiln-dried seeds with 80% power using a 0.05-level paired sample z-test if $\sigma_d = 70$ lb. For the paired t-test, calculation using Minitab shows that the power is actually

0.7839 using $n = 43$ per group, and $n = 45$ is needed to guarantee at least 80% power (the exact power is 0.8028). ■

14.4 CHAPTER NOTES

14.4.1 Proof of Theorem 14.1 (Gauss–Markov Theorem)

Let $\widetilde{\boldsymbol{\beta}}$ be any other linear unbiased estimator of $\boldsymbol{\beta}$. Write $\widetilde{\boldsymbol{\beta}} = [(\boldsymbol{X}'\boldsymbol{X})^{-1}\boldsymbol{X}' + \boldsymbol{A}]\boldsymbol{y} = \widehat{\boldsymbol{\beta}} + \boldsymbol{A}\boldsymbol{y}$, where $\boldsymbol{A}$ is any $(p+1) \times N$ matrix of constants (so that $\widetilde{\boldsymbol{\beta}}$ is a linear function of $\boldsymbol{y}$). For $\widetilde{\boldsymbol{\beta}}$ to be unbiased, we must have

$$\mathrm{E}(\widetilde{\boldsymbol{\beta}}) = \mathrm{E}(\widehat{\boldsymbol{\beta}}) + A\mathrm{E}(\boldsymbol{y}) = \boldsymbol{\beta} + \boldsymbol{A}\boldsymbol{X}\boldsymbol{\beta} \quad \text{for all } \boldsymbol{\beta}.$$

Hence $\boldsymbol{A}\boldsymbol{X} = \boldsymbol{O}$.

Next,

$$\begin{aligned}
\mathrm{Cov}(\widetilde{\boldsymbol{\beta}}) &= \mathrm{Cov}(\widehat{\boldsymbol{\beta}} + \boldsymbol{A}\boldsymbol{y}) \\
&= \mathrm{Cov}(\widehat{\boldsymbol{\beta}}) + \mathrm{Cov}(\boldsymbol{A}\boldsymbol{y}) + 2\mathrm{Cov}(\widehat{\boldsymbol{\beta}}, \boldsymbol{A}\boldsymbol{y}) \\
&= \sigma^2(\boldsymbol{X}'\boldsymbol{X})^{-1} + A\mathrm{Cov}(\boldsymbol{y})A' + 2\mathrm{Cov}((\boldsymbol{X}'\boldsymbol{X})^{-1}\boldsymbol{X}'\boldsymbol{y}, \boldsymbol{A}\boldsymbol{y}) \\
&= \sigma^2(\boldsymbol{X}'\boldsymbol{X})^{-1} + \sigma^2\boldsymbol{A}\boldsymbol{A}' + 2(\boldsymbol{X}'\boldsymbol{X})^{-1}\boldsymbol{X}'\mathrm{Cov}(\boldsymbol{y})A' \\
&= \sigma^2(\boldsymbol{X}'\boldsymbol{X})^{-1} + \sigma^2\boldsymbol{A}\boldsymbol{A}' + 2\sigma^2(\boldsymbol{X}'\boldsymbol{X})^{-1}\boldsymbol{X}'A' \\
&= \sigma^2(\boldsymbol{X}'\boldsymbol{X})^{-1} + \sigma^2\boldsymbol{A}\boldsymbol{A}' \quad (\text{since } \boldsymbol{X}'A' = \boldsymbol{O}').
\end{aligned}$$

Hence $\mathrm{Cov}(\widetilde{\boldsymbol{\beta}}) - \mathrm{Cov}(\widehat{\boldsymbol{\beta}}) = \sigma^2\boldsymbol{A}\boldsymbol{A}'$, which is positive semidefinite. Therefore for any two linear unbiased estimators $\boldsymbol{c}'\widetilde{\boldsymbol{\beta}}$ and $\boldsymbol{c}'\widehat{\boldsymbol{\beta}}$ of a parametric function $\boldsymbol{c}'\boldsymbol{\beta}$, we have

$$\begin{aligned}
\mathrm{Var}(\boldsymbol{c}'\widetilde{\boldsymbol{\beta}}) - \mathrm{Var}(\boldsymbol{c}'\widehat{\boldsymbol{\beta}}) &= \sigma^2\boldsymbol{c}'[(\boldsymbol{X}'\boldsymbol{X})^{-1} + \boldsymbol{A}\boldsymbol{A}']\boldsymbol{c} - \sigma^2\boldsymbol{c}'(\boldsymbol{X}'\boldsymbol{X})^{-1}\boldsymbol{c} \\
&= \sigma^2\boldsymbol{c}'\boldsymbol{A}\boldsymbol{A}'\boldsymbol{c} \geq 0
\end{aligned}$$

since $\boldsymbol{A}\boldsymbol{A}'$ is positive semidefinite.

14.4.2 Proof of Theorem 14.2

We can write

$$\begin{aligned}
\mathrm{SS}_{\mathrm{reg}} &= ||\widehat{\boldsymbol{y}} - \overline{\boldsymbol{y}}||^2 \\
&= \widehat{\boldsymbol{y}}'\widehat{\boldsymbol{y}} - \overline{\boldsymbol{y}}'\overline{\boldsymbol{y}} \\
&= \widehat{\boldsymbol{\beta}}'\boldsymbol{X}'\boldsymbol{X}\widehat{\boldsymbol{\beta}} - N\overline{y}^2 \\
&= \widehat{\boldsymbol{\beta}}'\boldsymbol{X}'\boldsymbol{X}(\boldsymbol{X}'\boldsymbol{X})^{-1}\boldsymbol{X}'\boldsymbol{y} - N\overline{y}^2 \\
&= \widehat{\boldsymbol{\beta}}'\boldsymbol{X}'\boldsymbol{y} - N\overline{y}^2.
\end{aligned}$$

This proves (14.32).

The distribution of SS_{reg} follows directly from Lemma A.2 by noting that SS_{reg} is a quadratic form $\boldsymbol{y}'[\boldsymbol{H} - (1/N)\boldsymbol{J}]\boldsymbol{y}$, where $\boldsymbol{H} - (1/N)\boldsymbol{J}$ is an idempotent matrix of rank p. Independence of SS_{reg} and SS_e (whose distribution was derived in Section 14.2.2) follows from the orthogonality of the projection matrices $\boldsymbol{H} - (1/N)\boldsymbol{J}$ and $\boldsymbol{I} - \boldsymbol{H}$.

Note that, under H_0, $\boldsymbol{\beta} = (\beta_0, 0, \ldots, 0)'$. Therefore $\boldsymbol{X\beta} = \beta_0 \mathbf{1}$ because the first column of $\boldsymbol{X}$ equals $\mathbf{1}$. Substituting this in (14.34) we get

$$\lambda^2 = \frac{\beta_0^2 \mathbf{1}'[\boldsymbol{H} - (1/N)\boldsymbol{J}]\mathbf{1}}{\sigma^2} = 0$$

since $\boldsymbol{H}\mathbf{1} = \mathbf{1}$ from Exercise 14.11 and $(1/N)\boldsymbol{J}\mathbf{1} = \mathbf{1}$, as can be easily verified. Therefore $\mathbf{1}'[\boldsymbol{H} - (1/N)\boldsymbol{J}]\mathbf{1} = 0$.

14.5 CHAPTER SUMMARY

(a) A linear model $E(y) = \beta_0 + \beta_1 x_1 + \cdots + \beta_p x_p$ is linear in the unknown parameters β_j's; the x_j's may be nonlinear functions of each other or of themselves. A model is nonlinear if it is a nonlinear function of the β_j's. An additive $N(0, \sigma^2)$ random error is generally assumed in a linear model.

(b) A common method for estimating the β_j's in a linear model is the LS method. Geometrically, finding the LS estimator of $\boldsymbol{\beta}$ involves finding the projection $\widehat{\boldsymbol{y}} = \boldsymbol{X}\widehat{\boldsymbol{\beta}}$, called the fitted vector, of the observed response vector $\boldsymbol{y}$ on to the the column space of the model matrix $\boldsymbol{X}$, called the estimation space. The dimension of the estimation space is the column rank r of $\boldsymbol{X}$, where $r = p + 1$ if $\boldsymbol{X}$ is full column rank. The residual vector $\widehat{\boldsymbol{e}} = \boldsymbol{y} - \widehat{\boldsymbol{y}}$ lies in the error space, which is orthogonal to the estimation space and has dimension $\nu = N - r$, called the error d.f. The fitted and residual vectors are useful for model diagnostics.

(c) The Gauss–Markov theorem states that among all linear (in $\boldsymbol{y}$) unbiased estimators of any linear parametric function $\boldsymbol{c}'\boldsymbol{\beta}$, the LS estimator $\boldsymbol{c}'\widehat{\boldsymbol{\beta}}$ has the minimum variance.

(d) The LS estimator $\widehat{\boldsymbol{\beta}}$ has an MVN distribution with mean vector $\boldsymbol{\beta}$ and covariance matrix $\sigma^2(\boldsymbol{X}'\boldsymbol{X})^{-1}$. Furthermore, $\widehat{\boldsymbol{\beta}}$ is independent of $s^2 = MS_e$, which is an unbiased estimator of σ^2 and has a $\sigma^2\chi^2_\nu/\nu$ distribution. These facts are used to derive confidence intervals and hypothesis tests on a single linear parametric function of $\boldsymbol{\beta}$ based on the t-distribution or on several functions based on the F-distribution. The latter is a test of the general linear hypothesis and has an ellipsoidal acceptance region. This test can also be derived using the extra sum of squares method, which expresses the F-statistic as a ratio of MS_{H_0} to MS_e.

(e) The test of the overall null hypothesis that $\beta_1 = \cdots = \beta_p = 0$ is a special case of the F-test of the general linear hypothesis. The associated decomposition of SS_{tot} into SS_{H_0} and SS_e along with the respective d.f. gives the ANOVA identity (14.30).

(f) The power of the F-test of the general linear hypothesis involves a noncentral F-distribution; its n.c.p. is a measure of the deviation of the alternative hypothesis from the null hypothesis. Sample size calculations can be done to guarantee a specified power requirement using charts or programs based on this distribution.

EXERCISES

Section 14.1 (Basic Linear Model and Least Squares Estimation)

14.1 Suppose we want to fit a straight line through the origin to the data $\{(x_i, y_i),\ i = 1, 2, \ldots, N\}$, that is, fit the model $y_i = \beta x_i + e_i,\ i = 1, 2, \ldots, N$. Write this model in the matrix notation spelling out the $\boldsymbol{X}$ matrix and $\boldsymbol{\beta}$ and $\boldsymbol{y}$ vectors.

14.2 Consider the two-sample problem from Section 2.2. Write a linear model in the form (14.1) for the data $y_{11}, y_{12}, \ldots, y_{1n_1}$ i.i.d. $N(\mu_1, \sigma^2)$ and $y_{21}, y_{22}, \ldots, y_{2n_2}$ i.i.d. $N(\mu_2, \sigma^2)$.

14.3 Let $f(\boldsymbol{x}) : \mathcal{R}^n \to \mathcal{R}^1$ be a scalar-valued function of $\boldsymbol{x} = (x_1, x_2, \ldots, x_n)'$ which is differentiable in each of its arguments. Define the vector derivative

$$\frac{df(\boldsymbol{x})}{d\boldsymbol{x}} = \left(\frac{\partial f(\boldsymbol{x})}{\partial x_1}, \frac{\partial f(\boldsymbol{x})}{\partial x_2}, \ldots, \frac{\partial f(\boldsymbol{x})}{\partial x_n}\right)'.$$

(a) Let $f(\boldsymbol{x}) = \boldsymbol{a}'\boldsymbol{x} = \sum_{i=1}^n a_i x_i$ where $\boldsymbol{a} = (a_1, a_2, \ldots, a_n)'$ is a vector of constants. Show that

$$\frac{df(\boldsymbol{x})}{d\boldsymbol{x}} = \boldsymbol{a}.$$

(b) Let $f(\boldsymbol{x}) = \boldsymbol{x}'\boldsymbol{A}\boldsymbol{x} = \sum_{i=1}^n \sum_{j=1}^n a_{ij} x_i x_j$ where $\boldsymbol{A}$ is an $n \times n$ symmetric matrix with elements $a_{ij} = a_{ji}$. Show that

$$\frac{df(\boldsymbol{x})}{d\boldsymbol{x}} = 2\boldsymbol{A}\boldsymbol{x}.$$

14.4 Refer to Exercise 14.1.

(a) Find the LS estimator of the slope coefficient β using the matrix approach.

(b) Describe the model and error spaces. What are their dimensions?

14.5 Consider the simple linear regression model (2.15). Suppose that we center the x_i's before fitting the model. In other words, we fit the model $y_i = \beta_0 + \beta_1 x_i^* + e_i$ $(1 \le i \le N)$ where $x_i^* = x_i - \overline{x}$.

(a) How does this transformation simplify the calculation of the LS estimators?

(b) Show that $\widehat{\beta_0} = \overline{y}$ and $\widehat{\beta_1} = S_{xy}/S_{xx}$.

14.6 In the linear model (14.1), suppose that $\boldsymbol{X}$ is a full-rank square matrix $(N = p + 1)$. Show that the LS estimator of $\boldsymbol{\beta}$ is given by solving a square system of equations $\boldsymbol{X\beta} = \boldsymbol{y}$, so $\widehat{\boldsymbol{\beta}} = \boldsymbol{X}^{-1}\boldsymbol{y}$.

14.7 Suppose that the parameters in the model (14.1) are required to satisfy the following constraints $\beta_p = \beta_1 + 2\beta_2$ and $\beta_{p-1} = 2\beta_1 - 3\beta_3$. Assume that the original $\boldsymbol{X}$ matrix is full-column rank.

(a) What are the basis vectors of the model space taking into account these two constraints?

(b) What are the dimensions of the model and error spaces?

(c) More generally, if there are q linearly independent constraints on the β's, what are the dimensions of the model and error spaces?

14.8 Show that $\boldsymbol{H} = \boldsymbol{X}(\boldsymbol{X}'\boldsymbol{X})^{-1}\boldsymbol{X}'$ and $\boldsymbol{I} - \boldsymbol{H}$ are projection matrices. Further show that the spaces onto which $\boldsymbol{H}$ and $\boldsymbol{I} - \boldsymbol{H}$ project any vector $\boldsymbol{y} \in \mathcal{R}^N$ are orthogonal, that is, $\boldsymbol{Hy}$ and $(\boldsymbol{I} - \boldsymbol{H})\boldsymbol{y}$ are orthogonal.

14.9 Show that $(\boldsymbol{I} - \boldsymbol{H})\boldsymbol{X}$ is a null matrix and hence $\boldsymbol{X}'\widehat{\boldsymbol{e}}$ is a null vector.

14.10 Show that $\text{rank}(\boldsymbol{H}) = \text{tr}(\boldsymbol{H}) = p + 1$ and $\text{rank}(\boldsymbol{I} - \boldsymbol{H}) = \text{tr}(\boldsymbol{I} - \boldsymbol{H}) = N - (p + 1)$.

14.11 Let $\boldsymbol{J}$ be an $N \times N$ matrix of all 1's and let $\boldsymbol{H}$ be the hat matrix.

(a) Show that $\boldsymbol{H1} = \boldsymbol{1}$, that is, the entries of $\boldsymbol{H}$ sum to 1 along any row and, because of the symmetry of $\boldsymbol{H}$, also along any column. Hence show that $\boldsymbol{JH} = \boldsymbol{J}$.

(b) Show by direct multiplication that $(1/N)\boldsymbol{J}$, $\boldsymbol{H} - (1/N)\boldsymbol{J}$, and $\boldsymbol{I} - \boldsymbol{H}$ are mutually orthogonal.

14.12 Refer to Exercise 14.1. Show that the entries of $\boldsymbol{H}$ are

$$h_{ij} = \frac{x_i x_j}{\sum x_i^2} \qquad (1 \le i, j \le N).$$

14.13 Show that the entries of $\boldsymbol{H}$ for the simple linear regression model (2.15) are

$$h_{ij} = \frac{1}{N} + \frac{(x_i - \overline{x})(x_j - \overline{x})}{S_{xx}} \qquad (1 \leq i, j \leq N).$$

14.14 Show that for a parametric function $\boldsymbol{c}'\boldsymbol{\beta}$ to be estimable $\boldsymbol{c}$ must be in the row space of $\boldsymbol{X}$.

14.15 Consider the model $y_{ij} = \mu + \alpha_i + e_{ij}$ $(1 \leq i \leq a, 1 \leq j \leq n_i)$ for a general one-way layout setting with no side conditions on either μ or the α_i. Show that the only linear parametric functions of the α_i that are estimable are contrasts of the form $\sum_{i=1}^{a} c_i \alpha_i$ where $\sum_{i=1}^{a} c_i = 0$.

14.16 Consider the one-way layout model $y_{ij} = \mu_i + e_{ij}$ $(1 \leq i \leq a, 1 \leq j \leq n_i)$, where $a = 3$ and $n_1 = 2, n_2 = 3, n_3 = 4$. Show that the hat matrix for this design equals

$$\boldsymbol{H} = \begin{bmatrix} \frac{1}{2} & \frac{1}{2} & 0 & 0 & 0 & 0 & 0 & 0 & 0 \\ \frac{1}{2} & \frac{1}{2} & 0 & 0 & 0 & 0 & 0 & 0 & 0 \\ 0 & 0 & \frac{1}{3} & \frac{1}{3} & \frac{1}{3} & 0 & 0 & 0 & 0 \\ 0 & 0 & \frac{1}{3} & \frac{1}{3} & \frac{1}{3} & 0 & 0 & 0 & 0 \\ 0 & 0 & \frac{1}{3} & \frac{1}{3} & \frac{1}{3} & 0 & 0 & 0 & 0 \\ 0 & 0 & 0 & 0 & 0 & \frac{1}{4} & \frac{1}{4} & \frac{1}{4} & \frac{1}{4} \\ 0 & 0 & 0 & 0 & 0 & \frac{1}{4} & \frac{1}{4} & \frac{1}{4} & \frac{1}{4} \\ 0 & 0 & 0 & 0 & 0 & \frac{1}{4} & \frac{1}{4} & \frac{1}{4} & \frac{1}{4} \\ 0 & 0 & 0 & 0 & 0 & \frac{1}{4} & \frac{1}{4} & \frac{1}{4} & \frac{1}{4} \end{bmatrix}.$$

What will be the form of $\boldsymbol{H}$ for arbitrary a and the n_i's?

14.17 In the model (14.1) suppose that the covariance matrix of $\boldsymbol{e}$ is $\sigma^2 \boldsymbol{W}^{-1}$, where $\boldsymbol{W}$ is an arbitrary known symmetric positive-definite matrix. To apply the LS method to this model, assume the following result: There exists an $N \times N$ nonsingular matrix $\boldsymbol{P}$ such that $\boldsymbol{P}'\boldsymbol{P} = \boldsymbol{W}$ and $\boldsymbol{P}\boldsymbol{W}^{-1}\boldsymbol{P}' = \boldsymbol{I}$. Make a nonsingular (one-to-one) transformation of the model as follows:

$$\boldsymbol{P}\boldsymbol{y} = \boldsymbol{P}\boldsymbol{X}\boldsymbol{\beta} + \boldsymbol{P}\boldsymbol{e}.$$

Let $\boldsymbol{y}^* = \boldsymbol{P}\boldsymbol{y}$, $\boldsymbol{X}^* = \boldsymbol{P}\boldsymbol{X}$, and $\boldsymbol{e}^* = \boldsymbol{P}\boldsymbol{e}$.

(a) Show that $\mathrm{Cov}(\boldsymbol{e}^*) = \sigma^2 \boldsymbol{I}$ and hence the LS method can be applied to the transformed model $\boldsymbol{y}^* = \boldsymbol{X}^*\boldsymbol{\beta} + \boldsymbol{e}^*$.

(b) Show that the LS estimator of $\boldsymbol{\beta}$ for the transformed model equals

$$\widehat{\boldsymbol{\beta}}_{\text{GLS}} = (\boldsymbol{X}'\boldsymbol{W}\boldsymbol{X})^{-1}\boldsymbol{X}'\boldsymbol{W}\boldsymbol{y}.$$

This is called the **generalized least squares (GLS)** estimator.

(c) Show that $\widehat{\boldsymbol{\beta}}_{\text{GLS}}$ minimizes $Q = (\boldsymbol{y} - \boldsymbol{X}\boldsymbol{\beta})'\boldsymbol{W}(\boldsymbol{y} - \boldsymbol{X}\boldsymbol{\beta})$.

14.18 Refer to the previous exercise and consider a special case where $\boldsymbol{W}$ is diagonal with diagonal entries $w_i > 0$. The GLS estimator is referred to as the **weighted least squares (WLS)** estimator in this case. An example is when each y_i is the average of n_i i.i.d. observations with a common variance σ^2 so that $w_i = n_i$. Give the explicit form of the objective function Q that is minimized.

14.19 Refer to the previous exercise. Show that the WLS estimator of β when fitting a straight line through the origin $y_i = \beta x_i + e_i \quad (1 \le i \le N)$ is

$$\widehat{\beta}_{\text{WLS}} = \frac{\sum_{i=1}^{N} w_i x_i y_i}{\sum_{i=1}^{N} w_i x_i^2}.$$

14.20 Refer to Exercise 14.18. Show that the WLS estimators of β_0 and β_1 when fitting a straight line $y_i = \beta_0 + \beta_1 x_i + e_i \quad (1 \le i \le N)$ are

$$\widehat{\beta}_{0,\text{WLS}} = \overline{y}_w - \overline{x}_w \widehat{\beta}_{1,\text{WLS}}, \quad \widehat{\beta}_{1,\text{WLS}} = \frac{\sum_{i=1}^{N} w_i (x_i - \overline{x}_w)(y_i - \overline{y}_w)}{\sum_{i=1}^{N} w_i (x_i - \overline{x}_w)^2},$$

where $\overline{x}_w$ and $\overline{y}_w$ are the weighted averages given by

$$\overline{x}_w = \frac{\sum_{i=1}^{N} w_i x_i}{\sum_{i=1}^{N} w_i} \quad \text{and} \quad \overline{y}_w = \frac{\sum_{i=1}^{N} w_i y_i}{\sum_{i=1}^{N} w_i}.$$

Section 14.2 (Confidence Intervals and Hypothesis Testing)

14.21 Refer to Exercise 14.1. Find $\text{Var}(\widehat{\beta})$ from $\boldsymbol{V} = (\boldsymbol{X}'\boldsymbol{X})^{-1}$.

14.22 Show that the residual vector $\widehat{\boldsymbol{e}}$ is distributed as

$$\widehat{\boldsymbol{e}} \sim \text{MVN}(\boldsymbol{0}, \sigma^2(\boldsymbol{I} - \boldsymbol{H})).$$

14.23 Show the following results.

(a) $\widehat{\boldsymbol{\beta}}$ and $\widehat{\boldsymbol{e}}$ are uncorrelated, that is, $\text{Cov}(\widehat{\boldsymbol{\beta}}, \widehat{\boldsymbol{e}})$ is a null matrix. Hence $\widehat{\boldsymbol{\beta}}$ and $\widehat{\boldsymbol{e}}$ are independently distributed. (*Hint*: Both $\widehat{\boldsymbol{\beta}}$ and $\widehat{\boldsymbol{e}}$ are linear transforms of the observation vector $\boldsymbol{y}$.)

(b) $\widehat{\boldsymbol{y}}$ and $\widehat{\boldsymbol{e}}$ are uncorrelated, that is, $\mathrm{Cov}(\widehat{\boldsymbol{y}}, \widehat{\boldsymbol{e}})$ is a null matrix. Hence $\widehat{\boldsymbol{y}}$ and $\widehat{\boldsymbol{e}}$ are independently distributed.

(c) $\widehat{\boldsymbol{\beta}}$ and SS_e (and hence $s^2 = \mathrm{MS}_e$) are independently distributed.

14.24 Refer to Exercises 14.17 and 14.19. Show the following results.

(a) $\mathrm{E}(\widehat{\boldsymbol{\beta}}_{\mathrm{GLS}}) = \boldsymbol{\beta}$ and $\mathrm{Cov}(\widehat{\boldsymbol{\beta}}_{\mathrm{GLS}}) = \sigma^2(\boldsymbol{X}'\boldsymbol{W}\boldsymbol{X})^{-1}$.

(b) $\mathrm{Var}(\widehat{\beta}_{\mathrm{WLS}}) = \sigma^2 / \sum_{i=1}^{N} w_i x_i^2$.

14.25 Suppose that the true linear model is

$$\boldsymbol{y} = \boldsymbol{X}_1\boldsymbol{\beta}_1 + \boldsymbol{X}_2\boldsymbol{\beta}_2 + \boldsymbol{e},$$

where $\boldsymbol{e} \sim \mathrm{MVN}(\boldsymbol{0}, \sigma^2\boldsymbol{I})$, $\boldsymbol{X}_1 : N \times p_1$, $\boldsymbol{\beta}_1 : p_1 \times 1$, $\boldsymbol{X}_2 : N \times p_2$, $\boldsymbol{\beta}_2 : p_2 \times 1$. But we wrongly fit the model $\boldsymbol{y} = \boldsymbol{X}_1\boldsymbol{\beta}_1 + \boldsymbol{e}$ and calculate the estimators of $\boldsymbol{\beta}_1$ and σ^2 as

$$\widehat{\boldsymbol{\beta}}_1 = (\boldsymbol{X}_1'\boldsymbol{X}_1)^{-1}\boldsymbol{X}_1'\boldsymbol{y}, \qquad s^2 = \frac{(\boldsymbol{y} - \boldsymbol{X}_1\widehat{\boldsymbol{\beta}}_1)'(\boldsymbol{y} - \boldsymbol{X}_1\widehat{\boldsymbol{\beta}}_1)}{N - p_1}$$

$$= \frac{\boldsymbol{y}'[\boldsymbol{I} - \boldsymbol{X}_1(\boldsymbol{X}_1'\boldsymbol{X}_1)^{-1}\boldsymbol{X}_1']\boldsymbol{y}}{N - p_1}.$$

(a) Show that, in general, $\widehat{\boldsymbol{\beta}}_1$ is a biased estimator with

$$\mathrm{Bias}(\widehat{\boldsymbol{\beta}}_1) = (\boldsymbol{X}_1'\boldsymbol{X}_1)^{-1}\boldsymbol{X}_1'\boldsymbol{X}_2\boldsymbol{\beta}_2.$$

Under what condition on $\boldsymbol{X}_1$ and $\boldsymbol{X}_2$ is $\widehat{\boldsymbol{\beta}}_1$ unbiased?

(b) Show that, in general, s^2 is a biased estimator with

$$\mathrm{Bias}(s^2) = \frac{1}{N - p_1}\boldsymbol{\beta}_2'\boldsymbol{X}_2'[\boldsymbol{I} - \boldsymbol{X}_1(\boldsymbol{X}_1'\boldsymbol{X}_1)^{-1}\boldsymbol{X}_1']\boldsymbol{X}_2\boldsymbol{\beta}_2.$$

Is the bias positive or negative? Under what condition on $\boldsymbol{\beta}_2$ is s^2 unbiased?

14.26 Consider the general linear model $\mathrm{E}(\boldsymbol{y}) = \boldsymbol{X}\boldsymbol{\beta}$ and $\mathrm{Cov}(\boldsymbol{y}) = \sigma^2\boldsymbol{W}^{-1}$, where $\boldsymbol{W}^{-1}$ is a symmetric positive-definite matrix. Prove the Gauss–Markov theorem for this model, namely, that $\widehat{\boldsymbol{\beta}}_{\mathrm{GLS}} = (\boldsymbol{X}'\boldsymbol{W}\boldsymbol{X})^{-1}\boldsymbol{X}'\boldsymbol{W}\boldsymbol{y}$ is the best linear unbiased estimator.

14.27 Assume the standard linear model (14.1) and let $\widehat{\boldsymbol{\beta}}$ be the usual LS estimator of $\boldsymbol{\beta}$. Let $\boldsymbol{\theta} = \boldsymbol{C}\boldsymbol{\beta}$, where $\boldsymbol{C}$ is a given $q \times (p+1)$ full row rank matrix. Use the Lagrangian multiplier method to show that the LS estimator of $\boldsymbol{\beta}$ under the constraint $\boldsymbol{\theta} = \boldsymbol{C}\boldsymbol{\beta} = \boldsymbol{\theta}_0$, where $\boldsymbol{\theta}_0$ is a given vector, equals

$$\widehat{\boldsymbol{\beta}}_0 = \widehat{\boldsymbol{\beta}} - \boldsymbol{V}\boldsymbol{C}'(\boldsymbol{C}\boldsymbol{V}^{-1}\boldsymbol{C}')^{-1}(\widehat{\boldsymbol{\theta}} - \boldsymbol{\theta}_0),$$

where $\boldsymbol{V} = (\boldsymbol{X}'\boldsymbol{X})^{-1}$ and $\widehat{\boldsymbol{\theta}} = \boldsymbol{C}\widehat{\boldsymbol{\beta}}$.

14.28 Show that the F-statistic (14.25) is invariant under a nonsingular transformation $\boldsymbol{\phi} = \boldsymbol{M}\boldsymbol{\theta}$, where $\boldsymbol{M}$ is any $q \times q$ nonsingular matrix. Thus, if we test the hypothesis $H_0 : \boldsymbol{\phi} = \boldsymbol{\phi}_0$ versus $H_1 : \boldsymbol{\phi} \neq \boldsymbol{\phi}_0$, where $\boldsymbol{\phi}_0 = \boldsymbol{M}\boldsymbol{\theta}_0$, then the test is equivalent to (14.25) for $H_0 : \boldsymbol{\theta} = \boldsymbol{\theta}_0$ versus $H_1 : \boldsymbol{\theta} \neq \boldsymbol{\theta}_0$ since the F-statistics are identical.

14.29 Use the result from Exercise 14.27 and the extra sum of squares method to show the result (14.27): $\mathrm{SS}_{H_0} = (\widehat{\boldsymbol{\theta}} - \boldsymbol{\theta}_0)'(\boldsymbol{C}\boldsymbol{V}\boldsymbol{C}')^{-1}(\widehat{\boldsymbol{\theta}} - \boldsymbol{\theta}_0)$.

14.30 Suppose we have two parameters θ_1 and θ_2, and their LS estimators $\widehat{\theta_1}$ and $\widehat{\theta_2}$ are independently distributed as $\widehat{\theta_1} \sim \mathrm{N}(\theta_1, \sigma_1^2)$ and $\widehat{\theta_2} \sim \mathrm{N}(\theta_2, \sigma_2^2)$. Write the formulas for $100(1-\alpha)\%$ confidence ellipses for the two cases (i) $\sigma_1^2 = \sigma_2^2 = 1$ and (ii) $\sigma_1^2 = 1, \sigma_2^2 = 4$. Comment on the shapes of the two ellipses.

14.31 Consider case (i) of the previous exercise, but now assume that $\widehat{\theta_1}$ and $\widehat{\theta_2}$ have a correlation coefficient ρ. Write the formula for $100(1-\alpha)\%$ confidence ellipse. Assume $\widehat{\theta_1} = \widehat{\theta_2} = 0$, $\nu = \infty$, and $1 - \alpha = 0.95$ and plot two ellipses for $\rho = 0.3$ and 0.7. Comment on the shapes of the two ellipses. For which ellipse is the projected confidence interval on θ_1 longer?

Section 14.3 (Power of F-Test)

14.32 For the one-way layout design, use the rule given in Section 14.3 to show that, under H_1: Not all μ_i's are equal,

$$\mathrm{E}(\mathrm{MS}_{H_0}) = \sigma^2 + \frac{\sum_{i=1}^{a} n_i \alpha_i^2}{a-1},$$

where $\alpha_i = \mu_i - \overline{\mu}$ and $\overline{\mu} = \sum_{i=1}^{a} n_i \mu_i / \sum_{i=1}^{a} n_i$. Further, show that the F-statistic has a noncentral F-distribution with n.c.p. equal to

$$\lambda^2 = \frac{\sum_{i=1}^{a} n_i \alpha_i^2}{\sigma^2}.$$

14.33 Consider the problem of testing $H_0 : \mu = 0$ versus $H_1 : \mu \neq 0$ based on a random sample $y_1, y_2, \ldots, y_n$ from a $\mathrm{N}(\mu, \sigma^2)$ distribution. The α-level t-test rejects H_0 if

$$|t| = \frac{|\overline{y}|\sqrt{n}}{s} > t_{n-1,\alpha/2},$$

where $\overline{y}$ and s are the sample mean and standard deviation of the data.

(a) What is the n.c.p. of the noncentral t-distribution (see Section A.4) of the t-statistic when $\mu \neq 0$?

(b) Using the Pearson–Hartley chart in Figure 14.4 or a suitable software, find the power of this test for $\alpha = 0.05, n = 16$, and $\mu/\sigma = 0.5$.

14.34 Compare the power of the t-test calculated in part (b) of the previous exercise with the power of the corresponding z-test, which assumes that σ is known. The power formula for the two-sided z-test can be readily obtained by modifying the formula (2.6) for the one-sided test, namely,

$$\pi(\mu) = \Phi\left(-z_{\alpha/2} + \frac{\mu\sqrt{n}}{\sigma}\right) + \Phi\left(-z_{\alpha/2} - \frac{\mu\sqrt{n}}{\sigma}\right).$$

14.35 Refer to Example 14.3 about testing $H_0 : \beta_1 = 0$ versus $H_1 : \beta_1 \neq 0$ in simple linear regression. What is the n.c.p. of the t-statistic for testing H_0?

APPENDIX A

Vector-Valued Random Variables and Some Distribution Theory

In this appendix we give a brief overview of the distribution theory of vector-valued random variables, focusing on the case when this distribution is multivariate normal (MVN). Many of these results are used throughout the text, but especially in Chapter 14 on linear models. Excellent references for more thorough expositions of the topics covered in this appendix are the books by Christensen (1996), Rencher (2000), and Searle (1971).

The mean vector and covariance matrix of a random vector are defined in Section A.1. The "sandwich formula" for the covariance matrix of a linear transform of a random vector is derived in Section A.2. The MVN distribution is introduced in Section A.3 followed by chi-square, F- and t-distributions in Section A.4. Next, Section A.5 gives some results on the distributions of quadratic forms of MVN random vectors. This section also states Cochran's theorem, which shows that idempotent quadratic forms are independently chi-square distributed. These results are useful in the analysis of variance of balanced designs in which the sums of squares due to different terms in the linear model are idempotent quadratic forms and hence independently chi-square distributed under the normality assumption. Section A.6 introduces the multivariate t-distribution which is useful in the multiple comparisons problems discussed in Chapter 4. Finally, Section A.7 gives a brief review of some multivariate normal sampling distribution theory.

Statistical Analysis of Designed Experiments: Theory and Applications By Ajit C. Tamhane

A.1 MEAN VECTOR AND COVARIANCE MATRIX OF RANDOM VECTOR

Let $\boldsymbol{x} = (x_1, x_2, \ldots, x_m)'$ be a random vector where $x_1, x_2, \ldots, x_m$ are jointly distributed r.v.'s with means $\mathrm{E}(x_i) = \mu_i$, variances

$$\mathrm{Var}(x_i) = \mathrm{E}[(x_i - \mu_i)^2] = \sigma_{ii} = \sigma_i^2,$$

and covariances

$$\mathrm{Cov}(x_i, x_j) = \mathrm{E}[(x_i - \mu_i)(x_j - \mu_j)] = \sigma_{ij}.$$

The **mean vector** of $\boldsymbol{x}$ equals

$$\boldsymbol{\mu} = \mathrm{E}(\boldsymbol{x}) = \begin{bmatrix} \mathrm{E}(x_1) \\ \mathrm{E}(x_2) \\ \vdots \\ \mathrm{E}(x_m) \end{bmatrix} = \begin{bmatrix} \mu_1 \\ \mu_2 \\ \vdots \\ \mu_m \end{bmatrix}.$$

The **covariance matrix** of $\boldsymbol{x}$ (denoted by $\boldsymbol{\Sigma} = \mathrm{Cov}(\boldsymbol{x})$) [1] is an $m \times m$ matrix with diagonal elements, $\mathrm{Var}(x_i) = \sigma_{ii} = \sigma_i^2$ and off-diagonal elements $\mathrm{Cov}(x_i, x_j) = \sigma_{ij}$:

$$\boldsymbol{\Sigma} = \begin{bmatrix} \sigma_{11} & \sigma_{12} & \cdots & \sigma_{1m} \\ \sigma_{21} & \sigma_{22} & \cdots & \sigma_{2m} \\ \vdots & \vdots & \ddots & \vdots \\ \sigma_{m1} & \sigma_{m2} & \cdots & \sigma_{mm} \end{bmatrix}.$$

Since $\mathrm{Cov}(x_i, x_j) = \mathrm{Cov}(x_j, x_i)$, we have $\sigma_{ij} = \sigma_{ji}$ for all $i \neq j$; therefore $\boldsymbol{\Sigma}$ is a symmetric matrix. Furthermore, $\boldsymbol{\Sigma}$ is a **positive semidefinite matrix** , that is, for all vectors $\boldsymbol{a} = (a_1, a_2, \ldots, a_m)'$, we have $\boldsymbol{a}'\boldsymbol{\Sigma}\boldsymbol{a} \geq 0$. In fact, if the r.v.'s x_i's are linearly independent, that is, if there is no vector $\boldsymbol{a} \neq \boldsymbol{0}$ such that $\boldsymbol{a}'\boldsymbol{x} = \sum a_i x_i$ equals a constant, then $\boldsymbol{\Sigma}$ is a **positive definite matrix**; in other words, $\boldsymbol{a}'\boldsymbol{\Sigma}\boldsymbol{a} > 0$ for all nonnull vectors $\boldsymbol{a}$.

It is easy to see that $\boldsymbol{\Sigma}$ can be expressed as

$$\boldsymbol{\Sigma} = \mathrm{E}\left(\begin{bmatrix} x_1 - \mu_1 \\ x_2 - \mu_2 \\ \vdots \\ x_m - \mu_m \end{bmatrix} \begin{bmatrix} x_1 - \mu_1, & x_2 - \mu_2, & \cdots, & x_m - \mu_m \end{bmatrix} \right)$$

$$= \mathrm{E}[(\boldsymbol{x} - \boldsymbol{\mu})(\boldsymbol{x} - \boldsymbol{\mu})'] = \mathrm{E}(\boldsymbol{x}\boldsymbol{x}') - \boldsymbol{\mu}\boldsymbol{\mu}'. \quad \text{(A.1)}$$

[1] We use the same notation Cov to denote different types of covariances, e.g., $\mathrm{Cov}(\boldsymbol{x})$ denotes the covariance matrix of a random vector $\boldsymbol{x}$, $\mathrm{Cov}(x_i, x_j)$ denotes the covariance of two scalar r.v.'s x_i and x_j, and $\mathrm{Cov}(\boldsymbol{x}, \boldsymbol{y})$ denotes the covariance matrix between two random vectors $\boldsymbol{x}$ and $\boldsymbol{y}$, i.e., the matrix of covariances between the r.v.'s x_i and the r.v.'s y_j.

A useful fact from elementary probability is that if the r.v.'s x_i and x_j are statistically independent, then $\text{Cov}(x_i, x_j) = 0$, but the converse is not always true. However, if x_i and x_j are jointly normally distributed (see Section A.3), then mutual independence is both necessary and sufficient for $\text{Cov}(x_i, x_j) = 0$.

A.2 COVARIANCE MATRIX OF LINEAR TRANSFORMATION OF RANDOM VECTOR

It is readily shown that if a, b, c, and d are constants $(a, b \neq 0)$, x and y are r.v.'s, and $u = ax + c$ and $v = by + d$, then

$$\text{Cov}(u, v) = ab\text{Cov}(x, y) \qquad \text{and} \qquad \text{Corr}(u, v) = \pm\text{Corr}(x, y),$$

where the sign is + if $ab > 0$ and the sign is − if $ab < 0$. Thus, if the r.v.'s x and y are linearly transformed, then the additive constants c and d have no effect on the covariance; furthermore, the multiplicative constants a and b have no effect on the correlation except through the sign of their product.

More generally, let $\boldsymbol{x} = (x_1, x_2, \ldots, x_m)'$ and $\boldsymbol{y} = (y_1, y_2, \ldots, y_n)'$ be two random vectors and $\boldsymbol{a} = (a_1, a_2, \ldots, a_m)'$ and $\boldsymbol{b} = (b_1, b_2, \ldots, b_n)'$ be two vectors of constants. Let

$$u = \sum_{i=1}^{m} a_i x_i = \boldsymbol{a}'\boldsymbol{x} \qquad \text{and} \qquad v = \sum_{j=1}^{n} b_j y_j = \boldsymbol{b}'\boldsymbol{y}.$$

If $\boldsymbol{\Omega} = \text{Cov}(\boldsymbol{x}, \boldsymbol{y})$ denotes the $m \times n$ covariance matrix between $\boldsymbol{x}$ and $\boldsymbol{y}$ whose elements are $\omega_{ij} = \text{Cov}(x_i, y_j)$ $(1 \leq i \leq m, 1 \leq j \leq n)$, then

$$\text{Cov}(u, v) = \text{Cov}(\boldsymbol{a}'\boldsymbol{x}, \boldsymbol{b}'\boldsymbol{y}) = \sum_{i=1}^{m}\sum_{j=1}^{n} a_i b_j \text{ Cov}(x_i, y_j) = \boldsymbol{a}'\boldsymbol{\Omega}\boldsymbol{b}. \tag{A.2}$$

A special case of the above formula is obtained by putting $v = u$:

$$\text{Var}(\boldsymbol{a}'\boldsymbol{x}) = \sum_{i=1}^{m}\sum_{j=1}^{m} a_i a_j \text{ Cov}(x_i, x_j) = \boldsymbol{a}'\boldsymbol{\Sigma}\boldsymbol{a}. \tag{A.3}$$

Note that $\text{Var}(\boldsymbol{a}'\boldsymbol{x}) = \boldsymbol{a}'\boldsymbol{\Sigma}\boldsymbol{a} \geq 0$ and equals zero iff $\boldsymbol{a}'\boldsymbol{x}$ equals a constant. This shows that $\boldsymbol{\Sigma}$ is positive semidefinite and is in fact positive definite if the r.v.' s x_i are linearly independent.

Suppose that $\boldsymbol{x}$ is an $m \times 1$ random vector and $\boldsymbol{A}$ is a $p \times m$ matrix of constants. Let $\boldsymbol{u} = \boldsymbol{A}\boldsymbol{x}$. It is readily shown that

$$\text{E}(\boldsymbol{A}\boldsymbol{x}) = \boldsymbol{A}\text{E}(\boldsymbol{x}) = \boldsymbol{A}\boldsymbol{\mu}. \tag{A.4}$$

Then using (A.1), it follows that

$$\begin{aligned}\mathrm{Cov}(\boldsymbol{u}) &= \mathrm{Cov}(\boldsymbol{A}\boldsymbol{x})\\ &= \mathrm{E}[(\boldsymbol{A}\boldsymbol{x} - \boldsymbol{A}\boldsymbol{\mu})(\boldsymbol{A}\boldsymbol{x} - \boldsymbol{A}\boldsymbol{\mu})']\\ &= \mathrm{E}[\boldsymbol{A}(\boldsymbol{x} - \boldsymbol{\mu})(\boldsymbol{x} - \boldsymbol{\mu})'\boldsymbol{A}']\\ &= \boldsymbol{A}\mathrm{E}[(\boldsymbol{x} - \boldsymbol{\mu})(\boldsymbol{x} - \boldsymbol{\mu})']\boldsymbol{A}'\\ &= \boldsymbol{A}\boldsymbol{\Sigma}\boldsymbol{A}'. \end{aligned} \tag{A.5}$$

This is known as the **sandwich formula**, which generalizes (A.3).

A.3 MULTIVARIATE NORMAL DISTRIBUTION

The random vector $\boldsymbol{x} = (x_1, x_2, \ldots, x_m)'$ has an MVN distribution with mean vector $\boldsymbol{\mu}$ and covariance matrix $\boldsymbol{\Sigma}$ [denoted by $\boldsymbol{x} \sim \mathrm{MVN}(\boldsymbol{\mu}, \boldsymbol{\Sigma})$] if the joint p.d.f. of $\boldsymbol{x} = (x_1, x_2, \ldots, x_m)'$ is given by[2]

$$f(\boldsymbol{x}) = \frac{1}{(2\pi)^{m/2}|\boldsymbol{\Sigma}|^{1/2}} \exp\left\{-\tfrac{1}{2}(\boldsymbol{x} - \boldsymbol{\mu})'\boldsymbol{\Sigma}^{-1}(\boldsymbol{x} - \boldsymbol{\mu})\right\}, \tag{A.6}$$

where $|\boldsymbol{\Sigma}|$ denotes the determinant of $\boldsymbol{\Sigma}$. In the above it is assumed that $\boldsymbol{\Sigma}$ is invertible or equivalently there are no linear dependencies among the x_i's. We will only consider this nonsingular case of the MVN distribution.

The marginal p.d.f. of each component r.v. x_i can be shown to be $\mathrm{N}(\mu_i, \sigma_i^2)$. If $\boldsymbol{\Sigma}$ is a diagonal matrix, $\boldsymbol{\Sigma} = \mathrm{diag}(\sigma_1^2, \sigma_2^2, \ldots, \sigma_m^2)$, that is, if $\mathrm{Cov}(x_i, x_j) = \sigma_{ij} = 0$ for all $i \neq j$, and thus the x_i's are uncorrelated, then the joint p.d.f. (A.6) factors into the product of the marginal p.d.f.'s:

$$f(\boldsymbol{x}) = \prod_{i=1}^{m} \frac{1}{\sqrt{2\pi}\sigma_i} \exp\left\{-\frac{1}{2\sigma_i^2}(x_i - \mu_i)^2\right\}.$$

Therefore the x_i's are independent and are distributed as $\mathrm{N}(\mu_i, \sigma_i^2)$. The converse of this result is immediate. Therefore if $\boldsymbol{x} = (x_1, x_2, \ldots, x_m)'$ is MVN distributed, then the x_i's are independent if and only if they are uncorrelated.

The following is a useful property of the MVN distribution: If $\boldsymbol{x} \sim \mathrm{MVN}(\boldsymbol{\mu}, \boldsymbol{\Sigma})$, then any fixed (nonrandom) nonsingular linear transformation of $\boldsymbol{x}$ also has an MVN distribution. Specifically, let $\boldsymbol{A}$ be a $p \times m$ fixed matrix with linearly independent rows. Then $\boldsymbol{u} = \boldsymbol{A}\boldsymbol{x} = (u_1, u_2, \ldots, u_p)'$ has an MVN distribution of dimension p with the mean vector and covariance matrix given by

$$\mathrm{E}(\boldsymbol{u}) = \boldsymbol{A}\boldsymbol{\mu} \qquad \text{and} \qquad \mathrm{Cov}(\boldsymbol{u}) = \boldsymbol{A}\boldsymbol{\Sigma}\boldsymbol{A}'$$

[2]All density functions are generically denoted by $f(\cdot)$.

using Eq. (A.4) and (A.5). Two cases of this result are of particular interest:

(a) Since $\boldsymbol{\Sigma}$ is invertible, it can be shown that there exists an $m \times m$ symmetric matrix, say $\boldsymbol{P}$, such that $\boldsymbol{P\Sigma P'} = \boldsymbol{I}$ and $\boldsymbol{P'P} = \boldsymbol{\Sigma}^{-1}$. Then

$$\boldsymbol{z} = \boldsymbol{P}(\boldsymbol{x} - \boldsymbol{\mu}) \tag{A.7}$$

is MVN with

$$\text{E}(\boldsymbol{z}) = \boldsymbol{0} \qquad \text{and} \qquad \text{Cov}(\boldsymbol{z}) = \boldsymbol{P\Sigma P'} = \boldsymbol{I},$$

that is, $z_1, z_2, \ldots, z_m$ are i.i.d. N(0, 1) r.v.'s. Thus (A.7) can be viewed as a standardizing transformation.

(b) Let $\boldsymbol{a} = (a_1, a_2, \ldots, a_m)'$ be a vector of constants. Then $u = \boldsymbol{a}'\boldsymbol{x} = \sum a_i x_i$ is univariate normal with

$$\text{E}(u) = \boldsymbol{a}'\boldsymbol{\mu} = \sum_{i=1}^{m} a_i \mu_i \qquad \text{and} \qquad \text{Var}(u) = \boldsymbol{a}'\boldsymbol{\Sigma a} = \sum_{i=1}^{m}\sum_{j=1}^{m} a_i a_j \sigma_{ij}.$$

Consider an MVN vector $\boldsymbol{z} = (z_1, z_2, \ldots, z_m)'$ such that the z_i's are N(0, 1) r.v.'s and $\text{Corr}(z_i, z_j) = \rho_{ij}$. In some multiple comparison and selection problems studied in Chapter 4, we need the upper α critical points of $\max_{1 \le i \le m} z_i$ and $\max_{1 \le i \le m} |z_i|$. For arbitrary correlations, these critical points can be computed using the SAS-IML program of Genz and Bretz (1999). In Tables C.6 and C.7, which are for the critical points of the multivariate t-distribution, we have tabulated these critical points (the entries for $\nu = \infty$) for the special case of equal correlation: $\rho_{ij} = \rho = \frac{1}{2}$ for selected values of $m = a, \rho, \alpha$.

A.4 CHI-SQUARE, F-, AND t-DISTRIBUTIONS

Let $z_1, z_2, \ldots, z_m$ be i.i.d. N(0, 1) r.v.'s. Then the r.v.

$$u = \boldsymbol{z}'\boldsymbol{z} = z_1^2 + z_2^2 + \cdots + z_m^2$$

has a **central chi-square distribution** with m d.f.. We denote this as $u \sim \chi^2_m$. The p.d.f. of u is given by

$$f(u) = \frac{1}{2^{m/2}\Gamma(m/2)} u^{m/2-1} e^{-u/2} \qquad \text{for } u \ge 0.$$

It can be shown that

$$\text{E}(\chi^2_m) = m \qquad \text{and} \qquad \text{Var}(\chi^2_m) = 2m.$$

Usually, we drop the prefix "central." We denote the upper α critical point of the chi-square distribution with m d.f. by $\chi^2_{m,\alpha}$. These critical points are tabulated in Table C.3.

To evaluate the powers of ANOVA F-tests we need the distribution of the sum of squares of independent normal r.v.'s with nonzero means. Toward this end, let $z_1, z_2, \ldots, z_m$ be independent r.v.'s with $z_i \sim \mathrm{N}(\delta_i, 1)$ where the δ_i's are arbitrary constants. Then the r.v.

$$u = z'z = z_1^2 + z_2^2 + \cdots + z_m^2$$

has a **noncentral chi-square distribution** with m d.f. and n.c.p.

$$\lambda^2 = \sum_{i=1}^{m} \delta_i^2 = \boldsymbol{\delta}'\boldsymbol{\delta}, \tag{A.8}$$

where $\boldsymbol{\delta} = (\delta_1, \delta_2, \ldots, \delta_m)'$. We write $u \sim \chi^2_m(\lambda^2)$. Note that the n.c.p. depends on the δ_i's only through $\sum \delta_i^2$. Thus, for example, a $\boldsymbol{\delta}$ vector with $\delta_1 = \cdots = \delta_m = \lambda/\sqrt{m}$ and another $\boldsymbol{\delta}$ vector with $\delta_1 = \lambda, \delta_2 = \cdots = \delta_m = 0$ both result in the same n.c.p. λ^2. The p.d.f. of the $\chi^2_m(\lambda^2)$ r.v. is given by

$$f(u) = e^{-\lambda^2/2} \sum_{k=0}^{\infty} \frac{(\lambda^2/2)^k}{k!} \left[\frac{u^{m/2+k-1} e^{-u/2}}{2^{m/2+k}\Gamma(m/2+k)} \right] \quad \text{for } u \geq 0.$$

Observe that if we put $w_k = e^{-\lambda^2/2}(\lambda^2/2)^k/k!$, then $\sum_{k=0}^{\infty} w_k = 1$, so the above p.d.f. is the weighted mixture of the p.d.f.'s of central χ^2_{m+2k} r.v.'s with these Poisson weights for $k = 0, \ldots, \infty$. If $\lambda^2 = 0$, then we have the central chi-square distribution, in which case we omit the n.c.p. and write $u \sim \chi^2_m$ as already indicated. The expected value and the variance of a noncentral chi-square r.v. are given by

$$\mathrm{E}[\chi^2_m(\lambda^2)] = m + \lambda^2 \quad \text{and} \quad \mathrm{Var}[\chi^2_m(\lambda^2)] = 2m + 4\lambda^2. \tag{A.9}$$

Let $u \sim \chi^2_m$ and $v \sim \chi^2_n$ be independently distributed. Then the ratio

$$w = \frac{u/m}{v/n} \sim \frac{\chi^2_m/m}{\chi^2_n/n}$$

is said to have a **central F-distribution** with m and n d.f., and we write $w \sim F_{m,n}$. Generally, we drop the prefix "central." The p.d.f. of an $F_{m,n}$ r.v. is given by

$$f(w) = \frac{\Gamma[(m+n)/2]}{\Gamma(m/2)\Gamma(n/2)} \left(\frac{m}{n}\right)^{m/2} w^{m/2-1} \left(1 + \frac{m}{n}w\right)^{-(m+n)/2} \quad \text{for } w \geq 0.$$

The lower α critical point of an $F_{m,n}$ r.v. is denoted by $f_{m,n,1-\alpha}$ and the upper α critical point by $f_{m,n,\alpha}$. The two are related by the formula

$$f_{m,n,1-\alpha} = \frac{1}{f_{n,m,\alpha}}.$$

As an example, $f_{5,10,0.05} = 3.33$ and $f_{10,5,0.95} = 1/3.33 = 0.300$. Hence only the upper α critical points are tabulated in Table C.4.

If $u \sim \chi^2_m(\lambda^2)$, $v \sim \chi^2_n$, and the two are independent, then the distribution of the ratio $w = (u/m)/(v/n)$ is called a **noncentral F-distribution** with n.c.p. λ^2 and d.f. m and n. We write $w \sim F_{m,n}(\lambda^2)$. The p.d.f. of an $F_{m,n}(\lambda^2)$ r.v. is the weighted mixture of the p.d.f.'s of central $F_{m+2k,n}$ r.v.'s with the same Poisson weights w_k for $k = 0, \ldots, \infty$ that appear in the formula for the p.d.f. of the noncentral chi-square distribution. If $\lambda^2 = 0$, then we have the central F-distribution; in this case we denote $F_{m,n}(0)$ simply by $F_{m,n}$.

A t r.v. with n d.f. is defined as

$$t_n = \frac{z}{\sqrt{u/n}},$$

where $z \sim \mathrm{N}(0, 1)$ and $u \sim \chi^2_n$ are independent. The p.d.f. of a t_n r.v. is given by

$$f(t) = \frac{\Gamma[(n+1)/2]}{\sqrt{\pi n}\,\Gamma(n/2)} \left(1 + \frac{t^2}{n}\right)^{-(n+1)/2}, \qquad -\infty < t < \infty.$$

Since $z^2 \sim \chi^2_1$, it follows that $t^2_n \sim F_{1,n}$. The percentiles of the two distributions are thus related by

$$t^2_{n,\alpha/2} = f_{1,n,\alpha}. \tag{A.10}$$

Analogous to noncentral F, we define a **noncentral t-distribution** with n d.f. and n.c.p. λ as the distribution of the r.v.

$$t_n(\lambda) = \frac{z + \lambda}{\sqrt{u/n}}.$$

Note that λ may be positive or negative. Furthermore, since $(z + \lambda)^2 \sim \chi^2_1(\lambda^2)$, it follows that

$$t^2_n(\lambda) \sim F_{1,n}(\lambda^2).$$

A.5 DISTRIBUTIONS OF QUADRATIC FORMS

A **quadratic form** in $\boldsymbol{x}$ is defined as

$$\boldsymbol{x}'\boldsymbol{A}\boldsymbol{x} = \sum_i a_{ii}x_i^2 + \sum_{i \neq j} a_{ij}x_i x_j,$$

where $\boldsymbol{x}$ is an m-dimensional random vector and $\boldsymbol{A}$ is an $m \times m$ matrix with constant elements a_{ij}. It is easy to show that $\boldsymbol{A}$ can be assumed to be symmetric without loss of generality. We will make this assumption from now on. In the following lemmas we give results about joint distributions of quadratic forms in $\boldsymbol{x}$.

Lemma A.1 *If* $E(\boldsymbol{x}) = \boldsymbol{\mu}$ *and* $Cov(\boldsymbol{x}) = \sigma^2\boldsymbol{V}$, *then*

$$E(\boldsymbol{x}'\boldsymbol{A}\boldsymbol{x}) = \boldsymbol{\mu}'\boldsymbol{A}\boldsymbol{\mu} + \sigma^2\, tr(\boldsymbol{A}\boldsymbol{V}).$$

Proof. We have

$$\begin{aligned}
\mathrm{E}(\boldsymbol{x}'\boldsymbol{A}\boldsymbol{x}) &= \mathrm{E}[\mathrm{tr}(\boldsymbol{A}\boldsymbol{x}\boldsymbol{x}')] \\
&= \mathrm{tr}[A\mathrm{E}(\boldsymbol{x}\boldsymbol{x}')] \quad (\text{since } \boldsymbol{A} \text{ is constant and } \mathrm{E}[\mathrm{tr}(\cdot)] = \mathrm{tr}[\mathrm{E}(\cdot)]) \\
&= \mathrm{tr}[\boldsymbol{A}(\boldsymbol{\mu}\boldsymbol{\mu}' + \sigma^2\boldsymbol{V})] \quad (\text{using } (A.1) \text{ with } \boldsymbol{\Sigma} = \sigma^2\boldsymbol{V}) \\
&= \boldsymbol{\mu}'\boldsymbol{A}\boldsymbol{\mu} + \sigma^2\mathrm{tr}(\boldsymbol{A}\boldsymbol{V}).
\end{aligned}$$

This proves the lemma. ■

Lemma A.2 *Let* $\boldsymbol{x} \sim MVN(\boldsymbol{\mu}, \sigma^2\boldsymbol{I})$. *Then* $\boldsymbol{x}'\boldsymbol{A}\boldsymbol{x}$ *is distributed as* σ^2 *times a noncentral chi-square r.v. with* n *d.f. and n.c.p.* $\lambda^2 = \boldsymbol{\mu}'\boldsymbol{A}\boldsymbol{\mu}/\sigma^2$ *[denoted as* $\boldsymbol{x}'\boldsymbol{A}\boldsymbol{x} \sim \sigma^2\chi^2_n(\lambda^2)$*] if and only if* $\boldsymbol{A}$ *is an idempotent matrix of rank* $n \leq m$.

Proof. Since $\boldsymbol{A}$ is symmetric, there exists an $m \times m$ orthogonal matrix $\boldsymbol{P}$ (i.e., $\boldsymbol{P}'\boldsymbol{P} = \boldsymbol{P}\boldsymbol{P}' = \boldsymbol{I}$) such that $\boldsymbol{P}'\boldsymbol{A}\boldsymbol{P}$ is a diagonal matrix with diagonal entries equal to the eigenvalues of $\boldsymbol{A}$. Since $\boldsymbol{A}$ is idempotent of rank n, it has n eigenvalues equal to 1 and $m - n$ eigenvalues equal to 0. Therefore $\boldsymbol{P}'\boldsymbol{A}\boldsymbol{P}$ can be written as

$$\boldsymbol{P}'\boldsymbol{A}\boldsymbol{P} = \begin{bmatrix} \boldsymbol{I}_{n\times n} & \boldsymbol{O}_{n\times(m-n)} \\ \boldsymbol{O}_{(m-n)\times n} & \boldsymbol{O}_{(m-n)\times(m-n)} \end{bmatrix}.$$

Let $\boldsymbol{y} = \boldsymbol{P}'\boldsymbol{x}$. It follows that $\boldsymbol{y}$ is MVN with $\mathrm{E}(\boldsymbol{y}) = \boldsymbol{P}'\boldsymbol{\mu}$ and $\mathrm{Cov}(\boldsymbol{y}) = \sigma^2\boldsymbol{P}'\boldsymbol{I}\boldsymbol{P} = \sigma^2\boldsymbol{P}'\boldsymbol{P} = \sigma^2\boldsymbol{I}$. Thus the y_i's are independent $\mathrm{N}(\delta_i, \sigma^2)$ r.v.'s where δ_i is the ith element of $\boldsymbol{\delta} = \boldsymbol{P}'\boldsymbol{\mu}$. Furthermore,

$$\boldsymbol{x}'\boldsymbol{A}\boldsymbol{x} = \boldsymbol{x}'\boldsymbol{P}\boldsymbol{P}'\boldsymbol{A}\boldsymbol{P}\boldsymbol{P}'\boldsymbol{x} = \boldsymbol{y}'\boldsymbol{P}'\boldsymbol{A}\boldsymbol{P}\boldsymbol{y} = \boldsymbol{y}'\begin{bmatrix} \boldsymbol{I}_{n\times n} & \boldsymbol{O}_{n\times(m-n)} \\ \boldsymbol{O}_{(m-n)\times n} & \boldsymbol{O}_{(m-n)\times(m-n)} \end{bmatrix}\boldsymbol{y} = \sum_{i=1}^{n} y_i^2.$$

Hence $\boldsymbol{x}'\boldsymbol{A}\boldsymbol{x}$ has a noncentral chi-square distribution with n d.f. and n.c.p.

$$\lambda^2 = \frac{\sum_{i=1}^{n}\delta_i^2}{\sigma^2} = \frac{\boldsymbol{\mu}'\boldsymbol{A}\boldsymbol{\mu}}{\sigma^2}.$$

Note that using formula (A.9) for the expectation of a noncentral chi-square r.v., we get

$$\mathrm{E}(\boldsymbol{x}'\boldsymbol{A}\boldsymbol{x}) = \sigma^2\mathrm{E}[\chi_n^2(\lambda^2)] = \sigma^2(\lambda^2 + n) = \boldsymbol{\mu}'\boldsymbol{A}\boldsymbol{\mu} + \sigma^2\mathrm{tr}(\boldsymbol{A}),$$

which agrees with the result of Lemma A.1. In the above we have used the fact that $n = \mathrm{rank}(\boldsymbol{A}) = \mathrm{tr}(\boldsymbol{A})$. The "only if" part of the proof is left to the reader. ■

Corollary A.1 *If $\boldsymbol{\mu} = \boldsymbol{0}$, then $\boldsymbol{x}'\boldsymbol{A}\boldsymbol{x}$ has a central chi-square distribution with n d.f.*

Corollary A.2 *If Cov($\boldsymbol{x}$) $= \sigma^2\boldsymbol{V}$, then Lemma A.2 holds if and only if $\boldsymbol{A}\boldsymbol{V}$ is an idempotent matrix of rank n.*

Proof. See Searle (1971, p. 57).

Lemma A.3 *Let $\boldsymbol{x} \sim MVN(\boldsymbol{\mu}, \sigma^2\boldsymbol{V})$ and let $\boldsymbol{A}$ and $\boldsymbol{B}$ be two symmetric $m \times m$ fixed matrices. The quadratic forms $\boldsymbol{x}'\boldsymbol{A}\boldsymbol{x}$ and $\boldsymbol{x}'\boldsymbol{B}\boldsymbol{x}$ are independent if and only if $\boldsymbol{A}\boldsymbol{V}\boldsymbol{B} = \boldsymbol{O}$.*

Proof. See Searle (1971, p. 59). ■

Lemma A.4 *Let $\boldsymbol{x} \sim MVN(\boldsymbol{\mu}, \sigma^2\boldsymbol{V})$, and let $\boldsymbol{A}$ be a symmetric $m \times m$ constant matrix and $\boldsymbol{B}$ be an $n \times m$ constant matrix. Then $\boldsymbol{x}'\boldsymbol{A}\boldsymbol{x}$ and $\boldsymbol{B}\boldsymbol{x}$ are independent if and only if $\boldsymbol{B}\boldsymbol{V}\boldsymbol{A} = \boldsymbol{O}$.*

Proof. See Searle (1971, p. 59). ■

Lemma A.5 *Let $\boldsymbol{x} \sim MVN(\boldsymbol{\mu}, \boldsymbol{\Sigma})$, where $\boldsymbol{\Sigma}$ is a positive-definite matrix. Then*

$$(\boldsymbol{x} - \boldsymbol{\mu})'\boldsymbol{\Sigma}^{-1}(\boldsymbol{x} - \boldsymbol{\mu}) \sim \chi_m^2.$$

More generally, if $\boldsymbol{\mu}_0 \neq \boldsymbol{\mu}$, then

$$(\boldsymbol{x} - \boldsymbol{\mu}_0)'\boldsymbol{\Sigma}^{-1}(\boldsymbol{x} - \boldsymbol{\mu}_0) \sim \chi_m^2(\lambda^2) \quad \textit{with} \quad \lambda^2 = (\boldsymbol{\mu} - \boldsymbol{\mu}_0)'\boldsymbol{\Sigma}^{-1}(\boldsymbol{\mu} - \boldsymbol{\mu}_0).$$

Proof. Make the standardizing transformation (A.7), $\boldsymbol{z} = \boldsymbol{P}(\boldsymbol{x} - \boldsymbol{\mu}) \sim$ MVN$(\boldsymbol{0}, \boldsymbol{I})$, and note that

$$(\boldsymbol{x} - \boldsymbol{\mu})'\boldsymbol{P}'\boldsymbol{P}(\boldsymbol{x} - \boldsymbol{\mu}) = \boldsymbol{z}'\boldsymbol{z} = \sum_{i=1}^{m} z_i^2 \sim \chi_m^2.$$

To prove the second part, make the transformation $z = P(x - \mu_0)$. Then $z \sim \text{MVN}(P(\mu - \mu_0), I)$, that is, the z_i are independent normal with unit variances and the mean vector of z is $P(\mu - \mu_0)$. Therefore,

$$(x - \mu_0)'\Sigma^{-1}(x - \mu_0) = z'z \sim \chi^2_m(\lambda^2)$$

with

$$\lambda^2 = (\mu - \mu_0)'P'P(\mu - \mu_0) = (\mu - \mu_0)'\Sigma^{-1}(\mu - \mu_0).$$

This proves the lemma. ■

The following theorem is known as **Cochran's theorem**.

Theorem A.1 *Let $y \sim MVN(\mu, \sigma^2 I)$ and let $y'y = y'Iy = y'A_1y + \cdots + y'A_ky$, where y is n-dimensional and the A_i are idempotent (projection) matrices with rank$(A_i) = r_i$. Then the quadratic forms $y'A_iy$ are independently distributed as $\sigma^2\chi^2_{r_i}(\lambda_i^2)$ with n.c.p.'s $\lambda_i^2 = \mu'A_i\mu/\sigma^2$ if and only if $\sum r_i = n$.*

Proof. From the properties of idempotent matrices listed in Section 14.1.1, we know that $\sum r_i = n$ if and only if $A_1 + \cdots + A_k = I$, where each A_i is idempotent or if and only if $A_iA_j = O$ for all $i \neq j$. But A_i is idempotent implies that $y'A_iy \sim \sigma^2\chi^2_{r_i}(\lambda_i^2)$ with $\lambda_i^2 = \mu'A_i\mu/\sigma^2$ using Lemma A.2. Also, using Lemma A.3, $A_iA_j = O$ for all $i \neq j$ implies that the quadratic forms $y'A_iy$ are independent. ■

Cochran's theorem underlies the ANOVA decompositions of orthogonal designs. For example, the following lemma gives an application of Cochran's theorem for the ANOVA decomposition used in balanced two-way layouts.

Lemma A.6 *Let $\{x_{ijk}\ (1 \le i \le a, 1 \le j \le b, 1 \le k \le n)\}$ be a vector of dimension $N = abn$ whose elements are arranged in the lexicographic order. Define the "bar" and "dot" notations as in Chapter 5, that is, a bar indicates the average and a dot indicates the subscript over which the average is taken. Then the five N-vectors (with their elements arranged in the same lexicographic order)*

$$\{\bar{x}_{\cdots}\}, \{\bar{x}_{i\cdot\cdot} - \bar{x}_{\cdots}\}, \{\bar{x}_{\cdot j\cdot} - \bar{x}_{\cdots}\}, \{\bar{x}_{ij\cdot} - \bar{x}_{i\cdot\cdot} - \bar{x}_{\cdot j\cdot} + \bar{x}_{\cdots}\}, \{x_{ijk} - \bar{x}_{ij\cdot}\}$$

are orthogonal to each other and represent projections of the vector $\{x_{ijk}\}$ into five mutually orthogonal subspaces, say S_1, S_2, S_3, S_4, and S_5, of dimensions $1, a-1, b-1, (a-1)(b-1)$, and $ab(n-1)$, respectively. If the x_{ijk} are independent normal with a constant variance, then these vectors are mutually independent.

Proof. Follows by checking that the dot products between all pairs of the above five vectors are zero, and $\{x_{ijk}\}$ is their sum. ■

Corollary A.3 *We have*

$$\sum_{i=1}^{a}\sum_{j=1}^{b}\sum_{k=1}^{n} x_{ijk}^2 = N\overline{x}_{...}^2 + bn\sum_{i=1}^{a}(\overline{x}_{i..} - \overline{x}_{...})^2 + an\sum_{j=1}^{b}(\overline{x}_{.j.} - \overline{x}_{...})^2$$

$$+ n\sum_{i=1}^{a}\sum_{j=1}^{b}(\overline{x}_{ij.} - \overline{x}_{i..} - \overline{x}_{.j.} + \overline{x}_{...})^2$$

$$+ \sum_{i=1}^{a}\sum_{j=1}^{b}\sum_{k=1}^{n}(x_{ijk} - \overline{x}_{ij.})^2.$$

If the x_{ijk} are independent normal with a common variance, then the terms on the right-hand side are independent chi-squared (in general, noncentral) with d.f. $1, a-1, b-1, (a-1)(b-1)$, and $ab(n-1)$, respectively, and the term on the left-hand side is chi-squared (in general, noncentral) with d.f. N.

Corollary A.4 *Let U and V be two subspaces formed by summing any subset of disjoint orthogonal subspaces $S_1, \ldots, S_5$, for example, $U = S_2$ and $V = S_3 \oplus S_4$ (where $\oplus$ represents the vector space formed of sums of vectors in the two subspaces). Then any vector in U is orthogonal to any vector in V, and under the given distributional assumptions, the two vectors are independent. For the given example, the vector $\{\overline{x}_{i..} - \overline{x}_{...}\}$ in U and the vector*

$$\{\overline{x}_{.j.} - \overline{x}_{...}\} + \{\overline{x}_{ij.} - \overline{x}_{i..} - \overline{x}_{.j.} + \overline{x}_{...}\} = \{\overline{x}_{ij.} - \overline{x}_{i..}\}$$

in V are independent if the x_{ijk} are mutually independent.

A.6 MULTIVARIATE t-DISTRIBUTION

Let $(z_1, z_2, \ldots, z_m)$ have a standard MVN distribution with correlation matrix $\{\rho_{ij}\}$ [i.e., $\text{Corr}(z_i, z_j) = \rho_{ij}$]. Let $u \sim \chi^2_\nu$ be distributed independently of $(z_1, z_2, \ldots, z_m)$. Define r.v.'s

$$t_1 = \frac{z_1}{\sqrt{u/\nu}}, t_2 = \frac{z_2}{\sqrt{u/\nu}}, \ldots, t_m = \frac{z_m}{\sqrt{u/\nu}}.$$

Then the joint distribution of $(t_1, t_2, \ldots, t_m)$ is known as the m-variate t-distribution with correlation matrix $\{\rho_{ij}\}$ and d.f. ν. The marginal distribution of each t_i is univariate Student's t with ν d.f. As $\nu \to \infty$, the distribution of $(t_1, t_2, \ldots, t_m)$ converges to that of $(z_1, z_2, \ldots, z_m)$.

In some multiple comparison and selection problems we need the upper α critical points of $\max_{1 \le i \le m} t_i$ and $\max_{1 \le i \le m} |t_i|$. For arbitrary correlations, these critical points can be computed using the SAS-IML program of Genz and Bretz (1999). The one-sided critical points are denoted by $t_{m,\nu,\{\rho_{ij}\},\alpha}$, which satisfy the equation

$$\mathrm{P}\left\{\max_{1\le i\le m} t_i \le t_{m,\{\rho_{ij}\},\alpha}\right\} = \mathrm{P}\{t_i \le t_{m,\{\rho_{ij}\},\alpha}\ (1 \le i \le m)\} = 1-\alpha.$$

The two-sided critical points are denoted by $|t|_{m,\{\rho_{ij}\},\alpha}$, which satisfy the equation

$$\mathrm{P}\left\{\max_{1\le i\le m} |t_i| \le |t|_{m,\{\rho_{ij}\},\alpha}\right\} = \mathrm{P}\{|t_i| \le |t|_{m,\{\rho_{ij}\},\alpha}\ (1 \le i \le m)\} = 1-\alpha.$$

These critical points are tabulated in Tables C.6 and C.7, respectively, for the special case of equal correlation: $\rho_{ij} = \rho = 0.5$ for selected values of $m,\ \nu,\ \alpha$.

A.7 MULTIVARIATE NORMAL SAMPLING DISTRIBUTION THEORY

The m-variate normal distribution was defined in Section A.3. Now consider n i.i.d. observations $\boldsymbol{x}_1, \boldsymbol{x}_2, \dots, \boldsymbol{x}_n$ from an MVN$(\boldsymbol{\mu}, \boldsymbol{\Sigma})$ distribution where $\boldsymbol{x}_i = (x_{i1}, x_{i2}, \dots, x_{im})'$. The sample mean vector and the sample covariance matrix are defined as

$$\overline{\boldsymbol{x}} = \frac{1}{n}\sum_{i=1}^{n}\boldsymbol{x}_i, \quad \boldsymbol{S} = \frac{1}{n-1}\sum_{i=1}^{n}(\boldsymbol{x}_i - \overline{\boldsymbol{x}})(\boldsymbol{x}_i - \overline{\boldsymbol{x}})' = \frac{1}{n-1}\left[\sum_{i=1}^{n}\boldsymbol{x}_i\boldsymbol{x}_i' - n\overline{\boldsymbol{x}}\overline{\boldsymbol{x}}'\right].$$

These sample statistics have the following properties:

(a) $\overline{\boldsymbol{x}}$ is an unbiased estimator of $\boldsymbol{\mu}$ and $\boldsymbol{S}$ is an unbiased estimator of $\boldsymbol{\Sigma}$.
(b) $\overline{\boldsymbol{x}}$ and $\boldsymbol{S}$ are distributed independently of each other.
(c) $\overline{\boldsymbol{x}} \sim$ MVN$\left(\boldsymbol{\mu}, \frac{1}{n}\boldsymbol{\Sigma}\right)$.
(d) Let $\boldsymbol{A} = (n-1)\boldsymbol{S} = \sum_{i=1}^{n}(\boldsymbol{x}_i - \overline{\boldsymbol{x}})(\boldsymbol{x}_i - \overline{\boldsymbol{x}})'$. For positive-definite $\boldsymbol{A}$, the p.d.f. of $\boldsymbol{A}$ is given by

$$f(\boldsymbol{A}) = \frac{|\boldsymbol{A}|^{(\nu-m-1)/2}\exp\left[-\frac{1}{2}\,\mathrm{tr}(\boldsymbol{\Sigma}^{-1}\boldsymbol{A})\right]}{2^{m\nu/2}\pi^{m(m-1)/4}|\boldsymbol{\Sigma}|^{\nu/2}\prod_{i=1}^{m}\Gamma[(\nu+1-i)/2]}.$$

This is called the **Wishart distribution** and is denoted by $W_m(\nu, \boldsymbol{\Sigma})$, where $\nu = n-1$ is the d.f. For the univariate case when $m = 1$ and $\boldsymbol{\Sigma} = \sigma^2$, this distribution reduces to the χ^2_ν distribution.

The following lemma is an extension of Lemma A.5 to the case where the covariance matrix is estimated. The proof of this lemma can be found in Anderson (1958).

Lemma A.7 *Let $\boldsymbol{x} \sim MVN(\boldsymbol{\mu}, \boldsymbol{\Sigma})$, where $\boldsymbol{\Sigma}$ is a positive definite matrix. Further let $\boldsymbol{S}$ be an $m \times m$ positive definite sample covariance matrix with ν d.f. independent of $\boldsymbol{x}$. Then the quadratic form*

$$T^2 = (\boldsymbol{x} - \boldsymbol{\mu})'\boldsymbol{S}^{-1}(\boldsymbol{x} - \boldsymbol{\mu}) \sim \frac{\nu m}{\nu - m + 1} F_{m,\nu-m+1}.$$

This extension forms the basis of **Hotelling's T^2-test** discussed in Section 13.2.1 on the mean vector of an MVN distribution or for comparing the mean vectors of two MVN distributions. Note that if $\nu \to \infty$, then $\boldsymbol{S} \to \boldsymbol{\Sigma}$ with probability 1, and the distribution of the quadratic form approaches $mF_{m,\infty} = \chi^2_m$. Thus we obtain the result of Lemma A.5.

APPENDIX B

Case Studies

Three case studies are described in this appendix. The first two are taken from student projects in my design of experiments class. The third is taken from an experiment reported by Phadke (1986). The purpose of these case studies is to illustrate complexities of practical experiments even in relatively simple settings. Hopefully, they will also help the students and instructors to get an idea of the level of difficulty expected in a class project and how the report should be written. (It should be noted that the accounts presented here are more concise than a typical report, which is about 20 pages long, and contains more detailed data analysis including treatment of outliers, model diagnostics, and different data transformations.)

B.1 CASE STUDY 1: EFFECTS OF FIELD STRENGTH AND FLIP ANGLE ON MRI CONTRAST

B.1.1 Introduction

This project was done by Nicole Campbell and Lan Ge in my Spring 2006 Design of Experiments class. The goal of the project was to study a method for magnetic resonance imaging (MRI), called blood oxygen level dependent (BOLD) imaging, which achieves tissue contrast without administration of an exogenous agent. Tissues containing oxygenated blood appear bright on BOLD images, while areas of deoxygenated blood appear dark. Applications of BOLD imaging include functional brain imaging and cardiovascular assessment.

Two key factors that affect the contrast between oxygenated and deoxygenated blood are flip angle and field strength. Flip angle refers to the degree to which magnetic moments in the tissues are tipped after application of radio-frequency (RF) pulse. RF tipping is essential for image creation. Theoretically, a high flip angle results in a greater signal and therefore greater image contrast. However,

Statistical Analysis of Designed Experiments: Theory and Applications By Ajit C. Tamhane

a higher flip angle deposits more energy in the tissue, which is not desirable; therefore it is of interest to know if a similar contrast can be achieved with a lower flip angle. The source of MR images is related to the alignment of magnetic moments with an external magnetic field. The most common field strength of a clinical scanner is 1.5 Tesla (T). For reference, 1 T $=$ 10,000G and the magnetic field of earth is less than 1 G. Recently, 3.0-T scanners have become available. Theoretically, there should be a twofold increase in contrast with 3.0 T versus 1.5 T. The trade-off is increased sensitivity to small disturbances that can distort the field.

B.1.2 Design

The experiment used the technique of inducing large differences in the oxygenation level of the muscle tissue of the lower leg of a subject by applying pressure to the upper leg. An inflatable pressure cuff placed around the subject's upper leg restricts the blood flow to the lower leg and generates deoxygenated tissue (ischemia). Release of the pressure generates elevated level of oxygen (reactive hypermia). The restrictions on regions of interest (ROI) placement (caused by the leg size and the need to avoid blood vessels) resulted in a different number of ROIs for each subject, causing the design to be highly imbalanced.

The response variable was "contrast," defined as

$$\text{Contrast} = y = \frac{S_{\text{hypermia}} - S_{\text{ischemia}}}{\sigma_{\text{noise}}},$$

where S_{hypermia} and S_{ischemia} are the average signal intensities of a given ROI during reactive hypermia and ischemia, respectively, and σ_{noise} is the standard deviation of the background noise, measured from an ROI of air outside the leg. This contrast measure was used to normalize the results from different subjects.

Two scanners, one with field strength 1.5 T and the other with field strength 3.0 T, were used. In each scanner, three flip angles (30°, 60°, and 90°) could be set, resulting in six treatment combinations. Three subjects participated in the study who were tested under all six treatment combinations. All measurements on a particular subject were carried out in the same scanner before switching to the other scanner. The order of the scanner selection and the flip angle for the given scanner were randomized. Thus this is a split-plot design with the scanner (the field strength) as the whole-plot factor (labeled A) and the flip angle as the subplot factor (labeled B). These two treatment factors are fixed, while the subject factor (labeled C) is random. Finally, it should be noted that the difference in the two field strengths is confounded with the difference in the two scanners, which differed in many respects, including age and hardware. The data for the experiment are shown in Table B.1.

Table B.1 Contrast Data for BOLD MRI Experiment

Flip Angle	Subject	Field Strength 1.5T	3T
30°	1	0.570, 0.221, 0.804 ($n = 3$)	2.904, 3.221, 3.939, 3.333, 2.728, 3.380, 3.798, 2.764, 4.226, 2.074 ($n = 10$)
	2	1.113 ($n = 1$)	2.488, 2.359, 3.537, 4.455 ($n = 4$)
	3	0.302, −0.160, 2.217 ($n = 3$)	5.170, 3.736, −0.435, −0.738, −0.208, 0.089 ($n = 6$)
60°	1	1.229, 0.088, 4.753 ($n = 3$)	4.987, 5.047, 3.564, 6.194, 5.585, 3.922, 5.382, 5.353, 6.058, 4.667 ($n = 10$)
	2	2.753, 2.460 ($n = 2$)	2.174, 1.684, 3.735, 5.887 ($n = 4$)
	3	3.762, 2.724, 3.213 ($n = 3$)	7.646, 4.985, 5.240, 6.248, 7.108, 7.934 ($n = 6$)
90°	1	2.356, 3.447, 7.265 ($n = 3$)	8.652, 5.968, 4.598, 6.668, 7.268, 5.224, 4.867, 3.788, 6.046, 6.223 ($n = 10$)
	2	4.153, 4.771 ($n = 2$)	4.176, 2.216, 3.551, 9.0156 ($n = 4$)
	3	5.030, 5.494, 7.332 ($n = 3$)	12.851, 9.187, 5.288, 7.668, 6.822, 9.475 ($n = 6$)

B.1.3 Data Analysis

The data analysis is complicated by highly unbalanced data and is at best approximate. However, there is no easy way to fix this, after the fact, given that the cell sample sizes range from 1 to 10 (which highlights the need to design the experiment so that it is balanced). We use the same model from Exercise 12.17 modified for unequal sample sizes:

$$y_{ijk\ell} = \mu + \alpha_i + \gamma_k + (\alpha\gamma)_{ik} + e_{ik}^{(1)} + \beta_j + (\alpha\beta)_{ij} + e_{ijk}^{(2)} + e_{ijk\ell}$$
$$(1 \le i \le a, 1 \le j \le b, 1 \le k \le c, 1 \le \ell \le n_{ijk}).$$

Minitab does not have the capability to analyze an unbalanced split-plot design; however, it does have the capability to analyze an unbalanced nested design. This capability can be used if the blocking effect due to subjects can be ignored, in which case the whole-plot treatment structure can be regarded as a completely randomized design with subjects nested in the whole-plot factor, field strength.

```
Analysis of Variance for Contrast, using Adjusted SS for Tests

Source                 DF   Seq SS   Adj SS   Adj MS      F      P
Field                   1   59.888   51.117   51.11 7   9.54  0.026 x
Angle                   2  217.344  147.231   73.615  11.13  0.003 x
Subject(Field)          4   25.029   24.923    6.231   0.79  0.561 x
Field*Angle             2    3.180    2.185    1.093   0.17  0.850 x
Angle*Subject(Field)    8   63.020   63.020    7.878   2.94  0.007
Error                  65  173.925  173.925    2.676
Total                  82  542.387
x Not an exact F-test.
S = 1.63578   R-Sq = 67.93%   R-Sq(adj) = 59.55%
Expected Mean Squares, using Adjusted SS

   Source                 Expected Mean Square for Each Term
1  Field                  (6) + 3.2432 (5) + 9.7297 (3) + Q[1, 4]
2  Angle                  (6) + 3.2698 (5) + Q[2, 4]
3  Subject(Field)         (6) + 4.3000 (5) + 12.9000 (3)
4  Field*Angle            (6) + 3.2698 (5) + Q[4]
5  Angle*Subject(Field)   (6) + 4.3187 (5)
6  Error                  (6)

Error Terms for Tests, using Adjusted SS

   Source                 Error DF  Error MS  Synthesis of Error MS
1  Field                      5.19     5.357  0.7542 (3) + 0.2458 (6)
2  Angle                      9.82     6.614  0.7571 (5) + 0.2429 (6)
3  Subject(Field)             8.02     7.855  0.9957 (5) + 0.0043 (6)
4  Field*Angle                9.82     6.614  0.7571 (5) + 0.2429 (6)
5  Angle*Subject(Field)      65.00     2.676  (6)
```

Display B.1 Minitab ANOVA output for MRI BOLD data under assumption of nonsignificant subject effect.

As seen in Section 12.4, the whole-plot error is then $SS_{C(A)}$ while the subplot error is $SS_{BC(A)}$. These sums of squares cannot be calculated using the formulas $SS_{C(A)} = SS_C + SS_{AC}$ and $SS_{BC(A)} = SS_{BC} + SS_{ABC}$, where the component sums of squares are taken from the ANOVA table computed by treating the design as a three-way crossed layout because of the nonorthogonality in the design. The ANOVA computed by making the assumption of no subject effects is shown in Display B.1. Notice that all the F-tests, except that for the angle–subject(field) interaction (which we use to estimate the subplot error and hence its F-test is to be ignored), are approximate. We see that the subject(field) effect is nonsignificant ($p = 0.561$) and the corresponding sum of squares may be used to estimate the whole-plot error [which is partially contaminated by the subplot error, as can be seen from the E(MS) expressions]. Thus the subject effect may be ignored and Display B.1 gives the ANOVA of the data that takes into account the split-plot nature of the design.

We see that only the field and angle main effects are significant; their interaction is nonsignificant ($p = 0.850$). Due to the unbalanced nature of the design, the synthetic error term used for testing the field main effect is obtained by pooling about 75% of the whole-plot error [i.e., the subject(field) mean-square error] and 25% of the replication error. Similarly, the synthetic error term used for testing the angle main effect is obtained by pooling about 75% of the subplot error [i.e., the angle–subject(field) mean square error] and 25% of the replication error. If the design were balanced, then only the former error term would be used in each case with no contribution from the latter. The normal and fitted-values plots for the residuals in Figure B.1 are quite satisfactory and the assumptions of normality and homoscedasticity do not appear to be seriously violated.

B.1.4 Results

The subject main effect and all interactions are nonsignificant. The main-effects plots are shown in Figure B.2. The mean contrast values for the field strength are 2.785 for 1.5 T and 4.656 for 3.0 T, which represents a 1.67 times increase when the field strength is doubled. We also see that the contrast increases nearly

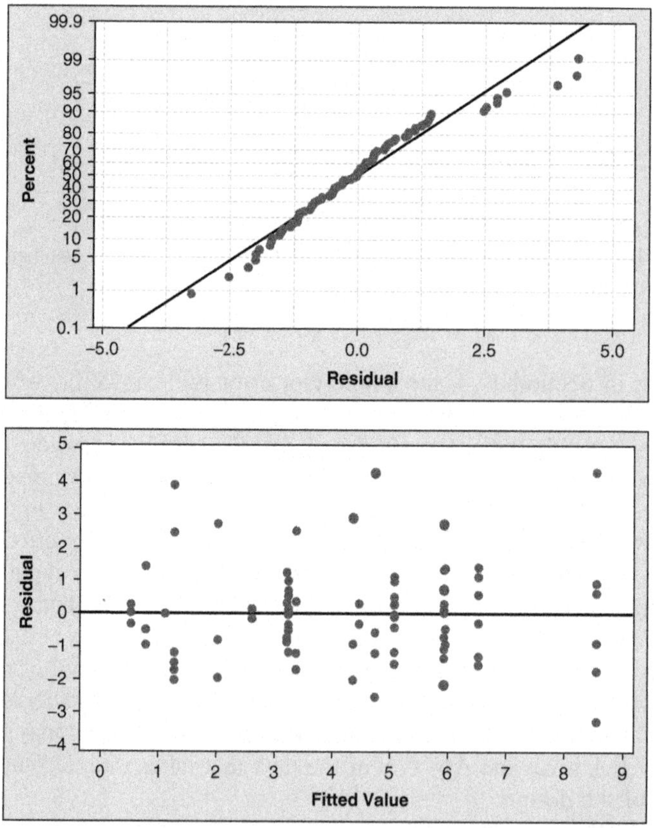

Figure B.1 Normal (top) and fitted-value (bottom) plots of residuals from MRI BOLD data.

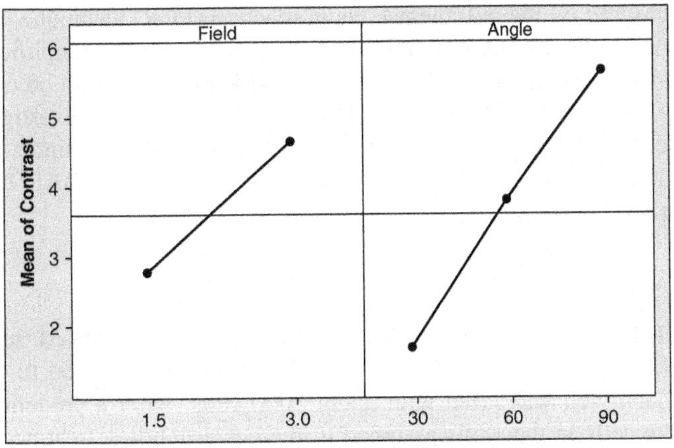

Figure B.2 Main-effects plots for MRI BOLD data.

linearly with the flip angle. Given the high significance of this effect and the nonsignicance of the field–angle interaction, it does not seem possible to use a lower flip angle to achieve a higher contrast, as was hoped. Thus both the field strength and flip angle must be set at the highest levels to maximize the image contrast.

B.2 CASE STUDY 2: GROWING STEM CELLS FOR BONE IMPLANTS

B.2.1 Introduction

This project was done by Yi Wu and Yuan Liao in my Spring 2006 Design of Experiments class. The goal of the project was to study the effects of three factors listed in Table B.2 on the growth of stem cells into mature bone cells for bone implantation.

One indicator of mature bone cells (response variable) is elevated alkaline phosphatase (ALP) gene expression—the more the ALP is expressed, the more the stem cells have grown into mature bone cells. ALP gene expression level

Table B.2 Treatment Factors for Stem Cell Growth Experiment

Factor	Levels
Growth factor concentration (A)	0, 15, and 30 ng/ml
Growth media (B)	Regular, osteogenic[a]
Incubating days (C)	3, 6, and 9 days

[a]Osteogenic media contains dexamethasone as an additive, which is supposed to induce bone differentiation. For simplicity of analysis the factor levels have been changed slightly so that they are equispaced.

was measured by the polymerase chain reaction (PCR) technique. This technique allows ALP gene to be amplified many times. Given the amplification constant (called the cycle number), the amount of ALP molecules can be calculated. This amount is also affected by the number of cells used. The investigators used the so-called GAPDH gene as an indicator of cell quantities since the number of GAPDH molecules is well correlated with cell quantities. GAPDH was used as a covariate.

B.2.2 Design

Each of the 18 factorial treatment combinations was replicated three times for a total of $N = 54$ observations, which were randomly assigned to the wells on a plate. Each well was filled with about 3000 cells. After a predefined number of days, the cells in the wells assigned to the corresponding treatment combination were destroyed and all of their contained molecules were released. The ALP and GAPDH molecules were then amplified by PCR. This process was repeated at 3, 6, and 9 days. The data are shown in Table B.3.

B.2.3 Data Analysis

Denote by $z_{ijk\ell}$ the value of ALP and $x_{ijk\ell}$ the value of the covariate GAPDH for the ℓth replicate of the treatment combination $(A = i, B = j, C = k)$ $(i = 1,\ 2,\ 3, j = 1,\ 2, k = 1,\ 2,\ 3, \ell = 1, 2, 3)$. Further let $\bar{z}_{ijk}$ and s_{ijk} be the mean and SD of the $z_{ijk\ell}$ $(\ell = 1, 2, 3)$. As a first step, a suitable transformation of ALP was sought. For this purpose, as suggested in Section 3.5, a plot of $\ln s_{ijk}$ versus $\ln \bar{z}_{ijk}$ was made (shown in Figure B.3). The plot is quite linear, suggesting a power transformation. The least squares fitted line has slope $\alpha = 1.558$. Hence $\lambda = 1 - \alpha = -0.558 \approx -0.5$. Therefore the transformation $y_{ijk\ell} = 1/\sqrt{z_{ijk\ell}}$ was employed.

For this transformed response the following linear model was fitted:

$$y_{ijk\ell} = \mu + \alpha_i + \beta_j + \gamma_k + (\alpha\beta)_{ij} + (\alpha\gamma)_{ik} + (\beta\gamma)_{jk} + (\alpha\beta\gamma)_{ijk} + \delta x_{ijk\ell} + e_{ijk\ell},$$

where the various terms have the usual meanings and side constraints and δ is the common (assumed) slope coefficient for the covariate. The ANOVA computed using Minitab is shown in Display B.2. All the effects are highly significant ($p < 0.05$) except the three-factor interaction, which has a p-value of 0.074. The normal and fitted-values plots of residuals are shown in Figure B.4. These plots show no significant violation of the normality or homoscedasticity assumptions.

B.2.4 Results

The parameter estimates are shown in Display B.3. The main-effect and interaction plots are shown in Figures B.5 and B.6. We see that all three factors have

Table B.3 Stem Cell Growth Data

No.	Growth Factor	Media	Incubation (days)	GAPDH	ALP	No.	Growth Factor	Media	Incubation (Days)	GAPDH	ALP
1	0	Regular	3	23.6	0.0095	28	0	Osteogenic	9	24.0	0.1045
2	0	Regular	3	23.5	0.0091	29	0	Osteogenic	9	23.8	0.0541
3	0	Regular	3	24.1	0.0091	30	0	Osteogenic	9	23.8	0.0826
4	15	Regular	3	24.0	0.0138	31	15	Osteogenic	9	23.8	0.0685
5	15	Regular	3	24.5	0.0087	32	15	Osteogenic	9	24.3	0.0752
6	15	Regular	3	23.9	0.0120	33	15	Osteogenic	9	23.7	0.0788
6	30	Regular	3	24.4	0.0167	34	30	Osteogenic	9	24.4	0.0826
8	30	Regular	3	23.8	0.0202	35	30	Osteogenic	9	24.2	0.1045
9	30	Regular	3	24.0	0.0222	36	30	Osteogenic	9	23.6	0.2115
15	0	Osteogenic	3	24.9	0.0211	37	0	Regular	6	23.0	0.0126
11	0	Osteogenic	3	24.7	0.0222	38	0	Regular	6	24.3	0.0087
12	0	Osteogenic	3	24.8	0.0202	39	0	Regular	6	23.7	0.0100
13	15	Osteogenic	3	25.0	0.0159	40	15	Regular	6	23.0	0.0184
14	15	Osteogenic	3	24.6	0.0167	41	15	Regular	6	23.3	0.0232
15	15	Osteogenic	3	24.4	0.0159	42	15	Regular	6	23.3	0.0192
16	30	Osteogenic	3	24.1	0.0232	43	30	Regular	6	22.9	0.0255
17	30	Osteogenic	3	23.3	0.0428	44	30	Regular	6	22.7	0.0280
18	30	Osteogenic	3	23.7	0.0280	45	30	Regular	6	23.1	0.0267
19	0	Regular	9	22.8	0.0145	46	0	Osteogenic	6	23.1	0.0541
20	0	Regular	9	22.2	0.0167	47	0	Osteogenic	6	22.7	0.0997
21	0	Regular	9	23.2	0.0120	48	0	Osteogenic	6	22.3	0.1148
22	15	Regular	9	21.9	0.0267	49	15	Osteogenic	6	25.1	0.0232
23	15	Regular	9	22.5	0.0211	50	15	Osteogenic	6	26.1	0.0175
24	15	Regular	9	22.8	0.0202	51	15	Osteogenic	6	25.8	0.0175
25	30	Regular	9	24.9	0.0152	52	30	Osteogenic	6	23.9	0.0355
26	30	Regular	9	24.0	0.0211	53	30	Osteogenic	6	23.3	0.0541
27	30	Regular	9	24.9	0.0167	54	30	Osteogenic	6	23.4	0.0685

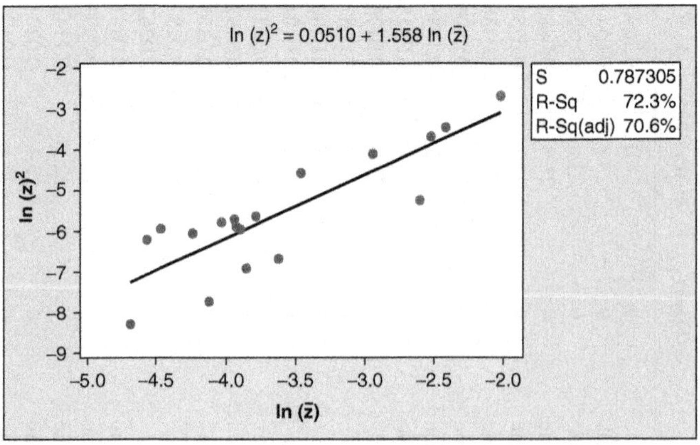

Figure B.3 Fitted-line plot of $\ln s_{ijk}$ versus $\ln \bar{z}_{ijk}$ for stem cell growth data (z = ALP).

Source	DF	Seq SS	Adj SS	Adj MS	F	P
Growth Factor	2	22.919	26.238	13.119	82.16	0.000
Media	1	119.841	96.422	96.422	603.90	0.000
Incubation	2	55.416	17.905	8.953	56.07	0.000
Growth Factor*Media	2	27.697	20.784	10.392	65.09	0.000
Growth Factor*Incubation	4	9.562	2.132	0.533	3.34	0.020
Media*Incubation	2	9.564	12.366	6.183	38.72	0.000
Growth Factor*Media*Incubation	4	14.280	1.498	0.375	2.35	0.074
GAPDH	1	8.251	8.251	8.251	51.68	0.000
Error	35	5.588	5.588	0.160		
Total	53	273.118				

sS = 0.399582   R-Sq = 97.95%   R-Sq(adj) = 96.90%

Display B.2 Minitab ANOVA output for stem cell growth data.

negative main effects at their high levels; also two-factor interaction effects are negative at high levels. The three-factor interaction, being nonsignificant, may be ignored. Recall that our goal is to maximize ALP and the response analyzed here is inversely proportional to the square root of ALP. Therefore the best treatment combination is when all three factors are set at their high levels: growth factor $(A) = 30$ ng/ml, incubation days $(B) = 9$, and media (C) = osteogenic.

To write the fitted model, define the following indicator variables:

$$(x_{11}, x_{12}) = \begin{cases} (1, 0) & \text{if } A = 0 \text{ ng/ml}, \\ (0, 1) & \text{if } A = 15 \text{ ng/ml}, \\ (-1, -1) & \text{if } A = 30 \text{ ng/ml}, \end{cases}$$

$$x_2 = \begin{cases} 1 & \text{if } B = \text{regular}, \\ -1 & \text{if } B = \text{osteogenic}, \end{cases}$$

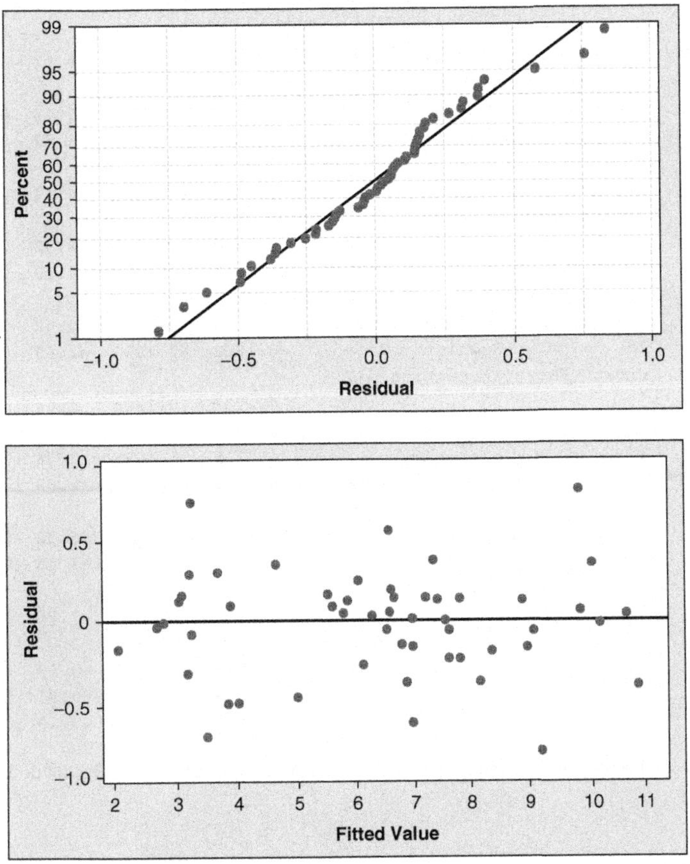

Figure B.4 Normal (top) and fitted-value (bottom) plots of residuals for stem cell growth data.

$$(x_{31}, x_{32}) = \begin{cases} (1, 0) & \text{if } C = 3 \text{ days}, \\ (0, 1) & \text{if } C = 6 \text{ days}, \\ (-1, -1) & \text{if } C = 9 \text{ days}. \end{cases}$$

Then the prediction equation up to two-factor interactions (rounding all coefficients from Display B.3 to three decimal places) is given by

$$\begin{aligned} \widehat{y} = & -23.348 + 0.790x_{11} + 0.145x_{12} + 1.874x_2 + 0.851x_{31} + 0.016x_{32} \\ & + 0.838x_{11}x_2 - 0.186x_{12}x_2 - 0.174x_{11}x_{31} + 0.058x_{11}x_{32} + 0.312x_{12}x_{31} \\ & -0.414x_{12}x_{32} - 0.530x_2x_{31} - 0.113x_2x_{32} + 1.258\text{GAPDH}. \end{aligned}$$

If all three factors are chosen at their high levels, then all x's equal -1. Also, assume GAPDH equals its average, which is 23.798. Substituting these values

Term			Coef	SE Coef	T	P
Constant			-23.348	4.165	-5.61	0.000
Growth Facto						
0			0.79012	0.08560	9.23	0.000
15			0.14487	0.08463	1.71	0.096
Media						
Regular			1.87414	0.07626	24.57	0.000
Incubation						
3d			0.8513	0.1023	8.32	0.000
6d			0.01564	0.08358	0.19	0.853
Growth Facto*Media						
0	Regular		0.83760	0.07887	10.62	0.000
15	Regular		-0.1862	0.1101	-1.69	0.100
Growth Facto*Incubation						
0	3d		-0.1739	0.1206	-1.44	0.158
0	6d		0.0578	0.1150	0.50	0.618
15	3d		0.3115	0.1088	2.86	0.007
15	6d		-0.4139	0.1537	-2.69	0.011
Media*Incubation						
Regular	3d		-0.53035	0.07887	-6.72	0.000
Regular	6d		-0.11287	0.07740	-1.46	0.154
Growth Facto*Media*Incubation						
0	Regular	3d	0.2591	0.1320	1.96	0.058
0	Regular	6d	0.0024	0.1690	0.01	0.989
15	Regular	3d	-0.0235	0.1283	-0.18	0.855
15	Regular	6d	-0.1105	0.1320	-0.84	0.408
GAPDH			1.2581	0.1750	7.19	0.000

Display B.3 Minitab parameter estimates output for stem cell growth data.

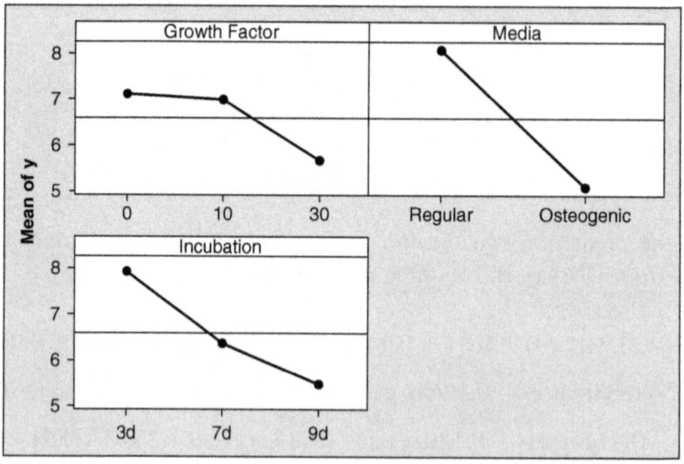

Figure B.5 Main-effects plots for stem cell growth data.

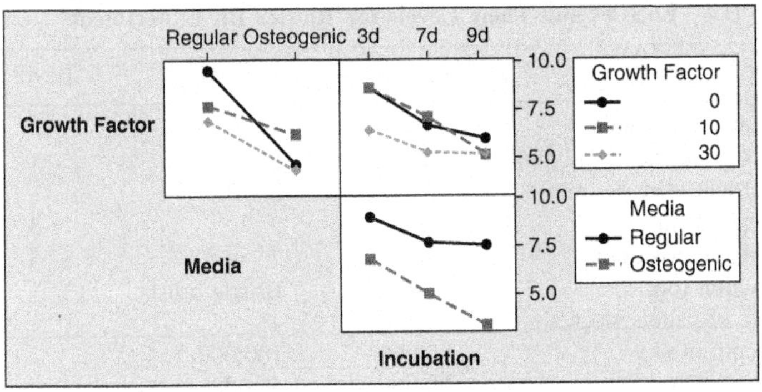

Figure B.6 Interaction plots for stem cell growth data.

in the above equation, we obtain $\widehat{y} = 2.705$ and hence $\widehat{z} = 1/(2.705)^2 = 0.137$. This is close to $\bar{z} = 0.133$ observed at this particular setting of the three factors.

B.3 CASE STUDY 3: ROUTER BIT EXPERIMENT

B.3.1 Introduction

Phadke (1986) discussed an experiment to study a routing process used to cut 8 × 4-in. printed wiring boards from 18 × 24-in. panels. The goal of the experiment was to increase the life of the router bit before it gets dull (which causes excessive dust formation and consequent expensive cleaning operation to smooth the edges of the boards) or fails. The routing machine had four spindles all of which were synchronized in their horizontal and vertical feeds and rotational speeds. Each spindle cut a separate stack of panels of $\frac{1}{16}$ in. thickness. Three or four panels could be cut simultaneously. Cutting more panels simultaneously increases the production rate but may shorten the bit life. Other design factors included three factors related to suction used to remove dust created due to cutting; they were suction pressure with higher pressure removing more dust, type of suction foot (solid ring or bristle brush), and depth of slot precut in backup panels used to provide air passage for dust to accumulate beneath the panels being cut. In addition, four types of router bits were used. Since all four spindle positions must be used in the production process, it is not a design factor. All other factors were set at two levels. The factors and their levels are summarized in Table B.4.

B.3.2 Design

Before designing the experiment, it is important to list which interactions are of interest. Four interactions were deemed to be of interest: BG, BJ, CJ, and GJ.

Table B.4 Factors and Their Levels for Router Bit Experiment

Factor	Level 1	Level 2	Level 3	Level 4
A: Suction (in. of Hg)	1	2^a		
B: Horizontal feed (in./min)	60^a	80		
C: Vertical feed (in./min)	10^a	50		
D: Type of bit	1	2	3	4^a
E: Spindle positions	1	2	3	4
F: Suction foot	Solid ring	Bristle brush[a]		
G: No. of panels stacked	3	4^a		
H: Depth of slot	$60{,}000^a$	100,000		
J: Speed (rpm)	30,000	$40{,}000^a$		

[a]Levels currently in use.

Source: Phadke (1986, Table I). Courtesy of AT&T Archives and History Center.

Thus the following model was postulated:

$$\begin{aligned}\text{E(Lifetime)} &= \beta_0 + \beta_A x_A + \beta_B x_B + \beta_C x_C + \beta_F x_F + \beta_G x_G \\ &\quad + \beta_H x_H + \beta_J x_J + \beta_{BG} x_B x_G + \beta_{BJ} x_B x_J + \beta_{CJ} x_C x_J \\ &\quad + \beta_{GJ} x_G x_J + \sum_{j=1}^{3} \beta_{Dj} x_{Dj} + \sum_{j=1}^{3} \beta_{Ej} x_{Ej},\end{aligned} \tag{B.1}$$

where the notation is (for $i = A, B, C, F, G, H, J$)

$$x_i = \begin{cases} -1 & \text{if } i\text{th factor is at low level ,} \\ +1 & \text{if } i\text{th factor is at high level} \end{cases}$$

and $(i = D,\ E)$

$$x_{ij} = \begin{cases} -1 & \text{if } i\text{th factor is at level 4,} \\ +1 & \text{if } i\text{th factor is at level } j,\ j = 1, 2, 3, \\ 0 & \text{if } i\text{th factor is at level } k,\ k \neq j, k = 1, 2, 3. \end{cases}$$

A total of 32 runs were made. The design was constructed as follows. Take any OA$(16, 2^{15}, 2)$ and assign two-level factors A–C and F–H to its six columns and factors D and E to six other columns using the method of replacement. Care must be taken so that the column corresponding to the interaction BG is not assigned to any factor. This uses up a total of 13 columns leaving 2 columns unassigned. The resulting $2^{6-4}4^2$ design is then crossed with two levels of J resulting in 32 runs, that is, the 16-run design is replicated for $J = 1$ and $J = 2$. Crossing by J allows estimation of the main effect of J and of all two-factor interactions involving J clear of all other two-factor interactions; see Example 8.2.

Phadke (1986) used Taguchi's $L_{16}(2^{15})$ array as the OA$(16, 2^{15}, 2)$ for assigning the columns; this array is shown in Table B.5 (1 and 2 are changed to −

and +, respectively, to be consistent with the notation followed in the book). The columns in this array are labeled with pseudofactors A', B', C', D' and their interactions, so that when the actual factors $A-H$ are assigned to these columns, their alias relationships can be easily derived.

The two-level factors are assigned to the columns of this orthogonal array as follows:

$$A \to A', B \to B', C \to -A'B', F \to -B'D', G \to -A'B'C'D', H \to -A'C'.$$

The four-level factors are assigned as follows:

$$D = 1, 2, 3, 4 \leftrightarrow (A'B'C', B'C'D') = (-, -), (-, +), (+, -), (+, +),$$

$$E = 1, 2, 3, 4 \leftrightarrow (C', D') = (-, -), (-, +), (+, +), (+, -).$$

The resulting design is shown in Table B.6. Therefore the interaction columns $(A'B'C')(B'C'D') = A'D'$ and $C'D'$ are left unassigned. Note that since D and E are both categorical factors, it is not very useful to partition their effects into contrasts of the type (9.13) since the goal here is to select the best level of D (bit type) and compare the levels of E (spindle position) with each other to see if any positions result in significantly better/worse bit life.

During each run the machine was stopped after every 100 in. of horizontal (in the $X - Y$ plane) movement of the router bit to inspect the amount of dust. If this amount was more than some threshold or if the bit broke, then the bit was regarded as failed. The lifetime was recorded as the midpoint of the last interval before failure was observed. If the bit had not failed by 1700 in. of movement, then that run was stopped and the lifetime was recorded as 1750 in. The data are shown in Table B.6.

Note that the data are interval censored if the bit failed before 1700 in. and right censored if it did not. In fact, 8 of 32 observations are right censored. Thus a correct analysis of the data would require appropriate survival methods not covered in this book. For sake of simplicity, the data are treated as uncensored and normally distributed.

The runs were arranged in groups of four as follows. For each group, the rotational speed, the horizontal feed, the vertical feed, and the suction pressure were kept the same. The four runs within each group correspond to the four spindles on which four different types of bit are mounted. From the description of the experiment it appears that the runs were not randomized. In particular, it is stated that the suction pressure (factor A) was difficult to change, so the design appears to be a split-plot design with the suction pressure as the whole plot and the other factors as subplots. Furthermore, of necessity, all four spindles are used together with the same speed and feeds. Therefore spindles form a sub-subplot with even smaller error. In the following analyses these issues are ignored.

Table B.5 L_{16} **Orthogonal Array**

	Column														
Run	A'	B'	C'	D'	$-A'B'$	$-A'C'$	$-A'D'$	$-B'C'$	$-B'D'$	$-C'D'$	$A'B'C'$	$A'C'D'$	$A'B'D'$	$B'C'D'$	$-A'B'C'D'$
1	−	−	−	−	−	−	−	−	−	−	−	−	−	−	−
2	−	−	−	+	−	−	+	−	+	+	−	+	+	+	+
3	−	−	+	−	−	+	−	+	−	+	+	+	−	+	+
4	−	−	+	+	−	+	+	+	+	−	+	−	+	−	−
5	−	+	−	−	+	−	−	+	+	−	+	−	+	+	+
6	−	+	−	+	+	−	+	+	−	+	+	+	−	−	−
7	−	+	+	−	+	+	−	−	+	+	−	+	+	−	−
8	−	+	+	+	+	+	+	−	−	−	−	−	−	+	+
9	+	−	−	−	+	+	+	−	−	−	+	+	+	+	+
10	+	−	−	+	+	+	−	−	+	+	+	−	−	−	−
11	+	−	+	−	+	−	+	+	−	+	−	−	+	−	−
12	+	−	+	+	+	−	−	+	+	−	−	+	−	+	+
13	+	+	−	−	−	+	+	+	+	−	−	+	−	+	−
14	+	+	−	+	−	+	−	+	−	+	−	−	+	−	+
15	+	+	+	−	−	−	+	−	+	+	+	−	−	−	+
16	+	+	+	+	−	−	−	−	−	−	+	+	+	+	−

Table B.6 Design and Data for Router Bit Experiment

Factor											
								$J = 1$		$J = 2$	
A	B	C	D	E	F	G	H	Run	Lifetime[a]	Run	Lifetime[a]
−	−	−	1	1	−	−	−	1	3.5	17	17.5
−	−	−	2	2	+	+	−	2	0.5	18	0.5
−	−	−	3	4	−	+	+	3	0.5	19	0.5
−	−	−	4	3	+	−	+	4	17.5	20	17.5
−	+	+	3	1	+	+	−	5	0.5	21	0.5
−	+	+	4	2	−	−	−	6	2.5	22	17.5
−	+	+	1	4	+	−	+	7	0.5	23	14.5
−	+	+	2	3	−	+	+	8	0.5	24	0.5
+	−	+	4	1	−	+	+	9	17.5	25	17.5
+	−	+	3	2	+	−	+	10	2.5	26	3.5
+	−	+	2	4	−	−	−	11	0.5	27	17.5
+	−	+	1	3	+	+	−	12	3.5	28	3.5
+	+	−	2	1	+	−	+	13	0.5	29	0.5
+	+	−	1	2	−	+	+	14	2.5	30	3.5
+	+	−	4	4	+	+	−	15	0.5	31	0.5
+	+	−	3	3	−	−	−	16	3.5	32	17.5

[a]The data are in hundreds of inches of movement of the bit in the $X - Y$ plane until the bit fails. The data are interval censored in 100-in. intervals and the recorded value is the midpoint of the last interval. If the lifetime exceeded 1700 in., then the value is recorded as 1750 in.

Source: Phadke (1986, Table II). Courtesy of AT&T Archives and History Center.

B.3.3 Data Analysis

The analysis of variance computed using Minitab is shown in Display B.4. The main effects are plotted in Figure B.7 and the interactions are plotted in Figure B.8. From these outputs we see that the following are significant at the 10% level: the main effects B, D, F, G, J and the interaction GJ. To maximize the bit life, the optimum settings are B (horizontal feed) low, D (bit) type 4, F (suction foot) low, G (stack height) low, and J (rotation speed) high.

The estimated coefficients of all terms included in the model (B.1) along with their standard errors, t-statistics, and p-values are shown in Display B.5. The coefficients for the four-level factors D and E are the deviations from the overall mean. Due to the orthogonal nature of the design, these deviations sum to zero for each factor. Therefore we can deduce the coefficients for levels 4 of D and E as $D4 = -(0.188 - 3.313 - 2.313) = 5.438$ and $E4 = -(1.313 - 1.813 + 2.063) = -1.563$. Retaining the terms significant at the 0.10 level, the prediction model is

$$\widehat{\text{Lifetime}} = 5.938 - 1.813x_B - 3.313x_{D1} - 1.750x_F - 2.625x_G + 2.375x_J - 2.312x_G x_J \quad \text{(B.2)}$$

Source	DF	Seq SS	Adj SS	Adj MS	F	P
A	1	0.00	0.00	0.00	0.00	1.000
B	1	105.13	105.13	105.13	4.07	0.063
C	1	8.00	8.00	8.00	0.31	0.587
D	3	367.38	367.38	122.46	4.74	0.017
E	3	93.63	93.63	31.21	1.21	0.343
F	1	98.00	98.00	98.00	3.80	0.072
G	1	220.50	220.50	220.50	8.54	0.011
H	1	3.12	3.12	3.12	0.12	0.733
J	1	180.50	180.50	180.50	6.99	0.019
B*G	1	4.50	4.50	4.50	0.17	0.683
B*J	1	4.50	4.50	4.50	0.17	0.683
C*J	1	10.13	10.13	10.13	0.39	0.541
G*J	1	171.13	171.13	171.13	6.63	0.022
Error	14	361.38	361.38	25.81		
Total	31	1627.88				

S = 5.08060   R-Sq = 77.80%   R-Sq(adj) = 50.84%

Display B.4 Minitab ANOVA output for router bit data.

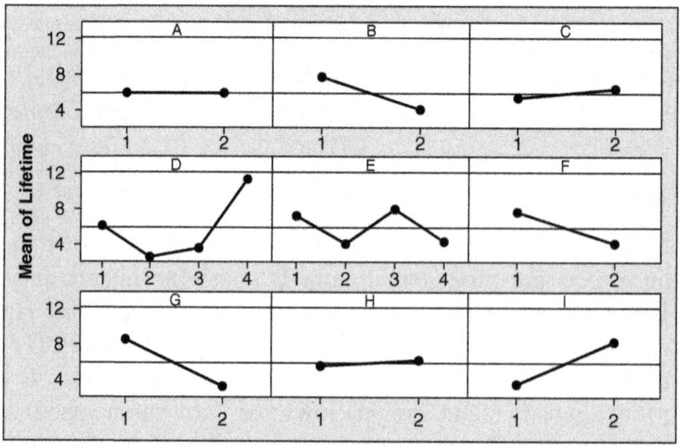

Figure B.7 Main-effects plot for router bit data.

B.3.4 Results

From the ANOVA output and the main-effects plots the following conclusions can be drawn:

(a) The D (bit type), G (no. of panels stacked), and J (rotational speed) main effects and the GJ interaction are significant at the 0.05 level. The B

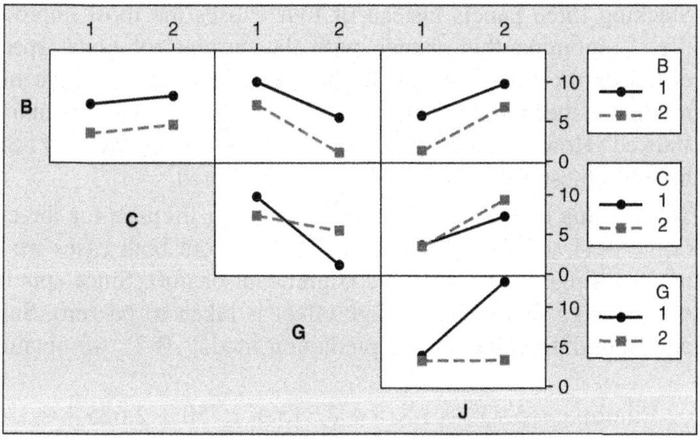

Figure B.8 Interaction effects plot for router bit data.

Term	Coef	SE Coef	T	P
Constant	5.938	0.8981	6.61	0.000
A	0.000	0.8981	0.00	1.000
B	-1.813	0.8981	-2.02	0.063
C	0.500	0.8981	0.56	0.587
D1	0.188	1.556	0.12	0.906
D2	-3.313	1.556	-2.13	0.051
D3	-2.313	1.556	-1.49	0.159
E1	1.313	1.556	0.84	0.413
E2	-1.813	1.556	-1.17	0.263
E3	2.063	1.556	1.33	0.206
F	-1.750	0.8981	-1.95	0.072
G	-2.625	0.8981	-2.92	0.011
H	0.313	0.8981	0.35	0.733
J	2.375	0.8981	2.64	0.019
B*G	-0.375	0.8981	-0.42	0.683
B*J	0.375	0.8981	0.42	0.683
C*J	0.563	0.8981	0.63	0.541
G*J	-2.312	0.8981	-2.57	0.022

Display B.5 Estimated coefficients of terms included in model (B.1) for router bit lifetime data.

(horizontal feed) and F (suction foot) main effects are significant at the 0.10 level.

(b) The currently used bit type 4 is the best. There are no significant differences between the spindle positions (E).

(c) The current horizontal feed is better, but the suction foot should be changed from the current bristle brush to the solid ring.

(d) Suction pressure seems to have no main effect. One may keep it at its present setting of 2 in. of Hg.

(e) Stacking three panels instead of four causes the most improvement in bit life. Combining this change with the current rotational speed of 40,000 enhances the bit life much further because of the positive main effect of rotational speed and its large negative interaction with the number of panels stacked. However, the production rate is lowered by 25%. So a trade-off between these two options must be considered.

(f) Toward this end we calculate the predicted lifetime for three-panel stacks ($x_G = -1$) and four-panel stacks ($x_G = 1$). In both cases we assume optimum settings of other factors (significant or not). Since spindle position is not a design factor, its average effect is taken to be zero. Substituting the corresponding values in the prediction model (B.2), we obtain

$$\widehat{\text{Lifetime}} = 5.938 + 1.813 + 3.313 + 1.750 + 2.625 + 2.375 + 2.312$$
$$= 20.126 \quad \text{for } x_G = -1$$

and

$$\widehat{\text{Lifetime}} = 5.938 + 1.813 + 3.313 + 1.750 - 2.625 + 2.375 - 2.312$$
$$= 10.252 \quad \text{for } x_G = +1.$$

Thus if the number of stacked panels is changed from the current setting of four panels to three panels, keeping the other factors at their optimum settings, then the production rate drops by 25% but the average bit life almost doubles from 1025.2 to 2012.6 in. The cost of replacing a failed bit and the cost of lost production must be factored into these calculations to choose the better of the two options.

The estimated mean lifetime of 2012.6 in. at the optimum settings may seem high since the highest data value is only 1750 in. However, 25% of the data are right censored, so it may not be unreasonable to expect that at the optimum settings the mean lifetime may exceed 1750 in. significantly. [Wu and Hamada (2000) take the highest data value as 1700 in. and, using a somewhat different model but the same optimum settings, estimate the lifetime to be 1793.25 in.] A confirmatory experiment should be performed to estimate the actual mean lifetime.

APPENDIX C

Statistical Tables

Statistical Analysis of Designed Experiments: Theory and Applications By Ajit C. Tamhane

Table C.1 Standard Normal c.d.f. $\Phi(z) = P(Z \leq z)$

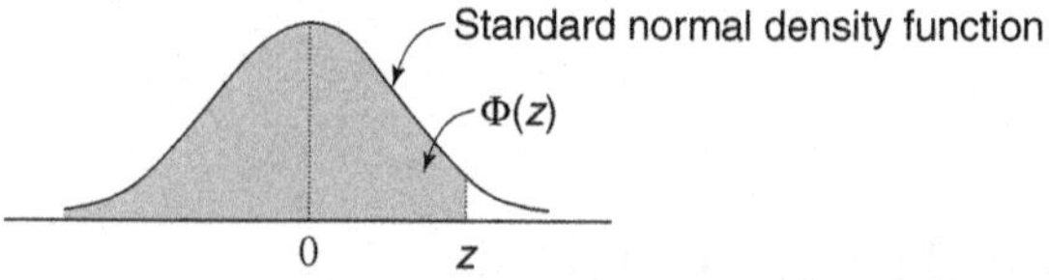

z	0.00	0.01	0.02	0.03	0.04	0.05	0.06	0.07	0.08	0.09
−3.4	0.0003	0.0003	0.0003	0.0003	0.0003	0.0003	0.0003	0.0003	0.0003	0.0002
−3.3	0.0005	0.0005	0.0005	0.0004	0.0004	0.0004	0.0004	0.0004	0.0004	0.0003
−3.2	0.0007	0.0007	0.0006	0.0006	0.0006	0.0006	0.0006	0.0005	0.0005	0.0005
−3.1	0.0010	0.0009	0.0009	0.0009	0.0008	0.0008	0.0008	0.0008	0.0007	0.0007
−3.0	0.0013	0.0013	0.0013	0.0012	0.0012	0.0011	0.0011	0.0011	0.0010	0.0010
−2.9	0.0019	0.0018	0.0017	0.0017	0.0016	0.0016	0.0015	0.0015	0.0014	0.0014
−2.8	0.0026	0.0025	0.0024	0.0023	0.0023	0.0022	0.0021	0.0021	0.0020	0.0019
−2.7	0.0035	0.0034	0.0033	0.0032	0.0031	0.0030	0.0029	0.0028	0.0027	0.0026
−2.6	0.0047	0.0045	0.0044	0.0043	0.0041	0.0040	0.0039	0.0038	0.0037	0.0036
−2.5	0.0062	0.0060	0.0059	0.0057	0.0055	0.0054	0.0052	0.0051	0.0049	0.0048
−2.4	0.0082	0.0080	0.0078	0.0075	0.0073	0.0071	0.0069	0.0068	0.0066	0.0064
−2.3	0.0107	0.0104	0.0102	0.0099	0.0096	0.0094	0.0091	0.0089	0.0087	0.0084
−2.2	0.0139	0.0136	0.0132	0.0129	0.0125	0.0122	0.0119	0.0116	0.0113	0.0110
−2.1	0.0179	0.0174	0.0170	0.0166	0.0162	0.0158	0.0154	0.0150	0.0146	0.0143
−2.0	0.0228	0.0222	0.0217	0.0212	0.0207	0.0202	0.0197	0.0192	0.0188	0.0183
−1.9	0.0287	0.0281	0.0274	0.0268	0.0262	0.0256	0.0250	0.0244	0.0239	0.0233
−1.8	0.0359	0.0352	0.0344	0.0336	0.0329	0.0322	0.0314	0.0307	0.0301	0.0294
−1.7	0.0446	0.0436	0.0427	0.0418	0.0409	0.0401	0.0392	0.0394	0.0375	0.0367
−1.6	0.0548	0.0537	0.0526	0.0516	0.0505	0.0495	0.0485	0.0475	0.0465	0.0455
−1.5	0.0668	0.0655	0.0643	0,0630	0.0618	0.0606	0.0594	0.0582	0.0571	0.0559
−1.4	0.0808	0.0793	0.0778	0.0764	0.0749	0.0735	0.0722	0.0708	0.0694	0.0681
−1.3	0.0968	0.0951	0.0934	0.0918	0.0901	0.0885	0.0869	0.0853	0.0838	0.0823
−1.2	0.1151	0.1131	0.1112	0.1093	0.1075	0.1056	0.1038	0.1020	0.1003	0.0985
−1.1	0.1357	0.1335	0.1314	0.1292	0.1271	0.1251	0.1230	0.1210	0.1190	0.1170
−1.0	0.1587	0.1562	0.1539	0.1515	0.1492	0.1469	0.1446	0.1423	0.1401	0.1379
−0.9	0.1841	0.1814	0.1788	0.1762	0.1736	0.1711	0.1685	0.1660	0.1635	0.1611
−0.8	0.2119	0.2090	0.2061	0.2033	0.2005	0.1977	0.1949	0.1922	0.1894	0.1867
−0.7	0.2420	0.2389	0.2358	0.2327	0.2296	0.2266	0.2236	0.2206	0.2177	0.2148
−0.6	0.2743	0.2709	0.2676	0.2643	0.2611	0.2578	0.2546	0.2514	0.2483	0.2451
−0.5	0.3085	0.3050	0.3015	0.2981	0.2946	0.2912	0.2877	0.2843	0.2810	0.2776
−0.4	0.3446	0.3409	0.3372	0.3336	0.3300	0.3264	0.3228	0.3192	0.3156	0.3121
−0.3	0.3821	0.3783	0.3745	0.3707	0.3669	0.3632	0.3594	0.3557	0.3520	0.3483
−0.2	0.4207	0.4168	0.4129	0.4090	0.4052	0.4013	0.3974	0.3936	0.3897	0.3859
−0.1	0.4602	0.4562	0.4522	0.4483	0.4443	0.4404	0.4364	0.4325	0.4286	0.4247
−0.0	0.5000	0.4960	0.4920	0.4880	0.4840	0.4801	0.4761	0.4721	0.4681	0.4641

Table C.1 Standard Normal c.d.f. $\Phi(z) = P(Z \le z)$ (*Continued*)

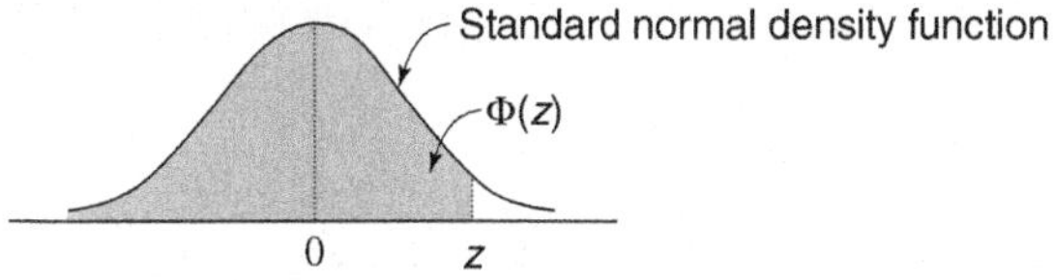

z	0.00	0.01	0.02	0.03	0.04	0.05	0.06	0.07	0.08	0.09
0.0	0.5000	0.5040	0.5080	0.5120	0.5160	0.5199	0.5239	0.5279	0.5319	0.5359
0.1	0.5398	0.5438	0.5478	0.5517	0.5557	0.5596	0.5636	0.5675	0.5714	0.5753
0.2	0.5793	0.5832	0.5871	0.5910	0.5948	0.5987	0.6026	0.6064	0.6103	0.6141
0.3	0.6179	0.6217	0.6255	0.6293	0.6331	0.6368	0.6406	0.6443	0.6480	0.6517
0,4	0.6554	0.6591	0.6628	0.6664	0.6700	0.6736	0.6772	0.6808	0.6844	0.6879
0.5	0.6915	0.6950	0.6985	0.7019	0.7054	0.7088	0.7123	0.7157	0.7190	0.7224
0.6	0.7257	0.7291	0.7324	0.7357	0.7389	0.7422	0.7454	0.7486	0.7517	0.7549
0.7	0.7580	0.7611	0.7642	0.7673	0.7704	0.7734	0.7764	0.7794	0.7823	0.7852
0.8	0.7881	0.7910	0.7939	0.7967	0.7995	0.8023	0.8051	0.8078	0.8106	0.8133
0.9	0.8159	0.8186	0.8212	0.8238	0.8264	0.8289	0.8315	0.8340	0.8365	0.8389
1.0	0.8413	0.8438	0.8461	0.8485	0.8508	0.8531	0.8554	0.8577	0.8599	0.8621
1.1	0.8643	0.8665	0.8686	0.8708	0.8729	0.8749	0.8770	0.8790	0.8810	0.8830
1.2	0.8849	0.8869	0.8888	0.8907	0.8925	0.8944	0.8962	0.8980	0.8997	0.9015
1.3	0.9032	0.9049	0.9066	0.9082	0.9099	0.9115	0.9131	0.9147	0.9162	0.9177
1.4	0.9192	0.9207	0.9222	0.9236	0.9251	0.9265	0.9278	0.9292	0.9306	0.9319
1.5	0.9332	0.9345	0.9357	0.9370	0.9382	0.9394	0.9406	0.9418	0.9429	0.9441
1.6	0.9452	0.9463	0.9474	0.9484	0.9495	0.9505	0.9515	0.9525	0.9535	0.9545
1.7	0.9554	0.9564	0.9573	0.9582	0.9591	0.9599	0.9608	0.9616	0.9625	0.9633
1.8	0.9641	0.9649	0.9656	0.9664	0.9671	0.9678	0.9686	0.9693	0.9699	0.9706
1.9	0.9713	0.9719	0.9726	0.9732	0.9738	0.9744	0.9750	0.9756	0.9761	0.9767
2.0	0.9772	0.9778	0.9783	0.9788	0.9793	0.9798	0.9803	0.9808	0.9812	0.9817
2.1	0.9821	0.9826	0.9830	0.9834	0.9838	0.9842	0.9846	0.9850	0.9854	0.9857
2.2	0.9861	0.9864	0.9868	0.9871	0.9875	0.9878	0.9881	0.9884	0.9887	0.9890
2.3	0.9893	0.9896	0.9898	0.9901	0.9904	0.9906	0.9909	0.9911	0.9913	0.9916
2.4	0.9918	0.9920	0.9922	0.9925	0.9927	0.9929	0.9931	0.9932	0.9934	0.9936
2.5	0.9938	0.9940	0.9941	0.9943	0.9945	0.9946	0.9948	0.9949	0.9951	0.9952
2.6	0.9953	0.9955	0.9956	0.9957	0.9959	0.9960	0.9961	0.9962	0.9963	0.9964
2.7	0.9965	0.9966	0.9967	0.9968	0.9969	0.9970	0.9971	0.9972	0.9973	0.9974
2.8	0.9974	0.9975	0.9976	0.9977	0.9977	0.9978	0.9979	0.9979	0.9980	0.9981
2.9	0.9981	0.9982	0.9982	0.9983	0.9984	0.9984	0.9985	0.9985	0.9986	0.9986
3.0	0.9987	0.9987	0.9987	0.9988	0.9988	0.9989	0.9989	0.9989	0.9990	0.9990
3.1	0.9990	0.9991	0.9991	0.9991	0.9992	0.9992	0.9992	0.9992	0.9993	0.9993
3.2	0.9993	0.9993	0.9994	0.9994	0.9994	0.9994	0.9994	0.9995	0.9995	0.9995
3.3	0.9995	0.9995	0.9995	0.9996	0.9996	0.9996	0.9996	0.9996	0.9996	0.9997
3.4	0.9997	0.9997	0.9997	0.9997	0.9997	0.9997	0.9997	0.9997	0.9997	0.9998

Table C.2 Critical Values $t_{\nu,\alpha}$ for the t-Distribution

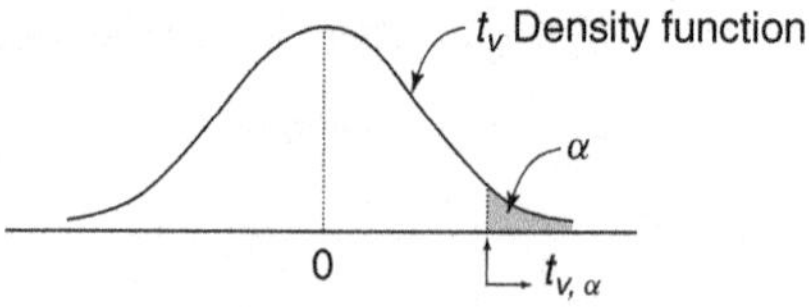

ν	α = .10	.05	.025	.01	.005	.001	.0005
1	3.078	6.314	12.706	31.821	63.657	318.31	636.62
2	1.886	2.920	4.303	6.965	9.925	22.326	31.598
3	1.638	2.353	3.182	4.541	5.841	10.213	12.924
4	1.533	2.132	2.776	3.747	4.604	7.173	8.610
5	1.476	2.015	2.571	3.365	4.032	5.893	6.869
6	1.440	1.943	2.447	3.143	3.707	5.208	5.959
7	1.415	1.895	2.365	2.998	3.499	4.785	5.408
8	1.397	1.860	2.306	2.896	3.355	4.501	5.041
9	1.383	1.833	2.262	2.821	3.250	4.297	4.781
10	1.372	1.812	2.228	2.764	3.169	4.144	4.587
11	1.363	1.796	2.201	2.718	3.106	4.025	4.437
12	1.356	1.782	2.179	2.681	3.055	3.930	4.318
13	1.350	1.771	2.160	2.650	3.012	3.852	4.221
14	1.345	1.761	2.145	2.624	2.977	3.787	4.140
15	1.341	1.753	2.131	2.602	2.947	3.733	4.073
16	1.337	1.746	2.120	2.583	2.921	3.686	4.015
17	1.333	1.740	2.110	2.567	2.898	3.646	3.965
18	1.330	1.734	2.101	2.552	2.878	3.610	3.922
19	1.328	1.729	2.093	2.539	2.861	3.579	3.883
20	1.325	1.725	2.086	2.528	2.845	3.552	3.850
21	1.323	1.721	2.080	2.518	2.831	3.527	3.819
22	1.321	1.717	2.074	2.508	2.819	3.505	3.792
23	1.319	1.714	2.069	2.500	2.807	3.485	3.767
24	1.318	1.711	2.064	2.492	2.797	3.467	3.745
25	1.316	1.708	2.060	2.485	2.787	3.450	3.725
26	1.315	1.706	2.056	2.479	2.779	3.435	3.707
27	1.314	1.703	2.052	2.473	2.771	3.421	3.690
28	1.313	1.701	2.048	2.467	2.763	3.408	3.674
29	1.311	1.699	2.045	2.462	2.756	3.396	3.659
30	1.310	1.697	2.042	2.457	2.750	3.385	3.646
40	1.303	1.684	2.021	2.423	2.704	3.307	3.551
60	1.296	1.671	2.000	2.390	2.660	3.232	3.460
120	1.289	1.658	1.980	2.358	2.617	3.160	3.373
∞	1.282	1.645	1.960	2.326	2.576	3.090	3.291

Source: Reprinted with permission of Pearson Education, Inc.

Table C.3 Critical Values $\chi^2_{\nu,\alpha}$ for Chi-Square Distribution

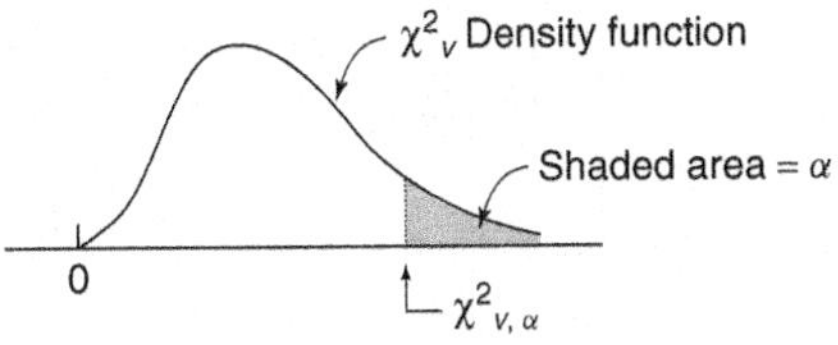

ν	α = .995	.99	.975	.95	.90	.10	.05	.025	.01	.005
1	0.000	0.000	0.001	0.004	0.016	2.706	3.843	5.025	6.637	7.882
2	0.010	0.020	0.051	0.103	0.211	4.605	5.992	7.378	9.210	10.597
3	0.072	0.115	0.216	0.352	0.584	6.251	7.815	9.348	11.344	12.837
4	0.207	0.297	0.484	0.711	1.064	7.779	9.488	11.143	13.277	14.860
5	0.412	0.554	0.831	1.145	1.610	9.236	11.070	12.832	15.085	16.748
6	0.676	0.872	1.237	1.635	2.204	10.645	12.592	14.440	16.812	18.548
7	0.989	1.239	1.690	2.167	2.833	12.017	14.067	16.012	18.474	20.276
8	1.344	1.646	2.180	2.733	3.490	13.362	15.507	17.534	20.090	21.954
9	1.735	2.088	2.700	3.325	4.168	14.684	16.919	19.022	21.665	23.587
10	2.156	2.558	3.247	3.940	4.865	15.987	18.307	20.483	23.209	25.188
11	2.603	3.053	3.816	4.575	5.578	17.275	19.675	21.920	24.724	26.755
12	3.074	3.571	4.404	5.226	6.304	18.549	21.026	23.337	26.217	28.300
13	3.565	4.107	5.009	5.892	7.041	19.812	22.362	24.735	27.687	29.817
14	4.075	4.660	5.629	6.571	7.790	21.064	23.685	26.119	29.141	31.319
15	4.600	5.229	6.262	7.261	8.547	22.307	24.996	27.488	30.577	32.799
16	5.142	5.812	6.908	7.962	9.312	23.542	26.296	28.845	32.000	34.267
17	5.697	6.407	7.564	8.682	10.085	24.769	27.587	30.190	33.408	35.716
18	6.265	7.015	8.231	9.390	10.865	25.989	28.869	31.526	34.805	37.156
19	6.843	7.632	8.906	10.117	11.651	27.203	30.143	32.852	36.190	38.580
20	7.434	8.260	9.591	10.851	12.443	28.412	31.410	34.170	37.566	39.997
21	8.033	8.897	10.283	11.591	13.240	29.615	32.670	35.478	38.930	41.399
22	8.643	9.542	10.982	12.338	14.042	30.813	33.924	36.781	40.289	42.796
23	9.260	10.195	11.688	13.090	14.848	32.007	35.172	38.075	41.637	44.179
24	9.886	10.856	12.401	13.848	15.659	33.196	36.415	39.364	42.980	45.558
25	10.519	11.523	13.120	14.611	16.473	34.381	37.652	40.646	44.313	46.925
26	11.160	12.198	13.844	15.379	17.292	35.563	38.885	41.923	45.642	48.290
27	11.807	12.878	14.573	16.151	18.114	36.741	40.113	43.194	46.962	49.642
28	12.461	13.565	15.308	16.928	18.939	37.916	41.337	44.461	48.278	50.993
29	13.120	14.256	16.147	17.708	19.768	39.087	42.557	45.772	49.586	52.333
30	13.787	14.954	16.791	18.493	20.599	40.256	43.773	46.979	50.892	53.672
31	14.457	15.655	17.538	19.280	21.433	41.422	44.985	48.231	52.190	55.000
32	15.134	16.362	18.291	20.072	22.271	42.585	46.194	49.480	53.486	56.328
33	15.814	17.073	19.046	20.866	23.110	43.745	47.400	50.724	54.774	57.646
34	16.501	17.789	19.806	21.664	23.952	44.903	48.602	51.966	56.061	58.964
35	17.191	18.508	20.569	22.465	24.796	46.059	49.802	53.203	57.340	60.272
36	17.887	19.233	21.336	23.269	25.643	47.212	50.998	54.437	58.619	61.581
37	18.584	19.960	22.105	24.075	26.492	48.363	52.192	55.667	59.891	62.880
38	19.289	20.691	22.878	24.884	27.343	49.513	53.384	56.896	61.162	64.181
39	19.994	21.425	23.654	25.695	28.196	50.660	54.572	58.119	62.420	65.473
40[a]	20.706	22.164	24.433	26.509	29.050	51.805	55.758	59.342	63.691	66.766

[a] For $\nu > 40$, $x^2_{\nu,\alpha} \simeq \nu\left(1 - \frac{2}{9\nu} + z_\alpha\sqrt{\frac{2}{9\nu}}\right)^3$.

Source: Reprinted with permission of Pearson Education, Inc.

Table C.4 Critical Values $f_{\nu_1,\nu_2,\alpha}$ for F-Distribution

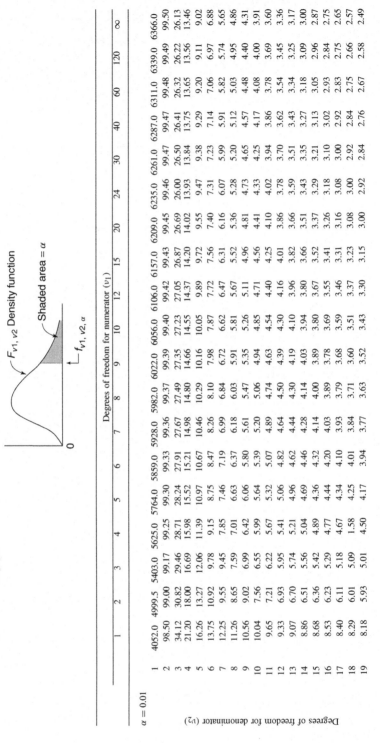

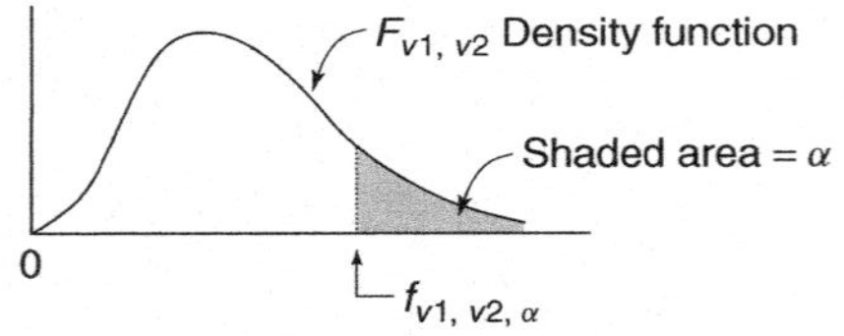

$\alpha = 0.01$

Degrees of freedom for denominator (ν_2)	Degrees of freedom for numerator (ν_1) 1	2	3	4	5	6	7	8	9	10	12	15	20	24	30	40	60	120	∞
1	4052.0	4999.5	5403.0	5625.0	5764.0	5859.0	5928.0	5982.0	6022.0	6056.0	6106.0	6157.0	6209.0	6235.0	6261.0	6287.0	6311.0	6339.0	6366.0
2	98.50	99.00	99.17	99.25	99.30	99.33	99.36	99.37	99.39	99.40	99.42	99.43	99.45	99.46	99.47	99.47	99.48	99.49	99.50
3	34.12	30.82	29.46	28.71	28.24	27.91	27.67	27.49	27.35	27.23	27.05	26.87	26.69	26.00	26.50	26.41	26.32	26.22	26.13
4	21.20	18.00	16.69	15.98	15.52	15.21	14.98	14.80	14.66	14.55	14.37	14.20	14.02	13.93	13.84	13.75	13.65	13.56	13.46
5	16.26	13.27	12.06	11.39	10.97	10.67	10.46	10.29	10.16	10.05	9.89	9.72	9.55	9.47	9.38	9.29	9.20	9.11	9.02
6	13.75	10.92	9.78	9.15	8.75	8.47	8.26	8.10	7.98	7.87	7.72	7.56	7.40	7.31	7.23	7.14	7.06	6.97	6.88
7	12.25	9.55	9.45	7.85	7.46	7.19	6.99	6.84	6.72	6.62	6.47	6.31	6.16	6.07	5.99	5.91	5.82	5.74	5.65
8	11.26	8.65	7.59	7.01	6.63	6.37	6.18	6.03	5.91	5.81	5.67	5.52	5.36	5.28	5.20	5.12	5.03	4.95	4.86
9	10.56	9.02	6.99	6.42	6.06	5.80	5.61	5.47	5.35	5.26	5.11	4.96	4.81	4.73	4.65	4.57	4.48	4.40	4.31
10	10.04	7.56	6.55	5.99	5.64	5.39	5.20	5.06	4.94	4.85	4.71	4.56	4.41	4.33	4.25	4.17	4.08	4.00	3.91
11	9.65	7.21	6.22	5.67	5.32	5.07	4.89	4.74	4.63	4.54	4.40	4.25	4.10	4.02	3.94	3.86	3.78	3.69	3.60
12	9.33	6.93	5.95	5.41	5.06	4.82	4.64	4.50	4.39	4.30	4.16	4.01	3.86	3.78	3.70	3.62	3.54	3.45	3.36
13	9.07	6.70	5.74	5.21	4.96	4.62	4.44	4.30	4.19	4.10	3.96	3.82	3.66	3.59	3.51	3.43	3.34	3.25	3.17
14	8.86	6.51	5.56	5.04	4.69	4.46	4.28	4.14	4.03	3.94	3.80	3.66	3.51	3.43	3.35	3.27	3.18	3.09	3.00
15	8.68	6.36	5.42	4.89	4.36	4.32	4.14	4.00	3.89	3.80	3.67	3.52	3.37	3.29	3.21	3.13	3.05	2.96	2.87
16	8.53	6.23	5.29	4.77	4.44	4.20	4.03	3.89	3.78	3.69	3.55	3.41	3.26	3.18	3.10	3.02	2.93	2.84	2.75
17	8.40	6.11	5.18	4.67	4.34	4.10	3.93	3.79	3.68	3.59	3.46	3.31	3.16	3.08	3.00	2.92	2.83	2.75	2.65
18	8.29	6.01	5.09	1.58	4.25	4.01	3.84	3.71	3.60	3.51	3.37	3.23	3.08	3.00	2.92	2.84	2.75	2.66	2.57
19	8.18	5.93	5.01	4.50	4.17	3.94	3.77	3.63	3.52	3.43	3.30	3.15	3.00	2.92	2.84	2.76	2.67	2.58	2.49

20	8.10	5.85	4.94	4.43	4.10	3.87	3.70	3.56	3.46	3.37	3.23	3.09	2.94	2.86	2.78	2.69	2.61	2.52	2.42
21	8.02	5.78	4.87	4.37	4.04	3.81	3.64	3.51	3.40	3.31	3.17	3.03	2.88	2.80	2.72	2.64	2.55	2.46	2.36
22	7.95	5.72	4.81	4.31	3.99	3.76	3.59	3.45	3.35	3.26	3.12	2.98	2.83	2.75	2.67	2.58	2.50	2.40	2.31
23	7.88	5.66	4.76	4.26	3.94	3.71	3.54	3.41	3.30	3.21	3.07	2.93	2.78	2.70	2.62	2.54	2.45	2.35	2.26
24	7.82	5.61	4.72	4.22	3.90	3.67	3.50	3.36	3.26	3.17	3.03	2.89	2.74	2.66	2.58	2.49	2.40	2.31	2.21
25	7.77	5.57	4.68	4.18	3.85	3.63	3.46	3.32	3.22	3.13	2.99	2.85	2.70	2.62	2.54	2.45	2.36	2.27	2.17
26	7.72	5.53	4.64	4.14	3.82	3.59	3.42	3.29	3.18	3.09	2.96	2.81	2.66	2.58	2.50	2.42	2.33	2.23	2.13
27	7.68	5.49	4.60	4.11	3.78	3.56	3.39	3.26	3.15	3.06	2.93	2.78	2.63	2.55	2.47	2.38	2.29	2.20	2.10
28	7.64	5.45	4.57	4.07	3.75	3.53	3.36	3.23	3.12	3.03	2.90	2.75	2.60	2.52	2.44	2.35	2.26	2.17	2.06
29	7.60	5.42	4.54	4.04	3.73	3.50	3.33	3.20	3.09	3.00	2.87	2.73	2.57	2.49	2.41	2.33	2.23	2.14	2.03
30	7.56	5.39	4.51	4.02	3.70	3.47	3.30	3.17	3.07	2.98	2.84	2.70	2.55	2.47	2.39	2.30	2.21	2.11	2.01
40	7.31	5.18	4.31	3.83	3.51	3.29	3.12	2.99	2.89	2.80	2.66	2.52	2.37	2.29	2.20	2.11	2.02	1.92	1.80
60	7.08	4.98	4.13	3.65	3.34	3.12	2.95	2.82	2.72	2.63	2.50	2.35	2.20	2.12	2.03	1.94	1.84	1.73	1.60
120	6.85	4.79	3.95	3.48	3.17	2.96	2.79	2.66	2.56	2.47	2.34	2.19	2.03	1.95	1.86	1.76	1.66	1.53	1.38
∞	6.63	4.61	3.78	3.32	3.02	2.80	2.64	2.51	2.41	2.32	2.18	2.04	1.88	1.79	1.70	1.51	1.47	1.32	1.00

Table C.4 Critical Values $f_{\nu_1,\nu_2,\alpha}$ for F-Distribution (*Continued*)

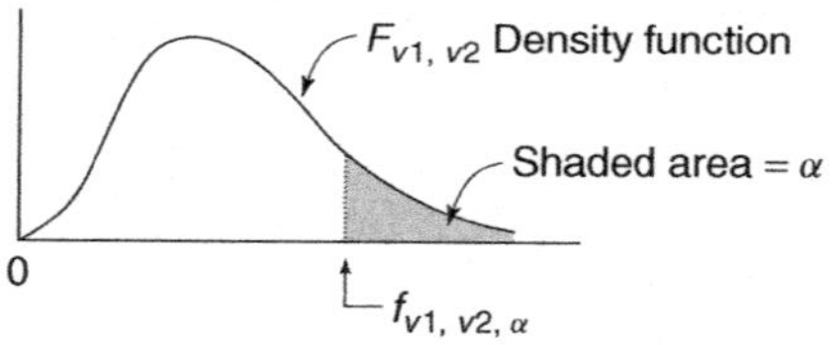

$\alpha = 0.05$

Degrees of freedom for numerator (ν_1); Degrees of freedom for denominator (ν_2)

ν_2	1	2	3	4	5	6	7	8	9	10	12	15	20	24	30	40	60	120	∞
1	161.4	199.5	215.7	224.6	230.2	234.0	236.8	238.9	240.5	241.9	243.9	245.9	248.0	249.1	250.1	251.1	252.2	253.3	254.3
2	18.51	19.00	19.16	19.25	19.30	19.33	19.35	19.37	19.38	19.40	19.41	19.43	19.45	19.45	19.46	19.47	19.48	19.49	19.50
3	10.13	9.55	9.28	9.12	9.01	8.94	8.89	8.85	8.81	8.79	8.74	8.70	8.66	8.64	8.62	8.59	8.57	8.55	8.53
4	7.71	6.94	6.59	6.39	6.26	6.16	6.09	6.04	6.00	5.96	5.91	5.86	5.80	5.77	5.75	5.72	5.69	5.66	5.63
5	6.61	5.79	5.41	5.19	5.05	4.95	4.88	4.82	4.77	4.74	4.68	4.62	4.56	4.53	4.50	4.46	4.43	4.40	4.36
6	5.99	5.14	4.76	4.53	4.39	4.28	4.21	4.15	4.10	4.06	4.00	3.94	3.87	3.84	3.81	3.77	3.74	3.70	3.67
7	5.59	4.74	4.35	4.12	3.97	3.87	3.79	3.73	3.68	3.64	3.57	3.51	3.44	3.41	3.38	3.34	3.30	3.27	3.23
8	5.32	4.46	4.07	3.84	3.69	3.58	3.50	3.44	3.39	3.35	3.28	3.22	3.15	3.12	3.08	3.04	3.01	2.97	2.93
9	5.12	4.26	3.86	3.63	3.48	3.37	3.29	3.23	3.18	3.14	3.07	3.01	2.94	2.90	2.86	2.83	2.79	2.75	2.71
10	4.96	4.10	3.71	3.48	3.33	3.22	3.14	3.07	3.02	2.98	2.91	2.85	2.77	2.74	2.70	2.66	2.62	2.58	2.54
11	4.84	3.98	3.59	3.36	3.20	3.09	3.01	2.95	2.90	2.85	2.79	2.72	2.65	2.61	2.57	2.53	2.49	2.45	2.40
12	4.75	3.89	3.49	3.26	3.11	3.00	2.91	2.85	2.80	2.75	2.69	2.62	2.54	2.51	2.47	2.43	2.38	2.34	2.30
13	4.67	3.81	3.41	3.18	3.03	2.92	2.83	2.77	2.71	2.67	2.60	2.53	2.46	2.42	2.38	2.34	2.30	2.25	2.21
14	4.60	3.74	3.34	3.11	2.96	2.85	2.76	2.70	2.65	2.60	2.53	2.46	2.39	2.35	2.31	2.27	2.22	2.18	2.13
15	4.54	3.68	3.29	3.06	2.90	2.79	2.71	2.64	2.59	2.54	2.49	2.40	2.33	2.29	2.25	2.20	2.16	2.11	2.07
16	4.49	3.63	3.24	3.01	2.85	2.74	2.66	2.59	2.54	2.49	2.42	2.35	2.28	2.24	2.19	2.15	2.11	2.06	2.01
17	4.45	3.59	3.20	2.96	2.81	2.69	2.61	2.55	2.49	2.45	2.38	2.31	2.23	2.19	2.15	2.10	2.06	2.01	1.96
18	4.41	3.55	3.16	2.93	2.77	2.66	2.58	2.51	2.46	2.41	2.34	2.27	2.19	2.15	2.11	2.06	2.02	1.97	1.92
19	4.38	3.52	3.13	2.90	2.74	2.63	2.54	2.48	2.42	2.38	2.31	2.23	2.16	2.11	2.07	2.03	1.98	1.93	1.88

20	4.35	3.49	3.10	2.87	2.71	2.60	2.51	2.45	2.39	2.35	2.28	2.20	2.12	2.08	2.04	1.99	1.95	1.90	1.84
21	4.32	3.47	3.07	2.84	2.68	2.57	2.49	2.42	2.37	2.32	2.25	2.18	2.10	2.05	2.01	1.96	1.92	1.87	1.81
22	4.30	3.44	3.05	2.82	2.66	2.55	2.46	2.40	2.34	2.30	2.23	2.15	2.07	2.03	1.98	1.94	1.89	1.84	1.78
23	4.28	3.42	3.03	2.80	2.64	2.53	2.44	2.37	2.32	2.27	2.20	2.13	2.05	2.01	1.96	1.91	1.86	1.81	1.76
24	4.26	3.40	3.01	2.78	2.62	2.51	2.42	2.36	2.30	2.25	2.18	2.11	2.03	1.98	1.94	1.89	1.84	1.79	1.73
25	4.24	3.39	2.99	2.76	2.60	2.49	2.40	2.34	2.28	2.24	2.16	2.09	2.01	1.96	1.92	1.87	1.82	1.77	1.71
26	4.23	3.37	2.98	2.74	2.59	2.47	2.39	2.32	2.27	2.22	2.15	2.07	1.99	1.95	1.90	1.81	1.80	1.75	1.69
27	4.21	3.35	2.96	2.73	2.57	2.46	2.37	2.31	2.25	2.20	2.13	2.06	1.97	1.93	1.88	1.84	1.79	1.73	1.67
28	4.20	3.34	2.95	2.71	2.56	2.45	2.36	2.29	2.24	2.19	2.12	2.04	1.96	1.91	1.87	1.82	1.77	1.71	1.65
29	4.18	3.33	2.93	2.70	2.55	2.43	2.35	2.28	2.22	2.18	2.10	2.03	1.94	1.90	1.85	1.91	1.75	1.70	1.64
30	4.17	3.32	2.92	2.69	2.53	2.42	2.33	2.27	2.21	2.16	2.09	2.01	1.93	1.89	1.84	1.79	1.74	1.68	1.62
40	4.09	3.23	2.84	2.61	2.45	2.34	2.25	2.18	2.12	2.08	2.00	1.92	1.84	1.79	1.74	1.69	1.64	1.59	1.51
60	4.00	3.15	2.76	2.53	2.37	2.25	2.17	2.10	2.04	1.99	1.92	1.84	1.75	1.70	1.65	1.59	1.53	1.47	1.39
120	3.92	3.07	2.68	2.45	2.29	2.17	2.09	2.02	1.96	1.91	1.81	1.75	1.66	1.61	1.55	1.55	1.43	1.35	1.25
∞	3.84	3.00	2.60	2.37	2.21	2.10	2.01	1.94	1.88	1.83	1.75	1.67	1.57	1.52	1.46	1.39	1.32	1.22	1.00

Table C.4 Critical Values $f_{\nu_1,\nu_2,\alpha}$ for F-Distribution (*Continued*)

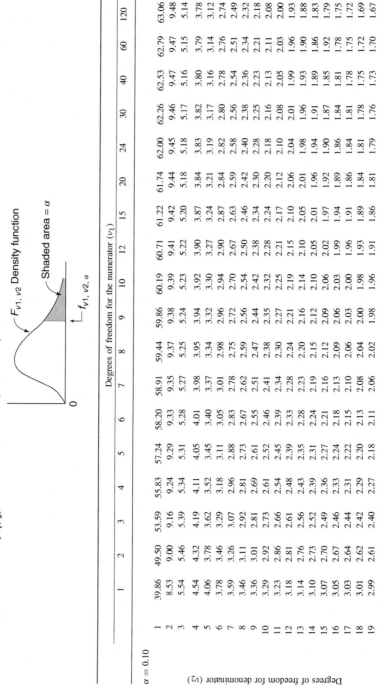

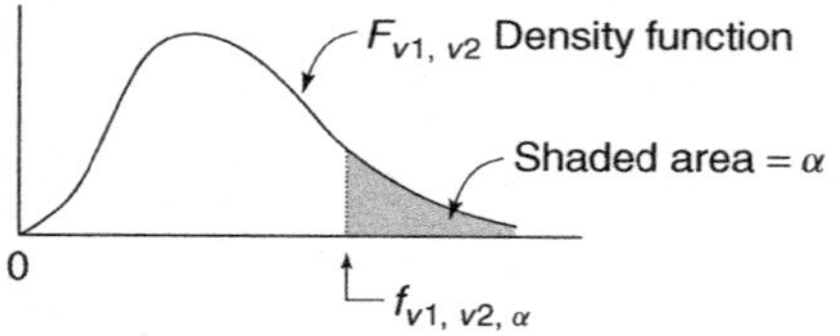

$\alpha = 0.10$

Degrees of freedom for denominator (ν_2)	Degrees of freedom for the numerator (ν_1): 1	2	3	4	5	6	7	8	9	10	12	15	20	24	30	40	60	120	∞
1	39.86	49.50	53.59	55.83	57.24	58.20	58.91	59.44	59.86	60.19	60.71	61.22	61.74	62.00	62.26	62.53	62.79	63.06	63.33
2	8.53	9.00	9.16	9.24	9.29	9.33	9.35	9.37	9.38	9.39	9.41	9.42	9.44	9.45	9.46	9.47	9.47	9.48	9.49
3	5.54	5.46	5.39	5.34	5.31	5.28	5.27	5.25	5.24	5.23	5.22	5.20	5.18	5.18	5.17	5.16	5.15	5.14	5.13
4	4.54	4.32	4.19	4.11	4.05	4.01	3.98	3.95	3.94	3.92	3.90	3.87	3.84	3.83	3.82	3.80	3.79	3.78	3.76
5	4.06	3.78	3.62	3.52	3.45	3.40	3.37	3.34	3.32	3.30	3.27	3.24	3.21	3.19	3.17	3.16	3.14	3.12	3.10
6	3.78	3.46	3.29	3.18	3.11	3.05	3.01	2.98	2.96	2.94	2.90	2.87	2.84	2.82	2.80	2.78	2.76	2.74	2.72
7	3.59	3.26	3.07	2.96	2.88	2.83	2.78	2.75	2.72	2.70	2.67	2.63	2.59	2.58	2.56	2.54	2.51	2.49	2.47
8	3.46	3.11	2.92	2.81	2.73	2.67	2.62	2.59	2.56	2.54	2.50	2.46	2.42	2.40	2.38	2.36	2.34	2.32	2.29
9	3.36	3.01	2.81	2.69	2.61	2.55	2.51	2.47	2.44	2.42	2.38	2.34	2.30	2.28	2.25	2.23	2.21	2.18	2.16
10	3.29	2.92	2.73	2.61	2.52	2.46	2.41	2.38	2.35	2.32	2.28	2.24	2.20	2.18	2.16	2.13	2.11	2.08	2.06
11	3.23	2.86	2.66	2.54	2.45	2.39	2.34	2.30	2.27	2.25	2.21	2.17	2.12	2.10	2.08	2.05	2.03	2.00	1.97
12	3.18	2.81	2.61	2.48	2.39	2.33	2.28	2.24	2.21	2.19	2.15	2.10	2.06	2.04	2.01	1.99	1.96	1.93	1.90
13	3.14	2.76	2.56	2.43	2.35	2.28	2.23	2.20	2.16	2.14	2.10	2.05	2.01	1.98	1.96	1.93	1.90	1.88	1.85
14	3.10	2.73	2.52	2.39	2.31	2.24	2.19	2.15	2.12	2.10	2.05	2.01	1.96	1.94	1.91	1.89	1.86	1.83	1.80
15	3.07	2.70	2.49	2.36	2.27	2.21	2.16	2.12	2.09	2.06	2.02	1.97	1.92	1.90	1.87	1.85	1.92	1.79	1.76
16	3.05	2.67	2.46	2.33	2.24	2.18	2.13	2.09	2.06	2.03	1.99	1.94	1.89	1.86	1.84	1.81	1.78	1.75	1.72
17	3.03	2.64	2.44	2.31	2.22	2.15	2.10	2.06	2.03	2.00	1.96	1.91	1.86	1.84	1.81	1.78	1.75	1.72	1.09
18	3.01	2.62	2.42	2.29	2.20	2.13	2.08	2.04	2.00	1.98	1.93	1.89	1.84	1.81	1.78	1.75	1.72	1.69	1.66
19	2.99	2.61	2.40	2.27	2.18	2.11	2.06	2.02	1.98	1.96	1.91	1.86	1.81	1.79	1.76	1.73	1.70	1.67	1.63

20	2.97	2.59	2.38	2.25	2.16	2.09	2.04	2.00	1.96	1.94	1.89	1.84	1.79	1.77	1.74	1.71	1.68	1.64	1.61
21	2.96	2.57	2.36	2.23	2.14	2.08	2.02	1.98	1.95	1.92	1.87	1.83	1.78	1.75	1.72	1.69	1.66	1.62	1.59
22	2.95	2.56	2.35	2.22	2.13	2.06	2.01	1.97	1.93	1.90	1.86	1.81	1.76	1.73	1.70	1.67	1.64	1.60	1.57
23	2.94	2.55	2.34	2.21	2.11	2.05	1.99	1.95	1.92	1.89	1.84	1.80	1.74	1.72	1.69	1.66	1.62	1.59	1.55
24	2.93	2.54	2.33	2.19	2.10	2.04	1.98	1.94	1.91	1.88	1.83	1.78	1.73	1.70	1.67	1.64	1.61	1.57	1.53
25	2.92	2.53	2.32	2.18	2.09	2.02	1.97	1.93	1.89	1.87	1.82	1.77	1.72	1.69	1.66	1.63	1.59	1.56	1.52
26	2.91	2.52	2.31	2.17	2.08	2.01	1.96	1.92	1.88	1.86	1.81	1.76	1.71	1.68	1.65	1.61	1.58	1.54	1.50
27	2.90	2.51	2.30	2.17	2.07	2.00	1.95	1.91	1.87	1.85	1.80	1.75	1.70	1.67	1.64	1.60	1.57	1.53	1.49
28	2.89	2.50	2.29	2.16	2.06	2.00	1.94	1.90	1.87	1.84	1.79	1.74	1.69	1.66	1.63	1.59	1.56	1.52	1.48
29	2.89	2.50	2.28	2.15	2.06	1.99	1.93	1.89	1.86	1.83	1.78	1.73	1.68	1.65	1.62	1.58	1.55	1.51	1.47
30	2.88	2.49	2.28	2.14	2.03	1.98	1.93	1.88	1.85	1.82	1.77	1.72	1.67	1.64	1.61	1.57	1.54	1.50	1.46
40	2.84	2.44	2.23	2.09	2.00	1.93	1.87	1.83	1.79	1.76	1.71	1.66	1.61	1.57	1.54	1.51	1.47	1.42	1.38
60	2.79	2.39	2.18	2.04	1.95	1.87	1.82	1.77	1.74	1.71	1.66	1.60	1.54	1.51	1.48	1.44	1.40	1.35	1.29
120	2.75	2.35	2.13	1.99	1.90	1.82	1.77	1.72	1.68	1.65	1.60	1.55	1.48	1.45	1.41	1.37	1.32	1.26	1.19
∞	2.71	2.30	2.08	1.94	1.85	1.77	1.72	1.67	1.63	1.60	1.55	1.49	1.42	1.38	1.34	1.30	1.24	1.17	1.00

Source: Reprinted with permission of Pearson Education, Inc.

Table C.5 Studentized Range Critical Values $q_{k,\nu,\alpha}$

	k								
ν	2	3	4	5	6	7	8	9	10
$\alpha = 0.05$									
5	3.635	4.602	5.218	5.673	6.033	6.330	6.582	6.802	6.995
10	3.151	3.877	4.327	4.654	4.912	5.124	5.305	5.461	5.599
15	3.014	3.674	4.076	4.367	4.595	4.782	4.940	5.077	5.198
20	2.950	3.578	3.958	4.232	4.445	4.620	4.768	4.986	5.008
30	2.888	3.486	3.845	4.102	4.302	4.464	4.602	4.720	4.824
40	2.858	3.442	3.791	4.039	4.232	4.389	4.521	4.635	4.735
60	2.829	3.399	3.737	3.977	4.163	4.314	4.441	4.550	4.646
120	2.800	3.356	3.685	3.917	4.096	4.241	4.363	4.468	4.560
∞	2.772	3.314	3.633	3.858	4.030	4.170	4.286	4.387	4.474
$\alpha = 0.10$									
5	2.850	3.717	4.264	4.664	4.979	5.238	5.458	5.648	5.816
10	2.563	3.270	3.704	4.018	4.264	4.465	4.636	4.783	4.913
15	2.479	3.140	3.540	3.828	4.052	4.235	4.390	4.524	4.641
20	2.439	3.078	3.462	3.736	3.950	4.124	4.271	4.398	4.510
30	2.400	3.017	3.386	3.648	3.851	4.016	4.155	4.275	4.381
40	2.381	2.988	3.349	3.605	3.803	3.963	4.099	4.215	4.317
60	2.363	2.959	3.312	3.562	3.755	3.911	4.042	4.155	4.254
120	2.344	2.930	3.276	3.520	3.707	3.859	3.987	4.096	4.191
∞	2.326	2.902	3.240	3.478	3.661	3.808	3.931	4.037	4.129

Source: Excerpted from H. L. Harter (1960), "Table of range and Studentized range," *Annals of Mathematical Statistics*, **31**, 1122–1147.

Table C.6 One-Sided Multivariate t Critical Points $t_{k,\nu,\alpha}$ for Common Correlation, $\rho = 0.5$

		k								
ν	α	2	3	4	5	6	7	8	9	10
5	.01	3.900	4.211	4.429	4.597	4.733	4.846	4.944	5.030	5.106
	.05	2.440	2.681	2.848	2.976	3.078	3.163	3.236	3.300	3.356
	.10	1.873	2.094	2.245	2.359	2.451	2.527	2.592	2.649	2.699
10	.01	3.115	3.314	3.453	3.559	3.644	3.715	3.777	3.830	3.878
	.05	2.151	2.338	2.466	2.562	2.640	2.704	2.759	2.807	2.849
	.10	1.713	1.899	2.024	2.119	2.194	2.256	2.309	2.355	2.396
15	.01	2.908	3.080	3.198	3.289	3.362	3.422	3.474	3.520	3.560
	.05	2.067	2.239	2.356	2.444	2.515	2.573	2.623	2.667	2.705
	.10	1.665	1.840	1.959	2.047	2.118	2.716	2.225	2.268	2.306
20	.01	2.813	2.972	3.082	3.166	3.233	3.289	3.337	3.378	3.416
	.05	2.027	2.192	2.304	2.389	2.456	2.512	2.559	2.601	2.637
	.10	1.642	1.813	1.927	2.013	2.081	2.137	2.185	2.227	2.263
25	.01	2.758	2.911	3.016	3.095	3.159	3.212	3.258	3.298	3.333
	.05	2.004	2.165	2.274	2.356	2.422	2.476	2.522	2.562	2.598
	.10	1.629	1.796	1.909	1.993	2.060	2.115	2.162	2.202	2.238
30	.01	2.723	2.871	2.973	3.050	3.111	3.163	3.207	3.245	3.279
	.05	1.989	2.147	2.255	2.335	2.399	2.453	2.498	2.537	2.572
	.10	1.620	1.786	1.897	1.980	2.046	2.100	2.146	2.186	2.222
35	.01	2.698	2.843	2.942	3.018	3.078	3.128	3.171	3.209	3.242
	.05	1.978	2.135	2.241	2.320	2.384	2.436	2.481	2.519	2.554
	.10	1.614	1.778	1.888	1.971	2.036	2.090	2.135	2.175	2.210
40	.01	2.680	2.822	2.920	2.994	3.053	3.103	3.145	3.181	3.214
	.05	1.970	2.125	2.230	2.309	2.372	2.424	2.468	2.506	2.540
	.10	1.609	1.772	1.882	1.964	2.028	2.082	2.127	2.167	2.201
50	.01	2.655	2.794	2.889	2.962	3.019	3.068	3.109	3.145	3.176
	.05	1.959	2.112	2.216	2.294	2.356	2.407	2.450	2.488	2.521
	.10	1.603	1.764	1.873	1.954	2.018	2.071	2.116	2.155	2.189
100	.01	2.0605	2.738	2.829	2.898	2.953	2.999	3.038	3.072	3.102
	.05	1.938	2.087	2.188	2.263	2.324	2.373	2.415	2.452	2.484
	.10	1.590	1.749	1.885	1.935	1.998	2.050	2.094	2.132	2.166
200	.01	2.581	2.711	2.800	2.867	2.921	2.966	3.004	3.037	3.006
	.05	1.927	2.074	2.174	2.249	2.308	2.357	2.398	2.434	2.466
	.10	1.583	1.741	1.847	1.926	1.988	2.039	2.083	2.121	2.154
∞	.01	2.558	2.685	2.772	2.837	2.889	2.933	2.970	3.002	3.031
	.05	1.916	2.062	2.160	2.234	2.292	2.340	2.381	2.417	2.448
	.10	1.577	1.734	1.838	1.916	1.978	2.029	2.072	2.109	2.148

Source: Excerpted from R. E. Bechhofer and C. W. Dunnett (1988), "Percentage points of multivariate t-distribution," *Selected Tables in Mathematical Statistics*, **11**, Providence, RI: American Mathematical Society.

Table C.7 Two-Sided Multivariate t Critical Points $|t|_{k,\nu,\alpha}$ for Common Correlation, $\rho = 0.5$

		k								
ν	α	2	3	4	5	6	7	8	9	10
5	.01	4.627	4.975	5.219	5.406	5.557	5.683	5.792	5.887	5.971
	.05	3.030	3.293	3.476	3.615	3.727	3.821	3.900	3.970	4.032
	.10	2.433	2.669	2.832	2.956	3.055	3.137	3.207	3.268	3.322
10	.01	3.531	3.739	3.883	3.994	4.084	4.159	4.223	4.279	4.329
	.05	2.568	2.759	2.890	2.990	3.070	3.137	3.194	3.244	3.288
	.10	2.149	2.335	2.463	2.559	2.636	2.700	2.755	2.802	2.844
15	.01	3.253	3.426	3.547	3.639	3.713	3.776	3.829	3.875	3.917
	.05	2.439	2.610	2.727	2.816	2.887	2.946	2.997	3.041	3.080
	.10	2.066	2.238	2.355	2.443	2.514	2.572	2.622	2.665	2.703
20	.01	3.127	3.285	3.395	3.479	3.547	3.603	3.651	3.694	3.731
	.05	2.379	2.540	2.651	2.735	2.802	2.857	2.905	2.946	2.983
	.10	2.027	2.192	2.304	2.388	2.455	2.511	2.559	2.600	2.636
25	.01	3.055	3.205	3.309	3.388	3.452	3.505	3.551	3.591	3.626
	.05	2.344	2.500	2.607	2.688	2.752	2.806	2.852	2.891	2.927
	.10	2.004	2.165	2.274	2.356	2.421	2.476	2.522	2.562	2.597
30	.01	3.009	3.154	3.254	3.330	3.391	3.442	3.486	3.524	3.558
	.05	2.321	2.474	2.578	2.657	2.720	2.772	2.817	2.856	2.890
	.10	1.989	2.147	2.254	2.335	2.399	2.452	2.498	2.537	2.572
35	.01	2.976	3.118	3.215	3.289	3.349	3.398	3.441	3.478	3.511
	.05	2.305	2.455	2.558	2.635	2.697	2.748	2.792	2.830	2.864
	.10	1.978	2.135	2.240	2.320	2.383	2.436	2.480	2.519	2.553
40	.01	2.952	3.091	3.186	3.259	3.317	3.366	3.408	3.444	3.476
	.05	2.293	2.441	2.543	2.619	2.680	2.731	2.774	2.812	2.845
	.10	1.970	2.125	2.230	2.309	2.372	2.424	2.468	2.506	2.540
50	.01	2.920	3.054	3.147	3.218	3.274	3.321	3.362	3.397	3.428
	.05	2.276	2.422	2.522	2.597	2.657	2.707	2.749	2.786	2.819
	.10	1.959	2.112	2.216	2.294	2.355	2.407	2.450	2.488	2.521
100	.01	2.856	2.983	3.071	3.137	3.191	3.235	3.273	3.306	3.335
	.05	2.244	2.385	2.481	2.554	2.611	2.659	2.700	2.735	2.767
	.10	1.938	2.087	2.188	2.263	2.323	2.373	2.415	2.452	2.484
200	.01	2.825	2.949	3.034	3.098	3.150	3.193	3.230	3.262	3.291
	.05	2.228	2.367	2.461	2.532	2.589	2.636	2.676	2.711	2.741
	.10	1.927	2.074	2.174	2.249	2.308	2.357	2.398	2.434	2.466
∞	.01	2.794	2.915	2.998	3.060	3.110	3.152	3.188	3.219	3.246
	.05	2.212	2.349	2.442	2.511	2.567	2.613	2.652	2.686	2.716
	.10	1.916	2.062	2.160	2.234	2.292	2.340	2.381	2.417	2.488

Source: Excerpted from R. E. Bechhofer and C. W. Dunnett (1988), "Percentage points of multivariate t-distribution," *Selected Tables in Mathematical Statistics*, **11**, (Providence, RI: American Mathematical Society.)

Table C.8 Studentized Maximum Critical Values $M_{k,\nu,\alpha}$ (One-sided Multivariate t Critical Values $t_{k,\nu,\rho,\alpha}$ with $\rho = 0$)

		k								
ν	α	2	3	4	5	6	7	8	9	10
5	.01	3.997	4.387	4.671	4.896	5.081	5.239	5.376	5.497	5.606
	.05	2.532	2.840	3.062	3.234	2.376	3.495	3.599	3.690	3.772
	.10	1.969	2.256	2.459	2.616	2.744	2.851	2.944	3.026	3.098
10	.01	3.161	3.394	3.560	3.690	3.796	3.886	3.963	4.032	4.094
	.05	2.211	2.439	2.598	2.720	2.820	2.903	2.976	3.039	3.096
	.10	1.787	2.018	2.178	2.300	2.398	2.481	2.552	2.614	2.670
15	.01	2.942	3.138	3.276	3.382	3.469	3.542	3.606	3.662	3.712
	.05	2.120	2.326	2.467	2.576	2.663	2.737	2.800	2.856	2.905
	.10	1.732	1.947	2.095	2.207	2.297	2.372	2.436	2.492	2.542
20	.01	2.842	3.021	3.147	3.243	3.322	3.388	3.444	3.494	3.539
	.05	2.076	2.271	2.405	2.507	2.590	2.659	2.718	2.770	2.816
	.10	1.706	1.914	2.055	2.162	2.248	2.320	2.381	2.434	2.481
25	.01	2.785	2.955	3.073	3.164	3.238	3.299	3.353	3.400	3.441
	.05	2.051	2.239	2.369	2.468	2.547	2.613	2.670	2.720	2.764
	.10	1.691	1.894	2.032	2.136	2.220	2.289	2.348	2.400	2.446
30	.01	2.748	2.912	3.026	3.113	3.184	3.243	3.294	3.338	3.378
	.05	2.034	2.219	2.346	2.442	2.519	2.584	2.639	2.687	2.730
	.10	1.681	1.881	2.017	2.119	2.201	2.269	2.327	2.378	2.422
35	.01	2.722	2.881	2.992	3.077	3.146	3.203	3.253	3.296	3.334
	.05	2.022	2.204	2.329	2.424	2.499	2.563	2.617	2.664	2.706
	.10	1.674	1.872	2.006	2.107	2.188	2.255	2.312	2.362	2.406
40	.01	2.703	2.859	2.968	3.051	3.118	3.174	3.222	3.264	3.302
	.05	2.014	2.194	2.317	2.410	2.485	2.547	2.600	2.647	2.688
	.10	1.668	1.865	1.998	2.098	2.178	2.244	2.301	2.350	2.393
50	.01	2.676	2.829	2.934	3.015	3.080	3.134	3.180	3.221	3.258
	.05	2.001	2.179	2.300	2.391	2.465	2.526	2.578	2.623	2.663
	.10	1.661	1.855	1.987	2.085	2.164	2.229	2.285	2.333	2.376
100	.01	2.625	2.769	2.869	2.944	3.005	3.056	3.100	3.138	3.171
	.05	1.978	2.150	2.267	2.355	2.425	2.483	2.533	2.577	2.615
	.10	1.647	1.837	1.965	2.061	2.137	2.200	2.254	2.301	2.342
200	.01	2.600	2.740	2.837	2.910	2.969	3.018	3.060	3.097	3.130
	.05	1.966	2.135	2.250	2.337	2.405	2.463	2.511	2.554	2.591
	.10	1.639	1.827	1.954	2.049	2.124	2.186	2.239	2.285	2.325
∞	.01	2.575	2.712	2.806	2.877	2.934	2.981	3.022	3.057	3.089
	.05	1.955	2.121	2.234	2.319	2.386	2.442	2.490	2.531	2.568
	.10	1.632	1.818	1.943	2.036	2.111	2.172	2.224	2.269	2.309

Source: Excerpted from R. E. Bechhofer and C. W. Dunnett (1988), "Tables of percentage points of multivariate student t distributions," *Selected Tables in Mathematical Statistics*, **11**, 1–371.

Table C.9 Studentized Maximum Modulus Critical Values $|M|_{k,\nu,\alpha}$ (Two-Sided Multivariate t Critical Values $|t|_{k,\nu,\rho,\alpha}$ for $\rho = 0$)

		k								
ν	α	2	3	4	5	6	7	8	9	10
5	.01	4.700	5.106	5.397	5.625	5.812	5.969	6.106	6.226	6.334
	.05	3.091	3.399	3.619	3.789	3.928	4.044	4.145	4.233	4.312
	.10	2.491	2.769	2.965	3.116	3.239	3.341	3.430	3.507	3.576
10	.01	3.567	3.801	3.968	4.098	4.205	4.295	4.373	4.441	4.503
	.05	2.609	2.829	2.983	3.103	3.199	3.281	3.351	3.412	3.467
	.10	2.193	2.410	2.562	2.678	2.771	2.850	2.918	2.977	3.029
15	.01	3.279	3.472	3.608	3.714	3.800	3.872	3.935	3.990	4.040
	.05	2.474	2.669	2.805	2.909	2.994	3.065	3.126	3.180	3.227
	.10	2.107	2.305	2.443	2.548	2.633	2.704	2.765	2.818	2.865
20	.01	3.149	3.323	3.446	3.540	3.617	3.682	3.738	3.787	3.831
	.05	2.411	2.594	2.721	2.819	2.897	2.963	3.020	3.070	3.114
	.10	2.065	2.255	2.386	2.486	2.567	2.634	2.691	2.742	2.786
25	.01	3.075	3.239	3.354	3.442	3.514	3.574	3.626	3.672	3.713
	.05	2.374	2.551	2.673	2.766	2.841	2.904	2.959	3.006	3.048
	.10	2.041	2.226	2.353	2.450	2.528	2.592	2.648	2.697	2.740
30	.01	3.027	3.185	3.295	3.379	3.447	3.505	3.554	3.598	3.637
	.05	2.350	2.522	2.641	2.732	2.805	2.866	2.918	2.964	3.005
	.10	2.025	2.207	2.331	2.426	2.502	2.565	2.620	2.667	2.709
35	.01	2.994	3.147	3.253	3.335	3.401	3.457	3.504	3.546	3.584
	.05	2.333	2.502	2.619	2.708	2.779	2.839	2.890	2.935	2.975
	.10	2.014	2.193	2.316	2.409	2.484	2.546	2.599	2.646	2.687
40	.01	2.969	3.119	3.223	3.303	3.367	3.421	3.468	3.508	3.545
	.05	2.321	2.488	2.602	2.690	2.760	2.819	2.869	2.913	2.952
	.10	2.006	2.183	2.305	2.397	2.470	2.532	2.584	2.630	2.671
50	.01	2.935	3.080	3.181	3.258	3.320	3.372	3.417	3.456	3.491
	.05	2.304	2.467	2.580	2.665	2.734	2.791	2.840	2.883	2.921
	.10	1.994	2.169	2.289	2.379	2.452	2.512	2.564	2.609	2.648
100	.01	2.869	3.006	3.100	3.172	3.229	3.278	3.319	3.356	3.388
	.05	2.270	2.427	2.535	2.616	2.682	2.736	2.783	2.823	2.859
	.10	1.971	2.141	2.257	2.345	2.414	2.473	2.522	2.565	2.604
200	.01	2.838	2.970	3.061	3.130	3.186	3.232	3.272	3.307	3.338
	.05	2.253	2.407	2.513	2.592	2.656	2.709	2.755	2.794	2.829
	.10	1.960	2.128	2.242	2.328	2.396	2.453	2.502	2.544	2.582
∞	.01	2.806	2.934	3.022	3.089	3.143	3.188	3.226	3.260	3.289
	.05	2.236	2.388	2.491	2.569	2.631	2.683	2.727	2.766	2.800
	.10	1.949	2.114	2.226	2.311	2.378	2.434	2.481	2.523	2.560

Source: Excerpted from R. E. Bechhofer and C. W. Dunnett (1988), "Tables of percentage points of multivariate student t distributions," *Selected Tables in Mathematical Statistics*, **11**, 1–371.

Table C.10 Critical Constants $c_\alpha(m-1, a-1, N-a)$ for Wilks' Λ Statistic for the Test of No Treatment–Time Interaction ($\alpha = 0.05$)

	$m=4$				$m=5$				$m=6$			
	a				a				a			
$N-a$	3	4	5	6	3	4	5	6	3	4	5	6
10	0.243	0.164	0.0864	0.117	0.155	0.091	0.057	0.038	0.092	0.046	0.026	0.015
15	0.405	0.309	0.243	0.195	0.314	0.219	0.159	0.119	0.239	0.152	0.102	0.0703
20	0.515	0.419	0.348	0.293	0.431	0.329	0.257	0.205	0.359	0.256	0.188	0.142
25	0.591	0.500	0.430	0.374	0.516	0.415	0.340	0.283	0.449	0.343	0.268	0.213
30	0.648	0.563	0.495	0.439	0.580	0.483	0.409	0.349	0.519	0.415	0.337	0.277
40	0.724	0.651	0.591	0.539	0.668	0.583	0.513	0.455	0.617	0.522	0.446	0.384
60	0.808	0.752	0.704	0.661	0.767	0.700	0.643	0.592	0.729	0.652	0.587	0.531
80	0.853	0.808	0.769	0.733	0.821	0.766	0.718	0.675	0.791	0.727	0.672	0.623
100	0.881	0.844	0.810	0.780	0.854	0.809	0.768	0.730	0.830	0.776	0.728	0.685

For $a = 2$, use $F = \left(\frac{N-m}{m-1}\right)\left(\frac{1-\Lambda}{\Lambda}\right) \sim F_{m-1,N-m}$.

For $m = 2$, use $F = \left(\frac{N-a}{a-1}\right)\left(\frac{1-\Lambda}{\Lambda}\right) \sim F_{a-1,N-a}$.

For $m = 3$, use $F = \left(\frac{N-a-1}{a-1}\right)\left(\frac{1-\sqrt{\Lambda}}{\sqrt{\Lambda}}\right) \sim F_{2(a-1),2(N-a-1)}$.

Source: Adapted from Timm (1975), Table IX. Reprinted by permission of Dr. Neil H. Timm.

Answers to Selected Exercises

Chapter 1

1.1 (a) Experimental. (b) Experimental. (c) Observational. (d) Experimental. (e) Observational.

1.2 (a) Treatment factors: bit size, rotational speed, feed rate; noise factors: coolant temperature, possibly others such as tool wear. (b) Number of levels of each factor 2; number of treatment combinations 8. (c) Experimental units: aluminum pieces, number of replicates 3. (d) Crossed.

1.3 (a) Treatment factors: fillers and filling heads. Fillers are fixed; filling heads are random. Filling heads are nested within fillers. (b) Experimental units: bottles. Number of replicates 5.

1.9 (a) Treatment factors: operating condition and water temperature. (b) Treatments: There are four treatments obtained by combining normal/severe operating condition and $175^\circ/150^\circ$ water temperature. (c) Blocking factors: plant. (d) Experimental units: pipes. (e) Replications: Within each plant there are two replications. (f) Randomization: The runs appear to be randomized.

1.10 (a) 0.243. (b) 12.

Chapter 2

2.1 (a) The box plot shows a fairly symmetric distribution. The normal plot shows no significant deviation from normality (AD statistic 0.204, $p = 0.863$).

(b) [5.3639, 5.5320].

2.2 (a) The box plot suggests a right-skewed distribution. The normal plot is not linear and the AD statistic is significant indicating lack of normality.

(b) Because of the outliers and nonnormality, a CI based on the normality assumption will not be accurate.

2.4 32.

2.5 (a) Matched pairs. (b) Matched pairs. (c) Independent samples. (d) Independent samples.(e) Independent samples.

2.7 (a) $F = 1.664$, $p = 0.177$. Homoscedasticity assumption is consistent with the data. (b) $t = 0.615$, $p = 0.539$. Difference on the log-scale is not significant.

2.8 (a) $F = 2.778$ ($p = 0.169$).

(b) Pooled variances $t = -2.164$ ($p = 0.050$).

(c) Separate variances $t = -1.825$ with 5.48 d.f. ($p = 0.129$). The two results are very different. Do not assume homoscedasticity.

2.9 $t = 2.14$, $p = 0.025$.

2.14 (a) The regression equation is Distance $= -70.7 + 4.14 \times$ speed. The residuals plot is quadratic; also variance increases with speed (heteroscedastic).

(b) $F_{\text{lof}} = 0.37$ with $p = 0.776$. No significant lack of fit is indicated. This result does not agree with the residuals plot.

(c) The regression equation is Distance $= 1.6 + 0.0517 \times \text{speed}^2$. The residuals plot does not show any pattern but shows heteroscedasticity.

(d) The kinetic energy of the car is proportional to its speed^2. This kinetic energy = braking energy = braking force × stopping distance. Hence stopping distance is proportional to speed^2.

2.17 (a) $\text{Corr}(x_1, x_2) = 0.913$, $\text{Corr}(x_1, y) = 0.971$, $\text{Corr}(x_2, y) = 0.904$.

(b) The regression equation is $\widehat{y} = -2.606 + 0.192x_1 + 0.341x_2$. Only x_1 is significant because x_1 and x_2 are highly correlated.

(c) $\widehat{\beta_1^*} = 0.875$, $\widehat{\beta_2^*} = 0.105$.

(d) $\widehat{\beta_1} = (s_y/s_{x_1})\widehat{\beta_1^*} = (1.501/6.830)0.875 = 0.192$, $\widehat{\beta_2} = (s_y/s_{x_2})\widehat{\beta_2^*} = (1.501/0.461)0.105 = 0.341$.

(e) $\widehat{y} = -2.68 + 0.213x_1$.

Chapter 3

3.3 $F = 14.73$ $(p = 0.000)$.

3.5 (a) The side-by-side box plot suggests significant differences between groups with HBSC having the highest and HBSS having the lowest level.

(b) The ANOVA shows highly significant differences between groups $(F = 50.0, p = 0.001)$.

3.6 The differences between the dose groups are significant $(F = 4.55, p = 0.001)$.

3.9 $B = 21.571$ $(p = 0.003)$.

3.10 For the raw data, Bartlett statistic is $B = 32.881(p = 0.000)$ and Levene statistic is $L = 1.605(p = 0.157)$. For the log-transformed data, Bartlett statistic is $B = 7.599(p = 0.369)$ and Levene statistic is $L = 0.557(p = 0.787)$. The homoscedasticity assumption is suspect for the raw data but is satisfied for the log-transformed data.

3.12 $F^* = 4.55, \nu_1 = 3.07, \nu_2 = 19.24$, with $p = 0.0139$. The test result is much less significant than that obtained under the homoscedasticity assumption in Exercise 3.6.

3.13 (a) $B = 10.457$ $(p = 0.015)$; the homoscedasticity hypothesis is rejected. (b) $F^* = 52.133, \nu_1 = 2.377, \nu_2 = 49.307$. Highly significant $(p = 0.000)$.

3.14 (a) The normal plot is satisfactory, but the plot against fitted values shows decreasing variance with the mean.

(b) Bartlett test: $B = 13.24, p = 0.004$; Levene test: $L = 3.12, p = 0.035$. Both tests reject the homoscedasticity hypothesis at $\alpha = 0.05$.

(c) $F^* = 2.994, \nu_1 = 2.56, \nu_2 = 38.22$ $(p = 0.000)$.

3.21 With $n = 13$ subjects per drug, power $= 0.9239$. With $n = 10$ subjects per drug, power $= 0.8205$.

3.22 (a) The fitted cubic orthogonal polynomial is

$$\widehat{y} = 13.925 - 0.685\xi_1(z) + 0.225\xi_2(z) + 0.005\xi_3(z),$$

where

$$\xi_1(z) = 2z, \xi_2(z) = z^2 - \tfrac{5}{4}, \qquad \xi_3(z) = \tfrac{10}{3}\left(z^3 - \tfrac{41}{20}z\right),$$

and $z = (x - 375)/50$. The t-statistics of the estimated regression coefficients are $t_1 = 11.417, t_2 = 1.679, t_3 = 0.083$. Only $\widehat{\beta}_1$ is significant at the 0.05 level. The relationship is essentially linear.

(b) $SS_1 = 37.538, SS_2 = 0.810, SS_3 = 0.002$, which add up to $SS_{trt} = 38.350$.

3.27 (a) $t = 3.212$ ($p = 0.002$).

(b) $\widehat{\mu}_1 = 20.471, \widehat{\mu}_2 = 11.011, SE(\widehat{\mu}_1 - \widehat{\mu}_2) = 6.583$.

(c) $t = 1.437$ ($p = 0.155$).

(d) Different results are obtained in (a) and (c) because the means are not adjusted for the covariate in (a) whereas they are in (b).

3.30 $\widehat{y}_{2,15} = 1.924, \widehat{e}_{2,15} = -0.624, SE(\widehat{e}_{2,15}) = 0.311, \widehat{e}^*_{2,15} = -2.01$.

Chapter 4

4.6 Bonferroni: 3.030; LSD: 2.042; Tukey: 2.900.

4.7 The three procedures identify the following pairs of stations as significantly different from each other:

LSD	3 & 6, 3 & 5, 3 & 1, 3 & 4, 2 & 6, 2 & 5
Bonferroni	3 & 6, 2 & 6
Tukey	3 & 6, 2 & 6

4.9

Comparison	Tukey 95% SCI	Bonferroni 95% SCI
HBS vs. HBSC	[−2.664, −0.676]	[−2.693, −0.647]
HBS vs. HBSS	[0.937, 2.899]	[0.908, 2.928]
HBSC vs. HBSS	[2.713, 4.463]	[2.688, 4.488]

All three groups are significantly different using $\alpha = 0.05$ according to both procedures.

4.13 (a) $\rho_{12} = 0.4264$.

(b) Drug A vs. control: $[-1.259, 2.559]$; drug B vs. control: $[0.837, 4.491]$. Drug B is significantly different from the control.

4.14 (a) Control vs. dose 1: $[-0.194, 0.899]$; control vs. dose 2: $[0.033, 0.957]$; control vs. dose 3: $[-0.118, 0.806]$. Only the control vs. dose 2 difference is significant.

(b) Control vs. dose 1: [−0.671, 0.253]; control vs. dose 2: [−0.860, 0.064]; control vs. dose 3: [−0.718, 0.205]. All three differences are nonsignificant.

4.18 (a) No multiplicity adjustment is needed. The t-statistic for the contrast: $t = 5.332$ ($p = 0.000$).

(b) Multiplicity adjustment is needed. Significant PTU effect at all doses except the zero dose.

4.19 (a) No PTU contrast: 0.5038; PTU contrast: −0.3502; difference: 0.8540; contrast vector for the difference: $(+1, -1, -1, +1, -1, +1, +1, -1)$.

(b) 0.3813.

(c) Scheffé procedure. The estimated contrast is not significant at $\alpha = 0.05$.

4.24 $n = 25$.

4.25 $N = 27$.

4.26 (a) Glues 1, 2, 7, and 8 are selected in the subset.

(b) The 95% Hsu SCIs for comparing each glue with the best glue are as follows: Glue 1: [−512.85,218.85]; Glue 2: [−218.85,512.85]; Glue 3: [−894.85,0]; Glue 4: [−1717.85,0]; Glue 5: [−979.85,0]; Glue 6: [−909.85,0]; Glue 7: [−639.85,91.85]; Glue 8: [−664.85,66.85]. If the CI falls below 0, then that glue is excluded from the subset. Thus, glues 3, 4, 5, and 6 are excluded from the subset.

4.27

Variety	A	B	C	D	E	F	G
ℓ_i	−32.242	−10.642	−14.242	−20.342	−10.442	−23.742	−20.842
u_i	0	10.442	6.842	0.742	10.642	0	0.242

Varieties A and F are nonbest and all others are candidates for best. There is no clear winner since no variety has the lower limit, $\ell_i = 0$ for its CI with the best.

Chapter 5

5.3 (a) Significant differences between both the temperatures ($F = 7.11$, $p = 0.005$) and coils ($F = 1295.36$, $p = 0.000$).

(b) Four differences are significant at $\alpha = 0.10$: 22 vs. 24, 22 vs. 25, 23 vs. 24, 23 vs. 25.

5.4 (a) Both positions ($F = 8.70$, $p = 0.000$) and batches have significant differences ($F = 12.25$, $p = 0.000$).

(b) Positions 2 and 3 are different from positions 4 and 1, and positions 6 and 8 are different from position 1.

5.5 (a) Both factors are highly significant ($F = 66.63$, $p = 0.000$ and $F = 19.65$, $p = 0.000$).

(b) Counter differs from Terma but not Fotobalk.

5.6 BIB design gives less precise estimate since not all treatments are compared in each block. This follows from

$$\frac{\mathrm{Var}_{\mathrm{RB}}(\widehat{\alpha}_i - \widehat{\alpha}_j)}{\mathrm{Var}_{\mathrm{BIB}}(\widehat{\alpha}_i - \widehat{\alpha}_j)} = \frac{a(c-1)}{c(a-1)} < 1.$$

5.7 The condition is

$$\frac{\sigma^2_{\mathrm{BIB}}}{\sigma^2_{\mathrm{RB}}} < \frac{a(c-1)}{c(a-1)}.$$

5.10 (a) Both shape and plate have significant effects ($F = 11.96$, $p = 0.010$ and $F = 10.90$, $p = 0.012$).

(b) All pairwise differences are significant except shape 2 vs. shape 4.

5.11 (a) $\widetilde{\alpha}_1 = -0.19, \widetilde{\alpha}_2 = 0.87, \widetilde{\alpha}_3 = -0.14, \widetilde{\alpha}_4 = -0.54$.

(b) $\widehat{\alpha}_1^* = 0.281, \widehat{\alpha}_2^* = -0.157, \widehat{\alpha}_3^* = 0.108, \widehat{\alpha}_4^* = -0.233$. They are close to the $\widehat{\alpha}_i$.

(c) $\mathrm{SE}(\widehat{\alpha}_i^*) = 0.194$.

5.12 (a) There are significant differences between pressures ($F = 29.90$, $p = 0.000$).

(b) 550 psi differs from the rest, 475 psi differs from 325 psi and 250 psi, and 400 psi differs from 325 psi.

(c) No significant run-to-run variation.

5.15 (a) Only the position effect is statistically significant ($F = 9.22$, $p = 0.006$).

(b) Position δ differs from the other three positions, which do not differ significantly from each other.

5.21 (a) Tire differences are not significant ($F = 2.44$, $p = 0.115$).

(b) Tire differences are significant ($F = 7.96$, $p = 0.007$).

(c) Tire differences are significant ($F = 11.42$, $p = 0.0075$).

5.22 There are significant differences between batches ($F = 8.60$, $p = 0.002$) and yogurts ($F = 20.21$, $p = 0.000$).

5.23 (a) Only the pattern effect is significant ($F = 12.52$, $p = 0.002$). Concave pattern takes significantly less time to assemble.

(b) The normality assumption is satisfied; the homoscedasticity assumption less so, but there is no gross violation of the assumption.

Chapter 6

6.6 The PTU main effect is significant ($F = 28.40$, $p = 0.000$). The dose main effect is nonsignificant ($F = 0.10$, $p = 0.960$), but the PTU-dose interaction is significant ($F = 3.83$, $p = 0.015$), so cannot conclude that there is no dose effect.

6.7 (a) $\widehat{\alpha}_1 = -5, \widehat{\alpha}_2 = -8, \widehat{\alpha}_3 = 13$ and $\widehat{\beta}_1 = -10, \widehat{\beta}_2 = 8, \widehat{\beta}_3 = 4, \widehat{\beta}_4 = 2, \widehat{\beta}_5 = -4$.

(b) The height main effect is almost significant ($F = 4.42$, $p = 0.051$); the position effect is nonsignificant ($F = 1.03$, $p = 0.449$).

6.8 (a) The interaction plot indicates presence of interaction.

(b) $F_{AB} = 1.361$ (nonsignificant). The reason for nonsignificance could be that the interaction is not multiplicative (check interaction residuals).

6.10 (a) $\text{SS}_A + \text{SS}_B + \text{SS}_{AB} + \text{SS}_e = 57.222 \neq \text{SS}_{\text{tot}} = 60.765$. The ANOVA identity does not hold. Only the alcohol–base interaction is significant ($F = 5.98$, $p = 0.016$).

(b) $\widehat{\overline{\mu}}_{\cdot 1} = 89.77, \widehat{\overline{\mu}}_{\cdot 2} = 90.87,\ \text{SE}(\widehat{\overline{\mu}}_{\cdot 1}) = \text{SE}(\widehat{\overline{\mu}}_{\cdot 2}) = 0.4908$.

(c) $t = -1.584, t^2 = F = 2.51$.

(d) $\overline{y}_{\cdot 1} = 89.73, \overline{y}_{\cdot 2} = 90.70, \text{SE}(\overline{y}_{\cdot 1}) = \text{SE}(\overline{y}_{\cdot 2}) = 0.4716$.

(e) $t = -1.454, t^2 = 2.114 \neq F = 2.51$.

6.11 (a) The communications main effect appears significant but the worry main effect does not. The interactions also appear significant.

(b) The graphical plot results are confirmed by the ANOVA (communication main effect: $F = 6.54$, $p = 0.003$; worry main effect: $F = 3.27$, $p = 0.078$; interaction: $F = 4.02$, $p = 0.025$).

(c) Satisfaction decreases more with the level of communication if worry is negative than if worry is positive. On the average, satisfaction decreases with communication, but there is no significant change with worry.

6.12 (a) $\widehat{\overline{\mu}}_{1\cdot} = 7.5, \widehat{\overline{\mu}}_{2\cdot} = 5.5, \widehat{\overline{\mu}}_{3\cdot} = 5.5$.

(b) Tukey 95% SCIs: low vs. medium: [0.266, 3.734]; low vs. high: [0.526, 3.474]; medium vs. high: [−1.710, 1.710].

Chapter 7

7.3 $\widehat{A} = 12.936$, $\widehat{B} = 17.374$, $\widehat{C} = -0.066$, $\widehat{AB} = -2.884$, $\widehat{AC} = -0.594$, $\widehat{BC} = 2.744$, $\widehat{ABC} = 8.566$.

7.6 (a) Only the A and B main effects are significant at $\alpha = 0.10$ ($F_A = 5.10$, $p_A = 0.054$ and $F_B = 9.21$, $p_B = 0.016$).

(b) $F_A = 6.18$, $p_A = 0.027$ and $F_B = 11.15$, $p_B = 0.005$.

(c) Both the fitted-values plot and the normal plot of the residuals are satisfactory.

7.10 (a) Let A and B be quantitative factors and C be qualitative. The design matrix is

$$\boldsymbol{D} = \begin{array}{c} \begin{matrix} A & B & C \end{matrix} \\ \begin{bmatrix} -1 & -1 & -1 \\ +1 & -1 & -1 \\ -1 & +1 & -1 \\ +1 & +1 & -1 \\ -1 & -1 & +1 \\ +1 & -1 & +1 \\ -1 & +1 & +1 \\ +1 & +1 & +1 \\ 0 & 0 & -1 \\ 0 & 0 & -1 \\ 0 & 0 & +1 \\ 0 & 0 & +1 \end{bmatrix} \end{array}.$$

(b) Straightforward check.

(c) 2 d.f.

7.12 (a) The main effects of the factors must be positive. A similar statement about the interactions cannot be made.

(b) The main-effect estimates have the expected positive signs ($\widehat{A} = 0.036$, $\widehat{B} = 0.123$, $\widehat{C} = 0.087$, $\widehat{D} = 0.114$). The half-normal plot shows that the four main effects and interactions $\widehat{BC}$ and $\widehat{BD}$ are significant at $\alpha = 0.05$.

(c) The main-effects model is

$$\hat{y} = -4.383 + 0.018x_1 + 0.061x_2 + 0.044x_3 + 0.057x_4$$
$$-0.011x_2x_3 - 0.011x_2x_4.$$

If all factors are set at high level, then $\hat{y} = -4.225$. Hence the predicted response rate is $\hat{p} = 1.4414\%$. The observed response rate when all factors are at high level is 1.44%.

7.13 PSE = 0.0107. Only $t_A = 3.3184$, $t_B = 1.4896$, $t_C = 8.1495$, $t_D = 10.6605$ are significant at $\alpha = 0.05$.

7.14 (a) Check that $\boldsymbol{X}'\boldsymbol{X} = 12\boldsymbol{I}$. (b) $\widehat{\beta}_0 = 8$, $\widehat{\beta}_1 = -4$. (c) $\text{Var}(\widehat{\beta}_j) = \sigma^2/12$ for $j = 0, 1, 2, 3$. (d) $\widehat{\sigma}^2 = \text{MS}_e = 2$, d.f. = 4. (e) $t_1 = -9.80$, significant at $\alpha = 0.05$.

7.15 (a) The main effects B, D and interactions AB, AC, AD, and BC are significant at $\alpha = 0.05$. The half-normal plot of the effects shows the same effects to be significant. The prediction model after omitting the nonsignificant terms is

$$\widehat{y} = 366 + 5.375x_2 - 6.625x_4 - 2.875x_1x_2 - 3.750x_1x_3 + 4.375x_1x_4 + 4.625x_2x_3.$$

(b) $t = 2.37$ $(p = 0.064)$.

7.16 Block 1: $\{(1), abc\}$; Block 2: $\{b, ac\}$; Block 3: $\{ab, c\}$; Block 4: $\{a, bc\}$.

7.17 Block 1: $\{(1), ab, ac, bc, ad, bd, cd, abcd\}$; Block 2: $\{a, b, c, d, abc, abd, acd, bcd\}$.

7.18 Block 1: $\{(1), bc, ad, abcd, abe, ace, bde, cde\}$; Block 2: $\{a, abc, d, bcd, be, ce, abde, acde\}$; Block 3: $\{b, c, abd, acd, ae, de, bcde, abce\}$; Block 4: $\{ab, ac, bd, cd, e, bce, ade, abcde\}$.

7.19 (a) ABC interaction.

(b) The main effects B (the amount of ammonium chloride) and C (the unit used) are significant ($t = 4.32$, $p = 0.005$ and $t = -5.06$, $p = 0.002$). The block (lot) effect is significant ($F = 24.62$, $p = 0.001$) but is confounded with ABC.

(c) The quality of ammonium chloride does not matter but the quantity does. The two units differ significantly. All two-way interactions are nonsignificant.

Chapter 8

8.1 The complete defining relation: $I = ABCE = BCDF = ADEF$. Resolution = IV. The alias structure:

$$AB = CE,\ AC = BE,\ AD = EF,\ AE = BC = DF,$$
$$AF = DE,\ BD = CF,\ BF = CD.$$

8.2 The complete defining relation: $I = ABE = ACDF = BCDEF$. The alias structure up to two-factor interactions: $A = BE, B = AE, E = AB, AC = DF, AD = CF$, and $AF = CD$. Resolution = III.

8.3 No. Yes. $ABCE$ is a word in the generator; $ABDE$ is not.

8.5 (a) 32 runs. (b) $A = BD = CE, F = GH, AF = BDF = CEF = AGH$.

8.6 (a) Generator: $I = ABD$. Resolution = III. (b) Generator $I = \pm ABCD$ gives a design of resolution IV.

8.7 (a) $I = ABCE = BCDF = ACDG = ABDH$.

(b) Resolution = IV. The design can be projected onto maximum four factors to get a full factorial.

(c)
```
Alias Structure (up to order 2)

A*B + C*E + D*H + F*G
A*C + B*E + D*G + F*H
A*D + B*H + C*G + E*F
A*E + B*C + D*F + G*H
A*F + B*G + C*H + D*E
A*G + B*F + C*D + E*H
A*H + B*D + C*F + E*G
```

(d) Half-normal plot shows two significant effects at $\alpha = 0.10$: B = wing length and C = body length.

(e) Use high B and low C to maximize the flight time. The maximum flight time = 3.325 sec.

8.10 Assign $D \rightarrow E'$ and $E \rightarrow D'$.

8.11 Assign $E \rightarrow F'$ and $F \rightarrow E'$.

8.14 The effect estimates are $\widehat{A} = -1.17$, $\widehat{B} = -0.50$, $\widehat{C} = -2.50$, $\widehat{D} = +0.50$, $\widehat{E} = -3.83$, $\widehat{F} = +14.83$, $\widehat{G} = -7.83$, $\widehat{H} = -1.17$, $\widehat{J} = -10.50$, $\widehat{K} = +2.83$, Error = -0.50.

The normal plot of the effects identifies only the main effects F, G, and J as significant at $\alpha = 0.05$.

8.19 OA$(12, 2^{11}, 2)$, Strength = 2.

8.20 (a) $A = BD = CE = FG, C = AE = BF = DG, E = AC = DF = BG$. Hierarchical principle.

(b) Yes. Yes.

(c) Since $E = -AC$ in the foldover design.

(d) No.

8.21 (a) The complete defining relation: $I = ABCF = ABDG = BCDEH = CDFG = ADEFH = ACEGH = BEFGH$ from which the alias structure up to second-order interactions can be deduced.

(b) Fold over factors C and D.

(c) The complete defining relation of the combined design is $I = CDFG = BCDEH$ and has resolution IV.

Chapter 9

9.2 (a) Both main effects (growth factor: $F = 38.04$, $p = 0.000$; incubation: $F = 13.85$, $p = 0.000$) and their interaction ($F = 6.99$, $p = 0.001$) are highly significant.

(b) All the effects are significant at the 0.05 level except the A quadratic main effect and AB quadratic–quadratic interaction.

(c) $$\hat{y} = 8.082 - 1.307\,\xi_1(z_A) + 0.177\xi_2(z_A) - 0.734\xi_1(z_B) + 0.198\xi_2(z_B) + 0.611\xi_1(z_A)\xi_1(z_B) + 0.290\xi_1(z_A)\xi_2(z_B) + 0.328\xi_2(z_A)\xi_2(z_B).$$

9.3 (a)

$$\begin{array}{c} \begin{matrix} A & B & C & D \end{matrix} \\ \begin{bmatrix} 0 & 0 & 0 & 0 \\ 0 & 1 & 0 & 1 \\ 0 & 0 & 1 & 1 \\ 0 & 2 & 0 & 2 \\ 0 & 0 & 2 & 2 \\ 0 & 1 & 1 & 2 \\ 0 & 1 & 2 & 0 \\ 0 & 2 & 1 & 0 \\ 0 & 2 & 2 & 1 \end{bmatrix} \end{array}, \quad \begin{array}{c} \begin{matrix} A & B & C & D \end{matrix} \\ \begin{bmatrix} 1 & 0 & 0 & 1 \\ 1 & 1 & 0 & 2 \\ 1 & 0 & 1 & 2 \\ 1 & 2 & 0 & 0 \\ 1 & 0 & 2 & 0 \\ 1 & 1 & 1 & 0 \\ 1 & 1 & 2 & 1 \\ 1 & 2 & 1 & 1 \\ 1 & 2 & 2 & 2 \end{bmatrix} \end{array}, \quad \begin{array}{c} \begin{matrix} A & B & C & D \end{matrix} \\ \begin{bmatrix} 2 & 0 & 0 & 2 \\ 2 & 1 & 0 & 0 \\ 2 & 0 & 1 & 0 \\ 2 & 2 & 0 & 1 \\ 2 & 0 & 2 & 1 \\ 2 & 1 & 1 & 1 \\ 2 & 1 & 2 & 2 \\ 2 & 2 & 1 & 2 \\ 2 & 2 & 2 & 0 \end{bmatrix} \end{array}.$$

(b) The alias relations are $A = AB^2C^2D = BCD^2$, $B = A^2BC^2D = ACD^2$, $C = A^2B^2CD = ABD^2$, $D = ABC = ABCD$.

9.4 (a) 121, (b) 9, 13.

9.5 (a) The complete defining relation is

$$I = ABD^2 = AB^2E^2 = ADE = BDE^2.$$

The design is as follows:

Run	A	B	C	D	E	Run	A	B	C	D	E	Run	A	B	C	D	E
1	0	0	0	0	0	10	1	0	0	1	1	19	2	0	0	2	2
2	0	0	1	0	0	11	1	0	1	1	1	20	2	0	1	2	2
3	0	0	2	0	0	12	1	0	2	1	1	21	2	0	2	2	2
4	0	1	0	1	2	13	1	1	0	2	0	22	2	1	0	0	1
5	0	1	1	1	2	14	1	1	1	2	0	23	2	1	1	0	1
6	0	1	2	1	2	15	1	1	2	2	0	24	2	1	2	0	1
7	0	2	0	2	1	16	1	2	0	0	2	25	2	2	0	1	0
8	0	2	1	2	1	17	1	2	1	0	2	26	2	2	1	1	0
9	0	2	2	2	1	18	1	2	2	0	2	27	2	2	2	1	0

(b) The aliases are as follows:

$$\begin{aligned}
A &= AB^2D = ABE = AD^2E^2 = ABDE^2 \\
&= BD^2 = BE = DE = AB^2D^2E, \\
B &= AB^2D^2 = AE^2 = ABDE = BD^2E = AD^2 \\
&= ABE^2 = AB^2DE = DE^2, \\
C &= ABCD^2 = AB^2CE^2 = ACDE = BCDE^2 = ABC^2D^2 \\
&= AB^2C^2E^2 = AC^2DE = BC^2DE^2, \\
D &= AB = AB^2DE^2 = AD^2E = BD^2E^2 = ABD \\
&= AB^2D^2E^2 = AE = BE^2, \\
E &= ABD^2E = AB^2E = ADE^2 = BD \\
&= ABD^2E^2 = AB^2E = AD = BDE.
\end{aligned}$$

9.7 (a) There are five such triples:

$$(A, B, AB), (C, D, CD), (AC, BD, ABCD),$$
$$(BC, ABD, ACD), (ABC, AD, BCD).$$

(b) Assign

$$X \to (A, B, AB), Y \to (C, D, CD), Z \to (AC, BD, ABCD),$$
$$T \to BC, U \to ABD, V \to ABC, W \to AD.$$

The design is

Run	T	U	V	W	X	Y	Z
1	1	0	0	1	0	0	3
2	1	1	1	0	1	0	2
3	0	1	1	0	2	0	1
4	0	0	0	0	3	0	0
5	0	0	1	1	0	1	2
6	0	1	0	0	1	1	3
7	1	1	0	1	2	1	0
8	1	0	1	0	3	1	1
9	1	1	0	0	0	2	1
10	1	0	1	1	1	2	0
11	0	0	1	0	2	2	3
12	0	1	0	1	3	2	2
13	0	1	1	0	0	3	0
14	0	0	0	1	1	3	1
15	1	0	0	0	2	3	2
16	1	1	1	1	3	3	3

9.8 (a)

$$\widehat{A}_L = 1.768,\ \widehat{B}_L = 19.445,\ \widehat{C}_L = -6.718,\ \widehat{D}_L = 13.789$$

and

$$\widehat{E}_L = -16.286,\ \widehat{E}_Q = 2.475,\ \widehat{E}_C = -9.961.$$

(b) The design is saturated (seven main effects estimated with eight observations). The main-effects plot confirms that the A and C main effects are the smallest; the B, D, and E main effects are larger in comparison.

9.9 The main effects of factors A, C, and D are significant at the 0.05 level; the main effect of B is also borderline significant. Interactions AD, AF, and CD are significant. Factor E (gun 7) is inactive. The final prediction equation is

$$\begin{aligned}\widehat{y} = {} & 1.732 - 0.085A + 0.071C + 0.565D_L - 0.038D_Q \\ & -0.081AD_L + 0.045AF_Q + 0.106CD_L - 0.073CF_L.\end{aligned}$$

Chapter 10

10.4

$$\begin{bmatrix} -1 & -1 & 0 & 0 \\ 1 & -1 & 0 & 0 \\ -1 & 1 & 0 & 0 \\ 1 & 1 & 0 & 0 \\ -1 & 0 & -1 & 0 \\ 1 & 0 & -1 & 0 \\ -1 & 0 & 1 & 0 \\ 1 & 0 & 1 & 0 \\ -1 & 0 & 0 & -1 \\ 1 & 0 & 0 & -1 \\ -1 & 0 & 0 & 1 \\ 1 & 0 & 0 & 1 \end{bmatrix}, \begin{bmatrix} 0 & -1 & -1 & 0 \\ 0 & 1 & -1 & 0 \\ 0 & -1 & 1 & 0 \\ 0 & 1 & 1 & 0 \\ 0 & -1 & 0 & -1 \\ 0 & 1 & 0 & -1 \\ 0 & -1 & 0 & 1 \\ 0 & 1 & 0 & 1 \\ 0 & 0 & -1 & -1 \\ 0 & 0 & 1 & -1 \\ 0 & 0 & -1 & 1 \\ 0 & 0 & 1 & 1 \end{bmatrix}.$$

10.5 (a) Only the quadratic terms are significant at $\alpha = 0.05$. The hierarchical model omitting the interaction terms is

$$\begin{aligned}\widehat{y} = 10.5062 + 0.0924x_1 - 0.0544x_2 + 0.1206x_3 \\ -2.8681x_1^2 - 2.5996x_2^2 - 2.9499x_3^2.\end{aligned}$$

(b) The model is essentially quadratic. Since the quadratic terms have negative signs, all three factor levels must be increased in order to minimize the particle size.

10.6 (a) After omitting run 7, the second-order model fit is

$$\begin{aligned}\widehat{y} &= 2.1838 + 1.1531x_1 + 0.8103x_2 - 0.0259x_3 + 2.7140x_1^2 \\ &\quad - 0.0410x_2^2 - 0.1215x_3^2 + 1.5277x_1x_2 - 0.2208x_1x_3 \\ &\quad + 0.1418x_2x_3.\end{aligned}$$

Factors A and B are active. The refitted model is

$$\begin{aligned}\widehat{y} = 2.0767 + 1.1852x_1 + 0.7782x_2 \\ +2.7157x_1^2 - 0.0394x_2^2 + 1.5824x_1x_2.\end{aligned}$$

(b) The eigenvalues of $\widehat{\boldsymbol{B}}$ are 2.9268 and -0.2504. So the stationary point is a saddle point. Its coordinates are $(-0.2218, -0.2514)$ on coded scale and $(4.9278, 18.4916)$ on raw scale.

10.7 (a) The second-order fit is

$$\hat{y} = 2.0244 + 1.9308x_1 - 1.3288x_2 + 1.4838x_1^2 + 0.7743x_2^2 - 0.6680x_1x_2.$$

(b) The eigenvalues of $\widehat{\boldsymbol{B}}$ are 1.6163 and 0.6418. The stationary point is a minimum. The coordinates of the stationary point are $X_1 = 4.870$, $X_2 = 23.837$. The minimum predicted value of the response is 1.1104.

10.8 Only the A, B, C main effects and AB interaction are significant ($F_A = 10.021$, $p_A = 0.000$; $F_B = 9.724$, $p_B = 0.000$; $F_C = 2.431$, $p_C = 0.041$; $F_{AB} = 3.978$, $p_{AB} = 0.004$). Omitting the nonsignificant terms the fitted response model is $\hat{y} = 42.512 + 16.9x_1 + 16.4x_2 + 4.1x_3 + 7.5x_1x_2$.

10.9 (a) Omitting the nonsignificant terms the fitted response model is

$$\hat{y} = 370.833 + 5.083x_2 - 6.083x_4 - 1.792x_1^2 - 2.292x_3^2 - 2.875x_1x_2 - 3.750x_1x_3 + 4.375x_1x_4 + 4.625x_2x_3 - 2.125x_3x_4.$$

(b) The eigenvalues of $\widehat{\boldsymbol{B}}$ are −4.4065, −3.5624, −0.9528, 3.5037. The stationary point is a saddle point.

10.11 $\beta_i = \beta_0' + \beta_i' + \beta_{ii}'$, $\beta_{ij} = \beta_{ij}' - \beta_{ii}' - \beta_{jj}'$.

10.13 (a) $\hat{y} = 103.16x_1 + 104.54x_2 + 103.98x_3 + 97.91x_4$.

(b) $\hat{y} = 104.10(x_1 + x_2 + x_3) + 97.77x_4$. Advantages: (i) simplified explanation, (ii) can make a single contour plot.

(c) No, because none of the data points are close to pure blends. So it is not known if the model holds at or near those points.

10.15 (a) $\hat{y} = 3.426x_1 + 4.540x_2 + 4.622x_3 - 7.068x_1x_2 - 6.305x_1x_3 - 5.677x_2x_3 + 52.200x_1x_2x_3$.

(b) $x_1 = 0.579275$, $x_2 = 0.420725$, $x_3 = 0$, $\hat{y} = 2.17205$.

10.16 Sony-US proportions: grade A: 0.4371; grade B: 0.3156; grade C: 0.1646; fraction defective: 0.0827.

Sony-Japan proportions: grade A: 0.6827; grade B: 0.2718; grade C: 0.0428; fraction defective: 0.0027.

Sony-Japan has a much higher proportion in grade A and much smaller proportion in grade C compared to Sony-USA. Also, the Sony-Japan fraction defective is much smaller than the Sony-USA fraction defective.

10.17 (a)

Run	$\overline{y}$	SN
1	17.525	24.025
2	19.475	25.522
3	19.025	25.335
4	20.125	25.904
5	22.825	26.908
6	19.225	25.326
7	19.850	25.711
8	18.338	24.852
9	21.200	26.152

(b)

A	$\overline{y}$	SN	B	$\overline{y}$	SN	C	$\overline{y}$	SN	D	$\overline{y}$	SN
1	18.675	24.961	1	19.167	25.213	1	18.363	24.734	1	20.520	25.695
2	20.730	26.046	2	20.210	25.761	2	20.267	25.859	2	19.517	25.520
3	19.796	25.572	3	19.817	25.604	3	20.570	25.985	3	19.163	25.364

(c) To maximize $\overline{y}$ choose A_2, B_2, C_3, and D_1. To maximize SN choose A_2, B_2, C_2, and D_1. The graphical analysis is not valid if interactions are present.

(d) A_2 and D_1 are best choices for robust performances with respect to variations in G and E, respectively. Final choice: A_2, B_2, C_2, D_1. If cheaper thin wall is preferred, then use A_2, B_1, C_2, D_1.

Chapter 11

11.1 (c) $\widehat{\rho} = 0.295$, (d) $[0.0157, 0.7441]$.

11.6 (a) Significant variability among cables ($F = 9.07$, $p = 0.000$).
(b) $\widehat{\sigma}^2_{\text{wire}} = 26.53, \widehat{\sigma}^2_{\text{cable}} = 17.83$.

11.7 (a) Significant variability among seams ($F = 8.09$, $p = 0.000$).
(b) $\widetilde{n} = 8.369, \widehat{\sigma}^2_{\text{Seam}} = 0.1030, \widehat{\sigma}^2_e = 0.1215$.

11.10 Only the part–setup interaction is significant ($F = 522.94$, $p = 0.000$). The variance component estimates are: part: 0.00043; setup: 0.00234; part–setup: 0.01031; error: 0.00006. The major contributor: part–setup interaction.

11.11 $\widehat{\sigma}^2_{\text{Gage}} = 126,216, \widehat{\sigma}^2_{\text{Part}} = 145,246$. Measurement system is not capable.

11.12 (a) The heads are not significantly different ($F = 1.32$, $p = 0.298$), but the periods are ($F = 7.79$, $p = 0.000$).

(b) Shift is the most significant factor ($F = 39.44$, $p = 0.025$). The night shift is causing the variability.

11.13 (a) Only the interaction is significant ($F = 16.28$, $p = 0.000$).

(b) $\widehat{\sigma}^2_{\text{Technician}} = -0.00257, \widehat{\sigma}^2_{\text{Instrument} \times \text{Technician}} = 0.00963, \widehat{\sigma}^2_e = 0.00189.$

11.15

$$
\begin{aligned}
\mathrm{E(MS}_A) &= \sigma_e^2 + bn\sigma_{ACD}^2 + bcn\sigma_{AD}^2 + bcdnQ_A,\\
\mathrm{E(MS}_{AB}) &= \sigma_e^2 + n\sigma_{ABCD}^2 + cn\sigma_{ABD}^2 + dn\sigma_{ABC}^2 + cdnQ_{AB},\\
\mathrm{E(MS}_{AC}) &= \sigma_e^2 + bn\sigma_{ACD}^2 + bdn\sigma_{AC}^2,\\
\mathrm{E(MS}_{ABC}) &= \sigma_e^2 + n\sigma_{ABCD}^2 + dn\sigma_{ABC}^2,\\
\mathrm{E(MS}_{CD}) &= \sigma_e^2 + n\sigma_{ABCD}^2 + an\sigma_{BCD}^2 + bn\sigma_{ACD}^2 + abn\sigma_{CD}^2.
\end{aligned}
$$

11.16 $\widehat{\sigma}^2_{\text{Gage}} = 0.0059, \widehat{\sigma}^2_{\text{Spring}} = 0.1416, \widehat{\sigma}_{\text{Gage}}/\widehat{\sigma}_{\text{Spring}} = 0.204.$ Measurement system is not capable.

Chapter 12

12.4 (a) The differences between the interventions are not significant ($F = 2.64$, $p = 0.218$).

(b) $\widehat{\sigma}^2_{\text{Class(Intervention)}} = 162.6, \widehat{\sigma}^2_e = 580.6.$

12.9 (a) The analysts show a lack of consistency at $\alpha = 0.05$ ($F = 3.10$, $p = 0.045$) and the samples show a lack of consistency at $\alpha = 0.10$ ($F = 1.85$, $p = 0.096$).

(b) $\widehat{\sigma}^2_{\text{Lab}} = 0.006, \widehat{\sigma}^2_{\text{Analyst}} = 0.007, \widehat{\sigma}^2_{\text{Sample}} = 0.003, \widehat{\sigma}^2_e = 0.023.$

12.10 The site effect cannot be tested since no error estimate is available (because $n = 1$). The lot and wafer effects are significant ($F = 6.114$, $p = 0.000$ and $F = 2.775$, $p = 0.000$). Variance component estimates are $\widehat{\sigma}^2_{\text{Lots}} = 30.344, \widehat{\sigma}^2_{\text{Wafers}} = 7.591, \widehat{\sigma}^2_{\text{Sites}} = 38.489$. To minimize variability, focus on sites and lots.

12.13 (a) Only the analyst and mouse(medium) main effects are significant ($F = 35.44$, $p = 0.000$ and $F = 14.45$, $p = 0.000$).

(b) $\widehat{\sigma}^2_{\text{Mouse(Medium)}} = 153.911, \widehat{\sigma}^2_{\text{Analyst} \times \text{Mouse(Medium)}} = 34.342.$

12.18 (a) Pretreat is the whole plot treatment and stain is the subplot treatment.

(b) Pretreat is significant at $\alpha = 0.05$ ($F = 13.49$, $p = 0.002$), but stain is not ($F = 1.53$, $p = 0.254$).

(c) Stain is now significant at $\alpha = 0.05$ ($F = 6.976$, $p = 0.006$), whereas pretreat is not ($F = 3.926$, $p = 0.186$). The subplot error is much smaller while the whole-plot error is larger than the error used in the analysis treating the design as a CR design.

12.19 The varietal differences are not significant ($F = 0.654$, $p = 0.541$). The differences between the dates of cutting are significant ($F = 3.357$, $p = 0.000$).

12.20 (a) The differences $|D_1 - D_2|$, $|D_1 - D_3|$, $|D_2 - D_3|$, and $|D_2 - D_4|$ are significant.

(b) The differences $|D_1 - D_2|$ and $|D_1 - D_3|$ are significant.

Chapter 13

13.2 (a) Cholesterol levels decrease in both groups with some large fluctuations from week to week.

(b) Mauchly's statistic: $\chi^2 = 29.881$ with 14 d.f. ($p = 0.008$).

(c) Group: $F = 0.00$, $p = 0.966$; week: $F = 1.81$, $p = 0.117$; group $\times$ week: $F = 0.47$, $p = 0.800$. GG adjustment factor $\widehat{\varepsilon} = 0.6645$, HF adjustment factor $\widetilde{\varepsilon} = 0.8419$. With the HF adjustment to d.f., the week and group $\times$ week effects are even more nonsignificant.

(d) There are no differences between boys and girls with regard to how their cholesterol levels change with time. The time effect (decreasing cholesterol levels) is probably not shown to be significant due to large weekly fluctuations.

13.3 (a) Temperature difference increases steeply with time in the amiodarone group. The time effect is mainly linear.

(b) Mauchly's statistic: $\chi^2 = 1.72$ with 5 d.f. ($p = 0.886$).

(c) Treatment: $F = 19.40$, $p = 0.000$; time: $F = 9.27$, $p = 0.000$; treatment $\times$ time: $F = 2.90$, $p = 0.021$.

(d) Yes.

13.4 Besides treatments, only the linear effect of time ($F = 25.03$, $p = 0.000$) and the linear component of treatment $\times$ time interaction ($F = 6.72$, $p = 0.003$) are significant.

13.5 (a) The $\%CO_2$ evolution increases with time in the control group and decreases in the 0.24 group; in the other two groups, it first increases and then decreases.

(b) Mauchly's statistic: $\chi^2 = 11.52$ with 5 d.f. ($p = 0.042$).

(c) Moisture: $F = 11.15$, $p = 0.003$; days: $F = 1.22$, $p = 0.324$; moisture $\times$ days: $F = 1.28$, $p = 0.296$. The days and moisture $\times$ days effects will be even less significant if the HF adjustment ($\widetilde{\varepsilon} = 0.7276$) is applied.

(d) Soil aeration level (as measured by $\%CO_2$ evolution) increased in control soil (with 0.24 moisture and no sludge) while it decreased in soil with 0.24 moisture and sludge added; also aeration level was the highest in the latter soil on all days. Thus sludge seems to make the most difference for soil with 0.24 moisture, but not for the other two soils with 0.26 and 0.28 moisture.

13.6 (a) The dental measurements increase with age in both groups; the increase is mainly linear. The plots are roughly parallel, indicating nonsignificant interaction between groups and age.

(b) Mauchly's statistic: $\chi^2 = 32.31$ with 2 d.f. Reject the sphericity assumption at $\alpha = 0.10$.

(c) Group: $F = 9.25$, $p = 0.005$; age: $F = 36.32$, $p = 0.000$; group $\times$ age: $F = 2.18$, $p = 0.097$.

13.9 Group: $F = 9.25$, $p = 0.005$; age: $T^2 = 123.17$, $F = 37.77$, $p = 0.000$; group $\times$ age: Wilks's $\Lambda = 0.7274 > c_{0.05}(3, 1, 21) = 0.6694$ (not significant at $\alpha = 0.05$). The results match with those from univariate analysis in Exercise 13.6.

13.10 Treatment: $F = 19.40$, $p = 0.000$; time: $T^2 = 25.081$, $F = 6.967$, $p = 0.008$; treatment $\times$ time: Wilks's $\Lambda = 0.308 < c_{0.05}(3, 2, 12) = 0.3157$ (significant at $\alpha = 0.05$). The results match with those from univariate analysis in Exercise 13.3.

13.11 Moisture: $F = 11.15$, $p = 0.003$; days: $T^2 = 6.975$, $F = 1.744$, $p = 0.257$; treatment $\times$ time: Wilks's $\Lambda = 0.2138 > c_{0.05}(3, 3, 8) = 0.0989$ (not significant at $\alpha = 0.05$). The results match with those from univariate analysis in Exercise 13.5

Chapter 14

14.1 $\boldsymbol{y} = (y_1, y_2, \ldots, y_N)'$, $\boldsymbol{X} = (x_1, x_2, \ldots, x_N)'$, and $\boldsymbol{\beta} = \beta$.

14.7 (b) Dimensions of model and error spaces are $p - 1$ and $N - p + 1$. (c) More generally, the dimensions are $p + 1 - q$ and $N - (p + 1 - q)$.

14.12

$$\boldsymbol{H} = \boldsymbol{X}(\boldsymbol{X}'\boldsymbol{X})^{-1}\boldsymbol{X}' = \frac{1}{\sum x_i^2}\boldsymbol{X}\boldsymbol{X}' = \frac{1}{\sum x_i^2}\{x_i x_j\}.$$

14.21 $\text{Var}(\widehat{\beta}) = \sigma^2 / \sum x_i^2$.

14.30 (i) The formula is

$$(\widehat{\theta}_1 - \theta_1)^2 + (\widehat{\theta}_2 - \theta_2)^2 \leq 2 f_{2,\infty,\alpha} = \chi^2_{2,\alpha},$$

which is a circle. (ii) The formula is

$$\frac{(\widehat{\theta}_1 - \theta_1)^2}{1} + \frac{(\widehat{\theta}_2 - \theta_2)^2}{4} \leq 2 f_{2,\infty,\alpha} = \chi^2_{2,\alpha},$$

which is an ellipse with the major (θ_2) axis twice as long as the minor (θ_1) axis.

14.33 (a) $\lambda^2 = n\mu^2/\sigma^2$. (b) 0.962.

14.34 0.9793.

References

Abraham, B., Chipman, H., and Vijayan, K. (1999), "Some risks in the construction and analysis of supersaturated designs," *Technometrics*, **41**, 135–141.

Adam, C. G. (1987), "Instrument panel process development," *5th Symposium on Taguchi Methods*, Romulus, MI: American Supplier Institute, 93–106.

Adamec, E., and Burdick, R. K. (2003), "Confidence intervals for a discrimination ratio in a gauge R & R study with three random factors," *Quality Engineering*, **15**, 383–389.

Addelman, S. (1962), "Orthogonal main-effect plans for asymmetrical factorial experiments," *Technometrics*, **4**, 21–46.

Anderson, T. W. (1958), *An Introduction to Multivariate Statistical Analysis*, New York: Wiley.

Anderson, V. L., and McLean, R. A. (1974), *Design of Experiments: A Realistic Approach*, New York: Marcel Dekker.

Bae, S., and Shoda, M. (2005), "Statistical optimization of culture conditions for bacterial cellulose production using Box–Behnken design," available: www.interscience.wiley.com.

Bates, D. M., and Watts, D. G. (1988), *Nonlinear Regression Analysis and Its Applications*, New York: Wiley.

Bechhofer, R. E. (1954), "A single-sample multiple decision procedure for ranking means of normal populations with known variances," *Annals of Mathematical Statistics*, **25**, 16–39.

Bechhofer, R. E., and Dunnett, C. W. (1988), "Tables of percentage points of multivariate student t distributions," *Selected Tables in Mathematical Statistics*, **11**, 1–371.

Bechhofer, R. E., Dunnett, C. W., and Sobel, M. (1954), "A two-sample multiple-decision procedure for ranking means of normal populations with a common unknown variance," *Biometrika*, **41**, 170–176.

Benjamini, Y., and Hochberg, Y. (1995), "Controlling the false discovery rate: A practical and powerful approach to multiple testing," *Journal of the Royal Statistical Society, Ser. B*, **57**, 289–300.

Best, D. J., and Rayner, J. C. W. (1987), "Welch's approximate solution for the Behrens-Fisher problem," *Technometrics*, **29**, 205–210.

Bechhofer, R. E., Santner, T. J., and Goldsman, D. M. (1995), *Design and Analysis of Experiments for Statistical Selection, Screening and Multiple Comparisons*, New York: Wiley.

Batra, P. K., and Parsad, R. (date unknown), available: http://www.iasri.res.in/iasriwebsite/DESIGNOFEXAPPLICATION/Electronic-book/index.htm.

Bechhofer, R. E., and Tamhane, A. C. (1981), "Incomplete block designs for comparing treatments with a control: General theory," *Technometrics*, **23**, 45–57.

Bisgaard, S. (1993), "Iterative analysis of data from two-level factorials," *Quality Engineering*, **6**, 319–330.

Bisgaard, S., and Fuller, H. T. (1994–1995), "Accommodating interaction effects in two-level fractional factorials: An alternative to linear graphs," *Quality Engineering*, **7**, 71–87.

Bisgaard, S., Fuller, H. T., and Barrios, E. (1996), "Two-level factorials run as split plot experiments," *Quality Engineering*, **8**, 705–708.

Bishop, T. A., and Dudewicz, E. J. (1978), "Exact analysis of variance with unequal variances: Test procedures and tables," *Technometrics*, **20**, 419–430.

Bjerke, F., Aastveit, A. H., Stroup, W. W., Kirkhaus, B., and Naes, T. (2004), "Design and analysis of storing experiments: A case study," *Quality Engineering*, **16**, 591–611.

Bliss, C. I. (1967), *Statistics in Biology, vol. 1*, New York: McGraw Hill.

Bose, R. C., Shrikhande, S. S., and Parker, E. T. (1960), "Some further results on the construction of mutually orthogonal Latin squares and the falsity of Euler's conjecture," *Canadian Journal of Mathematics*, **12**, 189–203.

Bowerman, B., and O'Connell, R. T. (1997), *Applied Statistics: Improving Business Processes*, Chicago: Richard D. Irwin.

Box, G. E. P. (1954), "Some theorems on quadratic forms applied in the study of analysis of variance problems, I. Effect of inequality of variance in the one-way classification," *Annals of Mathematical Statistics*, **25**, 290–302.

Box, G. E. P. (1988), "Signal to noise ratios, performance criteria and transformations," *Technometrics*, **30**, 1–17.

Box, G. E. P. (1992), "Teaching engineers experimental design with a paper helicopter," *Quality Engineering*, **4**, 453–459.

Box, G. E. P. (1993), "How to get lucky," *Quality Engineering*, **5**, 517–524.

Box, G. E. P. (1996), "Split plot experiments," *Quality Engineering*, **8**, 515–520.

Box, G. E. P. (1999), "Statistics as a catalyst to learning by scientific method, Part II—A discussion," *Journal of Quality Technology*, **31**, 16–29.

Box, G. E. P., and Behnken, D. W. (1960), "Some new three-level designs for the study of quantitative variables," *Technometrics*, **2**, 455–475.

Box, G. E. P., and Bisgaard, S. (1988), "Statistical tools for improving designs," *Mechanical Engineering*, **110**, 32–40.

Box, G. E. P., and Bisgaard, S. (1993), "What can you find out from twelve experimental runs," *Quality Engineering*, **4**, 663–668.

Box, G. E. P., and Cox, D. R. (1964), "An analysis of transformations (with discussion)," *Journal of the Royal Statistical Society, Series B*, **26**, 211–252.

Box, G. E. P., and Draper, N. R. (2007), *Response Surfaces, Mixtures, and Ridge Analyses*, 2nd ed., New York: Wiley.

Box, G. E. P., Hunter, J. S., and Hunter, W. G. (1978), *Statistics for Experimenters*, 1st ed., New York: Wiley.

Box, G. E. P., Hunter, J. S., and Hunter, W. G. (2005), *Statistics for Experimenters*, 2nd ed., New York: Wiley.

Box, G. E. P., and Jones, S. (1992a), "Designing products that are robust to the environment," *Total Quality Management*, **3**, 265–282.

Box, G. E. P., and Jones, S. (1992b), "Split-plot designs for robust product experimentation," *Journal of Applied Statistics*, **19**, 3–26.

Box, G. E. P., and Jones, S. (2000), "Split plots for robust product and process experimentation," *Quality Engineering*, **13**, 127–134.

Box, G. E. P., and Liu, P. (1999), "Statistics as a catalyst to learning by scientific method," *Journal of Quality Technology*, **31**, 1–15.

Box, G. E. P., and Meyer, R. D. (1985), "Some new ideas in the analysis of screening designs," *Journal of Research of the National Bureau of Standards*, **90**, 495–502.

Box, G. E. P., and Wilson, K. G. (1951), "On the experimental attainment of optimum conditions," *Journal of the Royal Statistical Society, Series B*, **13**, 1–45.

Box, G. E. P., et al. (2006), *How to Improve Anything*, New York: Wiley.

Brown, M. B., and Forsythe, A. B. (1974), "The ANOVA and multiple comparisons for data with heterogeneous variances," *Biometrics*, **30**, 719–724.

Byrne, D. M., and Taguchi, S. (1987), "The Taguchi approach to parameter design," *Quality Progress*, 19–26.

Cheng, S. W., and Wu, C. F. J. (2001), "Factor screening and response surface exploration (with discussion)," *Statistica Sinica*, **11**, 553–604.

Chiao, C-H., and Hamada, M. (2001), "Analyzing experiments with correlated multiple tests," *Journal of Quality Technology*, **33**, 451–465.

Christensen, R. (1996), *Plane Answers to Complex Questions: The Theory of Linear Models*, 2nd ed., New York: Springer-Verlag.

Cole, J. W. L., and Grizzle, J. E. (1966), "Applications of multivariate analysis of variance to repeated measurements experiments," *Biometrics*, **22**, 810–828.

Conover, W. J. (1999), *Practical Nonparametric Statistics*, 3rd ed., New York: Wiley.

Cornell, J. A. (1990a), *Experiments with Mixtures: Designs, Models, and the Analysis of Mixture Data*, 2nd ed., New York: Wiley.

Cornell, J. A. (1990b), "Mixture experiments," Chapter 7 in *Statistical Design and Analysis of Industrial Experiments* (Ed. S. Ghosh), New York: Marcel Dekker, pp. 175–209.

Crowder, M. J. (1995), "Keep timing the tablets: Statistical analysis of pill dissolution rates," *Applied Statistics*, **45**, 323–334.

D'Agostino, R. B., and Stephens, M. A. (1986), *Goodness-of-Fit Techniques*, New York: Marcel Dekker.

Daniel, C. (1959), "Use of half-normal plots in interpreting factorial two-level experiments," *Technometrics*, **1**, 311–341.

Darwin, C. (1876), *The Effect of Cross- and Self-Fertilization in the Vegetable Kingdom*, 2nd ed., London: John Murray.

Davies, O. L. (1963), *The Design and Analysis of Industrial Experiments*, Edinburgh and London: Oliver and Boyd, New York: Hafner.

Draper, N. R., and Smith, H. (1998), *Applied Regression Analysis*, 3rd ed., New York: Wiley.

Duncan, D. B. (1955), "Multiple range and multiple F tests," *Biometrics*, **11**, 1–42.

Dunn, O. J., and Clark, V. A. (1974), *Applied Statistics: Analysis of Variance and Regression*, New York: Wiley-Interscience.

Dunnett, C. W. (1955), "A multiple comparison procedure for comparing several treatments with a control," *Journal of the American Statistical Association*, **50**, 1096–1121.

Dunnett, C. W. (1970), "Multiple comparisons," in *Statistics in Endocrinology* (Eds. J. W. McArthur and T. Colton), Cambridge: MIT Press, pp. 79–103.

Dunnett, C. W., and Tamhane, A. C. (1991), "Step-down multiple tests for comparing treatments with a control in unbalanced one-way layouts," *Statistics in Medicine*, **10**, 939–947.

Elmore, R. T., Hettmansperger, T. P., and Xuan, F. (2004), "The sign statistic, one-way layouts and mixture models," *Statistical Science*, **19**, 579–587.

Fabian, V. (1962), "On multiple decision methods for ranking population means," *Annals of Mathematical Statistics*, **33**, 248–254.

Federer, W. T., and McCulloch, C. E. (1984), "Multiple comparisons procedures for some split plot and split block designs," in *Design of Experiments: Ranking and Selection (Essays in Honor of Robert E. Bechhofer)* (Eds. T. J. Santner and A. C. Tamhane), New York: Marcel Dekker, pp. 7–22.

Fisher, A. C., and Wallenstein, S. (1994), "Crossover designs in medical research," in *Statistics in the Pharmaceutical Industry*, 2nd ed. (Ed. C. R. Buncher and J-Y. Tsay), New York: Marcel Dekker, pp. 193–206.

Fisher, L. D., and van Belle, G. (1993), *Biostatistics: A Methodology for the Health Sciences*, New York: Wiley.

Fisher, R. A. (1935), *The Design of Experiments*, Edinburgh and London: Oliver and Boyd.

Fisher, R. A., and Yates, F. (1953), *Statistical Tables for Biological, Agricultural and Medical Research*, Edinburgh and London: Oliver and Boyd.

Fisher-Box, J. (1978), *R. A. Fisher: The Life of a Scientist*, New York: Wiley.

Fries, A., and Hunter, W. G. (1980), "Minimum aberration 2^{k-p} designs," *Technometrics*, **22**, 601–608.

Genz, A., and Bretz, F. (1999), "Numerical computation of multivariate t probabilities with application to power calculation of multiple contrasts," *Journal of Statistical Computation and Simulation*, **63**, 361–378.

Gibbons, J. D. (1998), "Nonparametric statistics," Chapter 12 in *Handbook of Statistical Methods for Engineers and Scientists*, 2nd ed. (Ed. H. M. Wadsworth), New York: McGraw-Hill.

Gibbons, J. D., Olkin, I., and Sobel, M. (1977), *Selecting and Ordering Populations: A New Statistical Methodology*, New York: Wiley.

Goldsman, D., Nelson, B. L., and Schmeiser, B. (1991), "Methods for selecting the best system," in *Proceedings of the 1991 Winter Simulation Conference* (Eds. B. L. Nelson, W. D. Kelton, and G. M. Clark), Piscataway, N.J.: Institute of Electrical and Electronics Engineers, Inc., pp. 177–186.

Gonzalez-de la Parra, M., and Rodriguez-Loaiza, P. (2003), "Application of analysis of means (ANOM) to nested designs for improving the visualization and understanding of the sources of variation of chemical and pharmaceutical processes," *Quality Engineering*, **15**, 663–670.

Gorrill, M. J., Rinehart, J. S., Tamhane, A. C., and Gerrity, M. (1991), "Comparison of the hamster sperm motility assay to the mouse one-cell and two-cell embryo bioassays as quality control tests for in vitro fertilization," *Fertility and Sterility*, **55**, 345–354.

Gosset, W. S. (1908), "The probable error of a mean," *Biometrika*, **6**, 1–25.

Graybill, F. A. (1961), *An Introduction to Linear Statistical Models*, Vol. 1, New York: McGraw-Hill.

Greenhouse, S. W., and Geisser, S. (1959), "On methods in the analysis of profile data," *Psychometrika*, **24**, 95–112.

Grubbs, F. E. (1973), "Errors of measurement, precision, accuracy, and the statistical comparison of measuring instruments," *Technometrics*, **15**, 53–66.

Gupta, S. S. (1956), "On a decision rule for a problem in ranking means," Mimeo Series No. 150, Institute of Statistics, University of North Carolina, Chapel Hill, NC.

Gupta, S. S. (1965), "On some multiple decision (selection and ranking) rules," *Technometrics*, **7**, 225–245.

Gupta, S. S., and Panchapakesan, S. (1979), *Multiple Decision Procedures: Theory and Methodology of Selecting and Ranking Populations*, New York: Wiley.

Hald, A. (1952), *Statistical Theory with Engineering Applications*, New York: Wiley, p. 434.

Hale-Bennett, C., and Lin, D. K. J. (1997), "From SPC to DOE: A case study at Meco, Inc.," *Quality Engineering*, **9**, 489–502.

Halvorsen, K. T. (1991), "Value splitting involving more factors," Chapter 6 in *Fundamentals of Exploratory Analysis of Variance* (Eds. D. C. Hoaglin, F. Mosteller, and J. W. Tukey), New York: Wiley.

Hamada, M., and Wu, C. F. J. (1992), "Analysis of designed experiments with complex aliasing," *Journal of Quality Technology*, **24**, 130–137.

Hamaker, H. (1955), "Experimental design in industry," *Biometrics*, **11**, 257–286.

Hancock, W., Zayko, M., Autio, M., and Ponagajba, D. (1997), "Analysis of components of variation in automotive stamping processes," *Quality Engineering*, **10**, 115–124.

Hand D. J., Daly F., Lunn A. D., McConway K. J., and Ostrowski E. (1994), *A Handbook of Small Data Sets*, London: Chapman & Hall.

Hansotia, B. J. (1990), "Sample size and design of experiment issues in testing offers," *Journal of Direct Marketing*, **4**, 15–25.

Harper, D., Kosbe, M., and Peyton, L. (1987), "Optimization of Ford Taurus wheel cover balance (by design of experiments—Taguchi method)," *5th Symposium on Taguchi Methods*, Romulus, MI: American Supplier Institute, 527–539.

Hayter, A. J. (1984), "A proof of the conjecture that the Tukey-Kramer multiple comparisons procedure is conservative," *Annals of Statistics*, **12**, 61–75.

Hedayat, A., and Wallis, W. D. (1978), "Hadamard matrices and their applications," *Annals of Statistics*, **6**, 1184–1238.

Hicks, C. R., and Turner, K. V. (1999), *Fundamental Concepts in the Design of Experiments*, 5th ed., New York: Oxford University Press.

Hirotsu, C. (1991), "An approach to comparing treatments based on repeated measures," *Biometrika*, **78**, 583–594.

Hoaglin, D. C., Mosteller, F., and Tukey, J. W. (1991), *Fundamentals of Exploratory Analysis of Variance*, New York: Wiley.

Hochberg, Y., and Tamhane, A. C. (1987), *Multiple Comparison Procedures*, New York: Wiley.

Hocking, R. R. (1985), *The Analysis of Linear Models*, Monterey, CA: Brooks/Cole.

Hocking, R. R. (1996), *Methods and Applications of Linear Models*, New York: Wiley.

Hollander, M., and Wolfe, D. A. (1998), *Nonparametric Statistical Methods*, 2nd ed., New York: Wiley.

Holm, S. (1979), "A simple sequentially rejective multiple test procedure," *Scandinavian Journal of Statistics*, **6**, 65–70.

Houf, R. E., and Berman, D. B. (1988), "Statistical analysis of power module thermal test equipment performance," *IEEE Transactions on Components, Hybrids and Manufacturing Technology*, **II**, 516–520.

Hsu, J. C. (1981), "Simultaneous confidence intervals for all distances from the 'best,'" *Annals of Statistics*, **9**, 1026–1034.

Hsu, J. C. (1984), "Constrained two-sided simultaneous confidence intervals for multiple comparisons with the best," *Annals of Statistics*, **12**, 1136–1144.

Hsu, J. C. (1996), *Multiple Comparisons: Theory and Methods*, London: Chapman and Hall.

Hunter, G. B., Hodi, F. S., and Eager, T. W. (1982), "High-cycle fatigue of weld repaired cast Ti-6Al-4V," *Metallurgical Transactions*, **13A**, 1589–1594.

Huynh, H., and Feldt, L. S. (1970), "Conditions under which mean square ratios in repeated measures designs have exact F distributions," *Journal of the American Statistical Association*, **65**, 1582–1589.

Huynh, H., and Feldt, L. S. (1976), "Estimation of Box correction for degrees of freedom from sample data in randomized block and split-plot designs," *Journal of Educational Statistics*, **1**, 69–82.

Jensen, C. R. (2002), "Variance component calculations: Common methods and misapplications in the semiconductor industry," *Quality Engineering*, **14**, 645–657.

John, P. W. M. (1998), *Statistical Design and Analysis of Experiments*, Classics Edition, Philadelphia: SIAM.

Johnson, R. A., and Wichern, D. W. (2002), *Applied Multivariate Statistical Analysis*, 5th ed., Upper Saddle River, NJ: Prentice-Hall.

Kacker, R. N., and Tsui, K. (1990), "Interaction graphical aids for planning experiments," *Journal of Quality Technology*, **22**, 1–14.

Khuri, A. I., and Cornell, J. A. (1996), *Response Surfaces: Designs and Analyses*, 2nd ed., New York: Marcel Dekker.

Kleinbaum, D. G., Kupper, L. L., and Muller, K. E. (1988), *Applied Regression Analysis and Other Multivariable Methods*, 2nd ed., Boston: PWS-Kent.

Kligo, M. B. (1988), "An application of fractional factorial experimental designs," *Quality Engineering*, **1**, 45–54.

Kuehl, R. O. (2000), *Design of Experiments: Statistical Principles of Research Design and Analysis*, 2nd ed., Pacific Grove, CA: Duxbury.

Lenth, R. V. (1989), "Quick and easy analysis of unreplicated factorials," *Technometrics*, **33**, 469–473.

Lin, D. K. J. (1993), "A new class of supersaturated designs," *Technometrics*, **35**, 28–31.

Lin, D. K. J. (1995), "Generating systematic supersaturated designs," *Technometrics*, **37**, 213–225.

Marcus, R., Peritz, E., and Gabriel, K. R., (1976), "On closed testing procedures with special reference to ordered analysis of variance," *Biometrika*, **63**, 655–660.

Mason, R. L., Gunst, R. F., and Hess, J. L. (2003), *Statistical Design and Analysis of Experiments*, 2nd ed., New York: Wiley.

Mauchly, J. W. (1940), "Significance test for sphericity of a normal n-variate distribution," *Annals of Mathematical Statistics*, **11**, 204–209.

McKenzie, J. D., Jr., and Goldman, R. (1999), *The Student Edition of MINITAB*, Reading, MA: Addison-Wesley.

Mee, R. W., and Peralta, M. (2000), "Semifolding 2^{k-p} designs," *Technometrics*, **42**, 122–143.

Mehrotra, D. V. (1997a), "Improving the Brown-Forsythe solution to the generalized Behrens-Fisher problem," *Communications in Statistics, Ser. A*, **26**, 1139–1145.

Mehrotra, D. (1997b), "ANOVA with unequal variances—Correcting a popular strategy," talk presented at Joint Statistical Meeting/American Statistical Association, Anaheim, CA.

Michaels, S. E. (1964), "The usefulness of experimental designs (with discussion)," *Applied Statistics*, **6**, 133–138.

Miller, R. G., Jr. (1981), *Simultaneous Statistical Inference*, 2nd ed., Heidelberg and Berlin: Springer-Verlag.

Moen, R. D., Nolan, T. W., and Provost, L. P. (1999), *Quality Improvement through Planned Experimentation*, 2nd ed., New York: McGraw-Hill.

Montgomery, D. C. (1997), *Introduction to Statistical Quality Control*, 3rd ed., New York: Wiley.

Montgomery, D. C. (2005), *Design and Analysis of Experiments*, 6th ed., New York: Wiley.

Montgomery, D. C., Peck, E., and Vining, G. G. (2006), *Introduction to Linear Regression Analysis*, 4th ed., New York: Wiley-Interscience.

Montgomery, D. C., and Runger, G. C. (1996), "Foldovers of 2^{k-p} resolution IV experimental designs," *Journal of Quality Technology*, **28**, 446–450.

Morris, V. M., Hargreaves, C., Overall, K., Marriott, P. J., and Hughes, J. G. (1997), "Optimization of the capillary electrophoresis separation of ranitidine and related compounds," *Journal of Chromatography*, **766**, 245–254.

Mosteller, F., Fienberg, S. E., and Rourke, R. E. K. (1983), *Beginning Statistics with Data Analysis*, Reading, MA: Addison-Wesley.

Myers, R. H., and Montgomery, D. C. (2002), *Response Surface Methodology*, 2nd ed., New York: Wiley.

Nachtsheim, C. J. (1987), "Tools for computer-aided design of experiments," *Journal of Quality Technology*, **19**, 132–160.

Natrella, M. G. (1963), *Experimental Statistics*, National Bureau of Standards, Handbook 91, Washington, DC: U.S. Department of Commerce.

Nelson, P. R. (1993), "Additional uses for the analysis of means and extended tables of critical values," *Technometrics*, **35**, 61–71.

Nelson, P. R. (1998), "Design and analysis of experiments," Chapter 15 in *Handbook of Statistical Methods for Engineers and Scientists*, 2nd ed. (Ed. H. M. Wadsworth), New York: McGraw-Hill.

Nguyen, N-K (1996), "An algorithmic approach to constructing supersaturated designs," *Technometrics*, **38**, 69–73.

Pearson, E. S., and Hartley, H. O. (1951), "Charts of the power function of the analysis of variance tests, derived from the non-central F-distribution," *Biometrika*, **38**, 112–130.

Pellicane, P. (1990), "Behavior of rubber-based elastomeric construction adhesive in wood joint," *Journal of Testing and Evaluation*, **18**, 256–264.

Peterson, I. (2005), "Sudoku math," available: http://www.sciencenews.org/articles/20050618/mathtrek.asp.

Phadke, M. S. (1986), "Design optimization case studies," *AT&T Technical Journal*, **65**, 51–68.

Phadke, M. S. (1989), *Quality Engineering Using Robust Design*, Englewood Cliffs, NJ: Prentice-Hall.

Pignatiello, J. J., and Ramberg, J. S. (1985), "Discussion of 'Off-line quality control, parameter design, and the Taguchi method,' by R. N. Kackar," *Journal of Quality Technology*, **17**, 198–206.

Pitman, E. J. G. (1937), "Significance tests which may be applied to samples from any populations," *Biometrika*, **29**, 322–335.

Plackett, R. L., and Burman, J. P. (1946), "The design of optimum multifactorial experiments," *Biometrika*, **33**, 305–325.

Potcner, K. J., and Kowlaski, S. M. (2004), "How to analyze a split-plot experiment," *Quality Progress*, 67–74.

Potthoff, R. F., and Roy, S. N. (1964), "A generalized multivariate analysis of variance model useful especially for growth curve problems," *Biometrika*, **51**, 313–326.

Rencher, A. C. (2000), *Linear Models in Statistics*, New York: Wiley.

Rice, J. A. (1988), *Mathematical Statistics and Data Analysis*, Pacific Grove, CA: Wadsworth & Brooks/Cole.

Roberts, R. M. (1989), *Serendipity: Accidental Discoveries in Science*, New York: Wiley.

Rothman, K. J. (1990), "No adjustments are needed for multiple comparisons," *Epidemiology*, **1**, 43–46.

Roy, R. (1990), *A Primer on the Taguchi Method*, New York: Van Nostrand Rinehold.

Roy, R. (2006), "Clutch plate rust inhibition process optimization study," Nutek, Inc. Report, available: http://nutek-us.com/csex-105.pdf.

Roy, S. N. (1953), "On a heuristic method of test construction and its use in multivariate analysis," *Annals of Mathematical Statistics*, **24**, 220–238.

Ryan, B. F., and Joiner, B. L. (2001), *MINITAB Handbook*, 4th ed., Belmont, CA: Duxbury.

Ryan, T. A. (1960), "Significance tests for multiple comparisons of proportions, variances and other statistics," *Psychological Bulletin*, **57**, 318–328.

Salsburg, D. (2001), *The Lady Tasting Tea: How Statistics Revolutionized Science in the Twentieth Century*, New York: W. H. Freeman.

Sanders, D., Leitnaker, M. G., and McLean, R. A. (2001–2002), "Randomized complete block designs in industrial studies," *Quality Engineering*, **14**, 1–8.

Scheffé, H. (1953), "A method for judging all contrasts in the analysis of variance," *Biometrika*, **40**, 87–104.

Scheffé, H. (1958), "Experiments with mixtures," *Journal of the Royal Statistical Society, Ser. B*, **21**, 344–360.

Scheffé, H. (1959), *The Analysis of Variance*, New York: Wiley.

Scheffé, H. (1963), "The simplex-centroid design for experiments with mixtures," *Journal of the Royal Statistical Society, Ser. B*, **25**, 235–263.

Searle, S. R. (1971), *Linear Models*, New York: Wiley.

Searle, S. R., Casella, G., and McCulloch, C. E. (1992), *Variance Components*, New York: Wiley.

Simpson, J., Olsen, A., and Eden, J. C. (1975), "A Bayesian analysis of a multiplicative treatment effect in weather modification," *Technometrics*, **17**, 161–166.

Snee, R. D. (1973), "Techniques for the analysis of mixture data," *Technometrics*, **15**, 517–528.

Snee, R. D. (1985a), "Graphical display of results of three-treatment randomized block experiments," *Applied Statistics*, **34**, 71–77.

Snee, R. D. (1985b), "Computer-aided design of experiments: Some practical experiences," *Journal of Quality Technology*, **17**, 222–236.

Spurrier, J. D., and Solorzano, E. (2004), "Multiple comparisons with more than one control," in *Recent Developments in Multiple Comparison Procedures* (Ed. Y. Benjamini, F. Bretz, and S. Sarkar), Beachwood, OH: Lecture Notes-Monograph Series **47**, Institute of Mathematical Statistics, pp. 119–128.

Steel, R. G. D., and Torrie, J. H. (1980), *Principles and Procedures of Statistics: A Biometrical Approach*, 2nd ed., New York: McGraw-Hill.

Taguchi, G. (1987), *Systems of Experimental Design*, vol. 1, Unipub, New York: Kraus International Publications.

Taguchi, G., and Wu, Y. (1979), *Introduction to Off-Line Quality Control*, Dearborn, MI: American Supplier Institute.

Tamhane, A. C., and Dunlop, D. D. (2000), *Statistics and Data Analysis: From Elementary to Intermediate*, Upper Saddle River, NJ: Prentice-Hall.

Tamhane, A. C., and Logan, B. R. (2004), "Finding the maximum safe dose for heteroscedastic data," *Journal of Biopharmaceutical Statistics*, **14**, 843–856.

Tarim, T. B., Kuntman, H. H., and Ismail, M. (1998), "Statistical design techniques for yield enhancement of low voltage CMOS VLSI," *International Symposium on Circuits and Systems (ISCAS'98)*, Monterey, CA, **2**, 331–334.

Tarry, G. (1901), "Le probleme de 36 officiers," *Compte Rendu de le'association Francaise pour l'avancement de Science Naturel*, **2**, 170–203.

Timm, N. H. (1975), *Multivariate Analysis with Applications in Education and Psychology*, Belmont, CA: Brooks/Cole.

Timmer, D. H. (2000), "An analysis of the reliability of an incandescent light bulb," *Quality Engineering*, **13**, 299–305.

Tukey, J. W. (1949), "One degree of freedom for non-additivity," *Biometrics*, **5**, 232–242.

Tukey, J. W. (1953), "The problem of multiple comparisons," manuscript, Princeton University, Princeton, NJ.

Weber, D. C., and Skillings, J. H. (2000), *A First Course in the Design of Experiments: A Linear Models Approach*, Baton Rouge, LA: CRC Press.

Westfall, P. H., Tobias, R. D., Rom, D., Wolfinger, R. D., and Hochberg, Y. (1999), *Multiple Comparisons and Multiple Test Using the SAS System*, Cary, NC: SAS Institute.

Westfall, P. H., and Young, S. S. (1993), *Resampling-Based Multiple Testing*, New York: Wiley.

Wilcox, R. R. (2005), *Introduction to Robust Estimation and Hypothesis Testing*, New York: Academic.

Williams, E. J. (1959), *Regression Analysis*, New York: Wiley.

Wu, C. F. J., and Chen, Y. (1992), "A graphical aided method for planning two-level experiments when certain interactions are important," *Technometrics*, **34**, 162–175.

Wu, C. F. J., and Hamada, M. (2000), *Experiments: Planning, Analysis, and Parameter Design Optimization*, New York: Wiley.

Xu, H., Cheng, S. W., and Wu, C. F. J. (2004), "Optimal projective three-level designs for factor screening and interaction detection," *Technometrics*, **46**, 280–292.

Yang, C-H., Lin, S-M., and Wen, T-C. (1995), "Application of statistical experimental strategies to the process optimization of waterborne polyurethane," *Polymer Engineering and Science*, **35**, 722–729.

Yates, F. (1937), *The Design and Analysis of Factorial Experiments*, Bulletin 35, Imperial Bureau of Social Science, England: Hafner (Macmillan).

Yates, F. (1940), "The recovery of interblock information in balanced incomplete block designs," *Annals of Eugenics*, **10**, 317–325.

Index

WILEY SERIES IN PROBABILITY AND STATISTICS

ESTABLISHED BY WALTER A. SHEWHART AND SAMUEL S. WILKS

Editors: *David J. Balding, Noel A. C. Cressie, Garrett M. Fitzmaurice, Iain M. Johnstone, Geert Molenberghs, David W. Scott, Adrian F. M. Smith, Ruey S. Tsay, Sanford Weisberg*
Editors Emeriti: *Vic Barnett, J. Stuart Hunter, Jozef L. Teugels*

The ***Wiley Series in Probability and Statistics*** is well established and authoritative. It covers many topics of current research interest in both pure and applied statistics and probability theory. Written by leading statisticians and institutions, the titles span both state-of-the-art developments in the field and classical methods.

Reflecting the wide range of current research in statistics, the series encompasses applied, methodological and theoretical statistics, ranging from applications and new techniques made possible by advances in computerized practice to rigorous treatment of theoretical approaches.

This series provides essential and invaluable reading for all statisticians, whether in academia, industry, government, or research.

† ABRAHAM and LEDOLTER · Statistical Methods for Forecasting
AGRESTI · Analysis of Ordinal Categorical Data
AGRESTI · An Introduction to Categorical Data Analysis, *Second Edition*
AGRESTI · Categorical Data Analysis, *Second Edition*
ALTMAN, GILL, and McDONALD · Numerical Issues in Statistical Computing for the Social Scientist
AMARATUNGA and CABRERA · Exploration and Analysis of DNA Microarray and Protein Array Data
ANDĚL · Mathematics of Chance
ANDERSON · An Introduction to Multivariate Statistical Analysis, *Third Edition*
* ANDERSON · The Statistical Analysis of Time Series
ANDERSON, AUQUIER, HAUCK, OAKES, VANDAELE, and WEISBERG · Statistical Methods for Comparative Studies
ANDERSON and LOYNES · The Teaching of Practical Statistics
ARMITAGE and DAVID (editors) · Advances in Biometry
ARNOLD, BALAKRISHNAN, and NAGARAJA · Records
* ARTHANARI and DODGE · Mathematical Programming in Statistics
* BAILEY · The Elements of Stochastic Processes with Applications to the Natural Sciences
BALAKRISHNAN and KOUTRAS · Runs and Scans with Applications
BALAKRISHNAN and NG · Precedence-Type Tests and Applications
BARNETT · Comparative Statistical Inference, *Third Edition*
BARNETT · Environmental Statistics
BARNETT and LEWIS · Outliers in Statistical Data, *Third Edition*
BARTOSZYNSKI and NIEWIADOMSKA-BUGAJ · Probability and Statistical Inference
BASILEVSKY · Statistical Factor Analysis and Related Methods: Theory and Applications
BASU and RIGDON · Statistical Methods for the Reliability of Repairable Systems
BATES and WATTS · Nonlinear Regression Analysis and Its Applications
BECHHOFER, SANTNER, and GOLDSMAN · Design and Analysis of Experiments for Statistical Selection, Screening, and Multiple Comparisons

*Now available in a lower priced paperback edition in the Wiley Classics Library.
†Now available in a lower priced paperback edition in the Wiley–Interscience Paperback Series.

BELSLEY · Conditioning Diagnostics: Collinearity and Weak Data in Regression
† BELSLEY, KUH, and WELSCH · Regression Diagnostics: Identifying Influential Data and Sources of Collinearity
BENDAT and PIERSOL · Random Data: Analysis and Measurement Procedures, *Third Edition*
BERRY, CHALONER, and GEWEKE · Bayesian Analysis in Statistics and Econometrics: Essays in Honor of Arnold Zellner
BERNARDO and SMITH · Bayesian Theory
BHAT and MILLER · Elements of Applied Stochastic Processes, *Third Edition*
BHATTACHARYA and WAYMIRE · Stochastic Processes with Applications
BILLINGSLEY · Convergence of Probability Measures, *Second Edition*
BILLINGSLEY · Probability and Measure, *Third Edition*
BIRKES and DODGE · Alternative Methods of Regression
BISWAS, DATTA, FINE, and SEGAL · Statistical Advances in the Biomedical Sciences: Clinical Trials, Epidemiology, Survival Analysis, and Bioinformatics
BLISCHKE AND MURTHY (editors) · Case Studies in Reliability and Maintenance
BLISCHKE AND MURTHY · Reliability: Modeling, Prediction, and Optimization
BLOOMFIELD · Fourier Analysis of Time Series: An Introduction, *Second Edition*
BOLLEN · Structural Equations with Latent Variables
BOLLEN and CURRAN · Latent Curve Models: A Structural Equation Perspective
BOROVKOV · Ergodicity and Stability of Stochastic Processes
BOULEAU · Numerical Methods for Stochastic Processes
BOX · Bayesian Inference in Statistical Analysis
BOX · R. A. Fisher, the Life of a Scientist
BOX and DRAPER · Response Surfaces, Mixtures, and Ridge Analyses, *Second Edition*
* BOX and DRAPER · Evolutionary Operation: A Statistical Method for Process Improvement
BOX and FRIENDS · Improving Almost Anything, *Revised Edition*
BOX, HUNTER, and HUNTER · Statistics for Experimenters: Design, Innovation, and Discovery, *Second Editon*
BOX, JENKINS, and REINSEL · Time Series Analysis: Forcasting and Control, *Fourth Edition*
BOX and LUCEÑO · Statistical Control by Monitoring and Feedback Adjustment
BRANDIMARTE · Numerical Methods in Finance: A MATLAB-Based Introduction
† BROWN and HOLLANDER · Statistics: A Biomedical Introduction
BRUNNER, DOMHOF, and LANGER · Nonparametric Analysis of Longitudinal Data in Factorial Experiments
BUCKLEW · Large Deviation Techniques in Decision, Simulation, and Estimation
CAIROLI and DALANG · Sequential Stochastic Optimization
CASTILLO, HADI, BALAKRISHNAN, and SARABIA · Extreme Value and Related Models with Applications in Engineering and Science
CHAN · Time Series: Applications to Finance
CHARALAMBIDES · Combinatorial Methods in Discrete Distributions
CHATTERJEE and HADI · Regression Analysis by Example, *Fourth Edition*
CHATTERJEE and HADI · Sensitivity Analysis in Linear Regression
CHERNICK · Bootstrap Methods: A Guide for Practitioners and Researchers, *Second Edition*
CHERNICK and FRIIS · Introductory Biostatistics for the Health Sciences
CHILÈS and DELFINER · Geostatistics: Modeling Spatial Uncertainty
CHOW and LIU · Design and Analysis of Clinical Trials: Concepts and Methodologies, *Second Edition*
CLARKE · Linear Models: The Theory and Application of Analysis of Variance
CLARKE and DISNEY · Probability and Random Processes: A First Course with Applications, *Second Edition*

*Now available in a lower priced paperback edition in the Wiley Classics Library.
†Now available in a lower priced paperback edition in the Wiley–Interscience Paperback Series.

* COCHRAN and COX · Experimental Designs, *Second Edition*
CONGDON · Applied Bayesian Modelling
CONGDON · Bayesian Models for Categorical Data
CONGDON · Bayesian Statistical Modelling
CONOVER · Practical Nonparametric Statistics, *Third Edition*
COOK · Regression Graphics
COOK and WEISBERG · Applied Regression Including Computing and Graphics
COOK and WEISBERG · An Introduction to Regression Graphics
CORNELL · Experiments with Mixtures, Designs, Models, and the Analysis of Mixture Data, *Third Edition*
COVER and THOMAS · Elements of Information Theory
COX · A Handbook of Introductory Statistical Methods
* COX · Planning of Experiments
CRESSIE · Statistics for Spatial Data, *Revised Edition*
CSÖRGŐ and HORVÁTH · Limit Theorems in Change Point Analysis
DANIEL · Applications of Statistics to Industrial Experimentation
DANIEL · Biostatistics: A Foundation for Analysis in the Health Sciences, *Eighth Edition*
* DANIEL · Fitting Equations to Data: Computer Analysis of Multifactor Data, *Second Edition*
DASU and JOHNSON · Exploratory Data Mining and Data Cleaning
DAVID and NAGARAJA · Order Statistics, *Third Edition*
* DEGROOT, FIENBERG, and KADANE · Statistics and the Law
DEL CASTILLO · Statistical Process Adjustment for Quality Control
DEMARIS · Regression with Social Data: Modeling Continuous and Limited Response Variables
DEMIDENKO · Mixed Models: Theory and Applications
DENISON, HOLMES, MALLICK and SMITH · Bayesian Methods for Nonlinear Classification and Regression
DETTE and STUDDEN · The Theory of Canonical Moments with Applications in Statistics, Probability, and Analysis
DEY and MUKERJEE · Fractional Factorial Plans
DILLON and GOLDSTEIN · Multivariate Analysis: Methods and Applications
DODGE · Alternative Methods of Regression
* DODGE and ROMIG · Sampling Inspection Tables, *Second Edition*
* DOOB · Stochastic Processes
DOWDY, WEARDEN, and CHILKO · Statistics for Research, *Third Edition*
DRAPER and SMITH · Applied Regression Analysis, *Third Edition*
DRYDEN and MARDIA · Statistical Shape Analysis
DUDEWICZ and MISHRA · Modern Mathematical Statistics
DUNN and CLARK · Basic Statistics: A Primer for the Biomedical Sciences, *Third Edition*
DUPUIS and ELLIS · A Weak Convergence Approach to the Theory of Large Deviations
EDLER and KITSOS · Recent Advances in Quantitative Methods in Cancer and Human Health Risk Assessment
* ELANDT-JOHNSON and JOHNSON · Survival Models and Data Analysis
ENDERS · Applied Econometric Time Series
† ETHIER and KURTZ · Markov Processes: Characterization and Convergence
EVANS, HASTINGS, and PEACOCK · Statistical Distributions, *Third Edition*
FELLER · An Introduction to Probability Theory and Its Applications, Volume I, *Third Edition,* Revised; Volume II, *Second Edition*
FISHER and VAN BELLE · Biostatistics: A Methodology for the Health Sciences
FITZMAURICE, LAIRD, and WARE · Applied Longitudinal Analysis
* FLEISS · The Design and Analysis of Clinical Experiments

*Now available in a lower priced paperback edition in the Wiley Classics Library.
†Now available in a lower priced paperback edition in the Wiley–Interscience Paperback Series.

FLEISS · Statistical Methods for Rates and Proportions, *Third Edition*
† FLEMING and HARRINGTON · Counting Processes and Survival Analysis
FULLER · Introduction to Statistical Time Series, *Second Edition*
† FULLER · Measurement Error Models
GALLANT · Nonlinear Statistical Models
GEISSER · Modes of Parametric Statistical Inference
GELMAN and MENG · Applied Bayesian Modeling and Causal Inference from Incomplete-Data Perspectives
GEWEKE · Contemporary Bayesian Econometrics and Statistics
GHOSH, MUKHOPADHYAY, and SEN · Sequential Estimation
GIESBRECHT and GUMPERTZ · Planning, Construction, and Statistical Analysis of Comparative Experiments
GIFI · Nonlinear Multivariate Analysis
GIVENS and HOETING · Computational Statistics
GLASSERMAN and YAO · Monotone Structure in Discrete-Event Systems
GNANADESIKAN · Methods for Statistical Data Analysis of Multivariate Observations, *Second Edition*
GOLDSTEIN and LEWIS · Assessment: Problems, Development, and Statistical Issues
GREENWOOD and NIKULIN · A Guide to Chi-Squared Testing
GROSS, SHORTLE, THOMPSON, and HARRIS · Fundamentals of Queueing Theory, *Fourth Edition*
GROSS, SHORTLE, THOMPSON, and HARRIS · Solutions Manual to Accompany Fundamentals of Queueing Theory, *Fourth Edition*
* HAHN and SHAPIRO · Statistical Models in Engineering
HAHN and MEEKER · Statistical Intervals: A Guide for Practitioners
HALD · A History of Probability and Statistics and their Applications Before 1750
HALD · A History of Mathematical Statistics from 1750 to 1930
† HAMPEL · Robust Statistics: The Approach Based on Influence Functions
HANNAN and DEISTLER · The Statistical Theory of Linear Systems
HARTUNG, KNAPP, and SINHA · Statistical Meta-Analysis with Applications
HEIBERGER · Computation for the Analysis of Designed Experiments
HEDAYAT and SINHA · Design and Inference in Finite Population Sampling
HEDEKER and GIBBONS · Longitudinal Data Analysis
HELLER · MACSYMA for Statisticians
HINKELMANN and KEMPTHORNE · Design and Analysis of Experiments, Volume 1: Introduction to Experimental Design, *Second Edition*
HINKELMANN and KEMPTHORNE · Design and Analysis of Experiments, Volume 2: Advanced Experimental Design
HOAGLIN, MOSTELLER, and TUKEY · Fundamentals of Exploratory Analysis of Variance
* HOAGLIN, MOSTELLER, and TUKEY · Exploring Data Tables, Trends and Shapes
* HOAGLIN, MOSTELLER, and TUKEY · Understanding Robust and Exploratory Data Analysis
HOCHBERG and TAMHANE · Multiple Comparison Procedures
HOCKING · Methods and Applications of Linear Models: Regression and the Analysis of Variance, *Second Edition*
HOEL · Introduction to Mathematical Statistics, *Fifth Edition*
HOGG and KLUGMAN · Loss Distributions
HOLLANDER and WOLFE · Nonparametric Statistical Methods, *Second Edition*
HOSMER and LEMESHOW · Applied Logistic Regression, *Second Edition*
HOSMER, LEMESHOW, and MAY · Applied Survival Analysis: Regression Modeling of Time-to-Event Data, *Second Edition*
† HUBER and RONCHETTI · Robust Statistics, *Second Edition*

*Now available in a lower priced paperback edition in the Wiley Classics Library.
†Now available in a lower priced paperback edition in the Wiley–Interscience Paperback Series.

HUBERTY · Applied Discriminant Analysis
HUBERTY and OLEJNIK · Applied MANOVA and Discriminant Analysis, *Second Edition*
HUNT and KENNEDY · Financial Derivatives in Theory and Practice, *Revised Edition*
HURD and MIAMEE · Periodically Correlated Random Sequences: Spectral Theory and Practice
HUSKOVA, BERAN, and DUPAC · Collected Works of Jaroslav Hajek—with Commentary
HUZURBAZAR · Flowgraph Models for Multistate Time-to-Event Data
IMAN and CONOVER · A Modern Approach to Statistics
† JACKSON · A User's Guide to Principle Components
JOHN · Statistical Methods in Engineering and Quality Assurance
JOHNSON · Multivariate Statistical Simulation
JOHNSON and BALAKRISHNAN · Advances in the Theory and Practice of Statistics: A Volume in Honor of Samuel Kotz
JOHNSON and BHATTACHARYYA · Statistics: Principles and Methods, *Fifth Edition*
JOHNSON and KOTZ · Distributions in Statistics
JOHNSON and KOTZ (editors) · Leading Personalities in Statistical Sciences: From the Seventeenth Century to the Present
JOHNSON, KOTZ, and BALAKRISHNAN · Continuous Univariate Distributions, Volume 1, *Second Edition*
JOHNSON, KOTZ, and BALAKRISHNAN · Continuous Univariate Distributions, Volume 2, *Second Edition*
JOHNSON, KOTZ, and BALAKRISHNAN · Discrete Multivariate Distributions
JOHNSON, KEMP, and KOTZ · Univariate Discrete Distributions, *Third Edition*
JUDGE, GRIFFITHS, HILL, LÜTKEPOHL, and LEE · The Theory and Practice of Econometrics, *Second Edition*
JUREČKOVÁ and SEN · Robust Statistical Procedures: Aymptotics and Interrelations
JUREK and MASON · Operator-Limit Distributions in Probability Theory
KADANE · Bayesian Methods and Ethics in a Clinical Trial Design
KADANE AND SCHUM · A Probabilistic Analysis of the Sacco and Vanzetti Evidence
KALBFLEISCH and PRENTICE · The Statistical Analysis of Failure Time Data, *Second Edition*
KARIYA and KURATA · Generalized Least Squares
KASS and VOS · Geometrical Foundations of Asymptotic Inference
† KAUFMAN and ROUSSEEUW · Finding Groups in Data: An Introduction to Cluster Analysis
KEDEM and FOKIANOS · Regression Models for Time Series Analysis
KENDALL, BARDEN, CARNE, and LE · Shape and Shape Theory
KHURI · Advanced Calculus with Applications in Statistics, *Second Edition*
KHURI, MATHEW, and SINHA · Statistical Tests for Mixed Linear Models
KLEIBER and KOTZ · Statistical Size Distributions in Economics and Actuarial Sciences
KLUGMAN, PANJER, and WILLMOT · Loss Models: From Data to Decisions, *Third Edition*
KLUGMAN, PANJER, and WILLMOT · Solutions Manual to Accompany Loss Models: From Data to Decisions, *Third Edition*
KOTZ, BALAKRISHNAN, and JOHNSON · Continuous Multivariate Distributions, Volume 1, *Second Edition*
KOVALENKO, KUZNETZOV, and PEGG · Mathematical Theory of Reliability of Time-Dependent Systems with Practical Applications
KOWALSKI and TU · Modern Applied U-Statistics
KROONENBERG · Applied Multiway Data Analysis
KVAM and VIDAKOVIC · Nonparametric Statistics with Applications to Science and Engineering

*Now available in a lower priced paperback edition in the Wiley Classics Library.
†Now available in a lower priced paperback edition in the Wiley–Interscience Paperback Series.

LACHIN · Biostatistical Methods: The Assessment of Relative Risks
LAD · Operational Subjective Statistical Methods: A Mathematical, Philosophical, and Historical Introduction
LAMPERTI · Probability: A Survey of the Mathematical Theory, *Second Edition*
LANGE, RYAN, BILLARD, BRILLINGER, CONQUEST, and GREENHOUSE · Case Studies in Biometry
LARSON · Introduction to Probability Theory and Statistical Inference, *Third Edition*
LAWLESS · Statistical Models and Methods for Lifetime Data, *Second Edition*
LAWSON · Statistical Methods in Spatial Epidemiology
LE · Applied Categorical Data Analysis
LE · Applied Survival Analysis
LEE and WANG · Statistical Methods for Survival Data Analysis, *Third Edition*
LEPAGE and BILLARD · Exploring the Limits of Bootstrap
LEYLAND and GOLDSTEIN (editors) · Multilevel Modelling of Health Statistics
LIAO · Statistical Group Comparison
LINDVALL · Lectures on the Coupling Method
LIN · Introductory Stochastic Analysis for Finance and Insurance
LINHART and ZUCCHINI · Model Selection
LITTLE and RUBIN · Statistical Analysis with Missing Data, *Second Edition*
LLOYD · The Statistical Analysis of Categorical Data
LOWEN and TEICH · Fractal-Based Point Processes
MAGNUS and NEUDECKER · Matrix Differential Calculus with Applications in Statistics and Econometrics, *Revised Edition*
MALLER and ZHOU · Survival Analysis with Long Term Survivors
MALLOWS · Design, Data, and Analysis by Some Friends of Cuthbert Daniel
MANN, SCHAFER, and SINGPURWALLA · Methods for Statistical Analysis of Reliability and Life Data
MANTON, WOODBURY, and TOLLEY · Statistical Applications Using Fuzzy Sets
MARCHETTE · Random Graphs for Statistical Pattern Recognition
MARDIA and JUPP · Directional Statistics
MASON, GUNST, and HESS · Statistical Design and Analysis of Experiments with Applications to Engineering and Science, *Second Edition*
McCULLOCH, SEARLE, and NEUHAUS · Generalized, Linear, and Mixed Models, *Second Edition*
McFADDEN · Management of Data in Clinical Trials, *Second Edition*
* McLACHLAN · Discriminant Analysis and Statistical Pattern Recognition
McLACHLAN, DO, and AMBROISE · Analyzing Microarray Gene Expression Data
McLACHLAN and KRISHNAN · The EM Algorithm and Extensions, *Second Edition*
McLACHLAN and PEEL · Finite Mixture Models
McNEIL · Epidemiological Research Methods
MEEKER and ESCOBAR · Statistical Methods for Reliability Data
MEERSCHAERT and SCHEFFLER · Limit Distributions for Sums of Independent Random Vectors: Heavy Tails in Theory and Practice
MICKEY, DUNN, and CLARK · Applied Statistics: Analysis of Variance and Regression, *Third Edition*
* MILLER · Survival Analysis, *Second Edition*
MONTGOMERY, JENNINGS, and KULAHCI · Introduction to Time Series Analysis and Forecasting
MONTGOMERY, PECK, and VINING · Introduction to Linear Regression Analysis, *Fourth Edition*
MORGENTHALER and TUKEY · Configural Polysampling: A Route to Practical Robustness
MUIRHEAD · Aspects of Multivariate Statistical Theory

*Now available in a lower priced paperback edition in the Wiley Classics Library.
†Now available in a lower priced paperback edition in the Wiley–Interscience Paperback Series.

MULLER and STOYAN · Comparison Methods for Stochastic Models and Risks
MURRAY · X-STAT 2.0 Statistical Experimentation, Design Data Analysis, and Nonlinear Optimization
MURTHY, XIE, and JIANG · Weibull Models
MYERS, MONTGOMERY, and ANDERSON-COOK · Response Surface Methodology: Process and Product Optimization Using Designed Experiments, *Third Edition*
MYERS, MONTGOMERY, and VINING · Generalized Linear Models. With Applications in Engineering and the Sciences
† NELSON · Accelerated Testing, Statistical Models, Test Plans, and Data Analyses
† NELSON · Applied Life Data Analysis
NEWMAN · Biostatistical Methods in Epidemiology
OCHI · Applied Probability and Stochastic Processes in Engineering and Physical Sciences
OKABE, BOOTS, SUGIHARA, and CHIU · Spatial Tesselations: Concepts and Applications of Voronoi Diagrams, *Second Edition*
OLIVER and SMITH · Influence Diagrams, Belief Nets and Decision Analysis
PALTA · Quantitative Methods in Population Health: Extensions of Ordinary Regressions
PANJER · Operational Risk: Modeling and Analytics
PANKRATZ · Forecasting with Dynamic Regression Models
PANKRATZ · Forecasting with Univariate Box-Jenkins Models: Concepts and Cases
* PARZEN · Modern Probability Theory and Its Applications
PEÑA, TIAO, and TSAY · A Course in Time Series Analysis
PIANTADOSI · Clinical Trials: A Methodologic Perspective
PORT · Theoretical Probability for Applications
POURAHMADI · Foundations of Time Series Analysis and Prediction Theory
POWELL · Approximate Dynamic Programming: Solving the Curses of Dimensionality
PRESS · Bayesian Statistics: Principles, Models, and Applications
PRESS · Subjective and Objective Bayesian Statistics, *Second Edition*
PRESS and TANUR · The Subjectivity of Scientists and the Bayesian Approach
PUKELSHEIM · Optimal Experimental Design
PURI, VILAPLANA, and WERTZ · New Perspectives in Theoretical and Applied Statistics
† PUTERMAN · Markov Decision Processes: Discrete Stochastic Dynamic Programming
QIU · Image Processing and Jump Regression Analysis
* RAO · Linear Statistical Inference and Its Applications, *Second Edition*
RAUSAND and HØYLAND · System Reliability Theory: Models, Statistical Methods, and Applications, *Second Edition*
RENCHER · Linear Models in Statistics
RENCHER · Methods of Multivariate Analysis, *Second Edition*
RENCHER · Multivariate Statistical Inference with Applications
* RIPLEY · Spatial Statistics
* RIPLEY · Stochastic Simulation
ROBINSON · Practical Strategies for Experimenting
ROHATGI and SALEH · An Introduction to Probability and Statistics, *Second Edition*
ROLSKI, SCHMIDLI, SCHMIDT, and TEUGELS · Stochastic Processes for Insurance and Finance
ROSENBERGER and LACHIN · Randomization in Clinical Trials: Theory and Practice
ROSS · Introduction to Probability and Statistics for Engineers and Scientists
ROSSI, ALLENBY, and McCULLOCH · Bayesian Statistics and Marketing
† ROUSSEEUW and LEROY · Robust Regression and Outlier Detection
* RUBIN · Multiple Imputation for Nonresponse in Surveys
RUBINSTEIN and KROESE · Simulation and the Monte Carlo Method, *Second Edition*
RUBINSTEIN and MELAMED · Modern Simulation and Modeling
RYAN · Modern Engineering Statistics

*Now available in a lower priced paperback edition in the Wiley Classics Library.
†Now available in a lower priced paperback edition in the Wiley–Interscience Paperback Series.

RYAN · Modern Experimental Design
RYAN · Modern Regression Methods, *Second Edition*
RYAN · Statistical Methods for Quality Improvement, *Second Edition*
SALEH · Theory of Preliminary Test and Stein-Type Estimation with Applications
* SCHEFFE · The Analysis of Variance
SCHIMEK · Smoothing and Regression: Approaches, Computation, and Application
SCHOTT · Matrix Analysis for Statistics, *Second Edition*
SCHOUTENS · Levy Processes in Finance: Pricing Financial Derivatives
SCHUSS · Theory and Applications of Stochastic Differential Equations
SCOTT · Multivariate Density Estimation: Theory, Practice, and Visualization
† SEARLE · Linear Models for Unbalanced Data
† SEARLE · Matrix Algebra Useful for Statistics
† SEARLE, CASELLA, and McCULLOCH · Variance Components
SEARLE and WILLETT · Matrix Algebra for Applied Economics
SEBER · A Matrix Handbook For Statisticians
† SEBER · Multivariate Observations
SEBER and LEE · Linear Regression Analysis, *Second Edition*
† SEBER and WILD · Nonlinear Regression
SENNOTT · Stochastic Dynamic Programming and the Control of Queueing Systems
* SERFLING · Approximation Theorems of Mathematical Statistics
SHAFER and VOVK · Probability and Finance: It's Only a Game!
SILVAPULLE and SEN · Constrained Statistical Inference: Inequality, Order, and Shape Restrictions
SMALL and McLEISH · Hilbert Space Methods in Probability and Statistical Inference
SRIVASTAVA · Methods of Multivariate Statistics
STAPLETON · Linear Statistical Models
STAPLETON · Models for Probability and Statistical Inference: Theory and Applications
STAUDTE and SHEATHER · Robust Estimation and Testing
STOYAN, KENDALL, and MECKE · Stochastic Geometry and Its Applications, *Second Edition*
STOYAN and STOYAN · Fractals, Random Shapes and Point Fields: Methods of Geometrical Statistics
STREET and BURGESS · The Construction of Optimal Stated Choice Experiments: Theory and Methods
STYAN · The Collected Papers of T. W. Anderson: 1943–1985
SUTTON, ABRAMS, JONES, SHELDON, and SONG · Methods for Meta-Analysis in Medical Research
TAKEZAWA · Introduction to Nonparametric Regression
TAMHANE · Statistical Analysis of Designed Experiments: Theory and Applications
TANAKA · Time Series Analysis: Nonstationary and Noninvertible Distribution Theory
THOMPSON · Empirical Model Building
THOMPSON · Sampling, *Second Edition*
THOMPSON · Simulation: A Modeler's Approach
THOMPSON and SEBER · Adaptive Sampling
THOMPSON, WILLIAMS, and FINDLAY · Models for Investors in Real World Markets
TIAO, BISGAARD, HILL, PEÑA, and STIGLER (editors) · Box on Quality and Discovery: with Design, Control, and Robustness
TIERNEY · LISP-STAT: An Object-Oriented Environment for Statistical Computing and Dynamic Graphics
TSAY · Analysis of Financial Time Series, *Second Edition*
UPTON and FINGLETON · Spatial Data Analysis by Example, Volume II: Categorical and Directional Data
† VAN BELLE · Statistical Rules of Thumb, *Second Edition*

*Now available in a lower priced paperback edition in the Wiley Classics Library.
†Now available in a lower priced paperback edition in the Wiley–Interscience Paperback Series.

VAN BELLE, FISHER, HEAGERTY, and LUMLEY · Biostatistics: A Methodology for the Health Sciences, *Second Edition*
VESTRUP · The Theory of Measures and Integration
VIDAKOVIC · Statistical Modeling by Wavelets
VINOD and REAGLE · Preparing for the Worst: Incorporating Downside Risk in Stock Market Investments
WALLER and GOTWAY · Applied Spatial Statistics for Public Health Data
WEERAHANDI · Generalized Inference in Repeated Measures: Exact Methods in MANOVA and Mixed Models
WEISBERG · Applied Linear Regression, *Third Edition*
WELSH · Aspects of Statistical Inference
WESTFALL and YOUNG · Resampling-Based Multiple Testing: Examples and Methods for *p*-Value Adjustment
WHITTAKER · Graphical Models in Applied Multivariate Statistics
WINKER · Optimization Heuristics in Economics: Applications of Threshold Accepting
WONNACOTT and WONNACOTT · Econometrics, *Second Edition*
WOODING · Planning Pharmaceutical Clinical Trials: Basic Statistical Principles
WOODWORTH · Biostatistics: A Bayesian Introduction
WOOLSON and CLARKE · Statistical Methods for the Analysis of Biomedical Data, *Second Edition*
WU and HAMADA · Experiments: Planning, Analysis, and Parameter Design Optimization
WU and ZHANG · Nonparametric Regression Methods for Longitudinal Data Analysis
YANG · The Construction Theory of Denumerable Markov Processes
YOUNG, VALERO-MORA, and FRIENDLY · Visual Statistics: Seeing Data with Dynamic Interactive Graphics
ZACKS · Stage-Wise Adaptive Designs
ZELTERMAN · Discrete Distributions—Applications in the Health Sciences
* ZELLNER · An Introduction to Bayesian Inference in Econometrics
ZHOU, OBUCHOWSKI, and McCLISH · Statistical Methods in Diagnostic Medicine

*Now available in a lower priced paperback edition in the Wiley Classics Library.
†Now available in a lower priced paperback edition in the Wiley–Interscience Paperback Series.

CPSIA information can be obtained
at www.ICGtesting.com
Printed in the USA
BVOW06*2319191017
498117BV00013BA/559/P

9 780471 750437